Electronic Circuit Fundamentals

Electronic Circuit Fundamentals

WALTER J. WEIR

Algonquin College of Applied Arts and Technology
Nepean, Ontario, Canada

PRENTICE-HALL, INC., *Englewood Cliffs, New Jersey 07632*

Library of Congress Cataloging-in-Publication Data

WEIR, WALTER J.
Electronic circuit fundamentals.

Includes index.
1. Electric circuits. 2. Electronic circuits. I. Title.
TK454.W46 1987 621.3815'3 86-30435
ISBN 0-13-250036-1

Editorial/production supervision: *Claudia Citarella and Jane Bonnell*
Cover design: *Diane Saxe*
Manufacturing buyer: *Margaret Rizzi*

Printed in the United States of America

10 9 8 7 6 5 4 3 2 1

ISBN 0-13-250036-1 025

Prentice-Hall International (UK) Limited, *London*
Prentice-Hall of Australia Pty. Limited, *Sydney*
Prentice-Hall Canada, Inc., *Toronto*
Prentice-Hall Hispanoamericana, S.A., *Mexico*
Prentice-Hall of India Private Limited, *New Delhi*
Prentice-Hall of Japan, Inc., *Tokyo*
Prentice-Hall of Southeast Asia Pte. Ltd., *Singapore*
Editora Prentice-Hall do Brasil, Ltda., *Rio de Janeiro*

Contents

3 OHM'S LAW 33

4 SERIES CIRCUITS 46

5 PARALLEL CIRCUITS 58

6 ENERGY, WORK, AND POWER 69

7 COMBINATION CIRCUITS 82

23 CAPACITANCE 330

24 *CR* AND *L/R* TIME-CONSTANT CIRCUIT 352

25 CAPACITIVE REACTANCE 376

26 IMPEDANCE IN THE SERIES AND PARALLEL *RC* CIRCUIT 388

27 SERIES RESONANT CIRCUIT 403

28 PARALLEL RESONANT CIRCUIT 425

Preface

ELECTRONIC CIRCUIT FUNDAMENTALS is designed for students taking technicican-level electronics courses at community colleges, technical schools, and secondary schools. The goal of the text is for students to comprehend fully the principles behind each fundamental circuit concept. To attain this goal, the text blends an easy going, person-to-person style of presentation with what I think you'll agree to be a very practical system for reinforcing the learning process.

Each chapter is divided into small building blocks of related concepts. These conceptual blocks consistently build upon preceding material, and are promptly followed by engaging and challenging practice questions. There are 750 practice questions altogether—considerably more than in other texts of this nature.

At least two examples are given for each circuit concept developed. These examples are not one-sided, but rather explore various methods of arriving at a given solution. Examples are founded on the practicality of the circuit, such as EIA resistor values and applications of the circuit. Formulas and problems are calculator-oriented wherever possible.

Review Questions (more than 500 in all) appear at the end of each chapter. These questions require students to either write an essay or solve a problem and should be of great assistance in summarizing the chapter.

Self-Examination Questions (more than 750) also appear at the end of each chapter. Students can use these to evaluate their progress—chapter by chapter. The answers to each Self-Examination Question can be found in the Instructor's Manual.

ORGANIZATION OF THE TEXT

As noted above, topics are presented in small blocks that consistently build on preceding material; new concepts are introduced only after that material has been thoroughly covered. The coverage of power in a dc circuit (Chapter 6) and in an ac circuit (Chapter 30) is a prime example of this style of presentation. Note that power and power rating of series and parallel resistors are presented only after series and parallel circuits have been thoroughly explored.

The chapter sequencing allows you greater flexibility to tailor the book to your specific curriculum:

Chapters 1 through 13 cover dc circuit concepts. The emphasis on Chapters 1 through 8 is critical regardless of the technical program; however, chapters such as

Direct Current Motors (Chapter 11), Batteries (Chapter 12), and Conductors and Insulators (Chapter 13) may be de-emphasized. Whatever the program requirements, this transition is easily accomplished since each of these three chapters is independent of Chapters 1 through 8.

Chapters 14 and 15 cover magnetism and electromagnetism, respectively, and are prerequisites for the following chapters on ac generation and ac circuits. In most community colleges and secondary schools, physics is usually taught concurrently with the electronics programs. The difficulty lies in synchronizing the study of magnetics with the study of electronics theory.

Chapters 16 through 31 cover ac generation and circuit theory. Such chapters as Filters (Chapter 29), Ac Power (Chapter 30), Ac Meters (Chapter 31), and Complex Numbers for Ac Circuits (Chapter 32) could be de-emphasized, since these chapters are independent of Chapters 16 through 28.

ELECTRON FLOW VERSUS CONVENTIONAL CIRCUIT FLOW

There are many reasons to introduce students to electron current flow. Perhaps the most important being Rutherford's atomic model, which shows that free electrons are the prime movers or carriers in a conductor. It is most certain that electrons flow in a vacuum tube or compose an electron beam in a cathode ray tube.

For semiconductor theory we require the electron current and conventional current flows. Thus, both concepts should be understood, but the most applicable should be introduced and used in a basic course. To this end, the electron flow direction has been used in this text with reminders that conventional flow produces identical results.

SUPPLEMENTARY MATERIAL

Lab Manual: Electronic Circuit Fundamentals

An integrative Lab Manual containing 37 experiments is available. Each experiment is suitable for a two-hour lab session, and requires students to prepare a pre-lab or pre-experiment report.

Instructor's Solutions Manual to the Lab Manual

This Instructor's Manual provides a convenient and quick reference to the Lab Manual. It also provides detailed, worked-out solutions to each pre-lab experiment, including typical measured results for each experiment.

Instructor's Solutions Manual to the Text

This Instructor's Manual provides a convenient and quick reference to the **Electronic Circuit Fundamentals** text. It also contains worked-out solutions to each end-of-chapter Self-Examination Question.

Transparency Masters

A set of transparency masters of more than 150 illustrations, waveform diagrams, and tables taken from the text is available.

Fundamentals of Mathematics: Outline and Review Problems for Electronic Circuit Fundamentals

This mathematics review manual is available for students who wish to refresh or polish their mathematics skills.

Acknowledgments

My sincere appreciation and thanks are extended to all the people who helped to make this textbook possible. To my reviewers: Edward Troyan, Lehigh Community College, Pennsylvania; and Jim Everett, PCAVTS, Plathe City, Missouri.

To Allen Bradley Canada Ltd, General Radio Inc., Eveready Battery Company, Duracell Inc., Canadian Standards Association, Weston Instruments, Hammond Manufacturing, and many other companies who contributed information and illustrations for the textbook.

To all the staff at Prentice-Hall: Len Rosen, former editor in Canada, who cajoled me into writing this book; Elio Ennamorati, editor in Canada; Gregory Burnell, editor; Alice Barr, editor; Jane Bonnell, editorial production; Claudia Citarella, editorial production; Maureen Lopez, editorial production; Melissa Halverstadt, marketing manager; and many others at Prentice-Hall, Inc., Englewood Cliffs, N.J.

To my wife Anne, who stood by patiently; Deirdre, who typed the manuscript; Catherine, who typed the Self-Examination Solutions; Karen, who humoured me; and Gregory, who co-authored the Mathematics Review Supplement and assisted in the Solutions Manual.

Thank you all.

Walter J. Weir

Electronic Circuit Fundamentals

1

Electricity

INTRODUCTION

To define electricity precisely is impossible since scientists are just beginning to probe the basic structure of matter, the atom. However, a broad definition can be applied by stating that *electricity* is that invisible energy inherent in nature which, when released by nature, takes a form such as a lightning discharge, and when released by man, takes the form of an electrical current flow either through a conductor to operate an electrical device such as a motor, lamp, or toaster, or through an electronic device such as a semiconductor or vacuum tube to do work in the form of heat, light, mechanical work, communication computation, and data processing.

Our present-day understanding of electricity is that all matter is made up of particles of electricity. Thus you and everyone else is made up of particles of electricity since all human beings are made up of matter. Matter, in turn, is made up of particles called molecules, and the molecules are made up of atoms. The atoms are made up of positive, negative, and neutral particles of electricity. Like any other law of nature, electricity has specific effects that can be measured and observed. These effects can be applied to a circuit to explain as well as predict the action of the circuit. To predict circuit action and verify the prediction, it is necessary to have a thorough understanding of basic dc and ac circuits.

MAIN TOPICS IN CHAPTER 1

OBJECTIVES

After studying Chapter 1, the student should be able to

1. Master the fundamental concepts of the structure of matter.
2. Master theoretical concepts of current flow in solids, liquids, and gases.
3. Interpret the difference between a conductor, an insulator, and a semiconductor.

1-1 STRUCTURE OF MATTER

Although the environment around us looks as if it is composed of mostly solid material, actually, it is the other way around. Matter occupies only a minute amount of space; therefore, anything that occupies space and has weight is considered to be matter.

Elements

If we were to take a closer look at matter, we would find that it contains two ingredients. The principal ingredient is called an *element*. There are some 105 elements to date, ranging from basic hydrogen to newly discovered elements such as nobelium 102 and lawrencium 103. Other recent discoveries have pushed the number of elements to 107 (see Table 1-1).

Compounds

The other ingredient that makes up matter is called a *compound*. Water, alcohol, salt, and steel, to name just a few examples, are compounds.

Elements versus Compounds

What is the difference between an element and a compound? To understand the difference, we need to magnify matter under an electron microscope. With a magnification factor of 100,000 times it is possible to see that matter is made up of molecules. Figure 1-1 shows a molecule of water; note that it is made up of smaller bits of matter called atoms.

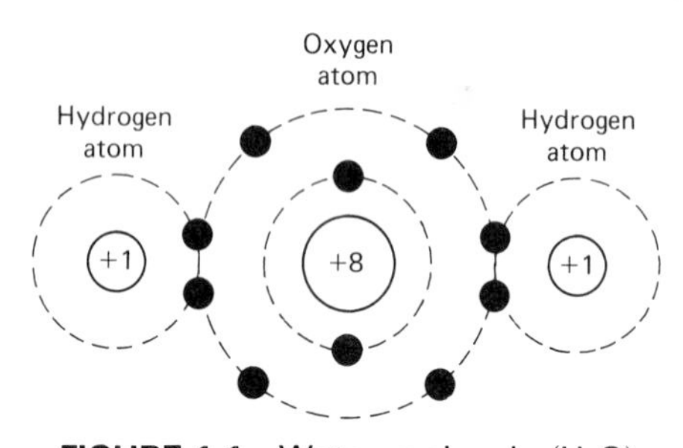

FIGURE 1-1 Water molecule (H_2O).

Now we can elaborate on the difference between an element and a compound. Shown in Fig. 1-2 is the most basic element, the hydrogen atom. Hydrogen and *all elements are composed of one type of atom*. Since an atom cannot be divided, we cannot make gold from lead as many ancient alchemists tried to do. All elements have their own special types of atoms, such as the hydrogen atom, which has one electron, and carbon, which has six electrons.

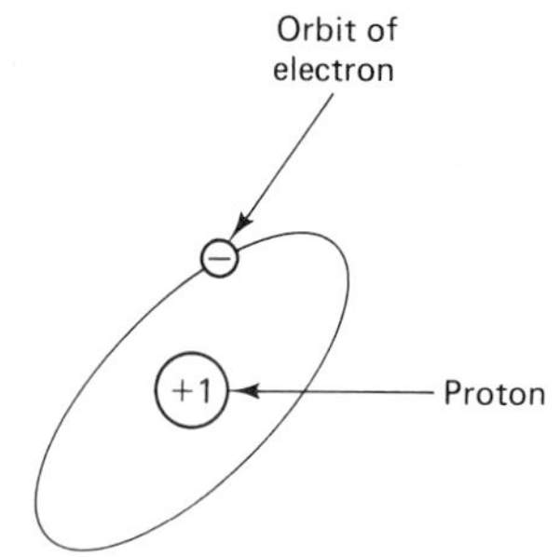

FIGURE 1-2 Hydrogen atom.

Compounds such as salt, water, alcohol, and gasoline have *two or more types of atoms joined chemically*. The molecule of water shown in Fig. 1-1 has two hydrogen atoms joined to an oxygen atom. "Joining" is the key here since a compound can be made only if the atoms are joined through a chemical bond. We would obtain a *mixture* if we simply combined (mixed), say, flour and water to make a dough. To form a compound, there must be a chemical reaction to bind two or more atoms together.

1-2 ATOM STRUCTURE

The generally accepted model of all atoms is the theory proposed by Niels Bohr in 1913. Although subsequently replaced by an abstract description of greater accuracy and usefulness to scientists, it remains in the minds of many as the mental image of the atom. This image, shown in Fig. 1-3, is that of a spherical nucleus consisting of protons and neutrons, with electrons orbiting about the nucleus in various energy levels or shells. If there are six protons in the nucleus of the atom, there will be six electrons in the shells or orbits of the atom. Each positively charged proton is able to attract and hold a negatively charged electron in orbit about the nucleus.

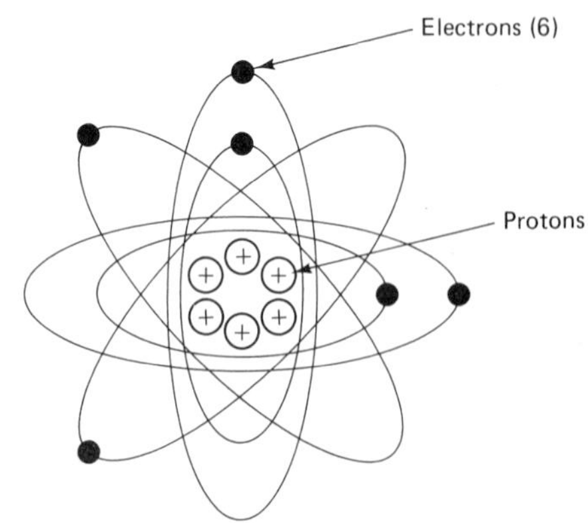

FIGURE 1-3 Representative drawing of a carbon atom.

TABLE 1-1 Electronic structure of atoms

Element	Atomic No.	SHELL K	L	M	N	O
Hydrogen	1	1				
Helium	2	2				
Lithium	3	2	1			
Beryllium	4	2	2			
Boron	5	2	3			
Carbon	6	2	4			
Nitrogen	7	2	5			
Oxygen	8	2	6			
Fluorine	9	2	7			
Neon	10	2	8			
Sodium	11	2	8	1		
Magnesium	12	2	8	2		
Aluminium	13	2	8	3		
Silicon	14	2	8	4		
Phosphorus	15	2	8	5		
Sulfur	16	2	8	6		
Chlorine	17	2	8	7		
Argon	18	2	8	8		
Potassium	19	2	8	8	1	
Calcium	20	2	8	8	2	
Scandium	21	2	8	9	2	
Titanium	22	2	8	10	2	
Vanadium	23	2	8	11	2	
Chromium	24	2	8	13	1	
Manganese	25	2	8	13	2	
Iron	26	2	8	14	2	
Cobalt	27	2	8	15	2	
Nickel	28	2	8	16	2	
Copper	29	2	8	18	1	
Zinc	30	2	8	18	2	
Gallium	31	2	8	18	3	
Germanium	32	2	8	18	4	
Arsenic	33	2	8	18	5	
Selenium	34	2	8	18	6	
Bromine	35	2	8	18	7	
Krypton	36	2	8	18	8	
Rubidium	37	2	8	18	8	1
Strontium	38	2	8	18	8	2
Yttrium	39	2	8	18	9	2
Zirconium	40	2	8	18	10	2
Niobium	41	2	8	18	12	1
Molybdenum	42	2	8	18	13	1
Technetium	43	2	8	18	13	2
Ruthenium	44	2	8	18	15	1
Rhodium	45	2	8	18	16	1
Palladium	46	2	8	18	18	
Silver	47	2	8	18	18	1
Cadmium	48	2	8	18	18	2
Indium	49	2	8	18	18	3
Tin	50	2	8	18	18	4
Antimony	51	2	8	18	18	5
Tellurium	52	2	8	18	18	6

Element	Atomic No.	SHELL K	L	M	N	O	P	Q
Iodine	53	2	8	18	18	7		
Xenon	54	2	8	18	18	8		
Cesium	55	2	8	18	18	8	1	
Barium	56	2	8	18	18	8	2	
Lanthanium	57	2	8	18	18	9	2	
Cerium	58	2	8	18	20	8	2	
Praseodymium	59	2	8	18	21	8	2	
Neodymium	60	2	8	18	22	8	2	
Promethium	61	2	8	18	23	8	2	
Samarium	62	2	8	18	24	8	2	
Europium	63	2	8	18	25	8	2	
Gadolinium	64	2	8	18	26	9	1	
Terbium	65	2	8	18	26	9	2	
Dysprosium	66	2	8	18	28	8	2	
Holmium	67	2	8	18	29	8	2	
Erbium	68	2	8	18	30	8	2	
Thulium	69	2	8	18	31	8	2	
Ytterbium	70	2	8	18	32	8	2	
Lutetium	71	2	8	18	32	9	2	
Hafnium	72	2	8	18	32	10	2	
Tantalum	73	2	8	18	32	11	2	
Tungsten	74	2	8	18	32	12	2	
Rhenium	75	2	8	18	32	13	2	
Osmium	76	2	8	18	32	14	2	
Iridium	77	2	8	18	32	15	2	
Platinum	78	2	8	18	32	17	1	
Gold	79	2	8	18	32	18	1	
Mercury	80	2	8	18	32	18	2	
Thallium	81	2	8	18	32	18	3	
Lead	82	2	8	18	32	18	4	
Bismuth	83	2	8	18	32	18	5	
Polonium	84	2	8	18	32	18	6	
Astatine	85	2	8	18	32	18	7	
Radon	86	2	8	18	32	18	8	
Francium	87	2	8	18	32	18	8	1
Radium	88	2	8	18	32	18	8	2
Actinium	89	2	8	18	32	18	9	2
Thorium	90	2	8	18	32	18	10	2
Protactium	91	2	8	18	32	20	9	2
Uranium	92	2	8	18	32	21	9	2
Neptunium	93	2	8	18	32	23	8	2
Plutonium	94	2	8	18	32	24	8	2
Americium	95	2	8	18	32	25	8	2
Curium	96	2	8	18	32	25	9	2
Berkelium	97	2	8	18	32	25	10	2
Californium	98	2	8	18	32	27	9	2
Einsteinium	99	2	8	18	32	27	10	2
Fermium	100	2	8	18	32	29	9	2
Mendelevium	101	2	8	18	32	29	10	2
Nobelium (?)	102	2	8	18	32	32	8	2
Lawrencium	103	2	8	18	32	32	9	2
	104	2	8	18	32	32	10	2

Note: Two or more forms of an element differing from each other in weight of atoms or mass are called isotopes. The following discoveries may be added to the elements in Table 1-1.

	Atomic No.
Kurchatovium, USSR, 1964, mass #260	104
Rutherfordium, USA, 1969, mass #257	104
Nielsbohrium, USSR, 1970, mass # ?	105
Hahnium, USA, 1970, mass #260	105
No name, USA, 1974, mass #263	106
No name, USSR, 1974, mass ?	106
No name, USSR, 1976, mass #263	107

Source: Canadian Electronic Teaching Aids, 452 Kraft Road, Fort Erie, Ontario, Canada, Data Sheet No. DAO–1.

The Electron

Figure 1-3 is a representative drawing of an atom whose electrons orbit about the nucleus. "Electron" comes from the Greek word for "amber," which is *elektron*. The Greeks observed the phenomenon of charging or "electrifying" bodies by rubbing pieces of amber with fur or cloth (similar to combing your hair) and then noting the attractive force between the amber (comb) and bits of paper or wool. Although the phenomenon could be reproduced, the concept of the electron was unknown.

The attraction force observed by the Greeks is now understood to be between two unlike charges, the electron and the proton. The proton and electron are considered to have an electrostatic force or field surrounding each. The electrostatic force of the electron is considered by convention to be *negative* and flows from the electron as shown in Fig. 1-4. The electron is the workhorse of both electricity and electronics.

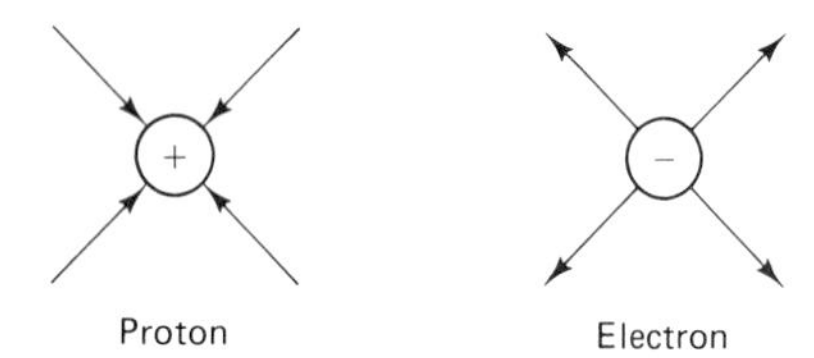

FIGURE 1-4 Electrostatic lines of force.

The Proton

By convention, the proton is assigned the opposite electrostatic field or force from the electron and is given a *positive* sign (Fig. 1-4). The proton is very massive, 1840 times heavier than the electron, and is the stable part of the atom. The *neutron*, a *neutral* particle of electricity, is not important for our purposes and will not be considered here.

Law of Electrostatic Field

Any two electrons will repel each other, by the *law of electrostatic action*. This law shows that two like charges or fields repel each other and two unlike charges attract each other.

This effect is easily demonstrated by charging a silk-rubbed glass rod and a fur-rubbed ebonite rod (see Fig. 1-5). In the case of the glass rod rubbed with silk, the silk removes electrons from the surface atoms of the glass so that the glass is charged positively. The ebonite rod is rubbed with fur so that the fur atoms give up electrons and these are left behind on the rod to charge it negatively. Touching a neutral pith ball with the glass rod will charge the ball positively. Electrons from the pith ball will be attracted by the positive charge on the glass rod and flow to the rod. The ball, having been depleted of electrons, is charged positively. Both the glass rod and the ball now have a positive charge and repel each other as shown in Fig. 1-5(a). If the negatively charged ebonite rod is brought close to but not touching the positively charged ball, the ball will be attracted to the rod as shown in Fig. 1-5(b).

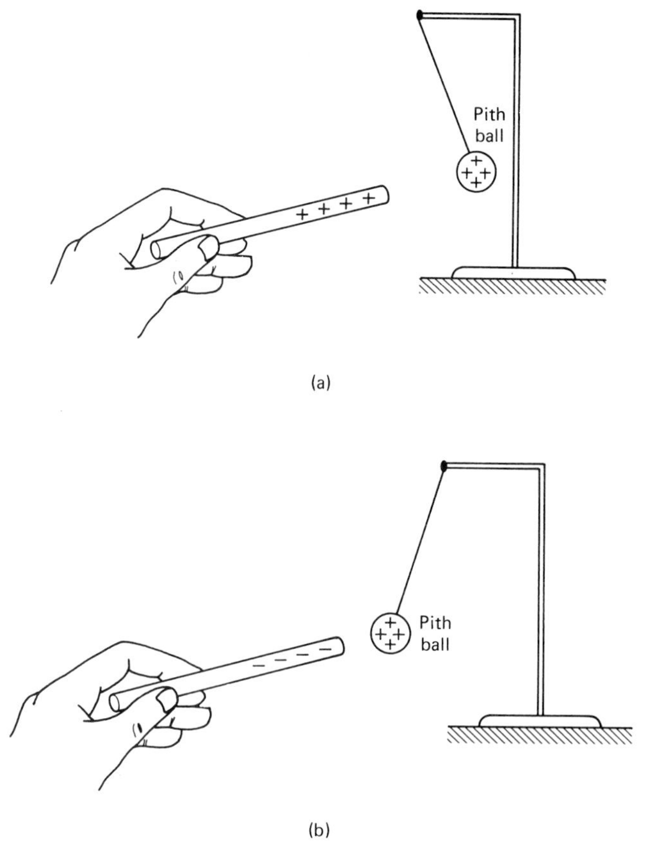

FIGURE 1-5 (a) Positively charged silk-rubbed glass rod repels positively charged pith ball. (b) Negatively charged fur-rubbed ebonite rod attracts positively charged pith ball.

PRACTICE QUESTIONS 1-1

Choose the most suitable word, phrase, or value to complete each statement.

1. Generally, anything that occupies space and has weight is called (*an electron, a proton, a neutron, matter, an atom*).
2. Two basic "building blocks" that make up all matter are called (*electrons and protons, protons and neutrons, elements and atoms, elements and compounds*).

3. The smallest particle of any substance is called a/an (*molecule, ion, atom, element, compound*).
4. All matter is made of particles of electricity in the form of (*atoms and molecules; protons and neutrons; electrons, protons, and neutrons*).
5. A positive particle of electricity is called a/an (*neutron, proton, electron, ion, atom*).
6. A negative particle of electricity is called a/an (*neutron, proton, electron, ion, atom*).
7. Lines of force about an electron by convention "flow" (*away from, toward*) the electron.
8. If two unlike charges are placed close to but not touching each other, lines of force will flow from the ________ (polarity) -charged body to the ________ (polarity) -charged body.
9. When placed close to but not touching each other, like charges will be (*repelled, attracted*).
10. When a negatively charged body is placed close to but not touching a neutral body, a ________ (polarity) charge is induced on that body.
11. An ebonite rod rubbed with fur will be charged ________ (polarity).
12. A glass rod rubbed with silk will be charged ________ (polarity).

1-3 SHELLS OR ENERGY LEVELS

The atom shown in Fig. 1-3 can be visualized as successive layers of an orange peel. The outermost shell of the atom has a definite order of 18 electrons, by the laws of nature. Similarly, the next shell or energy level has eight electrons, and the innermost orbit has two electrons.

The successive shells or energy levels are called *K, L, M, N, O, P,* and *Q* at increasing outward distance from the nucleus. Each shell has its maximum number of electrons for atomic stability as shown in Fig. 1-6.

Table 1-2 shows some examples of elements and their respective maximum number of electrons in the outermost orbit. Note that the maximum number of electrons that can be found in the *K* shell is two. After the third shell the maximum number that any other shell can hold when it is the outermost shell is eight electrons.

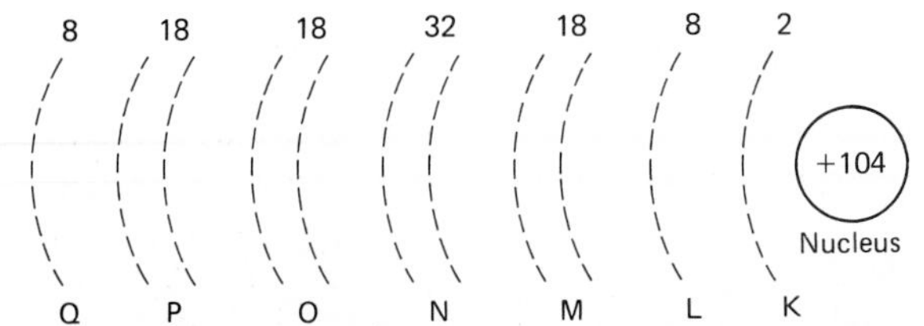

FIGURE 1-6 Successive shells and subshells of an atom.

TABLE 1-2

Shell	Maximum Electrons in Outermost Shell	Element	Atomic Number
K	2	Helium	2
L	8	Neon	10
M	8	Argon	18
N	8	Krypton	36
O	8	Xenon	54
P	8	Radon	86

The maximum number of electrons in a filled inner shell is equal to $2n^2$, where n is the shell number in sequential outward order from the nucleus. For example, the maximum number of electrons in the first shell is $2 \times 1^2 = 2$, in the second shell $2 \times 2^2 = 8$, in the third shell $2 \times 3^2 = 18$, and so on. These values apply only to an inner shell that is filled with its maximum number of electrons.

The elements shown in Table 1-2 all have eight electrons in their outermost shell and all share a remarkable property: They are almost totally inert; that is, they have no chemical properties or, under normal conditions, they do not interact with any other types of atoms. With the exception of these inert gases, all the other elements interact with one another.

Atomic Number

The *atomic number* represents the number of electrons or protons in the atom of an element. Looking at Table 1-1 and summing the electrons in each shell will give the atomic number of the element. For example, copper has four shells of 2, 8, 18, and 1 electron, for a total of 29 electrons, or an atomic number of 29.

Electron Valence

The number of electrons in the outermost orbit of an atom is the *electron valence* value. Table 1-1 shows that germanium has a valence of four, copper has a valence of one, and antimony has a valence of five. Since the maximum number of electrons that the outermost shell can hold is eight, the number of electrons required to fill the shell is called a *negative valence*. Copper, then, can be considered to have of valence of +1 or −7. Similarly, germanium and silicon have a valence of +4 or −4.

Positive and Negative Ions

Whether an electrical current flows through a solid or through a liquid, the principles are similar: an ion flow in one direction and an electron flow in the opposite direction. An understanding of ions will provide a better operational appraisal of cells, neon glow lamps, fluorescent lamps, and electroplating.

Neutral Atom. An atom whose negative orbital electron charge balances the positive proton charge of the nucleus is called a *neutral atom.*

Positive Ion. An atom that loses an electron(s) from its outermost orbital shell becomes a *positive ion* since the positive charge of the protons is now greater than the negative charge of the electrons, as shown in Fig. 1-7.

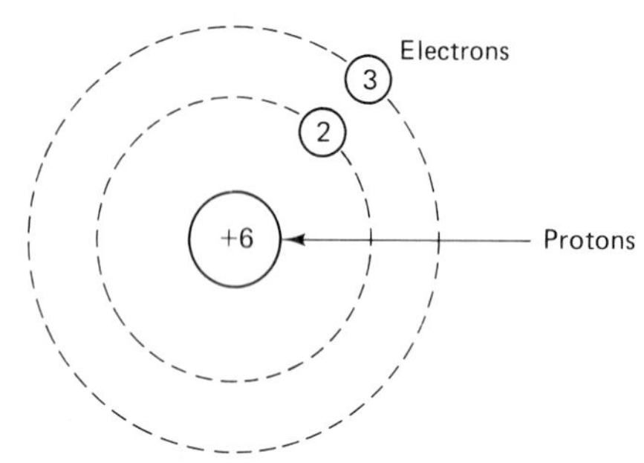

FIGURE 1-7 Positive ion.

Negative Ion. An atom that gains an electron(s) in its outermost orbital shell becomes a *negative ion* since the negative charge of the electron shells is now greater than the positive charge of the protons.

1-4 CURRENT FLOW IN LIQUIDS

Luigi Galvani, a professor of anatomy at the University of Bologna, Italy, discovered by chance that a dead frog's legs would contract when his scalpel touched the nerves at the same time that a spark was being drawn from a nearby electric (friction) machine. Much later, further experimentation led him to attach a short length of copper to the frog's spinal cord and an iron plate to the feet. The spasm occurred when he completed the circuit by touching the copper to the iron.

Volta, a professor at the University of Pavia, repeated Galvani's experiment with a frog and noted that the muscles and nerves of a dead frog would twitch when touched by two dissimilar metals that were separated by a piece of paper soaked in salt water. To increase this effect, Volta created a pile of alternating disks of copper and zinc separated by paper soaked in an acid solution. This first rudimentary cell, known as the *voltaic pile,* is depicted in Fig. 1-8.

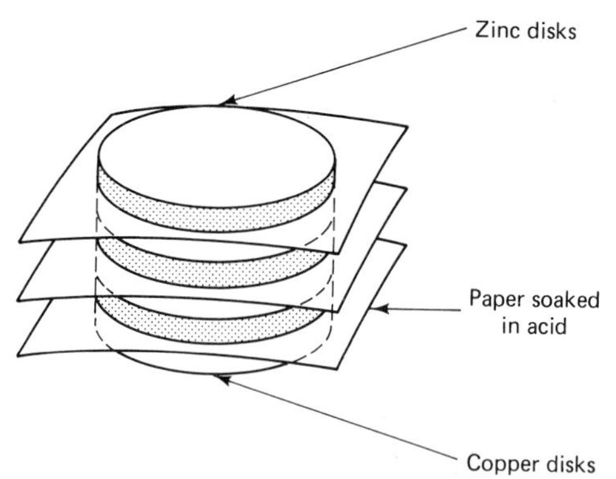

FIGURE 1-8 Voltaic pile.

Volta improved the voltaic pile by submerging two dissimilar metals in a glass cup. Such an arrangement, known as a *voltaic cell,* is shown in Fig. 1-9. The acid solution, which will sustain an ion current, is called an *electrolyte.* Volta was able to obtain a potential difference of charge between the two electrodes and thus an electric charge flow by chemical action. The chemical action of the acid on the two metal electrodes causes separation or dislocation of the molecules. The zinc electrode accumulates an excess of electrons and becomes negatively charged, while the copper electrode loses electrons and becomes positively charged. As a result of this chemical action, a *potential difference* (pd) or *voltage* (named after Volta) is developed between the two electrodes, and an electron current will flow from the zinc electrode through the lamp to the copper electrode. The current flow through the electrolyte is termed *convection current flow.*

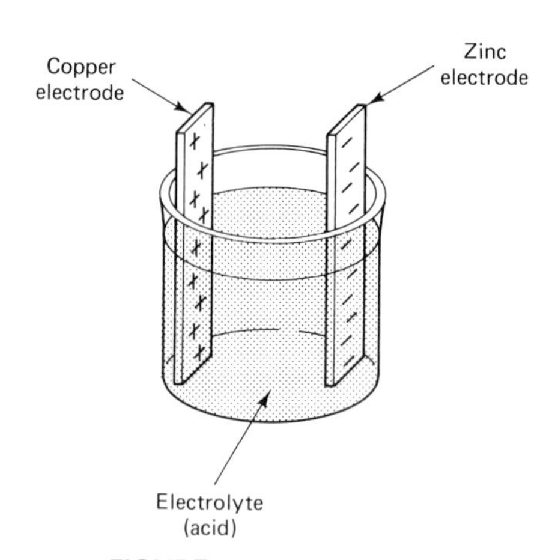

FIGURE 1-9 Voltaic cell.

The schematic symbol for a voltaic cell is shown in Fig. 1-10. By convention the negative electrode is represented by a short vertical line, and the positive electrode is represented by a longer vertical line. The negative or positive charge symbol is also usually printed adjacent to each electrode, indicating its polarity, as shown in Fig. 1-10.

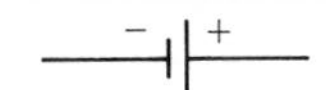

FIGURE 1-10 Schematic symbol for a voltaic cell.

1-5 CURRENT FLOW IN GASES

Inert gases such as neon have eight electrons in their outermost orbit, and therefore they do not interact with any other atom. Inert gases can be made to conduct an ion current provided that the voltage is sufficiently high to force ionization of the low-density gas.

The graphical representation of Fig. 1-11 shows the characteristic of a typical neon lamp. Under initial conditions the gas acts as an insulator with a resistance of about 1000 megohms (MΩ). If the voltage applied across the two electrodes is increased in value, some of the electrons will be stripped from the neon gas atoms and accelerated toward the positive terminal of the source. These electrons will be given sufficient velocity by the higher voltage to dislodge other electrons and form new positive ions. The positive ions are attracted to the negative electrode, and in their travel they help to dislodge other electrons. This process is referred to as *avalanche current*. It can be seen in Fig. 1-11 that the forward voltage must be high enough to reach the ignition point (i.e., where the gas glows). The glow is due to the electrons giving up their bonding energy. For the typical neon lamp the ignition point is about 110 volts (V). Once the gas is ionized the voltage required to maintain ionization is much lower and is typically 85 V. A lower voltage or deionization voltage will stop ionization and extinguish the lamp. The flow of current in a gas is the result of positive ions flowing in one direction and negative electrons in the opposite direction. This flow is termed *convection current*.

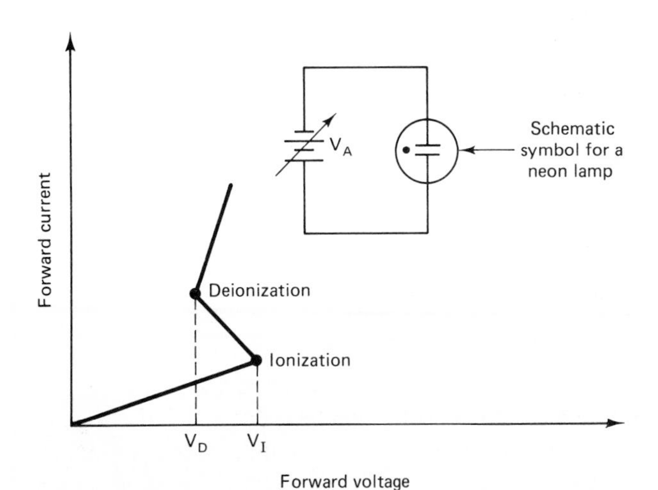

FIGURE 1-11 Voltage and current characteristics of a neon lamp.

1-6 ELECTRON FLOW IN SOLIDS

Having looked at the copper atom, we know that it has a free or valence electron in its outermost shell. In a copper conductor the valence electrons will move randomly within the conductor with no net movement in any particular direction, as shown in Fig. 1-12.

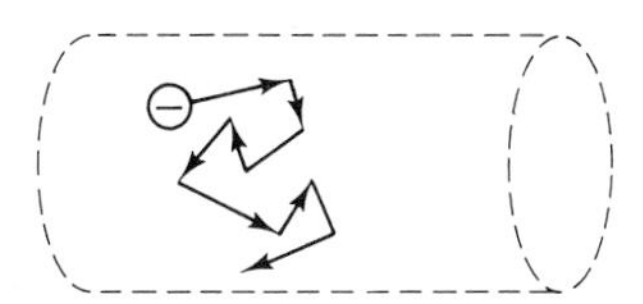

FIGURE 1-12 Random motion of electrons.

With a voltage source connected to the copper conductor the net movement of the electrons will be toward the positive terminal and away from the negative terminal by the laws of electrostatic attraction and repulsion, as shown in Fig. 1-13. The valence electrons form a cloud of free electrons, leaving behind positively charged copper atoms or positive ions. This cloud of electrons gives up some of its electrons as electrons recombine with other atoms and picks up some electrons that have been dislodged by the drifting cloud. The net result is a negative charge or *conduction current* flowing in one direction along the copper atomic structure and positive ions flowing in the opposite direction. The effective charge flows at the speed of light, 300×10^6 metres per second (m/s).

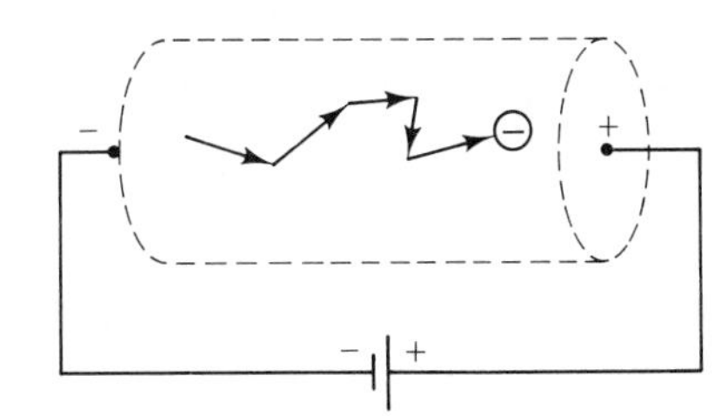

FIGURE 1-13 Electron drift.

PRACTICE QUESTIONS 1-2

Choose the most suitable word, phrase, or value to complete each statement. Questions may have more than one correct answer.

1. The orbits of planetary electrons are called (*rings, shells, valence, levels, energy levels*).
2. For a chemically stable atom the number of electrons in the valence bond is considered to be (*4, 2, 16, 32, 8*).
3. The atomic number for each element designates the number of (*electrons, protons, neutrons, shells, free electrons*) in the atom of that element.

4. The value that gives the number of electrons in an incomplete outermost shell or energy level is called the atomic (*number, valence, weight, energy*).
5. The number of outer orbit electrons is considered to be a (*neutral, positive, negative*) valence, as these electrons are in addition to the stable shells.
6. The number of electrons needed to complete the outermost ring of an atom is considered to be a (*neutral, positive, negative*) valence, as these electrons are required to complete the shell.
7. An atom that has gained an electron in its valence band is termed a/an (*element, molecule, ion, positive ion*).
8. Substances in which electrons spontaneously leave their atoms and become free negative charges are termed (*semiconductors, insulators, conductors*).
9. When a source of potential difference is connected to the solid conductor, the free electrons drift toward the (*negative charge, positive charge*).
10. A motion or flow of free electrons in a common direction through a conductor is called (*hole flow, electron flow, current flow, ion flow*).
11. To obtain a current flow through the conductor, a (*pressure, pd, resistance, friction, voltage*) is applied to force the free electrons to move in a common direction.
12. An electrical current-conducting solution is called (*an acid ion solution, an electrolyte, electolytic, conducting*).
13. Current flow through the electrolyte is in the form of a movement of (*electrons, ions, protons, molecules, atoms*).
14. The positive ions will travel toward the ________ (polarity) electrode where they will be neutralized.
15. A gas will behave as an electrical insulator until the gas becomes (*positive, ion free, ionized*).
16. A gas such as neon will conduct an ion current when a high voltage is applied; this current flow is called a/an (*electron current, ionized current, convection current*).

1-7 CONDUCTORS, INSULATORS, AND SEMICONDUCTORS

Conductors

A material that is a good conductor of electricity contains a large number of electrons free to move about the material, such as copper, discussed in Section 1-6. Most of the metals have this property, but copper is used extensively since it is more economical and easily drawn or formed into wire. Silver is a better conductor than copper since it has about 5% more free electrons per unit volume than copper. However, silver is very expensive. Aluminum is being used to a greater extent as copper becomes more expensive. Aluminum is not quite as good a conductor as copper, since it contains only about 60% as many free electrons per unit volume as copper. The approximate number of free electrons per cubic centimetre and the percentage per cubic centimetre for copper, silver, and aluminum are shown in Table 1-3.

Insulators

An insulator is a very poor conductor. Insulators are characterized as materials whose electrons are tightly bound to their nuclei. There are no electrons that are free to move about in the atomic structure of insulators; therefore, materials such as glass, mica, ceramic, plastic, and oil are good insulators and serve to prevent current flow. In practice it is impossible to construct a perfect insulator, since there are always some impurities in any material, and given sufficient applied voltage, the insulator will conduct electric current. Materials that do not conduct an electric current are also referred to as *dielectrics*.

Semiconductors

Semiconductors are materials that lie halfway between conductors and insulators. Carbon, germanium, and silicon are good examples of semiconductors. Heat or electrical energy will cause the semiconductor to conduct electricity. Note in Table 1-4 that the semiconductors

TABLE 1-3

Conductor	Atomic Number	Electron Valence	Approx. Number of Free Electrons (cm^3)	Approx. % per cm^3
Copper	29	+1	8.5×10^{22}	94
Silver	47	+1	9.0×10^{22}	100
Aluminum	13	+3	5.1×10^{22}	57

TABLE 1-4

Semiconductor	*Atomic Number*	*Electron Valence*
Carbon	6	±4
Silicon	14	±4
Germanium	32	±4

have four valence electrons. This means that each adjoining atom shares four valence electrons to complete its outermost shell or ring with eight electrons. In this way the electrons are tightly bound to each atom, and it takes a greater amount of energy to dislodge one or more electrons from the valence band. Semiconductors are used to construct diodes and transistors.

SUMMARY

An element is indivisible and is made of one type of atom.

A compound represents two or more different types of atoms joined chemically.

An atom has planetary electrons with protons and neutrons forming the nucleus.

An electron is a negative particle of electricity.

A proton is a positive particle of electricity.

A neutron is a neutral particle of electricity.

Like electrical charges repel.

Unlike electrical charges attract.

Shells: Atoms have successive shells of electrons, termed *K, L, M, N, O, P,* and *Q*, at increasing outward distance from the nucleus.

Filled inner shells have a maximum number of electrons equal to $2n^2$.

Inert gases have a full outermost shell with eight electrons.

The atomic number represents the number of electrons or protons in an atom.

The electron valence represents the number of electrons in the outermost ring and is considered to be positive valence.

A negative valence represents the number of electrons that must be added to the outermost ring to complete its chemical complement of eight electrons.

A positive ion is an atom that has been depleted of one or more electrons from its valence shell.

A negative ion is an atom that gains one or more electrons in its valence shell.

Electron flow in solids is called conduction current flow.

Ion flow through an electrolyte is called convection current flow. Inert gases can be made to conduct an ion current provided that the voltage is sufficiently high enough.

A conductor of electricity is a material that has a large number of free electrons free to move about the atomic structure of the material.

An insulator is a very poor conductor and is characterized by materials whose atomic electrons are tightly bound to their nuclei.

Semiconductors are materials that lie halfway between conductors and insulators.

Germanium, silicon, and carbon are semiconductors.

Silver is the best conductor, but copper is the most economical.

Insulators include mica, glass, air, plastic, rubber, ceramic, and oil.

REVIEW QUESTIONS

1-1. Explain why matter is made up of particles of electricity.

1-2. Explain two differences between an electron and a proton.

1-3. State the law of electric charges.

1-4. What factors determine the size of force between charged bodies?

1-5. What number of electrons constitute a filled inner shell?

1-6. What does "atomic number" specify?

1-7. Given a valence of ± 4, explain the reason for the two signs and the number four.

1-8. Define a positive and a negative ion.

1-9. What is a free electron?

1-10. Define conduction current flow.

1-11. Define convection current flow.

1-12. How many free electrons would a perfect insulator have?

1-13. How is it possible for free electrons to be moving randomly within a conductor with no net movement in any particular direction? How does an applied voltage change the situation?

1-14. What constitutes a current flow in liquids?

1-15. What constitutes a current flow in gases?

1-16. Name five conductors and five insulators.

1-17. What two types of energy must be applied to a semiconductor before it conducts a current?

SELF-EXAMINATION

Choose the most suitable word, phrase, or value to complete each statement. Questions may have more than one correct answer.

1-1. An element is made up of indivisible particles of matter called (*electrons, protons, atoms, neutrons, molecules*).

1-2. In nature there are approximately ________ (number) elements.

1-3. A compound is made up of two or more (*electrons, protons, atoms, neutrons, molecules*).

1-4. Alcohol, gasoline, water, and oil are (*compounds, elements*).

1-5. Two or more atoms joined together chemically form (*a compound, an element, two atoms, a molecule, a proton*).

1-6. An atom has negatively charged particles of electricity orbiting about its nucleus called (*protons, neutrons, electrons, mesons, molecules*).

1-7. The nucleus of an atom has particles of electricity called (*protons, neutrons, electrons, mesons, molecules*).

1-8. Two like-charged particles of electricity will (*repel, neutralize, attract*) each other.

1-9. Two unlike-charged particles of electricity will (*repel, neutralize, attract*) each other.

1-10. The electrons orbiting about an atom are "layered" in successive order from the nucleus. These layers are called (*orbits, energy layers, shells, energy levels*).

1-11. For a chemically stable atom the number of electrons found in the outermost ring will be (*2, 4, 6, 8, 16*).

1-12. The atomic number for any element or atom can be determined by the number of (*neutrons, shells, rings, protons, electrons*).

1-13. The atomic number for germanium is ________.

1-14. Germanium will have ________ valence electrons.
(number)

1-15. A sodium atom that loses one electron becomes a/an (*molecule, ion, positive ion, negative ion*).

1-16. Spontaneous ionization in the interatomic spaces constitutes an electrical current flow in a (*gas, liquid, conductor*).

1-17. To have a net movement of conduction current in a conductor, a/an (*charge, energy, potential difference, positive charge*) must be applied.

1-18. A solution that will sustain an ion current is defined as a/an (*electrolytic, acid, electrolyte, voltaic*).

1-19. The current flow through an electrolyte is termed (*electron, convention, convection, conduction*).

1-20. A voltage of (*50 V, 100 V, 110 V, 150 V*) is needed to ionize neon gas.

1-21. Once the neon gas is ionized, current flow through the gas is the result of (*negative ions, electrons, positive ions, neutral ions*).

1-22. A material that has more free electrons per unit volume is a good (*insulator, semiconductor, conductor*).

1-23. The two most common conductors are ________ and ________.

1-24. Materials whose electrons are tightly bound to their nuclei are (*conductors, semiconductors, insulators*).

1-25. Materials that have valence electrons sharing the orbit or shell of an adjoining atom can generally be considered to be (*conductors, semiconductors, insulators*).

ANSWERS TO PRACTICE QUESTIONS

1-1

1. matter
2. elements and compounds
3. atom
4. electrons, protons, and neutrons
5. proton
6. electron
7. away from
8. negative to positive
9. repelled
10. positive
11. negative
12. positive

1-2

1. shells or energy levels
2. eight
3. electrons or protons
4. valence
5. positive
6. negative
7. ion
8. conductors
9. positive charge
10. electron flow
11. pd or voltage
12. electrolyte
13. ions
14. negative
15. ionized
16. convection current

2

Electrical Circuit Characteristics

INTRODUCTION

In Chapter 1 we have seen the need to have a potential difference to produce an electric current. As depicted in Fig. 1–9, the positive potential difference of the dry cell or voltaic pile will attract the cloud of free electrons in the wire. The acceleration of the cloud of electrons toward the positive terminal causes collisions between other copper atoms. At the point of collision the electrons decelerate rapidly and undergo a change in direction. Once again the electrons accelerate toward the positive terminal of the source but soon collide again. This process is repeated and the electrons travel along the conductor, alternately undergoing acceleration due to the applied potential difference or voltage and deceleration due to collisions with the fixed copper atom. There are two factors of consequence due to the electron collisions: heat released by the copper atom whenever a collison takes place, and an "interatomic friction" or opposition to current flow. This opposition to electric current is the property called electrical *resistance* and is a property common to other materials. As a consequence of resistance, energy in the form of heat is released by the copper atom and lost to the air.

In this chapter we give our attention to resistance in a simple circuit as well as factors governing the resistance of a conductor and manufactured resistors.

MAIN TOPICS IN CHAPTER 2

2-1 An Electrical Circuit (Hydraulic Analogy)
2-2 Simple Circuit and Electrical Units
2-3 Factors Governing the Resistance of a Wire
2-4 Specific Resistance of a Material
2-5 Resistors
2-6 Resistance-Value Selection
2-7 Measuring Resistance with an Ohmmeter

OBJECTIVES

After studying Chapter 2, the student should be able to

1. Master the theoretical concept of a simple electrical circuit.
2. Apply electrical units to measure voltage, current, and resistance in an electrical circuit.
3. Define the factors affecting resistance of a conductor.
4. Define the difference between thin film, thick film, carbon composition, and wire-wound resistors.
5. Use the color code for resistors.
6. Use an ohmmeter to measure resistor values.

2-1 AN ELECTRICAL CIRCUIT (HYDRAULIC ANALOGY)

Shown in Fig. 2-1(a) is a closed water circuit or loop made up of a water pump driven by an electric motor to produce a source of water pressure. The pump produces a pressure to force or circulate the water through the furnace coils, where it is heated, then to the room radiators, where the heat is dissipated.

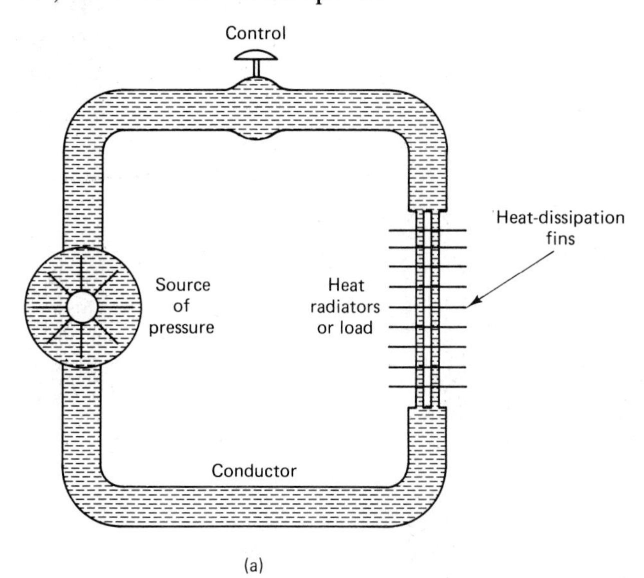

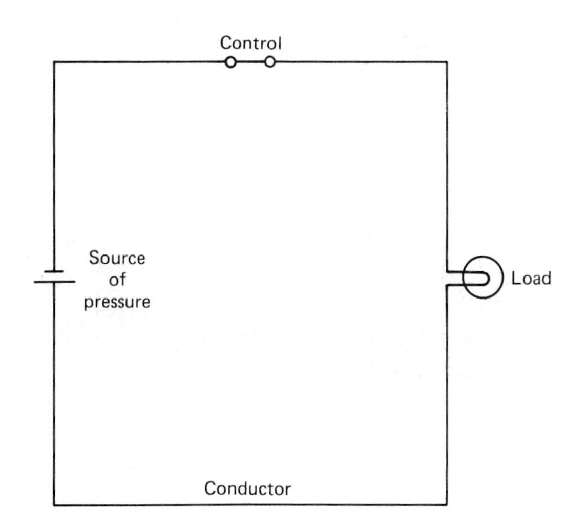

FIGURE 2-1 (a) Water circuit. (b) Electrical circuit.

In the electrical circuit [Fig. 2-1(b)], a dry cell (flashlight battery) provides the source of electrical pressure. The potential difference, pd, produced by the dry cell forces the electrons through the conductor(s) in the circuit, then through the lamp. The current flowing through the lamp heats the tungsten filament with resultant light and heat energy dissipated by the lamp. *The device that uses the effect of electrical current flow is termed a load.*

Control of the flow in the water circuit is accomplished by a simple water valve that closes or opens the path for the water flow. Similarly, the switch in the electrical circuit opens the path to stop the electrical current flow and closes the path to complete the circuit and allow current to flow. *A device such as a switch; a relay, which is a remotely controlled switch; or a thermostat, which is a heat-actuated switch, controls the flow of electricity by opening and closing the circuit.*

Source of Pressure

In both circuits it requires a source of pressure (see Table 2-1) to cause an effect: "water current" and an electrical current flow, respectively. The pressure in the water circuit is measured in pascal, newtons per square metre, or pounds per square inch. The pressure or potential difference in the electrical circuit is measured in volts.

TABLE 2-1

Term	*Water Circuit*	*Electrical Circuit*
Pressure	Newtons/m^2 or lb/in.^{2}	Potential difference or voltage
Quantity of flow	Litres or gallons	Coulomb charge
Rate of current flow	Litres per second Gallons per second	Coulombs per second or ampere
Opposition to current flow	Friction	Resistance

Both water flow and electrical current flow require a conductor: a pipe or conduit for water flow, and copper wire for electrical current flow. The water flow would be measured in litres or gallons. The electrical flow is measured in terms of an electrical charge called a coulomb. One coulomb is a charge of 6.25×10^{18} electrons.

Rate of Current Flow

Since the current flow is moving around the circuit in some given time period, it must be stated as a *rate of flow*. In the water circuit the rate of flow can be stated as litres per second or gallons per second. In the electrical circuit the rate of current flow is coulombs per second, a unit called the ampere. One ampere is equal to 1 coulomb flowing past a point in a circuit in 1 second.

The water and electrical current are forced by the source of pressure to flow through the load and then return to the source of pressure. Note that neither the water nor the electrical current is "used up."

Opposition to Current Flow

In the case of the water circuit the opposition to the flow of water stems from the friction of the water flow against the walls of the conducting pipes. The pipes have a certain diameter as well as length, which adds to the overall friction or opposition. The result of this opposition is a loss in pressure that the water pump or source of pressure must continually make up.

As stated in the introduction, the electron current is accelerated by the applied voltage toward the positive terminal. In accelerating toward the positive terminal, the electrons collide with the fixed copper ions. The collisions decelerate the electrons that undergo a change in direction at the same time. The change of direction is interrupted by the acceleration due to the positive attraction of the voltage. This continual acceleration, collision, with resultant change in direction and acceleration, is the "atomic friction" or opposition to the electric current flow in a conductor.

The "atomic friction" in the tungsten filament of the load or lamp adds greatly to the overall opposition to the electrical current flow and is called resistance. Resistance is a property common to most materials.

Current Flow Convention

A study of dc circuits as well as transistor and rectifier circuits necessitates that a direction must be established for an electrical current. Establishing the direction of the current flow will enable us to explain circuit characteristics such as voltage polarity, to trace a circuit for troubleshooting, to connect polarized capacitors, to understand how semiconductor devices work, and so on.

Historically, before Niels Bohr discovered the atom in 1913, early experimenters assumed that the current in a wire was caused by a movement of positive charges. With Bohr's atomic theory and Rutherford's discovery of the proton, the current flow through a conductor can be established as a negative charge.

Conventional Current Direction. This theory is widely used in modern semiconductor theory and circuit concepts. For dc circuit concepts it is sufficient to know that a conventional current is a positive ion charge. Therefore, there will be a drift of charges from a positive point to a negative point, as shown in Fig. 2-2(a).

Electron Current Direction. The electron current flow convention presupposes a flow of negative charges from a negative point to a positive point, as shown in Fig. 2-2(b).

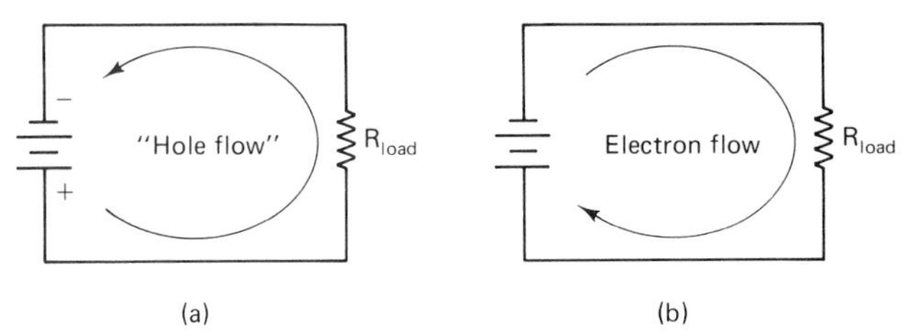

FIGURE 2-2 (a) Conventional current flow. (b) Electron current flow.

> *Note:* The current direction used throughout the text will be electron current flow. Students of electronics must be able to apply both concepts and understand that the results are identical.

2-2 SIMPLE CIRCUIT AND ELECTRICAL UNITS

Emf or Potential Difference

A constant potential difference is developed by a battery because a battery consists a number of cells connected to give a greater voltage than once cell could give. This potential difference is the force that causes the electron current to move through the conductor and through the load. The potential of a cell was given the name *electromotive force* or emf. The unit of measurement of potential difference was named the volt in honor of Alessandro Volta. The symbol for voltage is V, as shown in Table 2-2. The capital letter E, derived from "electromotive force," is also used as a symbol for voltage.

The emf of a cell is considered to be a measure of the amount of potential available to move a quantity of charge. In the familiar English system of units:

> The potential difference between two points is defined as 1 volt when 0.738 foot-pound (ft-lb) of work is necessary to move 1 coulomb of charge from one point to the other.

The unit of work (or energy) in the metre-kilogram-second (mks) system of units is the joule. The joule is equal to 0.738 ft-lb of work, so that

> One volt is the potential difference between two points when 1 joule of work is required to move 1 coulomb of charge from one point to the other.

TABLE 2-2

Term	Electrical Term	Electrical Unit	Symbol
Pressure	emf or potential difference	Volt (V)	V
Quantity of flow	Coulomb charge	6.25×10^{18} electrons (C)	Q
Rate of current flow	Coulombs/second	Ampere (A)	I
Opposition to current flow	Resistance	Ohm (Ω)	R
Ease of current flow	Conductance	Siemens (S)	G

Coulomb Charge

The unit for measuring the quantity of electric charge is the coulomb, named after the French engineer and scientist Charles-Augustin Coulomb (1736–1806). The unit of charge is derived from the charge present on one electron. Scientists have now determined with some accuracy that the charge on one electron is 1.60×10^{-19} coulomb. If the reciprocal of this value is taken, then 1 coulomb of charge is

$$1 \text{ C} = 6.25 \times 10^{18} \text{ electrons}$$

Since charge can be thought of as a quantity of electrons, the symbol Q is used to designate a charge measured in coulombs (see Table 2-2).

Coulombs per Second

With 1 volt applied we now have a charge of electrons flowing through the circuit. If a certain point was taken in the circuit and the electrons passing that point were counted to be 6.25×10^{18} or 1 coulomb every second, this would represent an electric charge rate flow of 1 C per second.

A convenient unit of measurement for coulombs per second was named the ampere in honor of the French physicist André Marie Ampère (1775–1836).

> An ampere is an electric current of 1 coulomb of electric charge passing a point in a circuit in 1 second.

The symbol for the ampere is I, from the French word *intensité* (of electron flow). Based on this definition, electric current can be expressed as

$$I \text{ (amperes)} = \frac{Q \text{ (coulombs)}}{t \text{ (second)}} \qquad (2\text{-}1)$$

EXAMPLE 2-1

An electrical current of 3 C flows past a certain point in a circuit in $\frac{1}{4}$ s. Calculate the current flowing in amperes.

Solution: Applying equation (2-1) gives

$$I \text{ (amperes)} = \frac{Q \text{ (coulombs)}}{t \text{ (second)}}$$

$$I = \frac{3 \text{ C}}{0.25 \text{ s}} = 12 \text{ A}$$

Resistance

Opposition to current flow is the property known as resistance. The resistance of a conductor was deduced by Georg Simon Ohm (1787–1854). Ohm set up an experiment with a number of cells connected so that the voltage of each cell aided the voltage of the others. He then connected a wire to one end of a cell terminal and the free end of the wire alternately to each cell terminal. He discovered that the current would double when the free end of the wire was connected from the first cell to the second and would triple when the free end was connected to the third cell.

From this observation and others using different-size conductors, he determined that the value of the emf divided by the current produced was equal to a constant. The constant proposed by Ohm is a measure of the electric resistance of a circuit. This relationship is known as *Ohm's law*. The unit of resistance is the ohm, and the symbol for resistance is R (see Table 2-2). The abbreviation for ohms is the Greek capital letter omega (Ω).

Since the volt and the ampere have been defined previously, the ohm can now be defined.

> One ohm is the electric resistance when a potential difference of 1 volt produces a current flow of 1 ampere through the resistance. The symbol for resistance is —⋀⋀⋀—

Conductance

Resistance was defined as the opposition offered to current flow. *Conductance* is the opposite of resistance. Conductance (symbol G) is a measure of the ease with which a circuit or resistor allows current flow. The lower the resistance, the higher will be its conductance. The SI unit for conductance is the siemens (S) named after Ernst von Siemens.

Expressed mathematically, conductance is the reciprocal of resistance, or

$$G = \frac{1}{R} \quad \text{siemens} \tag{2-2}$$

Also,

$$G = \frac{I}{V} \quad \text{siemens} \tag{2-2a}$$

Conversely, resistance will be the reciprocal of conductance by transposition of equation (2-2), or

$$R = \frac{1}{G} \quad \text{ohms} \tag{2-3}$$

PRACTICE QUESTIONS 2-1

Choose the most suitable word or phrase to complete each statement.

1. The copper conductors in an electrical circuit conduct (*pressure, flow, resistance, current flow, electrical current flow*).
2. Before an electrical current can flow in a circuit, a source of (*current, resistance, pressure, emf*) must be applied to the circuit.
3. Control of the current flow through a circuit is by means of a switch. When the circuit is broken by the switch, the circuit is said to be a/an (*closed circuit, short circuit, open circuit*).
4. Referring to Question 3, when the circuit is completed by the switch, the circuit is said to be a/an (*closed circuit, short circuit, open circuit*).
5. The rate of current flow in an electrical circuit is termed (*rate of flow, coulombs, coulombs per second, volts per second*).
6. Opposition to electrical current flow is referred to as (*mhos, ohms, amperes, siemens, resistance, volts*).
7. In comparison to the conductors, the load offers (*some, none, greater, infinite*) opposition to current flow.
8. The unit of potential difference is called (*an ampere, an ohm, emf, a volt, a volta*).
9. When 1 joule of work is required to move 1 coulomb of charge from one point to another, a potential difference of (*1 V, 10 V, 6.25 × 10¹⁸ V, 0.5 V*) is set up between the two points.
10. The symbol for the coulomb is ________.
11. The *intensité* of electron flow is measured in the unit called (*volt, ampere, ohm, mho*).
12. The symbol for the ampere is ________.
13. Coulombs per second is defined as (*current flow, charge flow, amperes, volts, ohms*).
14. The unit of resistance is called the ________.
15. The symbol for resistance is ________.
16. The abbreviation for the ohm is the Greek capital letter ________.
17. The measure of ease with which a circuit allows current to flow is termed (*resistance, resistivity, conductance, inductance*).
18. The unit of conductance is called the ________.
19. Conductance is the reciprocal of ________.
20. The siemens is a unit of (*resistance, resistivity, conductance, inductance*).

2-3 FACTORS GOVERNING THE RESISTANCE OF A WIRE

As discussed previously, the opposition to electric current flow in a material is called resistance. Nearly all materials exhibit resistance to current flow under ordinary conditions. A few materials have almost zero resistance when cooled to absolute zero. The opposition to the electrical current flow through a conductor is called the *ohmic resistance,* the term "ohmic" coming from the definition of an ohm given previously. The ohmic resistance is dependent on four factors:

1. Length of the conductor
2. Cross-sectional area of the conductor
3. Type of material
4. Temperature

Resistance is most undesirable in the case of electrical wiring circuits. Resistors used for electronic circuits depend on these factors as well as heater elements, toaster elements, lamp filaments, and other devices.

Length of the Conductor

The resistance of a test conductor can be determined by connecting a specified length and cross-sectional area conductor to a fixed voltage source as shown in Fig. 2-3(a). With the fixed voltage, a resultant current (I) will flow. If the length of the same conductor is doubled, the current would be reduced by one-half, as shown in Fig. 2-3(b). Since the applied voltage is constant, the resistance of the conductor must be doubled. Repeating the measurements for any length of conductor, it would be found that the resistance of the conductor is directly proportional to its length. This relationship exists for all electrical-current-conducting materials. In summary, the resistance of a conductor is directly proportional to its length, or

$$R \propto L$$

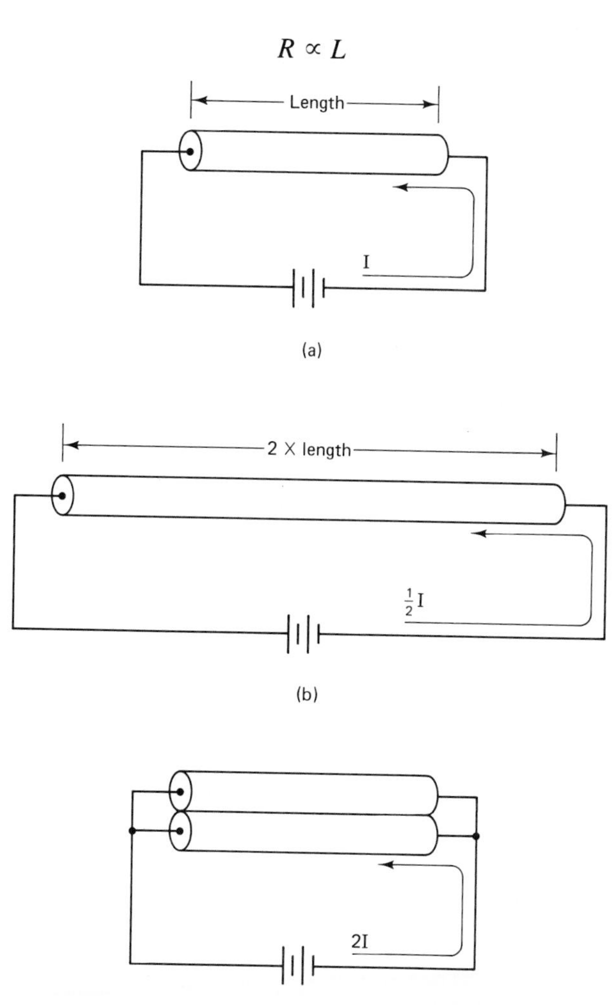

FIGURE 2-3 Factors governing resistance of a conductor.

Cross-Sectional Area

The effect of cross-sectional area on the resistance of a conductor can be determined similarly. The length of the conductor in Fig. 2-3(c) is the same original length L, but the cross-sectional area is doubled (two conductors connected together), with the result that the current is doubled. Since the applied voltage is the same, the resistance of the conductor must have reduced by one-half. If these measurements are repeated as the cross-sectional area is increased, it will be found that the resistance of the conductor is inversely proportional to its cross-sectional area, or

$$R \propto \frac{1}{A}$$

The experimental results can be summarized by stating that the resistance of any material is directly proportional to its length and inversely proportional to its cross-sectional area. In equation format,

$$R = \frac{L}{A} \quad \text{ohms} \tag{2-4}$$

Type of Material

As discussed in the introduction to Chapter 2, the resistance of copper conductor is due to the collisions between drifting electrons and fixed atoms. The copper conductor gains heat energy as a result of the collisions. Other materials behave similarly and heat is released to a greater or less extent depending on the material. Silver and copper, the best conductors, produce the least heating effect with a given cross-sectional area. High-resistance materials such as tungsten and nickel-iron alloys are used for resistive heating elements such as lamp filaments, electric heater, and stove elements because they produce a lot of heat. It is apparent from this that all materials have different resistive characteristics. This resistive characteristic is discussed in Section 2-4.

Temperature Coefficient of Resistance

This is a general term that refers to a number usually used as a multiplier to correct a value such as resistance for a change in temperature. The temperature coefficient for several materials is shown in Table 2-3.

Positive Temperature Coefficient. Looking at Table 2-3 shows that the resistance of the materials listed increases by the factors shown for each 1-degree rise in temperature, with the exception of carbon. When an increase in temperature causes an increase in resis-

TABLE 2-3
Temperature coefficient of materials

Material	*Temperature Coefficient, αR (Ω/°C) (Reference Temp. 20°C)*
Aluminum	0.0039
Brass	0.0020
Carbon	−0.0005
Copper	0.00393
Gold	0.0034
Nichrome	0.0004
Silver	0.0038
Tungsten	0.0045
Constantan	0.000008

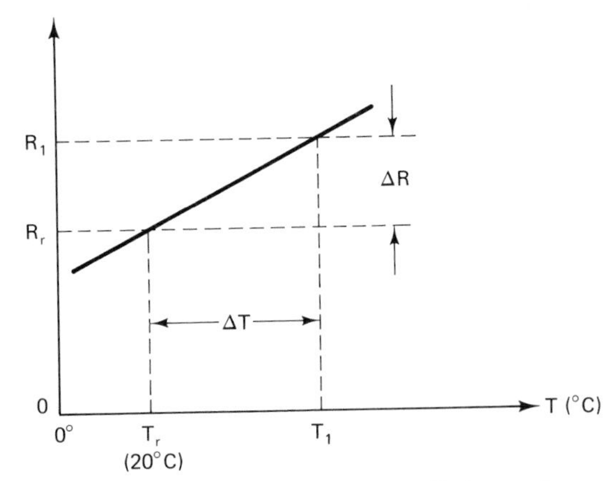

FIGURE 2-4 Temperature coefficient of resistance.

tance (i.e., resistance and temperature change in the same direction), the material is said to have a positive temperature coefficient.

Negative Temperature Coefficient. When an increase in temperature causes a *decrease* in resistance (i.e., resistance and temperature change in opposite directions), the material is said to have a negative temperature coefficient. This is shown in Table 2-3, where the resistance of carbon decreases by 5×10^{-4} Ω/°C.

Zero Temperature Coefficient. Materials such as constantan and manganin exhibit an extremely small or no change in resistance with a change in temperature. As a practical example, it is necessary to correct capacitance for the temperature coefficient in turned FM and TV circuits. This is done by shunting a capacitor having a negative temperature coefficient across the tuned circuit to minimize or bring the circuit down to the zero temperature coefficient of capactiance. This prevents the channel from drifting when the receiver heats up.

The *temperature coefficient of resistance* is given the symbol α (Greek lowercase letter alpha) and it may be defined as the percentage or unit change in resistance per 1-°C change in temperature. Note that the reference temperature is 20°C. The temperature coefficient of resistance can be illustrated graphically as shown in Fig. 2-4. Let

R_r = resistance at a reference temperature T_r
R_1 = resistance at a higher temperature T_1
ΔT = change in temperature [*Note*: The Greek capital letter delta (Δ) means a change]
ΔR = change in resistance
αR = temperature coefficient of resistance at the reference temperature

In equation form, the temperature coefficient of resistance at some reference temperature T_r may be stated as

$$\alpha R = \frac{\Delta R/\Delta T}{R_r} \quad \text{ohms} \tag{2-5}$$

EXAMPLE 2-2

Given that the temperature of a length of No. 24 copper wire rises to a value of 25°C, where the resistance of the wire at 20°C was 13.09 Ω, and its new resistance is 19.63 Ω, find the temperature coefficient of copper.

Solution: Find ΔR, the change in resistance with the increase in temperature.

$$\Delta R = 19.63 - 13.09 = 6.54\ \Omega$$

Find ΔT:

$$\Delta T = 150°\text{C} - 20°\text{C} = 130°\text{C}$$

Therefore,

$$\alpha R = \frac{6.54\ \Omega/130°\text{C}}{13.09\ \Omega} = 0.0038\ \Omega/°\text{C}$$

A commonly encountered problem is to determine the resistance of a conductor at some temperature when ΔR and αR are specified for some reference temperature T_r. This type of problem can be solved by rewriting equation (2-5) and solving for R_1, as shown in equations (2-6) and (2-7).

$$\frac{\alpha R = R_1 - R_r/T_1 - T_r}{R_r} \tag{2-6}$$

Then

$$R_1 = R_r[1 + \alpha R(T_1 - T_r)] \qquad (2\text{-}7)$$

EXAMPLE 2-3

A tungsten wire has a 14-Ω resistance at 20°C. Calculate its resistance at 400°C.

Solution: Using equation (2-7) and the temperature coefficient from Table 2-3, we have

$$\begin{aligned} R_1 &= 14[1 + 0.0045(400°C - 20°C)] \\ &= 14[1 + (45 \times 10^{-4} \times 380°C)] \\ &= 14 \times 2.71 = 37.94\ \Omega \end{aligned}$$

2-4 SPECIFIC RESISTANCE OF A MATERIAL

Good conductors contain a large number of free electrons that become current carriers. These materials have low values of resistivity, as shown in Table 2-4. Good insulators have very few free electrons, so that materials such as glass, mica, and Bakelite have high resistivity. Semiconductors such as silicon and germanium have resistivities that are midway between those of conductors and insulators. Resistivity is measured under two conditions: the unit volume of a material and its temperature. The materials shown in Table 2-4 have resistivity or specific resistance in units of cubic centimetres. The symbol used for specific resistance is ρ (Greek lowercase letter rho) and the unit for ρ is the ohm-centimetre (Ω-cm).

TABLE 2-4
Resistivity of materials at 20°C

Material	*Resistivity, ρ (Ω-cm)*
Silver	1.5×10^{-6}
Copper	1.7×10^{-6}
Gold	2.4×10^{-6}
Aluminum	2.8×10^{-6}
Nichrome	1.0×10^{-4}
Carbon	3.5×10^{-3}
Germanium	4.5×10^{1}
Silicon	2.3×10^{5}
Glass	1.0×10^{14}
Mica	1.0×10^{16}
Bakelite	1.0×10^{12}

From Section 2-3 we learned that the resistance of a conductor was dependent on its length, cross-sectional area, and temperature. Formula (2-4) can now be finalized to enable calculation of the resistance of any length and cross-sectional area of a conductor. In equation form,

$$R = \rho\frac{L}{A} \quad \text{ohms} \qquad (2\text{-}8)$$

The units of resistivity given in Table 2-4 are Ω-cm. To keep the units consistent, the dimensions of the conductor must be in centimetres.

$$R(\Omega) = \rho(\Omega\text{-cm})\frac{L\ (\text{cm})}{A\ (\text{cm}^2)} \qquad (2\text{-}9)$$

EXAMPLE 2-4

Calculate the net resistance of 100 m of copper wire having a diameter of 5 mm.

Solution: The cross-sectional area must be in square centimetres; therefore,

$$\begin{aligned} A(\text{cm}^2) = \pi r^2 = \frac{\pi d^2}{4} &= \frac{\pi(5 \times 10^{-1})^2}{4} \\ &= 19.63 \times 10^{-2}\,\text{cm}^2 \end{aligned}$$

Using equation (2-9) and Table 2-4 yields

$$R = \rho\frac{L}{A} = 1.7 \times 10^{-6}\ (\Omega\text{-cm}) \times \frac{100 \times 10^2\ \text{cm}}{19.63 \times 10^{-2}\ \text{cm}^2} = 8.66 \times 10^{-2}\ \Omega$$

EXAMPLE 2-5

Find the resistance per metre of aluminum wire with a diameter of 2.5 mm.

Solution: The cross-sectional area must be in square centimetres therefore,

$$\begin{aligned} A(\text{cm}^2) = \frac{\pi d^2}{4} &= \frac{\pi(2.5 \times 10^{-1})^2}{4} \\ &= 4.9 \times 10^{-2}\ \text{cm}^2 \end{aligned}$$

Using equation (2-9) and Table 2-4 gives us

$$R = \rho\frac{L}{A} = 2.8 \times 10^{-6}(\Omega\text{-cm}) \times \frac{1 \times 100\,\text{cm}}{4.9 \times 10^{-2}\ \text{cm}^2} = 5.7 \times 10^{-3}\,\Omega/\text{m}$$

PRACTICE QUESTIONS 2-2

Choose the most suitable word or phrase to complete each statement.

1. The ohmic resistance of a conductor is directly

proportional to the (*diameter, radius, length, material*).

2. The ohmic resistance of a conductor is inversely proportional to the (*diameter, radius, length, material*).
3. With certain materials such as copper, an increase in temperature will (*increase, decrease*) the ohmic resistance.
4. A number used as a multiplier to correct a value such as resistance for a change in temperature is a __________.
5. When the ohmic resistance of a substance increases with an increase in temperature, the material has a (*positive, negative*) temperature coefficient.
6. The resistance of a length of copper wire is 5.9 Ω at 20°C. Determine the new resistance of the wire at 120°C.
7. The resistance of a length of nichrome wire is 1000 Ω at 20°C. After being submerged in a liquid for some time, the resistance falls to 880 Ω. Calculate the temperature of the liquid.
8. The resistance of a certain length of wire was measured and found to be 330 Ω at 20°C and 448.8 Ω at 100°C. Calculate the temperature coefficient and identify the material.
9. Calculate the resistance of a 200-m length of aluminum wire that has a diameter of 2 mm.
10. What length of nichrome wire having a diameter of $\frac{1}{2}$ mm must be used to obtain a resistance of 33 Ω?

2-5 RESISTORS

Having looked at the resistivity of various materials, we now turn our attention to resistance that is made artificially. Artificial resistance is in the form of a tubular resistor or resistors of other shapes. Resistors are used extensively in electronic circuits as voltage dividers and current limiters.

Carbon-Composition Resistors

The standard carbon-composition fixed resistor is considered to be the backbone of the electronic industry. A cross-sectional view of a carbon-composition resistor is shown in Fig. 2-5. A mixture of carbon and binders is

FIGURE 2-5 Hot-molded carbon-composition fixed resistor. (Courtesy of Allen-Bradley Canada Ltd.)

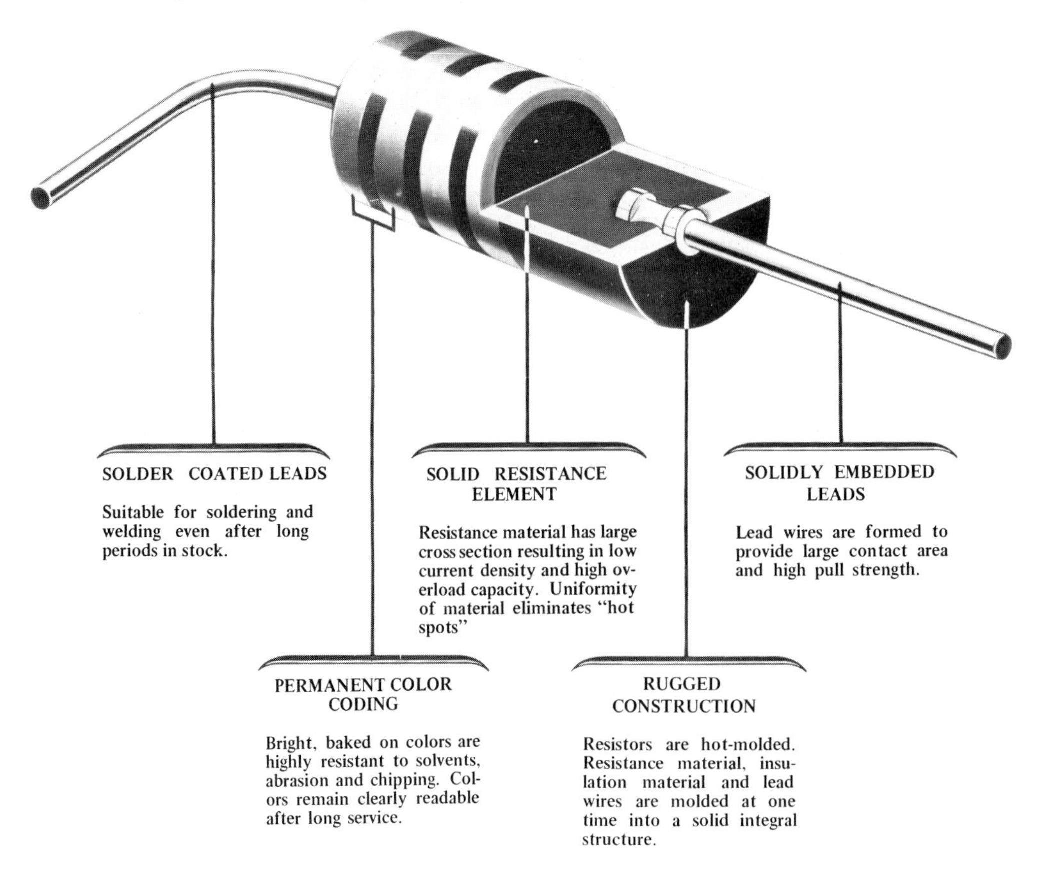

hot-molded by fully automated production machinery that uses high temperature and pressure to mold the insulation, resistance material, and heated lead wires together into a single dense structure.

Helpful hints for using composition resistors

1. Resistance change due to load life* can be minimized to a few percent in many thousands of hours by limiting the maximum operating surface temperature of the resistor under load to 100°C, typically achieved by operating at 70°C ambient air with 50% derating.

2. Resistance changes due to the heat of soldering is positive and may be permanent if the resistor has moisture present in its body. It can be greatly minimized if resistors are dry at the time of soldering.

Thin-Film Resistors

Thin-film resistors are ideally suited wherever precision, stability, reliability, low noise, or good high-frequency characteristics are prime considerations. Thin-film fixed resistors are defined by the electronics industry as those resistors whose resistance element is in the form of a film having a thickness in the order of 10^{-6} in. Because of the highly sophisticated manufacturing techniques used for these types of resistors, their electrical parameters can be closely controlled. This makes for high-precision, close-tolerance components.

Thick-Film Resistors

Thick-film resistors are defined by the electronics industry as those resistors whose resistance element is in the form of a film having a thickness greater than 10^{-6} in., or thick enough to be seen. Thick-film resistors are manufactured as precision and semiprecision low-power devices with tolerances of 1%, 2%, and 5%. The three major types of thick-film resistors are metal-oxide, metal-glaze, and bulk-property film resistors, which use only metal to conduct a current.

Manufacture of Film Resistors

There are seven basic steps in film-resistor manufacturing regardless of type: application of film, termination, rough sorting, spiraling (adjusting to final resistance range), application of protective media, marking, and testing. Most film resistors use a cylindrical ceramic substrate.

*The power or wattage rating of a resistor is determined by an industry standard test known as *load life*.

Carbon Film. Carbon film is "deposited" by cracking carbon-bearing gases at extremely high temperatures. First, ceramic substrates are heated to approximately 1100°C (2000°F) in an inert atmosphere or in an approximate vacuum. Then carbon-bearing gases are introduced near the hot substrates and the gas is "cracked." The carbon is deposited on the ceramics. The deposit time and the amount of carbon available in the gases determines the film thickness and the final resistance.

Metal Film. Metal film is obtained by condensing metal on a ceramic substance in a very high vacuum. The substrates are mounted in a vacuum chamber, where the metal, initially in wire or powdered form, is heated to evaporation temperature. The substrates are rotated for uniform exposure to the metal-vapor stream boiling off the heated metal source. The type of metal evaporated, usually nickel, chromium, and/or aluminum, and the condensed film thickness determine the blank resistance of the basic device.

The blank resistance has a low value of resistance. To increase the resistance a spiral is cut through the film down the length of the blank, which makes the resistive path long and narrow. The result is a resistance increase of up to several thousand times. The spiraling equipment has associated bridge circuitry that monitors the resistance value as the cut is being made. When the resistance reaches the desired value, the cut is stopped and the resistor is ejected from the machine. The resistor is now electrically complete.

The resistor is encapsulated with a hermetic seal enclosure and an organic enclosure. The hermetic seal enclosure is applied by painting, molding, or dipping. Typical materials used are silicones, epoxy, alkyds, and phenolics. The organically housed variety of film resistors are the most commonly used types. Figure 2-6 shows a cutaway view of a thick-film metal-glaze resistor.

FIGURE 2-6 Cutaway view of a thick-film resistor.

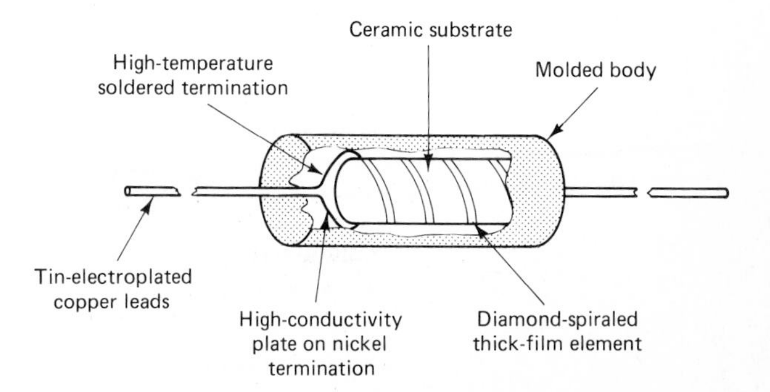

Carbon Film versus Metal Film. Table 2-5 summarizes three basic characteristics ranges of both deposited-carbon and metal-film resistors available from manufacturers.

Thin- and Thick-Film Chip Resistors

The continuing shift to microminiaturization of electronic circuitry and equipment has led to the thin- and thick-film chip resistors. An example of a thick-film chip resistor is shown in Fig. 2-7(a), with the comparative physical size (1.6 × 3.2 × 0.6 mm) shown in Fig. 2-7(b).

The conventional method of providing precision attentuation requirements in digital or analog instrumentation has been with voltage dividers assembled from discrete wire-wound or film resistors. Thin- and thick-film technology has made it possible to manufacture precision resistor networks either customized or made available in standard circuit configuration and resistance values such as shown in Fig. 2-8.

Ideally suited for use in any electronic equipment requiring close tolerances and low temperature coefficients, these thin- and thick-film networks are packaged in 8-, 14-, and 16-lead molded dual-in-line packages (DIPs) or single-in-line packages (SIPs), as shown in Fig. 2-8. Applications include ladder networks, digital multimeters, current summing applications, precision attenuators, A/D and DA converters, communication equipment, precision voltage dividers, telemetry equipment, and measurement bridges.

Wire-Wound Resistors

Although metal-film, especially bulk-property thick-film resistors, have almost entirely replaced the wire-wound resistor, many improvements have been made to the wire-wound resistor to enhance its applications.

Wire-wound resistors have always had the disadvantage that they are electrically similar to a coil and have some of the coil's characteristics, such as inductive and capacitive reactance with resultant phase shift. These factors limit the high-frequency use of wire-wound resistors. With new winding techniques and new metal alloys used as the resistance element, precision wire-wound resistors have extremely low stray inductance and capacitance.

Precision Wire-Wound Resistors. A precision wire-wound resistor is an accurately determined length of metal-alloy wire wrapped around a core and sealed to protect it from mechanical or environmental hazards. The resistance ranges from 0.001 Ω to 60 MΩ, the tolerance reaching 0.001%, and the temperature coefficient approaching 0.5 ppm/°C.

Figure 2-9 shows a cutaway view of a miniature precision wire-wound resistor. These resistors come in a variety of packages, ranging from miniature units for printed-circuit use to larger encased units that are used as accurate resistance standards. The ability of these devices to perform over the megahertz pulse frequency range while maintaining excellent tolerance over an extended temperature range make them useful for high-speed electronic devices such as computers.

2-6 RESISTANCE-VALUE SELECTION

Resistors are widely available in standard Electronics Industry Association (EIA) values. These are shown in Table 2-6. Note that the available values depend on the tolerance. A maximum of two significant figures is given; the actual resistor value is found by multiplying

TABLE 2-5

	Resistor Range	Temp. Coefficient (ppm/°C)	Resistor Tolerance (%)
Deposited carbon	1 Ω to 200 MΩ	−150 to −500	0.5 to 10
Metal film	1 Ω to 40 MΩ	3 to 150	0.01 to 1

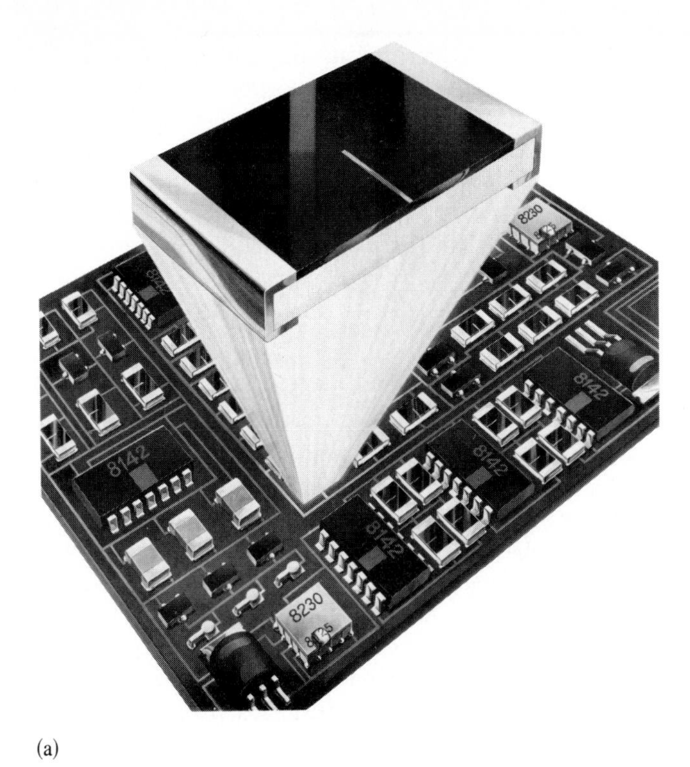

(a)

(b)

FIGURE 2-7 (a) Exploded view of a thick-film chip resistor. (Courtesy of Allen-Bradley Canada Ltd.) (b) Comparative size of thick-film chip resistors. (Courtesy of Allen-Bradley Canada Ltd.)

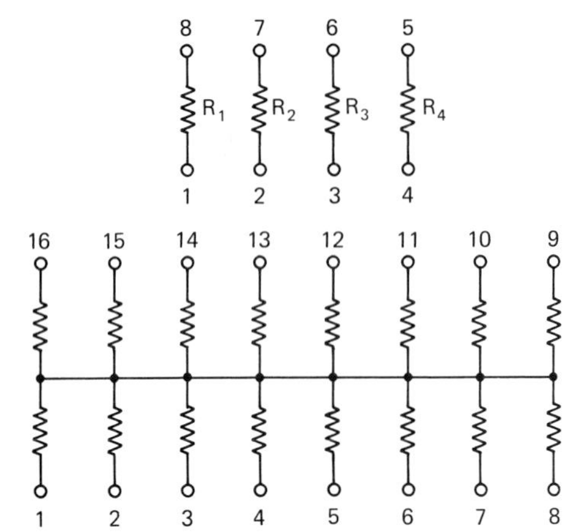

FIGURE 2-8 Schematic diagram of dual-in-line package, thin-film resistor network.

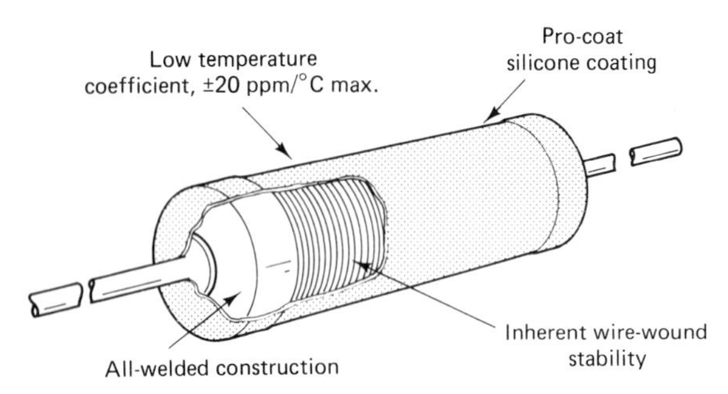

FIGURE 2-9 Precision wire-wound resistor.

the significant figures by any multiplier between 10^{-2} and 10^{6}. *When constructing circuits, you normally use the closest EIA standard-value resistor.*

Resistor Color Code

With the exception of precision resistors, where the resistance value is stamped on the body, most resistors have color bands to identify their resistance values. The color coding of resistance values by means of bands is shown in Table 2-7. For commercial applications resistors normally have four color bands. For military and aerospace applications the resistors have five bands. In each case the tolerance band is readily located since it is either silver or gold, and the first significant figure is the outermost band to the left (see Table 2-7) of the tolerance band. Resistors with four bands have their bands grouped toward one end or lead of the resistor. When located this way, the left outermost band is the first significant figure, the next band is the second significant figure, and the third band is the multiplier. The fourth band is the tolerance and is either silver or gold. The absence of a fourth band implies a 20% tolerance.

TABLE 2-6

Standard EIA resistor values[a] (ohms)

20%	*10%*	*5%*	*20%*	*10%*	*5%*	*20%*	*10%*	*5%*
10	10	10	100	100	100	1.0 k	1.0 k	1.0 k
		11			110			1.1 k
	12	13		120	130		1.2 k	1.3 k
15	15	15	150	150	150	1.5 k	1.5 k	1.5 k
		16			160			1.6 k
	18	18		180	180		1.8 k	1.8 k
		20			200			2.0 k
22	22	22	220	220	220	2.2 k	2.2 k	2.2 k
		24			240			2.4 k
	27	30		270	300		2.7 k	3.0 k
33	33	33	330	330	330	3.3 k	3.3 k	3.3 k
		36			360			3.6 k
	39	39		390	390		3.9 k	3.9 k
		43			430			4.3 k
47	47	47	470	470	470	4.7 k	4.7 k	4.7 k
		51			510			5.1 k
	56	56		560	560		5.6 k	5.6 k
		62			620			6.2 k
68	68	68	680	680	680	6.8 k	6.8 k	6.8 k
		75			750			7.5 k
	82	82		820	820		8.2 k	8.2 k
		91			910			9.1 k

[a]For higher values, add zeros to the basic numbers up to 22 MΩ, the maximum value.

TABLE 2-7 EIA resistor color code

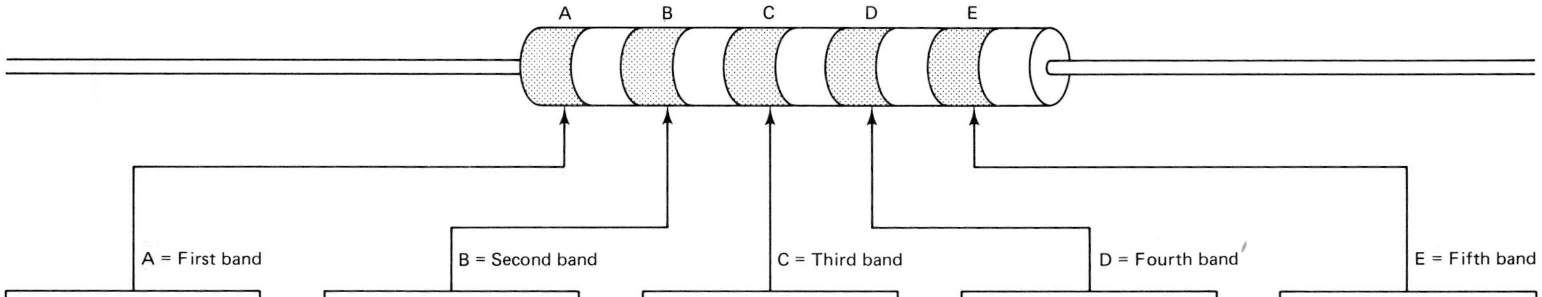

A B C D E

A = First band

First significant figure	
Color	Digit
Black	0
Brown	1
Red	2
Orange	3
Yellow	4
Green	5
Blue	6
Violet	7
Gray	8
White	9

B = Second band

Second significant figure	
Color	Digit
Black	0
Brown	1
Red	2
Orange	3
Yellow	4
Green	5
Blue	6
Violet	7
Gray	8
White	9

C = Third band

Multiplier	
Color	Multiplier
Black	1
Brown	10
Red	100
Orange	1,000
Yellow	10,000
Green	100,000
Blue	1,000,000
Silver	0.01
Gold	0.1

D = Fourth band

Tolerance	
Color	Tolerance
Silver	±10%
Gold	±5%

E = Fifth band

Failure rate level (%/1000 h)	
Color	Level
Brown	M = 1.0%
Red	P = 0.1%
Orange	R = 0.01%
Yellow	S = 0.001%

EXAMPLE 2-6

Given a resistor with the following color code, determine its resistance value and percentage tolerance.

First band: brown
Second band: red
Third band: orange
Fourth band: gold

Solution: From Table 2-7

First band: = 1
Second band: = 2
Third band: the multiplier 10^3 or 1000
Fourth band: the tolerance ±5%

Therefore, the resistor value is 12,000 Ω ±5% tolerance.

EXAMPLE 2-7

Given a resistor with the following color code, determine its resistance value and percentage tolerance.

First band: blue
Second band: gray
Third band: gold
Fourth band: gold

Solution: From Table 2-7

First band: = 6
Second band: = 8
Third band: a multiplier of 0.1
Fourth band: a tolerance of ±5%

Therefore, the resistor value is 68 × 0.1 = 6.8 Ω ±5% tolerance.

Implication of Tolerance

The tolerance of a resistor indicates minimum and maximum values for that specific resistor. A tolerance of 5% means that the color-code value of the resistor can be 5% higher or 5% lower. Thus a 5% tolerance means within ±5%.

EXAMPLE 2-8

A 1200-Ω resistor having a tolerance of 5% can be 0.05 × 1200 = 60 Ω higher or 60 Ω lower than the color-code value of 1200 Ω. This means that the 1200-Ω resistor has an actual resistance value of 1200 Ω or somewhere between 1140 and 1260 Ω. This is a matter of economics, since it is easier to manufacture a large quantity of resistors having a resistance value close to 1200 Ω than to manufacture resistors having a resistance of exactly 1200 Ω.

Failure Rate Level

Industrial specifications in general are modeled after various military specifications. Modifications include such things as different physical sizes, closer tolerance and temperature coefficients, load-life tests, different humidity test cycles, and reliability definitions.

Reliability, as defined by established tests, is expressed as "mean time between failures" or as "maximum failure rate in percent per 1000 hours" and is largely confined to military and aerospace applications. The fifth band is then applicable in these specialized circuits.

2-7 MEASURING RESISTANCE WITH AN OHMMETER

Resistance is measured by means of an ohmmeter. An ohmmeter has its own battery or internal power supply for the purpose of passing a current through the resistance to be measured.

This means that an ohmmeter must never be connected to a circuit in which a power supply is switched ON.

The use of an ohmmeter to measure the resistance of a component is illustrated in Fig. 2-10. The ohmmeter leads are connected to each side of the resistor. The polarity of the connection is not important. In the case of a component that forms part of a circuit, the circuit supply must be OFF, and one terminal of the component must be disconnected from the circuit before attempting to measure its resistance, as shown in Fig. 2-11.

Ohmmeters have a range switch that identifies the ohmmeter function of a multimeter. Typical ranges or multipliers are $R \times 1$, $R \times 10$, $R \times 100$, $R \times 10\text{k}$, and

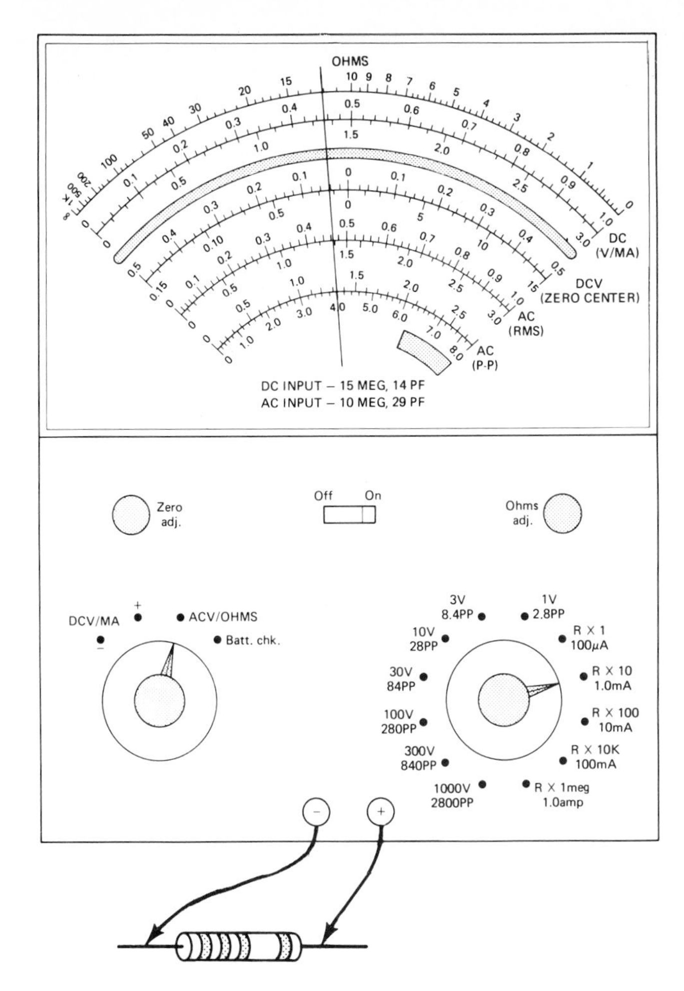

FIGURE 2-10 Using the ohmmeter function of a VOM.

$R \times 1$M. This means that a reading on the ohms scale of the meter is multiplied by the range factor, for example, by 10 if the switch is set to $R \times 10$, or by 1MΩ if the switch is set to $R \times 1$M. With multimeters (i.e., meters that also measure voltage and current), the function switch must be set for ohms (see Fig. 2-12).

The meter pointer moves across a calibrated scale. This scale is usually the uppermost scale of a multimeter and is identified by its label, simply OHMS. Note in Fig. 2-12 that the ohmmeter scale 0 Ω (0 position) is on the right-hand side of the scale. The left-hand side of the scale is marked ∞ or infinity, an extremely high value which would indicate an open circuit. Note that the length of the scale representing a 10-Ω change from 0 to 10 Ω is greater than the scale length from 10 to 20 Ω, which also represents a 10-Ω change in resistance. Moving from right to left, the ohmmeter scale becomes more and more crowded, which makes the scale nonlinear.

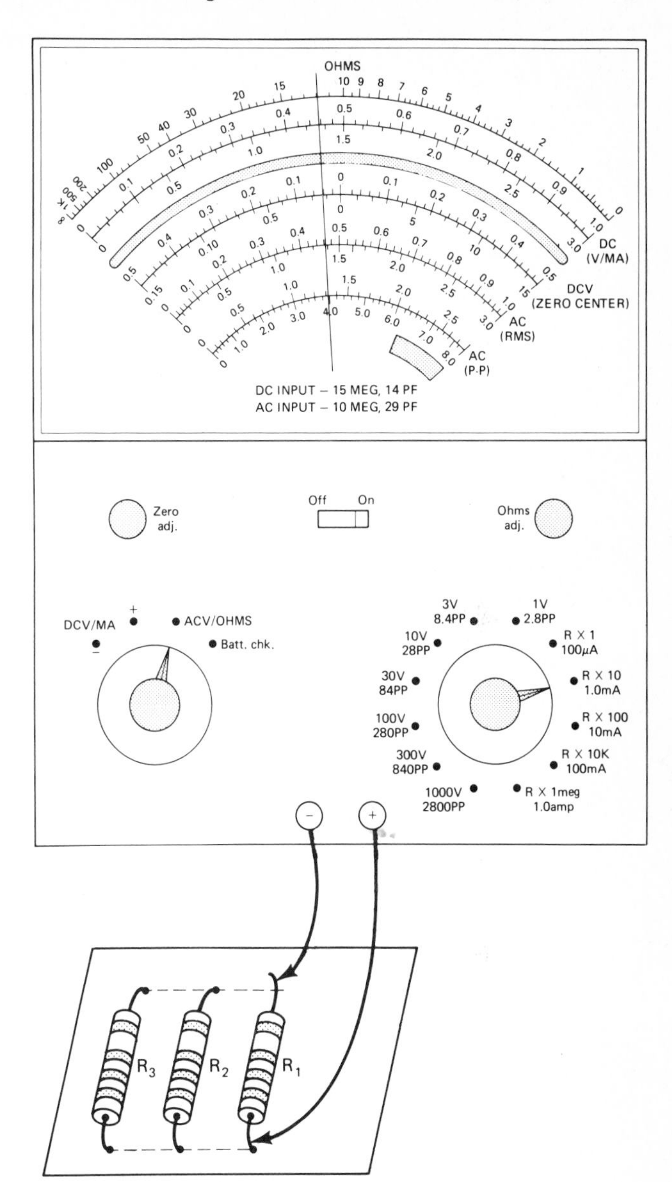

FIGURE 2-11 Disconnect one terminal lead to measure resistance of R_1.

Readying the Ohms Function to Measure Resistance

With the function switch set to OHMS or R, the range switch set to $R \times 1$, and the test leads plugged into the meter jacks, the test-lead tips are shorted together. The meter pointer will swing toward zero ohms. Since the internal battery voltage will govern the meter deflection to zero, a drop in this voltage due to use and aging will cause the pointer to stop short of zero ohms. To compensate for this, the ZERO OHMS adjust control should

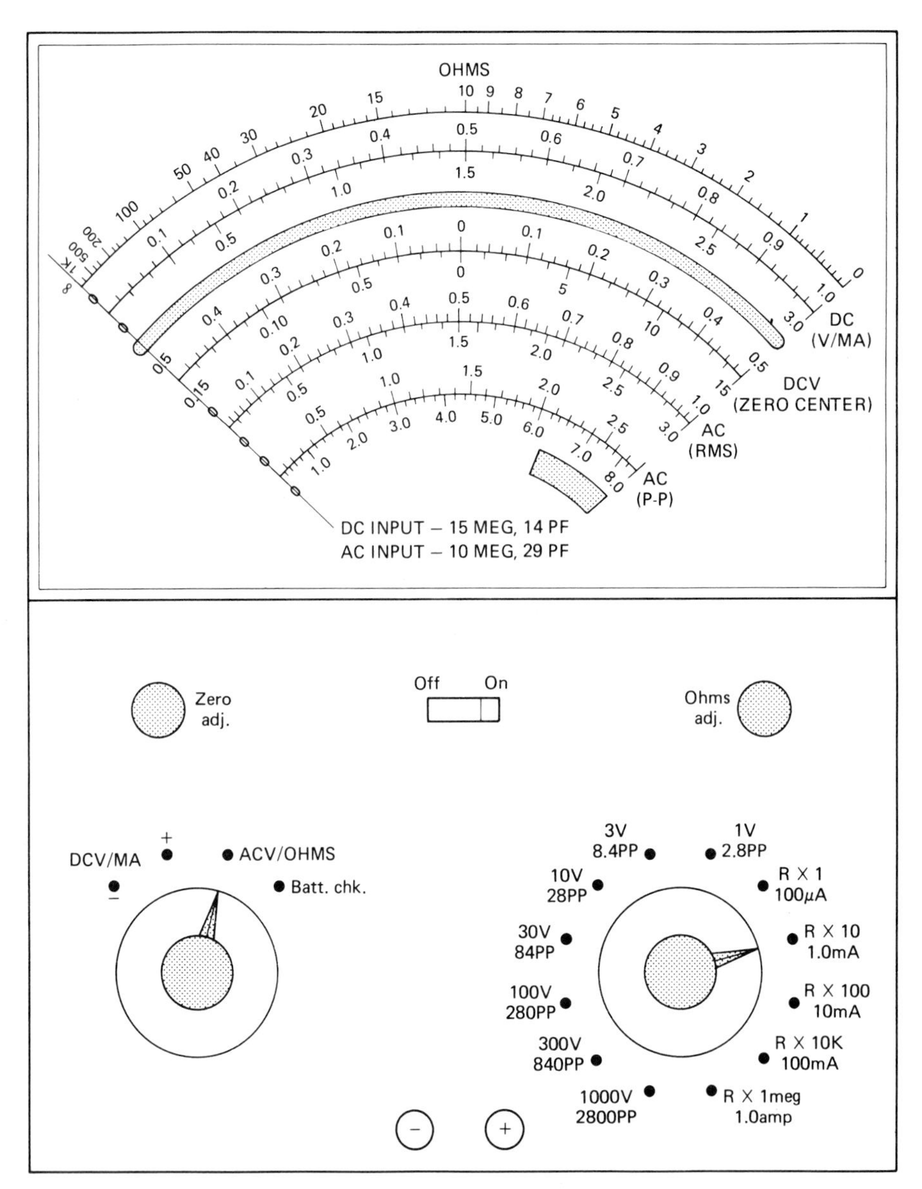

FIGURE 2-12

> Before measuring a resistance, the ohmmeter test leads must be shorted together and the ZERO OHMS control adjusted to zero ohms (0 Ω) for each range of the ohmmeter being used.

be adjusted until the pointer indicates exactly 0 Ω on the scale.

Since the ohms scale is very crowded on the left-hand end and expanded on the right-hand end, resistance readings are most accurate when the pointer is within the center scale. To this end the ohms range switch is used to enable a reading to be taken around midscale.

Continuity Testing with an Ohmmeter

An ohmmeter set on its lowest resistance range can be used as a continuity tester. As an example, a fuse can be tested for continuity (i.e., good versus blown fuse) by removing it from its fuse holder and applying the ohmmeter leads to each end. A blown fuse will show infinite resistance or an open circuit; a good fuse will show zero ohms resistance. A conductor and other devices can similarly be tested for continuity (i.e., for a complete path for current flow).

SUMMARY

An electrical circuit has a source of electrical pressure connected through insulated copper conductors to a load that uses the effect of electrical current flow.

An open circuit has an incomplete path for current flow.

A closed circuit completes the path for current flow.

A source of electrical pressure is a dry cell, and is measured in volts.

The rate of current flow is measured in amperes.

One ampere is a current flow of 1 coulomb per second past a point in the circuit.

Opposition to the flow of current is called resistance.

Conductance is a measure of the ease with which a circuit allows current flow, and is the reciprocal of resistance.

The ohmic resistance of a conductor is directly proportional to its length.

The ohmic resistance of a conductor is inversely proportional to its cross-sectional area.

A positive temperature coefficient means that the resistance of the material increases with temperature increase.

A negative temperature coefficient means that the resistance of the material decreases with temperature increase.

A zero temperature coefficient indicates that the resistance of the material remains constant with a change in temperature.

The temperature coefficient of resistance at room temperature can be stated as

$$\alpha R = \frac{\Delta R/\Delta T}{R_r} \tag{2-5}$$

where αR = temperature coefficient of resistance
ΔT = change in temperature
ΔR = change in resistance
R_r = resistance at reference temperature

To calculate the resistance of a conductor at some temperature when ΔR and αR are specified for a reference temperature T_r:

$$R_1 = R_r[1 + \alpha R(T_1 - T_r)] \tag{2-7}$$

The specific resistance of a material is for a cubic centimetre and the unit of resistivity is the ohm-centimetre.

The resistance of any material given its length, cross-sectional area, and resistivity can be calculated by

$$R\ (\Omega) = \rho\ (\Omega\text{-cm})\frac{L\ (\text{cm})}{A\ (\text{cm}^2)} \tag{2-9}$$

Carbon-composition resistors are composed of a mixture of carbon and binders, unit molded.

Thin-film resistors are made by depositing a thin layer of about 10^{-6} in. of carbon or metal film, are of high precision, and have low-noise, high-stability, high-reliability, and good high-frequency characteristics.

Thick-film resistors are made by depositing greater than 10^{-6} in. of carbon or metal film. They are manufactured as high-precision and semiprecision units, with low power and tolerances of 1%, 2%, and 5%.

Metal-film resistors are made by condensing nickel, chromium, and/or aluminum on a ceramic substrate. A spiral cut through the blank gives the desired resistance.

Precision wire-wound resistors are made by wrapping an accurately determined length of metal-alloy wire around a ceramic substrate core and are sealed to protect them from mechanical or environmental hazards.

Wire-wound resistors are available in the resistance range 0.001 Ω to 60 MΩ, with a tolerance reaching 0.001% and a temperature coefficient approaching 0.5 ppm/°C.

The EIA (Electronic Industries Association) has standardized the resistor color code to identify resistor values as well as the range of values normally available.

Resistance values are normally measured with an ohmmeter.

REVIEW QUESTIONS

2-1. List three components that make up a simple electrical circuit.

2-2. Explain briefly why an electrical source of pressure is needed to cause an electrical current flow.

2-3. What is the difference between a volt and a coulomb charge?

2-4. What is the difference between resistance and conductance?

2-5. List four factors that govern the resistance of a conductor.

2-6. Explain the effect on resistance of a certain conductor when:
- **(a)** Its length is doubled.
- **(b)** Its area is halved.

2-7. What is the difference between a negative, a positive, and a zero temperature coefficient?

2-8. Define resistivity. What is the resistivity $(\text{cm})^3$ of aluminum and copper?

2-9. Explain briefly the difference between composition resistors and film resistors.

2-10. What is the difference between metal-film and carbon-film resistors in terms of
- **(a)** Resistance range?
- **(b)** Resistance tolerance?

2-11. What advantages do wire-wound resistors offer?

2-12. What is the resistance tolerance of a resistor?

2-13. On an ohmmeter scale what would an open circuit indicate? A closed circuit? A short circuit?

2-14. Why must one end of a resistor be disconnected from a circuit before measuring its value with an ohmmeter?

2-15. Why must the power be turned off before measuring the resistance in a circuit?

2-16. A ohmmeter pointer can no longer be zero ohms adjusted. What remedy must be taken to correct this?

2-17. State the resistance value and tolerance for the following resistor color codes:
- **(a)** Brown, black, gold, gold
- **(b)** Yellow, violet, orange, silver
- **(c)** Gray, red, black, gold

SELF-EXAMINATION

Where applicable, choose the most suitable word, phrase, or value to complete each statement; or draw a schematic diagram where required.

2-1. Draw a simple circuit using a battery as a source connected to a resistance.

2-2. The battery provides a source of (*free electrons, ions, electrical pressure, conductance, emf*).

2-3. Electrical pressure is measured in the unit called (*ampere, ohm, siemens, coulomb, volt*).

2-4. Electrical current flow is measured in the unit called (*ampere, siemens, coulomb, volt, ohm*).

2-5. Opposition to the flow of an electrical current is measured in the unit called (*ampere, ohm, siemens, coulomb, volt*).

2-6. If the resistance of a conductor is halved, the area of the same conductor must have (*doubled, quadrupled, halved, remained constant*).

2-7. If the resistance of a conductor is doubled, the length of the same conductor must have (*doubled, quadrupled, halved, remained the same*).

2-8. With an increase in temperature the copper wire resistance will (*decrease, remain the same, increase*).

2-9. A material that has a negative temperature coefficient is (*copper, aluminum, silver, carbon, constantan.*)

2-10. A material that has an almost-zero temperature coefficient is (*copper, aluminum, silver, carbon, constantan*).

2-11. A copper wire has a 12-Ω resistance at 20°C. Calculate its resistance at 70°C.

2-12. What length of nichrome wire having a diameter of 0.15 mm must be used to obtain a resistance of 40 Ω?

2-13. Between 0 and 85°C, carbon-composition resistors have a temperature coefficient that is essentially (*negative, positive, zero, 150 to 500 ppm/°C*).

2-14. With a maximum ambient temperature of 70°C, composition resistors should be derated by a factor of ________.
(%)

2-15. Carbon-film resistors have a temperature coefficient that is (*negative, positive, zero, −150 to −500 ppm/°C*).

2-16. Metal-film resistors have a temperature coefficient that is (*negative, positive, zero, 3 to 150 ppm/°C*).

2-17. When measuring resistance with an ohmmeter, the ohmmeter range should be chosen so that the pointer is within the (*right, middle, left, uncrowded*) end of the scale.

2-18. Connecting an ohmmeter across a fuse that has blown will give a reading of (*0 Ω, 1 Ω, ∞ Ω*).

2-19. Connecting an ohmmeter from one end of a lead or wire to the other end of the wire will give a reading of (*0 Ω, 1 Ω, ∞ Ω*).

2-20. State one precaution that should be taken before measuring the value of a resistor connected in a circuit.

2-21. What is the resistance value for each resistor having the following color-code bands?
- **(a)** Yellow, violet, gold, gold
- **(b)** Brown, black, black, gold
- **(c)** Blue, gray, brown, silver
- **(d)** Brown, red, green, gold

ANSWERS TO PRACTICE QUESTIONS

2-1

1. electrical current flow
2. emf
3. open circuit
4. closed circuit
5. coulombs per second
6. resistance
7. greater
8. a volt
9. 1 V
10. C
11. ampere
12. A
13. amperes
14. ohm
15. R
16. omega
17. conductance
18. siemens
19. resistance
20. conductance

2-2

1. length
2. diameter
3. increase
4. temperature coefficient of resistance
5. positive
6. $R_1 = 8.2\ \Omega$
7. $\Delta T = 300°\text{C}$
   $\therefore T = 300°\text{C} + 20°\text{C} = 320°\text{C}$
8. $\alpha R = 0.0045\ \Omega/°\text{C}$ and is tungsten
9. $R = 1.78\ \Omega$
10. $L = 648$ cm

3

Ohm's Law

INTRODUCTION

Georg Simon Ohm, an instructor of mathematics and physics, published a pamphlet, "Mathematical Theory of the Galvanic Circuit," in 1827, which contains the basis of Ohm's law.

Ohm's basic idea was that each current is pushed along by a definite "intensity" or "pressure." The circuit opposes this pressure, the size of the opposition varying with the conductor's thickness, length, composition, and even temperature. In any circuit the volume or quantity of current flow is always directly proportional to the pressure.

In Ohm's time the theory was so revolutionary that only a few of his fellow scientists endorsed it. To quote Charles Wheatstone from one of his lectures in 1843: "Ohm's law is not yet generally understood and admitted, even by many persons engaged in original research."

Today, the relationship between the volt, ampere, and the ohm is indeed simple through the mathematical language of an equation that shows the direct proportion relationship of voltage and current and the inverse relationship between current and resistance.

MAIN TOPICS IN CHAPTER 3

3-1 Ohm's Law: Linear Device and Nonlinear Device
3-2 Multiple and Submultiple Electrical Units of Measurement
3-3 Measuring Current with a Multimeter
3-4 Measuring Voltage with a Multimeter

OBJECTIVES

After studying Chapter 3, the student should be able to

1. Differentiate between linear and nonlinear resistance characteristics.
2. Apply Ohm's law equations to solve simple circuit parameters.
3. Measure current and voltage using a multimeter to verify linear and nonlinear device characteristics through graphical analysis.
4. Measure current and voltage using a multimeter to verify Ohm's law in a simple circuit.

3-1 OHM'S LAW: LINEAR DEVICE AND NONLINEAR DEVICE

Linear Device

> For a linear device such as a resistor, the current is directly proportional to voltage.

In Chapter 2 we considered resistance to be the property of a conductor's size, length, composition, and temperature. In electronic circuits resistors are inserted into circuits to produce deliberate voltage drops for specific applications such as transistor biasing, shown in Fig. 3-1.

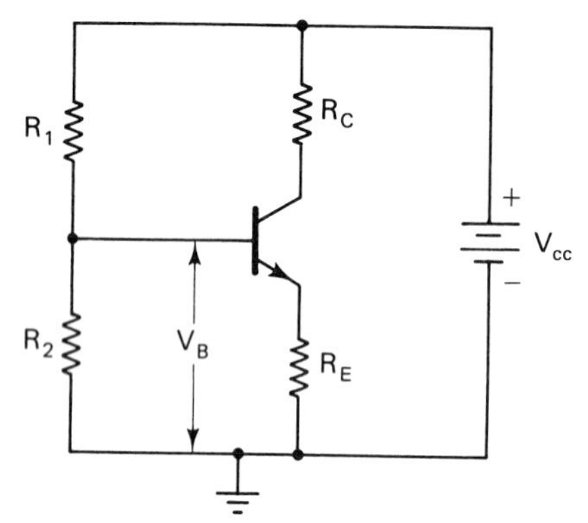

FIGURE 3-1 Resistors R_1 and R_2 provide a voltage divider bias V_B.

Composition-carbon, thin-film, thick-film, and wire-wound resistors operated within a specified temperature range are linear devices that obey Ohm's law; that is, the current will increase or decrease directly with an increase or decrease in voltage. This relationship can be plotted graphically as shown in Fig. 3-2.

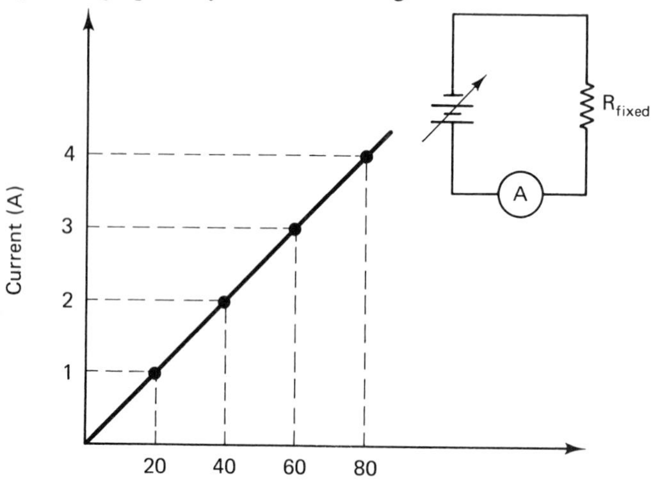

FIGURE 3-2 Current is directly proportional to voltage.

Figure 3-2 shows the direct proportion relationship of voltage to current or

$$I \propto V \tag{3-1}$$

If the voltage is doubled from 20 V to 40 V, the current in the circuit will double from 1 A to 2 A. Doubling the voltage again will double the previous current. In this manner a device such as a resistor operated within a specified temperature range can be considered as a linear device.

Similarly, the inverse relationship for Ohm's law says that the current will be inversely proportional to the resistance in the circuit. This relationship can be plotted graphically as shown in Fig. 3-3.

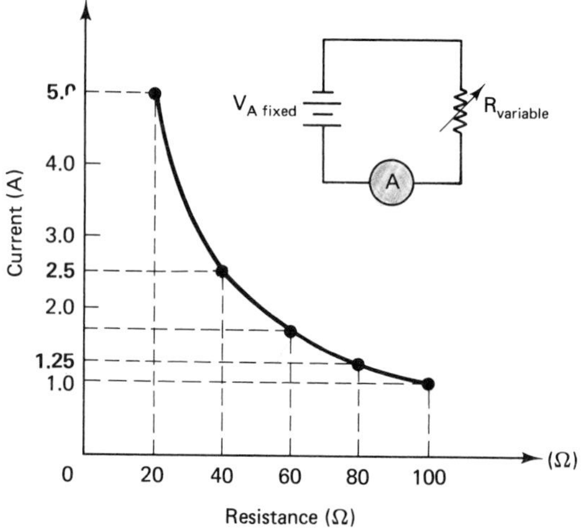

FIGURE 3-3 Current is inversely proportional to resistance.

For a linear device such as a resistor, the current is inversely proportional to the resistance. Figure 3-3 shows the inverse proportion relationship of current to resistance or

$$I \propto \frac{1}{R} \tag{3-2}$$

If, for example, your body contact resistance across a certain voltage is reduced by one-half due to contact pressure or an increased damp condition, the current through your body will double.

Ohm's Law Equation. From formulas (3-1) and (3-2) or $I \propto V$ and $I \propto 1/R$ *Ohm's law* can be written, in equation format,

$$I = \frac{V}{R} \tag{3-3}$$

where V is in volts
R is in ohms
I is in amperes

By cross-multiplying we obtain

$$V = IR \qquad \text{volts} \tag{3-4}$$

and by transposing terms we can find R:

$$R = \frac{V}{I} \qquad \text{ohms} \tag{3-5}$$

Given two known electrical values the third unknown can easily be determined by applying the appropriate equation, (3-3) to (3-5).

EXAMPLE 3-1

Find the required voltage to force a current flow of 0.01A through a 1000-Ω resistance.

Solution: From equation (3-4),

$$V = IR = 0.01 \text{ A} \times 1 \times 10^3 \text{ }\Omega = 10 \text{ V}$$

EXAMPLE 3-2

Find the hot resistance of the elements in a toaster if the current drawn is 10 A.

Solution: To find the resistance, we must have both the voltage and the current [See equation (3-5)]. But we know that the toaster plugs into a convenience outlet that supplies 110 V, so

$$R = \frac{V}{I} = \frac{110 \text{ V}}{10 \text{ A}} = 11 \text{ }\Omega$$

EXAMPLE 3-3

What is the value of current flowing through a human body if the contact voltage is 110 V and the contact resistance is 2000 Ω? Is this condition hazardous?

Solution: From equation (3-3),

$$I = \frac{V}{R} = \frac{110 \text{ V}}{2 \times 10^3 \text{ }\Omega} = 0.055 \text{ A}$$

A current of approximately 0.150 A causes death. Certainly, a current of 0.055 A is hazardous since it could stop your heart and/or lungs from functioning.

Conductance. Recall that conductance was the ability of a circuit to conduct a current flow. Since conductance is the reciprocal of resistance, conductance can be stated in terms of Ohm's law by substituting V/I for R, or

$$G = \frac{I}{R} = \frac{I}{V/I}$$

Then

$$G = \frac{I}{V} \qquad \text{siemens} \tag{3-6}$$

EXAMPLE 3-4

What is the conductance of the toaster in Example 3-2?

Solution: Using equation (3-6) will give the conductance.

$$G = \frac{I}{V} = \frac{10 \text{ A}}{110 \text{ V}} = 9.1 \times 10^{-2} \text{ S}$$

Nonlinear Device

A device whose voltage–current characteristics change with temperature has a nonlinear resistance response curve, as shown in Fig. 3-4. Such a device is a *thermistor*. Thermistors are essentially semiconductors that behave as "thermal resistors," resistors that have a negative temperature coefficient.

Figure 3-4(a) shows that for low values of applied voltage current is proportional to the applied voltage and behaves according to Ohm's law. As the voltage is increased, the increased current begins to raise the temperature of the device. An increase in temperature reduces the resistance and the current continues to increase more rapidly. At some maximum value this self-heating effect will reach a steady-state condition limited by the voltage source or more commonly by the voltage drop across a fixed resistor placed in series with the thermistor. Note that the thermistor has a negative temperature coefficient of resistance; therefore, its resistance decreases with temperature, as shown in Fig. 3-4(b).

In the case of the incandescent lamp the tungsten filament has a positive temperature coefficient, so its resistance increases with temperature and the current levels out to a steady-state value once the filament reaches its working resistance, as shown in Fig. 3-4(a).

For nonlinear devices such as the thermistors shown in Fig. 3-5, Ohm's law is applicable provided that the device is held in a steady-state condition. Ohm's law cannot predict a change in the thermistor's condition because the thermistor's resistance is temperature dependent.

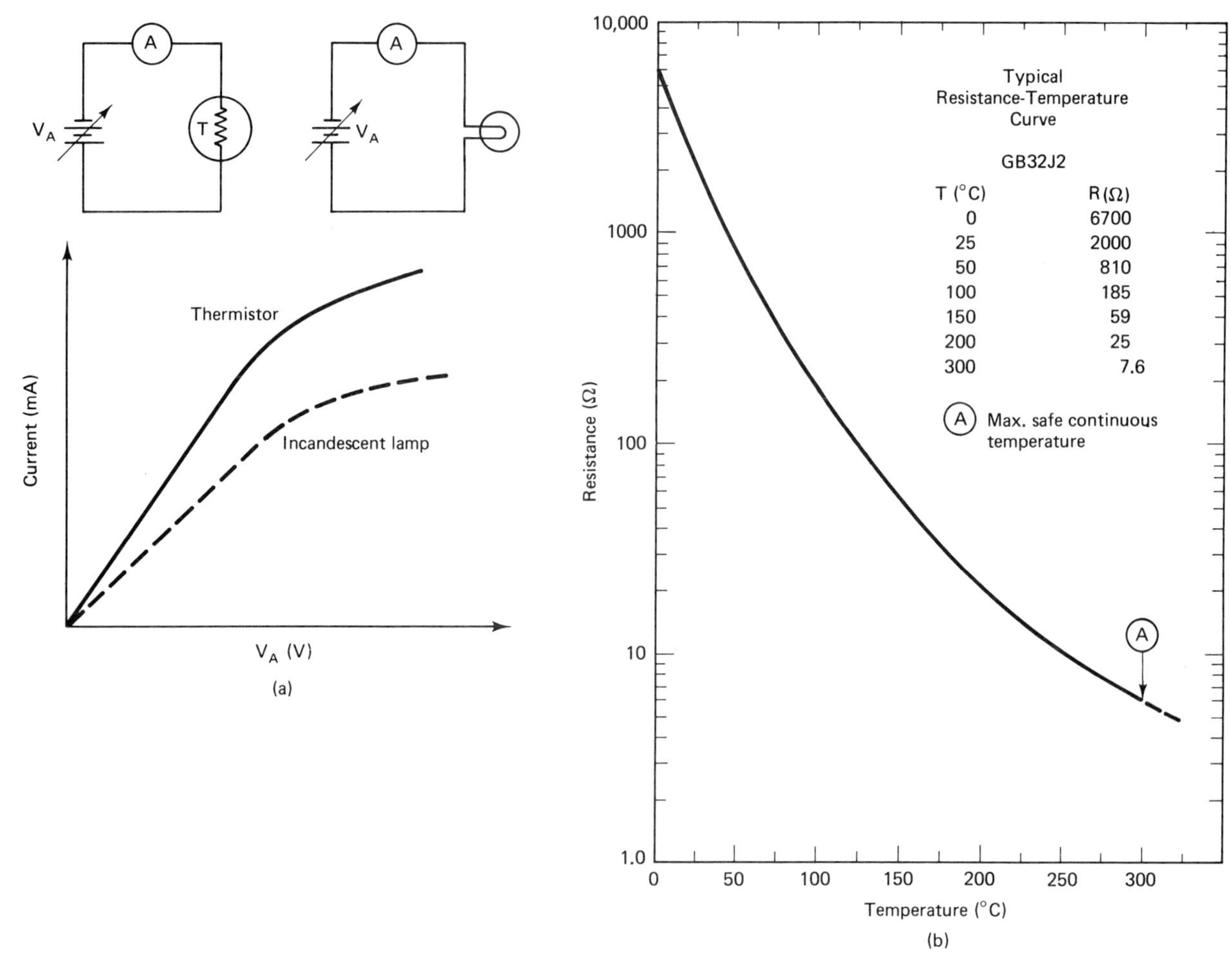

FIGURE 3-4 (a) Nonlinear *V–I* characteristics of thermistor and incandescent lamp. (b) Typical resistance–temperature curve for a thermistor. [Part (b) courtesy of Fenwal Electronics, Inc.]

PRACTICE QUESTIONS 3-1

Choose the most suitable word, phrase, or value to complete each statement.

1. A device such as a resistor that is operated within a specified temperature range is a (*linear, nonlinear*) device.
2. A device such as a thermistor or lamp filament does not obey Ohm's law and is a (*linear, nonlinear*) device.
3. Ohm's law tells us that doubling the voltage across a resistor will (*halve, triple, double, not change*) the current through the resistor.
4. Ohm's law tells us that halving the resistance in a circuit will (*halve, triple, double, not change*) the current in the circuit.
5. A thermistor will have a (*small, large, steady*) inrush or increase of current upon application of a voltage.
6. A 12-V battery will supply ________ amperes to a lamp having a hot resistance of 5.5 Ω.
7. What resistance is required to limit a current of 2.2 A with an applied voltage of 100 V?

3-2 MULTIPLE AND SUBMULTIPLE ELECTRICAL UNITS OF MEASUREMENT

The standard SI unit prefixes are shown in Table 3-1. The International System of Units (SI) is a rationalized and coherent system of units. It is a simplified system of measurement based on the metric system. The SI system has been developed and approved by countries that participate in the activities of the General Conference of Weights and Measures, the International Organization for Standardization, and the International Electrotechnical Commission. Canada and the United States are members of all these organizations. The SI units are rec-

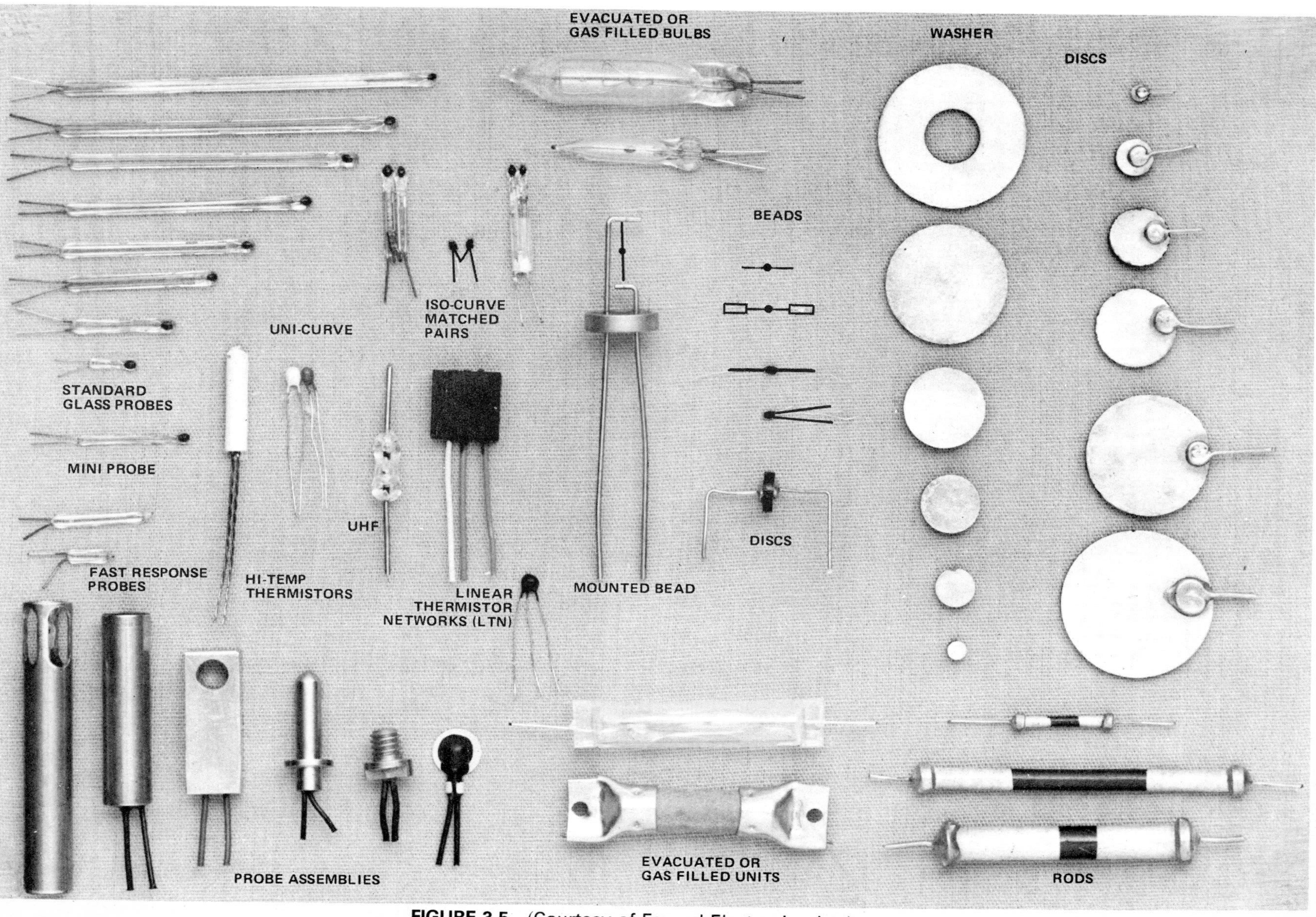

FIGURE 3-5 (Courtesy of Fenwal Electronics, Inc.)

TABLE 3-1
SI unit prefixes

Prefix	*Multiplier*	*Symbol*
tera	10^{12}	T
giga	10^{9}	G
mega	10^{6}	M
kilo	10^{3}	k
hecto	10^{2}	h
deka	10	da
deci	10^{-1}	d
centi	10^{-2}	c
milli	10^{-3}	m
micro	10^{-6}	μ
nano	10^{-9}	n
pico	10^{-12}	p
femto	10^{-15}	f
atto	10^{-18}	a

ommended for use in Canada and are endorsed by the Institute of Electrical and Electronics Engineers (IEEE).

In this chapter we concern ourselves only with the units for resistance and current. Multiples and submultiples of a basic unit such as resistance and current can be indicated by attaching a prefix to the unit as shown in Table 3-2.

EXAMPLE 3-5

Convert the following.
(a) 0.003 A to milliamperes
(b) 10 mA to amperes
(c) 0.0001 V to microvolts
(d) 100 μV to volts
(e) 5600 Ω to kilohms
(f) 5.6 kΩ to ohms

Solution:

(a) $$0.003\text{ A} = \frac{3}{1000}\text{ A} = 3 \times 10^{-3}\text{ A} = 3\text{ mA}$$

(b) $$10\text{ mA} = \frac{10}{1000}\text{ A} = 10 \times 10^{-3}\text{ A} = 0.010\text{ A}$$

(c) $$0.0001\text{ V} = \frac{100}{1{,}000{,}000}\text{ V} = 100 \times 10^{-6}\text{ V} = 100\ \mu\text{V}$$

(d) $$100\ \mu\text{V} = \frac{100}{1{,}000{,}000}\text{ V} = 100 \times 10^{-6}\text{ V} = 0.0001\text{ V}$$

(e) $$5600\ \Omega = \frac{5600}{1000} = 5600 \times 10^{-3}\text{ k}\Omega = 5.6\text{ k}\Omega$$

(f) $$5.6\text{ k}\Omega = 5.6 \times 1000\ \Omega = 5.6 \times 10^{3}\ \Omega = 5600\ \Omega$$

TABLE 3-2
SI unit prefixes

Value	*Scientific Notation*	*Prefix*	*Symbol*
1,000,000	10^{6}	mega	M
1,000	10^{3}	kilo	k
0.001	10^{-3}	milli	m
0.000,001	10^{-6}	micro	μ

PRACTICE QUESTIONS 3-2

Choose the most suitable word, phrase, or value to complete each statement.

1. 1 A expressed as milliamperes is ________.
2. 25 μA expressed as amperes is ________.
3. 300 μA shown as milliamperes is ________.
4. 0.000006 A as microamperes is ________.
5. 20 mV is 20,000 ________.
6. 1 V is ________ millivolts.
7. 1 μV is ________ millivolts.
8. 1 V is ________ microvolts.
9. 0.005 V is ________ millivolts.
10. There are ________ volts in 5 kV.
11. A 300-W lamp consumes ________ kilowatts.
12. 18,000 V is ________ kilovolts.
13. 0.0063 mV expressed as ________ would be 6.3.
14. A 50-kΩ resistor has ________ ohms resistance.
15. The resistance in ohms of a 3.3-MΩ resistor is ________ ohms.
16. In 600 $\mu\Omega$ we have ________ ohms.
17. 0.000000007 A as nanoamperes is ________.

3-3 MEASURING CURRENT WITH A MULTIMETER

Regardless of the type of multimeter that you use, the principles involved in reading the scale and connecting the meter into the circuit to read current or voltage are the same. Two basic settings must be made on any meter before it can be used for its proper function.

1. *Function switch.* The function switch provides the desired function that you select for the meter to do. For this case we require the meter to read current; therefore, the function switch is set to DCV/MA. The oblique stroke simply means "or" (see Fig. 3-6).

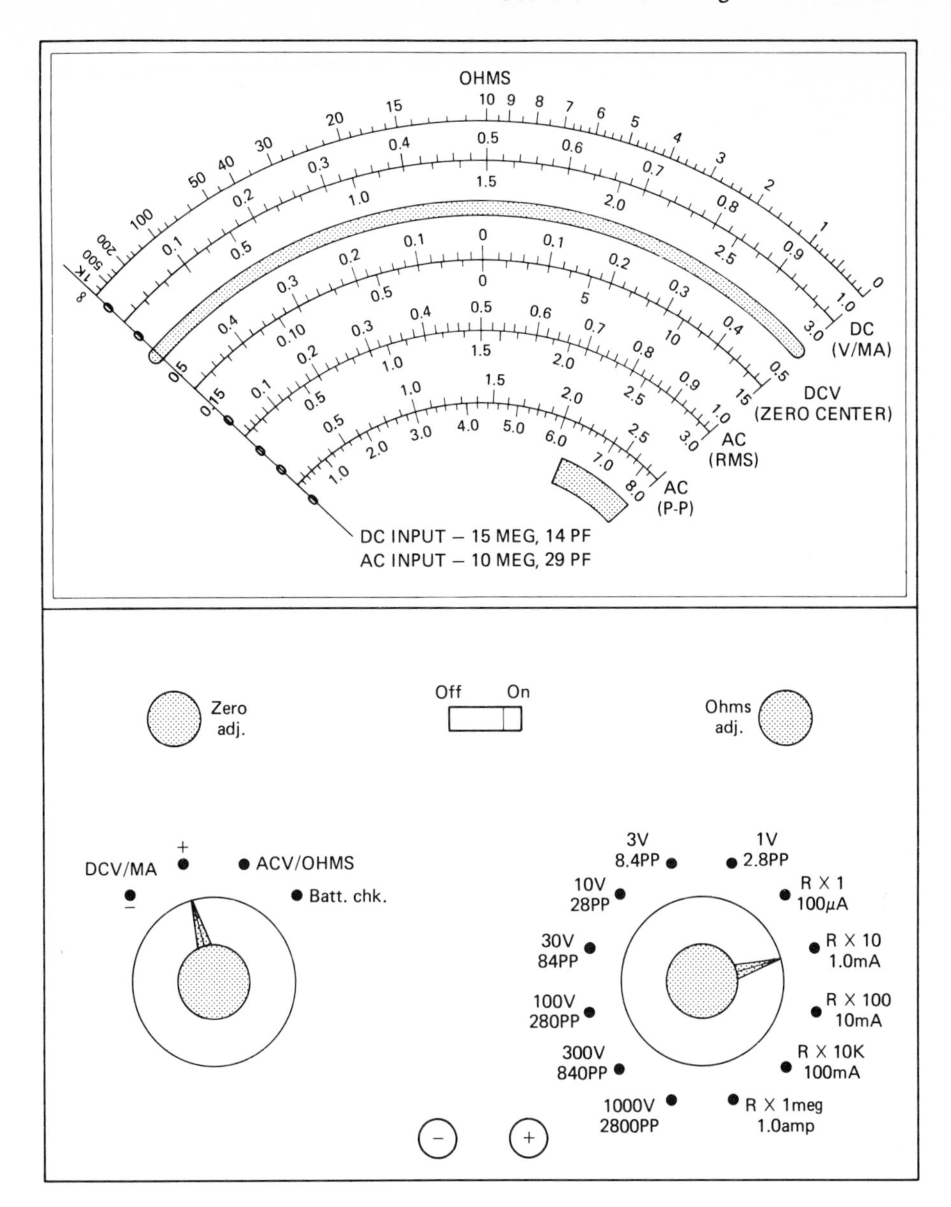

FIGURE 3-6 Electronic multimeter.

2. *Range switch.* The range switch selects the upper limit or full-scale current that the meter will be capable of measuring. With an unknown current to be measured, always select the highest current range to avoid damaging the meter pointer/meter, then switch to a lower range, until you get a mid- to full-scale deflection.

Note that the FET multimeter in Fig. 3-6 has the following current ranges: 100 μA, 1.0 mA, 10 mA, 100 mA, and 1.0 A. Now you are ready to connect the meter into the circuit to measure current.

Connecting the mA Meter into the Circuit

Since current flows in a conductor and through the device or resistor, the path for this current flow must be opened and the mA meter connected in series, that is, between the two points that were opened (see Fig. 3-7).

Polarity of the Meter Leads

In order for the meter pointer to deflect from zero to the full-scale reading, the polarity of the meter must be observed. Consider that the black lead is negative (for most meters), and therefore electrons must enter the black lead, flow through the meter, and exit by the red lead (see Fig. 3-7). The same principle applies regardless of where you connect the mA meter into the circuit.

Reading the mA Scales

For this FET multimeter the current ranges are combined on one scale, 0 to 1 (see Fig. 3-6). With the 100 μA each measurement is multiplied by 100 and the units

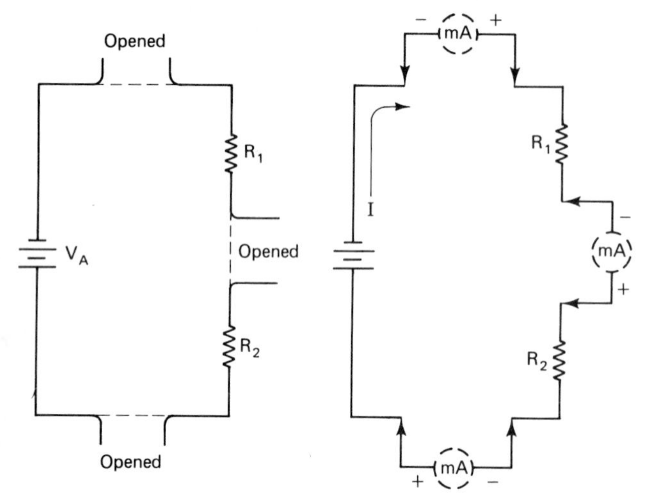

FIGURE 3-7 Multimeter connection.

are microamperes. The 1-mA range measurement is read directly. Similarly, each milliampere measurement is multiplied by the range number selected on the range switch.

3-4 MEASURING VOLTAGE WITH A MULTIMETER

For the milliammeter application there are two basic settings regardless of the make of multimeter before it can be used for its proper function.

1. *Function switch.* The function switch provides the desired function that you select the meter to do. Since we require the meter to read dc volts, the function switch is set to DCV/MA (see Fig. 3-6).
2. *Range switch.* The range switch selects the upper limit or full-scale voltage that the meter will be capable of measuring.

> With an unknown voltage to be measured, always select the highest voltage range, to avoid damaging the meter pointer/meter, then switch to a lower range until you get a mid- to full-scale deflection.

Observe (see Fig. 3-6) that this meter has two DCV scales, 0–1.0 and 0–3.0.

Connecting the Voltmeter across a Voltage

A potential difference or voltage drop exists between two points when one point has a higher potential than the other point. Such a potential difference exists across voltage sources, resistors, other types of loads, transistor elements, and capacitors. A voltmeter is always connected across two such points as shown in Fig. 3-8.

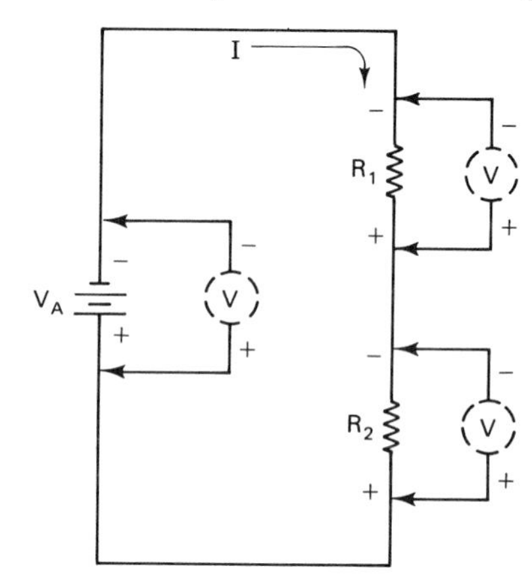

FIGURE 3-8 Voltmeter connection.

Polarity of the Meter Leads

It is necessary that the meter lead polarity be observed in order to get a deflection of the meter pointer to the right of the zero. As in the case of the milliampere meter, the black lead is negative and the red lead is positive. Observing the direction of electron current flow would make the end of the load resistor negative where electrons enter that end and positive where electrons leave that end. Therefore, the black meter lead connects to the negative side of the resistor or load and the red lead connects to the positive side of the resistor or load (see Fig. 3-8).

Reading the Voltage Scales

As shown in Fig. 3-6, there are two dc voltmeter scales, 1.0 V and 3.0 V. The position of the range switch must be observed and the appropriate multiplier applied for each reading on either scale. With the voltmeter function switch set on 3 V, the voltage reading is read directly from the 3-V scale. Range settings of 30 V and 300 V are also read off this 3-V scale; however, each reading must be multiplied by 10 and 100, respectively.

For the 1-V scale, the same procedure must be followed, so that a 1-V setting on the range switch will enable readings to be made directly on the 1-V scale. Range settings of 10 V, 100 V, and 1000 V are read on the 1-V scale. However, each reading must be multiplied by 10, 100, and 1000, respectively. Regardless of the manufacturing make or model of deflection meter, reading the various scales will be similar.

> Set the range switch, note the appropriate range scale, and apply the range multiplier to your meter reading.

Parallax Error When Reading a Meter Scale

Although this type of meter reading error is not significant for general applications, it should nevertheless be mentioned at this time.

Parallax error is the error introduced to a reading by observing the meter pointer from one side or the other. To overcome this type of reading error, a mirror is mounted alongside the meter scale. Reading errors are minimized by aligning the pointer with the mirror-image pointer. Precision meters have a mirror mounted alongside the scales.

PRACTICE QUESTIONS 3-3

1. Given that the meter shown in Fig. 3-9 is set and connected to measure current, determine the meter readings for the following additional range settings:

Range	*No. 1 mA Reading*	*No. 2 mA Reading*
100 μA		
1 mA		
10 mA		
100 mA		
1 A		

2. Given that the meter shown in Fig. 3-9 is set and connected to measure voltage, determine the meter readings for the following additional range settings:

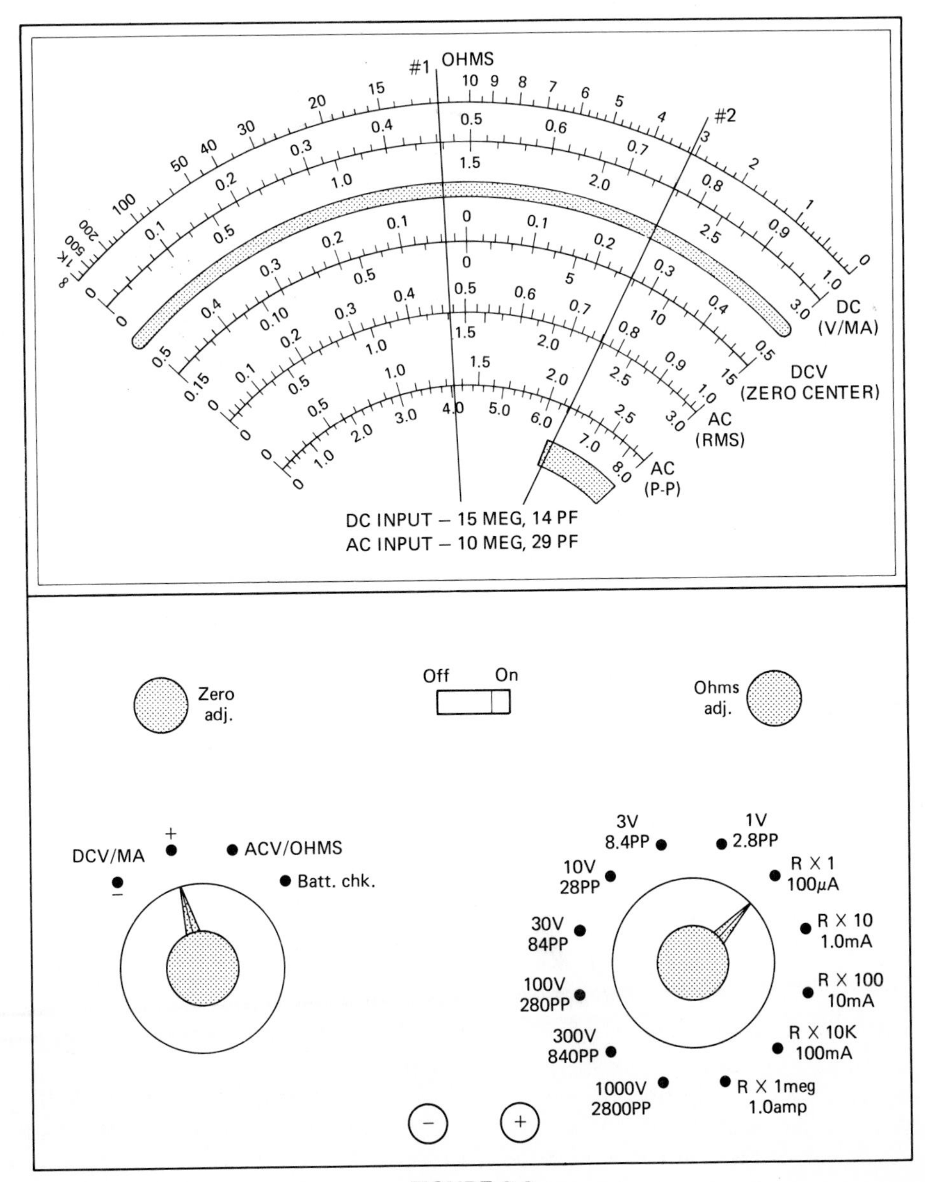

FIGURE 3-9

Range	No. 1 Volts Reading	No. 2 Volts Reading
1 V		
3 V		
10 V		
30 V		
100 V		
300 V		
1000 V		

SUMMARY

A device obeys Ohm's law when the current through it varies directly as the voltage applied.

A current through a linear device will vary directly (i.e., uniformly) to the applied voltage.

Resistors operated within their specified temperature range will behave as linear devices.

A current through a linear device will vary inversely to the resistance of the device, or $I \propto 1/R$.

Ohm's law states that the current is directly proportional to the applied voltage and inversely proportional to the resistance, or $I \propto V/R$.

A current through a nonlinear resistance device will vary nonuniformly to the applied voltage, and its value is dependent on temperature.

Nonlinear resistance devices are thermistors, semiconductors, and tungsten lamp filaments.

The multiple prefix for 1000 is kilo (k) and for 1,000,000 is mega (M).

The submultiple prefix for 1/1000 is milli (m) and for 1/1,000,000 is micro (μ).

A milliammeter (ammeter) measures current and therefore must be connected in series.

With an unknown current to be measured, always select the highest current range, to avoid damaging the meter pointer/meter, then a lower range until you get a mid- to full-scale deflection.

Regardless of the make or model of multimeter you use, the range switch setting will determine the scale you use and the multiplier for each meter reading.

The polarity of the mA leads must be observed in order to obtain left-to-right deflection of the pointer, where normally the black lead is negative and the red lead is positive.

A voltmeter measures a potential difference or voltage drop, therefore must be connected in parallel or across a resistor, device, or source.

The polarity of the voltmeter leads must be observed in order to obtain left-to-right deflection of the pointer, where normally the black lead is negative and the red lead is positive.

REVIEW QUESTIONS

3-1. State the relationship of current to voltage for a linear device.

3-2. State the relationship of current to voltage for a nonlinear device.

3-3. What factor governs the resistance of a nonlinear device?

3-4. When can resistors properly be called linear devices?

3-5. Why do semiconductors, thermistors, and tungsten lamp filaments have a nonlinear resistance characteristic?

3-6. With a fixed resistance the current doubles with double the voltage and triples with triple voltage. In your own words, state the law to support this statement.

3-7. With a fixed voltage, the current is halved with double the resistance or is doubled with half the resistance. In your own words, state the law that supports this statement.

3-8. Give the scientific notation value and symbol for milli, micro, kilo, and mega.

3-9. Give the two main multimeter settings used to measure a current of 65 mA. What scale on the meter shown in Figure 3-9 would be used, and what would the reading be?

3-10. Give the two main multimeter settings used to measure a voltage of 23 V. What scale on the meter shown in Figure 3-9 would be used, and what would the reading be?

3-11. Explain briefly what determines the polarity of a voltage across a resistor.

3-12. What is the effect on the meter reading of not observing the correct polarity of the mA meter or voltmeter leads with respect to the current or voltage being measured?

3-13. What might be the effect on the meter reading of trying to measure an unknown current or voltage on the low range of the meter?

3-14. It is necessary to measure the current through a resistor mounted on a printed-circuit board. Explain how the mA meter is connected.

3-15. Explain a simple method to determine the current through the resistor of Question 3-14.

SELF-EXAMINATION

Where applicable, choose the most suitable word, phrase, or value to complete each statement.

3-1. For a linear device such as a resistor the current is (*inversely, uniformly, directly, indirectly*) proportional to the voltage.

3-2. Ohm's law states that the ________ is directly proportional to the voltage applied.

3-3. For a linear device such as a resistor, the current is (*inversely, uniformly, directly, indirectly*) proportional to the resistance.

3-4. Ohm's law states that the current is inversely proportional to the ________.

3-5. Find the required voltage to force a current flow of 35 mA through a 5.6-kΩ resistor.

3-6. A current of 40 mA flows through a resistor that has an applied voltage of 6.7 V. What is its resistance?

3-7. A resistance of 8.2 kΩ is connected across a voltage of 25 V. Find the current through the resistance.

3-8. A thermistor's resistance is dependent on (*manufacturer, voltage, temperature, size*).

3-9. A tungsten filament of a lamp has a (*negative, positive*) temperature coefficient.

3-10. The resistance characteristic with temperature of a thermistor or tungsten filament is (*linear, uniform, nonlinear, directly proportional*).

Refer to Fig. 3–6 for Questions 3–11 to 3–17.

3-11. To what setting and which two controls must be set on the multimeter to measure a current of 18 mA?

3-12. On which scale is the current of Question 3-11 read?

3-13. What is the actual reading and the multiplier?

3-14. Which two controls must be set on the multimeter, and to what setting, to measure a voltage of 18 V?

3-15. On which scale is the voltage of Question 3-14 read?

3-16. What is the actual reading and the multiplier?

3-17. An unknown current is to be measured. To what range would you set the switch?

3-18. Redraw the circuit schematic shown in Fig. 3-10 to show how a mA meter is connected to measure the current through source V_2. Label the polarity of the leads.

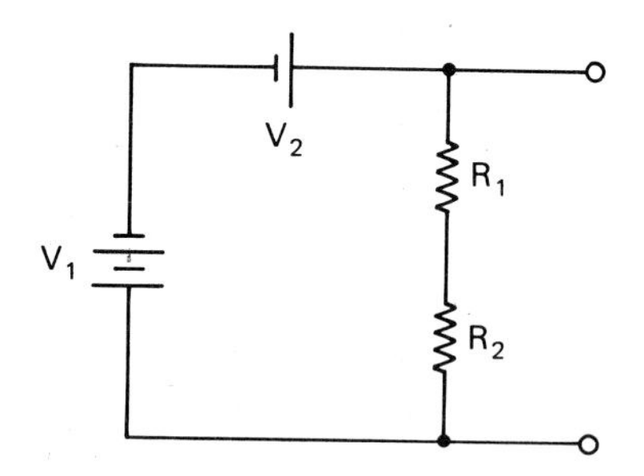

FIGURE 3-10 Circuit schematic.

3-19. Redraw the schematic of Fig. 3-10 to show how a voltmeter is connected to measure the voltage across R_2.

3-20. What is the effect on the meter pointer deflection with the lead polairty connected incorrectly?

ANSWERS TO PRACTICE QUESTIONS

3-1

1. linear
2. nonlinear
3. double
4. double
5. small
6. 2.2
7. 45.5 Ω

3-2

1. 1000 mA
2. 0.000025 A or 25×10^{-6} A
3. 0.3 mA
4. 6 μA
5. 20,000 μV
6. 1000
7. 0.001
8. 1,000,000
9. 5
10. 5000
11. 0.3
12. 18
13. μV
14. 50,000
15. 3.3×10^{6}
16. 600×10^{-6}
17. 7 nA

3-3

1. For the mA readings:

Range	*No. 1 mA Reading*	*No. 2 mA Reading*
100 μA	46 μA (approx.)	76 μA (approx.)
1 mA	0.46 mA	0.76 mA
10 mA	4.6 mA	7.6 mA
100 mA	46 mA	76 mA
1 A	0.46 A	0.76 A

2. For the voltage readings:

Range	*No. 1 Volts Reading*	*No. 2 Volts Reading*
1 V	0.46 V (approx.)	0.76 V (approx.)
3 V	1.4 V	2.3 V
10 V	4.6 V	7.6 V
30 V	14 V	23 V
100 V	46 V	76 V
300 V	140 V	230 V
1000 V	460 V	760 V

4

Series Circuits

INTRODUCTION

The first application of a series circuit connection was probably made by Alessandro Volta. Volta invented the first battery around 1800, by applying Galvani's discovery that two dissimilar pieces of metal (copper and zinc) separated by moist paper could produce electricity.

Volta combined a number of "pairs of metal" to make his famous "voltaic pile." The order was copper plate, moist paper, zinc plate moist paper, copper plate moist paper, and so on. What Volta did was to connect a number of cells in series so that the voltage produced by each cell aided the voltage of each of the other cells to provide a much larger voltage from the group.

Around 1809, Sir Humphry Davy in London was using a battery of 2000 cells, providing close to 3000 V. To this day a battery is produced in much the same manner. Individual voltage cells are connected in series to give an output voltage such as two 1.5-V D cells connected in series to give 3 V for a flashlight or six 2-V (approx.) cells to give 12 V (approx.) for a car battery.

Your study of the series circuit will help you to understand voltage, current, and resistance characteristics as well as applications of the series connection.

MAIN TOPICS IN CHAPTER 4

4-1 Series Circuit Characteristics
4-2 Ohm's Law for a Series Circuit
4-3 Voltage-Division Relationship to Resistance and to a Common Ground
4-4 Internal Resistance of a Source
4-5 Applications of a Series Circuit

OBJECTIVES

After studying Chapter 4, the student should be able to

1. Relate current voltage and resistance characteristics of a series circuit.
2. Apply Ohm's law equations to solve for series circuit voltage, current, and resistance.
3. Relate voltage drop proportionality to resistance proportionality about a series circuit.
4. Relate the internal resistance of a source(s) to series circuit characteristics.
5. Relate the series circuit to practical applications.

4-1 SERIES CIRCUIT CHARACTERISTICS

The simplest series type of circuit or connection is the flashlight battery. Two or three cells are connected in series to obtain a higher potential or voltage, as shown in Fig. 4-1. The increased voltage available from this circuit configuration enables the lamp to have more current flowing through it, with resultant extra light brightness from the flashlight.

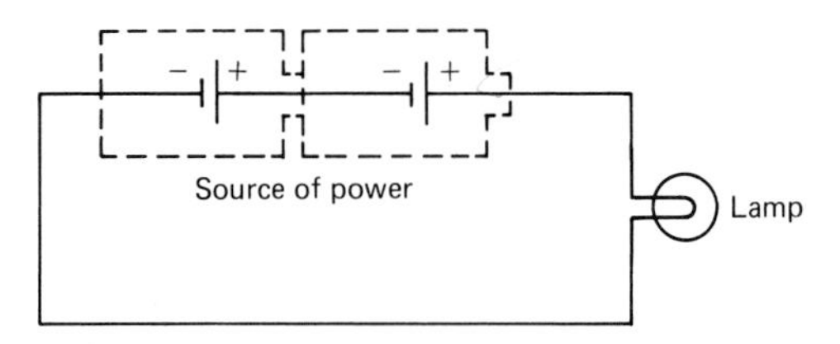

FIGURE 4-1 Schematic diagram of flashlight battery.

Current Path

> The series circuit has one path for current flow.

An electron current flows from the highest potential to the lowest so that the current in the circuit schematic of Fig. 4-2 will flow from the negative terminal of cell *A* through the lamp, to the positive terminal of cell *B*, out of the negative terminal of cell *B*, and back to the positive terminal of cell *A*.

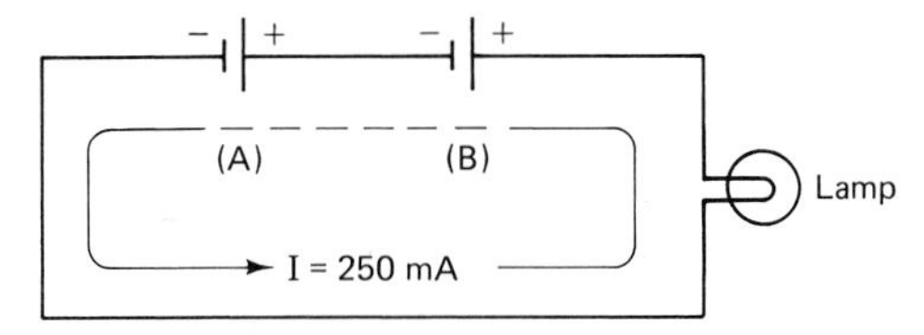

FIGURE 4-2 One path for current, which is the same in all parts of the circuit.

Removing one of the cells or the lamp from the circuit would open the path for the current, which would stop the current from flowing and the lamp would go out. Since the current flow can be interrupted anywhere along the circuit, the series connection or circuit has only one path for current flow.

Current

> The amount of current flow is the same in all parts of a series circuit.

EXAMPLE 4-1

As shown in Fig. 4-2, the amount of current flowing through the cells and the lamp will be the same, or 250 mA. In equation format,

$$I_T = I_A = I_B = I_L$$

or

$$I_T = I_1 = I_2 = I_3 \text{ etc.} \qquad \text{amperes} \quad (4\text{-}1)$$

Series-Aiding and Series-Opposing Voltages

A flashlight cell produces 1.5 V when "fresh." With the two cells connected in series as shown in Fig. 4-3, the flashlight battery now delivers 3 V. With three cells connected in series, the battery would deliver 4.5 V.

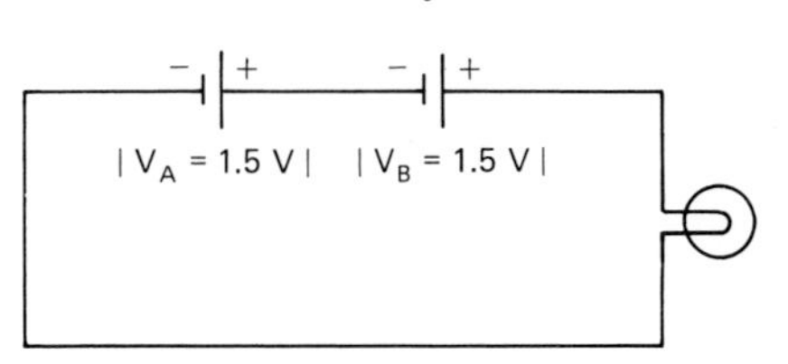

FIGURE 4-3 Voltage sources $V_A + V_B = 3$ V.

Series-Aiding Voltages. When two or more sources are connected from a negative terminal of one source to the positive terminal of the next source around the complete circuit or path so that a negative terminal and a positive terminal will be left as the battery voltage or source of power, they are said to be connected *series aiding*. As shown in Fig 4-4, the source voltages aid each other to produce an overall voltage source equal to the sum of the individual sources, in this case 4.5 V. Applications of the series-aiding connection are in batteries, transformer windings, and solar or photo cells particularly.

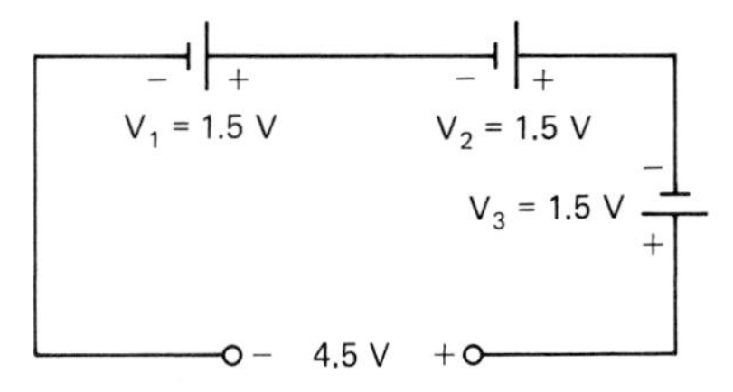

FIGURE 4-4 Series-aiding connection.

Series-Opposing Voltages. When two or more sources are connected from a negative terminal of one source to the negative terminal of the next source or from the positive terminal of one source to the positive terminal of the next source around the complete circuit or path so that either two negative terminals or two positive terminals are left as the source of power as shown in Fig. 4-5, they are said to be connected *series opposing*. The source voltages oppose each other, that is, they cancel each other. With two cells each producing 1.5 V, the voltages of each cancel and produce 0 V at the source terminals. If one source voltage is greater than the other, the largest voltage source will cancel the smaller voltage, leaving a difference voltage as the battery terminal voltage.

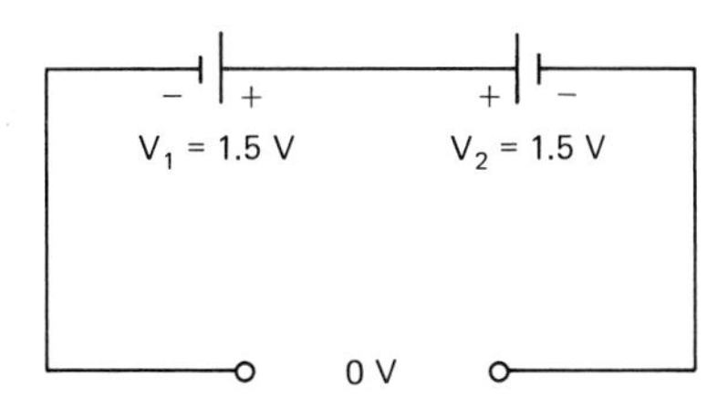

FIGURE 4-5 Series-opposing voltages.

EXAMPLE 4-2

(a) A 60-V and a 40-V dc source are connected series aiding. Show how they are connected, and the final voltage.

(b) Show how the two sources would be connected series opposing, and the final voltage.

Solution:

(a) The series-aiding circuit is shown in Fig 4-6(a); the final voltage is 100 V.

(b) The series-opposing circuit is shown in Fig. 4-6(b); the final voltage is 20 V.

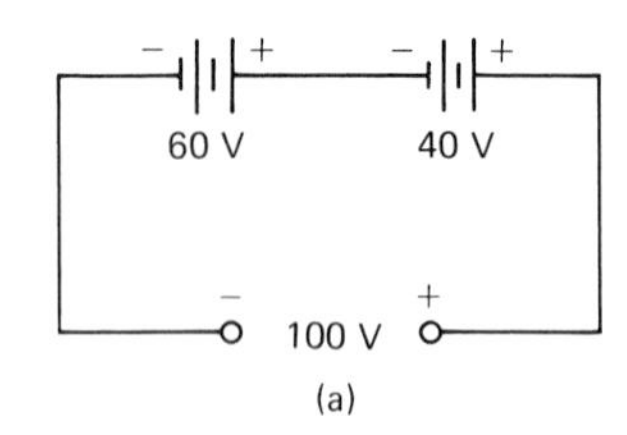

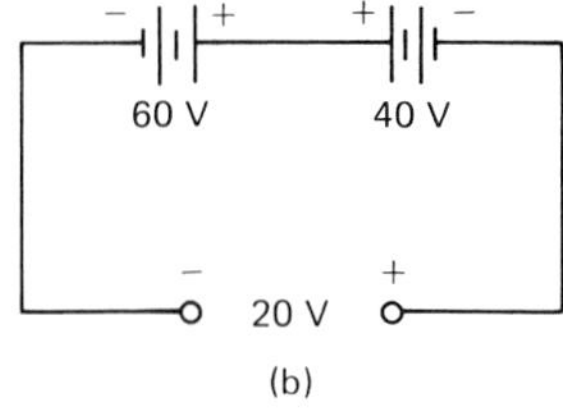

FIGURE 4-6 Circuit schematic for Example 4-2. (a) Series-aiding connection. (b) Series-opposing connection.

Voltage Drops in a Series Circuit

Power sources such as a cell, battery, generator, transformer, and solar cell all provide voltage and current. The voltage provided by the source forces the current through the loads. When current flows through a load or resistance, a voltage difference set up by the source causes a higher potential or voltage at one end of the resistor with respect to the other end of the resistor. This enables negative electrons to flow from a point that is negative to a point that is positive by comparison.

Figure 4-7 shows that A will be negative with respect to B and B will be negative with respect to C, and so on. Each resistor about the circuit of Fig. 4-7 will have a voltage drop determined by Ohm's law.

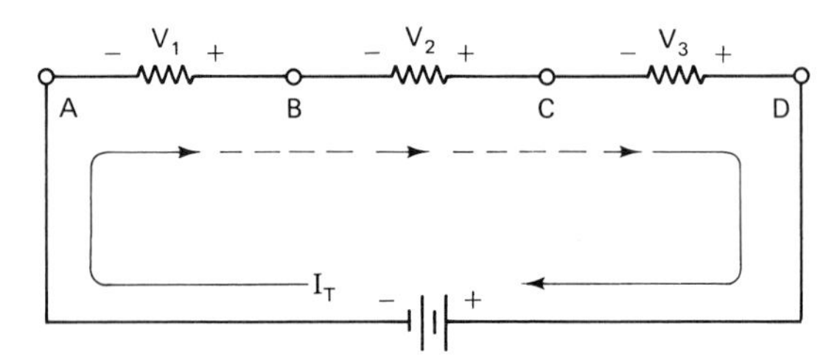

FIGURE 4-7 Polarity of voltage drops.

Polarity of Voltage Drops

As shown in Fig. 4-7, the electron current flows from the negative side of the dc source around the circuit and back to the positive side of the source. Where electrons enter a resistor or load, that end will be negative with respect to the other end. Where electrons leave a resistor or load, that end will be positive with respect to the other end of the resistor. The voltmeter leads must be connected so that the black lead of the voltmeter connects to the negative side of each load and the red lead to the positive side of each load, as shown in Fig. 4-8.

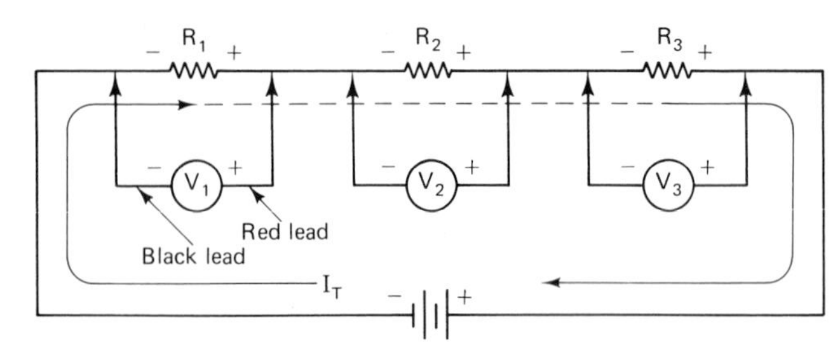

FIGURE 4-8 Voltmeter lead polarity.

In a series circuit, the sum of the voltage drops about a series circuit will be equal to the source voltage.

Whether there are two or more cells connected in series or two or more loads connected in series, the sum of the voltage drops about the circuit will always be equal to the source voltage. In equation form as shown in Fig. 4-9,

$$V_T = V_1 + V_2 + V_3 \quad \text{etc.} \qquad \text{volts} \qquad (4\text{-}2)$$

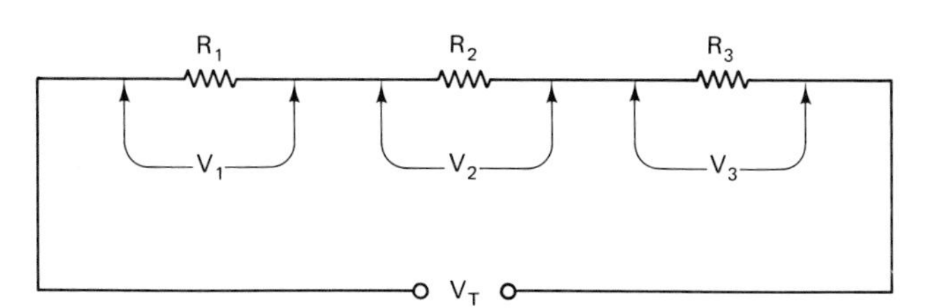

FIGURE 4-9 Sum of voltage drops will be equal to V_T.

EXAMPLE 4-3

Three resistors have voltage drops of 30 V, 20 V, and 60 V. What is the source voltage?

Solution: Using equation (4-2)* yields

$$V_T = 30\text{ V} + 20\text{ V} + 60\text{ V}$$
$$= 110\text{ V}$$

Resistance

The sum of the resistances about a series circuit will be equal to the total resistance that the source "sees."

Since there is only one path for the current to flow and the same current must flow through all the loads, the current must be forced through each resistance or load in turn, with a resultant voltage loss across each resistance or load. Although the current "meets" this opposition in turn, the source voltage "sees" an overall opposition to the current flow. This overall opposition will be equal to the sum of all the resistance in the current path. From equation (4-2),

$$V_T = V_1 + V_2 + V_3 \qquad (4\text{-}2)$$

Substituting IR for V, we have

$$I_T R_T = I_1 R_1 + I_2 R_2 + I_3 R_3 \quad \text{etc.}$$

Since the current is the same,

$$R_T = R_1 + R_2 + R_3 \quad \text{etc.} \qquad \text{ohms} \qquad (4\text{-}3)$$

*Note that the subscript T after the V indicates total, although this could imply source voltage, applied voltage, or generator voltage.

EXAMPLE 4-4

Three resistors, 1.2 kΩ, 2.2 kΩ, and 3.3 kΩ, are connected in series. What is the total resistance of the circuit?

Solution

$$R_T = 1.2\text{ k}\Omega + 2.2\text{ k}\Omega + 3.3\text{ k}\Omega$$
$$= 6.7\text{ k}\Omega$$

PRACTICE QUESTIONS 4-1

Choose the most suitable word, phrase, or value to complete each statement.

1. A circuit with two or more loads and one path for current flow is termed a/an (*open, closed, short, series, parallel*) circuit.
2. Electrons will always flow from the ________ potential to the lowest potential.
3. In a series circuit the current flowing from or back into the source will always be the same as the current flowing through the ________.
4. Two or more voltage sources connected negative to positive will give an applied voltage which will be the (*difference, sum, zero, maximum*) voltage of the sources.
5. Two or more voltage sources connected positive to positive will give an applied voltage which will be the (*difference, sum, zero, maximum*) voltage of the sources.
6. The polarity of one end of a resistor with electrons flowing into that end will be (*positive, negative, neutral, maximum*).
7. To cause electrons to flow through a resistor there must be a (*source, current source, voltage source, potential difference*) between the two ends of the resistor.
8. The negative lead of the voltmeter is normally ________ in color.
9. A deflection type of multimeter must be connected correctly as to polarity to ensure that the pointer deflects from zero to (*midscale, full scale, the right, the left*).
10. In a series circuit the (*difference, sum, source*) of the voltage drops will be equal to the source voltage.
11. What is the source voltage if the voltage drops around a series circuit are 12 V, 15 V, and 36 V?

12. In a series circuit the (*difference, sum, source*) of the resistance around the circuit will be equal to the total resistance.
13. Four resistors, 680 Ω, 1.5 kΩ, 390 Ω, and 2.7 kΩ, are connected in series. What is the total resistance of the circuit?
14. In a series circuit an open circuit means that the circuit is (*fully, partially, not at all*) functional.

4-2 OHM'S LAW FOR A SERIES CIRCUIT

Current

Since current is common in all parts of the circuit, Ohm's law can be applied directly to any load or resistor. Given that the values of the load resistances are known and the voltage across the resistor, the current can be found using equation (3-3), $I = V/R$.

EXAMPLE 4-5

Given $R_2 = 3.3$ kΩ and $V_2 = 6.6$ V, find the total current in the circuit shown in Fig 4-10.

Solution: The current is found by using equation (3-3).

$$I_2 = I_T = \frac{V_2}{R_2} = \frac{6.6 \text{ V}}{3.3 \text{ k}\Omega} = 2 \text{ mA}$$

The 2-mA current flows through each of the load resistors R_1, R_2, and R_3 from the source supply and back to the source supply. It is the total current in the circuit.

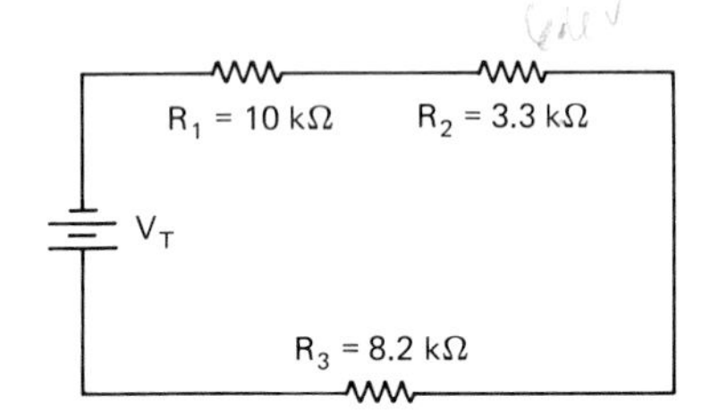

FIGURE 4-10 Series circuit schematic.

Voltage

Once the current is known in a series circuit, the voltage drops across each resistor can be found using Ohm's law, equation (3-4), $V = IR$. The voltage drops can then be summed to give the source or applied voltage.

EXAMPLE 4-6

From Example 4-5, find the voltage drop across R_1 and R_3. What is the source voltage?

Solution: Current is common through all parts of a series circuit and was found to be 2 mA, where $V_2 = 6.6$ V. Therefore,

$$V_1 = I_1R_1 = 2 \text{ mA} \times 10 \text{ k}\Omega = 20 \text{ V}$$

and

$$V_3 = I_3R_3 = 2 \text{ mA} \times 8.2 \text{ k}\Omega = 16.4 \text{ V}$$

The total or source voltage will be equal to the sum of the voltage drops, or

$$V_T = V_1 + V_2 + V_3$$
$$= 20 \text{ V} + 6.6 \text{ V} + 16.4 \text{ V} = 43 \text{ V}$$

The total or source voltage can also be determined using the Ohm's law equation, $V_T = I_TR_T$. For this equation we need to find the total resistance in the circuit from equation (3-5), $R = V/I$, so that

$$R_T = \frac{V_T}{I_T} = \frac{43 \text{ V}}{2 \text{ mA}} = 21.5 \text{ k}\Omega$$

and

$$V_T = I_TR_T$$
$$= 2 \text{ mA} \times 21.5 \text{ k}\Omega = 43 \text{ V}$$

Resistance

Individual resistance values can be calculated by using the equation $R = V/I$.

EXAMPLE 4-7

From the schematic in Example 4-5, we know that the current is 2 mA through each of the resistors. Having found the voltage drop across each resistor, we can verify the value of each, or

$$R_1 = \frac{V_1}{I_1} = \frac{20 \text{ V}}{2 \text{ mA}} = 10 \text{ k}\Omega$$

$$R_2 = \frac{V_2}{I_2} = \frac{6.6 \text{ V}}{2 \text{ mA}} = 3.3 \text{ k}\Omega$$

and

$$R_3 = \frac{V_3}{I_3} = \frac{16.4 \text{ V}}{2 \text{ mA}} = 8.2 \text{ k}\Omega$$

The total resistance can also be verified using

$$R_T = R_1 + R_2 + R_3$$

$$= 10\text{ k}\Omega + 3.3\text{ k}\Omega + 8.2\text{ k}\Omega = 21.5\text{ k}\Omega$$

The complete series circuit can be analyzed once the current is known or can be calculated. Find the current in the circuit by using either the voltage drop or resistance of any individual resistor and apply Ohm's law, $I = V/R$; or find or use the total voltage and resistance of the circuit and apply Ohm's law, $I = V_T/R_T$.

4-3 VOLTAGE-DIVISION RELATIONSHIP TO RESISTANCE AND TO A COMMON GROUND

The voltage across a specific resistor or resistance will be proportional to the ratio of that resistance and to the total resistance in the circuit.

Ohm's law equation $V = IR$ tells us that the voltage will be directly proportional to the resistance given that the current is constant. Since the current is constant in all parts of the series circuit, we can apply this direct relationship: From

$$V_1 = I_1 R_1 \qquad (1)$$

since $I_1 = I_T$, substitute V_T/R_T for I_1 in equation (1).

$$V_1 = \frac{V_T}{R_T} \times R_1$$

or

$$V_1 = \frac{R_1}{R_T} \times V_T \quad \text{volts} \qquad (4\text{-}4)$$

Similarly,

$$V_2 = \frac{R_2}{R_T} \times V_T \quad \text{etc.} \quad \text{volts}$$

EXAMPLE 4-8

Three resistors, 2.2 kΩ, 15 kΩ, and 33 kΩ, are connected in series to an applied voltage of 100 V. Find the voltage drop across each resistor.

Solution: The current is common; therefore, we can apply voltage-division formula (4-4), where

$$V_T = 100\text{ V}$$

$$R_T = R_1 + R_2 + R_3 = 2.2\text{ k}\Omega + 15\text{ k}\Omega + 33\text{ k}\Omega$$

$$= 50.2\text{ k}\Omega$$

$$V_1 = \frac{R_1}{R_T} \times V_T \qquad (4\text{-}4)$$

$$= \frac{2.2\text{ k}\Omega}{50.2\text{ k}\Omega} \times 100\text{ V} = 4.4\text{ V}$$

$$V_2 = \frac{R_2}{R_T} \times V_T = \frac{15\text{ k}\Omega}{50.2\text{ k}\Omega} \times 100 = 29.9\text{ V}$$

$$V_3 = \frac{R_3}{R_T} \times V_T = \frac{33\text{ k}\Omega}{50.2\text{ k}\Omega} \times 100$$

$$= 65.7\text{ V}$$

Verify: $V_T = 4.4\text{ V} + 29.9\text{ V} + 65.7\text{ V} = 100\text{ V}$.

PRACTICE QUESTIONS 4-2

1. Two equal lamps are connected in series. The current through lamp 1 is 250 mA; what is the total current in the circuit?
2. What is the current through lamp 2 of Question 1?
3. Assume that the lamps of Question 1 are connected to 12.6 V. What would the voltage drop be across each lamp?
4. Calculate the resistance of each lamp in Question 3.
5. Calculate the total resistance in the circuit of Question 3.
6. Three resistors, 1.8 kΩ, 4.7 kΩ, and 6.0 kΩ, are connected in series. The voltage across the 1.8 kΩ is 9 V. Find the source (i.e., total current) in the circuit.
7. Find the total source voltage in the circuit of Question 6.
8. Find the voltage drops across the 4.7-kΩ and the 6.0-kΩ resistors is in the circuit of Question 6.
9. Find the total resistance of the circuit of Question 6.
10. Verify the voltage drops across each resistor in the circuit of Question 6 using voltage-division formula (4-4).

4-4 INTERNAL RESISTANCE OF A SOURCE

There are many dc and ac sources of voltage and current. Some of these sources are the photocell, microphone, telephone, thermocouple, generator, and of course the battery. All of these sources have one thing in common—some form of internal resistance. This is an introduction to internal resistance of a source; therefore, we will look only at the internal resistance of a battery and its characteristics to the series circuit.

An Ideal Source

An ideal source is able to provide constant voltage under varying conditions of load current. The closest approach to this ideal source is the present state-of-the-art voltage and current electronic regulators which form the integral part of a dc source.

Internal Resistance of a Cell

Regardless of the type of cell, all cells rely on using the conversion of chemical energy into electrical energy. The chemical action, construction, and lead connections of the cell contribute to its internal resistance. This provides us with a source that is less than ideal. The schematic diagram of Fig. 4-11 shows an equivalent circuit of a dry cell with its internal resistance r in series.

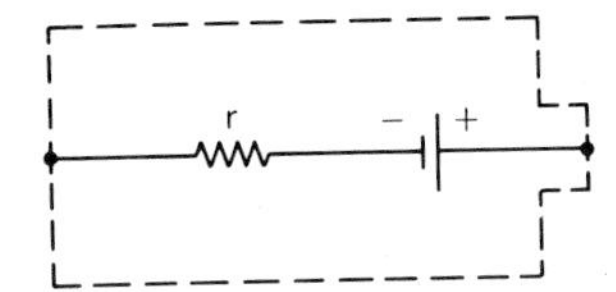

FIGURE 4-11 Internal resistance of cell/battery.

No-Load versus Full-Load Voltage

Figure 4-12 shows a voltmeter that draws negligible current connected across the terminals of the cell, which will indicate the no-load or open-circuit voltage of the cell or battery. This voltage will normally be the rated voltage of the cell or battery: for example, 1.5 V for a carbon cell, 9 V for an alkaline-manganese battery. Since there is no current flowing in the no-load or open circuit, there is zero voltage drop across the internal resistance of the source. Therefore, the battery is able to maintain its full-rated voltage.

With a load-connected or closed circuit across the battery terminals, current will flow and cause a voltage drop (Fig. 4-12) across the internal resistance. The amount of this voltage drop will be governed by the current drawn by the load, so that the final voltage available across the battery terminals will be less than the rated voltage.

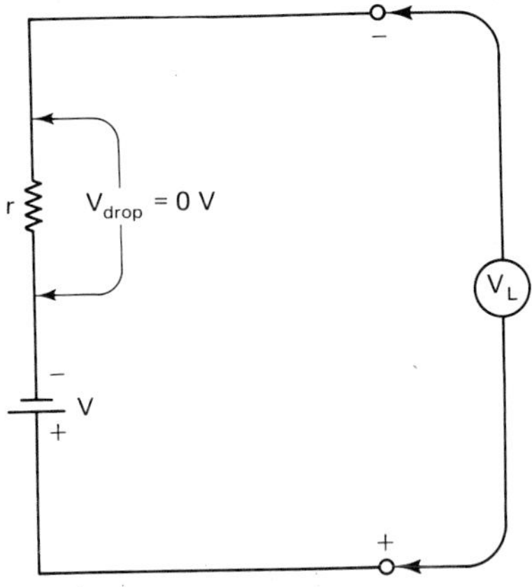

FIGURE 4-12 No load voltage is equal to cell or battery voltage.

Cell Battery Shelf Life and Aging

Since it is a chemical action that produces an electrical pressure and current, a cell's "chemistry" deteriorates over a period of time and use. This increases the internal resistance of the cell and the cell can no longer provide its rated voltage as well as its rated current. It is possible to recharge or revitalize the chemical reaction inside a cell.

Cells such as the carbon-zinc and alkaline-manganese cannot be recharged and are discarded. It is important to ensure that these types of cells have not been "sitting on a shelf" for an indefinite period since their chemical reaction will have greatly deteriorated. The next time you see a sale on batteries, be cautious—they might not have much life left in them due to their extended shelf life. Cells and battery voltages must be taken under load conditions in order to ascertain the true voltage being delivered to the load.

EXAMPLE 4-9

A 12.6-V battery having an internal resistance of 0.6 Ω is to be charged at the rate of 5 A from a dc power supply. To what voltage must the dc power supply be set?

Solution: The battery must have 12.6 V across it in order to be charged. The dc power source must supply 12.6 V plus the voltage drop across the internal resistance, which will be $V = IR = 5\text{ A} \times 0.6\ \Omega = 3\text{ V}$. The power supply must provide 12.6 V + 3 V or 15.6 V.

EXAMPLE 4-10

The no-load voltage of a portable radio battery is 9 V. The full-load voltage of the same battery is 7.5 V with 70 mA. Find the internal resistance of the battery.

> ***Solution:*** The difference between the no-load and full-load voltage is 9–7.5 V = 1.5 V. This 1.5 V is the voltage drop across the internal resistance of the source with 70 mA flowing through it. Using Ohm's law, $R = V/I$, we obtain 1.5/70 mA = 21.4 Ω for the internal resistance of the battery.

4-5 APPLICATIONS OF A SERIES CIRCUIT

The series circuit has many applications. Some of the more obvious applications are as follows:

Batteries. As discussed previously, cells are connected in series to provide a higher battery voltage.

Series lamps. These are lamps connected in series, especially for decorative purposes such as Christmas tree light sets.

Appliance motors. Appliance motors that are typically used in vacuum cleaners, mixers, blenders, lawn mowers, and so on, have a series type of motor. The motor armature windings are series connected through a commutator and brushes to the field winding and to the ac line cord.

Electronic circuits. Resistors are used in series and in parallel configuration, normally to provide a bias voltage or current for transistor and other semiconductor circuits, as shown in Fig. 4-13.

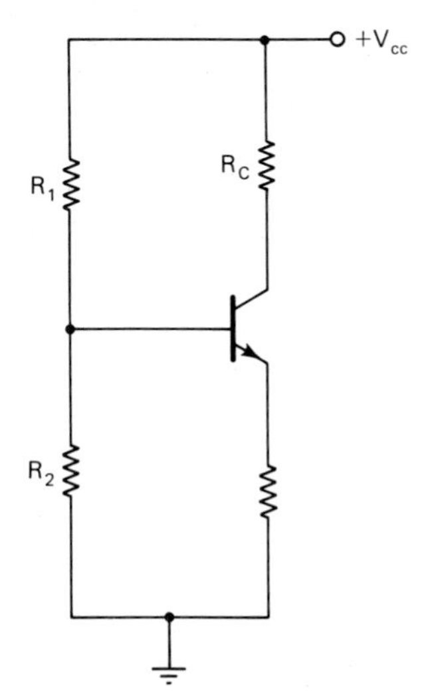

FIGURE 4-13 Resistors R_1 and R_2 provide a series circuit voltage divider.

Voltage dividers. Where two or more values of dc voltages and currents are required to supply an electronic circuit from one source of power, a voltage divider will provide these requirements. A *voltage divider* is a number of resistors connected in series across a power source. Figure 4-14 shows a simple voltage divider.

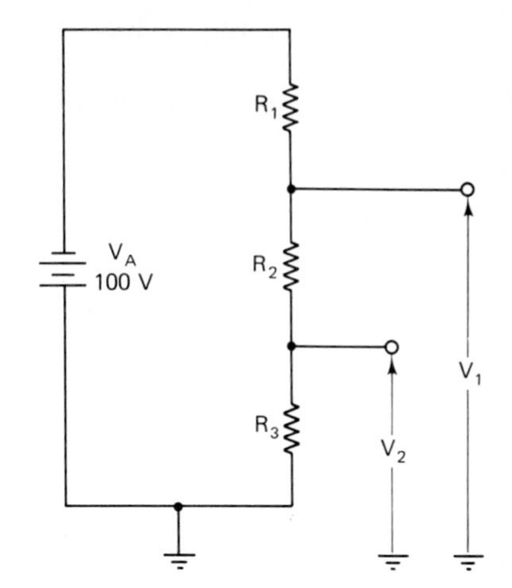

FIGURE 4-14 A simple voltage divider.

Open Circuits. The simplest method of detecting an open circuit in a series path is to take voltage readings across each load. The normal loads will give a zero voltage reading since there is no current flowing through them. The open-circuited load will give an applied voltage reading since the voltmeter completes the circuit through its internal resistance. This is shown in Fig. 4-15.

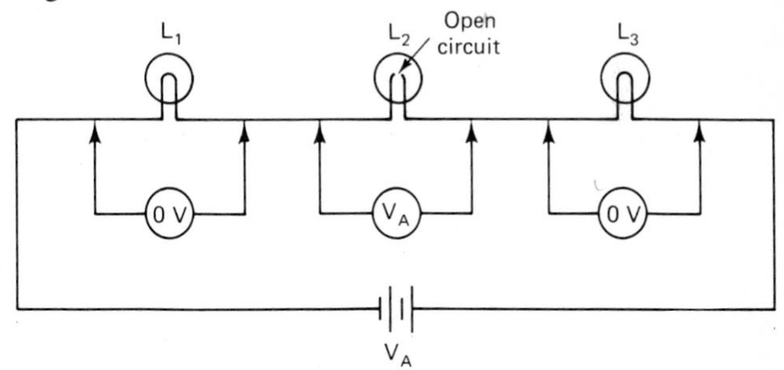

FIGURE 4-15 Full voltage V_A will be read by voltmeter across the open circuit of lamp 2.

To keep the remaining lamps operating, a thermistor is shunted or connected in parallel with each lamp in the circuit. With the lamp fully operating, the thermistor gets very little current and its resistance remains very high. Once the lamp burns out, a higher current flows through the thermistor. The current heats up the thermistor, its resistance drops, and the operating current is allowed to flow to the remaining lamps.

Short Circuits. A short circuit across a load will reduce the voltage across the load to zero volts. This throws the applied voltage across the remaining loads, with a resultant increase in voltage across each load. Ohm's law tells us that an increase in voltage will increase the current. The result of this increase in current would be a serious overload of the remaining lamps and burnout within a short period of time.

SUMMARY

The series circuit has two or more loads with one path for current flow.

The amount of current flow is the same in all parts of a series circuit, that is,

$$I_T = I_1 = I_2 = I_3 \quad \text{etc.} \qquad \text{amperes} \tag{4-1}$$

Series-aiding voltages connect positive to negative, always leaving positive and negative lead polarities.

Series-opposing voltages connect positive to positive or negative to negative, always leaving two similar lead polarities.

A voltage drop is the result of current flow through a resistance.

The polarity of the voltage drop will be negative where electrons enter the resistance and positive where electrons leave the resistance.

The sum of the voltage drops about a series circuit will be equal to the source voltage, or

$$V_T = V_1 + V_2 + V_3 \quad \text{etc.} \qquad \text{volts} \tag{4-2}$$

The sum of the resistance about a series circuit will be equal to the total resistance of the circuit, or

$$R_T = R_1 + R_2 + R_3 \quad \text{etc.} \qquad \text{ohms} \tag{4-3}$$

Ohm's law can be applied to any resistor in a series circuit to determine its voltage drop given or finding the current through it, that is, $V = IR$.

The voltage across a specific resistor or resistance will be proportional to the ratio of that resistance to the total resistance in the circuit.

Internal resistance is inherent in all types of voltage and current sources.

The internal resistance of a cell increases with age because the chemical reaction decreases.

With no current being drawn from a source, its internal resistance has no voltage drop across it, and therefore gives its rated voltage.

With a load connected to the source, the internal resistance drops the available voltage to the load.

Series connections are used to increase a battery voltage.

Appliance motors have armature and field coil windings connected in series.

Two or more resistors are connected in series to provide a voltage divider for electronic circuits.

An open circuit does not have a current flow.

A short circuit has little or no resistance, therefore will not limit the current flow and causes a very large increase in current flow.

REVIEW QUESTIONS

4-1. Briefly compare a simple circuit to a series circuit.

4-2. Why is the amount of current the same in all parts of a series circuit?

4-3. Given three dc sources, explain briefly how they are connected series aiding.

4-4. Given three dc sources, explain briefly how they are connected series opposing.

4-5. Electron current flows into the right-hand end of a resistor. What is its polarity with respect to its left end?

4-6. There is never a current loss, whereas there is a voltage loss across a resistor. Explain why.

4-7. What is the effect of voltage drops or losses about a series circuit with respect to the source voltage?

4-8. What two factors will determine any voltage drop or loss across a resistance?

4-9. The largest resistance in a series circuit will have the greatest voltage drop. Explain.

4-10. What is the effect of two or more resistances connected in a series circuit with respect to the net resistance that the source "sees"?

4-11. If three equal-value resistors are connected in series, what effect does this have on the voltage drop of each?

4-12. What causes the internal resistance of a cell?

4-13. Under what condition should a battery voltage be measured? Explain why.

4-14. What effect would an open circuit have on three loads connected in series?

4-15. What effect would a short circuit across one of the loads have on three loads connected in series?

SELF-EXAMINATION

Choose the most suitable word, phrase, or value to complete each statement.

4-1. Show how four 1.5-V cells are connected in series aiding.

4-2. What voltage rating must a flashlight lamp have with the four cells in Question 4-1?

4-3. You inadvertently purchase a replacement lamp rated at 3 V for the flashlight of Question 4-1. The result of the circuit would be that (*it would not work, the lamp would be too bright, the cells would burn out, the lamp would burn out*).

4-4. One of the four cells deteriorates rapidly and "goes dead." The result on the circuit would be that (*the lamp would burn out, the lamp would be dim, the lamp would not work*).

4-5. One of the four cells is placed in a reverse position (i.e. positive to positive) in the flashlight. The result on the circuit would be that (*it would not work, the lamp would burn out, the lamp would be dim*).

4-6. What would be the output voltage of the series cell connection in Question 4-5?

4-7. To find the current in a series circuit, Ohm's law can be applied to the (*source, loads, conductors, voltmeter*).

4-8. To have current flow through series-connected resistors, there must be (*a low resistance, a potential difference, emf, a voltage source*) across each resistor.

4-9. The ________ (polarity) lead of a voltmeter is always connected to the end of a resistor that has electrons flowing out of that end.

4-10. By convention the ________ (color) lead of a voltmeter is always considered to be negative.

Refer to Fig. 4-16 for Questions 4-11 to 4-24. Use the following values:

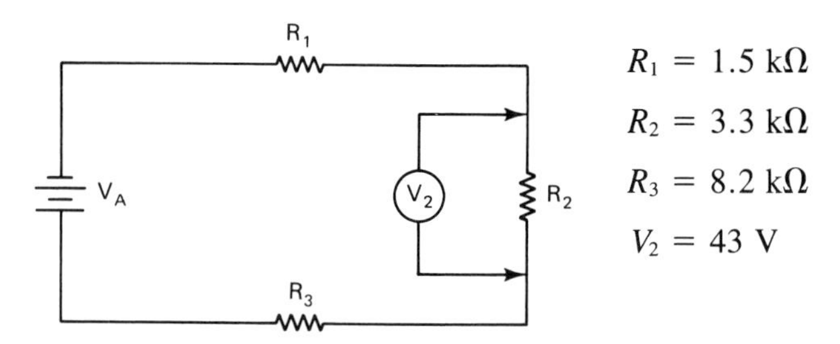

$R_1 = 1.5\ k\Omega$

$R_2 = 3.3\ k\Omega$

$R_3 = 8.2\ k\Omega$

$V_2 = 43\ V$

FIGURE 4-16 Series circuit schematic.

4-11. Calculate the value of I_2.

4-12. Calculate the value of I_T.

4-13. Calculate the value of V_1.

4-14. Calculate the value of V_3.

4-15. Calculate the value of V_A.

4-16. Find the total resistance in the circuit.

4-17. R_2 has a short circuit across it. Calculate I_T. (Assume the V_A of Question 4-15).

4-18. Find R_T with the short across R_2.

4-19. Assume the V_A of Question 4-15. Find V_1 with the short across R_2.

4-20. Calculate V_2 with the short across R_2.

4-21. R_1 is open circuited. Find the voltage across R_1 using a voltmeter. Assume the source voltage of Question 4-15.

4-22. Find the voltage across R_2 with R_1 open circuited.

4-23. Find the voltage across R_3 with R_1 open circuited.

4-24. Find the total resistance with R_1 open circuited.

Refer to Fig. 4-17 for Questions 4-25 to 4-27.

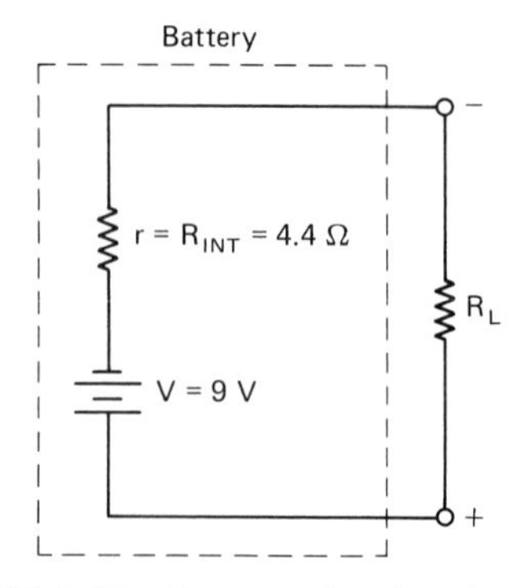

FIGURE 4-17 Battery circuit schematic.

4-25. What is the open-circuit voltage of this battery?

4-26. Given the load resistance is 90 Ω, what is the closed-circuit voltage of this battery?

4-27. The battery has deteriorated with use. The load now receives 8.2 V. What is the value of its internal resistance?

Refer to Fig. 4-18 for Questions 4-28 to 4-30. Use the following values:

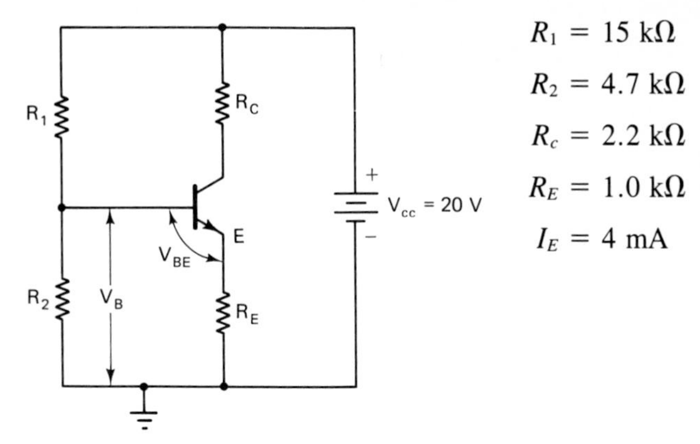

R_1 = 15 kΩ

R_2 = 4.7 kΩ

R_c = 2.2 kΩ

R_E = 1.0 kΩ

I_E = 4 mA

FIGURE 4-18 Series circuit schematic.

4-28. Calculate V_B.
4-29. Calculate V_E.
4-30. Calculate V_{BE}.

ANSWERS TO PRACTICE QUESTIONS

4-1

1. series
2. highest
3. components or load
4. sum
5. difference
6. negative
7. potential difference
8. black
9. the right
10. sum
11. 63 V
12. sum
13. 5.27 kΩ
14. not at all

4-2

1. 250 mA
2. 250 mA
3. 6.3 V
4. $R = 25.2\ \Omega$
5. $R_T = 50.4\ \Omega$
6. 5 mA
7. 62.5 V
8. 23.5 V across the 4.7 kΩ
   30 V across the 6.0 kΩ
9. $R_T = 12.5$ kΩ
10. $V_1 = \dfrac{R_1}{R_T} \times V_A = \dfrac{1.8\text{ k}\Omega}{12.5\text{ k}\Omega} \times 62.5\text{ V} = 9\text{ V}$

$$V_2 = \frac{R_2}{R_T} \times V_A = \frac{4.7\text{ k}\Omega}{12.5\text{ k}\Omega} \times 62.5\text{ V} = 23.5\text{ V}$$

$$V_3 = \frac{R_3}{R_T} \times V_A = \frac{6.0\text{ k}\Omega}{12.5\text{ k}\Omega} \times 62.5\text{ V} = 30\text{ V}$$

5

Parallel Circuits

INTRODUCTION

Unlike the series circuit connection, where the loads are dependent on each other for continuous operation, the parallel circuit loads operate independently. The parallel circuit is used exclusively in the home, commerce, and industry. This type of circuit enables you to plug in a table lamp and operate a radio from the same power outlet, independent of each other. Of course, you are limited to the number of appliances that you can connect in parallel by the size of the conductor, which in turn determines the size of the fuse. This is why the fuse or breaker "blows" when you plug a toaster and a frying pan into the same outlet.

The parallel circuit is especially useful in electronic circuits, where applications will be directed to resistors connected in parallel, meter shunts, speakers, and to parallel tuned circuits and many other parallel connections.

MAIN TOPICS IN CHAPTER 5

5-1 Parallel Circuit Characteristics
5-2 Ohm's Law for a Parallel Circuit
5-3 Current-Division Relationship to Parallel Resistance
5-4 Resistor Configurations to Give Net Resistance
5-5 Applications of Parallel Circuits

OBJECTIVES

After studying Chapter 5, the student should be able to

1. Relate current, voltage, and resistance characteristics of a parallel circuit.
2. Apply Ohm's law equations to solve for parallel circuit voltage, current, and resistance.
3. Relate current-division proportionality to resistance proportionality of parallel resistors.
4. Connect various resistor combinations in parallel to give a desired net resistance.
5. Relate the parallel circuit to practical applications.

5-1 PARALLEL CIRCUIT CHARACTERISTICS

In Chapter 4 we learned to connect cells in series to obtain higher voltages, as Sir Humphry Davy did using

2000 cells to provide 3000 V. If the same cells were now connected in parallel, we would obtain a total voltage of 1.5 V. But the total current available would now be 2000 times greater than that obtained from one cell. Each cell provides its own branch for current flow; therefore, each cell adds its available current to the total source current.

Current Path

> The parallel circuit has two or more branches for current flow.

Shown in Fig. 5-1 are three 1.5-V cells connected in parallel. The output voltage of the parallel combination or battery is 1.5 V, so what is the advantage, since only one cell is needed to produce 1.5 V? The advantage is in the total current available. With one cell being able to sustain a current of, say, 1 A, the parallel combination would be able to sustain a current of 3 A. Each cell has its own branch for current flow; therefore, a load connected to the combination would be able to "draw" current from the three cells as shown in Fig. 5-2. Note that each cell is independent. Removing one of the cells reduces the total available current from 3 A to 2 A.

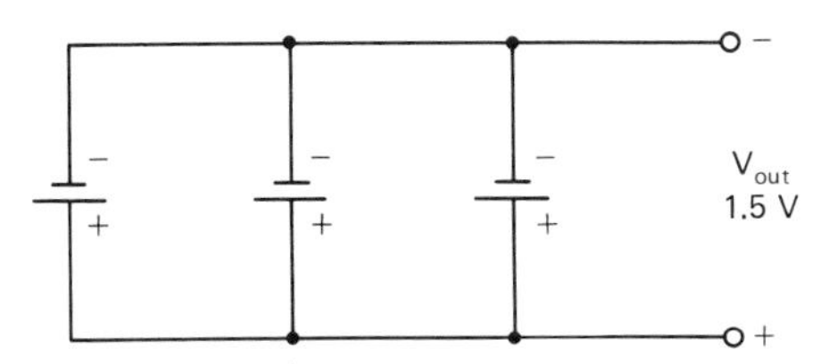

FIGURE 5-1 Cells connected in parallel.

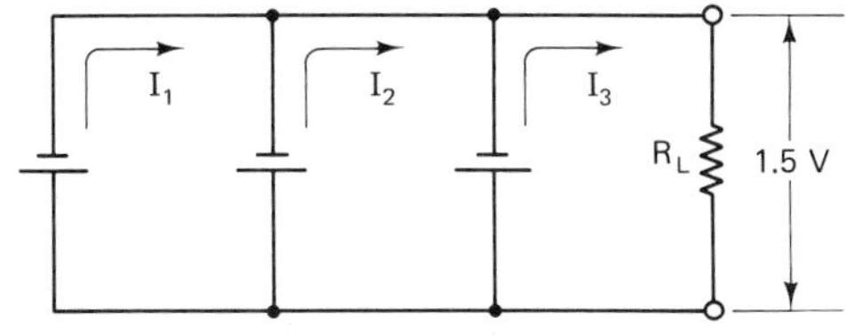

FIGURE 5-2 Current flow in a parallel circuit.

Current

> The sum of the currents in a parallel circuit is equal to the total current flow from the source or returning to the source.

In electronic circuits the ac/dc source provides the voltage and current with load resistors connected in parallel or some combination of series parallel. In the parallel circuit the load currents have independent branches for current flow, as shown in Fig. 5-3. The current flow from and returning back to the source will always be equal to the sum I_T of the load currents.

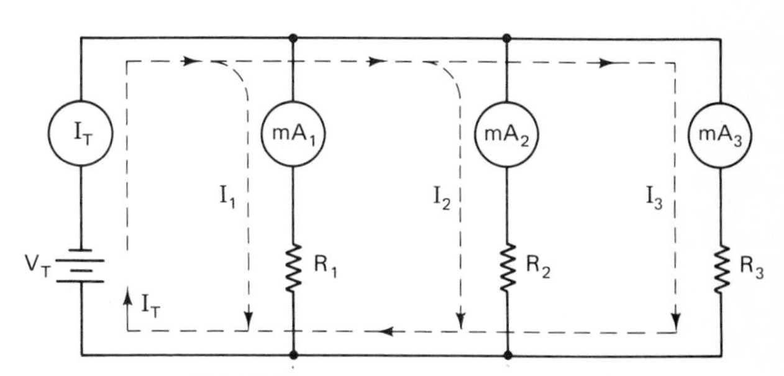

FIGURE 5-3 Parallel resistors.

Note that one end of a resistor or load must be disconnected from the parallel branch in order to connect a milliammeter in series with that respective load. The polarity of the milliammeter loads must be observed according to electron current flow. In equation form,

$$I_T = I_1 + I_2 + I_3 \text{ etc.} \qquad \text{amperes} \qquad (5\text{-}1)$$

EXAMPLE 5-1

Three resistors of Fig. 5-3 have branch currents of 10 mA, 8.2 mA, and 15 mA, respectively. What is the total current flowing from the source?

Solution: Using equation (5-1), we have

$$I_T = 10 + 8.2 + 15 = 33.2 \text{ mA}$$

EXAMPLE 5-2

The parallel circuit shown in Figure 5-4 has currents $I_1 = 10$ mA, $I_2 = 8.2$ mA, and, $I_3 = 15$ mA. With the same three resistors as in Example 5-1, what would milliammeters 1, 2, and 3 read?

Solution: The current through mA1 would be the sum of the I_2 and I_3 currents, or $mA_1 = I_2 + I_3 = 8.2$ mA $+ 15$ mA $= 23.2$ mA. The current through mA2 would be $I_3 = 15$ mA. The current through mA3 would be the sum of all three branch currents, or the total current, $I_T = I_1 + I_2 + I_3 = 10$ mA $+ 8.2$ mA $+ 15$ mA $= 33.2$ mA.

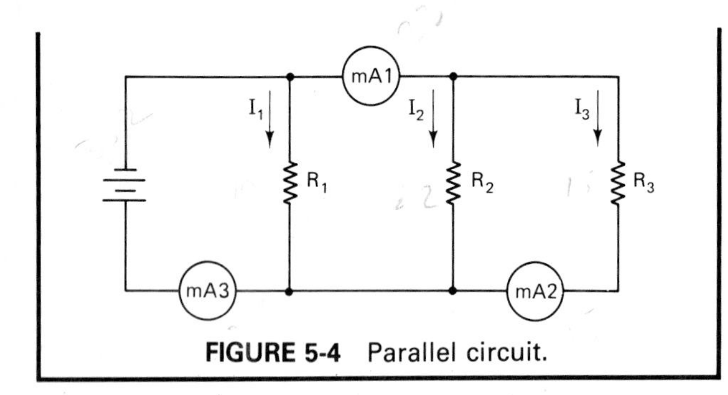

FIGURE 5-4 Parallel circuit.

Voltage

> The voltage is common to all branches or loads of a parallel circuit.

One side or end of each load in a parallel circuit is connected to the high-potential side of the source, while the other end or side is connected to the low-potential side of the source, as shown in Fig. 5-5.

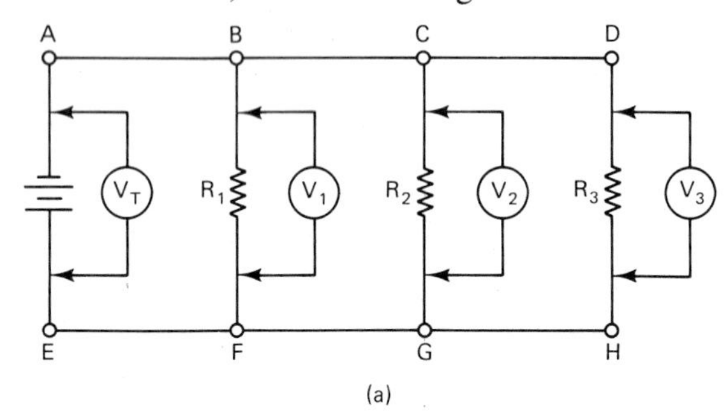

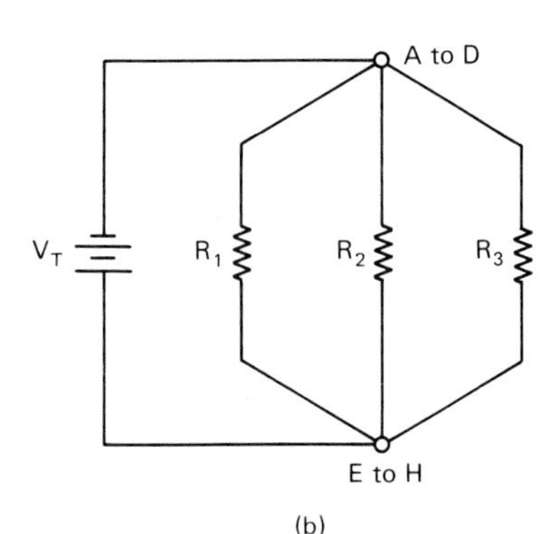

FIGURE 5-5 (a) Voltage is common in a parallel circuit. (b) Equivalent circuit.

Since all high-potential B, C, and D points are common to the source high-potential point A, and all the low-potential points F, G, and H are common to the source low-potential point E, the voltage of the source A to E will be common or the same to all the load resistors. In equation form,

$$V_T = V_1 = V_2 = V_3 \quad \text{etc.} \qquad \text{volts} \qquad (5\text{-}2)$$

EXAMPLE 5-3

Given that the voltage across R_2 in Fig. 5-5 is 100 V, what are the values of V_1, V_2, and V_T?

Solution: Voltage is common in all loads connected in parallel or

$$V_T = V_1 = V_2 = V_3 = 100 \text{ V}$$

Resistance

The individual resistance of each branch can be found by Ohm's law, $R = V/I$, where the voltage is common but the current is the separate load current of that respective resistor or branch. "Looking" into the circuit from the source are two or more branches for current flow, where the sum of the currents is equal to the total current, or

$$I_T = I_1 + I_2 + I_3 \quad \text{etc.} \qquad (5\text{-}1)$$

Substituting V/R for the current gives us

$$\frac{V_T}{R_T} = \frac{V_1}{R_1} + \frac{V_2}{R_2} + \frac{V_3}{R_3} \quad \text{etc.} \qquad (5\text{-}1a)$$

Since voltage is common, and dividing by V, equation (5-1a) can be written

$$\frac{1}{R_T} = \frac{1}{R_1} + \frac{1}{R_2} + \frac{1}{R_3} \quad \text{etc.} \qquad \text{ohms} \qquad (5\text{-}3)$$

Recall that conductance was discussed in Chapter 2. Conductance is a measure of a circuit's ability to allow current to flow and is the reciprocal of resistance, or $G = 1/R$. Substituting conductance for the reciprocal of the resistance in equation (5-3) gives

$$G_T = G_1 + G_2 + G_3 \quad \text{etc.} \qquad \text{siemens} \qquad (5\text{-}4)$$

EXAMPLE 5-4

What total resistance does the source "see" in the circuit of Fig. 5-4 given $R_1 = 10\text{ k}\Omega$, $R_2 = 12.2\text{ k}\Omega$, and $R_3 = 6.7\text{ k}\Omega$?

Solution: Using equation (5-3) yields

$$\frac{1}{R_T} = \frac{1}{10\text{ k}\Omega} + \frac{1}{12.2\text{ k}\Omega} + \frac{1}{6.7\text{ k}\Omega}$$

$$= \frac{270.7}{817.4}$$

Therefore,

$$R_T = \frac{817.4}{270.7} \cong 3\ \text{k}\Omega$$

Using a calculator

Step 1. Enter 10. The display shows 10.
Step 2. Press the 1/x key. The display shows 0.1.
Step 3. Press the + key. The display shows 0.1.
Step 4. Enter 12.2. The display shows 12.2.
Step 5. Press the 1/x key. The display shows 8.1967-02 or 0.0819672.
Step 6. Press the + key. The display shows 0.1819672.
Step 7. Enter 6.7. The display shows 6.7.
Step 8. Press the 1/x key. The display shows 0.1492537.
Step 9. Press the = key. The display shows 0.3312209.

The total conductance of the parallel circuit is found in step 9 and is in siemens. To find the total resistance, one more step is required.

Step 10. Press the 1/x key. The display shows 3.0191328.

Therefore, the total resistance of the circuit to four significant figures is 3.019 kΩ.

Total Resistance Product-over-Sum Method. The total resistance of two parallel resistors can be found using the product-over-sum method. From

$$\frac{1}{R_T} = \frac{1}{R_1} + \frac{1}{R_2} \quad \text{ohms} \qquad (5\text{-}3)$$

A common denominator can be found by multiplying R_1 and R_2, or

$$\frac{1}{R'_T} = \frac{R_1 + R_2}{R_1 \times R_2}$$

Invert to find R'_T:

$$R'_T = \frac{R_1 \times R_2}{R_1 + R_2} \quad \text{ohms} \qquad (5\text{-}5)$$

The product-over-sum method is applicable only for two parallel resistors at a time. If the total resistance of three or more resistors in parallel is to be found using the product-over-sum method, the total resistance of one pair must be found first, then this total resistance (R'_T) is used in conjunction with the remaining resistance to find the total resistance.

EXAMPLE 5-5

What is the total resistance of the three resistors in the circuit of Example 5-4 using the product-over-sum method?

Solution: Take any pair of resistors first, say R_3 and R_2:

$$R'_T = \frac{R_2 \times R_3}{R_2 + R_3} = \frac{12.2\ \text{k}\Omega \times 6.7\ \text{k}\Omega}{12.2\ \text{k}\Omega + 6.7\ \text{k}\Omega}$$

$$= \frac{81.7\ \text{k}\Omega}{18.9\ \text{k}\Omega} = 4.32\ \text{k}\Omega$$

Therefore,

$$R_T = \frac{R'_T \times R_1}{R'_T + R_1} = \frac{4.32\ \text{k}\Omega \times 10\ \text{k}\Omega}{4.32\ \text{k}\Omega + 10\ \text{k}\Omega}$$

$$= 3\ \text{k}\Omega$$

Note: The total resistance in a parallel circuit will always be smaller than the smallest branch resistance value.

The reason for this is that the lowest-value resistance or branch has the greatest current flow, therefore contributing the greatest proportion of conductance to the total conductance of the circuit. Adding the other conductance branches to the circuit increases the total conductance and decreases the net resistance.

PRACTICE QUESTIONS 5-1

Choose the most suitable word or phrase to complete each statement.

1. A circuit with two or more loads and two or more branches for current flow is termed a/an (*open, closed, short, series, parallel*) circuit.
2. Connecting loads in parallel will cause the source current to (*decrease, remain the same, increase*).
3. Removing loads in parallel will cause the source current to (*decrease, remain the same, increase*).

4. Current flow from the source and back to the source will be equal to the (*difference, sum*) of the load currents.
5. Loads connected in (*series, series–parallel, parallel*) will operate independently of each other.
6. Three 12-V batteries connected in parallel supply a load of lamps. Each battery is able to supply 25 A. How many lamps can be connected to the batteries if one lamp draws 1 A?
7. Three lamps are connected in parallel. Each lamp draws 1 A. What is the total current drawn from the source? Give your answer in equation form.
8. For the same circuit as in Question 7, one of the lamps burns out; what is the total current drawn from the source? Give your answer in equation form.
9. Given that a fuse or circuit breaker has a rating of 15 A, what is the maximum number of lamps (in Question 7) that can be connected in parallel?
10. Given that the lamps of Question 7 plug into the convenience outlet of your home, what is the voltage across each lamp?
11. Given that you accidentally plug the lamps of Question 7 into 220-V, what would the current be through each lamp? What is the total current from the 220-V source?
12. What is the voltage across each lamp of Question 7?
13. The three lamps of Question 7, of 110 V each, are connected in parallel. Calculate the total resistance.
14. Remove one of the lamps in Question 13. What is the total resistance of the circuit?
15. Remove two of the lamps in Question 13. What is the total resistance of the circuit?

5-2 OHM'S LAW FOR A PARALLEL CIRCUIT

Applying Ohm's law to each individual resistance connected in parallel will give its current, resistance, or voltage. Applying the calculated information to the characteristics of a parallel circuit as discussed in Section 5-2 will yield the necessary voltage, current, and resistance to solve the complete circuit.

EXAMPLE 5-6

In the parallel circuit shown in Fig. 5-6, the current that mA_2 reads is 3.6 mA, while mA_3 reads 7.4 mA. Find (a) V_T, (b) I_1, (c) R_3, (d) I_T, and (e) R_T.

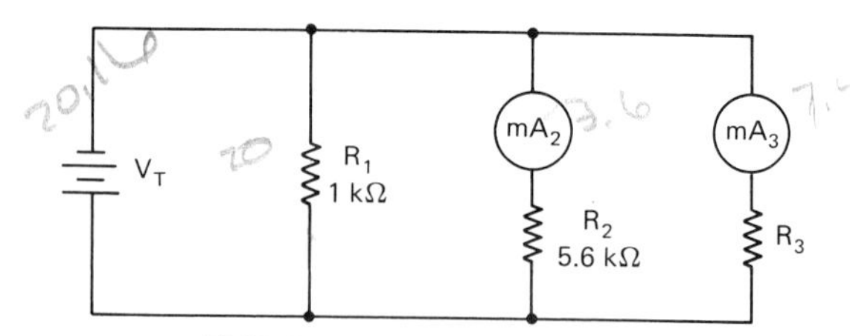

FIGURE 5-6 Parallel circuit.

Solution: Solving for the voltage across any one of the resistances will give the applied voltage from

$$V_T = V_1 = V_2 = V_3 \qquad (5\text{-}2)$$

(a) Resistor R_2 has a current of 3.6 mA flowing through it, so that

$$V_T = V_2 = I_2 R_2$$
$$= 3.6 \text{ mA} \times 5.6 \text{ k}\Omega \cong 20 \text{ V}$$

Therefore,

$$V_T = V_1 = V_2 = V_3 \cong 20 \text{ V}$$

(b) Solve for I_1: Since $V_1 = 20$ V, then

$$I_1 = \frac{V_1}{R_1} = \frac{20 \text{ V}}{1 \text{ k}\Omega} = 20 \text{ mA}$$

(c) Solve for R_3: Since $V_3 = 20$ V, then

$$R_3 = \frac{V_3}{I_3} = \frac{20 \text{ V}}{7.4 \text{ mA}} = 2.7 \text{ k}\Omega$$

(d) Solve for I_T: From equation (5-1),

$$I_T = I_1 + I_2 + I_3$$
$$= 20 \text{ mA} + 3.6 \text{ mA} + 7.4 \text{ mA}$$
$$= 31 \text{ mA}$$

(e) Solve for R_T:

$$R_T = \frac{V_T}{I_T} = \frac{20 \text{ V}}{31 \text{ mA}} = 645 \ \Omega$$

Verify R_T using the product-over-sum formula (5-5):

$$R_T' = \frac{R_3 \times R_2}{R_3 + R_2} = \frac{2.7 \times 5.6}{2.7 + 5.6} = 1.8 \text{ k}\Omega$$

$$R_T = \frac{R_T' \times R_1}{R_T' + R_1} = \frac{1.8 \times 1}{1.8 + 1} = 643 \ \Omega$$

Verify I_T using Ohm's law:

$$I_T = \frac{V_T}{R_T} = \frac{20 \text{ V}}{643 \ \Omega} = 31 \text{ mA}$$

EXAMPLE 5-7

For the parallel circuit shown in Fig. 5-7, mA_1 = 17 mA, mA_2 = 11 mA, mA_T = 65 mA, and V_3 = 67 V. Find (a) V_T, (b) R_1, (c) R_2, (d) R_3, and (e) R_T.

Solution: (a) The source voltage V_T is easily solved since the voltage V_3 is the parallel voltage of the circuit, from the equation

$$V_T = V_1 = V_2 = V_3 \quad \text{etc.} \qquad (5\text{-}2)$$

Therefore,

$$V_3 = V_T = 67 \text{ V}$$

(b), (c) Solve for R_1 and R_2: Since the voltage is common in all parts of a parallel circuit, $V_T = V_1 = V_2 = V_3 = 67$ V, the resistor values can be found using Ohm's law:

$$R_1 = \frac{V_T}{I_1} = \frac{67 \text{ V}}{17 \text{ mA}} = 3.9 \text{ k}\Omega$$

$$R_2 = \frac{V_T}{I_2} = \frac{67 \text{ V}}{11 \text{ mA}} = 6.1 \text{ k}\Omega$$

(d) To solve for R_3, the current I_3 is needed. From equation (5-1),

$$I_T = I_1 + I_2 + I_3$$

$$I_3 = I_T - (I_1 + I_2)$$

$$= 65 \text{ mA} - (17 \text{ mA} + 11 \text{ mA})$$

$$= 37 \text{ mA}$$

Therefore,

$$R_3 = \frac{V_T}{I_T} = \frac{67 \text{ V}}{37 \text{ mA}} = 1.8 \text{ k}\Omega$$

(e) Solve for R_T:

$$R_T = \frac{V_T}{I_T} = \frac{67 \text{ V}}{65 \text{ mA}} \cong 1.0 \text{ k}\Omega$$

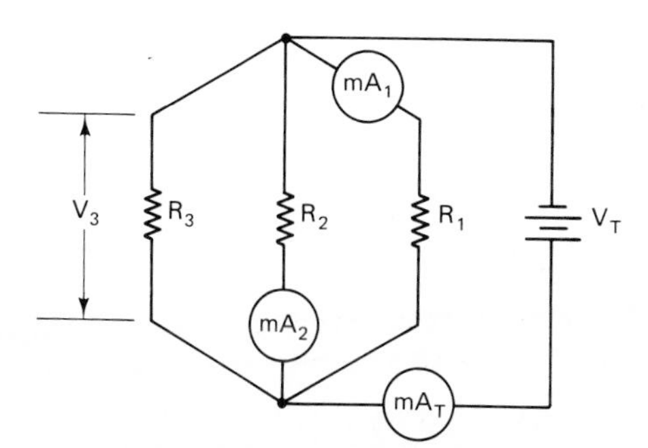

FIGURE 5-7 Parallel circuit.

5-3 CURRENT-DIVISION RELATIONSHIP TO PARALLEL RESISTANCE

If the voltage across a resistance is kept constant and the resistance is doubled, the current will be halved. If the resistance is halved, the current will be doubled. This relationship can be seen from Ohm's law, $V = IR$.

Current division for the parallel branch circuit is shown in Fig. 5-8. From Ohm's law, $V = IR$, where $V_1 = V_2 = V_T$, then

$$V_1 = I_T R_T \qquad (1)$$

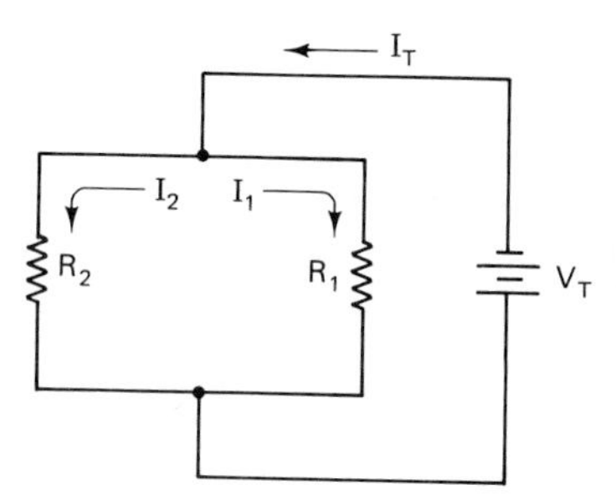

FIGURE 5-8 Current division for two parallel resistances.

Substitute product/sum for R_T in equation (1):

$$V_1 = I_T \frac{R_1 \times R_2}{R_1 + R_2} \qquad (2)$$

Substitute $I_1 R_1$ for V_1 in equation (2):

$$I_1 R_1 = I_T \frac{R_1 \times R_2}{R_1 + R_2} \qquad (3)$$

Solving for I_1 gives

$$I_1 = I_T \times \frac{R_2}{R_1 + R_2} \quad \text{amperes} \qquad (5\text{-}6a)$$

Similarly, the current I_2 can be found:

$$I_2 = I_T \times \frac{R_1}{R_1 + R_2} \qquad (5\text{-}6b)$$

EXAMPLE 5-8

Two resistors, 2.2 kΩ and 5.6 kΩ, are connected in parallel. The total current flow in this parallel branch is 49 mA. Find the current through each resistance.

Solution: Using equation (5-6) gives us

$$I_1 = I_T \times \frac{R_2}{R_1 + R_2}$$

$$= 49 \text{ mA} \times \frac{5.6 \text{ k}\Omega}{2.2 \text{ k}\Omega + 5.6 \text{ k}\Omega} \cong 35 \text{ mA}$$

and

$$I_2 = I_T \times \frac{R_1}{R_1 + R_2}$$

$$= 49 \text{ mA} \times \frac{2.2 \text{ k}\Omega}{2.2 \text{ k}\Omega + 5.6 \text{ k}\Omega} \cong 14 \text{ mA}$$

$$I_T = I_1 + I_2 = 35 + 14 = 49 \text{ mA}$$

PRACTICE QUESTIONS 5-2

Refer to the parallel circuit schematic Fig. 5-9 for Questions 1 to 7. Use the values

$$R_1 = 1.5 \text{ k}\Omega$$
$$R_2 = 5.1 \text{ k}\Omega$$
$$R_3 = 4.7 \text{ k}\Omega$$
$$V_2 = 30 \text{ V}$$

FIGURE 5-9 Circuit for Practice Questions 1 to 7.

1. Find V_T.
2. Find I_1.
3. Find I_2.
4. Find I_3.
5. Find I_T.
6. Find R_T.
7. Given *conventional current flow,* indicate all branches for current flow direction from the labeled schematic.

Refer to the parallel circuit schematic Fig. 5-10 for Questions 8 to 14. Use the values

S_1 is open circuited

$$R_1 = 3.3 \text{ k}\Omega$$
$$R_2 = 1.2 \text{ k}\Omega$$
$$R_3 = 6.8 \text{ k}\Omega$$
$$I_T = 24 \text{ mA}$$

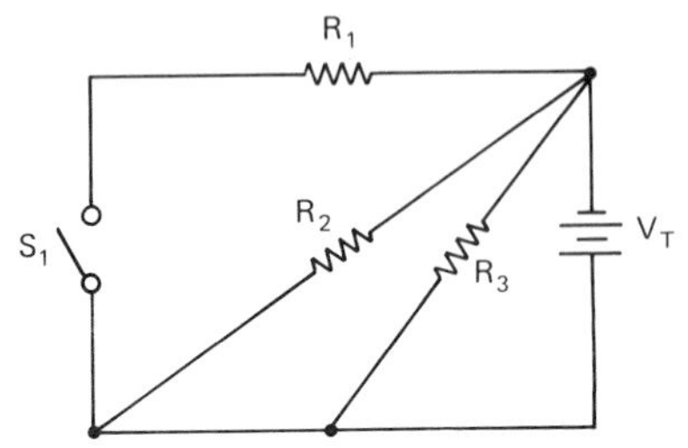

FIGURE 5-10 Circuit for Practice Questions 8 to 14.

8. Find I_2.
9. Find I_3.
10. Find V_T.
11. Find R_T.
12. Find I_1 with switch S_1 closed.
13. Find R_T with switch S_1 closed.
14. Find I_T with switch S_1 closed.

5-4 RESISTOR CONFIGURATIONS TO GIVE NET RESISTANCE

There are times, especially when constructing a prototype circuit, that you might require a certain value resistance other than an EIA standard value. Combining resistors in series or parallel (with limitation) will give a nonstandard resistance value.

Resistors of Equal Value in Parallel

The simplest combination of two or more resistors to give any specified value of resistance is the equal-value method. Combining equal-value resistors gives a total resistance that will be the value of one of the resistor values divided by the number of resistors connected in parallel, or

$$R_T = \frac{R}{N} \quad \frac{\text{resistance of one resistor}}{\text{number of resistors}} \qquad \text{ohms} \quad (5\text{-}7)$$

EXAMPLE 5-9

Three 10-kΩ resistors are connected in parallel. What is their net resistance?

Solution: Using equation (5-7) yields

$$R_T = \frac{R}{N} = \frac{10 \text{ k}\Omega}{3} = 3.3 \text{ k}\Omega$$

Notation for parallel resistors: A notation that is used to show resistors in parallel consists of two parallel lines (‖). The three resistors connected in parallel in Example 5-9 would be written as

$$R_T = R_1 \parallel R_2 \parallel R_3$$

Determining an Unknown Parallel Resistance

Combining equal-value resistors to obtain a desired value of resistance, although it is the simplest method, limits the possible value of equivalent resistance. A useful method to determine the value of a resistor to parallel with another value of resistor, to obtain a required value of resistance, is obtained by transposing the product-over-sum formula (5-5) to obtain

$$R_x = \frac{R \times R_T}{R - R_T} \quad \text{ohms} \qquad (5\text{-}8)$$

EXAMPLE 5-10

What resistance value is required in parallel with a 1.8-kΩ resistance to give an equivalent resistance of 1.4 kΩ? Select the closest 5% EIA-value resistor.

Solution: Using equation (5-8) yields

$$R_x = \frac{R \times R_T}{R - R_T} = \frac{1.8 \times 1.4}{1.8 - 1.4} = 6.3 \text{ k}\Omega$$

The closest 5% EIA-value resistor would be 6.2 kΩ.

5-5 APPLICATIONS OF PARALLEL CIRCUITS

The most common use of the parallel circuit is for lighting circuits as well as power circuits. You would most certainly want your lighting and power outlets to be independent of each other.

Not-so-obvious parallel (and series) connections are electric water heater, dryer, and stove elements. Both ac and dc motors have field coil windings and armature windings that are interconnected in parallel or in series or in combination series parallel.

In electronic circuits, resistors, capacitors, and inductors are connected in various series, parallel, or series–parallel combinations. It is essential to recognize the configuration of the connection so that a proper analysis of the circuit can be made.

SUMMARY

A parallel circuit has two or more branches for current flow.

The sum of the currents in a parallel circuit is equal to the total current flow, or

$$I_T = I_1 + I_2 + I_3 \text{ etc.} \quad \text{amperes} \qquad (5\text{-}1)$$

The voltage is common to all branches or loads of a parallel circuit, or

$$V_T = V_1 = V_2 = V_3 \text{ etc.} \quad \text{volts} \qquad (5\text{-}2)$$

The total conductance in a parallel circuit is the sum of the conductances, or

$$G_T = G_1 + G_2 + G_3 \text{ etc.} \quad \text{siemens} \qquad (5\text{-}4)$$

The total resistance in a parallel circuit is the reciprocal of the conductance, or

$$\frac{1}{R_T} = \frac{1}{R_1} + \frac{1}{R_2} + \frac{1}{R_3} \text{ etc.} \quad \text{ohms} \qquad (5\text{-}3)$$

For two parallel-connected resistors, the net resistance is given by

$$R_T' = \frac{R_1 \times R_2}{R_1 + R_2} \quad \text{ohms} \qquad (5\text{-}5)$$

The net resistance of a number of parallel-connected resistors is always less than the smallest individual resistors.

For two parallel-connected resistors with a total current I_T, the current-division relationship to resistance gives the current in each resistor by

$$I_1 = I_T \times \frac{R_2}{R_1 + R_2} \quad \text{and} \quad I_2 = I_T \times \frac{R_1}{R_1 + R_2} \tag{5-6}$$

The net resistance of a number of equal-value resistors connected in parallel is given by

$$R_T = \frac{R}{N} \quad \frac{\text{resistance of one } R}{\text{number of } R\text{'s}} \tag{5-7}$$

Given the net resistance of two resistors in parallel and the value of one resistor, the unknown value can be found using the formula

$$R_x = \frac{R \times R_T}{R - R_T} \quad \text{ohms} \tag{5-8}$$

REVIEW QUESTIONS

5-1. State two advantages of a parallel circuit over a series circuit.

5-2. Explain why the sum of the branch currents in a parallel circuit is equal to the total source current.

5-3. Explain why the voltage is common to all branches or loads of a parallel circuit.

5-4. The simple sum of resistance, as in the series circuit, gives net resistance, while the simple sum of conductance of branches in parallel gives net resistance. Explain the difference.

5-5. Explain a simple method of finding the net resistance of equal-value resistors connected in parallel.

5-6. Draw a schematic diagram to show three parallel resistors connected to a dc source with milliammeter symbols located in the appropriate place, to show current being measured through each resistor and the current *returning* to the source. Show the polarities of the leads.

5-7. The total resistance of two parallel resistors is 1075 Ω. One of the resistor values is 3.8 kΩ; find the other resistor value.

5-8. Two resistors, 1.8 kΩ and 5.6 kΩ, are connected in parallel to a dc source. The source current is 18 mA. Find the current through each resistor by using the current-division method.

SELF-EXAMINATION

Choose the most suitable word, phrase, or value to complete each statement.

5-1. A parallel circuit is characterized by having two or more branches for (*voltage, resistance, current, coulombs, cells*).

5-2. With two or more sources connected in parallel, a greater amount of (*voltage, resistance, current, coulombs, cells*) is available from the sources.

5-3. Given that a parallel circuit has three branch currents, two of the branch currents total 6.8 mA and the source current is 12 mA. What is the value of the third branch current?

5-4. In a parallel circuit the (*current, voltage, resistance, coulombs, cell*) is independent of each load and is the same as the source.

5-5. The simple sum of resistance connected in a ________ circuit configuration will give the total resistance of the circuit.

5-6. The sum of two or more branch conductances will give the total conductances of a ________ circuit configuration.

5-7. The total resistance in a parallel circuit will always be (*greater, smaller*) than the smallest branch resistance value.

5-8. Show the product-over-sum formula for the total resistance of two resistors.

Refer to the parallel circuit of Fig. 5-11 for Questions 5-9 to 5-20. Use the following values:

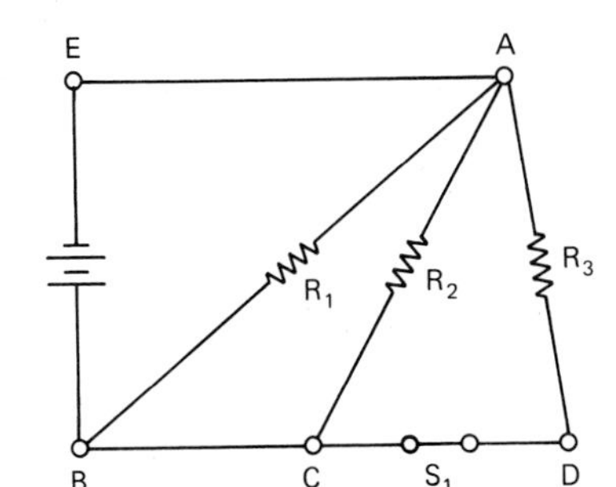

$R_1 = 10\ \text{k}\Omega$

$R_2 = 6.2\ \text{k}\Omega$

$R_3 = 15\ \text{k}\Omega$

$V_2 = 80\ \text{V}$

FIGURE 5-11 Parallel circuit schematic.

5-9. Given *conventional current flow*, indicate all branches for current flow direction from the labeled schematic.

5-10. Given that S_1 is closed, find I_1.

5-11. Given that S_1 is closed, find I_2.

5-12. Given that S_1 is closed, find I_3.

5-13. Find V_T.

5-14. Find R_T with S_1 closed.

5-15. Find I_T with S_1 open.

5-16. Find R_T with S_1 open.

5-17. Given that $R_1 = R_2 = R_3 = 15\ \text{k}\Omega$, find R_T using a simplified method. Show your work.

5-18. Show how to connect four 1-kΩ resistors to give a net resistance of 1 kΩ.

5-19. Two resistors in parallel have a combined resistance of 4.4 kΩ. One resistor's value is 15 kΩ. Find the value of the other resistor.

5-20. Explain why all lighting and appliance outlets must be connected in parallel.

ANSWERS TO PRACTICE QUESTIONS

5-1
1. parallel
2. increase
3. decrease
4. sum
5. parallel
6. 75 lamps

7. $I_T = I_1 + I_2 + I_3 = 3\text{ A}$
8. $I_T = I_1 + I_2 = 2\text{ A}$
9. 15 lamps
10. 110 to 120 V
11. 2 A to each lamp
    6.0 A to 3 lamps
12. 220 V
13. $R_T = \dfrac{V}{I_T} = \dfrac{110\text{ V}}{3\text{ A}} = 36.7\ \Omega$
14. $R_T = \dfrac{V}{I_T} = \dfrac{110\text{ V}}{2\text{ A}} = 55\ \Omega$
15. $R_T = \dfrac{V}{I_T} = \dfrac{110\text{ V}}{1\text{ A}} = 110\ \Omega$

5-2

1. $V_T = V_2 = 30\text{ V}$
2. $I_1 = 20\text{ mA}$
3. $I_2 = 5.9\text{ mA}$
4. $I_3 = 6.4\text{ mA}$
5. $I_T = 32.3\text{ mA}$
6. $R_T = 929\ \Omega$
7. A to H for R_1
   C to F for R_2
   D to E for R_3
8. $I_2 = 20.4\text{ mA}$
9. $I_3 = 3.6\text{ mA}$
10. $V_T = 24.5\text{ V}$
11. $R_T = 1.02\text{ k}\Omega$
12. $I_1 = 7.4\text{ mA}$
13. $R_T = 779\ \Omega$
14. $I_T = 31.5\text{ mA}$

6

Energy, Work, and Power

INTRODUCTION

Conserve energy; reduce your auto speed on the highway to conserve energy; reduce your heating thermostat setting. These platitudes sound familiar since the energy that is being conserved by following this advice is the nonrenewable energy provided by oil.

Stored energy takes many forms such as oil, gas, coal, food, and many others. In the electrical circuit the source provides energy in the form of a voltage and current flow. Once current flows, power is dissipated and work is done in such forms as heating a toaster element, lighting a lamp, or turning the armature of a motor. To understand energy and power is to understand why a resistor or a transistor heats up. Energy is stored in a capacitor in the form of an electric field or in an inductor in the form of a magnetic field. Since energy cannot be destroyed, it must be converted from one form of energy to another. A thorough understanding of energy and power will explain many phenomena, not only in this chapter, but with regard to other aspects of electronic components and circuits, as well.

MAIN TOPICS IN CHAPTER 6

OBJECTIVES

After studying Chapter 6, the student should be able to

1. Master work, energy, and power concepts.
2. Apply power equations to simple and combination circuit characteristics.
3. Apply power equations to identify voltage limits of power rated resistors.
4. Apply power equations to identify current limits of power-rated resistors.
5. Apply power equations to connect various combinations of series–parallel power-rated resistors to obtain desired voltage, current, and power rating of the combination.
6. Master the concept of maximum power transfer.

6-1 MECHANICAL WORK AND ENERGY

Scientific data indicate that energy cannot be created or destroyed and is constant in the universe. This leaves us with the principle that energy changes from one form to another form. As an example, the car battery converts chemical energy into electrical energy, and the battery is a form of stored work.

Work is done in the process of converting energy from one form to another. *Energy* is stored work. Mechanical work is quite obvious, as examples are seen every day. Moving an object through a distance requires a force with resultant work. Doing work obviously requires an expenditure of energy.

Work is done whenever a force is exerted over a distance (see Fig. 6-1):

$$\text{work} = \text{force} \times \text{distance}$$

$$W = F \times d \qquad (6\text{-}1)$$

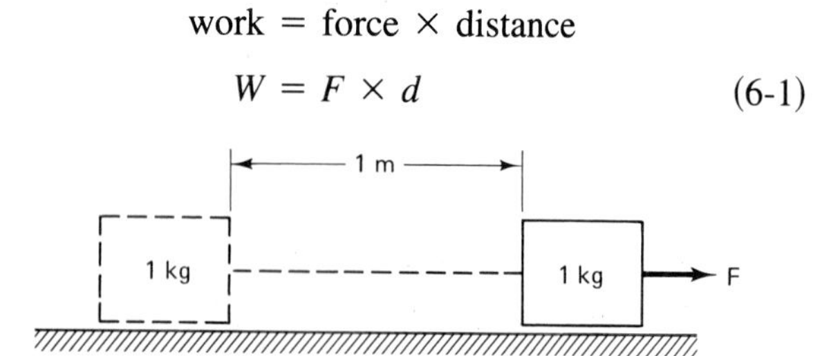

FIGURE 6-1 Work is done when a force is exerted over a distance.

The newton-metre is a unit of mechanical work and is equal to 1 joule. The joule is a standard unit of work. When a force of 1 newton acts on a 1 kilogram weight through a distance of 1 metre, 1 joule of work is done.

6-2 ELECTRICAL WORK AND ENERGY

The principle of work and energy can be applied to an electric charge since like charges repel and unlike charges attract. A force (volt) must be applied to a charge in order to move it through a certain distance.

One joule of electrical work is done when 1 coulomb of electrons is moved through a potential difference of 1 volt. In equation form,

$$W = VQ \quad \text{joules} \qquad (6\text{-}2)$$

where W = work done or energy expended, in joules
V = potential difference, in volts
Q = charge, in coulombs

Since $Q = It$, the following equation determines the amount of work done in a length of time t.

$$W = V \times I \times t \quad \text{joules} \qquad (6\text{-}3)$$

where W = work or energy expended, in joules
V = potential difference, in volts
I = current, in amperes
t = time, in seconds

6-3 ELECTRICAL POWER

Power is the rate at which work is done. As an example, when a cyclist applies energy to the pedals of a ten-speed bike, he or she has the option of selecting the rate at which the pedals are pushed or energy is applied to the chain sprocket and rear wheel. Thus the cyclist maintains a certain speed uphill, or against the wind on a level stretch. Power involves not only how much work is being done but also the length of time required to do the work. *Power* is the rate at which work is done, or

$$\text{power} = \frac{\text{energy}}{\text{time}} = \frac{\text{work done}}{\text{time taken to do the work}}$$

In the electrical circuit potential energy is supplied by a source of voltage. When a load such as a resistor is connected to the source, current flows through the resistance to produce heat. The heat energy released into the surrounding air cannot be returned to the source; therefore, the source must supply this energy. The rate at which the heat energy is released is the *power*. Whatever amount of power a load or loads require must be provided by the source; otherwise, it cannot maintain the voltage required to supply the current.

> Electrical power is the rate at which electrical energy is expended.

From equation (6-3),

$$\text{power} = \frac{\text{work}}{\text{time}} = \frac{V \times I \times t}{t} = \frac{\text{joules}}{\text{seconds}}$$

and 1 joule per second expended is equal to 1 watt of power, or

$$P = VI \quad \text{watts} \qquad (6\text{-}4)$$

The watt is the unit of electrical power named after James Watt, who devised a method of measuring his then-developed steam engine power to the average power that a horse could develop over the same interval. From definitions for 1 horsepower, where

$$1 \text{ hp} = 550 \text{ ft-lb/min}$$

$$1 \text{ W} = 0.7376 \text{ ft-lb/s}$$

one electrical horsepower is

$$1 \text{ hp} = \frac{550 \text{ ft-lb}}{0.7376 \text{ ft-lb}} = 746 \text{ W}$$

EXAMPLE 6-1

A toaster connected to the 120-V ac power source draws 10 A of current. What power do the toaster elements dissipate to toast the bread?

Solution: Using equation (6-4) gives us

$$P = VI$$

$$120 \text{ V} \times 10 \text{ A} = 1200 \text{ W} \quad \text{or} \quad 1.2 \text{ kW}$$

Note: 1 kW = 1 kilowatt = 1000 watts.

EXAMPLE 6-2

A certain resistor is connected across 150 V and has 10 mA flowing through it. What power does it dissipate in the form of heat energy?

Solution: Using equation (6-4) yields

$$P = VI$$

$$150 \text{ V} \times 10 \times 10^{-3} \text{ A} = 1.5 \text{ W}$$

The resistor dissipates 1.5 W of heat energy.

6-4 ELECTRICAL POWER EQUATIONS

There are two alternative power equations derived from the basic power equation (6-4).

I^2R Equation

Since $P = VI$ [equation (6-4)] and $V = IR$, [Ohm's law equation], substitute IR for V. Therefore,

$$P = IR \times I \quad \text{or} \quad I^2R \qquad \text{watts} \qquad (6\text{-}5)$$

This equation is very useful for calculating the power dissipation of resistors and the power loss of conductors.

EXAMPLE 6-3

A 10-kΩ resistor has 10 mA of current flowing through it. What is the power dissipated by the resistor?

Solution: Using equation (6-5) yields

$$P = I^2R$$

$$(10 \times 10^{-3} \text{ A})^2 \times 10 \times 10^3 \ \Omega = 1 \text{ W}$$

The resistor dissipates 1 W.

EXAMPLE 6-4

A certain length of copper has a 1.1-Ω resistance. If the current flowing through the conductor is 15 A, what is the power loss by the conductor? What form does this power loss take, and where does this loss go?

Solution: Using equation (6-5) gives us

$$P = I^2R$$

$$= (15 \text{ A})^2 \times 1.1 \ \Omega = 247.5 \text{ W}$$

The power loss of 247.5 W is in the form of heat and heats the conductor. The heat is dissipated into the surrounding air and is said to be lost since it does not return to the source.

V^2/R Equation

Since $P = VI$ [equation (6-4)] and $I = V/R$ [Ohm's law equation], substitute V/R for I. Therefore,

$$P = V \times \frac{V}{R} \quad \text{or} \quad \frac{V^2}{R} \qquad \text{watts} \qquad (6\text{-}6)$$

This equation is very useful for calculating the voltage across a resistor given its maximum power dissipation.

EXAMPLE 6-5

What is the maximum voltage that can be applied across a 6800-Ω resistor with a power rating of 2 W?

Solution: Using equation (6-6) gives us

$$P = \frac{V^2}{R}$$

Therefore

$$V = \sqrt{PR} = \sqrt{2 \text{ W} \times 6800 \ \Omega} = 117 \text{ V}$$

PRACTICE QUESTIONS 6-1

Choose the most suitable word, phrase, or value to complete each statement.

1. Work is done in the process of converting ________ from one form to another.
2. Coal, oil and gas all have stored ________.
3. When an applied force moves a mass through a distance, ________ is done.

4. The newton-metre is a unit of mechanical work and is equal to ________.
5. The basic unit of work in the metric system is the joule. When a force of 1 newton acts on a ________ weight through a distance of ________ 1 joule of work is done.
6. One ________ of work is done when 1 coulomb of electrons is moved through a potential difference of 1 volt.
7. State the answer to Question 6 in equation format.
8. The rate at which work is done is termed ________.
9. In the electrical circuit power is dissipated by a resistor in the form of ________ energy.
10. The power dissipated by the resistor must be made up by the ________.
11. One joule of heat energy dissipated per second is equal to 1 ________.
12. In the electrical circuit 1 W of power is equal to ________.
13. One horsepower is equal to ________ watts.
14. One kilowatt is equal to ________ watts.
15. One millwatt is equal to ________ watt.
16. Calculate the amount of current flowing through a 100-W 120-V lamp.
17. Calculate the voltage required to dissipate 200 W with a current flow of 2 A.
18. A certain length of copper conductor has a 2-V loss. If the current through the conductor was 15 A, what is the power loss?
19. The resistance of the secondary winding of a transformer is 0.5 Ω. If the secondary current is 5 A, find the power loss due to the internal resistance of the winding.
20. Calculate the voltage loss due to the internal resistance of the winding.

6-5 WATTAGE RATING OF RESISTORS

Resistors are rated according to maximum wattage, maximum voltage, and resistance range. The maximum ratings indicate those values that are not to be exceeded since the resistor will be damaged or destruct.

Wattage rating is the ability of a resistor to be able to dissipate heat into the surrounding air or environment. The need for a certain wattage depends on the voltage and current in the circuit; however, many factors determine the wattage rating of a resistor. The selected resistor is not necessarily that with the lowest wattage rating capable of handling the load. The life of a resistor is largely determined by the resistor operating temperature, which is a result of both the amount of power dissipated and the environmental heat encountered. Enclosures, higher ambient temperatures, the presence of nearby components also dissipating heat, and many other factors affect the wattage that a resistor should be required to dissipate in a particular application. Figure 6-2 shows the relative size of molded composition and carbon-film resistors for various wattage ratings.

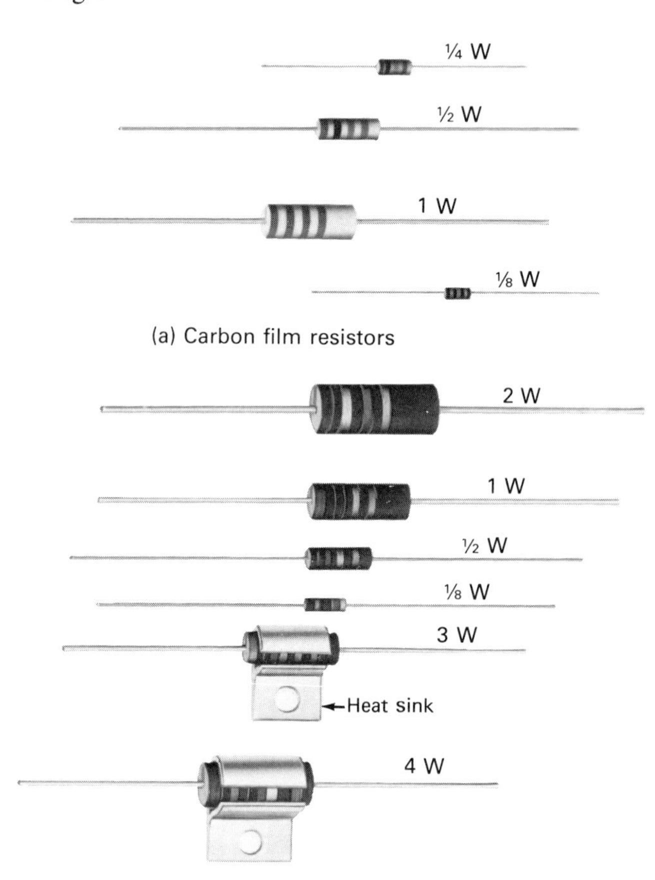

FIGURE 6-2 Relative size of molded composition and carbon-film resistors for various wattage ratings. (Courtesy of Allen-Bradley Canada Ltd.)

Derating Wattage Dissipation of Resistors

The general practice is to operate resistors at 50%, 25%, or another percentage of the actual *VI* watts. The actual watts multiplied by the product for a given condition will keep the resistor maximum temperature rise within the desired limits. Some of the conditions are discussed briefly below.

Ambient Temperature. The ambient temperature is the temperature of the surrounding air or environment. Ambient temperatures higher than standard restrict the permissible temperature rise for a given hot-spot temperature and therefore reduce the permissible wattage.

Enclosures. Enclosures limit the flow of heat from the resistor. Accurate estimation is very difficult except by comparison with similar known measured conditions. Calculated results give only approximate accuracy unless other factors are known precisely. As a guide, resistors in nonventilated enclosures should be derated 100% if the area is totally enclosed and 15 to 50% if the enclosure is of mesh construction.

Grouping. Resistors banks or groups will be affected by each other's heat dissipation. Their proximity to each other, contact or space allowed between the resistor and the circuit board, and environmental factors govern their wattage rating.

Air Cooling. Forcing air through an enclosure reduces the ambient temperature considerably. The required wattage rating of resistors can then be greatly reduced, as well as other components. Calculating the volume of cooling air is not within the scope of this book.

Soldering Precautions. Composition and film resistors are sensitive to thermal extremes because of the particular nature of the lead connection. Caution must be used when soldering or welding. This is extremely important in the case of dip soldering for assembly-line purposes.

Limited Temperature Rise. A less-than-maximum temperature rise is always desirable to protect adjacent components and to reduce change in resistance with load and resultant increase in reliability.

Other Application Considerations. Very high resistances, high voltage, radio-frequency or high-frequency pulse applications, and other military requirements require special consideration.

Resistor Specifications and Reliability

Many industrial specifications have been written by large users of resistors, which, in general, have been modeled after various military specifications. Modifications include such things as different physical sizes, closer tolerances and temperature coefficients, load-life tests, different humidity test cycles, and reliability definitions.

Reliability, not in a general sense, but as "established reliability," expressed as "mean time between failures" or as "maximum failure rate in percent per 1000 hours," is a concept still largely confined to certain military and aerospace applications. There are two military and industry standard designations for resistors used. The MIL-R-11 meets a minimal standard specification for resistance, wattage, and reliability. The MIL-R-39008 rating is an "established reliability." This is designated by the fifth color band. See Table 2-7.

The selection of resistors for given applications, when made with full knowledge and with proper allowances for anticipated conditions, can result in long and dependable service life.

6-6 VOLTAGE LIMITATION OF RESISTORS

Resistors are rated according to maximum wattage, maximum voltage, and resistance range. Consider for a moment the common insulated composition fixed resistance covered by MIL-R-11. It provides for power ratings of $\frac{1}{10}$, $\frac{1}{4}$, $\frac{1}{2}$, 1, and 2 W and a maximum continuous working voltage of 500 V for 1-W resistors, yet it covers resistances as high as 22 MΩ. According to the power law, $V = \sqrt{PR}$, 4700 V would have to be applied to a 22-MΩ resistor to dissipate 1 W. But the resistor is limited by a MIL specification to the application of 500 V.

Since $R = V^2/P$, it is apparent that the "critical-resistance" value is 250,000 Ω. It is the only resistance value at which both maximum voltage and maximum wattage can be tolerated. Values above 250,000 Ω (the critical resistance) are limited in wattage by the maximum voltage (500 V), and we find that the 22-MΩ resistor is only capable of slightly more than 0.01 W dissipation. In higher nominal resistance values, the maximum voltage rating becomes a factor, and at a resistance defined as the "critical value," the resistor operates at maximum voltage and maximum power at the same time.

At values above the critical-resistance value, the power is less than the maximum, because the voltage is at its specified limit. Below critical value, the voltage will be less because the power has reached its limit. In practice, the critical-resistance value is defined as the nearest standard value below the point at which maximum wattage is dissipated with the maximum rated voltage.

For carbon-composition resistors the critical resistance is reached for the following dc ratings:

$\frac{1}{10}$ W–150 V $\quad$ $\frac{1}{4}$ W–250 V

$\frac{1}{2}$ W–350 V $\quad$ 1 W–500 V

2 W–700 V

Carbon-film resistors are manufactured to provide a range of limiting or maximum voltage from 150 to

1000 V. Metal-film resistors' maximum voltage ranges from 150 to 500 V. Applications where higher-voltage limits are required necessitate the use of high-voltage resistors. In all cases, manufacturers' specifications must be consulted and critical resistance must be considered.

EXAMPLE 6-6

Calculate the critical-resistance value for a $\frac{1}{2}$-W composition resistor.

Solution: The maximum voltage rating for a $\frac{1}{2}$-W composition resistor is 350 V since

$$R = \frac{V^2}{P} = \frac{(350\ \text{V})^2}{\frac{1}{2}\ \text{W}} = 250\ \text{k}\Omega$$

6-7 WATTAGE RATING RESISTORS CONNECTED IN SERIES OR PARALLEL

A very important factor in the power rating or life of composition resistors or thin and thick-film resistors is the heat sink to which they are connected. As a rule of thumb, approximately three-fourths of the dissipated heat goes out through the lead wires and must have some place to go. The *heat sink* is the immediate body to which the resistor leads are soldered. Some examples of heat sinks include the copper conducting strip of a printed-circuit board, power transistors, and power rectifier diodes.

In the standard load-life test, the specifications call for "lightweight" terminals about an inch from the resistor body. If in actual application the resistor is connected to a heat sink with greater thermal capacity, the performance would be expected to be better than when lightweight terminals alone are used.

By the same reasoning, the life of the resistor would be shortened if it were connected to the terminal of a power diode or transistor, which would make it difficult for the resistor to get rid of its heat. For this reason a 4-W resistor cannot always be made from two 2-W resistors simply by connecting them in series or parallel without taking adequate precautions for removing their heat. The heat sink to which resistors are connected affects their rating. Resistors operated in series or parallel should be derated unless an adequate heat sink is provided.

Series Power Combination Connection

Assuming that the resistors have an adequate heat sink, the wattage rating can be increased by connecting two or more resistors in series if the maximum wattage or voltage rating of each resistor is considered and the critical resistance value of each is also considered.

EXAMPLE 6-7: Series Power Combination Connection

Calculate the power dissipation of each resistor in the circuit shown in Fig. 6-3, the total power, and the net resistance of the combination.

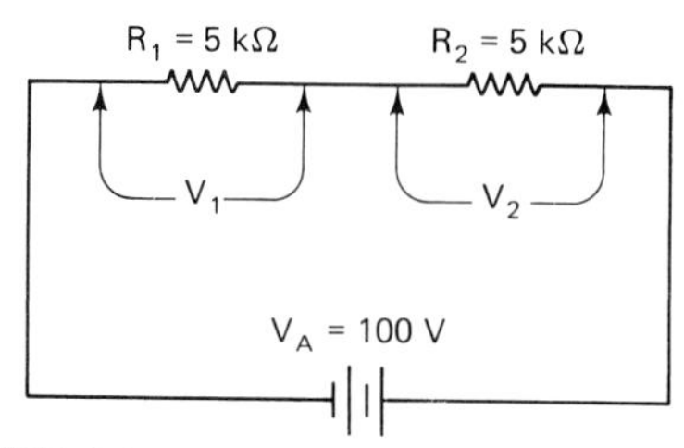

FIGURE 6-3 Power rating of series combination.

Solution: Since the applied voltage is known, find V_1 and V_2, where $V_1 = V_2$ since $R_1 = R_2$.

$$V_1 = V_2 = \frac{R_1}{R_1 + R_2} \times V_A$$
$$= \frac{5\ \text{k}\Omega}{5\ \text{k}\Omega + 5\ \text{k}\Omega} \times 100\ \text{V}$$
$$= \frac{5\ \text{k}\Omega}{10\ \text{K}\Omega} \times 100\ \text{V}$$
$$= 50\ \text{V}$$

Using formula (6-6) yields

$$P_1 = \frac{V^2}{R_1} = \frac{(50\ \text{V})^2}{5\ \text{k}\Omega} = \frac{1}{2}\ \text{W}$$

$$P_2 = \frac{V^2}{R_2} = \frac{(50\ \text{V})^2}{5\ \text{k}\Omega} = \frac{1}{2}\ \text{W}$$

Total wattage

$$P_T = P_1 + P_2 = 0.5\ \text{W} + 0.5\ \text{W} = 1\ \text{W}$$

Net resistance of the combination

$$R_T = R_1 + R_2 = 10\ \text{k}\Omega$$

Note that in the case of similar values, the power rating must be the same for each resistor.

EXAMPLE 6-8: Series Power Combination Connection

For each resistor in the circuit shown in Fig. 6-4

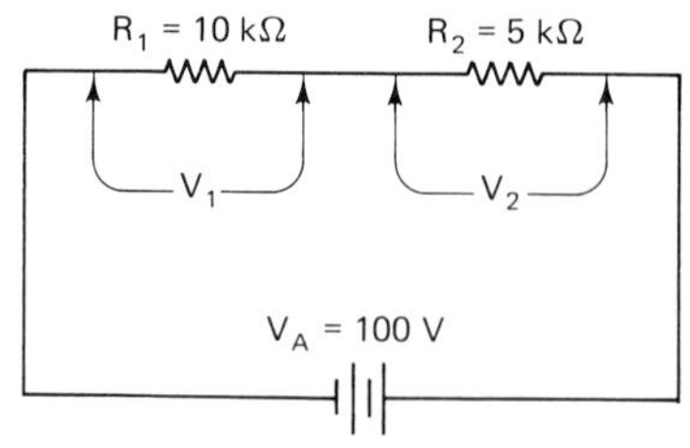

FIGURE 6-4 Highest-value series resistor dissipates greatest power.

calculate the power dissipation, total power, and net resistance of the combination.

Solution: Since the applied voltage is known, find V_1 and V_2.

$$V_1 = \frac{R_1}{R_1 + R_2} \times V_A = \frac{10\ \text{k}\Omega}{10\ \text{k}\Omega + 5\ \text{k}\Omega} \times 100$$

$$= 67\ \text{V}$$

$$V_2 = \frac{R_2}{R_1 + R_2} \times V_A = \frac{5\ \text{k}\Omega}{10\ \text{k}\Omega + 5\ \text{k}\Omega} \times 100$$

$$= 33\ \text{V}$$

Using formula (6-6) gives us

$$P_1 = \frac{V^2}{R_1} = \frac{(67\ \text{V})^2}{10\ \text{k}\Omega} = 0.45\ \text{W}$$

$$P_2 = \frac{V^2}{R_2} = \frac{(33\ \text{V})^2}{5\ \text{k}\Omega} = 0.22\ \text{W}$$

Total wattage

$$P_T = P_1 + P_2 = 0.45\ \text{W} + 0.22\ \text{W} = 0.67\ \text{W}$$

Net resistance of the combination

$$R_T = R_1 + R_2 = 15\ \text{k}\Omega$$

Note that the power dissipation is greatest by the *highest*-value resistor.

From Examples 6-7 and 6-8 it can be seen that two or more series-connected resistors will have a total power rating equal to the sum of the power rating of each provided that they are equal-value resistors.

If the resistors are not of equal value, the highest resistance value will dissipate the greater power; therefore, its power rating must be chosen as the working power rating and the remaining resistor with the same power rating will be more than adequate.

Parallel Power Combination Connection

Assuming that the resistors have an adequate heat sink, the wattage rating can be increased by connecting two or more resistors in parallel if the maximum wattage or voltage rating of each resistor is considered and the critical resistance value of each is also considered.

EXAMPLE 6-9: Parallel Power Combination Connection

For each resistor in the circuit of Fig. 6-5, calculate the power dissipation, total power, and net resistance of the combination.

Solution: Use formula (6-6), since the voltage is common in a parallel circuit.

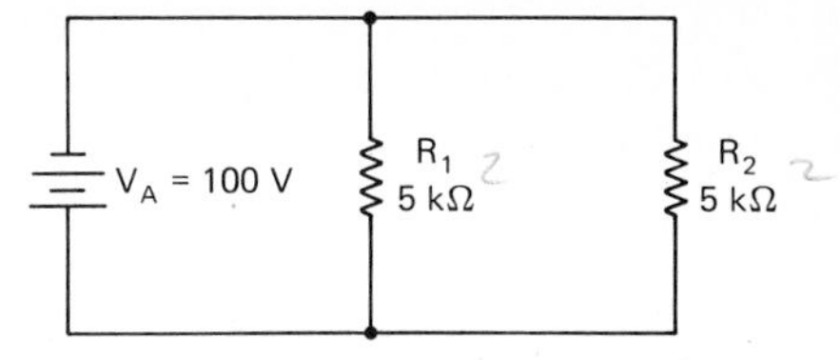

FIGURE 6-5 Power rating of parallel combination.

$$P_1 = \frac{V^2}{R_1} = \frac{(100\ \text{V})^2}{5\ \text{k}\Omega} = 2\ \text{W}$$

$$P_2 = \frac{V^2}{R_2} = \frac{(100\ \text{V})^2}{5\ \text{k}\Omega} = 2\ \text{W}$$

Total wattage

$$P_T = P_1 + P_2$$

$$= 2\ \text{W} + 2\ \text{W} = 4\ \text{W}$$

Net resistance of the combination

$$R_T = \frac{5\ \text{k}\Omega}{2}$$

$$= 2.5\ \text{k}\Omega$$

EXAMPLE 6-10: Parallel Power Combination Connection

For each resistor in the circuit of Fig. 6-6, calculate the power dissipation, total power, and net resistance of the combination.

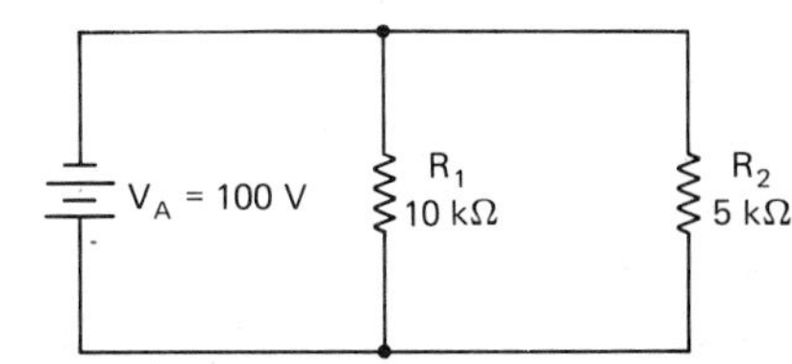

FIGURE 6-6 Lowest-value parallel resistor dissipates greatest power.

Solution: Use formula (6-6), since the voltage is common in a parallel circuit.

$$P_1 = \frac{V^2}{R_1} = \frac{(100\ \text{V})^2}{10\ \text{k}\Omega} = 1\ \text{W}$$

$$P_2 = \frac{V^2}{R_2} = \frac{(100\ \text{V})^2}{5\ \text{k}\Omega} = 2\ \text{W}$$

Total wattage

$$P_T = P_1 + P_2$$

$$= 1\ \text{W} + 2\ \text{W} = 3\ \text{W}$$

Net resistance of the combination

$$R_T = \frac{\text{product}}{\text{sum}}$$

$$= \frac{10\ \text{k}\Omega \times 5\ \text{k}\Omega}{10\ \text{k}\Omega + 5\ \text{k}\Omega} = 3.3\ \text{k}\Omega$$

Note that the power dissipation is greatest by the *lowest*-value resistor, as shown in Example 6-10. From Example 6-10 it can be seen that two or more parallel-connected resistors will have a total power rating equal to the sum of the power rating of each provided that they are equal-value resistors. If the resistors are not of equal value, the lowest resistance value will dissipate the greater power. Therefore, its power rating must be chosen as the working power rating and the remaining resistors with the same power rating will be more than adequate.

PRACTICE QUESTIONS 6-2

Choose the most suitable word, phrase, or value to complete each statement, or draw a schematic diagram where required.

1. State three characteristics by which resistors are rated.
2. The ability of a device to dissipate heat energy into the surrounding environment is termed the ________ rating of the device.
3. The operating temperature and the environmental temperature will largely determine the ________ expectancy of a resistor.
4. State three factors that determine the wattage rating and life of a resistor.
5. An industry standard test known as ________ determines the wattage rating of a resistor.
6. Resistor wattage ratings should be derated by a factor of ________ percent if the enclosure area is total.
7. Resistor wattage rating should be derated by a factor of ________ percent if the enclosure is mesh construction.
8. The ________ resistance is the value of a resistor at which both maximum voltage and maximum wattage can be tolerated.
9. Since the lead wires of a resistor conduct heat, the wattage rating of a resistor will be affected by the ________ ________ to which the leads are connected.
10. The wattage rating of resistors connected in series or parallel must be ________ unless an adequate heat sink is provided.
11. With two unequal resistors connected in series, the (*lowest, highest*)-value resistor will dissipate the greatest wattage.
12. Two resistors, 110 kΩ and 55 kΩ, are connected in series. For any applied voltage, which resistor will dissipate the highest power?
13. If each resistor in Question 12 has a 1-W rating, what is the highest voltage that can safely be applied?
14. If the 110-kΩ and the 55-kΩ 1-W resistors are connected in parallel, what is the maximum safe total current for the combination?
15. Show how you would connect 1-kΩ 1-W resistors to obtain a net resistance of 1 kΩ, 4 W.
16. A 22-kΩ resistor dissipates $\frac{1}{2}$ W in a certain circuit. If we increase the voltatge by 50%, the power dissipated will be ________ watts.
17. The highest permissible voltage that can be applied to a 18-kΩ 2-W resistor is ________ volts.
18. The maximum current that can be passed through a 10-kΩ 1-W resistor is ________ milliamperes.

6-8 MAXIMUM POWER TRANSFER THEOREM

For most cases of power transfers from sources such as amplifier to speaker, microphone to amplifier input, or transmitter output to antenna, it is required to maximize the power transfer from the output of one device to the input of another device, or vice versa. This is possible provided that the internal resistance or impedance of the device is taken into consideration.

Ideal Source versus a Practical Source

An introduction to an ideal source was given in Section 4-4, which was applied to cells and batteries and the effect of internal resistance on the voltage output of the battery. An ideal source does not have any internal resistance and is able to provide a constant voltage under varying conditions of load current. The maximum *VI* power put out by the source would be completely available to the load under any load current condition. This is shown graphically in Fig. 6-7, where the load power

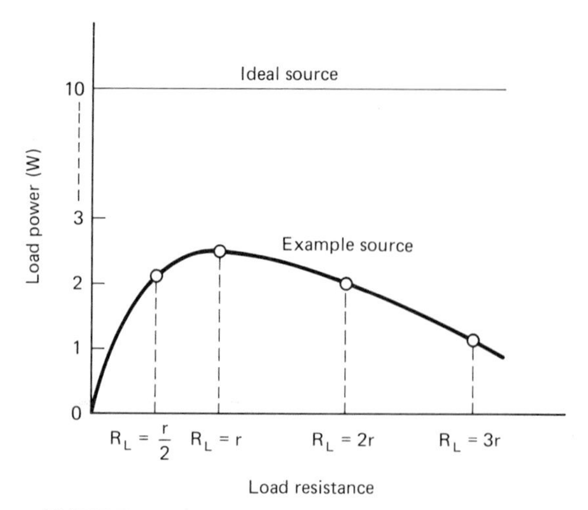

FIGURE 6-7 Graph of load power for various values of internal resistance, *r*.

provided by the ideal source is 100 percent regardless of load resistance.

Dc power sources such as batteries, solar cells, and dc power supplies, as well as ac sources such as amplifiers, microphones, function generators, and other sources of emf, have an internal resistance. Ac source internal resistance is termed *impedance*. Impedance is the total ac resistance due to capacitance, inductance, and resistance, and is discussed in Chapter 26.

The load power curve for the example source shown in Fig. 6-7 is far from ideal. Why is this so different? The internal resistance (r) shown in the schematic circuit to determine load power of an example source (see Fig. 6-8) will have a voltage loss due to the load current. This reduces the voltage and power transfer to the load.

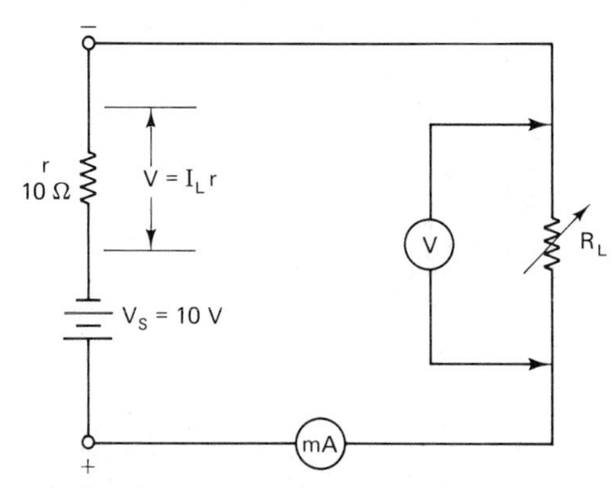

FIGURE 6-8 Circuit to determine load power of example circuit.

To show when maximum power transfer takes place, the power transfer to the load in the circuit of Fig. 6-8 can be calculated for values of R_L as follows: The load current I_L is calculated from the equation

$$I_L = \frac{V_S}{r + R_L} \quad \text{amperes}$$

The load currents for the various values of R_L are shown in Table 6-1. Next, the power in the load R_L is calculated from the equation $P = I^2R$.

TABLE 6-1

Load power calculations for the example source

Load R_L Value	*I_L (A)*	*P_L (W)*	*P_r (W)*
$\frac{r}{2}$	0.67	2.2	4.5
r	0.5	2.5	2.5
$2r$	0.33	2.2	1.1
$3r$	0.20	1.2	0.4

From the curve for the example source of Fig. 6-6, the maximum power transfer takes place when $R_L = r$. It is also readily determined from Table 6-1, which shows $P_L = 2.5$ W for this load and internal resistance condition. Note that both the load and the internal resistance are dissipating the same power, so that only 50% of the power is really available to the load. The remainder of the power from the source is dissipated by the source through its internal resistance. The maximum power transfer theorem states that maximum power will be delivered to the load when the load resistance is equal to the source resistance.

A practical example of source resistance or impedance is a speaker. Its impedance of 4 to 128 Ω must be matched to the output transistor(s) of an amplifier. Another example is a television antenna. Its terminal impedance must be matched by means of a special TV line called a 300 twin lead or a 75-Ω coaxial cable. Similarly, the output impedance of a transmitter power amplifier must be matched to the transmitting antenna. The voltage source impedance of devices such as a microphone, tape deck, phono input, tuner input, solar cell, thermocouple transducer, thermistor, and other types of transducers all have varying output impedance that must be matched to the input impedance of an amplifier.

6-9 POWER MEASUREMENT: THE WATTMETER

Rather than measure the current and voltage separately and from this measurement calculate the VI power, a wattmeter combines both functions to give a direct watt reading on its calibrated scale. The wattmeter has two separate coils, a current coil that measures the load current and the voltage coil that measures the voltage. The voltage coil is the moving coil coupled mechanically to the pointer. The torque applied to the pointer is proportional to the product of the load current and the current flowing through the voltage coil (see Fig. 6-9).

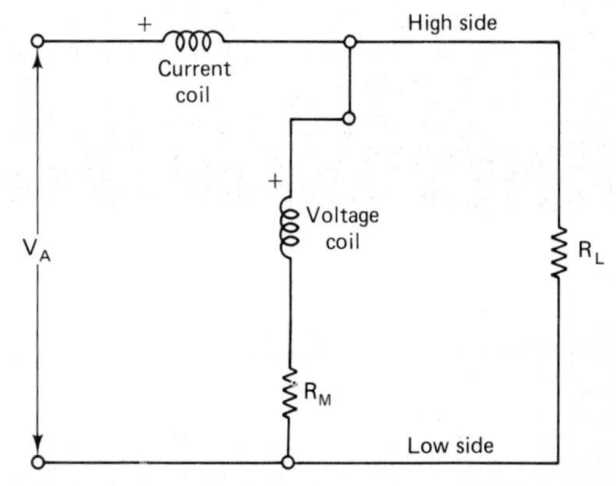

FIGURE 6-9 Wattmeter connection to measure load power.

If the wattmeter is connected to measure the load power (Fig. 6-9), the load current enters the + terminal of the current coil, and the + terminal of the potential coil connected to the higher-voltage side of the load. This connection ensures that the meter deflection is upscale. The wattmeter described briefly is known as an electrodynamometer and can be used to measure both dc and ac power.

SUMMARY

Energy cannot be created or destroyed; therefore, it can only be converted from one form to another form. *Work* is done in the process of converting energy from one form to another:

$$\text{work} = \text{force} \times \text{distance} \qquad \text{or} \qquad W = F \times d \tag{6-1}$$

where W is in joules
F is 1 N
d is 1 m

One joule of electrical work is done when 1 coulomb of electrons is moved through a potential difference of 1 volt.

$$W = VQ \qquad \text{joules} \tag{6-2}$$

where V is in volts
Q is charge in coulombs

The amount of work done in a length of time t is determined by

$$W = V \times I \times t \qquad \text{joules} \tag{6-3}$$

Power is the rate at which work is done:

$$\text{power} = \frac{\text{energy}}{\text{time}} = \frac{\text{work done}}{\text{time taken}}$$

Electrical power is the rate at which electrical energy is expended:

$$\text{power} = \frac{\text{work}}{\text{time}} = \frac{V \times I \times t}{t} = \frac{\text{joules}}{\text{seconds}}$$

The basic equation for electrical power is

$$P = VI \qquad \text{watts} \tag{6-4}$$

One electrical horsepower is

$$1 \text{ hp} = 746 \text{ W}$$

Two other power equations are:

$$P = I^2R \qquad \text{watts} \tag{6-5}$$

$$P = \frac{V^2}{R} \qquad \text{watts} \tag{6-6}$$

Resistors are rated according to maximum wattage, limiting voltage, and resistance range. *Wattage rating* is the ability of a resistor to dissipate heat.

Factors to consider when selecting the power rating of a resistor are ambient temperature, enclosures, grouping, air cooling, soldering precautions, limited temperature rise, and special applications.

The critical resistance value of a resistor is defined as the nearest standard value below the point at which maximum wattage is dissipated with maximum rated voltage.

The wattage rating of a resistor and its useful life depend on the heat sink to which the resistor is connected.

A heat sink will aid or prevent the removal of heat from a resistor through its leads.

Connecting two equal-value resistors in series will double their wattage rating.

The highest-value resistor in series will dissipate the greatest power.

The lowest-value resistor in parallel will dissipate the greatest power.

All voltage sources have some form of internal resistance. The internal resistance of ac sources is called impedance.

It is desirable and efficient to maximize the power transfer from the output of one device to the input of another device.

The maximum power transfer theorem states that maximum power will be delivered to the load when the load resistance is equal to the source resistance.

The output resistance or impedance of a device must be matched to the input impedance of another device to achieve maximum power transfer.

Examples of where output impedance must be matched to input impedance are: a microphone, a tape deck, a solar cell, and a thermocouple.

An electrodynamometer wattmeter measures dc and ac power directly.

REVIEW QUESTIONS

6-1. Define energy.

6-2. Define work.

6-3. State the principle of the conservation of energy law.

6-4. What is the standard unit of work?

6-5. Define electrical work.

6-6. Define electrical power.

6-7. What is the unit of electrical power?

6-8. Define an electrical horsepower.

6-9. State the basic power formula.

6-10. Calculate the current required by a 200-W 120-V lamp.

6-11. Calculate the wattage of 120-V 10-A toaster.

6-12. Calculate the power dissipation of a 56-kΩ resistor that has a 350-V drop.

6-13. List three factors by which resistors are rated.

6-14. List five factors that affect the wattage rating of a resistor.

6-15. Define the critical value of a resistance.

6-16. Explain why the life of a resistor can be increased or decreased when connected to a heat sink.

6-17. Explain which resistor of two unequal-value series resistors dissipates the greater power.

6-18. Explain which resistor of two unequal-value parallel resistors dissipates the greater power.

6-19. What is the basic difference between an ideal voltage source and a practical voltage source such as a battery, solar cell, amplifier, microphone, or other source of emf?

6-20. From Table 6-1, when does maximum power transfer take place?

6-21. Explain why only 50% of the total power available reaches the load at the maximum power transfer point.

6-22. Give three examples of when maximum power transfer is essential.

6-23. What are the function and advantage of a wattmeter?

SELF-EXAMINATION

Choose the most suitable word, phrase, or value to complete each statement.

6-1. What type of energy is converted to electrical energy in a car battery?

6-2. When energy is converted from one form to another, ________ is done.

6-3. One ________ of electrical work is done when 1 coulomb of electrons is moved through a potential difference of 1 volt.

6-4. The rate at which work is done is termed ________.

6-5. The rate at which electrical energy is expended is termed ________.

6-6. The unit of electrical power is the ________.

6-7. One horsepower is ________ watts.

6-8. What is the voltage across a device that dissipates 600 W with a current of 6 A?

6-9. A 220-V electric dryer dissipates 4 kW. Calculate its current.

6-10. Calculate the power rating of a 4.7-kΩ resistor with 100 mA flowing through it.

6-11. Resistors are rated according to maximum ________, limiting or maximum ________, and ________ range.

6-12. The life of a resistor is determined largely by the resistor operating ________.

6-13. The operating temperature of a resistor will be affected by five factors: ________, ________, ________, ________, and ________.

6-14. The wattage rating of a resistor must be derated by ________ percent for a total enclosure.

6-15. The wattage rating of a resistor will be limited by its maximum or limiting ________.

6-16. The resistance value at which both the maximum voltage and maximum wattage of a resistor can be tolerated is termed ________ ________.

6-17. Calculate the critical resistance of a carbon film resistor with a 150-V $\frac{1}{4}$-W rating.

6-18. A 33-kΩ resistor dissipates $\frac{1}{8}$ W in a circuit. If the voltage is increased 50%, the power dissipated will be ________ watts.

6-19. What standard wattage rating must the resistor of Question 6-18 have if it is derated by 50%?

6-20. A speaker has an impedance of 64 Ω. What must the amplifier impedance be to achieve maximum power transfer?

6-21. A wattmeter has two coils, a ________ coil and a ________.

6-22. The ________ coil of a wattmeter is coupled mechanically to the pointer.

6-23. The wattmeter must be connected so that the load current flows through the ________ coil.

6-24. The wattmeter must be connected so that the potential or voltage coil is connected across the ________.

ANSWERS TO PRACTICE QUESTIONS

6-1

1. energy
2. energy
3. work
4. 1 J
5. 1 kg, 1 m
6. joule
7. $W = VQ$ joules
8. power
9. heat
10. source
11. watt
12. 1 W = 1 V × 1 A
13. 746
14. 1000
15. 1/1000
16. 0.83 A
17. 100 V
18. 30 W loss
19. 12.5 W loss
20. 2.5 V

6-2

1. maximum wattage
   maximum voltage
   resistance
2. wattage
3. life
4. voltage
   current
   environmental temperature
5. the load-life test
6. 100
7. 15 to 50
8. critical
9. heat sink
10. derated
11. highest
12. 110 kΩ
13. 495 V
14. 7.3 mA
15.

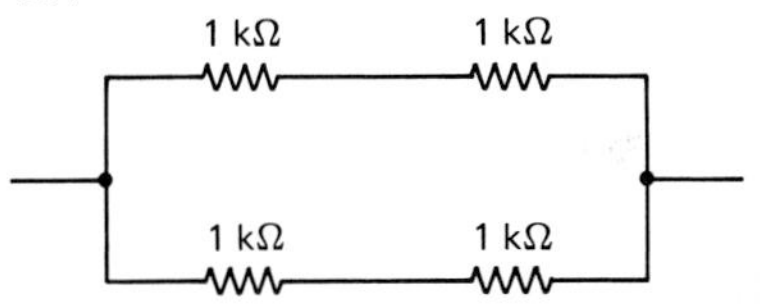

or

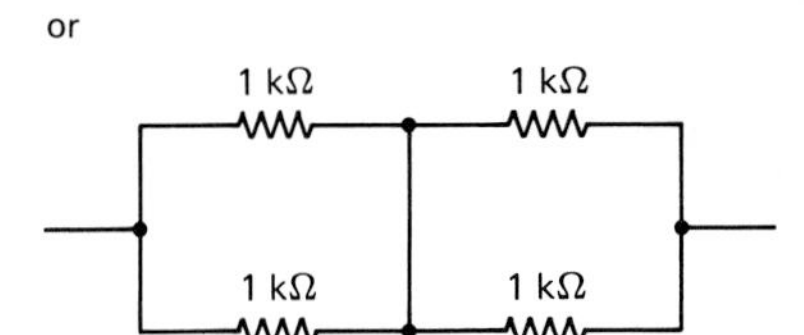

16. 1.125
17. 190
18. 10

7

Combination Circuits

INTRODUCTION

The three basic circuits studied in the previous chapters—the simple circuit, the series circuit, and the parallel circuit—provide all of the circuit characteristics and laws that can be applied to solve series–parallel combination circuits.

A combination circuit has passive components that are connected in series with other parallel-connected components. Important applications of combination circuits are for use in circuits such as a bridge circuit and a series voltage divider with parallel branches that have load voltages and currents. This allows a load or loads to operate at voltages different from the source voltage. In this chapter we focus on solving combination circuits. In later chapters we deal with the practical design of the circuits mentioned.

MAIN TOPICS IN CHAPTER 7

7-1 Series Resistance Branches Connected in Parallel
7-2 Parallel Resistance Branches Connected in Series
7-3 Series–Parallel Resistor Circuits
7-4 Jumbled Series–Parallel Resistance Circuits

OBJECTIVES

After studying Chapter 7, the student should be able to

1. Master combination circuits by applying series and parallel circuit laws as well as power equations to solve the circuit for all voltage, current, and resistance parameters.
2. Connect various resistor combinations in series–parallel to verify theoretical values of voltage, current, resistance, and power.

7-1 SERIES RESISTANCE BRANCHES CONNECTED IN PARALLEL

A series branch with two or more loads is connected in parallel with a similar branch as shown in Fig. 7-1(a). Solving for voltage drops, current, and resistance in such a circuit involves previous series and parallel circuit laws as well as Ohm's law. From the schematic of Fig. 7-1(a) the following observations can be made:

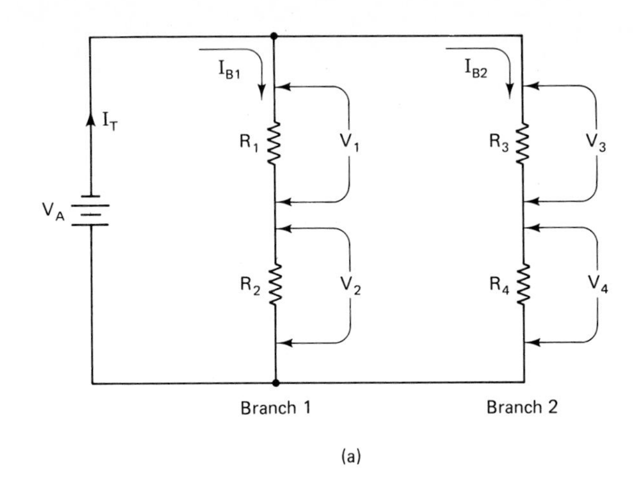

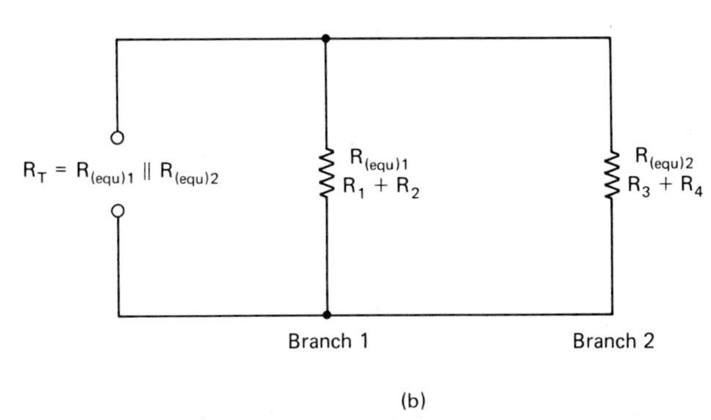

FIGURE 7-1 (a) Combination circuit schematic. (b) Equivalent circuit.

Voltage drops

1. The sum of the series voltage drops V_1 and V_2 will be equal to the total branch 1 voltage.
2. Similarly, the sum of the series voltage drops V_3 and V_4 will be equal to the branch 2 voltage.
3. Branches 1 and 2 are connected in parallel: therefore, their voltages are common to the applied voltage V_A.

Currents

1. Resistors R_1 and R_2 are connected in series; therefore, both have the same current, I_{B1}.
2. Similarly, resistors R_3 and R_4 are connected in series; therefore, they have the same current, I_{B2}.
3. Branches 1 and 2 are connected in parallel; therefore, the sum of the currents of each branch will equal the total line current I_T.

Resistance

1. Resistors R_1 and R_2 are connected in series; therefore, the equivalent resistance of branch 1 will be equal to the sum of R_1 and R_2 [see Fig. 7-1(b)].
2. Similarly R_3 and R_4 are connected in series; therefore, the equivalent resistance $R_{(equ)}$ of branch 2 will be equal to the sum of R_3 and R_4.
3. The parallel combination for the resistance of branch 1 and branch 2 will be equal to the total resistance of the circuit.

EXAMPLE 7-1

Solve for all voltage drops, currents, and total resistance in the circuit of Fig. 7-2.

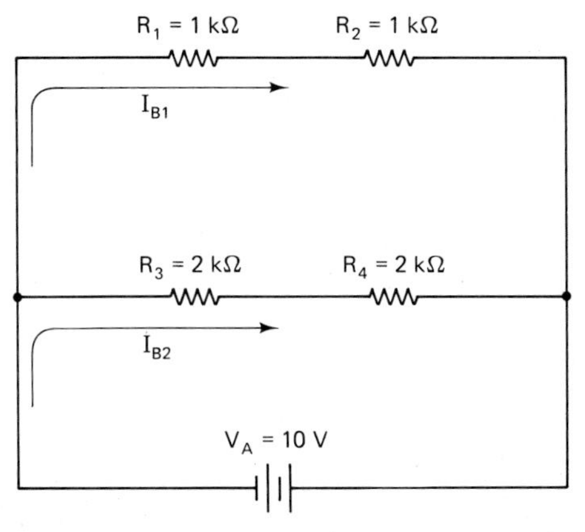

FIGURE 7-2 Combination for Example 7-1.

Solution: The voltage drops can be solved using Ohm's law; however, the current I_{B1} and I_{B2} must first be calculated. A second method, using resistance proportionality, can also be used to solve for the voltage drops.

Method 1

$$\text{Current } I_{B1} = \frac{V_{B1}}{R_T} = \frac{V_A}{R_1 + R_2}$$

$$= \frac{10\text{ V}}{1\text{ k}\Omega + 1\text{ k}\Omega} = \frac{10\text{ V}}{2\text{ k}\Omega} = 5\text{ mA}$$

Voltage $V_1 = V_2$ since both resistance values are equal.

$$V_1 = V_2 = IR = 5\text{ mA} \times 1\text{ k}\Omega = 5\text{ V}$$

$$\text{Current } I_{B2} = \frac{10\text{ V}}{2\text{ k}\Omega + 2\text{ k}\Omega} = \frac{10\text{ V}}{4\text{ k}\Omega} = 2.5\text{ mA}$$

Voltage $V_3 = V_4$ since the two resistance values are equal.

$$V_3 = V_4 = IR = 2.5\text{ mA} \times 2\text{ k}\Omega = 5\text{ V}$$

Method 2

$$V_1 = V_2 = \frac{R_1}{R_1 + R_2} \times V_A$$

$$= \frac{1\ \text{k}\Omega}{1\ \text{k}\Omega + 1\ \text{k}\Omega} \times 10\ \text{V} = \frac{1\ \text{k}\Omega}{2\ \text{k}\Omega} \times 10\ \text{V}$$

$$= 5\ \text{V}$$

and

$$V_3 = V_4 = \frac{R_3}{R_3 + R_4} \times V_A$$

$$= \frac{2\ \text{k}\Omega}{2\ \text{k}\Omega + 2\ \text{k}\Omega} \times 10\ \text{V} = 5\ \text{V}$$

Therefore,

$$V_3 = V_4 = 5\ \text{V}$$

Total current I_T will be equal to the sum of the branch currents, or

$$I_T = I_{B1} + I_{B2} = 5\ \text{mA} + 2.5\ \text{mA} = 7.5\ \text{mA}$$

Total resistance R_T will be equal to the parallel resistance of the two series branches, or

$$R_T = (R_1 + R_2) \parallel (R_3 + R_4)$$

$$= (1\ \text{k}\Omega + 1\ \text{k}\Omega) \parallel (2\ \text{k}\Omega + 2\ \text{k}\Omega)$$

$$= \frac{\text{product}}{\text{sum}} = \frac{2\ \text{k}\Omega \times 4\ \text{k}\Omega}{2\ \text{k}\Omega + 4\ \text{k}\Omega} = 1.33\ \text{k}\Omega$$

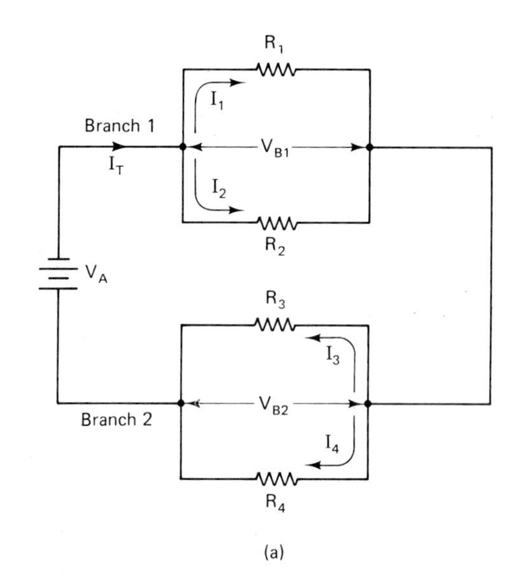

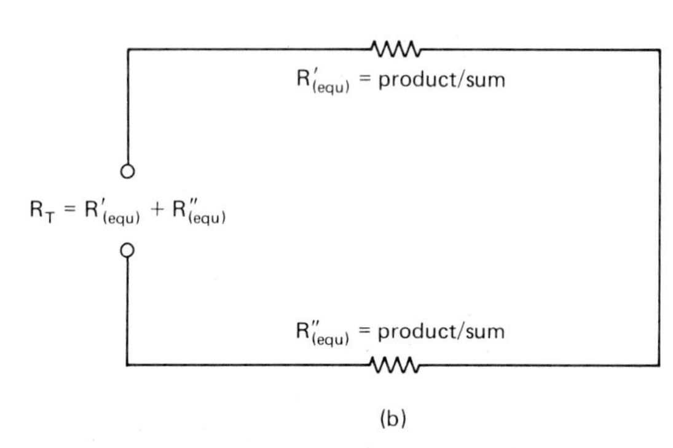

FIGURE 7-3 (a) Combination circuit schematic. (b) Equivalent circuit.

7-2 PARALLEL RESISTANCE BRANCHES CONNECTED IN SERIES

A parallel branch with two or more loads is connected in series with a similar branch as shown in Figure 7-3(a). Solving for voltage drops, current, and resistance, as in all cases, involves previous series and parallel circuit laws as well as Ohm's law. From the schematic of Fig. 7-3(a) the following observations can be made.

Voltage drops

1. R_1 and R_2 are connected in parallel; therefore, V_1 will be equal to V_2 or the parallel branch voltage V_{B1} will be the same across each parallel resistance.
2. Similarly, V_3 will be equal to V_4, or the parallel branch voltage V_{B2} will be the same across each parallel resistance.
3. Branches 1 and 2 are connected in series with each other; therefore, V_{B1} summed with V_{B2} will be equal to the source voltage V_A.

Currents

1. The line current I_T flows into branch 1 and divides (or splits) into two current values I_1 and I_2; therefore, the sum of I_1 and I_2 must be equal to the total line current I_T.
2. Similarly, the line current I_T flows into branch 2 and splits into two current values, I_3 and I_4; therefore, the sum of I_3 and I_4 must be equal to the total line current I_T.

Resistance

1. Resistors R_1 and R_2 are in parallel; therefore, their equivalent resistance must be the parallel resistance of branch 1.
2. Similarly, R_3 and R_4 are in parallel; therefore, their equivalent resistance must be the parallel resistance of branch 2.
3. The total resistance of the circuit is the sum of the equivalent resistance values of branch 1 and branch 2 [see Fig. 7-3(b)].

EXAMPLE 7-2

Solve for all voltage drops, currents, and total resistance in the circuit of Fig. 7-4.

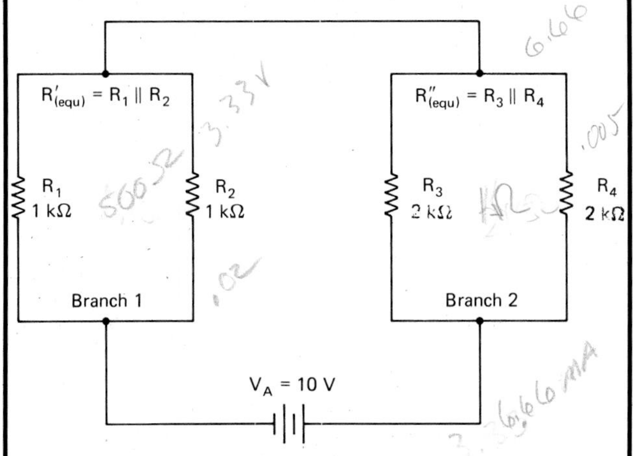

FIGURE 7-4 Combination circuit schematic.

Solution: To find the voltage drops about the circuit, the current must be known since $V = IR$, so that the parallel branches must be reduced to equivalent resistance and then the total current can be calculated.

$$R'_{(equ)} = R_1 \parallel R_2$$

$$= \frac{\text{product}}{\text{sum}} = \frac{1\ \text{k}\Omega \times 1\ \text{k}\Omega}{1\ \text{k}\Omega + 1\ \text{k}\Omega} = 500\ \Omega$$

or

$$R'_{(equ)} = \frac{R}{N} = \frac{1\ \text{k}\Omega}{2} = 500\ \Omega$$

$$R''_{(equ)} = R_3 \parallel R_4$$

$$= \frac{\text{product}}{\text{sum}} = \frac{2\ \text{k}\Omega \times 2\ \text{k}\Omega}{2\ \text{k}\Omega + 2\ \text{k}\Omega} = 1\ \text{k}\Omega$$

or

$$R''_{(equ)} = \frac{R}{N} = \frac{2\ \text{k}\Omega}{2} = 1\ \text{k}\Omega$$

$$R_T = R'_{(equ)} + R''_{(equ)} = 500\ \Omega + 1\ \text{k}\Omega = 1.5\ \text{k}\Omega$$

The total current can be calculated:

$$I_T = \frac{V_A}{R_T} = \frac{10\ \text{V}}{1.5\ \text{k}\Omega} = 6.67\ \text{mA}$$

Since 6.67 mA flows into parallel branches 1 and 2, the voltages about the circuit can be calculated using either of two methods.

Method 1. The voltages V_1 and V_2 are common; therefore,

$$V_1 = V_2 = I_T \times R'_{(equ)}$$

$$= 6.67\ \text{mA} \times 500\ \Omega = 3.33\ \text{V}$$

$$V_3 = V_4 = I_T \times R''_{(equ)}$$

$$= 6.67\ \text{mA} \times 1\ \text{k}\Omega = 6.67\ \text{V}$$

Method 2. $R'_{(equ)}$ is in series with $R''_{(equ)}$, so that the voltage drops across each can be calculated using the voltage-division method.

$$V_1 = V_2 = \frac{R'_{(equ)}}{R'_{(equ)} + R''_{(equ)}} \times V_A$$

$$= \frac{500\ \Omega}{500\ \Omega + 1\ \text{k}\Omega} \times 10\ \text{V} = 3.33\ \text{V}$$

$$V_3 = V_4 = \frac{R''_{(equ)}}{R'_{(equ)} + R''_{(equ)}} \times V_A$$

$$= \frac{1\ \text{k}\Omega}{500\ \Omega + 1\ \text{k}\Omega} \times 10\ \text{V} = 6.67\ \text{V}$$

The voltage-divider method can be used to calculate the voltage drops without the need for calculating the current; however, the currents must be known, so they will need to be calculated.

Method 1. The currents through R_1 and R_2 will be equal since resistor R_1 equals resistor R_2. This means that the current splits into one-half of the total current through each resistor, or

$$I_1 = I_2 = \frac{R_2}{R_1 + R_2} \times I_T$$

$$= \frac{1\ \text{k}\Omega}{1\ \text{k}\Omega + 1\ \text{k}\Omega} \times 6.67\ \text{mA} = 3.33\ \text{mA}$$

Method 2. Since the voltage across $R_1 \parallel R_2$ is 3.33 V, the current can be calculated using Ohm's law, or

$$I_1 = I_2 = \frac{V_1}{R_1} = \frac{3.33\ \text{V}}{1\ \text{k}\Omega} = 3.33\ \text{mA}$$

Also,

$$I_3 = I_4 = \frac{V_3}{R_3} = \frac{6.67\ \text{V}}{2\ \text{k}\Omega} = 3.33\ \text{mA}$$

7-3 SERIES–PARALLEL RESISTOR CIRCUITS

The circuit of Fig. 7-5(a) is more complex-looking than the previous circuits that were analyzed. If the current paths are traced, the circuit breaks down into two paral-

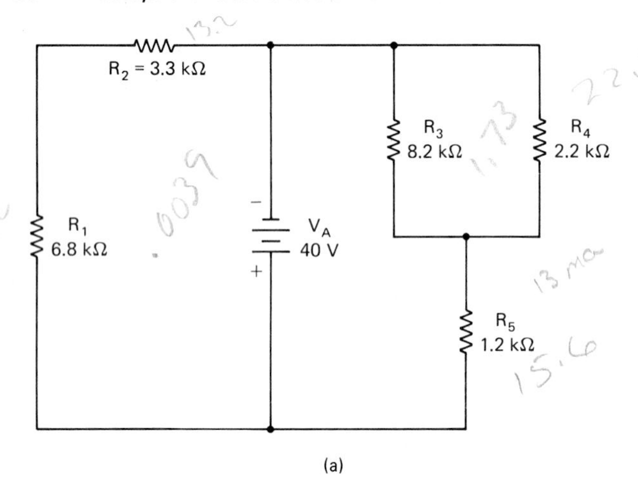

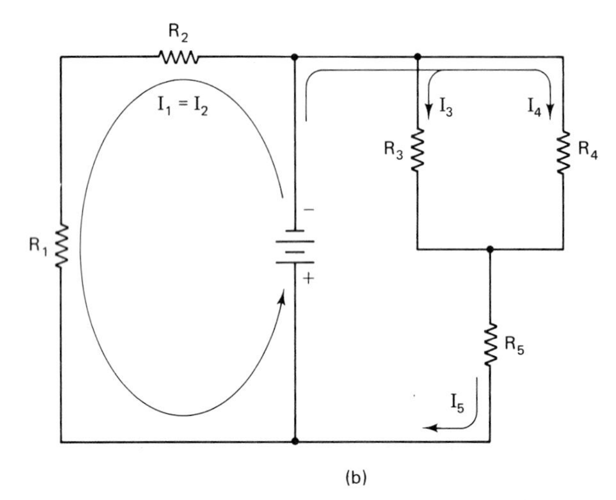

FIGURE 7-5 (a) Combination circuit schematic. (b) Current paths.

lel paths, as shown in Fig. 7-5(b). Solving for the current in each path provides the necessary data to solve for voltage drops about each current path.

From Fig. 7-3(b) the current through R_1 and R_2 is common since R_1 and R_2 are in series; therefore,

$$I_1 = I_2 = \frac{V_A}{R_T} = \frac{V_A}{R_1 + R_2}$$

$$= \frac{40 \text{ V}}{6.8 \text{ k}\Omega + 3.3 \text{ k}\Omega} \cong 4 \text{ mA}$$

The voltage drops across R_2 and R_1 can now be calculated.

$$V_2 = I_2R_2 = 4 \text{ mA} \times 3.3 \text{ k}\Omega \cong 13 \text{ V}$$

$$V_1 = I_1R_1 = 4 \text{ mA} \times 6.8 \text{ k}\Omega \cong 27 \text{ V}$$

An alternative solution for finding the voltage drops across R_2 and R_1 would be to use the voltage-division method, or

$$V_2 = \frac{R_2}{R_1 + R_2} \times V_A$$

$$= \frac{3.3 \text{ k}\Omega}{6.8 \text{ k}\Omega + 3.3 \text{ k}\Omega} \times 40 \text{ V} = 13 \text{ V}$$

$$V_1 = \frac{R_1}{R_1 + R_2} \times V_A$$

$$= \frac{6.8 \text{ k}\Omega}{6.8 \text{ k}\Omega + 3.3 \text{ k}\Omega} \times 40 \text{ V} \cong 27 \text{ V}$$

In the second current loop, currents I_3 and I_4 combine to form I_5. Ohm's law can be applied to solve for I_5, but the parallel branch must be reduced to an equivalent resistance, and from there the total resistance of the path can be found.

$$R_{(\text{equ})} = \frac{\text{product}}{\text{sum}} = \frac{R_3 \times R_4}{R_3 + R_4} = \frac{8.2 \text{ k}\Omega \times 2.2 \text{ k}\Omega}{8.2 \text{ k}\Omega + 2.2 \text{ k}\Omega}$$

$$= 1.7 \text{ k}\Omega$$

$$R_T = R_{(\text{equ})} + R_5 = 1.7 \text{ k}\Omega + 1.2 \text{ k}\Omega$$

$$= 2.9 \text{ k}\Omega$$

$$I_5 = \frac{V_A}{R_T} = \frac{40 \text{ V}}{2.9 \text{ k}\Omega} = 13.8 \text{ mA}$$

The voltage drops about the second current path can now be found:

$$V_5 = I_5R_5 = 13.8 \text{ mA} \times 1.2 \text{ k}\Omega = 16.6 \text{ V}$$

Since V_5 is in series with the parallel combination of R_3 and R_4, the voltage across R_3 and R_4 will be the difference between the applied voltage and V_5, or

$$V_3 = V_4 = V_A - V_5 = 40 \text{ V} - 16.6 \text{ V} = 23.4 \text{ V}$$

The current through R_3 and R_4 can now be solved:

$$I_3 = \frac{V_3}{R_3} = \frac{23.4 \text{ V}}{8.2 \text{ k}\Omega} = 2.9 \text{ mA}$$

$$I_4 = \frac{V_4}{R_4} = \frac{23.4 \text{ V}}{2.2 \text{ k}\Omega} = 10.6 \text{ mA}$$

An alternative solution for finding the voltage drops across $R_3 \parallel R_4$ and R_5 would be to use the voltage-division method, or

$$V_3 = V_4 = \frac{R_{(\text{equ})}}{R_{(\text{equ})} + R_5} \times V_A$$

$$= \frac{1.7 \text{ k}\Omega}{1.7 \text{ k}\Omega + 1.2 \text{ k}\Omega} \times 40 \text{ V} = 23.4 \text{ V}$$

$$V_5 = \frac{R_5}{R_{(equ)} + R_5} \times V_A$$

$$= \frac{1.2\ \text{k}\Omega}{1.7\ \text{k}\Omega + 1.2\ \text{k}\Omega} \times 40\ \text{V} = 16.6\ \text{V}$$

Another alternative solution for solving the second current loop would be to use the current-division method through R_3 and R_4, or

$$I_3 = \frac{R_4}{R_3 + R_4} \times I_T = \frac{2.2\ \text{k}\Omega}{8.2\ \text{k}\Omega + 2.2\ \text{k}\Omega} \times 13.8\ \text{mA}$$

$$= 2.9\ \text{mA}$$

$$I_4 = \frac{R_3}{R_3 + R_4} \times I_T = \frac{8.2\ \text{k}\Omega}{8.2\ \text{k}\Omega + 2.2\ \text{k}\Omega} \times 13.8\ \text{mA}$$

$$= 10.9\ \text{mA}$$

PRACTICE QUESTIONS 7-1

Choose the most suitable word, phrase, or value to complete each statement. Refer to Fig. 7-6 for Questions 1 to 10.

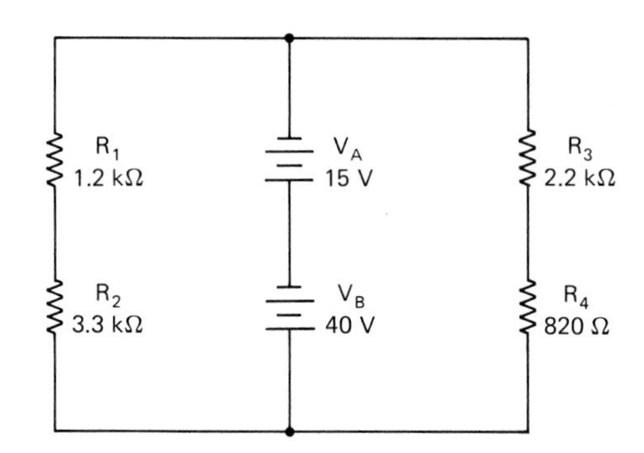

FIGURE 7-6 Combination circuit schematic.

1. Resistors R_1 and R_2 are in (*series, parallel*) with each other because they both have the same ________.
2. Resistors R_3 and R_4 are in (*series, parallel, series–parallel*) with R_1 and R_2.
3. Resistors R_1 and R_2 are in (*series, parallel, series–parallel*) with the source voltages V_A and V_B.
4. Resistors R_3 and R_4 are in (*series, parallel, series–parallel*) with the source voltages V_A and V_B.
5. Source voltages V_A and V_B are in (*series, parallel, series–parallel*) with each other.
6. The total applied voltage in this circuit will be ________.
7. Calculate the current through R_3 and R_4.
8. Calculate the current through R_1 and R_2.
9. Calculate the voltage drops across R_1 and R_2 using the voltage-division method.
10. Calculate the voltage drops across R_3 and R_4 using Ohm's law.

Refer to Fig. 7-7 for Questions 11 to 20.

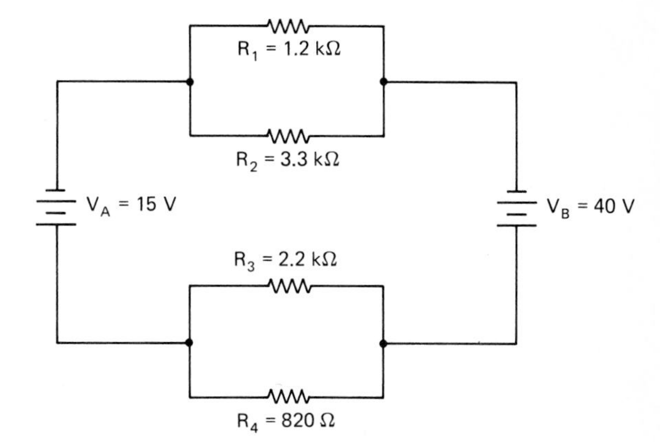

FIGURE 7-7 Combination circuit schematic.

11. Resistors R_1 and R_2 are in (*series, parallel*) with each other because they both have the same ________.
12. Resistors R_3 and R_4 are in (*series, parallel, series–parallel*) with R_1 and R_2.
13. Resistors R_1 and R_2 are in (*series, parallel, series–parallel*) with the source voltages V_A and V_B.
14. Resistors R_3 and R_4 are in (*series, parallel, series–parallel*) with the source voltages V_A and V_B.
15. Source voltage V_A and V_B are in (*series, parallel, series–parallel*) with each other.
16. The total applied voltage in this circuit will be ________ volts.
17. Calculate the current through R_3 and R_4.
18. Calculate the current through R_1 and R_2.
19. Calculate the current through R_3 and R_4 using the current-division method.
20. Calculate the voltage drop across R_1 and R_2 using the voltage-division method.

7-4 JUMBLED SERIES–PARALLEL RESISTANCE CIRCUITS

Not all circuit schematic diagrams will have the resistors neatly laid out in series and parallel. These cases involve rough schematics drawn from a specific printed-circuit board or chassis or perhaps a roughed-out design circuit. The technician must be able to "decipher" the rough schematic to analyze or perhaps construct the circuit.

The circuit of Fig. 7-8(a) can be redrawn so that the resistors in parallel or in series can be very obvious. This is done by taking the outermost current path and

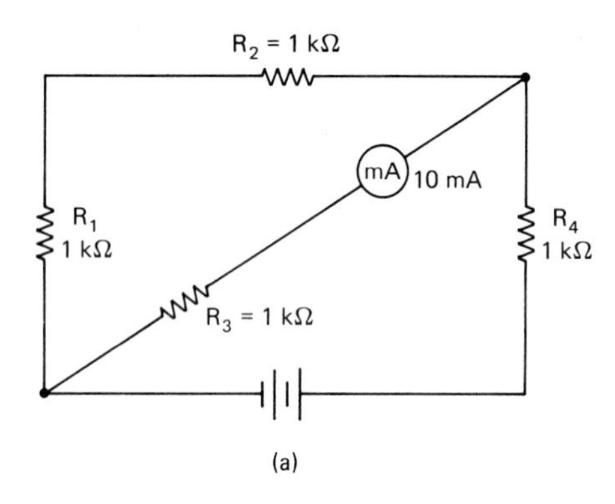

(a)

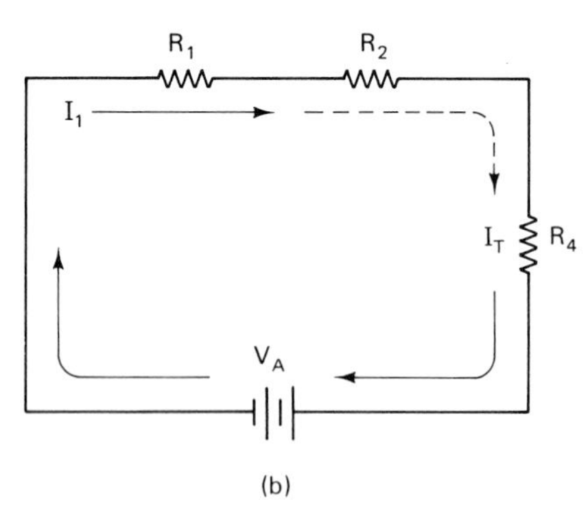

(b)

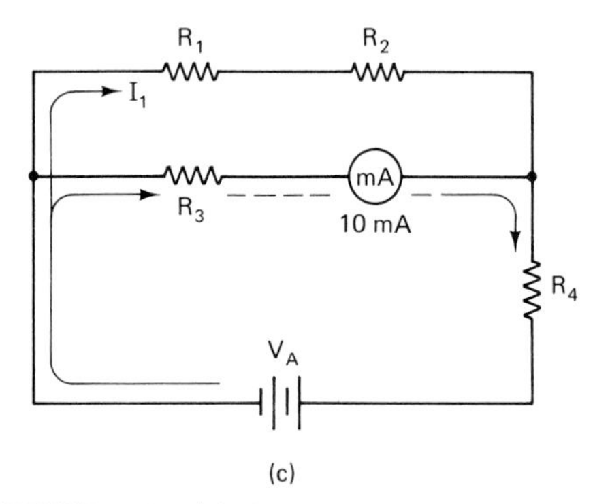

(c)

FIGURE 7-8 (a) Circuit schematic. (b) Simplified circuit (stage 1). (c) Simplified circuit (stage 2).

drawing in all the resistors, as shown in Fig. 7-8(b), stage 1.

The second current path can now be drawn in as shown in Fig. 7-8(c), stage 2. From this simplified schematic the following observations can be made:

1. The current through R_3 is shown to be 10 mA.
2. Current I_1 flows through R_2; therefore, R_1 and R_2 are in series with each other, and thus $I_1 = I_2$.
3. R_3 is across or in parallel with R_1 and R_2, so $V_1 + V_2 = V_3$.
4. The current through R_4 will be the sum of the I_3 and ($I_1 = I_2$) currents, or $I_4 = I_3 + I_1$

Solve for V_A:

$$V_1 + V_2 = V_3 = I_3R_3 = 10\text{ mA} \times 1\text{ k}\Omega = 10\text{ V}$$

$$I_1 = I_2 = \frac{V_3}{R_1 + R_2} = \frac{10\text{ V}}{1\text{ k}\Omega + 1\text{ k}\Omega} = 5\text{ mA}$$

The current flowing into R_4 will be the sum of I_1 and I_3:

$$I_4 = I_1 + I_3 = 5\text{ mA} + 10\text{ mA} = 15\text{ mA}$$

The voltage drop $V_4 = I_4R_4$.

$$V_4 = 15\text{ mA} \times 1\text{ k}\Omega = 15\text{ V}$$

and

$$V_A = V_3 + V_4 = 10\text{ V} + 15\text{ V} = 25\text{ V}$$

EXAMPLE 7-3

The circuit of Fig. 7-9(a) can be redrawn so that the resistors in parallel or in series can be very obvious. This is done by taking the outermost current path and drawing in all the resistors as shown in Fig. 7-9(b), stage 1.

The second current path can now be drawn in as shown in Fig. 7-9(c), stage 2, and finally, the third current path can be drawn in as shown in Fig. 7-9(d). Solving for V_A yields

$$\text{current } I_5 = I_4 = \frac{V_5}{R_5} = \frac{10\text{ V}}{1\text{ k}\Omega} = 10\text{ mA}$$

$$V_4 = I_4R_4 = 10\text{ mA} \times 1\text{ k}\Omega = 10\text{ V}$$

Also $R_4 = R_5$, then $V_4 = V_5 = 10$ V.

R_3 is in parallel with the series string of R_5 and R_4, so

$$V_3 = V_4 + V_5 = 10\text{ V} + 10\text{ V} = 20\text{ V}$$

and

$$I_3 = \frac{V_3}{R_3} = \frac{20\text{ V}}{1\text{ k}\Omega} = 20\text{ mA}$$

The current through R_2 will be the sum of I_3 and I_5 since R_2 is in series with the parallel branches R_3 and ($R_5 + R_4$).

$$I_2 = I_3 + I_5 = 20\text{ mA} + 10\text{ mA} = 30\text{ mA}$$

and

$$V_2 = I_2R_2 = 30\text{ mA} \times 1\text{ k}\Omega = 30\text{ V}$$

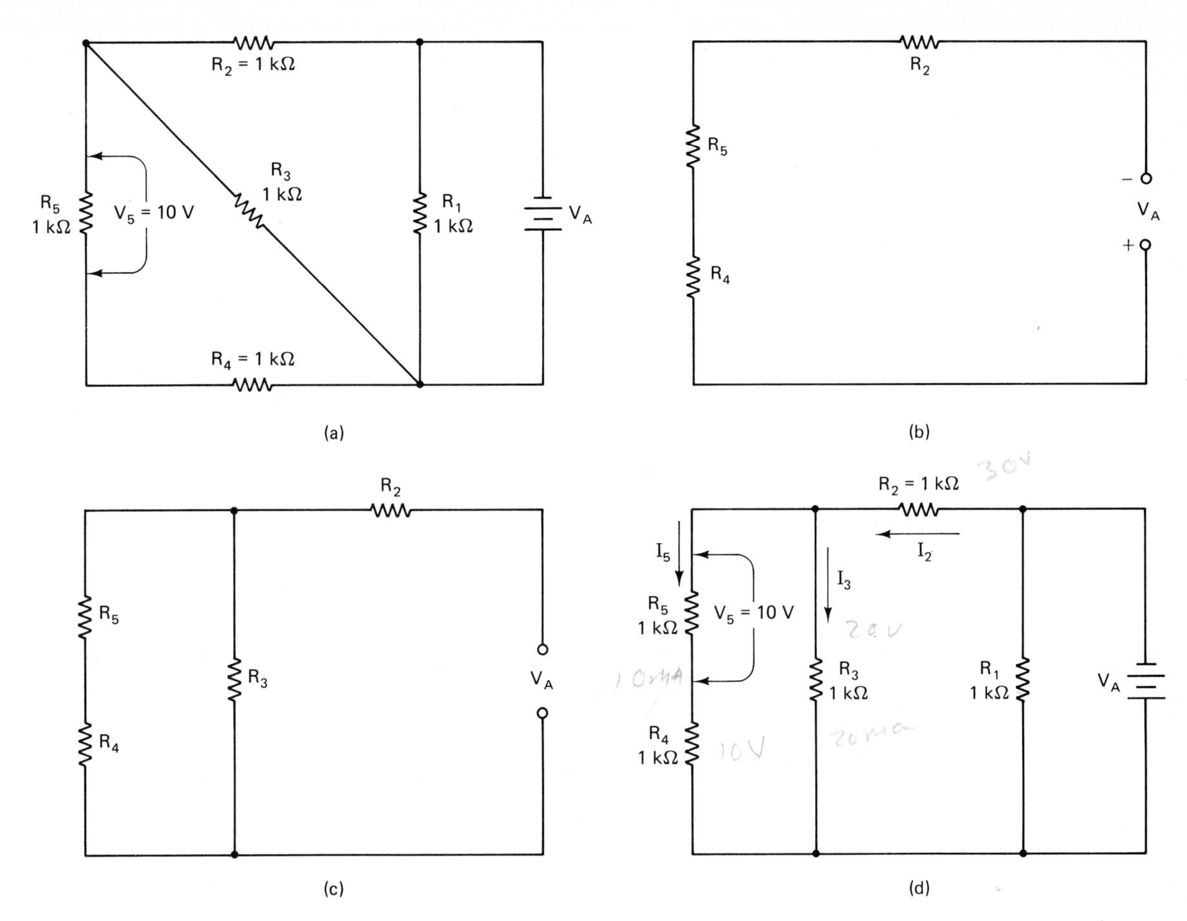

FIGURE 7-9 (a) Circuit schematic. (b) Simplified circuit (stage 1). (c) Simplified circuit (stage 2). (d) Simplified circuit.

V_2 is in series with the parallel combination of R_3 and R_4R_5; therefore,

$$V_T = V_2 + V_3 = 30\text{ V} + 20\text{ V} = 50\text{ V}$$

Finally, R_1 is shunted or in parallel with the source V_A, so its voltage drop will equal V_A or 50 V.

$$\text{Current } I_1 = \frac{V_1}{R_1} = \frac{V_A}{R_1} = \frac{50\text{ V}}{1\text{ k}\Omega} = 50\text{ mA}$$

Power Ratings of Resistors

Generally, the power rating of resistors that are connected in parallel can be conveniently calculated using either $P = V^2/R$ or $P = VI$, since voltage is common in a parallel circuit.

Similarly, the power rating of resistors that are connected in series can be conveniently calculated using $P = I^2R$ or $P = VI$ since current is common in a series circuit.

From the schematic of Fig. 7-9(d), the power rating of R_1 can easily be calculated since it is across the source voltage:

$$P_1 = \frac{V_1}{R_1} = \frac{V_A^2}{R_1} = \frac{(50\text{ V})^2}{1\text{ k}\Omega} = 2.5\text{ W}$$

Since R_2 is in series with the parallel group R_3 and $(R_5 + R_4)$, its power rating is

$$P_2 = I_2^2R_2 = (30\text{ mA})^2 \times 1\text{ k}\Omega = 0.9\text{ W} \quad \text{or} \quad 900\text{ mW}$$

The power rating of R_3 can be calculated with any of the three formulas since the voltage current and resistance are known.

$$P_3 = V_3I_3 = 20\text{ V} \times 20\text{ mA} = 400\text{ mW}$$

and

$$P_5 = \frac{V^2}{1\text{ k}\Omega} = \frac{(10\text{ V})^2}{1\text{ k}\Omega} = 0.1\text{ W} \quad \text{or} \quad 100\text{ mW}$$

$$P_4 = I^2R = (10\text{ mA})^2 \times 1\text{ k}\Omega = 100\text{ mW}$$

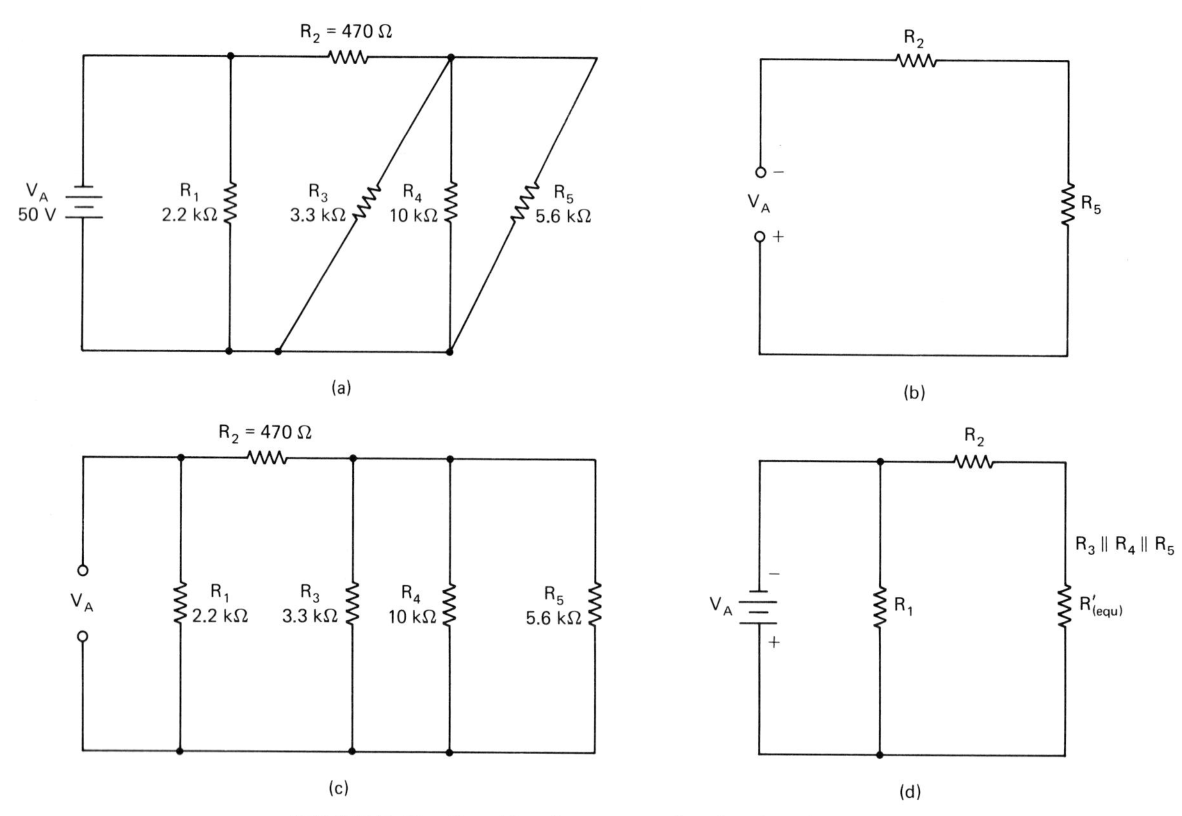

FIGURE 7-10 Combination networks circuit schematic.

EXAMPLE 7-4

The network circuit, Fig. 7-10(a), is redrawn to show the resistors that are obviously in series or parallel. This is done by taking the outermost current path and drawing in all the resistors as shown in Fig. 7-10(b), stage 1.

The second current path can now be drawn in as shown in Fig. 7-10(c), stage 2. This shows the resistors $R_3 \parallel R_4 \parallel R_5$. Finally, by reducing $R_3 \parallel R_4 \parallel R_5$ to an equivalent total resistance, the circuit is simplified sufficiently to enable an easy approach to the solution for voltage current and resistance.

Solution: We solve for the total resistance in the circuit. R_1 is in parallel with R_2 and $R'_{(equ)}$ [see Fig. 7-10(d)]:

$$R_T = R_1 \parallel (R_2 + R'_{(equ)})$$

$$R'_{(equ)} = \frac{\text{product}}{\text{sum}} \text{(two } R\text{'s at a time)}$$

$$R_{(equ)} = \frac{R_3 \times R_4}{R_3 + R_4} = \frac{3.3\ \text{k}\Omega \times 10\ \text{k}\Omega}{3.3\ \text{k}\Omega + 10\ \text{k}\Omega} = 2.5\ \text{k}\Omega$$

$$R'_{(equ)} = \frac{R_{(equ)} \times R_5}{R_{(equ)} + R_5} = \frac{2.5\ \text{k}\Omega \times 5.6\ \text{k}\Omega}{2.5\ \text{k}\Omega + 5.6\ \text{k}\Omega}$$

$$= 1.7\ \text{k}\Omega$$

$$R_T = 2.2\ \text{k}\Omega \parallel (470\ \Omega + 1.7\ \text{k}\Omega)$$

$$= 2.2\ \text{k}\Omega \parallel 2.2\ \text{k}\Omega \cong \frac{2.2\ \text{k}\Omega}{2} \cong 1.1\ \text{k}\Omega$$

The total circuit current can now be calculated:

$$I_T = \frac{V_A}{R_T} = \frac{50\ \text{V}}{1.1\ \text{k}\Omega} \cong 46\ \text{mA}$$

Resistor R_2 is in series with the parallel branches $R_3 \parallel R_4 \parallel R_5$. I_2 can easily be calculated since the total resistance of this current path is known and is connected across the source.

$$I_2 = \frac{V_A}{R_T} = \frac{V_A}{R_2 + R'_{(equ)}} = \frac{50\ \text{V}}{470\ \Omega + 1.7\ \text{k}\Omega}$$

$$= 23\ \text{mA}$$

The current through the parallel resistors can be calculated by Ohm's law, but first the voltage drop across the parallel group must be found.

$$V_3 = V_4 = V_5 = I_T \times R'_{(equ)}$$

$$= 23\ \text{mA} \times 1.7\ \text{k}\Omega \cong 39\ \text{V}$$

$$I_3 = \frac{V_3}{R_3} = \frac{39\ \text{V}}{3.3\ \text{k}\Omega} = 11.8\ \text{mA}$$

$$I_4 = \frac{V_4}{R_4} = \frac{39\text{ V}}{10\text{ k}\Omega} = 3.9\text{ mA}$$

$$I_5 = \frac{V_5}{R_5} = \frac{39\text{ V}}{5.6\text{ k}\Omega} = 7\text{ mA}$$

Summing I_3, I_4, and I_5 should prove the I_2 current in series with the parallel group:

$$I_2 = 11.8\text{ mA} + 3.9\text{ mA} + 7\text{ mA}$$
$$= 22.7\text{ mA} \cong 23\text{ mA}$$

The voltage drop across R_2 can easily be found:

$$V_2 = I_2R_2 = 23\text{ mA} \times 470\ \Omega = 10.8\text{ V}$$

Summing V_2 and $V_3 \parallel V_4 \parallel V_5$ should prove the total voltage V_A:

$$V_A = V_2 + V_3 = 10.8\text{ V} + 39\text{ V}$$
$$= 49.8\text{ V} \cong 50\text{ V}$$

Resistor R_1 is independent of any other resistances and is in shunt, or parallel with the source voltage V_A. Current I_1 can easily be found using Ohm's law:

$$I_1 = \frac{V_A}{R_1} = \frac{50\text{ V}}{2.2\text{ k}\Omega} = 22.7\text{ mA}$$

The total current I_2 can be proven by summing I_2 with I_1, or

$$I_T = I_1 + I_2 = 22.7\text{ mA} + 23\text{ mA}$$
$$= 45.7\text{ mA} \cong 46\text{ mA}$$

The power rating of each parallel resistor can be calculated by using the voltage-squared formula:

$$P_5 = \frac{V_5^2}{R_5} = \frac{(39\text{ V})^2}{5.6\text{ k}\Omega} = 0.27\text{ W} \quad\text{or}\quad 270\text{ mW}$$

$$P_4 = \frac{V_4^2}{R_4} = \frac{(39)^2}{10\text{ k}\Omega} = 0.15\text{ W} \quad\text{or}\quad 150\text{ mW}$$

$$P_3 = \frac{V_3^2}{R_3} = \frac{(39\text{ V})^2}{3.3\text{ k}\Omega} = 0.46\text{ W} \quad\text{or}\quad 460\text{ mW}$$

R_2 is in series, so the I^2R formula might be more convenient:

$$P_2 = I_2^2R_2 = (23\text{ mA})^2 \times 470\ \Omega$$
$$= 0.248\text{ W} \quad\text{or}\quad 248\text{ mW}$$

Finally, R_1 is shunted across the source; therefore, the V^2/R formula might be more convenient:

$$P_1 = \frac{V_1^2}{R_1} = \frac{(50\text{ V})^2}{2.2\text{ k}\Omega} = 1.1\text{ W}$$

SUMMARY

Solving for voltage drops, current, and resistance in a series–parallel circuit involves a knowledge of series and parallel circuit laws and Ohm's law.

Voltages

Series circuit: $V_A = V_1 + V_2 + V_3$

Parallel circuit: $V_A = V_1 = V_2 = V_3$

Current

Series circuit: $I_T = I_1 = I_2 = I_3$

Parallel circuit: $I_T = I_1 + I_2 + I_3$

Resistance

Series circuit: $R_T = R_1 + R_2 + R_3$

Parallel circuit: $R_T = \frac{\text{product}}{\text{sum}}$ or $R_T = \frac{R_1 \times R_2}{R_1 + R_2}$ or $\frac{1}{R_T} = \frac{1}{R_1} + \frac{1}{R_2} + \frac{1}{R_3}$

Voltage-division series circuit

$$V_1 = \frac{R_1}{R_1 + R_2,\text{ etc.}} \times V_A \quad\text{or}\quad V_2 = \frac{R_2}{R_1 + R_2,\text{ etc.}} \times V_A$$

Current-division parallel circuit

$$I_1 = \frac{R_2}{R_1 + R_2} \times I_T \quad \text{or} \quad I_2 = \frac{R_1}{R_1 + R_2} \times I_T$$

Simplifying a series–parallel circuit: Although there is more than one method to simplify a combination circuit, it is recommended that a beginning student use the current loop method, starting with an outermost loop as shown in the examples.

For series resistance the current is common, so a convenient power equation to use would be $P = I^2R$. For parallel resistance the voltage is common, so a convenient power equation to use would be $P = V^2/R$.

REVIEW QUESTIONS

7-1. Draw a schematic diagram to show two branch circuits connected in parallel with the source. Each branch circuit has three series resistances.

7-2. Explain how the three series resistances are reduced to an equivalent resistance.

7-3. Draw a schematic diagram to show two parallel resistance branches connected in series with the source. Each parallel branch has three resistors.

7-4. Explain how the three parallel resistances are reduced to an equivalent resistance.

7-5. Give a brief description of a simple method to redraw a combination circuit to visually enhance series and parallel components.

7-6. Draw a schematic diagram to show how four 2-kΩ, $\frac{1}{2}$-W resistors are connected to give a combined resistance of 2 kΩ at 2 W.

7-7. Two resistors are connected in series. The resistance of one is 3.3 kΩ; the resistance of the second is 200 Ω. If each resistor is rated at $\frac{1}{2}$ W, which resistor would dissipate the greatest power? Explain.

7-8. The two resistors of Question 7-7 are connected in parallel. Which resistor would dissipate the greatest power? Explain.

7-9. Of the three power equations, state the most convenient equation to use for resistors connected in parallel.

7-10. Of the three power equations, state the most convenient equation to use for resistors connected in series.

SELF-EXAMINATION

Refer to Fig. 7-11 for Questions 7-1 to 7-5.

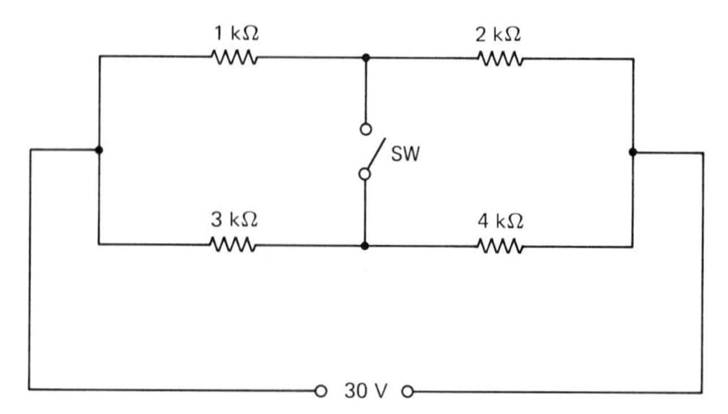

FIGURE 7-11 Combination circuit schematic.

7-1. What is the net resistance with the switch open?

7-2. What is the net resistance with the switch closed?

7-3. What is the voltage drop across the 2-kΩ resistor with the switch closed?

7-4. What is the current through the 3-kΩ resistor with the switch closed?

7-5. For maximum voltage across the 4-kΩ resistor, should the switch be open or closed?

Refer to Fig. 7-12 for Questions 7-6 to 7-10.

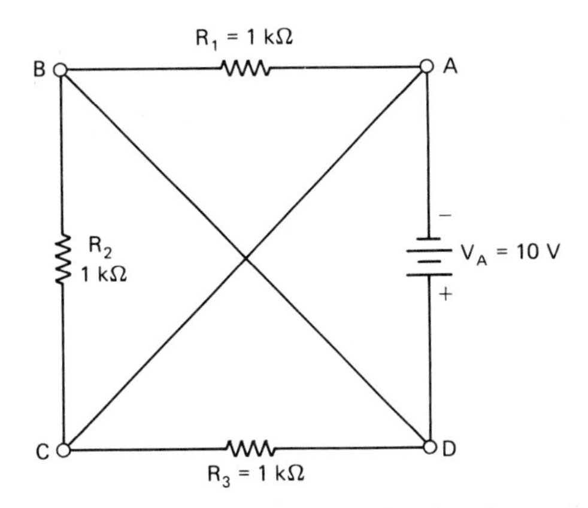

FIGURE 7-12 Combination circuit schematic.

7-6. Calculate the current through R_2.

7-7. What is the net resistance of this circuit?

7-8. What is the total current from the source?

7-9. What is the polarity of *B* relative to *C*?

7-10. How much current flows in the diagonal wire linking points *A* to *C*?

Refer to Fig. 7-13 for Questions 7-11 to 7-13.

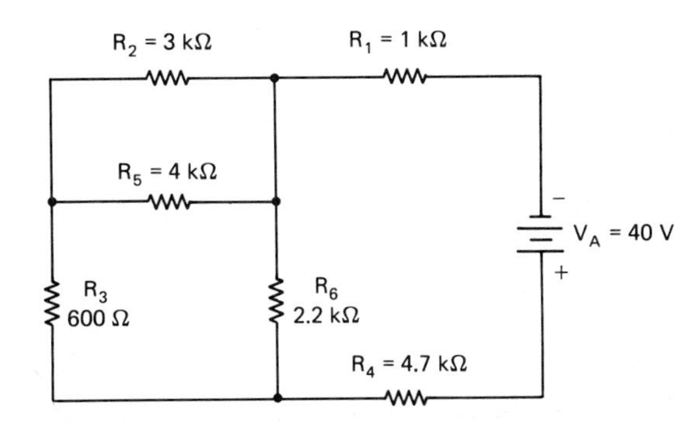

FIGURE 7-13 Combination circuit schematic.

7-11. Calculate the net resistance of the circuit.

7-12. Calculate the current through R_3.

7-13. Calculate the power rating of R_4.

Refer to Fig. 7-14 for Questions 7-14 to 7-16.

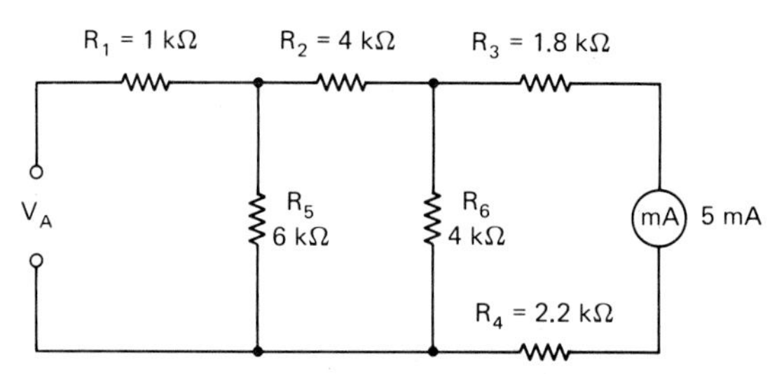

FIGURE 7-14 Combination circuit schematic.

7-14. Calculate the applied voltage V_A.

7-15. Calculate the total resistance in the circuit.

7-16. Calculate the total power dissipation of the circuit.

ANSWERS TO PRACTICE QUESTIONS

7-1

1. series, current
2. parallel
3. series
4. series
5. series
6. 55 V
7. 18.2 mA
8. 12.2 mA
9. $V_1 = \dfrac{R_1}{R_1 + R_2} \times V_A = 14.7$ V

   $V_2 = \dfrac{R_2}{R_1 + R_2} \times V_A = 40.3$ V
10. $V_3 = 40$ V

    $V_4 = 15$ V
11. parallel, voltage
12. series
13. series
14. series
15. series
16. 25
17. $I_3 = 4.6$ mA

    $I_4 = 12.3$ mA
18. $I_1 = 12.4$ mA

    $I_2 = 4.5$ mA
19. $I_3 = \dfrac{R_4}{R_3 + R_4} \times I_T = 4.6$ mA

    $I_4 = \dfrac{R_3}{R_3 + R_4} \times I_T = 12.4$ mA
20. $V_1 = \dfrac{R'_T}{R_T} \times V_A = 14.9$ V

    $V_1 = V_2 = 14.9$ V

8

Voltage Dividers

INTRODUCTION

A voltage divider is a series circuit that supplies a lower voltage or voltages to one or more loads from a single power source. There are many applications or needs for voltage dividers, some of which are a simple volume control to reduce or increase the audio input voltage from the detector circuit of a radio, or varying the gain of a microphone. Another need is for a simple voltage divider to supply base bias for transistor circuits. Voltage doublers, triplers, and so on, are also voltage dividers. Notably, the series voltage divider is used to supply various levels of voltage and current to two or more parallel loads from one source.

MAIN TOPICS IN CHAPTER 8

OBJECTIVES

After studying Chapter 8, the student should be able to

1. Master voltage-divider circuits by applying series and parallel circuit laws to design a loaded series voltage divider.
2. Master the voltage-divider concepts of a bridge circuit and relate the bridge circuit to simplified practical applications.
3. Relate positive and negative voltages to ground reference.
4. Master potentiometer uses, construction, and characteristics.

8-1 POTENTIOMETER AS A VOLTAGE DIVIDER

A simple fixed voltage divider is two resistances in series across a source. Figure 8-1 shows a simple voltage divider connected across an ac generator source, such as an audio-frequency generator. Its function is to provide a lower input voltage to a transistor amplifier from a higher generator output voltage. The ratio of voltage drops across R_1 to R_2 will be 9 to 1, or 90% of the generator output voltage will be dropped by R_1 and 10%

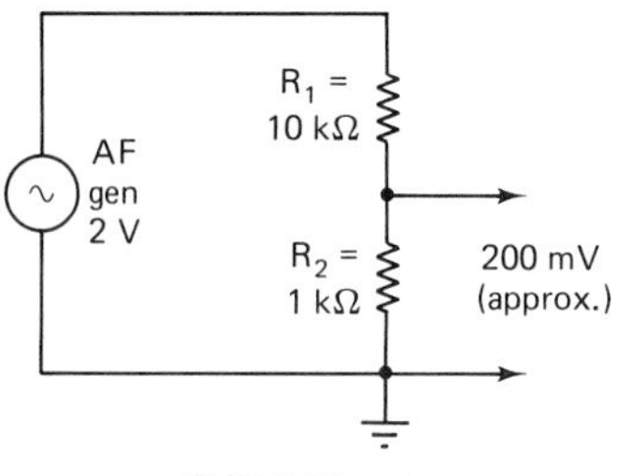

FIGURE 8-1

(approx.) of the voltage will actually be inputted into the transistor amplifier. This is useful in cases where the input voltage to the amplifier must be kept to low-level values of a few hundred millivolts.

A more practical method of reducing input voltages to a device is to use a variable resistance, specifically a potentiometer, as a voltage divider. Carbon variable resistors or potentiometers (pots) are constructed by depositing a carbon compound on a fiber disk. A slider contact on a rotating shaft varies the resistance as the arm shaft is rotated (see Fig. 8-2). Carbon pots have a maximum power rating of 2 W. For applications where greater power-dissipation ratings are needed, a wire-wound pot is used. Instead of carbon, a resistance wire is wound on a fiber or ceramic ring. Like the carbon pot, a slider contact on a rotating shaft varies the resistance as the arm shaft is rotated (see Fig. 8-3).

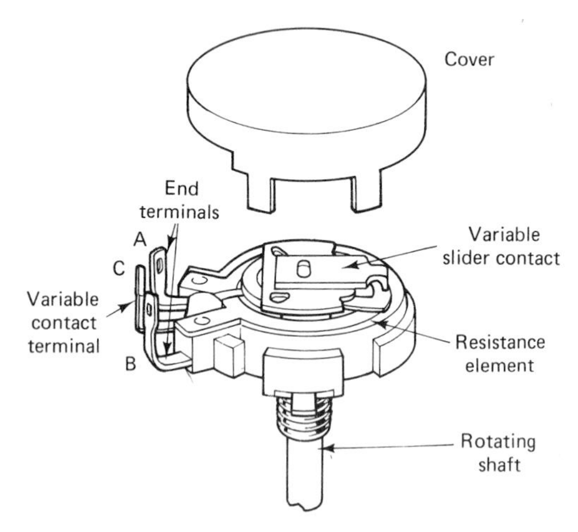

FIGURE 8-2 Carbon potentiometer.

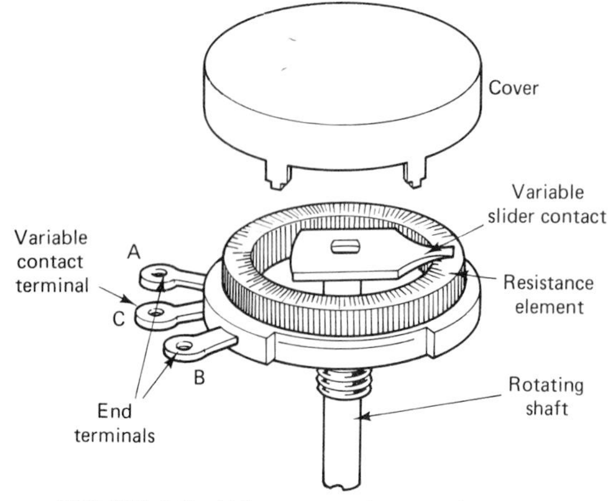

FIGURE 8-3 Wire-wound potentiometer.

The schematic symbol for a potentiometer is shown in Fig. 8-4 for each of the applications of a potentiometer used as a gain control for a microphone and for phono input.

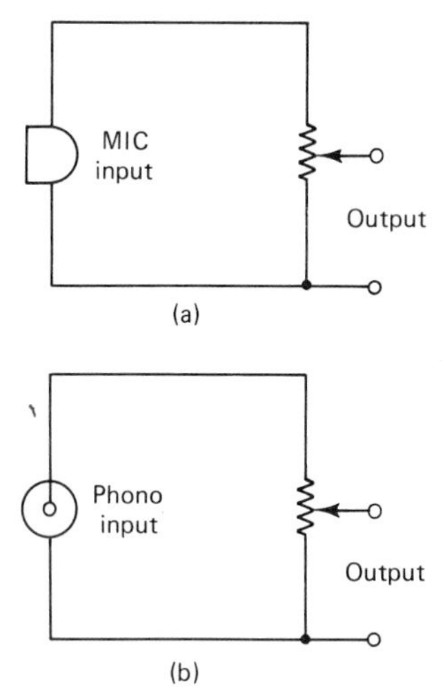

FIGURE 8-4 (a) Microphone gain control. (b) Phono gain control.

8-2 POTENTIOMETER TAPERS AND APPLICATIONS

Figure 8-5 shows the various potentiometer percentage resistance change with percentage shaft rotation tapers. Applications of these tapers are as follows;

A For a uniform resistance change with shaft rotation, the linear taper A potentiometer is selected. Applications include the ohmmeter zero adjust, FET voltmeter zero adjust, oscilloscope gain controls, and many other requirements where a change in resistance with shaft rotation ensures a uniform change in voltage.

B This is a modified log where approximately 20% of its resistance is at half of the shaft rotation. It is used in audio volume controls and tone control circuits.

C This is a semilog taper where 10% of its resistance is at half of the shaft rotation. It is also used in audio volume controls and tone control circuits.

H This is a tapped log curve used for tone controls with bass compensation.

P This is the reverse of B, used as a contrast control in television. A contrast control increases or decreases the black level in monochrome TV and increases or decreases the color level in color TV sets.

Q This is the reverse of C, used as a brightness control in television sets.

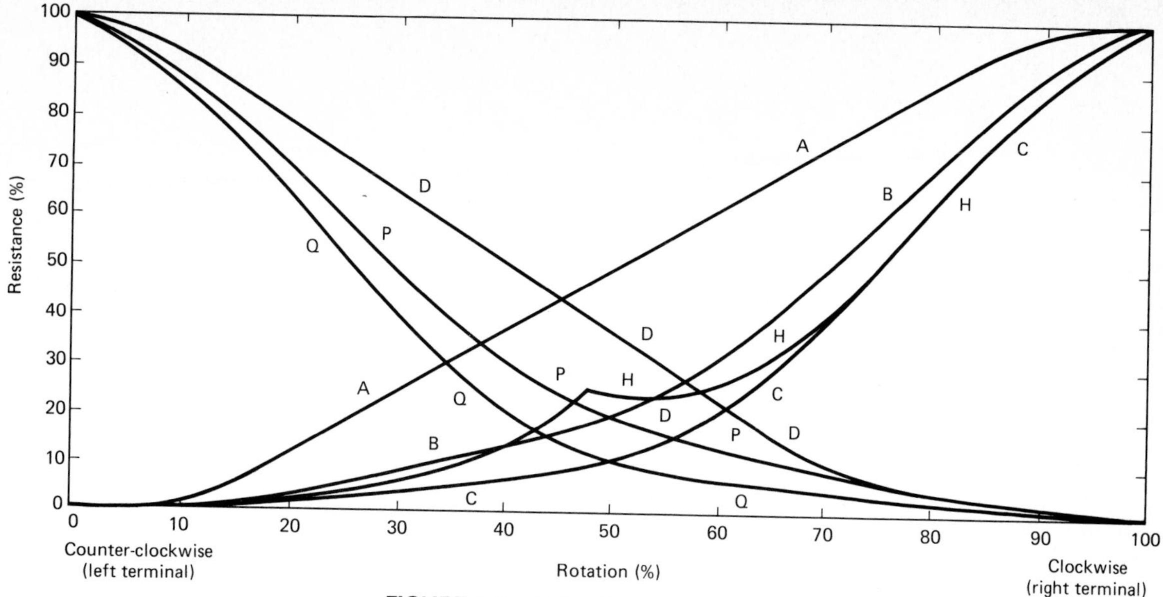

FIGURE 8-5 Potentiometer tapers.

8-3 BRIDGE CIRCUIT

One of the most significant circuits in electrical and electronic systems is the bridge circuit. Its importance ranges from industrial control circuits and navigational control circuits to measuring resistance, capacitance, and inductance. Sir Charles Wheatstone in about 1840 developed the theory behind the bridge circuit and its precision measuring ability. Today a Wheatstone bridge connotes a resistance, capacitance, or inductance precision measuring instrument.

The schematic of Fig. 8-6(a) shows a commonly used circuit configuration for a bridge circuit. Figure 8-6(b) is a simplified version of the same circuit. The symbol between points A and B represents a galvanometer. A galvanometer is an extremely sensitive meter capable of measuring currents as small as 100 picoamperes (pA). The galvanometer increases the sensitivity of the bridge circuit's null point. The pointer of the galvanometer is preset to midscale for a zero current reading. A zero current reading on the galvanometer midscale is called the null point and is obtained when the bridge is balanced or "nulled."

Balanced Bridge

When the bridge is said to be balanced, the voltage across R_1 will be exactly equal to the voltage across R_2, so that the current through the galvanometer is zero. Since $R_3 = R_4$ for a balanced condition, V_3 will also equal V_4, or $V_1 = V_2$ and $V_3 = V_4$.

R_1 and R_3 form a series voltage divider, as does R_2 and R_4, so that the ratio of V_1 to V_3 will be exactly equal to the ratio of V_2 to V_4, or

$$\frac{V_1}{V_3} = \frac{V_2}{V_4}$$

The ratios of the voltages can be written in terms of resistance ratios by substituting IR for voltages and canceling the currents since the currents are common in each arm of the bridge circuit [i.e., $I_1 = I_3$ in the left arm of Fig. 8-6(b) and $I_2 = I_4$ in the right arm]. Rewriting the voltages by substituting IR gives

$$\frac{I_1 R_1}{I_3 R_3} = \frac{I_2 R_2}{I_4 R_4}$$

The currents cancel to give

$$\frac{R_1}{R_3} = \frac{R_2}{R_4} \quad \text{ohms} \qquad (8\text{-}1)$$

The circuit of Fig. 8-6(a) can be used as a Wheat-

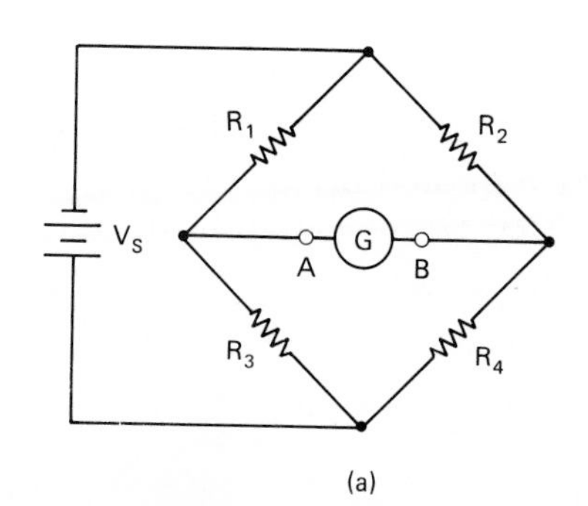

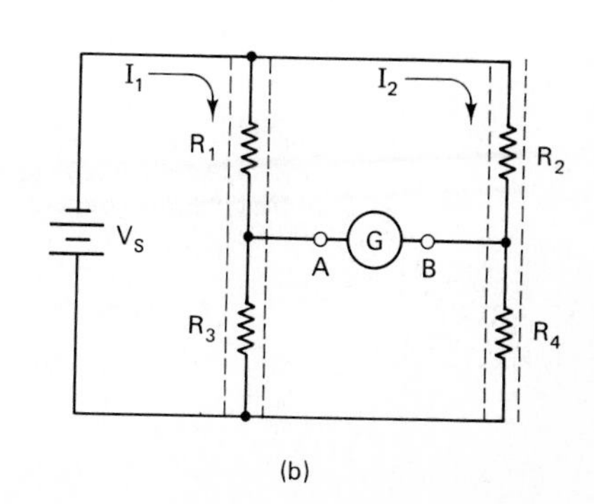

FIGURE 8-6 (a) Bridge circuit configuration. (b) Simplified bridge circuit.

stone bridge by substituting R_4 with a variable resistance R_V and using the terminals of R_3 as the unknown resistance, as shown in Fig. 8-7. Equation (8-1) can now be rewritten to give

$$\frac{R_1}{R_X} = \frac{R_2}{R_V}$$

Solving for R_X gives

$$R_X R_2 = R_1 R_V \quad \text{and} \quad R_X = \frac{R_1 R_V}{R_2} \quad \text{ohms} \quad (8\text{-}2)$$

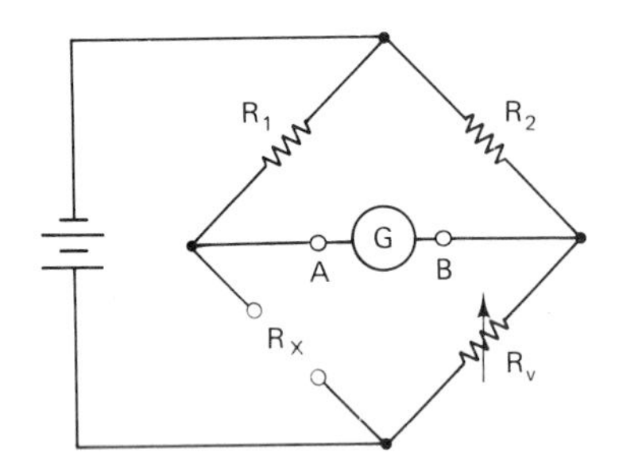

FIGURE 8-7 Wheatstone bridge.

A lab type of Wheatstone bridge has precision resistors R_1 and R_2 as well as variable resistor R_V. Figure 8-8 shows one type of precison Wheatstone bridge manufactured by GenRad, Inc.

EXAMPLE 8-1

In the Wheatstone bridge circuit of Figure 8-7, $R_1 = 1\ \text{k}\Omega$, $R_2 = 1.5\ \text{k}\Omega$, and $R_V = 3\ \text{k}\Omega$. Find R_X.

Solution: From the balanced bridge equation (8-2),

$$R_X = \frac{R_1 \times R_V}{R_2} = \frac{1\ \text{k}\Omega \times 3\ \text{k}\Omega}{1.5\ \text{k}\Omega} = 2\ \text{k}\Omega$$

EXAMPLE 8-2

In the Wheatstone bridge circuit of Fig. 8-7, $R_X = 3.3\ \text{k}\Omega$, $R_1 = 2\ \text{k}\Omega$, and $R = 2\ \text{k}\Omega$. Find the value of R_V.

Solution: Using the balanced bridge equation (8-2),

$$R_X R_2 = R_1 R_V$$

Therefore,

$$R_V = \frac{R_X R_2}{R_1}$$

$$= \frac{3.3\ \text{k}\Omega \times 2\ \text{k}\Omega}{2\ \text{k}\Omega} = 3.3\ \text{k}\Omega$$

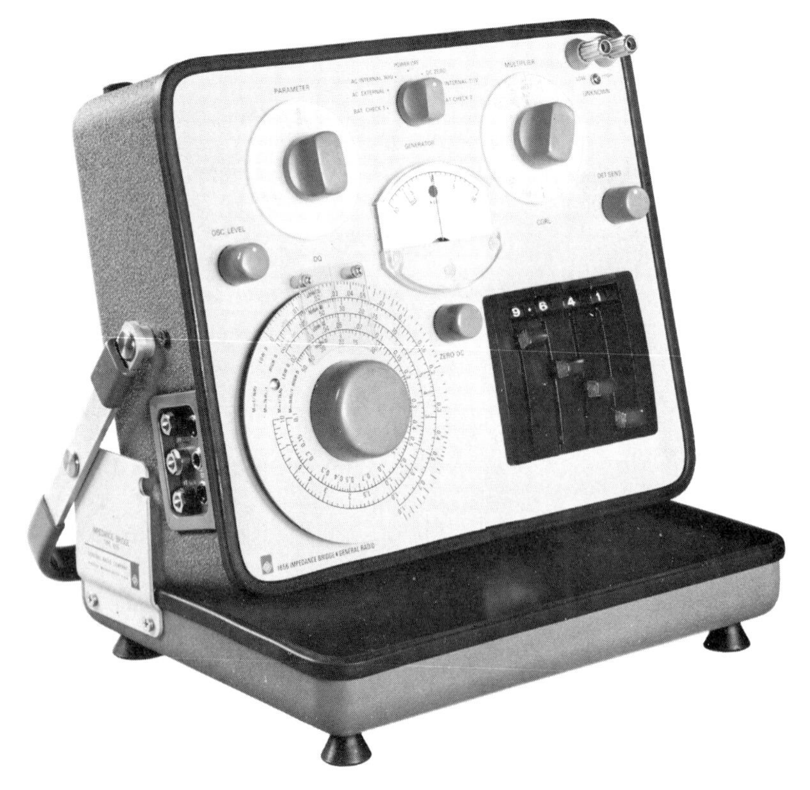

FIGURE 8-8 Courtesy of GenRad, Inc., Wheatstone bridge.

8-4 DESIGN OF A VOLTAGE DIVIDER WITH RESISTIVE LOADS

Voltage dividers are useful in obtaining different values of voltages from a single power supply. If two voltages, say 15 V and 5 V are required from a 24-V source, a voltage divider with two taps as shown in Fig. 8-9 would be designed.

With the loads not connected and the circuit complete through R_1, R_2, and R_3, a current called the bleeder current (I_B) will flow continuously from the source. With an ordinary voltage source the bleeder current helps to stabilize the source voltage under varying current load conditions. Because the bleeder current must be supplied by the source, the overall current requirement of the voltage-divider circuit must include the bleeder current. Generally, the bleeder current should not exceed 10% of the total current required by the loads.

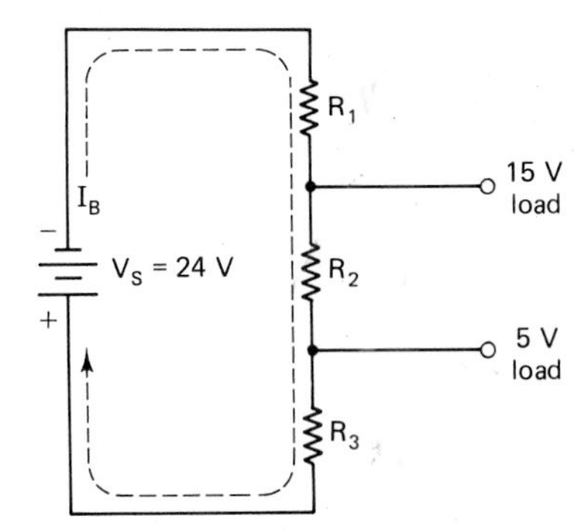

FIGURE 8-9 Two-tap voltage divider.

EXAMPLE 8-3: Designing a Voltage Divider

A design, regardless of its nature, requires some type of specifications. Specifications outline the requirements and limits of the desirable characteristics of a device. In this case the specifications are as follows.

Voltage-divider specifications

Voltage source: 24 V at 100 mA dc

Load 1 requirement: 15 V at 50 mA, ±5%

Load 2 requirement: 5 V at 30 mA, ±5%

Bleeder current: Not more than 10% of the total load current.

Checking to ensure that the power supply will meet the load and bleeder current requirements shows that the total current required from the source will be less than 100 mA.

$$I_T = I_{L1} + I_{L2} + I_B$$

where

$$I_B = 10\% \text{ of } (I_{L1} + I_{L2})$$

$$= \frac{10}{100} \times (50 \text{ mA} + 30 \text{ mA}) = 8 \text{ mA}$$

$$I_T = 50 \text{ mA} + 30 \text{ mA} + 8 \text{ mA} = 88 \text{ mA}$$

The 88 mA is under the maximum 100-mA rating of the power supply, so its current capacity is adequate.

Calculating the Resistance Values for R_1, R_2, and R_3. Each resistance value can be calculated using Ohm's law (i.e., $R = V/I$); however, the current and voltage drop across each resistor must be determined first.

Currents: The total current flowing through each of the voltage-divider resistors can easily be determined once their respective paths are drawn in, as shown in Fig. 8-10. The schematic shows

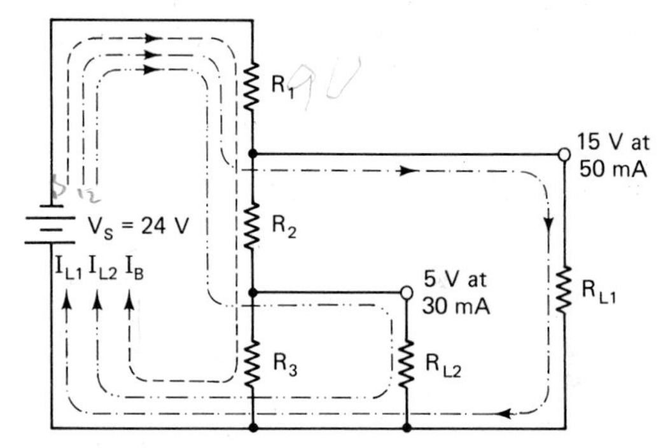

FIGURE 8-10 Voltage-divider currents.

that all three currents, I_{L1}, I_{L2}, and I_B, flow through R_1; therefore, I_1 can be summed:

$$I_1 = I_{L1} + I_{L2} + I_B = 50 \text{ mA} + 30 \text{ mA} + 8 \text{ mA}$$

$$= 88 \text{ mA}$$

The same schematic shows the currents through R_2 to be

$$I_2 = I_{L2} + I_B = 30 \text{ mA} + 8 \text{ mA} = 38 \text{ mA}$$

Finally, the current through R_3 is shown to be only I_B; therefore,

$$I_3 = I_B = 8 \text{ mA}$$

Voltages: Examining the schematic of Fig. 8-11 shows that the voltage across R_1 must be dropped from 24 V to 15 V, so V_1 is the difference between 24 V and 15 V, or $V_1 = 24 \text{ V} - 15 \text{ V} = 9 \text{ V}$.

The voltage across R_2 must be 10 V since R_2 must drop the voltage from 15 V to 5 V or $I_2 = 15 \text{ V} - 5 \text{ V} = 10 \text{ V}$. Since R_3 is in parallel

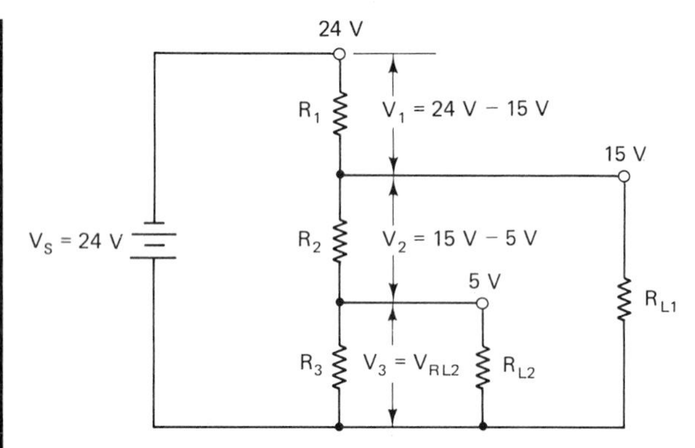

FIGURE 8-11 Voltage-divider voltage drops.

with R_{L2}, its voltage is equal to the load 2 voltage of 5 V, or $I_R = I_{L2} = 5$ V.

Putting the currents together with the voltages allows R_1, R_2, and R_3 to be solved.

$$R_1 = \frac{V_1}{I_1} = \frac{24\text{ V} - 15\text{ V}}{I_1 + I_{L2} + I_B}$$

$$= \frac{9\text{ V}}{50\text{ mA} + 30\text{ mA} + 8\text{ mA}} = 102\ \Omega$$

$$R_2 = \frac{V_2}{I_2} = \frac{15\text{ V} - 5\text{ V}}{I_{L2} + I_B} = \frac{10\text{ V}}{30\text{ mA} + 8\text{ mA}}$$

$$= 263\ \Omega$$

$$R_3 = \frac{V_3}{I_3} = \frac{5\text{ V}}{8\text{ mA}} = 625\ \Omega$$

EIA Standard Values for the Resistors. The odd values of calculated resistance are not available readily and it would be advantageous to use "off-the-shelf" EIA standard-value resistors since the load specifications have a ±5% tolerance. Referring to Table 2-6 gives the following EIA values for each resistor:

$102\ \Omega \longrightarrow$ EIA value of $100\ \Omega \pm 5\%$

$263\ \Omega \longrightarrow$ EIA value of $270\ \Omega \pm 5\%$

$625\ \Omega \longrightarrow$ EIA value of $620\ \Omega \pm 5\%$

Power Dissipation and Standard Power Rating Selection. The power rating for each resistor must be chosen to ensure that the voltage divider does not overheat and burn out. Off-the-shelf or standard power ratings are to be used after derating the calculated power dissipation of each resistor by 100% for a totally enclosed area.

Standard power ratings are normally $\frac{1}{10}$ W, $\frac{1}{8}$ W, $\frac{1}{4}$ W, $\frac{1}{2}$ W, 1 W, and 2 W for most composition resistors. The wattage rating of carbon film and metal film will vary with the manufacturer's specifications. For general applications the composition resistor power ratings can serve as a guide. For specific applications carbon-film and metal-film resistors with the appropriate wattage rating must be selected from the manufacturer's catalog.

$$P_1 = V_1 I_1 = 9\text{ V} \times 88\text{ mA} = 792\text{ mW}$$

$$P_1 \text{ derated by } 100\% = 792\text{ mW} \times 2 = 1.6\text{ W}$$

The closest standard power rating for this resistor is 2 W; therefore R_1 is a 100-Ω 2-W resistor.

$$P_2 = V_2 I_2 = 10\text{ V} \times 38\text{ mA} = 380\text{ mW}$$

$$P_2 \text{ derated by } 100\% = 380\text{ mW} \times 2 = 760\text{ mW}$$

The closest standard power rating for this resistor is 1 W; therefore R_2 is a 270-Ω 1-W resistor.

$$P_3 = V_3 I_3 = 5\text{ V} \times 8\text{ mA} = 40\text{ mW}$$

$$P_3 \text{ derated by } 100\% = 40\text{ mW} \times 2 = 80\text{ mW}$$

The closest standard power rating for this resistor is $\frac{1}{10}$ W; therefore, P_3 is a 620-Ω $\frac{1}{10}$-W resistor.

PRACTICE QUESTIONS 8-1

Choose the most suitable word, phrase, or value to complete each statement.

1. A form of variable voltage divider is called a ________.
2. Draw the schematic symbol for a potentiometer.
3. Carbon pots have a maximum power rating of ________ watts.
4. Two applications for a pot are ________ and ________.
5. ________ of a pot is the percentage resistance change with percentage shaft rotation.
6. Where a change in resistance with shaft rotation ensures a uniform change in voltage, a type ________ taper is used.
7. For audio volume control and tone control circuits, a type ________ or ________ taper is used.
8. For tone controls used as contrast controls in TV sets, a reverse of the ________ type taper is used.
9. The reverse of the ________ taper is used as a brightness control in television sets.
10. One of the most important circuits used in many control applications and measurements is called a ________ ________.

Refer to the schematic of Fig. 8-12 for Questions 11 to 13.

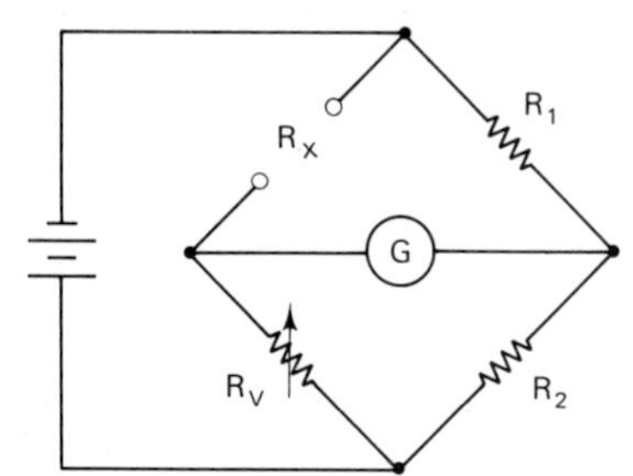

FIGURE 8-12 Wheatstone bridge.

11. Show the ratio of resistances for the branches of the bridge circuit.

12. Given that $R_1 = 10\ \text{k}\Omega$, $R_2 = 50\ \text{k}\Omega$, and $R_V = 3\ \text{k}\Omega$, find R_X.

13. Given that $R_X = 2700\ \Omega$, $R_1 = 10\ \text{k}\Omega$, $R_2 = 50\ \text{k}\Omega$, find R_V.

Refer to the schematic of Fig. 8-13 for questions 14 to 21.

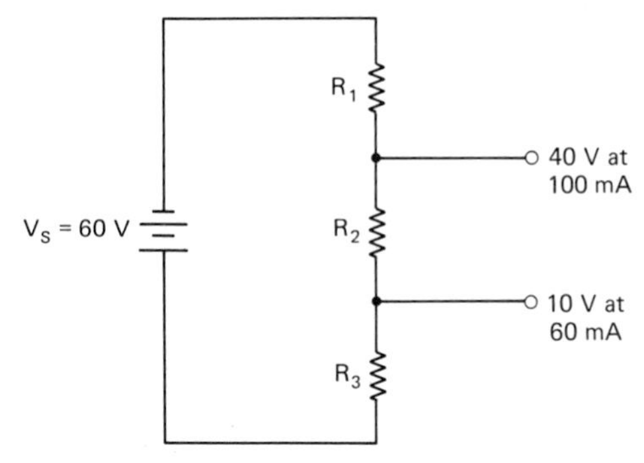

FIGURE 8-13 Voltage divider.

14. Calculate the bleeder current if it is not to exceed 10% of the total current required by the loads.

15. What is the minimum current rating of the source?

16. Find the standard EIA value for R_1.

17. Find the standard EIA value for R_2.

18. Find the standard EIA value for R_3.

19. Find the standard power-dissipation rating for R_1 (use 100% derating).

20. Find the standard power-dissipation rating for R_2 (use 100% derating).

21. Find the standard power-dissipation rating for R_3 (use 100% derating).

8-5 REFERENCING VOLTAGES TO CHASSIS GROUND

All electronic equipment, from instrumentation such as an oscilloscope to consumer appliances, plugs into an ac power line. The power line has two insulated conductors for 120 V plus an uninsulated ground wire. The ground wire and the neutral (white) conductor are both connected to the city water system, or special ground rods are driven into the ground to properly ground the two conductors. What is the purpose of grounding the conductors? This is discussed in more detail in a later chapter; we will only note here that grounding serves to reduce or minimize the danger of shock and of fire due to lightning strikes.

The printed-circuit board of electronic equipment is normally mounted on a metal chassis to give mechanical rigidity to the device. The housing of most devices is plastic; however, industrial lab and some consumer equipment has metal cabinets. The metal parts of the equipment, termed "hardware," are usually grounded to the ground side of the ac utility system.

Hardware or chassis ground is synonymous with the "ground" that is used as a return conductor for electronic circuits. The circuit return conductors or wire runs from one point to another point are eliminated by the use of the common ground bus or chassis. This reduces stray inductance and capacitance, especially in radio-frequency circuits.

All dc voltages used in transistor and logic circuits must be referenced with respect to a common point, which can be the chassis or common circuit ground. In this manner a positive voltage would be interpreted as positive with respect to ground. To reverse the polarity you simply reference the corresponding source polarity. The commonly used symbols for ground are shown in Fig. 8-14.

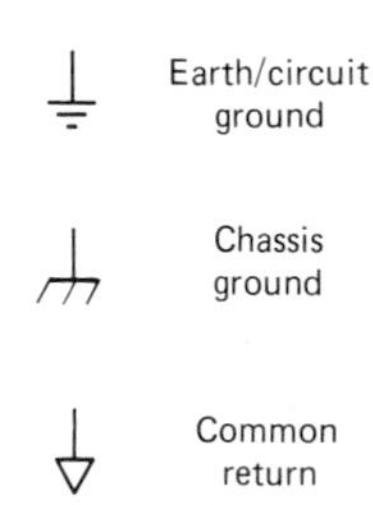

FIGURE 8-14 Ground symbols.

EXAMPLE 8-4

An *NPN* transistor circuit requires a positive dc V_{cc} bias voltage at the collector with respect to ground, and a positive base bias voltage, V_B, with respect to ground, as shown in Fig. 8-15(a). For a *PNP* transistor the V_{cc} polarity with respect to ground is reversed, as shown in Fig. 8-15(b).

EXAMPLE 8-5

An operational amplifier requires both a positive

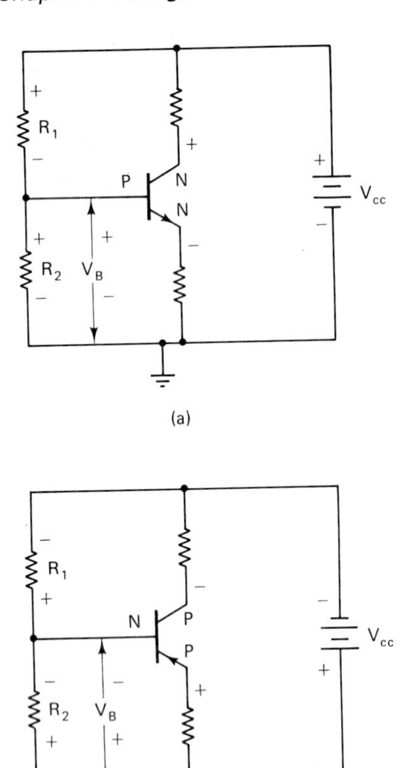

FIGURE 8-15 (a) NPN transistor biasing. (b) PNP transistor biasing.

and a negative voltage with a common ground from a single-source power supply. This is obtained by the connections shown in Fig. 8-16(a).

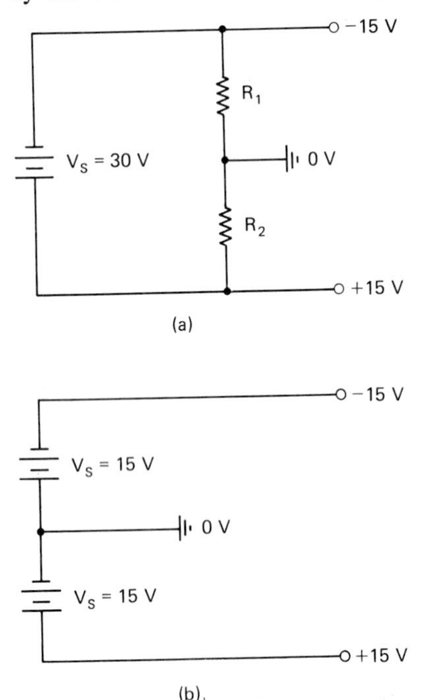

FIGURE 8-16 (a) Single-source system. (b) Dual-source system.

A dual-voltage source would provide the same voltages and polarity with respect to ground, as shown in Fig. 8-16(b).

8-6 VOLTAGE DIVIDER TO PROVIDE A POSITIVE AND A NEGATIVE VOLTAGE

Figure 8-17 shows the symbol for an operational amplifier. An operational amplifier is an integrated microcircuit (IC) that has high gain and other desirable characteristics. Note that it requires both a positive and a negative polarity voltage. A dual-source supply, as shown in Fig. 8-16(b), is normally used; however, a voltage divider can be designed as follows.

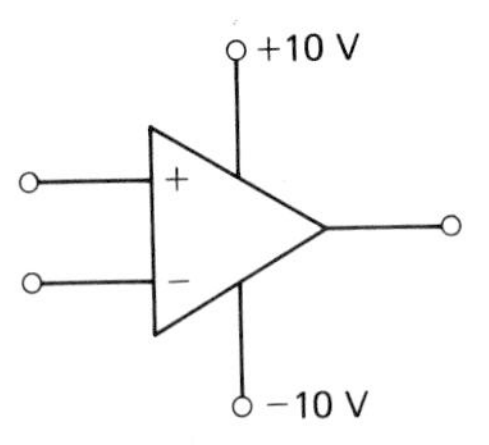

FIGURE 8-17 Operational amplifier symbol.

EXAMPLE 8-6

Specifications

Power source: Single dc source 30 V at 100 mA rating.

Op amp requirements: +15 V at 28 mA, −15 V at 28 mA.

Solution: Since the supply voltage is 30 V, it will need to be dropped by 15 V; therefore, two equal-value series resistors connected across a source with their center tap grounded will provide a positive and a negative voltage, as shown in Fig. 8-18. The bleeder current through R_1 and R_2 can be preset to 10% of the load current, or 3 mA.

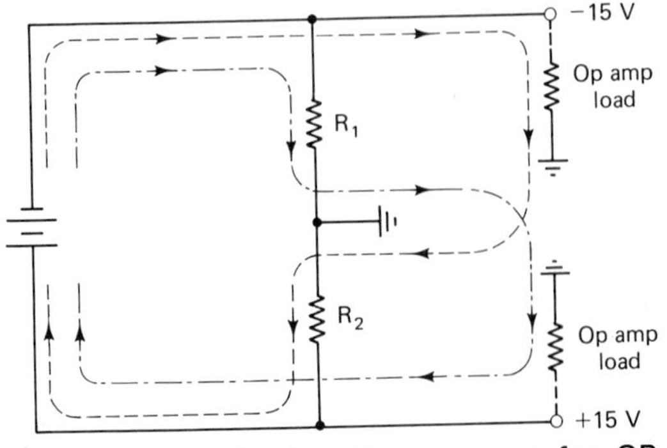

FIGURE 8-18 Dual voltage source for OP AMP supply.

With a 15-V drop across R_1 and a current of 28 mA plus the 3-mA bleeder current flowing through it, R_1 can be calculated:

$$R_1 = \frac{V_1}{I_1} = \frac{15 \text{ V}}{28 \text{ mA} + 3 \text{ mA}} = 484 \ \Omega$$

Standard EIA value for R_1: 470 Ω ± 5%

R_1 power rating:
$P = VI = 15 \text{ V} \times 31 \text{ mA} = 465 \text{ mW}$

Allowing 100% derating:
$P_1 = 2 \times 465 \text{ mW} = 0.9 \text{ W}$

Standard power rating for R_1: 1 W

Since R_2 voltage and currents are identical to those of R_1, its value and power rating are the same as those of R_1.

EXAMPLE 8-7

A certain circuit requires the following voltages and currents.

Specifications

Power source: Single dc source 35 V rated at 100 mA.

Load 1 requirement: +20 V at 30 mA.

Load 2 requirements: ±15 V at 21 mA (dual supply).

Draw a schematic and determine the standard EIA values and standard power rating (derating 100%) for each resistor used in a voltage divider to supply the foregoing load requirements. The bleeder current is 10% of the total load current.

Solution: The schematic for the circuit is shown in Fig. 8-19. The bleeder current is to be approximately 10% of the load currents:

$$I_B = 0.1 \times (30 \text{ mA} + 21 \text{ mA} + 21 \text{ mA})$$
$$= 7 \text{ mA}$$

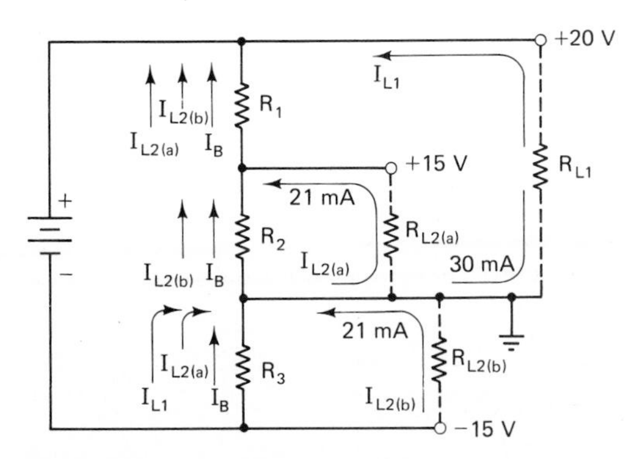

FIGURE 8-19 Voltage-divider circuit for Example 8-7.

Resistance and Wattage Values for R_3. Tracing the paths for the currents shows that load 1 and load 2(a) currents plus the bleeder current will flow through R_3. Therefore,

$$I_3 = I_{L1} + I_{L2(a)} + I_B$$
$$= 30 \text{ mA} + 21 \text{ mA} + 7 \text{ mA} = 58 \text{ mA}$$

R_3 is in parallel with the load $R_{L2(b)}$, so $V_3 = V_{L2(b)} = 15$ V. Therefore,

$$R_3 = \frac{V_3}{I_3} = \frac{15 \text{ V}}{58 \text{ mA}} = 258 \ \Omega$$

Closest EIA value for R_3 would be 270 Ω ± 5% (from Table 2-6)

Power rating of R_3: $P_3 = V_3 I_3 = 15 \text{ V} \times 58 \text{ mA} = 870 \text{ mW}$

Standard power rating for $R_3 = 2 \times 870 \text{ mW} = 2 \text{ W}$ (derate by 100%)

Resistance and Wattage Rating for R_2. The schematic diagram of Fig. 8-19 shows currents I_B and $I_{L2(b)}$ flowing through R_2. Therefore,

$$I_2 = I_B + I_{L2(b)} = 7 \text{ mA} + 21 \text{ mA} = 28 \text{ mA}$$

Since R_2 is in parallel with $R_{L2(a)}$, $V_2 = V_{L2(a)} = 15$ V.

$$R_2 = \frac{V_2}{I_2} = \frac{15 \text{ V}}{28 \text{ mA}} = 536 \ \Omega$$

Closest EIA value for R_2 would be 560 Ω ± 5% (from Table 2-6)

Power rating for R_2: $P_2 = V_2 I_2 = 15 \text{ V} \times 28 \text{ mA} = 420 \text{ mW}$

Standard power rating for R_2: $2 \times 420 \text{ mW} = 1 \text{ W}$

Resistance and Wattage Rating for R_1. The schematic diagram of Fig. 8-19 shows currents I_B, $I_{L2(a)}$, and $I_{L2(b)}$ flowing through R_1:

$$I_1 = I_B + I_{L2(a)} + I_{L2(b)}$$
$$= 7 \text{ mA} + 21 \text{ mA} + 21 \text{ mA}$$
$$= 49 \text{ mA}$$

R_1 must drop the voltage from 20 V to 15 V, so

$$V_1 = 20 \text{ V} - 15 \text{ V} = 5 \text{ V}$$

Therefore,

$$R_1 = \frac{V_1}{I_1} = \frac{5 \text{ V}}{49 \text{ mA}} = 102 \ \Omega$$

Closest EIA value for R_1 would be 100 Ω ± 5% (from Table 2-6)

Power rating for R_1: $P_1 = V_1 I_1 = 5\ \text{V} \times 49\ \text{mA} = 245\ \text{mW}$

Standard power rating for R_1: $2 \times 245\ \text{mW} = \frac{1}{2}\ \text{W}$ (derate by 100%)

Is the dc source maximum current rating sufficient to provide load and bleeder current?

$$I_T = I_B + I_{L1} + I_{L2(a)} + I_{L2(b)}$$
$$= 7\ \text{mA} + 30\ \text{mA} + 21\ \text{mA} + 21\ \text{mA}$$
$$= 79\ \text{mA}$$

The source provides 100 mA and has sufficient capacity since the maximum current is only 79 mA.

SUMMARY

A simple voltage divider is two resistors in series across a single source.

The ratio of voltage drops will be equal to the ratio of the resistance of the divider resistors.

A variable voltage divider is termed a potentiometer.

A potentiometer has two fixed terminals and one terminal that has a movable contact coupled to a shaft.

A potentiometer must be chosen with the applicable taper.

A potentiometer taper is the percentage resistance change with the percentage shaft rotation.

A linear taper means that the resistance change with shaft rotation is uniform.

A nonlinear taper such as a log or semilog taper has very little change in resistance with initial shaft rotation until the slider reaches approximately 50% of rotation, where the change increases rapidly.

The bridge circuit used for precision measuring of resistance, inductance, and capacitance is known as a Wheatstone bridge.

The bridge is said to be balanced when the voltage ratio of V_1 to V_3 is exactly equal to the ratio of V_2 to V_4 from Fig. 8-6(a).

Voltage ratio can be written in terms of resistance, or

$$\frac{R_1}{R_3} = \frac{R_2}{R_4} \tag{8-1}$$

Substituting R_2 for R_X and R_V for R_4 and solving for R_X gives

$$R_X = \frac{R_1 R_V}{R_3} \tag{8-2}$$

Voltage dividers provide two or more different values of voltages from a single source to supply two or more loads.

A voltage divider is designed from a set of specifications.

Specifications outline the requirements and limits of the desirable characteristics of a device.

Since a voltage divider is formed from a series string of resistors connected across a source, a bleeder current will flow through the string, normally not to exceed 10% of the total load current requirement.

Standard EIA resistance values and power ratings are used for resistors in the voltage divider.

Derating the power dissipation of a resistor means increasing its wattage rating by a factor of 15% (1.15× for open air) to 100% (2.0× for enclosed) or greater.

A common return point to electronic circuits is termed a ground.

When the chassis (metal or copper bus) is used as a common return to electronic circuits, it is termed a chassis ground.

When voltages, particularly dc voltages, are referenced to ground, all voltage readings are interpreted as being with respect to ground unless otherwise stated.

Commonly used symbols for ground are shown in Fig. 8-14.

Two equal-value series resistors connected across a source with their center tap grounded will provide a positive and a negative voltage (see Fig. 8-18).

REVIEW QUESTIONS

8-1. Draw a schematic of a simple voltage divider to reduce the output of a generator from 1 V to 100 mV. Show the value of the resistors.

8-2. Explain how a potentiometer can be used as a voltage divider.

8-3. List three applications of a potentiometer.

8-4. Explain what is meant by the "taper" of a potentiometer.

8-5. List the tapers required for the following applications.
- **(a)** Gain controls
- **(b)** Tone control with bass compensation
- **(c)** Audio volume and tone controls
- **(d)** TV contrast control
- **(e)** TV brightness control

8-6. For what purposes is a Wheatstone bridge used?

8-7. Under what circuit condition is a Wheatstone bridge "balanced"?

8-8. What is the value of the variable resistance of a Wheatstone bridge if R_X is 5.6 kΩ, R_1 is 10 kΩ, and R_2 = 1 kΩ of Fig. 8-12?

8-9. Draw a schematic diagram of a bridge circuit, utilizing a thermistor to measure temperature. Show how it would be possible to calibrate the "thermometer."

8-10. Explain the function of a voltage divider.

8-11. Give two reasons why a bleeder current is necessary in a voltage divider.

8-12. What are "specifications"?

8-13. Under what condition can EIA standard value resistors be substituted for the calculated resistance values of a voltage divider?

8-14. List the standard power ratings for most types of resistors.

8-15. Give two reasons why the electric utility grounds the neutral wire to a water system or to ground.

8-16. What is meant by the term "hardware" as applied to electronic equipment?

8-17. Why is a common bus or ground used in electronic circuits?

8-18. What polarity voltage with respect to ground does a *PNP* transistor need? An *NPN*?

8-19. What polarity voltage with respect to ground does an operational amplifier need?

8-20. Draw the schematic symbols for chassis ground, ground, and common return.

SELF-EXAMINATION

Choose the most suitable word, phrase, or value to complete each statement.

8-1. Refer to Fig. 8-1; what is the output voltage of the voltage divider if the audio-frequency generator is delivering 3.5 V?

8-2. The maximum power rating of carbon pots is ________ watts.

8-3. A potentiometer to be used as a gain control or as an ohmmeter zero ohms adjust must have a ________ taper.

8-4. A potentiometer to be used as a volume control or a tone control must have a ________ taper or a ________ taper.

8-5. A significant circuit used in industrial control, navigational control, and for measuring resistance, inductance, or capacitance is termed as ________ ________.

8-6. The purpose of the galvanometer in a bridge circuit is to (*increase, decrease*) the sensitivity response of the circuit at the null point.

8-7. For the bridge circuit to be balanced, the voltage the galvanometer reads must be ________.

8-8. Refer to Fig. 8-6(a); for the bridge circuit to be balanced, the ratio of R_2 to R_4 must be equal to the ratio of ________.

8-9. Refer to Fig. 8-7; which of the resistances is adjusted to balance the bridge?

8-10. If R_1 is 100 kΩ, R_2 is 1 kΩ, and R_V is 5.6 kΩ, what is the value of R_X?

Refer to Fig. 8-10 for Questions 8-11 to 8-17. With the source voltage at 35 V and the load requirements remaining the same, calculate the following values.

8-11. Find the bleeder current if it is not to exceed 10% of the total current required by the loads.

8-12. Calculate the EIA resistance value of R_1.

8-13. Calculate the EIA resistance value of R_2.

8-14. Calculate the EIA resistance value of R_3.

8-15. Calculate the standard power rating of R_1. Use 100% derating.

8-16. Repeat Question 8-15 for R_2.

8-17. Repeat Question 8-15 for R_3.

8-18. The metal parts of electronic equipment are termed ________ and are usually connected to the ________ wire of the ac utility system.

8-19. All ________ voltages used in transistor and logic circuits must be referenced with respect to a common point, which can be the chassis or a common circuit ground.

Refer to Fig. 8-20 for Questions 8-20 to 8-25.

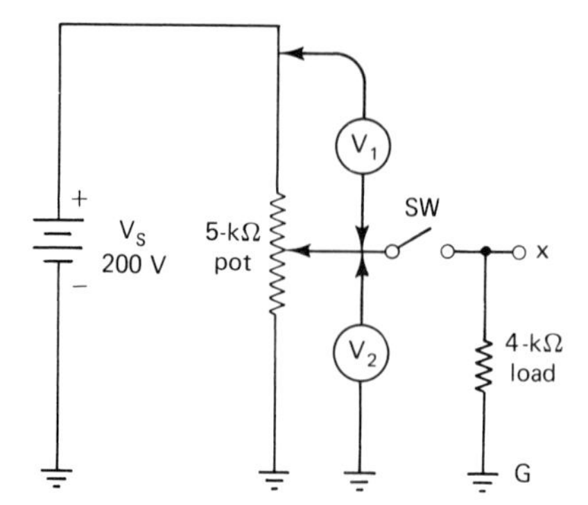

FIGURE 8-20 Potentiometer circuit.

8-20. With SW open circuited, what will V_1 read when V_2 reads 75 V?

8-21. (Switch open) What does V_1 read when the sliding tap is set 2 kΩ from the chassis or ground end?

8-22. What is the polarity of V_2 with respect to ground?

8-23. (Switch closed) If the sliding tap is set exactly halfway on the tapped resistor, what is the reading on V_2?

8-24. (Switch closed) What is the current drawn from the source when the sliding tap is set 1 kΩ from the tap end of the tapped resistor?

8-25. What will V_1 read given the conditions in Question 8-24?

8-26. Design a voltage divider to meet the following specifications: It is to provide ±20 V at 30 mA to two separate loads and 30 V at 30 mA to a third load. All resistors must be the nearest EIA 5% tolerance value. Derate the power dissipation ratings of the resistors by 100%. The dc supply has a rating of 50 V at 100 mA.

ANSWERS TO PRACTICE QUESTIONS

8-1

1. potentiometer
2. 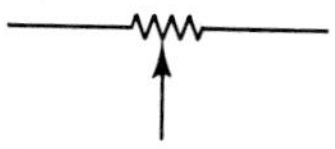
3. 2
4. gain control, ohmmeter zero adjust
5. taper
6. linear
7. modified log or semilog
8. modified log
9. semilog
10. bridge circuit
11. $\frac{R_x}{R_V} = \frac{R_1}{R_2}$
12. $R_x = 600\ \Omega$
13. $R_V = 13.5\ \text{k}\Omega$
14. 16 mA
15. 176 mA
16. 110 Ω
17. 390 Ω
18. 620 Ω
19. 7 W; use 10 W
20. 4.5 W; use 5 W
21. 0.3 W; use 1/2 W

9

Kirchhoff's Law

INTRODUCTION

Gustav Robert Kirchhoff was one of the great scientists who carried out studies in electricity. Kirchhoff, about 1847, formulated two basic circuit laws that govern the behavior of voltages around a closed circuit and currents at junctions of two or more components in a circuit. These laws are called Kirchhoff's voltage law and Kirchhoff's current law.

All the circuits in previous chapters were easily solved using Ohm's law. Many types of circuits have components connected in a more complex manner than just simply series, parallel, or series–parallel. One example is a circuit with two voltages applied in different branches shown in Fig. 9-6. Another example is the bridge circuit discussed in Chapter 8. These circuits are easily solved using Kirchhoff's laws applied to three circuit analysis methods. These methods are particularly useful in the analysis of circuits with two or more voltage or current sources, and are the *branch current, node voltage,* and *mesh current methods.*

MAIN TOPICS IN CHAPTER 9

9-1 Kirchhoff's Voltage Law
9-2 Kirchhoff's Current Law
9-3 Kirchhoff's Voltage Law and the Bridge Circuit
9-4 Branch Current Method for a Multisource Network
9-5 Node Voltage Method for a Multisource Network
9-6 Mesh Current Method for a Multisource Network

OBJECTIVES

After studying Chapter 9, the student should be able to

1. Master the concepts of Kirchhoff's voltage and current laws by applying these laws to bridge circuits and multisource networks.
2. Apply Kirchhoff's laws to solve complex circuits using the branch current method, the node voltage method, and the mesh current method.

9-1 KIRCHHOFF'S VOLTAGE LAW

The series circuit characteristics and laws outlined in Chapter 4 showed that the sum of the voltage drops about a series circuit must be equal to the source voltage. The schematic of Fig. 9-1 will serve to refresh and restate the voltage law for this circuit:

$$V_T = V_1 + V_2 + V_3 \qquad \text{volts} \qquad (4\text{-}2)$$

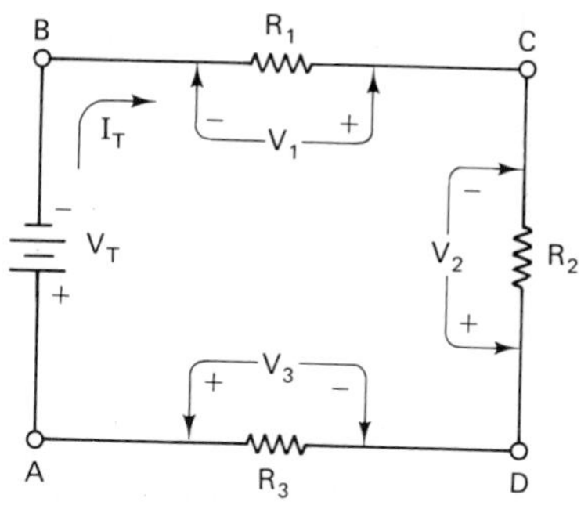

FIGURE 9-1 Series circuit.

The circuit of Fig. 9-1 can also be called a closed circuit, a complete circuit, or a closed loop, since current is the same or common to all parts of the circuit. *Kirchhoff's voltage law* states:

> Around any closed circuit, or a complete circuit, or a closed loop, the algebraic sum of the voltages will equal zero.

The term "algebraic" was avoided in the series circuit, but Kirchhoff stated his concept of the series circuit voltage law in a slightly different equation format, so the voltage sources and the voltage drops about the circuit of Fig. 9-1 must be summed algebraically.

To sum the voltages algebraically, the polarity of the voltage drops must first be established. For the circuit of Fig. 9-1, this means following a current path from a voltage source in the circuit and labeling the polarity of each component negative where electrons enter a component and positive where electrons leave a component. The polarities are shown in the schematic of Fig. 9-1.

Kirchhoff's voltage law can be applied to show that the algebraic sum of the voltages will be equal to zero by going around the circuit or loop from any starting point A, B, C, D of the series circuit (Fig. 9-1).

Starting at A

$$+V_T - V_1 - V_2 - V_3 = 0$$

Note that the first polarity encountered around the loop will be the algebraic symbol for that particular voltage. If the voltage drops are transposed in the algebraic equation, equation (4-2) is obtained.

Starting at B

$$-V_1 - V_2 - V_3 + V_T = 0$$

Once again, if the voltage drops are transposed in the algebraic equation, equation (4-2) is obtained. Kirchhoff's voltage law proves this series circuit equation, $V_T = V_1 + V_2 + V_3$.

The following are the general steps to apply to obtain a voltage loop equation:

1. Select a starting point for the loop, usually from the source.
2. Go around the circuit and assign the component polarity according to the electron current or conventional current flow.
3. Apply Kirchhoff's voltage law around the loop.

EXAMPLE 9-1

1. For the circuit of Fig. 9-2, the starting point for this example is A, using the electron current flow direction.
2. The appropriate polarities have been assigned according to the electron flow.
3. Applying Kirchhoff's voltage law yields

$$+V_B - V_2 + V_A - V_1 = 0$$

$$V_B = +V_2 - V_A + V_1$$

$$V_B = 10\text{ V} - 18\text{ V} + 14\text{ V} = 6\text{ V}$$

A similar Kirchhoff's voltage loop equation can be obtained by looping the voltages using the conventional current direction starting at B:

$$-V_B + V_1 - V_A + V_2 = 0$$

Therefore,

$$V_1 - V_A + V_2 = V_B$$

$$14\text{ V} - 18\text{ V} + 10\text{ V} = V_B = 6\text{ V}$$

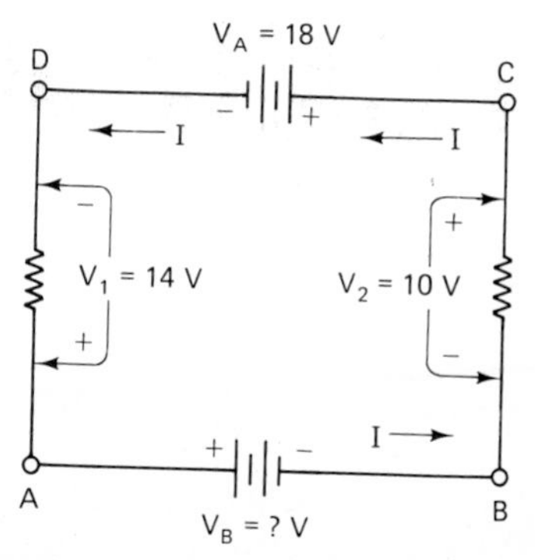

FIGURE 9-2 Two-source closed circuit.

9-2 KIRCHHOFF'S CURRENT LAW

The parallel circuit characteristics outlined in Chapter 5 showed that the sum of the currents in a parallel circuit was equal to the total current flowing from or back into the source. The schematic of Fig. 9-3(a) will serve to refresh and restate the current law for this circuit:

$$I_T = I_1 + I_2 \quad \text{etc.} \qquad \text{amperes} \tag{5-1}$$

The circuit of Fig. 9-3(a) can also be considered as having two current paths or two current loops. The total current entering the junction (or node) of R_1 and R_2 at A will be equal to the sum of the currents leaving the junction (or node) of R_1 and R_2 at B. *Kirchhoff's current law* states:

> The algebraic sum of the currents flowing into a junction (or node) must be equal to the algebraic sum of the currents flowing out of that point.

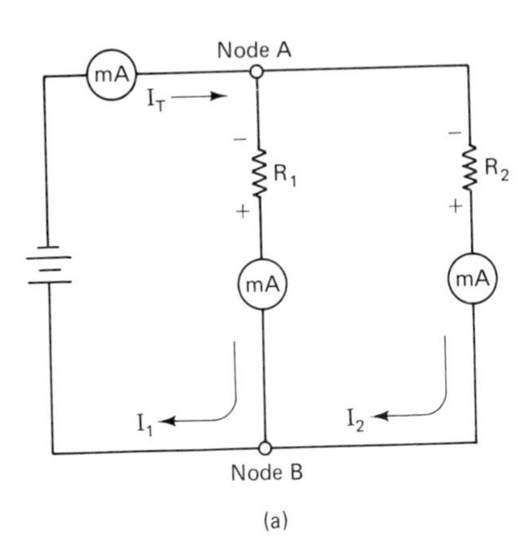

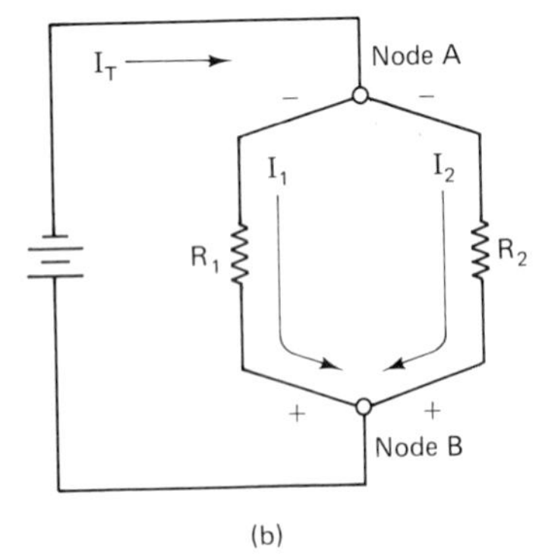

FIGURE 9-3 (a) Parallel branch currents. (b) Node currents.

The term "algebraic" was avoided in the parallel circuit, but as in the case of the voltage law, Kirchhoff stated the current law in the same equation format, so the currents in any circuit must be summed algebraically.

To sum the currents algebraically, the polarity of the currents must be established first. For the circuit of Fig. 9-3(a), this means following the current paths from a voltage source in the circuit and labeling the polarity at respective junctions or nodes, negative where current enters the junction and positive where current leaves the junction. The polarities are shown in Fig. 9-3(b).

Kirchhoff's current law can be applied to show that the algebraic sum of the currents entering a node will be equal to the algebraic sum of the currents leaving that point. The current entering node A is the total current I_T and is negative. The currents leaving at node B are I_1 and I_2 and are positive; therefore, Kirchhoff's current equation states that

$$-I_T + I_1 + I_2 = 0$$

Transposing I_T in the equation will give the familiar equation (5-1) of a parallel circuit, $I_T = I_1 + I_2$.

9-3 KIRCHHOFF'S VOLTAGE LAW AND THE BRIDGE CIRCUIT

The bridge circuit discussed in Chapter 8 lends itself to the application of Kirchhoff's voltage law. Using Kirchhoff's voltage law, the voltage A to B [see Fig. 9-4(a)] can be shown to be zero for a balanced bridge circuit condition or some other value for an unbalanced bridge circuit condition. The polarity A to B can also be determined using Kirchhoff's voltage law.

Balanced Bridge Condition

To solve the voltage loop equation for the voltage loop, the voltage drops must first be calculated.

$$V_1 = \frac{R_1}{R_1 + R_2} \times 6\text{ V} = \frac{2.2\text{ k}\Omega}{2.2\text{ k}\Omega + 6.6\text{ k}\Omega} \times 6\text{ V}$$
$$= 1.5\text{ V}$$

$$V_2 = \frac{R_2}{R_1 + R_2} \times 6\text{ V} = \frac{6.6\text{ k}\Omega}{2.2\text{ k}\Omega + 6.6\text{ k}\Omega} \times 6\text{ V}$$
$$= 4.5\text{ V}$$

$$V_3 = V_1 = 1.5\text{ V}$$

$$V_4 = V_2 = 4.5\text{ V}$$

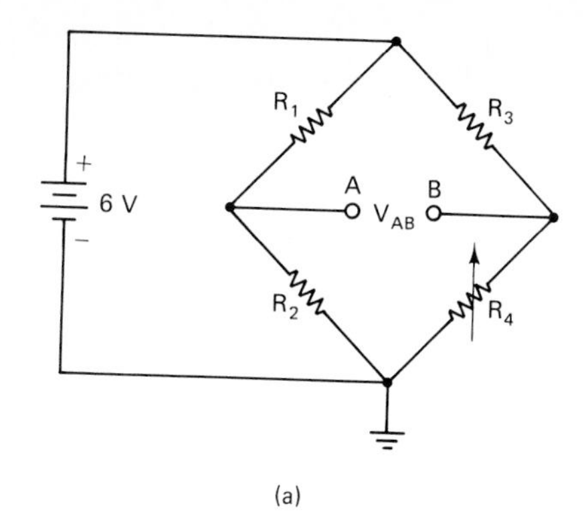

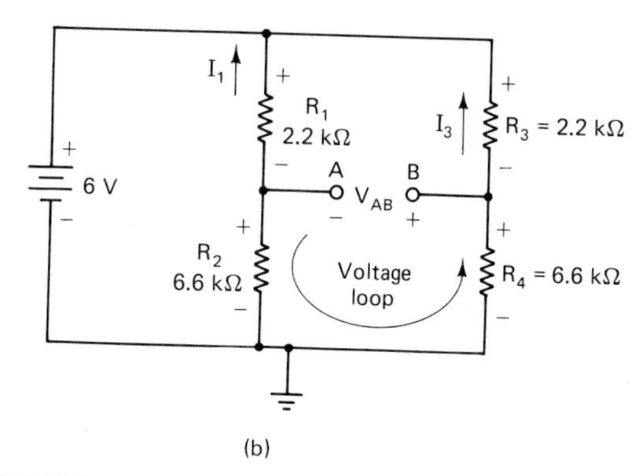

FIGURE 9-4 (a) Bridge circuit. (b) Simplified bridge circuit.

The next step is to go around the circuit and assign the component polarities according to electron or conventional current flow, as shown in the schematic of Fig. 9-4(b).

Finally, a voltage loop equation can be obtained by applying Kirchhoff's voltage law around the loop from A to B. Applying Kirchhoff's voltage law to the voltage loop beginning at A gives

$$+V_2 - V_4 \pm V_{AB}0 = 0$$

$$4.5\text{ V} - 4.5\text{ V} \pm V_{AB} = 0$$

Therefore,

$$V_{AB} = 0\text{ V}$$

The voltage V_{AB} is 0 V, which is the condition for a balanced bridge.

Unbalanced Bridge Condition

With R_4 changed to 8 kΩ in the schematic of Fig. 9-4(b), the bridge will be unbalanced (i.e., the voltage A to B will no longer be 0 V). Recalculating V_4 and applying Kirchhoff's voltage law will show the value of V_{AB} as well as its polarity at A since the loop equation was started at A.

Recalculate V_4:

$$V_4 = \frac{R_4}{R_3 + R_4} \times 6\text{ V} = \frac{8\text{ k}\Omega}{2.2\text{ k}\Omega + 8\text{ k}\Omega} \times 6\text{ V}$$
$$= 4.7\text{ V}$$

Apply Kirchhoff's voltage law:

$$+V_2 - V_4 \pm V_{AB} = 0$$
$$V_2 - V_4 = \pm V_{AB}$$
$$4.5\text{ V} - 4.7\text{ V} = -0.2\text{ V}$$

V_{AB} is negative by 0.2 V at A with respect to B.

EXAMPLE 9-2

With R_4 changed to 4 kΩ in the schematic of Fig. 9-4(b), the bridge will be unbalanced. What are the voltage and the polarity A to B?

Solution: Recalculate V_4 and apply Kirchhoff's voltage law to the voltage loop.

$$V_4 = \frac{R_4}{R_3 + R_4} \times 6\text{ V} = \frac{4\text{ k}\Omega}{2.2\text{ k}\Omega + 4\text{ k}\Omega} \times 6\text{ V}$$
$$= 3.9\text{ V}$$

$$+V_2 - V_4 \pm V_{AB} = 0$$
$$V_2 - V_4 = \pm V_{AB}$$
$$4.5\text{ V} - 3.9\text{ V} = 0.6\text{ V}$$

V_{AB} is positive by 0.6 V at A with respect to B.

The calculations for the voltage A to B of the bridge circuit show that the polarity A to B will change depending on the unbalanced resistive condition in one of its branches. For this reason the null point of the galvanometer pointer is zero centered at the midrange of its scale. For a negative polarity it is deflected to the left, and a positive polarity will cause it to be deflected to the right from its null position.

PRACTICE QUESTIONS 9-1

1. State Kirchhoff's voltage law.
2. What must first be established about a closed circuit, a complete circuit, or a closed loop before Kirchhoff's voltage law can be applied?
3. From Fig. 9-2, write the voltage loop equation by

starting counterclockwise at C. Repeat for the voltage loop equation starting at D.

4. From Fig. 9-5, write the voltage loop equation using the electron flow direction beginning at B and solve for V_B.

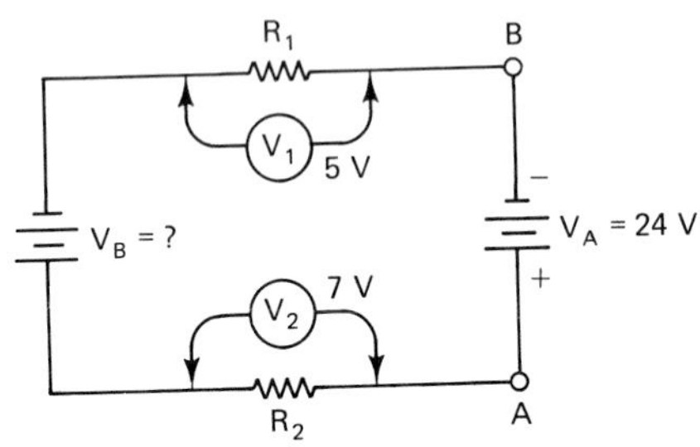

FIGURE 9-5 Circuit schematic for Practice Questions 9-1.

5. From Fig. 9-5 write the voltage loop equation using the conventional current flow direction beginning at B and solve for V_B.
6. From Fig. 9-4(b) write the loop equation for V_{AB} using the V_1 and V_3 voltage loop.
7. From Fig. 9-4(b) with R_3 changed to 3.3 kΩ, calculate the voltage drop across V_1 and V_3, then write the voltage loop equation and solve for V_{AB} and its polarity at A with respect to B.

9-4 BRANCH CURRENT METHOD FOR A MULTISOURCE NETWORK

All the circuits to this point have had a single voltage source and were solved simply using Ohm's law and Kirchhoff's voltage law. Circuits with two or more volage sources (see Fig. 9-6) must be solved using a combination of methods. In the branch current method the circuit currents are solved by obtaining the voltage loops for each source, substituting IR in the equation, then solving for the currents by substitution in the simultaneous equations. The general steps used in applying the branch current method are as follows:

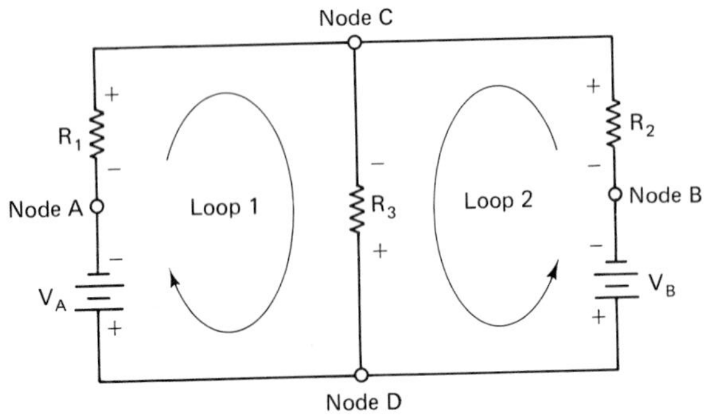

FIGURE 9-6 Multiple-source circuit.

1. Assign polarities to each resistor in the circuit by looping the current according to the electron or conventional current direction.
2. Apply Kirchhoff's voltage law around each closed loop or circuit and substitute IR for each voltage.
3. Apply Kirchhoff's current law to include all branch currents flowing into and out of circuit nodes.
4. Substitute all known numerical circuit values in each equation and rewrite equations.
5. Solve for one of the unknown currents in the simultaneous equations and substitute its value in an equation to solve for the remaining currents.

EXAMPLE 9-3

Use the branch current method to find each branch current in Fig. 9-7.

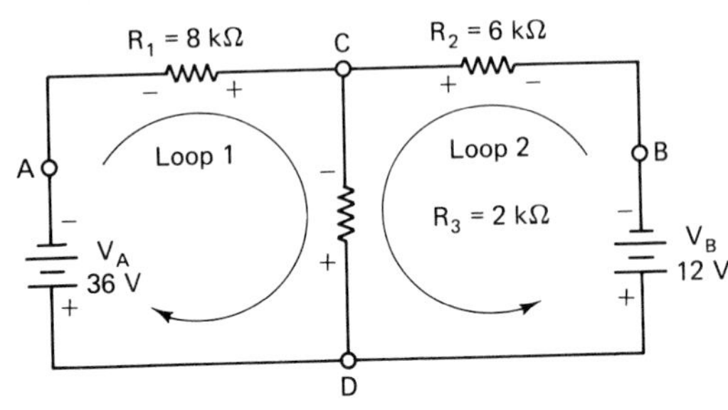

FIGURE 9-7 Kirchhoff's voltage and current application for a multisource circuit.

Solution

1. Assign polarities to each resistor in the circuit by looping the current according to the electron flow as shown in Fig. 9-7.
2. Apply Kirchhoff's voltage law around each closed loop or circuit.

(a) For loop 1 starting at A:

$$-V_1 - V_3 + V_A = 0$$

$$-I_1R_1 - I_3R_3 + V_A = 0$$

Therefore,

$$V_A = I_1R_1 + I_3R_3 \tag{1}$$

(b) For loop 2 starting at B:

$$-V_2 - V_3 + V_B = 0$$

$$-I_2R_2 - I_3R_3 + V_B = 0$$

Therefore,

$$V_B = I_2R_2 + I_3R_3 \tag{2}$$

3. Apply Kirchhoff's current law to include I_1 and I_2 flowing into and out of R_3 at nodes C and D.

$$-I_3 + I_1 + I_2 = 0$$

$$I_3 = I_1 + I_2 \quad (3)$$

4. Substitute all known numerical circuit values in each equation and rewrite the equations.

(a) For equation (1) (Note: The kΩ have been dropped since the current value will be in mA):

$$V_A = I_1R_1 + I_3R_3 \quad (1)$$

and

$$I_3 = I_1 + I_2 \quad (3)$$

$$V_A = I_1R_1 + (I_1 + I_2)R_3$$

$$36 = 8I_1 + 2I_1 + 2I_2$$

Gathering terms gives equation (4):

$$36 = 10I_1 + 2I_2 \quad (4)$$

(b) For equation (2),

$$V_B = I_2R_2 + I_3R_3 \quad (2)$$

and

$$I_3 = I_1 + I_2 \quad (3)$$

$$V_B = I_2R_2 + (I_1 + I_2)R_3$$

$$12 = 6I_2 + 2I_1 + 2I_2$$

Gathering terms gives equation (5):

$$12 = 2I_1 + 8I_2 \quad (5)$$

5. Solve for one of the unknown currents in the simultaneous equations and substitute its value in an equation to solve for the remaining currents.
To solve for I_1, subtract equation (5) from equation (4) by multiplying equation (4) by 8 and equation (5) by 2.

$$36 = 10I_1 + 2I_2 \quad \text{Eq. (4)} \times 8$$

$$12 = 2I_1 + 8I_2 \quad \text{Eq. (5)} \times 2$$

$$288 = 80I_1 + 16I_2 \quad (6)$$

$$24 = 4I_1 + 16I_2 \quad (7)$$

Subtract equation (7) from (6) by changing the signs of equation (7) and adding.

$$264 = 76I_1$$

$$I_1 = 3.47 \text{ mA}$$

Solve for I_2 by substituting 3.47 mA for I_1 in equation (4).

$$36 = (10 \times 3.47 \text{ mA}) + 2I_2$$

$$= 34.7 + 2I_2$$

$$I_2 = \frac{36 - 34.7}{2} = 0.65 \text{ mA}$$

and

$$I_3 = I_1 + I_2 = 3.47 \text{ mA} + 0.65 \text{ mA} = 4.12 \text{ mA}$$

Solve for the voltage drops around each loop.

$$V_1 = I_1R_1 = 3.47 \text{ mA} \times 8 \text{ k}\Omega = 27.8 \cong 28 \text{ V}$$

$$V_2 = I_2R_2 = 0.65 \text{ mA} \times 6 \text{ k}\Omega = 3.9 \text{ V} \cong 4 \text{ V}$$

$$V_3 = I_3R_3 = 4.12 \text{ mA} \times 2 \text{ k}\Omega = 8.2 \cong 8 \text{ V}$$

Checking the voltage drops about the two voltage loops by using Kirchhoff's voltage law gives:
Around loop 1 with V_A:

$$-V_1 - V_B + V_A = 0$$

$$-28 \text{ V} - 8 \text{ V} + 36 \text{ V} = 0$$

Around loop 2 with V_A:

$$-V_2 - V_3 + V_B = 0$$

$$-4 \text{ V} - 8 \text{ V} + 12 \text{ V} = 0$$

Check the current nodes C and D:

$$-I_3 + I_1 + I_2 = 0$$

$$-4.12 \text{ mA} + 3.47 \text{ mA} + 0.65 \text{ mA} = 0$$

PRACTICE QUESTION 9-2

1. Solve the two-source circuit of Fig. 9-8 for currents and voltages using Kirchhoff's voltage and current laws.

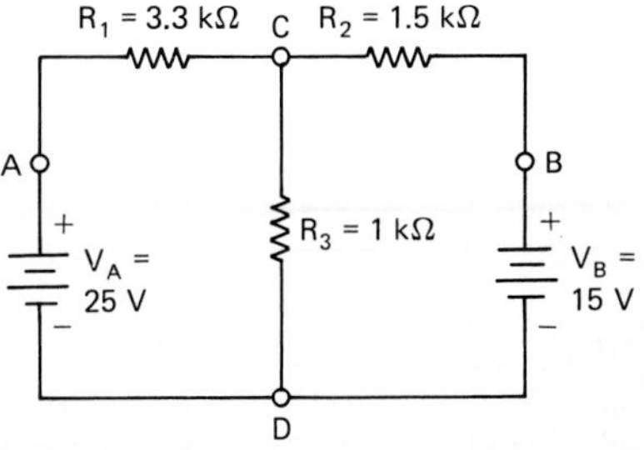

FIGURE 9-8 Circuit schematic for Practice Questions 9-2 and 9-3.

9-5 NODE VOLTAGE METHOD FOR A MULTISOURCE NETWORK

The node voltage method is based on finding the voltages at each node in the circuit using Kirchhoff's current law, keeping in mind that a node is the junction of two or more current paths. The general procedure for this method is as follows:

1. Label all the current nodes in the circuit (see Fig. 9-9).

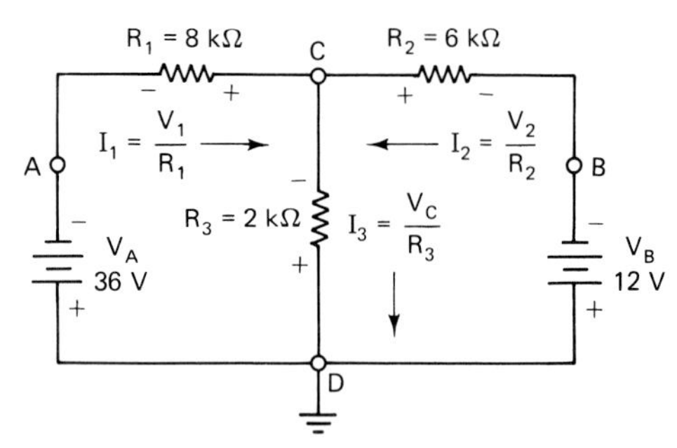

FIGURE 9-9 Node voltage circuit analysis for a multisource circuit.

2. Reference one node to ground so that all voltages will be taken with respect to ground.
3. Assign current directions at each node where the voltage is unknown, except at the node that is referenced to ground.
4. Apply Kirchhoff's current law to each node to obtain a current equation.
5. Assign voltage designations to each node where the voltage is unknown.
6. Express the currents in terms of circuit voltages with respect to the reference node in terms of Ohm's law.
7. Substitute voltage equations into the current equations of step 4 and insert the numerical circuit values of all components and sources of emf.
8. Gather like terms and solve for the node voltage. Use the node voltage to find the other voltage values.

EXAMPLE 9-4

Solve the two-source circuit of Fig. 9-9 using the node voltage method.

Solution

1. The current nodes in the circuit have been labeled *A*, *B*, *C*, and *D*.
2. Node *D* has been referenced to ground.
3. Current directions have been assigned using the electron current direction.
4. Kirchhoff's current equation for node *C* is

$$-I_3 + I_1 + I_2 = 0$$

Therefore,

$$I_3 = I_1 + I_2$$

5. Assign voltage designations to each node where the voltage is unknown. In this case the only unknown voltage is at node *C* since node *D* is the reference or zero voltage point. Node *C* is designated V_C.
6. Apply Ohm's Law and express the currents in terms of circuit voltages with respect to the reference node (ground).

$$I_1 = \frac{V_1}{R_1} = \frac{-V_A - V_C}{R_1}$$

$$I_2 = \frac{V_2}{R_2} = \frac{-V_B - V_C}{R_2}$$

$$I_3 = \frac{V_3}{R_3} = \frac{V_C}{R_3}$$

7. Substitute voltage equations into the current equation of node C, procedure (4), and solve for V_C.

$$-I_3 + I_1 + I_2 = 0$$

$$\frac{-V_C}{R_3} + \frac{-V_A - V_C}{R_1} + \frac{-V_B - V_C}{R_2} = 0$$

Insert numerical circuit values of all components and sources of emf. (*Note:* The kΩ have been dropped since currents will be in mA.)

$$\frac{-V_C}{2} + \frac{-36 - V_C}{8} + \frac{-12 - V_C}{6} = 0$$

Gathering like terms and solving for the node voltage gives

$$\frac{-V_C}{2} - \frac{V_C}{8} - \frac{V_C}{6} = \frac{36}{8} + \frac{12}{6}$$

$$\frac{-12V_C - 3V_C - 4V_C}{24} = \frac{108 + 48}{24}$$

$$-19V_C = 156$$

$$V_C = -8.2 \text{ V}$$

V_C node is -8.2 V with respect to D (i.e. ground). The other voltages V_1 and V_2 can now be calculated since

$$V_1 + V_3 = V_A$$

Therefore,

$$V_1 = V_A - V_3$$
$$= -36 \text{ V} - (-8.2 \text{ V}) = -27.8 \cong -28 \text{ V}$$

Similarly,

$$V_2 + V_3 = V_B$$

and

$$V_2 = -V_B - (-V_3) = -12 + 8.2 \text{ V} \cong -4 \text{ V}$$

PRACTICE QUESTION 9-3

1. Use the schematic of the circuit shown in Fig. 9-8, and solve for node C voltage with respect to ground D using the node voltage method. Solve for V_1 and V_2.

9-6 MESH CURRENT METHOD FOR A MULTISOURCE NETWORK

A mesh is a closed path or "frame" (analogous to a window frame), similar to the branch current loop. The difference is that the mesh current does not divide at a branch point. A mesh current is an assumed current to which Kirchhoff's voltage law is applied around each mesh or loop.

The closed paths or frames in the schematic of Fig. 9-10 are *a-b-e-d-a* and *b-c-f-e-b*. The outside path, *a-b-c-f-e-d-a*, is a loop but not a mesh.

Since a mesh current is assumed to flow without dividing, the mesh current A flows through R_1, R_3, and V_A. Mesh current B flows through R_2, V_B, and R_3. The general procedure used in applying the mesh current method is as follows:

1. Assume a clockwise direction to draw in a mesh current for each closed loop or frame. The clockwise direction should be consistent to all loops.
2. Label the loop currents for each frame I_A, I_B, and so on, or some other form of subscript(s).
3. Label the polarity of each resistor according to the assumed direction of loop current consistent to both loops negative to positive or positive to negative. Where a resistor has more than one loop current flowing through it, each loop voltage drop across that resistor may have the same or opposite polarity signs.
4. Apply Kirchhoff's voltage law to each loop following the assumed direction of the loop current.
   (a) Where the assumed resistor currents flow in the same direction, the total voltage drop will be the product of the resistance and the sum of the two loop currents.
   (b) Where the assumed resistor currents flow in opposite directions, the total voltage drop will be the product of the resistance and the difference of the two currents, *with the larger current assumed to be in the loop being traced.*
   (c) When the mesh current flows into the positive terminal of the source, the algebraic sign for that source voltage is positive. When the mesh current flows into the negative terminal of the source, the algebraic sign for that source voltage is negative.
5. Substitute the numerical circuit values in the equations.
6. Collect like terms for the mesh current equations.
7. Solve using simultaneous equations. The answers to each current will come out positive, indicating that the direction of that mesh current direction is the same as the assumed direction. The answer will be negative if the current is flowing opposite to the assumed mesh direction.

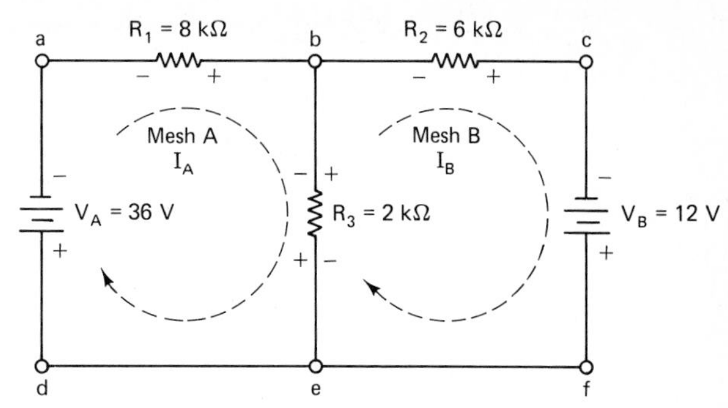

FIGURE 9-10 Mesh currents.

EXAMPLE 9-5

1. A clockwise direction for the mesh currents is shown for each loop in the Fig. 9-10 schematic.

2. The mesh currents have been labeled I_A and I_B.

3. The polarity of each resistor has been labeled according to the assumed direction of current consistent to both loops (i.e., clockwise). The polarity of R_3 is labeled according to each loop and has the opposite polarity for loop current I_B.

4. Applying Kirchhoff's voltage law to each loop gives

(a) Mesh *A*

$$-V_1 - V_3 + V_A = 0$$

$$V_A = V_1 + V_3$$

$$= I_1R_1 + I_3R_3$$

But $I_3 = I_A - I_B$. Note that the currents I_A and I_B flow in the opposite directions, with I_A the current loop being traced assumed to be the larger.

$$V_A = I_AR_1 + (I_A - I_B)R_3$$

$$36\text{ V} = 8I_A + 2I_A - 2I_B$$

$$36 = 10I_A - 2I_B \qquad (1)$$

(b) Mesh *B*

$$-V_2 - V_B - V_3 = 0$$

$$-V_B = V_2 + V_3$$

$$= I_2R_2 + I_3R_3$$

But $I_3 = I_B - I_A$. Note that the currents I_A and I_B flow in the opposite direction, with I_B the current loop being traced assumed to be the larger.

$$-V_B = I_BI_2 + (I_B - I_A)R_3$$

$$-12 = 6I_B + 2I_B - 2I_A$$

$$-12 = 8I_B - 2I_A \qquad (2)$$

5. The numerical circuit values have been substituted in equations (1) and (2).

6. Like terms have been collected and two simultaneous equations obtained [(1) and (2)].

7. Solve using simultaneous equations. Multiply equation (1) by 4, then add equation (2) and equation (1) to solve for I_A.

$$144 = 40I_A - 8I_B$$

$$-12 = -2I_A + 8I_B$$

$$132 = 38I_A$$

$$I_A = 3.47\text{ mA}$$

Substitute 3.74 mA for I_A in equation (1) to solve for I_B.

$$36 = (10\text{ k}\Omega \times 3.47\text{ mA}) - 2I_B$$

$$-I_B = 0.65\text{ mA}$$

The negative sign indicates that the current is flowing opposite the assumed loop current direction and

$$I_3 = I_A - I_B$$

$$= 3.47\text{ mA} - (-0.65\text{ mA}) = 4.1\text{ mA}$$

SUMMARY

Kirchhoff's voltage law states: Around any closed circuit, or a complete circuit, or a closed loop, the algebraic sum of the voltages will equal zero.

Kirchhoff's current law states: The algebraic sum of the current flowing into a junction or node must be equal to the algebraic sum of the currents flowing out of that point.

The balanced bridge circuit of Fig. 9-4(a) will have 0 V *A* to *B*.

Applying Kirchhoff's voltage law to the upper or lower voltage loop *A* to *B* will give the polarity *A* to *B* as well as the voltage *A* to *B* of the bridge circuit.

Procedure for solving branch currents of a multisource network.

1. Assign polarities to each resistor in the circuit by looping the current according to electron or conventional flow as shown in Fig. 9-7.
2. Apply Kirchhoff's voltage law around each closed loop or circuit and substitute *IR* for each voltage.

3. Apply Kirchhoff's current law to include all branch currents flowing into and out of circuit nodes.
4. Substitute all known numerical circuit values in each equation and rewrite equations.
5. Solve for one of the unknown currents in the simultaneous equations and substitute its value in an equation to solve for the remaining currents.

Procedure for solving voltages about a multisource network using the node voltage method:

1. Label all the current nodes in the circuit (see Figure 9-8).
2. Reference one node to ground so that all voltages will be taken with respect to ground.
3. Assign current directions at each node where the voltage is unknown, except at the node that is referenced to ground.
4. Apply Kirchhoff's current law to each node to obtain a current equation.
5. Assign voltage designations to each node where the voltage is unknown.
6. Apply Ohm's law and express the currents in terms of circuit voltages with respect to the reference node.
7. Substitute voltage equations into the current equations of step 4 and insert numerical circuit values of all components and sources of emf.
8. Gather like terms and solve for the node voltage. Use the node voltage to find the other voltage values.

A mesh is a closed path or "frame" to current flow. A mesh current direction is assigned and Kirchhoff's voltage law applied around each closed loop or frame.

The procedure used in applying the mesh current method to solve a multisource network is as follows:

1. Assume a clockwise direction to draw in a mesh current for each closed loop.
2. Label the loop currents for each "frame" with some form of subscript (e.g., I_A, I_B, etc., or I_1, I_2, etc.).
3. Label the polarity of each resistor according to the assumed direction of loop current consistent with both loops. Where a resistor has more than one loop current flowing through it, each loop voltage drop across that resistor may have the same or opposite polarity signs.
4. Apply Kirchhoff's voltage law to each loop following the assumed direction of the loop current.
   - **(a)** Where the assumed resistor currents flow in the same direction, the total voltage drop will be the product of the resistance and the sum of the two loop currents.
   - **(b)** Where the assumed resistor currents flow in opposite directions, the total voltage drop will be the product of the resistance and the difference of the two currents, *with the larger current assumed to be the loop being traced.*
   - **(c)** When the mesh current flows into the positive terminal of the source, the algebraic sign for that source voltage is positive. When the mesh current flows into the negative terminal of the source, that source voltage is negative.
5. Substitute the numerical circuit values in the equation.
6. Collect like terms for the mesh current equations.

7. Solve using simultaneous equations. The answers to each current will come out positive if that mesh current direction is the same as the assumed direction. The answer will be negative if the current is flowing opposite to the assumed mesh direction.

REVIEW QUESTIONS

9-1. State Kirchhoff's voltage law.

9-2. State Kirchhoff's current law.

9-3. Explain a method to show a balanced or an unbalanced bridge condition.

9.4. Explain why the bridge circuit galvanometer pointer is zero centered at mid-scale.

9-5. List the procedure in applying the branch current method to solve a complex resistive network.

9-6. List the procedure in applying the node voltage method to solve a complex resistive network.

9-7. List the procedure in applying the mesh current method to solve a complex resistive network.

SELF-EXAMINATION

9-1. **(a)** Use Kirchhoff's voltage law to calculate the value of V_{XY} in Fig. 9-11.
(b) What is the polarity at X with respect to Y?

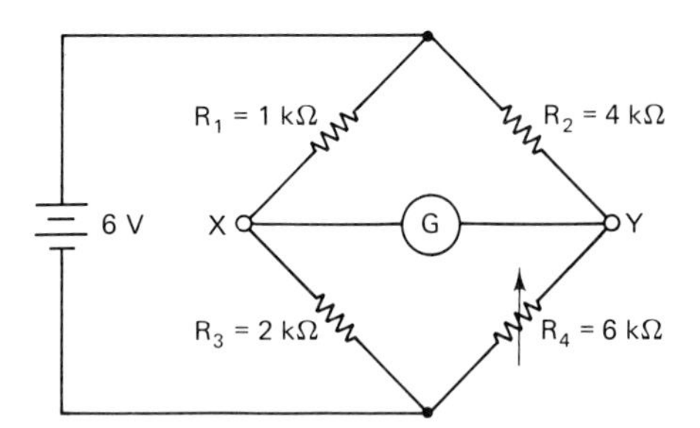

FIGURE 9-11 Bridge circuit.

9-2. Make the necessary changes to the circuit of Fig. 9-11 to obtain a balanced bridge condition. Prove your balanced bridge condition by applying Kirchhoff's voltage law.

9-3. Solve the currents and voltage drops in Fig. 9-12 by the branch current method.

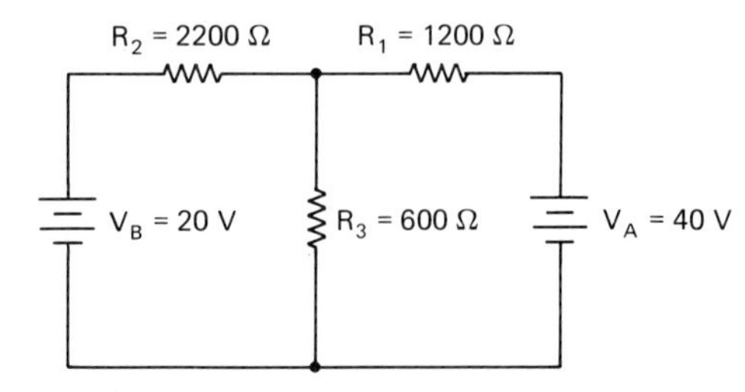

FIGURE 9-12 Branch currents circuit schematic.

9-4. Reverse the polarity of V_B in Fig. 9-13 and solve the currents and voltage drops by the branch current method.

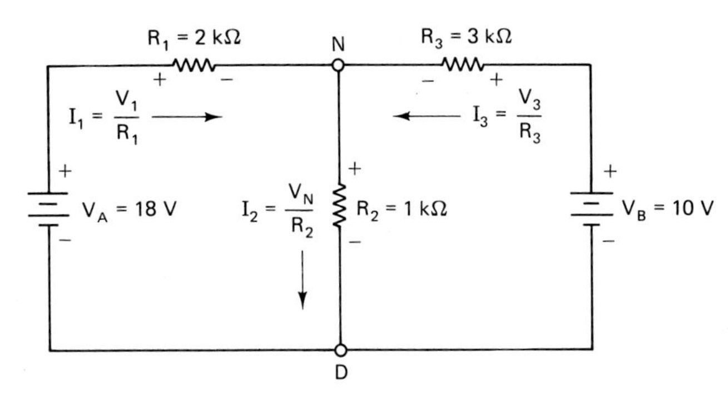

FIGURE 9-13 Node voltage circuit schematic.

9-5. Solve the currents and voltage drops in Fig. 9-13 by the node voltage method.

9-6. Solve all currents and voltage drops in Fig. 9-8 by the mesh current method.

ANSWERS TO PRACTICE QUESTIONS

9-1

1. Around any closed circuit, or a complete circuit, or a closed loop, the algebraic sum of the voltages will equal zero.
2. Polarities of voltage drops must be established before Kirchhoff's voltage law can be applied.
3. Figure 9-2 voltage loop equation starting counterclockwise at C:

$$V_A - V_1 + V_B - V_2 = 0$$

Starting at D:

$$-V_1 + V_B - V_2 + V_A = 0$$

4. Figure 9-5 voltage loop equation using electron current flow:

$$-V_1 - V_B - V_2 + V_A = 0$$
$$-V_1 - V_2 + V_A = V_B$$
$$-5\text{ V} - 7\text{ V} + 24\text{ V} = V_B$$
$$V_B = 12\text{ V}$$

5. Using conventional current flow, the voltage loop equation for Fig. 9-5 is

$$-24\text{ V} + 7\text{ V} + V_B + 5\text{ V} = 0$$
$$V_B = 24\text{ V} - 7\text{ V} - 5\text{ V}$$
$$= 12\text{ V}$$

6. Figure 9-4(b) loop equation for V_{AB}:

$$-V_1 + V_3 = 0$$

7. With R_3 changed to 3.3 kΩ:

$$V_1 = \frac{R_1}{R_1 + R_2} \times V_A = \frac{2.2\text{ k}\Omega}{2.2\text{ k}\Omega + 6.6\text{ k}\Omega} \times 6\text{ V} = 1.5\text{ V}$$

$$V_3 = \frac{R_3}{R_3 + R_4} \times V_A = \frac{3.3\text{ k}\Omega}{3.3\text{ k}\Omega + 6.6\text{ k}\Omega} \times 6\text{ V} = 2\text{ V}$$

Loop equation V_{AB}:

$$-1.5\text{ V} + 2\text{ V} = 0.5\text{ V}$$

Therefore, A is positive with respect to B.

9-2 1. Loop A (Fig. 9-8) gives

$$I_1R_1 + I_3R_3 = 25\text{ V} \tag{1}$$

Loop B gives

$$I_2R_2 + I_3R_3 = 15\text{ V} \tag{2}$$

Current node C to D gives

$$+I_1 + I_2 - I_3 = 0$$
$$I_1 + I_2 = I_3$$

Substitute $I_1 + I_2$ for I_3 and the circuit numerical values give

$$3.3I_1 + (I_1 + I_2)1.0 = 25$$
$$3.3I_1 + 1.0I_1 + 1.0I_2 = 25$$

Therefore,

$$4.3I_1 + 1.0I_2 = 25 \tag{3}$$

Similarly, for equation (2):

$$1.5I_2 + 1.0I_1 + 1.0I_2 = 15$$
$$1.0I_1 + 2.5I_2 = 15 \tag{4}$$

Solve for I_1 by multiplying equation (3) by 2.5 and equation (4) by 1, then subtract equation (4) from (3).

$$4.3I_1 + 1.0I_2 = 25 \times 2.5 \tag{3}$$
$$1.0I_1 + 2.5I_2 = 15 \times 1 \tag{4}$$
$$10.75I_1 + 2.5I_2 = 62.5$$
$$1.0I_1 + 2.5I_2 = 15$$
$$\overline{\qquad 9.75I_1 = 47.5 \qquad}$$
$$I_1 = 4.87\text{ mA}$$

Substitute 4.87 mA for I_1 in equation (4) and solve for I_2:

$$(1.0 \times 4.87 \text{ mA}) + 2.5I_2 = 15$$
$$15 - 4.87 = 2.5I_2$$
$$I_2 = 4.05 \text{ mA}$$

and

$$I_3 = I_1 + I_2 = 4.87 \text{ mA} + 4.05 \text{ mA} = 8.9 \text{ mA}$$

Solve for voltages:

$$V_1 = I_1R_1 = 4.87 \text{ mA} \times 3.3 \text{ k}\Omega \cong 16 \text{ V}$$
$$V_2 = I_2R_2 = 4.05 \text{ mA} \times 1.5 \text{ k}\Omega \cong 6 \text{ V}$$
$$V_3 = I_3R_3 = 8.9 \text{ mA} \times 1 \text{ k}\Omega \cong 9 \text{ V}$$

Voltage loop A:

$$V_1 + V_3 - 25 \text{ V} = 0$$
$$16 \text{ V} + 9 \text{ V} - 25 \text{ V} = 0$$

Voltage loop B:

$$V_2 + V_3 - 15 \text{ V} = 0$$
$$6 \text{ V} + 9 \text{ V} - 15 = 0$$

9-3 1. Kirchhoff's current equation for node C is

$$I_1 + I_2 - I_3 = 0$$

Therefore,

$$I_3 = I_1 + I_2$$

Assign a voltage designation to each node where the voltage is unknown, in this case node C, and express the currents in terms of voltages with respect to the reference node using Ohm's law.

$$I_1 = \frac{V_1}{R_1} = \frac{V_A - V_C}{R_1}$$
$$I_2 = \frac{V_2}{R_2} = \frac{V_B - V_C}{R_2}$$
$$I_3 = \frac{V_3}{R_3} = \frac{V_C}{R_3}$$

Substitute voltage equations into the current equation for node C.

$$I_1 + I_2 - I_3 = 0$$
$$\frac{V_A - V_C}{R_1} + \frac{V_B - V_C}{R_2} - \frac{V_C}{R_3} = 0$$

Substitute circuit values from the schematic:

$$\frac{25\text{ V} - V_C}{3.3\text{ k}\Omega} + \frac{15\text{ V} - V_C}{1.5\text{ k}\Omega} - \frac{V_C}{1\text{ k}\Omega} = 0$$

Solve for V_C:

$$\frac{25}{3.3\text{ k}\Omega} - \frac{V_C}{3.3\text{ k}\Omega} + \frac{15}{1.5\text{ k}\Omega} - \frac{V_C}{1.5\text{ k}\Omega} - \frac{V_C}{1\text{ k}\Omega} = 0$$

Gather like terms:

$$-\frac{V_C}{3.3\text{ k}\Omega} - \frac{V_C}{1.5\text{ k}\Omega} - \frac{V_C}{1\text{ k}\Omega} = -\frac{25}{3.3\text{ k}\Omega} - \frac{15}{1.5\text{ k}\Omega}$$

Therefore,

$$-\frac{1.5V_C}{4.95\text{ k}\Omega} - \frac{3.3V_C}{4.95\text{ k}\Omega} - \frac{4.95V_C}{4.95} = -\frac{37.5}{4.95} - \frac{49.5}{4.95}$$

$$-9.75V_C = -87 \quad \text{and} \quad V_C = 8.9\text{ V} \cong 9\text{ V}$$

Find V_1:

$$V_A = V_1 + V_3$$

Therefore,

$$V_1 = V_A - V_3 = 25 - 9 = 16\text{ V}$$

$$V_2 = V_B - V_3$$

$$= 15 - 9\text{ V} = 6\text{ V}$$

10

Network Theorems

INTRODUCTION

In Chapter 9 Kirchhoff's voltage and current laws were used to solve complex circuits where the simple application of Ohm's law was not practical. Although derived from Kirchhoff's laws, the network theorem provides a more powerful and shorter method to solve both ac and dc complex circuits. Network theorems reduce complex circuits to simple equivalent circuits that can be solved using the laws of series and parallel circuits.

MAIN TOPICS IN CHAPTER 10

10-1 Constant Voltage and Current Source
10-2 Superposition Theorem
10-3 Thévenin's Theorem
10-4 Thévenizing a Circuit with Two Voltage Sources
10-5 Norton's Theorem
10-6 Norton-to-Thévenin Conversion
10-7 Millman's Theorem

OBJECTIVE

After studying Chapter 10, the student should be able to

1. Master the theorems as shown and explained in this chapter by applying each to a multisource complex circuit, to solve for currents and voltages in these types of circuits.

10-1 CONSTANT VOLTAGE AND CURRENT SOURCE

Constant Voltage Source

Figure 10-1(a) shows the familiar symbol for an ideal dc voltage source. The ideal voltage source provides a constant fixed voltage regardless of the current required by the load resistance connected to its terminals.

Both dc and ac sources are far from ideal and have some internal resistance. The example of a dry cell or battery was given in Chapter 4, where the internal resistance increased as the chemical action producing the cell emf decreased. An ac transformer is another excellent

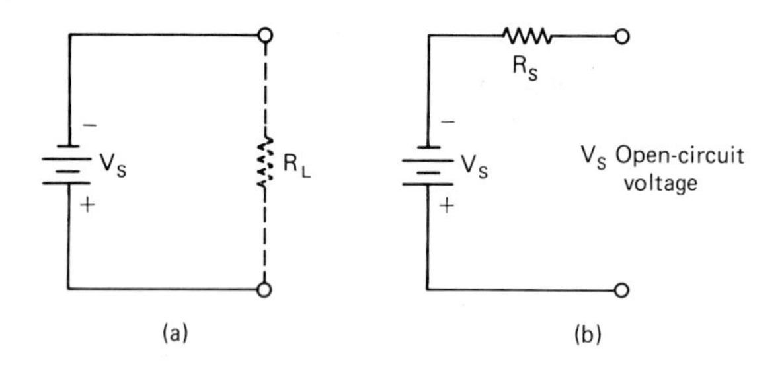

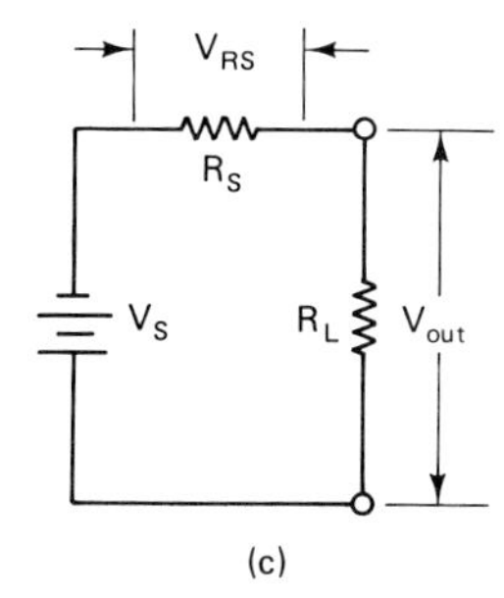

FIGURE 10-1 (a) Ideal voltage source. (b) Voltage source and its internal resistance R_S. (c) Loaded voltage source.

example where the internal resistance is due to the resistance of the winding. Figure 10-1(b) shows the internal resistance R_S in series with the source.

With no load connected to the terminal, the *open-circuit voltage* of the source will always be exactly the source voltage since no current flows through the source internal resistance and the voltage drop across the internal resistance will be zero.

When a load is connected across the terminals as shown in Fig. 10-1(c), current flows and a voltage loss across the internal resistance will reduce the output voltage since R_S and R_L become a voltage divider. The voltage across R_L can be determined by using the voltage-divider formula:

$$V_{out} = \frac{R_L}{R_S + R_L} \times V_S$$

If R_L is much greater than R_S, the source approaches ideal and most of the voltage will be across R_L since very little voltage will be lost across the internal resistance R_S. With a changing load the current from the source changes and the internal resistance voltage drop changes. The voltage-divider action causes a change in load voltage. The larger R_L is compared to R_S, the less change there will be in the output voltage. As a rule, R_L should be at least 10 times greater than R_S ($R_L \geq 10R_S$).

Example 10-1 shows the effect on the output voltage with a change in R_L.

When the load resistance is at least 10 times greater than the source resistance, the voltage source resistance is assumed to be zero for ideal source purposes and the total source voltage is applied across the load resistance.

EXAMPLE 10-1

Calculate the output voltage of the source given that R_L is 10 Ω, 50 Ω, 100 Ω, and R_S is 10 Ω, with a V_S of 10 V in the schematic diagram of Fig. 10-1(c).

Solution: For R_L = 10 Ω:

$$V_{out} = \frac{R_L}{R_S + R_L} \times V_S = \frac{10\ \Omega}{10\ \Omega + 10\ \Omega} \times 10\ \text{V}$$

$$= 5\ \text{V}$$

For R_L = 50 Ω:

$$V_{out} = \frac{50\ \Omega}{10\ \Omega + 50\ \Omega} \times 10\ \text{V} = 8.3\ \text{V}$$

For R_L = 100 Ω:

$$V_{out} = \frac{100\ \Omega}{10\ \Omega + 100\ \Omega} \times 10\ \text{V} = 9.1\ \text{V}$$

With R_L = 100 Ω or $R_L \geq 10R_S$, the output voltage is within 10% of the ideal source voltage.

Constant-Current Source

The symbol for an ideal current source is shown in Fig. 10-2(a). The *arrow indicates the direction of electron current flow* and I_S is the value of the source current. An ideal current source provides a fixed or constant value of current regardless of the load resistance value, since its internal resistance is infinitely large and acts almost like an open circuit. Constant-current sources are designed into many integrated circuits such as operational amplifiers, waveform generators, and digital-to-analog or analog-to-digital converters.

If R_S of Fig. 10-2(b) is at least 10 times greater than R_L, the current source will approach the ideal. A very simple method of providing a constant current is to

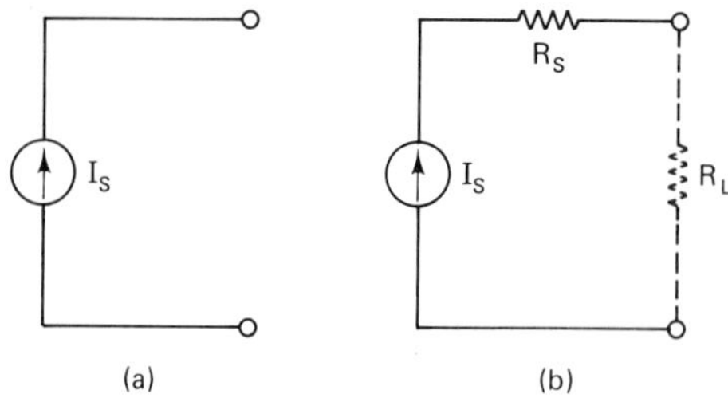

FIGURE 10-2 (a) Ideal current source. (b) Loaded current source.

add a large resistance in series with a source. Of course, the voltage must be increased since the voltage-divider action of R_S and R_L will cause most of the voltage to be dropped across R_S, which will be 10 times greater than R_L. This, however, provides constant-current regulation to the load.

The equivalent current source is shown in Fig. 10-3. Since the ideal current source is said to have infinite resistance, the source resistance is considered to be in parallel, or across the current source symbol. Once the load is connected, it too is in parallel, so the circuit acts as a current divider. Example 10-2 shows the effect on the load current when R_L is changed.

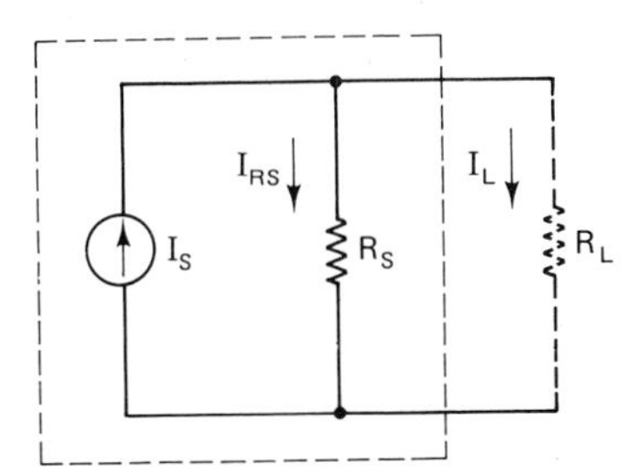

FIGURE 10-3 Equivalent current source.

EXAMPLE 10-2

Calculate the load current in Fig. 10-4 given that R_L is 10 Ω, 50 Ω, 100 Ω, and R_S is 1 kΩ, with an I_S of 10 mA.

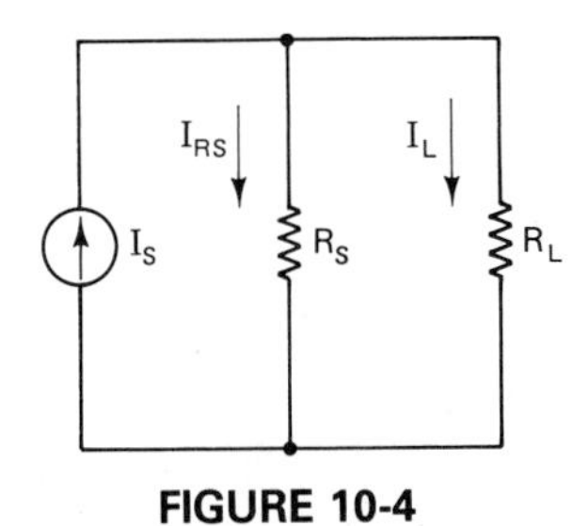

FIGURE 10-4

Solution: For $R_L = 10\ \Omega$:

$$I_L = \frac{R_S}{R_S + R_L} \times I_S = \frac{1000\ \Omega}{1000\ \Omega + 10\ \Omega} \times 10\text{ mA}$$

$$= 9.9\text{ mA}$$

For $R_L = 50\ \Omega$:

$$I_L = \frac{1000\ \Omega}{1000\ \Omega + 50\ \Omega} \times 10\text{ mA} = 9.5\text{ mA}$$

For $R_L = 100$:

$$I_L = \frac{1000\ \Omega}{1000\ \Omega + 100\ \Omega} \times 10\text{ mA} = 9.1\text{ mA}$$

With $R_S = 1000\ \Omega$ or $R_S \geq 10R_L$, the load current is within 10% of the ideal source current.

When the source resistance is 10 times greater than the load resistance, a constant-current source is assumed to have an infinite parallel source resistance, and all the source current is assumed to flow through the load.

Source Conversions

In complex circuit analysis (or other type of analysis) it is convenient to be able to convert from a voltage source to a current source, and vice versa. The equivalent voltage and current source is shown in Fig. 10-5. For both configurations of sources, the load current produced by the sources will be identical for the same load resistance and voltage.

Connecting a load resistance to each source and calculating the load current shows that for the voltage source V_S, the current I_S is the same as I_L. The load resistance and R_S act as a voltage divider as shown in Fig. 10-6(a); therefore,

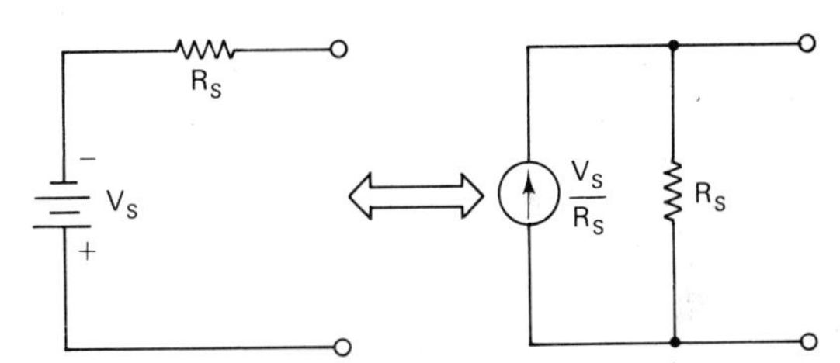

FIGURE 10-5 Conversion of voltage to current source.

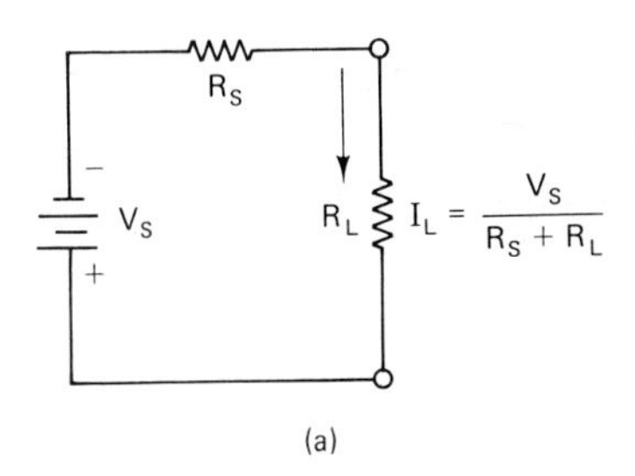

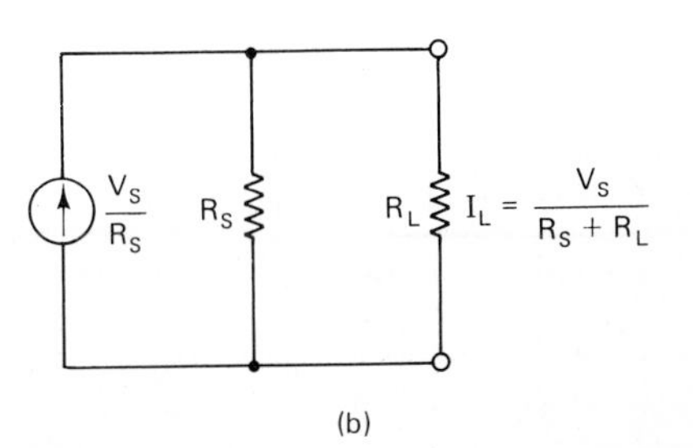

FIGURE 10-6 (a) Terminal equivalent voltage source. (b) Terminal equivalent current source.

$$I_L = \frac{V_S}{R_S + R_L}$$

For the current source V_S/R_S is substituted for I_S, and V_S is the same as V_L since they are considered to be in parallel, as shown in Fig. 10-6(b). By the current-division formula,

$$I_L = \frac{R_S}{R_S + R_L} \times I_S = \frac{R_S}{R_S + R_L} \times \frac{V_S}{R_S}$$

Therefore,

$$I_L = \frac{V_S}{R_S + R_L}$$

EXAMPLE 10-3: Converting a Voltage Source to a Current Source

Convert the voltage source in Fig. 10-7(a) to an equivalent current source.

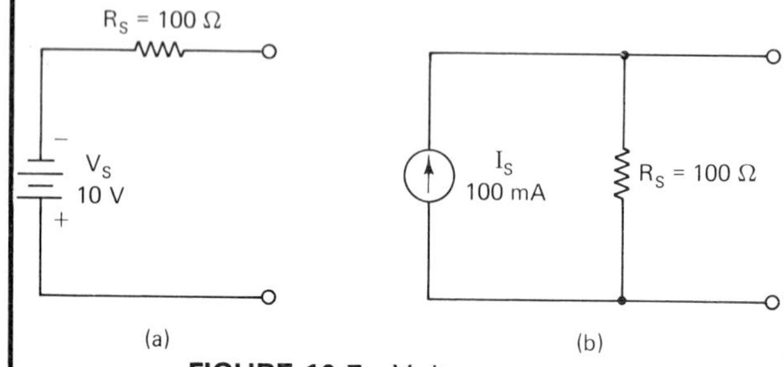

FIGURE 10-7 Voltage source.

Solution

$$I_S = \frac{V_S}{R_S} = \frac{10\text{ V}}{100\ \Omega} = 100\text{ mA}$$

$$R_S = 100\ \Omega$$

The equivalent current source is shown in Fig. 10-7(b).

EXAMPLE 10-4: Converting a Current Source to a Voltage Source

Convert the current source in Fig. 10-8(a) to an equivalent voltage source.

Solution

$$V_S = V_L = I_S R_S = 1\text{ mA} \times 10\text{ k}\Omega = 10\text{ V}$$

$$R_S = 10\text{ k}\Omega$$

The equivalent voltage source is shown in Fig. 10-8(b).

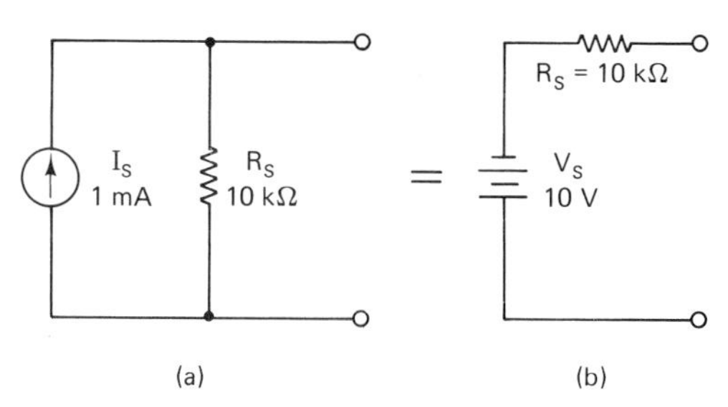

FIGURE 10-8 (a) Current source. (b) Equivalent voltage source.

PRACTICE QUESTIONS 10-1

Where applicable, choose the most suitable word or phrase to complete each statement.

1. Draw the symbol for an ideal current source.
2. The direction of current flow is shown by the __________ in the ideal current source symbol.
3. A voltage source can be termed a constant-voltage source when R_L is (*one-half, one times, two times, 10 times*) greater than R_S.
4. A current source can be termed a constant-current source when R_S is (*one-half, one times, two times, 10 times*) greater than R_L.
5. In Fig. 10-7(a), given that $V_S = 36$ V and $R_S = 720\ \Omega$, convert and draw the schematic diagram of the equivalent current source.
6. In Fig. 10-8(a), given that $I_S = 10$ mA and $R_S = 1\text{ k}\Omega$, convert and draw the schematic diagram of the equivalent voltage source.

10-2 SUPERPOSITION THEOREM

The superposition theorem provides another method to solve branch currents by applying Kirchhoff's current law to a multisource circuit. The *superposition theorem* can be stated:

> The current in each branch of a multisource circuit can be determined by finding the current in each branch produced by its source acting alone and with all other sources replaced by their internal resistance. The superimposed currents in each branch will be the algebraic sum of the individual source currents in that branch.

The general steps to follow in applying the superposition theorem are as follows:

1. Select one voltage source and replace all other sources with its internal resistance (i.e., a short circuit across the terminals).
2. Reduce the current source to zero by disconnecting its terminals; any internal parallel resistance remains.
3. Use the electron or conventional current flow to label the current direction. Find the total current in the selected branch due to the remaining voltage source.
4. Repeat steps 1, 2, and 3 for each source in turn until the branch current components have been calculated for all sources.
5. Sum all the individual current values algebraically by superimposing all branch current polarities and values over each component.

EXAMPLE 10-5

Find the branch currents of Fig. 10-9 by using the superposition theorem.

Solution

Step 1. Select voltage source V_A and short-circuit all other sources as shown in Fig. 10-10(a).

Step 2. Delete, since there are no current sources shown in this schematic.

Step 3. Use the electron or conventional current flow to label the current direction. Find the total current in the selected branch due to the remaining voltage source. The total current by Ohm's law in Fig. 10-10(a) is

$$I_A = I_1' = \frac{V_A}{R_T} = \frac{36\text{ V}}{9.5\text{ k}\Omega} = 3.79\text{ mA}$$

where

$$R_T = R_1 + (R_3 \| R_2)$$

$$= 8\text{ k}\Omega + \frac{2\text{ k}\Omega \times 6\text{ k}\Omega}{2\text{ k}\Omega + 6\text{ k}\Omega} = 9.5\text{ k}\Omega$$

Currents I_2' and I_3' can be solved by the current-division formula for parallel paths.

$$I_2' = \frac{R_3}{R_2 + R_3} \times I_T = \frac{2\text{ k}\Omega}{6\text{ k}\Omega + 2\text{ k}\Omega} \times 3.79\text{ mA}$$

$$= 0.948\text{ mA}$$

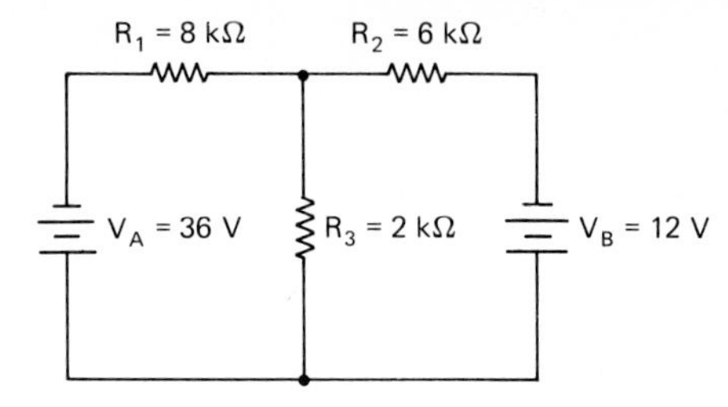

FIGURE 10-9 Circuit schematic for Example 10-5.

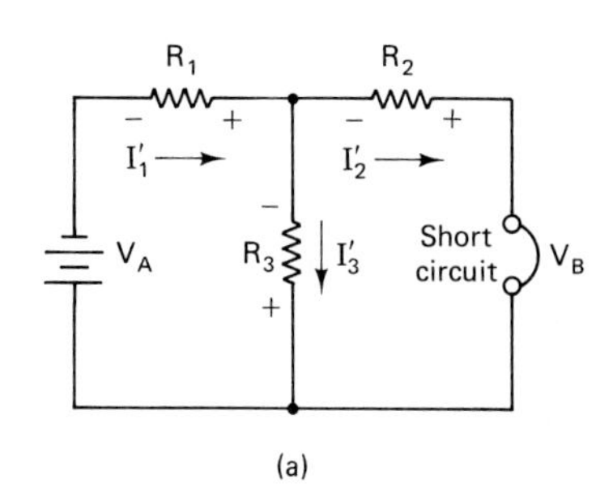

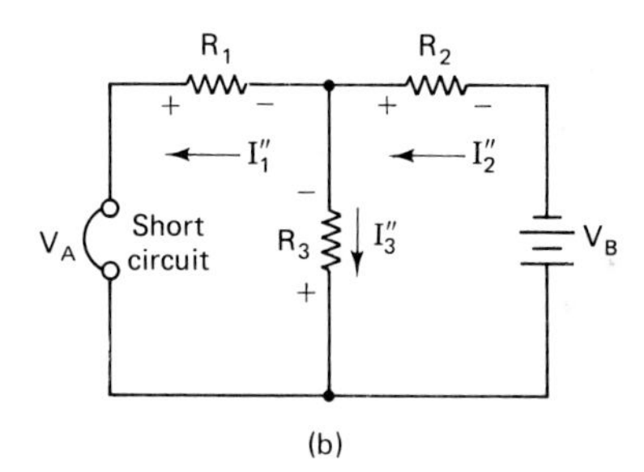

FIGURE 10-10

Similarly,

$$I_3' = \frac{R_2}{R_2 + R_3} \times I_T = \frac{6\text{ k}\Omega}{6\text{ k}\Omega + 2\text{ k}\Omega} \times 3.79\text{ mA}$$

$$= 2.84\text{ mA}$$

Step 4. Repeat steps 1 and 3 for source V_B. The total current by Ohm's law in Fig. 10-10(b) is

$$I_B = I_2'' = \frac{V_B}{R_T} = \frac{12\text{ V}}{7.6\text{ k}\Omega} = 1.58\text{ mA}$$

where

$$R_T = R_2 + (R_3 \| R_1)$$

$$= 6\text{ k}\Omega + \frac{2\text{ k}\Omega \times 8\text{ k}\Omega}{2\text{ k}\Omega + 8\text{ k}\Omega} = 7.6\text{ k}\Omega$$

Currents I_1'' and I_3'' can be solved by the current-division formula for parallel paths.

$$I_1'' = \frac{R_3}{R_1 + R_3} \times I_T = \frac{2\text{ k}\Omega}{8\text{ k}\Omega + 2\text{ k}\Omega} \times 1.58\text{ mA}$$

$$= 0.316\text{ mA}$$

$$I_3'' = \frac{R_1}{R_1 + R_3} \times I_T = \frac{8\text{ k}\Omega}{8\text{ k}\Omega + 2\text{ k}\Omega} \times 1.58\text{ mA}$$

$$= 1.26\text{ mA}$$

Step 5. Superimposing each branch current and summing each current algebraically gives

$$I_1 = -I_1' + I_1'' = -3.79\text{ mA} + 0.316\text{ mA}$$

$$= -3.47\text{ mA}$$

The negative polarity indicates that I_1 flows in the direction shown in Fig. 10-10(a).

$$I_2 = -I_2' + I_2'' = -0.948\text{ mA} + 1.58\text{ mA}$$

$$= 0.63\text{ mA}$$

The positive sign indicates that I_2 flows in the direction shown in Fig. 10-10(b).

$$I_3 = -I_3' - I_3'' = -2.84\text{ mA} - 1.26\text{ mA}$$

$$= -4.1\text{ mA}$$

The minus sign indicates that I_3 flows in the direction labeled on both diagrams.

EXAMPLE 10-6: Conventional Current Direction

Calculate the current through resistor R_3 for the circuit in Fig. 10-11 using the superposition theorem.

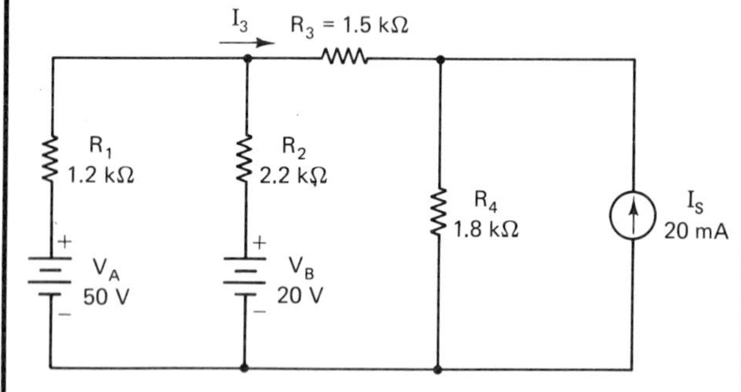

FIGURE 10-11 Circuit schematic for conventional current direction.

Solution: The circuit for the V_A source acting independently is shown in Fig. 10-12(a). The total current I_A is

$$I_A = \frac{V_A}{R_1 + [R_2 \parallel (R_3 + R_4)]}$$

$$= \frac{50\text{ V}}{1.2\text{ k}\Omega + [2.2\text{ k}\Omega \parallel (1.5\text{ k}\Omega + 1.8\text{ k}\Omega)]}$$

$$= 19.84\text{ mA}$$

Find I_3' using the current-divider formula for parallel paths.

$$I_3' = \frac{R_2}{R_2 + (R_3 + R_4)} \times I_A$$

$$= \frac{2.2\text{ k}\Omega}{2.2\text{ k}\Omega + 1.5\text{ k}\Omega + 1.8\text{ k}\Omega)} \times 19.84\text{ mA}$$

$$= 7.94\text{ ma}$$

The circuit for V_B source acting independently is shown in Fig. 10-12(b). The total current I_B is

$$I_B = \frac{V_B}{R_2 + [R_1 \parallel (R_3 + R_4)]}$$

$$= \frac{20\text{ V}}{2.2\text{ k}\Omega + [1.2\text{ k}\Omega \parallel (1.5\text{ k}\Omega + 1.8\text{ k}\Omega)]}$$

$$= 6.49\text{ mA}$$

Find I_3'' using the current-divider formula for parallel paths.

$$I_3'' = \frac{R_1}{R_1 + (R_3 + R_4)} \times I_B$$

$$= \frac{1.2\text{ k}\Omega}{1.2\text{ k}\Omega + (1.5\text{ k}\Omega + 1.8\text{ k}\Omega)} \times 6.49\text{ mA}$$

$$= 1.73\text{ mA}$$

The circuit for the current source acting independently is shown in Fig. 10-12(b). Find I_3''' using the current-divider formula for parallel paths.

$$I_3''' = \frac{R_4}{R_4 + R_3 + (R_1 \parallel R_2)} \times I_S$$

$$= \frac{1.8\text{ k}\Omega}{1.8\text{ k}\Omega + 1.5\text{ k}\Omega + (1.2\text{ k}\Omega \parallel 2.2\text{ k}\Omega)}$$

$$\times 20\text{ mA}$$

$$= 8.83\text{ mA}$$

Solving for I_3 by superimposing the currents and adding them algebraically gives

$$I_3 = I_3' + I_3'' - I_3'''$$

$$= 7.94\text{ mA} + 1.73\text{ mA} - 8.83\text{ mA}$$

$$= 0.84\text{ mA}$$

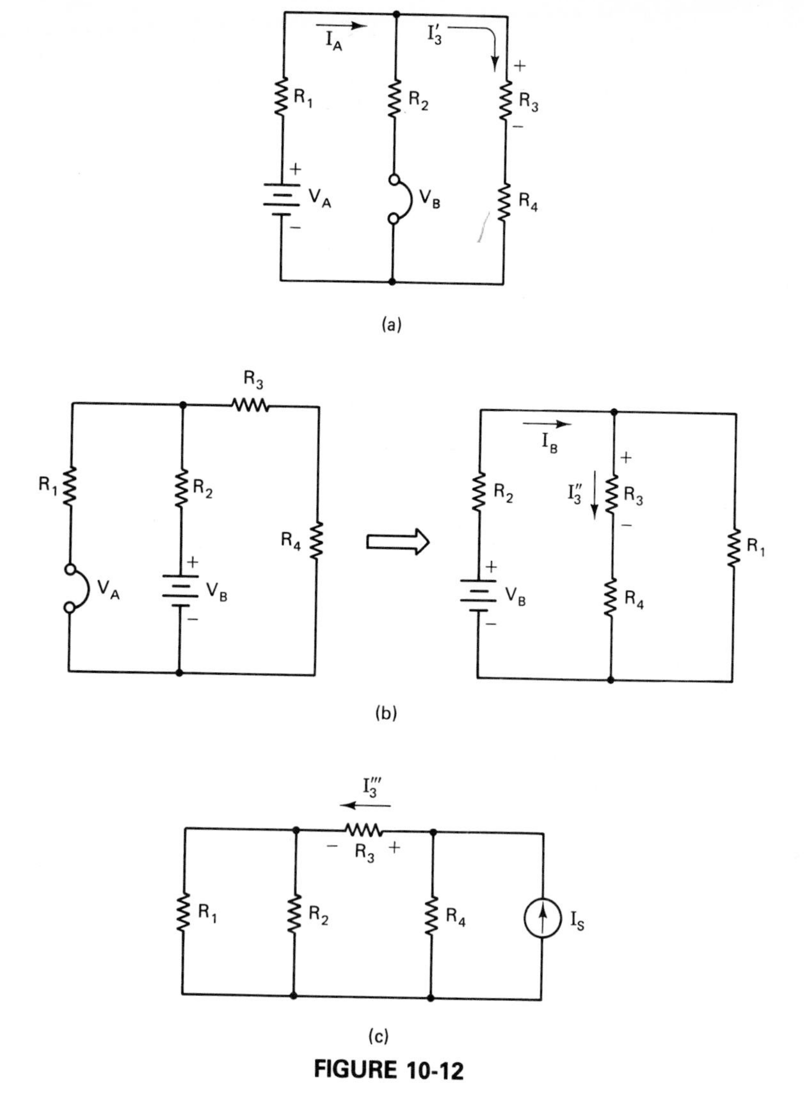

FIGURE 10-12

PRACTICE QUESTIONS 10-2

1. Use the superposition theorem to solve for the current through R_5 in the circuit of Fig. 10-13.
2. Convert the current source in the circuit of Fig. 10-14 to a voltage source and find the current through R_1 using the superposition theorem.

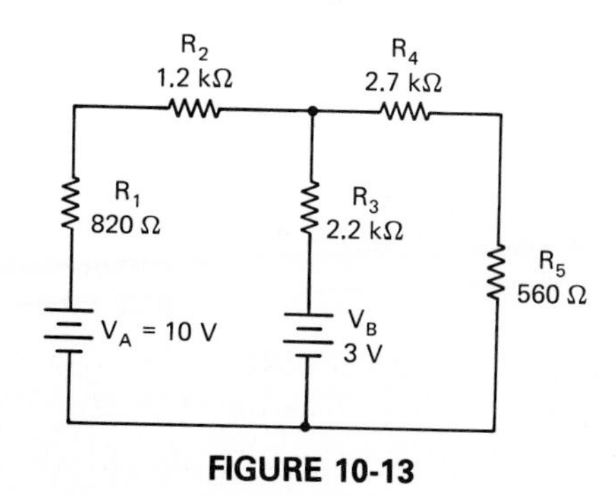

FIGURE 10-13

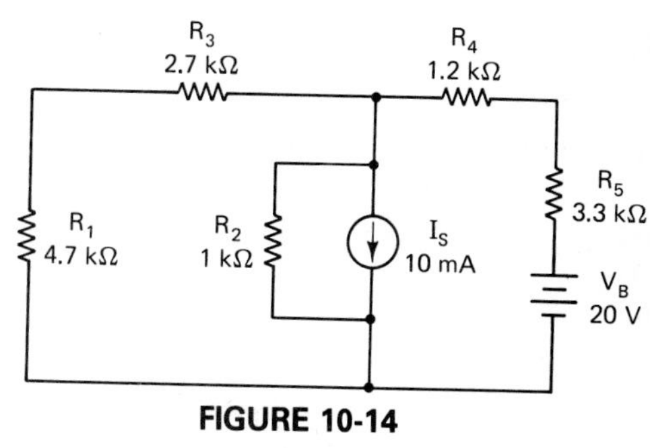

FIGURE 10-14

10-3 THÉVENIN'S THEOREM

Thévenin's theorem provides a very powerful method of analyzing a complex resistive circuit, particularly when repeated calculations are to be made for various load or single-component values.

Thévenin's theorem reduces a complex circuit containing emfs and linear components into a single

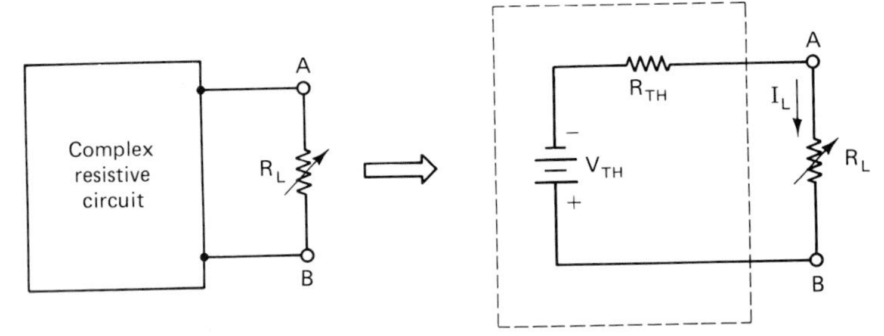

FIGURE 10-15 Thévenin's equivalent circuit.

equivalent series resistance, R_{TH}, and emf, V_{TH}, as shown in Fig. 10-15. Once the circuit is reduced, the load or single-component current can be calculated using Ohm's law.

Finding Thévenin's Voltage

The equivalent voltage V_{TH} is one part of the series Thévenin equivalent circuit and is the open-circuit voltage of the single component or load. This means that the single component or load must first be disconnected from the circuit and the voltage across the two points A to B in the circuit of Fig. 10-16 calculated by means of the series voltage-divider formula.

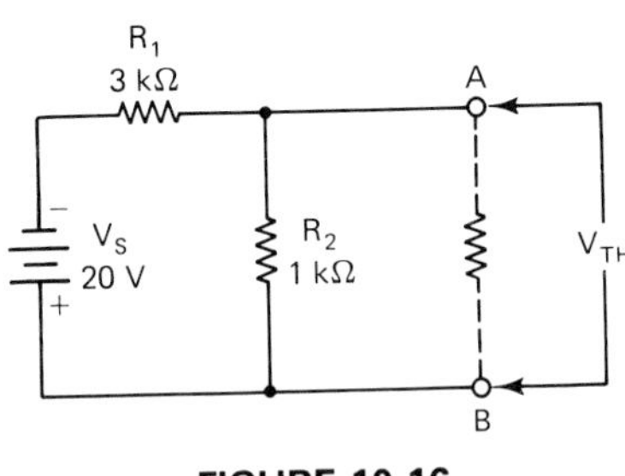

FIGURE 10-16

For the circuit in Fig. 10-16, the Thévenin voltage V_{TH} is in parallel with R_2. Finding V_2 will give V_{TH}. Using the voltage-divider formula gives

$$V_{TH} = V_2 = \frac{R_2}{R_1 + R_2} \times V_S$$

$$= \frac{1\ \text{k}\Omega}{1\ \text{k}\Omega + 3\ \text{k}\Omega} \times 20\ \text{V} = 5\ \text{V}$$

EXAMPLE 10-7

Find the Thévenin voltage, V_{TH}, for the load resistance, R_L, in the circuit of Fig. 10-17.

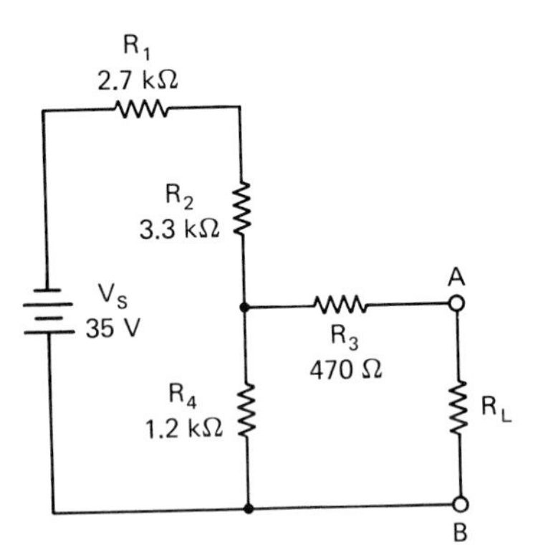

FIGURE 10-17 Circuit schematic for Example 10-7 and 10-9.

Solution: Disconnect R_L and calculate the open-circuit voltage at the load terminals, A to B, by applying the series voltage-divider formula. With the load resistor removed, there is no current flow through R_3 and no voltage drop across R_3, so that $V_{TH} = V_4$.

$$V_{TH} = V_4 = \frac{R_4}{R_4 + R_1 + R_2} \times V_S$$

$$= \frac{1.2\ \text{k}\Omega}{1.2\ \text{k}\Omega + 2.7\ \text{k}\Omega + 3.3\ \text{k}\Omega} \times 35\ \text{V}$$

$$= 5.8\ \text{V}$$

Finding the Thévenin Equivalent Resistance

The net resistance of the circuit looking into the terminals A and B must be found to complete the Thévenizing of the circuit. As defined by Thévenin's theorem, R_{TH} is the total resistance as seen from the output terminals of the load or single component in a given circuit, with all sources replaced by their internal resistance.

To find R_{TH}, redraw the circuit with each voltage source replaced by a short circuit in series with its internal resistance, and each current source replaced by an open circuit in parallel with its internal resistance.

EXAMPLE 10-8

Find R_{TH} for the circuit in the schematic of Fig. 10-16.

Solution: The circuit of Fig. 10-16 has been redrawn (see Fig. 10-18) with its voltage source replaced by a short circuit. Resistor R_1 is now in parallel with R_2, and the Thévenin equivalent resistance R_{TH} can be found by means of parallel circuit laws:

$$R_T = \frac{R_1 R_2}{R_1 + R_2} = \frac{3\text{ k}\Omega \times 1\text{ k}\Omega}{3\text{ k}\Omega + 1\text{ k}\Omega} = 750\ \Omega$$

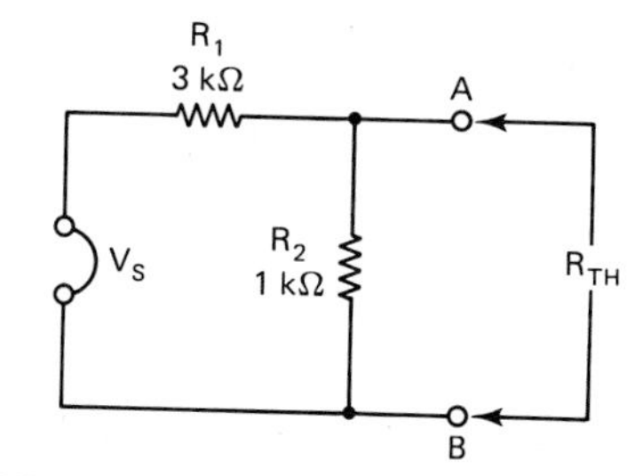

FIGURE 10-18 Circuit schematic for Example 10-8.

For the circuit of Fig. 10-16, V_{TH} was 5 V and R_{TH} is 750 Ω. The load current can now be solved by using Ohm's law since the load is in series with R_{TH} (see Fig. 10-19). Assume that R_L is 820 Ω; then the load current is

$$I_L = \frac{V_{TH}}{R_{TH} + R_L} = \frac{5\text{ V}}{750\ \Omega + 820\ \Omega} = 3.18\text{ mA}$$

If the load resistance value is changed, the load current can easily be solved by substituting the new load value into the formula.

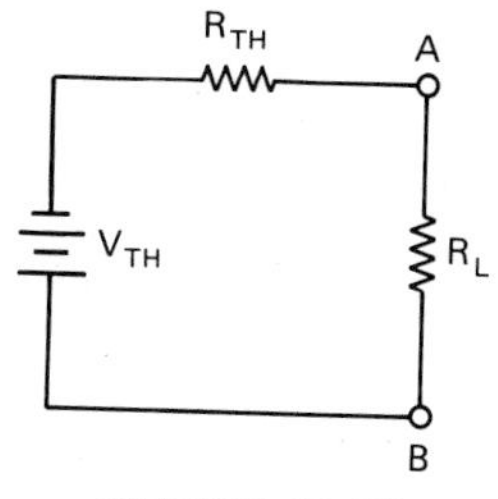

FIGURE 10-19

EXAMPLE 10-9

Find R_{TH} for the circuit in the schematic of Fig. 10-17.

Solution: The circuit of Fig. 10-17 has been redrawn with its voltage source replaced by a short circuit (see Fig. 10-20). Resistors $(R_1 + R_2)$ are in parallel with R_4. The Thévenin equivalent resistance can be calculated as follows:

$$R_{TH} = [(R_1 + R_2) \parallel R_4] + R_3$$

Using product/sum to find R_{TH} gives

$$R_{TH} = \frac{R_4(R_1 + R_2)}{R_4 + R_1 + R_2} + R_3$$

$$= \frac{1.2\text{ k}\Omega(2.7\text{ k}\Omega + 3.3\text{ k}\Omega)}{1.2\text{ k}\Omega + 2.7\text{ k}\Omega + 3.3\text{ k}\Omega} + 470\ \Omega$$

$$= 1\text{ k}\Omega + 470\ \Omega = 1.47\text{ k}\Omega$$

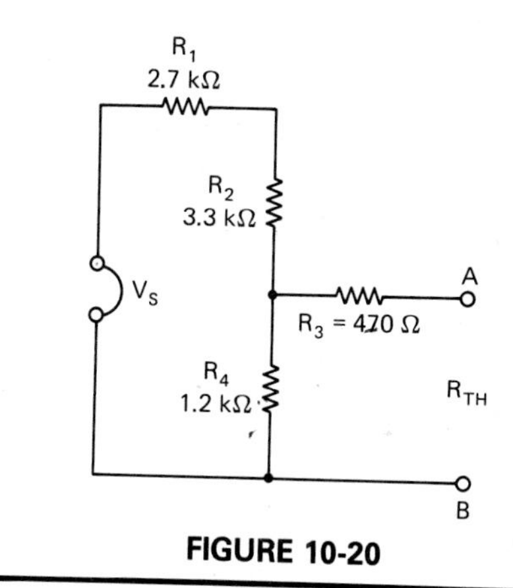

FIGURE 10-20

For the circuit of Fig. 10-17, V_{TH} was 5.8 V and R_{TH} is 1.47 kΩ. The load current can now be solved using Ohm's law since the load is in series with R_{TH} (see Fig. 10-21). Assume that the load resistance is 560 Ω; then the load current is

$$I_L = \frac{V_{TH}}{R_{TH} + R_L} = \frac{5.8\text{ V}}{1.47\text{ k}\Omega + 560\ \Omega} = 2.85\text{ mA}$$

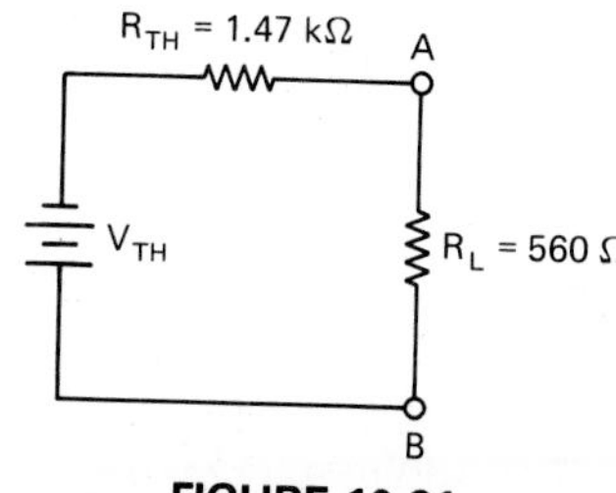

FIGURE 10-21

Reducing a circuit to its Thévenin equivalent of V_{TH} and R_{TH} is termed *Thévenizing the circuit*. The general steps to Thévenize a circuit are as follows:

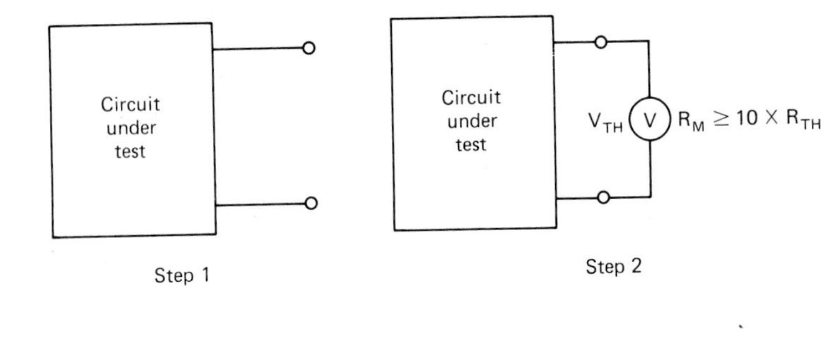

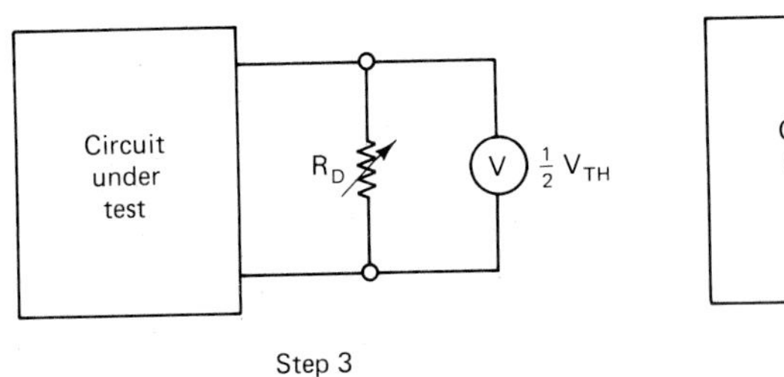

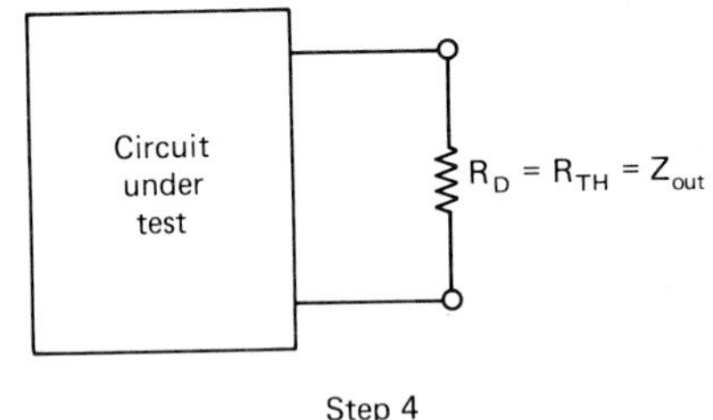

FIGURE 10-22 Steps to find the output impedance of any device or circuit. Step 1: Remove R_L; Step 2: measure open-circuit voltage V_{TH}; Step 3: connect R_D and adjust for $\frac{1}{2}V_{TH}$; Step 4: resistance of decade = $R_{TH} = Z_{out}$.

1. Remove the resistor or load for which you want to find the Thévenin equivalent circuit.
2. Calculate the open-circuit voltage V_{TH} across the two terminals.
3. Calculate the resistance R_{TH} seen between the two terminals with all voltage sources shorted and all current sources opened.
4. Reconnect the resistor or load, which is now in series with R_{TH} and V_{TH}. Its current can be calculated by using

$$I_L = \frac{V_{TH}}{R_{TH} + R_L}$$

Finding V_{TH} and R_{TH} by Measurement

Although Thévenin's theorem is basically a powerful method of analyzing theoretical circuits, it can also be applied to a practical circuit using simple measurement techniques. One application of this would be to measure the output impedance (resistance) of a transistor amplifier. The following steps illustrate the procedure to measure the internal resistance of any device or circuit (see Fig. 10-22).

1. Disconnect any output load from the circuit under test.
2. Measure the open-circuit voltage. To prevent loading the circuit, the meter's internal resistance must be at least 10 times the expected R_{TH}. Use an oscilloscope or a FET input type of VOM.
3. Connect a variable resistance or a decade resistance box to the output terminals and adjust its resistance for one-half of the open-circuit voltage V_{TH}.
4. The resistance of the decade box will now equal R_{TH}.

The limitation of this measurement technique is the ability of the circuit to provide the required current to the variable resistance without loading the circuit. The power rating of the variable resistance or the decade resistance box must also be considered.

PRACTICE QUESTIONS 10-3

1. What is the greatest advantage of reducing a circuit to a Thévenin equivalent circuit?
2. What specific formula or rule is used to find V_{TH}?
3. Explain how R_{TH} is determined.
4. Find the Thévenin equivalent circuit and calculate the load current in the circuit of Fig. 10-23.

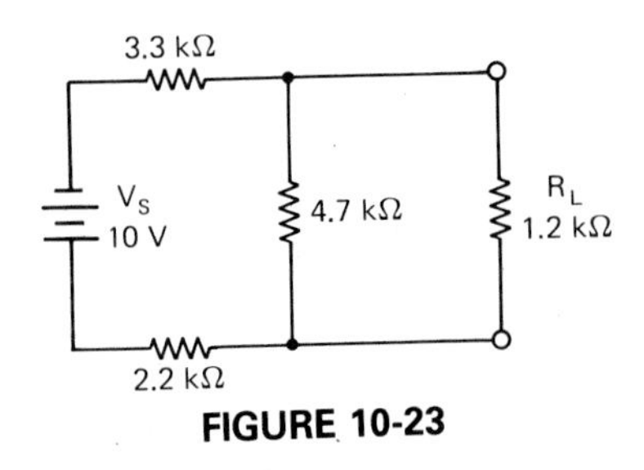

FIGURE 10-23

10-4 THÉVENIZING A CIRCUIT WITH TWO VOLTAGE SOURCES

To Thévenize the circuit of Fig. 10-24(a) requires that the load resistor be removed, and then V_{TH}, the open-circuit voltage A to B, is calculated as shown in Fig. 10-24(b).

To find V_{TH}, the polarities of the voltage drops must be shown in order to properly add them algebraically with V_A and V_B. This is obviously done by using the greater emf source as the "current mover" shown in Fig. 10-24(b). Thévenin's voltage V_{TH} can now be calculated.

The circuit in Fig. 10-24(b) is a simple series circuit with a current given by

$$I = \frac{V_A - V_B}{R_1 + R_2} = \frac{36 - 12}{8\text{ k}\Omega + 6\text{ k}\Omega} = \frac{24\text{ V}}{14\text{ k}\Omega} = 1.71\text{ mA}$$

The open-circuit voltage A and B can be found using either source, V_A or V_B.

Using V_A

$$V_{TH} = V_A - V_{R1}$$

and

$$V_{R1} = I_1R_1 = 1.71\text{ mA} \times 8\text{ k}\Omega = 13.7\text{ V}$$

Therefore,

$$V_{TH} = 36\text{ V} - 13.7\text{ V} = 22.3\text{ V}$$

Using V_B

$$V_{TH} = V_B + V_{R2}$$

and

$$V_{R2} = I_2R_2 = 1.71\text{ mA} \times 6\text{ k}\Omega = 10.3\text{ V}$$

Therefore,

$$V_{TH} = 12\text{ V} + 10.3\text{ V} = 22.3\text{ V}$$

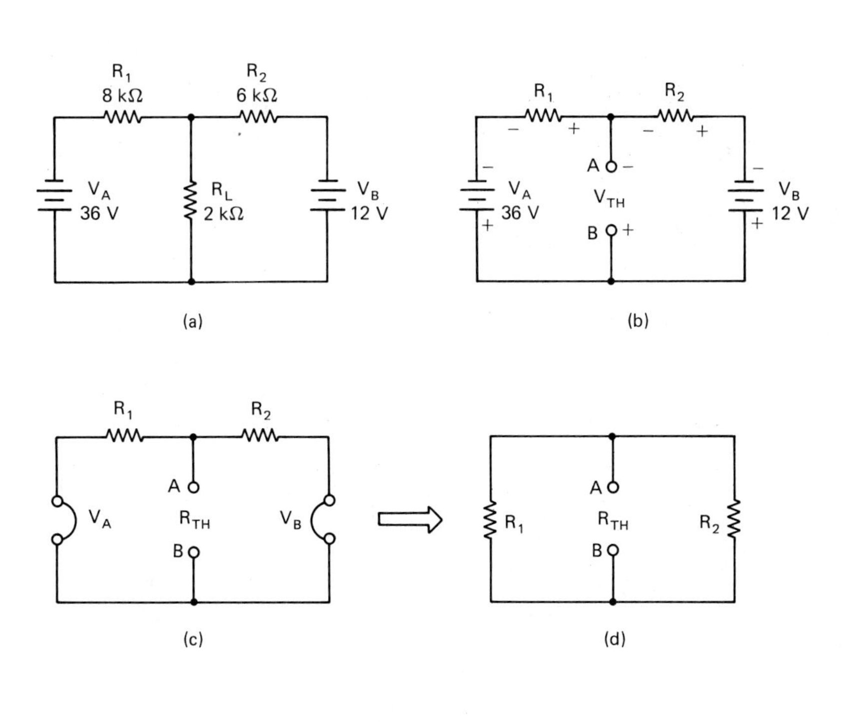

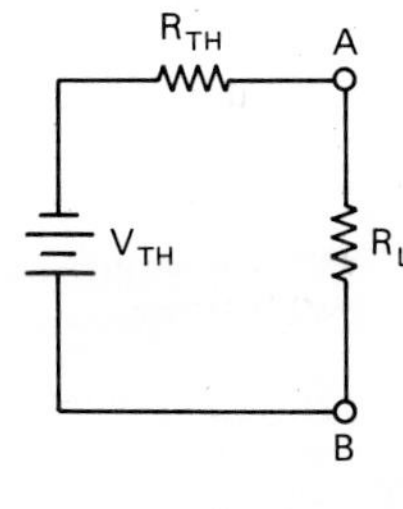

FIGURE 10-24 Thévenizing a two voltage source circuit.

The Thévenin equivalent resistance is found by shorting the voltage sources and finding the total resistance between A to B shown in Fig. 10-24(c) and (d).

$$R_{TH} = R_1 \parallel R_2 = \frac{8\ \text{k}\Omega \times 6\ \text{k}\Omega}{8\ \text{k}\Omega + 6\ \text{k}\Omega} = 3.43\ \text{k}\Omega$$

The current through the load resistor R_L [Fig. 10.24(e)] is obtained using Ohm's law.

$$I_L = \frac{V_{TH}}{R_{TH} + R_L} = \frac{22.3\ \text{V}}{3.43\ \text{k}\Omega + 2\ \text{k}\Omega} = 4.1\ \text{mA}$$

PRACTICE QUESTIONS 10-4

1. Find the current through R_3 using Thévenin's theorem in the circuit of Fig. 10-25.
2. Find the current if $R_3 = 4.7\ \text{k}\Omega$.
3. Explain why it is important to label the voltage polarities with respect to the greater emf source.

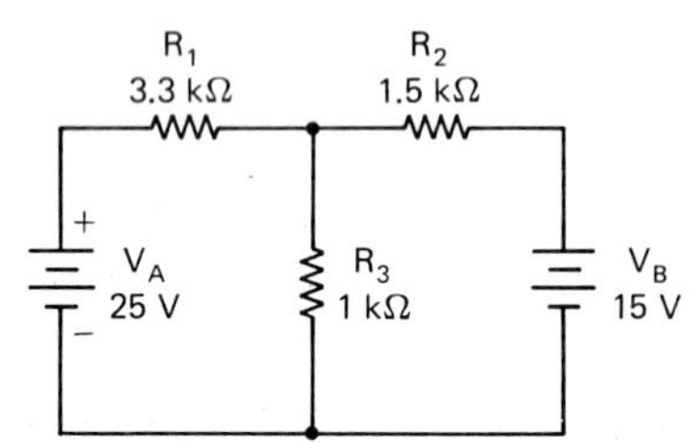

FIGURE 10-25 Circuit for Practice Questions 10-4.

10-5 NORTON'S THEOREM

Thévenizing a circuit reduced the circuit to an equivalent voltage source V_{TH} in series with R_{TH} and the load resistor. Norton's theorem reduces the circuit to an equivalent current source I_N in parallel with Norton's equivalent resistance R_N. Figure 10-26 shows the basic format of Norton's equivalent circuit.

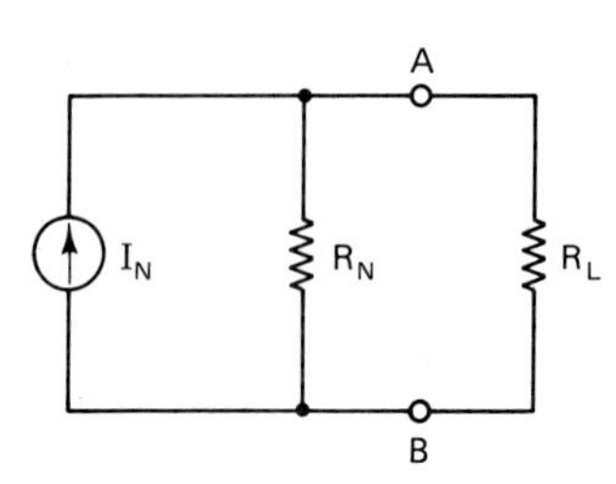

FIGURE 10-26 Norton's equivalent circuit.

Norton's equivalent current (I_N) is defined as the short-circuit current between two points in a circuit. Norton's equivalent resistance is identical in value to Thévenin's R_{TH} and is the total resistance between two terminals in a circuit with all sources replaced by their internal resistance.

Finding Norton's Equivalent Current

In the circuit of Fig. 10-27(a), R_L is replaced by a short circuit between the terminals of the load resistance (i.e., A to B) as shown in Fig. 10-27(b). I_N is the current flowing through the short circuit and is calculated from the following.

The total resistance seen by the source is first calculated from the circuit of Fig. 10-27(b).

$$\begin{aligned} R_T &= R_1 + (R_3 \parallel R_2) \\ &= 3\ \text{k}\Omega + (1\ \text{k}\Omega \parallel 1\ \text{k}\Omega) \\ &= 3.5\ \text{k}\Omega \end{aligned}$$

The total current from the source will be

$$I_T = \frac{V_S}{R_T} = \frac{20\ \text{V}}{3.5\ \text{k}\Omega} = 5.7\ \text{mA}$$

Applying the current-divider formula gives I_N and is

$$\begin{aligned} I_N &= \frac{R_3}{R_2 + R_3} \times I_T = \frac{1\ \text{k}\Omega}{1\ \text{k}\Omega + 1\ \text{k}\Omega} \times 5.7\ \text{mA} \\ &= 2.85\ \text{mA} \end{aligned}$$

The 2.85 mA is the value of Norton's current source.

Finding Norton's Equivalent Resistance

As stated previously, R_N is identical in value to R_{TH} and is found by shorting the sources and calculating the total resistance betweeen two terminals A and B of any circuit, as shown in Fig. 10-27(c). From Fig. 10-27(c), R_2 is in series with R_1 and R_3, which are in parallel with each other. Calculating R_N gives

$$\begin{aligned} R_N &= R_2 + (R_1 \parallel R_3) \\ &= 1\ \text{k}\Omega + \frac{3\ \text{k}\Omega \times 1\ \text{k}\Omega}{3\ \text{k}\Omega + 1\ \text{k}\Omega} = 1.75\ \text{k}\Omega \end{aligned}$$

The Norton equivalent circuit for the schematic shown in Fig. 10-27(a) is redrawn in Fig. 10-27(d).

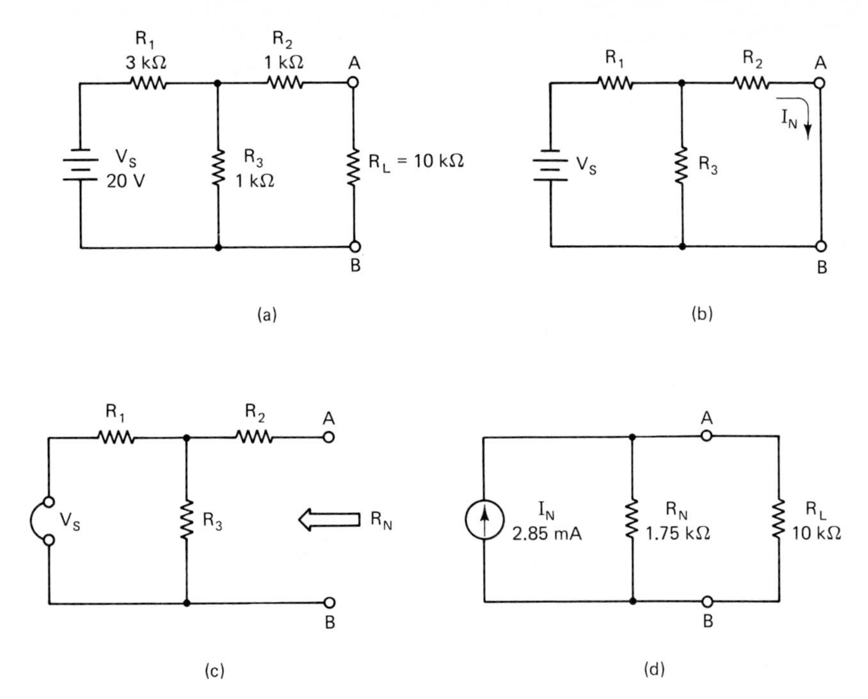

FIGURE 10-27 Finding Norton's equivalent circuit.

The load current can now be calculated by connecting the load resistance to terminals A to B. Using the current-division formula gives

$$I_L = \frac{R_N}{R_N + R_L} \times I_N$$

$$= \frac{1.75\ \text{k}\Omega}{1.75\ \text{k}\Omega + 10\ \text{k}\Omega} \times 2.85\ \text{mA} = 0.42\ \text{mA}$$

PRACTICE QUESTIONS 10-5

1. Find the current through R_3 using Norton's theorem in the circuit of Fig. 10-28.
2. Draw the Norton equivalent circuit and label the values of I_N and R_N obtained in Question 1.

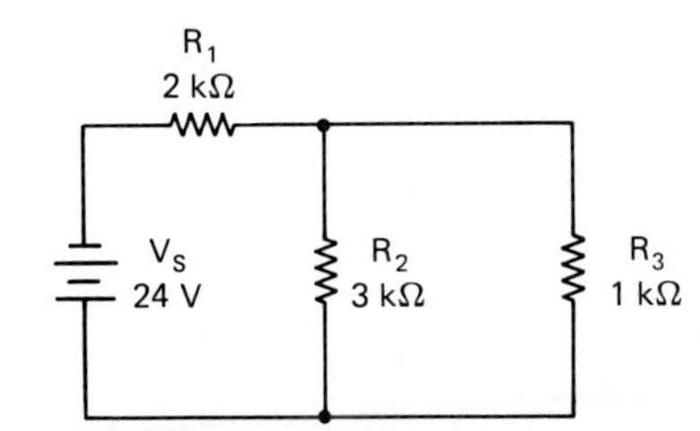

FIGURE 10-28 Circuit for Practice Questions 10-5.

10-6 NORTON-TO-THÉVENIN CONVERSION

In Section 10-7 we will be looking at Millman's theorem, which requires that a number of voltage sources be converted to current sources in order to solve a parallel multiple-source circuit. To convert from Norton's to Thévenin's, and vice versa, requires that you keep the basic principles of each theorem in mind, mainly that Thévenin's theorem specfies a series R_{TH} with the source voltage, while Norton's theorem specifies a parallel R_N with the current source. Conversion can then be as simple as using Ohm's law.

In both Thévenin's and Norton's theorems the equivalent resistance is calculated in the same manner. The sources are shorted and the total resistance is found by looking into the circuit from the terminals of the load resistance. Since both R_{TH} and R_N are calculated in the same manner,

$$R_N = R_{TH}$$

Converting Norton's equivalent current source to Thévenin's equivalent voltage source is a simple application of Ohm's law, $V = IR$. In this case

$$V_{TH} = I_N R_N$$

With these two basic formulas you can move from Norton's to Thévenin's theorem, and vice versa.

$$R_{TH} = R_N \longleftrightarrow R_N = R_{TH}$$

$$V_{TH} = I_N R_N \longleftrightarrow I_N = \frac{V_{TH}}{R_{TH}}$$

10-7 MILLMAN'S THEOREM

Millman's theorem is strictly applied to voltage sources in parallel or to a combination of voltage and current sources in parallel, as shown in Fig. 10-29(a). *Millman's theorem* states:

> A single current source can represent any number of parallel current sources provided that the individual currents are summed algebraically. The total parallel resistance with the individual current sources opened will be equal to the representative source resistance.

The general steps to follow in applying Millman's theorem are as follows:

1. Convert all voltage sources to current sources, and calculate each current as shown in Fig. 10-29(b).
2. Algebraically add all current sources to obtain the Norton equivalent current [see Fig. 10-29(c)].
3. Determine the total parallel sources resistance in order to obtain the equivalent resistance R_N [see Fig. 10-29(c)].
4. Convert the Norton equivalent circuit to the Thévenin equivalent circuit, as required [see Fig. 10-29(d)].

EXAMPLE 10-10

Use Millman's theorem to find the current through R_L and the voltage across R_L in the circuit of Fig. 10-29(a).

Solution

1. The voltage sources have been converted to current sources as shown in Fig. 10-29(b). Calculating each current gives

$$I_1 = \frac{V_1}{R_1} = \frac{40\text{ V}}{2\text{ k}\Omega} = 20\text{ mA}$$

$$I_2 = -15\text{ mA}$$

$$I_3 = \frac{V_3}{R_3} = \frac{30\text{ V}}{3\text{ k}\Omega} = 10\text{ mA}$$

(a)

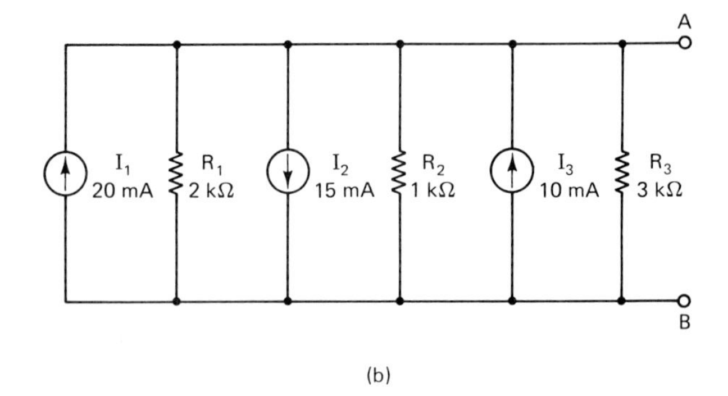

(b)

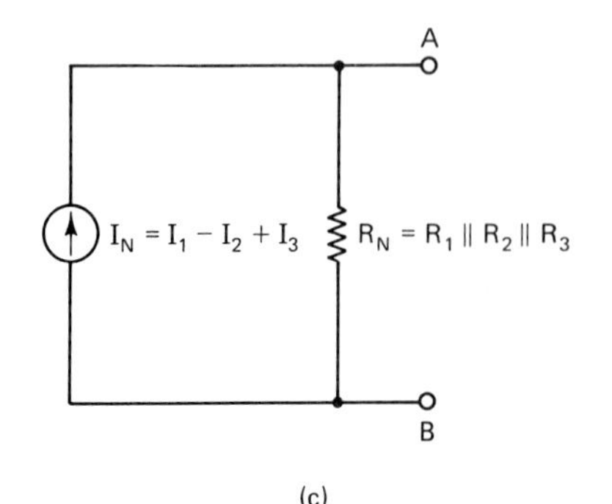

(c)

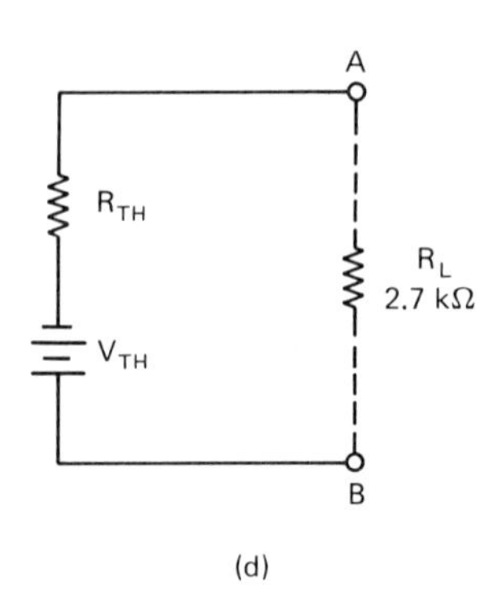

(d)

FIGURE 10-29 Circuit for Millman's theorem.

2. Adding the currents algebraically to obtain the Norton equivalent current gives

$$I_N = I_1 - I_2 + I_3$$
$$= 20 \text{ mA} - 15 \text{ mA} + 10 \text{ mA} = 15 \text{ mA}$$

3. The total parallel resistance of all the sources will give R_N [see Fig. 10-29(b) and (c)].

$$R_N = R_1 \parallel R_2 \parallel R_3$$
$$= 2 \text{ k}\Omega \parallel 1 \text{ k}\Omega \parallel 3 \text{ k}\Omega = 545 \ \Omega$$

4. Norton's equivalent current and resistance can now be converted to the Thévenin equivalent voltage and resistance, and the load voltage and current calculated.

$$R_{TH} = R_N = 545 \ \Omega$$
$$V_{TH} = I_N R_N = 15 \text{ mA} \times 545 \ \Omega = 8.2 \text{ V}$$

$$I_L = \frac{V_{TH}}{R_{TH} + R_L} = \frac{8.2 \text{ V}}{545 \ \Omega + 2700 \ \Omega} = 2.5 \text{ mA}$$

$$V_L = I_L R_L = 2.5 \text{ mA} \times 2.7 \text{ k}\Omega = 6.8 \text{ V}$$

PRACTICE QUESTIONS 10-6

1. To what type of circuit does Millman's theorem strictly apply?
2. State the four general steps to follow in applying Millman's theorem.
3. Solve for the load current and the load volage in the circuit of Fig. 10-30.

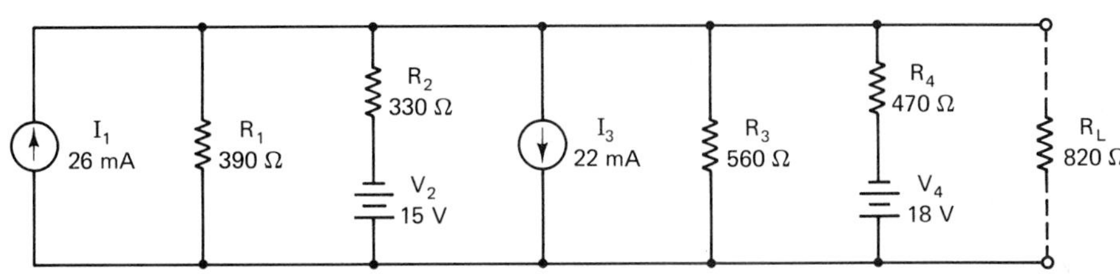

FIGURE 10-30 Circuit for Practice Questions 10-6.

SUMMARY

The ideal voltage source provides a constant fixed voltage regardless of the current required by the load resistance connected to its terminals.

A voltage source resistance is assumed to be zero when the load resistance is 10 times greater and its total voltage is within 10% of the ideal source voltage.

An ideal current source provides a fixed or constant current regardless of the load resistance. Its internal resistance is considered to be infinitely large.

All the source current is assumed to flow through the load when the source resistance is 10 times greater than the load resistance.

A current source can be converted to a voltage source by applying Ohm's law, $V_S = I_S R_S$, and connecting R_S in series.

A voltage source can be converted to a current source by applying Ohm's law, $I_S = V_S/R_S$, and connecting R_S in parallel.

The general steps to follow in applying the superposition theorem are as follows:

1. Select one voltage source and replace all other sources with its internal resistance (i.e., a short circuit across the terminals).
2. Reduce the current source to zero by disconnecting its terminals; any internal parallel resistance remains.
3. Use electron or conventional current flow to label the current direction in the selected branch due to the remaining voltage source.

4. Repeat steps 1, 2, and 3 for each source in turn until the branch current components have been calculated for all sources.
5. Sum all the individual current values algebraically by superimposing all branch current polarities and values over each component.

Thévenin's theorem allows repeated load current calculations to be made easily.

The Thévenin equivalent voltage, V_{TH}, is found by using the voltage-divider rule with the load resistance disconnected:

$$V_{TH} = \frac{R_2}{R_1 + R_2} \times V_S \qquad \text{(Fig. 10-16)}$$

The Thévenin equivalent resistance, R_{TH}, is found by shorting the source and calculating the resultant total resistance with the load disconnected.

The general steps to Thévenize a circuit are as follows:

1. Remove the resistor or load of which you want to find the Thévenin equivalent circuit.
2. Calculate the open-circuit voltage V_{TH} across the two terminals.
3. Calculate the resistance R_{TH} seen between the two terminals with all voltage sources shorted and all current sources opened.
4. Reconnect the resistor or load, which is now in series with R_{TH} and V_{TH}. The I_L current can be calculated by using

$$I_L = \frac{V_{TH}}{R_{TH} + R_L}$$

Norton's theorem reduces a circuit to an equivalent current source I_N in parallel with the Norton equivalent resistance, R_N.

The Norton equivalent current, I_N, is the short-circuit current with the load removed and a short circuit between the terminals of the load resistance. To find I_N, the total current in the current must first be calculated, then the current-divider formula applied:

$$I_N = \frac{R_3}{R_2 + R_3} \times I_T \qquad \text{[Fig. 10-27(b)]}$$

R_N is equal to V_{TH} and is found in the same manner. Short the voltage source and calculate the total resistance looking into the circuit from the disconnected load terminals.

Norton-to-Thévenin conversion, and vice versa:

$$R_{TH} = R_N \longleftrightarrow R_N = R_{TH}$$

$$V_{TH} = I_N R_N \longleftrightarrow I_N = \frac{V_{TH}}{R_{TH}}$$

Millman's theorem is applicable only to parallel voltage and current sources. General steps to follow in applying Millman's theorem:

1. Convert all voltage sources to current sources, and calculate each current.
2. Algebraically add all current sources to obtain the Norton equivalent current.
3. Determine the total parallel sources resistance in order to obtain the equivalent resistance R_N.

4. Convert the Norton equivalent circuit to the Thévenin equivalent circuit, as required.

REVIEW QUESTIONS

10-1. What are the characteristics of an ideal voltage source?

10-2. What are the characteristics of an ideal current source?

10-3. When does a practical voltage source approach ideal source characteristics?

10-4. When does a practical current source approach ideal source characteristics?

10-5. Explain how you would convert a voltage source to a current source.

10-6. Explain how you would convert a current source to a voltage source.

10-7. Give the five general steps to follow in applying the superposition theorem.

10-8. What is the advantage of applying Thévenin's theorem to any circuit?

10-9. List the four general steps to Thévenize a circuit.

10-10. List the four general steps to measure the output impedance of any device or circuit.

10-11. What are two limitations of the method in Question 10-10?

10-12. What is the difference between the Norton equivalent circuit and the Thévenin equivalent circuit?

10-13. Give the formulas to convert from the Norton to the Thévenin equivalent circuit.

10-14. Give the formulas to convert from the Thévenin to the Norton equivalent circuit.

10-15. To what type of circuit configuration is Millman's theorem strictly applied?

10-16. List the four general steps to follow in applying Millman's theorem.

SELF-EXAMINATION

10-1. Given a constant-current source of $I_S = 1.5$ A and $R_S = 10\ \Omega$, calculate the equivalent constant-voltage source V_S, R_S, and draw the schematic diagram.

10-2. Given a constant-voltage source of $V_S = 18$ V and $R_S = 1\ \text{k}\Omega$, calculate the equivalent constant-current source I_S, R_S, and draw the schematic diagram.

10-3. For the constant-current source in Question 10-1, what must be the load resistance to obtain a practical constant-voltage source? Calculate the load current using this load resistance value.

10-4. Find the current through R_3 and R_4 in the circuit of Fig. 10-31 using the superposition theorem.

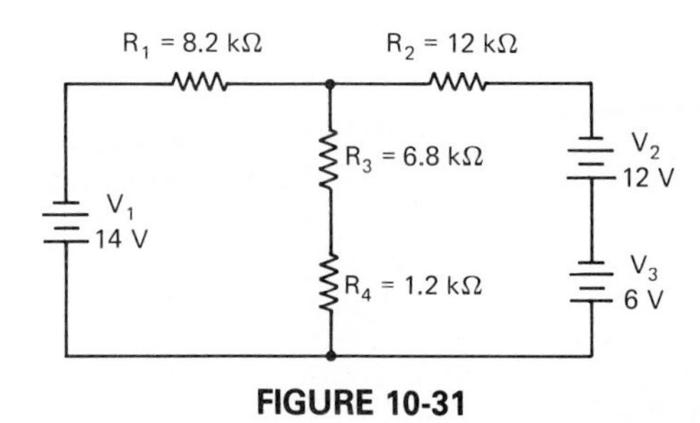

FIGURE 10-31

10-5. Find the current through R_3 and R_2 in the circuit of Fig. 10-32 using the superposition theorem.

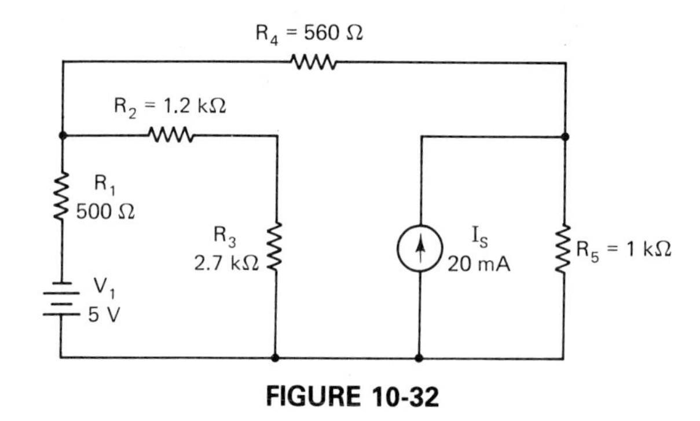

FIGURE 10-32

10-6. Find the current through R_L in the circuit of Fig. 10-33 using Thévenin's theorem.

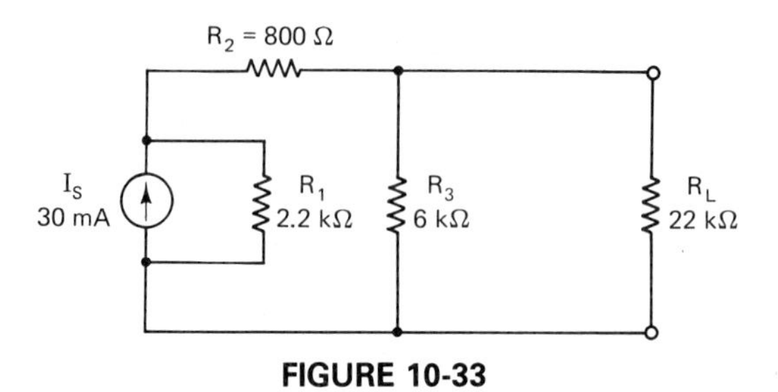

FIGURE 10-33

10-7. Find the current through R_3 and R_4 in the circuit of Fig. 10-34 using Thévenin's theorem.

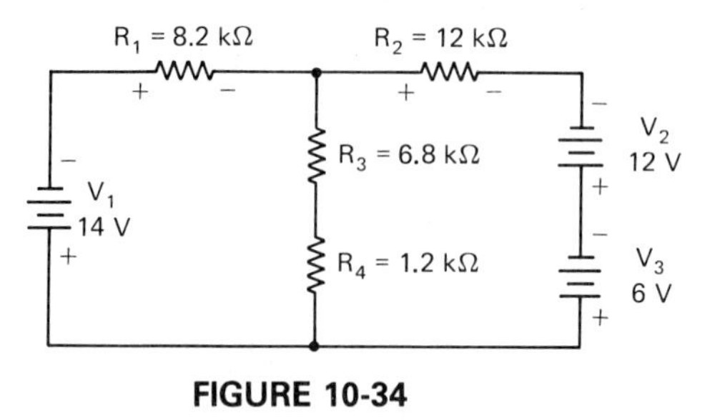

FIGURE 10-34

10-8. Calculate and draw the Norton equivalent circuit A to B for the circuit of Fig. 10-35.

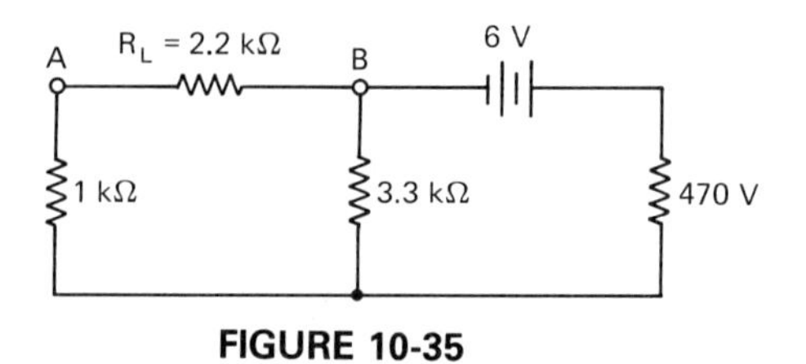

FIGURE 10-35

10-9. Calculate the load current in the circuit of Question 10-8.

10-10. Find the current through R_5 in the circuit of Fig. 10-36 using Norton's theorem.

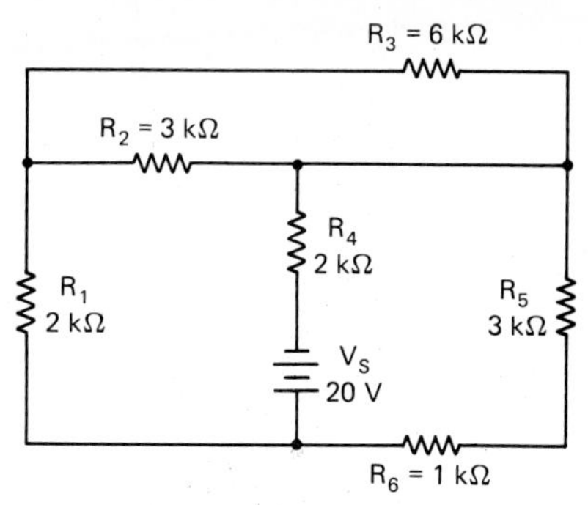

FIGURE 10-36

10-11. Convert the Norton equivalent circuit in Question 10-10 to the Thévenin equivalent circuit and calculate the current through the new value for R_5 of 5 kΩ.

10-12. Use Millman's theorem to reduce the circuit of Fig. 10-37 to an equivalent current source and find R_N and I_N.

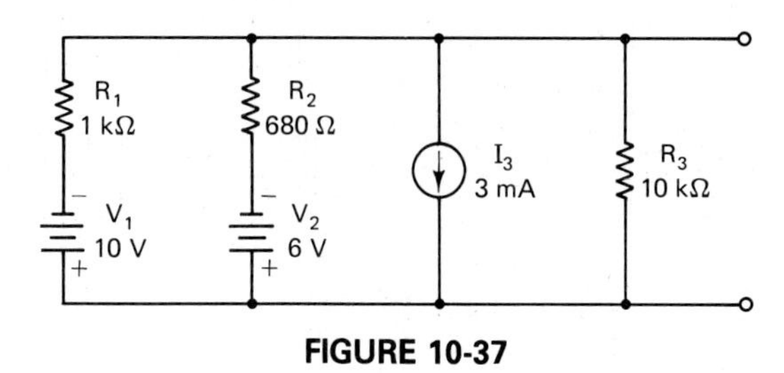

FIGURE 10-37

10-13. Use Millman's theorem to reduce the circuit of Fig. 10-38 to a single voltage source, and find the Thévenin V_{TH} and R_{TH}.

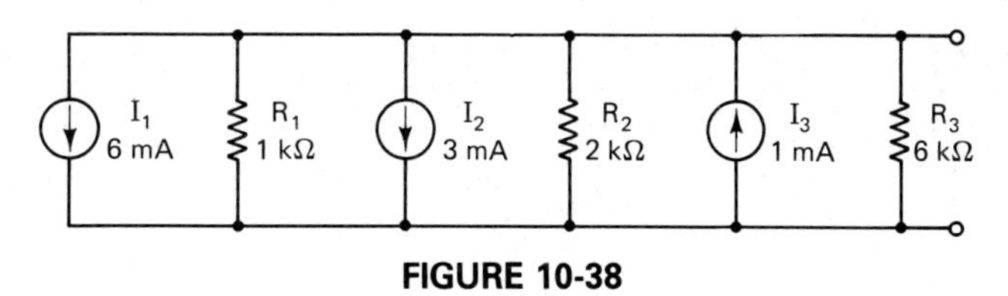

FIGURE 10-38

ANSWERS TO PRACTICE QUESTIONS

10-1
1. See Fig. 10-2(a)
2. arrowhead
3. 10 times
4. 10 times
5. $I_S = \dfrac{V_S}{R_S} = \dfrac{36\text{ V}}{720\ \Omega} = 50\text{ mA}$
   [see Fig. 10-7(b) for a similar diagram]
6. $V_S = I_S R_S = 10\text{ mA} \times 1\text{ k}\Omega = 10\text{ V}$
   [see Fig. 10-8(b) for a similar diagram]

10-2 1.

$$I_A = \frac{V_A}{R_1 + R_2 + [R_3 \parallel (R_4 + R_5)]}$$

$$= \frac{10 \text{ V}}{0.82 \text{ k}\Omega + 1.2 \text{ k}\Omega + [2.2 \text{ k}\Omega \parallel (2.7 \text{ k}\Omega + 0.56 \text{ k}\Omega)]} = 3 \text{ mA}$$

Find I_5 by the current-division formula.

$$I_5' = \frac{R_3}{R_3 + (R_4 + R_5)} \times I_A$$

$$= \frac{2.2 \text{ k}\Omega}{2.2 \text{ k}\Omega + (2.7 \text{ k}\Omega + 0.56 \text{ k}\Omega)} \times 3 \text{ mA} = 1.2 \text{ mA}$$

$$I_B = \frac{V_B}{R_3 + [(R_1 + R_2) \parallel (R_4 + R_5)]}$$

$$= \frac{3 \text{ V}}{2.2 \text{ k}\Omega + [(0.82 \text{ k}\Omega + 1.2 \text{ k}\Omega) \parallel (2.7 \text{ k}\Omega + 0.56 \text{ k}\Omega)]} = 0.87 \text{ mA}$$

Find I_5'' by the current-division formula.

$$I_5'' = \frac{R_1 + R_2}{(R_1 + R_2) + (R_4 + R_5)} \times I_B$$

$$= \frac{0.82 \text{ k}\Omega + 1.2 \text{ k}\Omega}{(0.82 \text{ k}\Omega + 1.2 \text{ k}\Omega) + (2.7 \text{ k}\Omega + 0.56 \text{ k}\Omega)} \times 0.87 \text{ mA} = 0.33 \text{ mA}$$

$$I_5 = I_5' + I_5'' = 1.2 \text{ mA} + 0.33 \text{ mA} = 1.53 \text{ mA}$$

2. Converting the current source to a voltage source gives

$$V_A = I_S R_2 = 10 \text{ mA} \times 1 \text{ k}\Omega = 10 \text{ V}$$

Then

$$I_A = \frac{V_A}{R_2 + [(R_1 + R_3) \parallel (R_4 + R_5)]}$$

$$= \frac{10 \text{ V}}{1 \text{ k}\Omega + [(4.7 \text{ k}\Omega + 2.7 \text{ k}\Omega) \parallel (1.2 \text{ k}\Omega + 3.3 \text{ k}\Omega)]} = 2.63 \text{ mA}$$

Use the current-division formula to find I_1'.

$$I_1' = \frac{R_4 + R_5}{R_4 + R_5 + R_3 + R_1} \times I_A$$

$$= \frac{1.2 \text{ k}\Omega + 3.3 \text{ k}\Omega}{1.2 \text{ k}\Omega + 3.3 \text{ k}\Omega + 2.7 \text{ k}\Omega + 4.7 \text{ k}\Omega} \times 2.63 \text{ mA} \cong 1.0 \text{ mA}$$

$$I_B = \frac{V_B}{(R_4 + R_5) + [R_2 \parallel (R_1 + R_3)]}$$

$$= \frac{20 \text{ V}}{1.2 \text{ k}\Omega + 3.3 \text{ k}\Omega + [1 \text{ k}\Omega \parallel (4.7 \text{ k}\Omega + 2.7 \text{ k}\Omega)]} = 3.72 \text{ mA}$$

Use the current-division formula to find I''_1:

$$I''_1 = \frac{R_2}{R_2 + R_1 + R_3} \times I_B$$

$$= \frac{1\ \text{k}\Omega}{1\ \text{k}\Omega + 4.7\ \text{k}\Omega + 2.7\ \text{k}\Omega} \times 3.72\ \text{mA} = 0.44\ \text{mA}$$

$$I_1 = I'_1 + I''_1 = 1.0\ \text{mA} + 0.44\ \text{mA} \cong 1.4\ \text{mA}$$

10-3

1. The greatest advantage of reducing a circuit to a Thévenin equivalent is that it allows the load current to be calculated easily with a change in load resistance.
2. Formula to find V_{TH} is the voltage-divider formula:

$$V_{\text{TH}} = \frac{R_x}{R_T} \times V_S$$

3. R_{TH} is obtained by shorting the source and calculating for net resistance *looking back* from the load terminals.
4. Remove the load resistance and short the source to find R_{TH}.

$$R_{\text{TH}} = (3.3\ \text{k}\Omega + 2.2\ \text{k}\Omega) \parallel 4.7\ \text{k}\Omega = 2.53\ \text{k}\Omega$$

$$V_{\text{TH}} = \frac{4.7\ \text{k}\Omega}{4.7\ \text{k}\Omega + 3.3\ \text{k}\Omega + 2.2\ \text{k}\Omega} \times 10\ \text{V} = 4.6\ \text{V}$$

$$I_L = \frac{V_{\text{TH}}}{R_{\text{TH}} + R_L} = \frac{4.6\ \text{V}}{2.5\ \text{k}\Omega + 1.2\ \text{k}\Omega} = 1.2\ \text{mA}$$

10-4

1. Removing R_3 and solving for the total current gives

$$I = \frac{V_A - V_B}{R_1 + R_2} = \frac{25\ \text{V} - 15\ \text{V}}{3.3\ \text{k}\Omega + 1.5\ \text{k}\Omega} = 2.08\ \text{mA}$$

Solving for V_{TH} by using V_A gives

$$V_{\text{TH}} = -V_{R1} + V_A$$

and

$$V_{R1} = I_1R_1 = 2.08\ \text{mA} \times 3.3\ \text{k}\Omega = 6.86\ \text{V}$$

$$V_{\text{TH}} = -6.86\ \text{V} + 25\ \text{V} = 18.1\ \text{V}$$

Solving for V_{TH} by using V_B gives

$$V_{\text{TH}} = +V_{R2} + V_B$$

and

$$V_{R2} = I_2R_2 = 2.08\ \text{mA} \times 1.5\ \text{k}\Omega = 3.1\ \text{V}$$

$$V_{\text{TH}} = 3.1\ \text{V} + 15\ \text{V} = 18.1\ \text{V}$$

$$R_{TH} = R_1 \parallel R_2 = \frac{3.3\text{ k}\Omega \times 1.5\text{ k}\Omega}{3.3\text{ k}\Omega + 1.5\text{ k}\Omega} = 1.03\text{ k}\Omega$$

The current through R_3 can now be found:

$$I_3 = \frac{V_{TH}}{R_{TH} + R_3} = \frac{18.1\text{ V}}{1.03\text{ k}\Omega + 1\text{ k}\Omega} = 8.9\text{ mA}$$

2. If $R_3 = 4.7\text{ k}\Omega$:

$$I_3 = \frac{V_{TH}}{R_{TH} + R_3} = \frac{18.1\text{ V}}{1.03\text{ k}\Omega + 4.7\text{ k}\Omega} = 3.16\text{ mA}$$

3. Voltage polarities must be labeled with respect to the greater emf source since the algebraic sum of the voltages will depend on the voltage drops in the direction of the current flow (i.e., from the highest potential to the lowest).

10-5 1. Disconnect R_3 and short its terminals A to B to solve for I_N:

$$I_N = \frac{V_S}{R_T} = \frac{24}{2\text{ k}\Omega} = 12\text{ mA}$$

Note that R_2 has been shorted out also. Disconnect the short and find R_N looking into terminals A to B. By shorting the source,

$$R_N = R_1 \parallel R_2 = 2\text{ k}\Omega \parallel 3\text{ k}\Omega = 1.2\text{ k}\Omega$$

Reconnect R_3 and find its current by using the current-divider formula.

$$I_3 = \frac{R_N}{R_N + R_2} \times I_N = \frac{1.2\text{ k}\Omega}{1.2\text{ k}\Omega + 1\text{ k}\Omega} \times 12\text{ mA} = 6.55\text{ mA}$$

2. The Norton equivalent circuit:

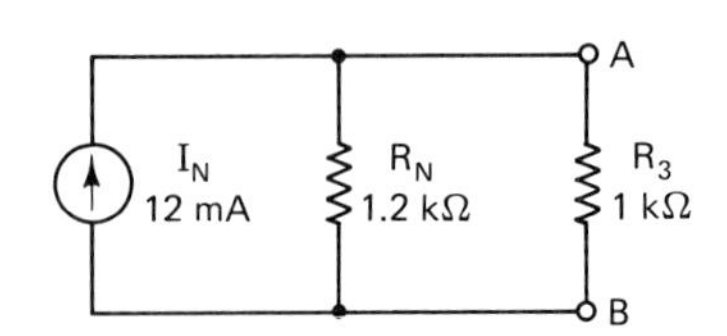

10-6 1. Applies strictly to parallel circuit.
2. See Section 10-7 for the steps to follow in applying Millman's theorem.
3. Converting all voltage sources to current sources in Fig. 10-30 and calculating the currents gives

$$I_1 = 26\text{ mA}$$

$$I_2 = \frac{V_2}{R_2} = \frac{15}{330\ \Omega} = 45.5\text{ mA}$$

$$I_3 = -22\text{ mA}$$

$$I_4 = \frac{V_4}{R_4} = \frac{18\text{ V}}{470\ \Omega} = 38.3\text{ mA}$$

Algebraically adding the current sources to find I_N gives

$$I_N = I_1 + I_2 - I_3 + I_4$$
$$= 26 \text{ mA} + 45.5 \text{ mA} - 22 \text{ mA} + 38.3 \text{ mA} = 87.8 \text{ mA}$$

The total parallel resistance of the sources gives R_N:

$$R_N = R_1 \parallel R_2 \parallel R_3 \parallel R_4$$
$$= 390\ \Omega \parallel 330\ \Omega \parallel 560\ \Omega \parallel 470\ \Omega = 105\ \Omega$$

Converting I_N and R_N into the Thévenin equivalent resistance and voltage gives

$$R_{\text{TH}} = R_N = 105\ \Omega$$
$$V_{\text{TH}} = I_N R_N = 87.8 \text{ mA} \times 0.105 \text{ k}\Omega = 9.2 \text{ V}$$

Calculating the load current and voltage gives

$$I_L = \frac{V_{\text{TH}}}{R_{\text{TH}} + R_L} = \frac{9.2 \text{ V}}{105\ \Omega + 820\ \Omega} = 9.95 \cong 10 \text{ mA}$$
$$V_L = I_L R_L = 10 \text{ mA} \times 0.82 \text{ k}\Omega \cong 8.2 \text{ V}$$

11

Direct-Current Meters

INTRODUCTION

To this point your concern was to be able to solve and measure dc currents and voltages about a circuit. In this chapter we investigate the circuit concepts of a d'Arsonval meter movement used as a multifunction meter or volt-ohm-milliammeter (VOM) to measure these currents and voltages.

The first form of a "moving-needle meter" was discovered by Hans C. Oersted in 1819. Oersted noted that a pivoted magnetic needle placed parallel to a non-current-carrying conductor would be deflected at a right angle to the same current-carrying conductor. The needle would turn in the opposite direction if the current was reversed.

Oersted's discovery initiated the development of the galvanometer by J. S. C. Schweigger in 1820. The galvanometer was progressively improved by such scientists as Lord Kelvin in 1858, C. V. Boys in 1890, and others. It was Arséne d'Arsonval in 1882 who introduced the first deflecting moving-coil galvanometer. Around 1884, Edward Weston greatly improved the d'Arsonval moving-coil galvanometer by adding pivoted sapphire bearings and spiral bronze torque springs as well as improving the permanent magnetic system of the meter. To this day most dc and ac analog meters use the basic permanent-magnet moving-coil or d'Arsonval galvanometer.

MAIN TOPICS IN CHAPTER 11

OBJECTIVES

After studying Chapter 11, the student should be able to

1. Master the basic concepts of a d'Arsonval meter movement and measure the internal resistance of any meter movement.
2. Apply Ohm's law to solve current shunts, voltmeter multipliers, and ohmmeter resistance values used in multiranges of a VOM.

3. Master the concept of meter accuracy and meter loading effect on the circuit being measured.
4. Have an understanding of the fundamental specifications of an analog dc VOM.

11-1 D'ARSONVAL MOVING-COIL GALVANOMETER

The moving-coil mechanism shown in Fig. 11-1 operates on the principle that a current-carrying conductor will move at a right angle to the magnetic flux lines set up by the permanent magnet. The moving coil has a number of turns of enamel insulated wire wound on a former, as shown in Fig. 11-2. The former is pivoted at its center on precision jeweled bearings. Current flows into the moving coil by means of leads soldered to two torque opposing springs, and from there leads are brought out to the meter terminals.

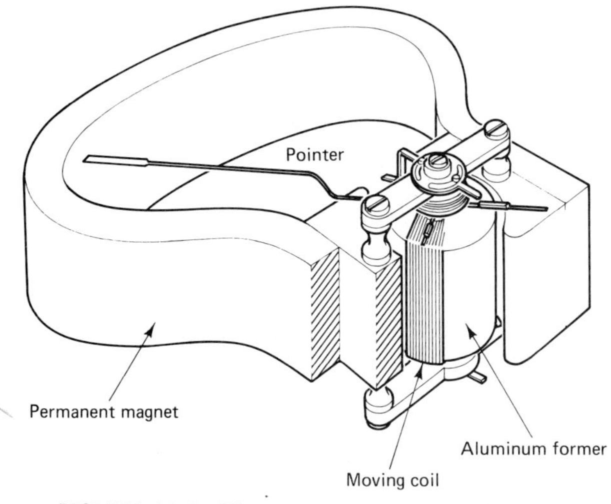

FIGURE 11-1 The permanent-magnet moving-coil mechanism. (Courtesy of Weston Instruments, Newark, NJ.)

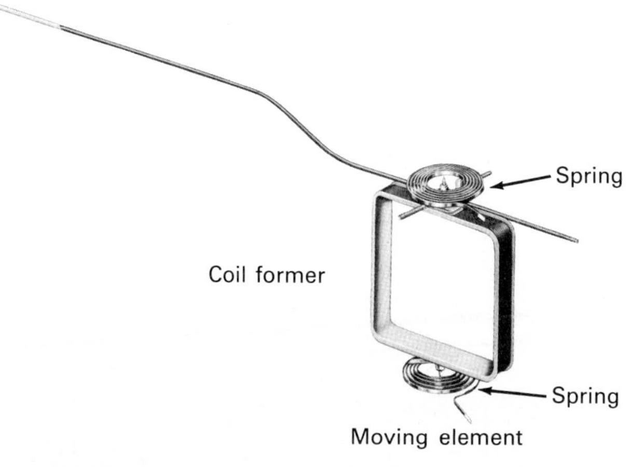

FIGURE 11-2 Moving coil. (Courtesy of Weston Instruments, Newark, NJ.)

Since the magnetic field is constant, the torque produced by the current-carrying coil and opposed by the springs will be proportional to the current flowing through the coil. The pointer mounted mechanically to the moving coil is deflected and its position along a calibrated scale is a measure of the current flow.

The aluminum former, shown in Fig. 11-1, is an enclosed cylinder mounted within the moving coil. Its purpose is to "damp out" the meter needle oscillations as the meter needle swings to indicate the measured value. Current flowing through the moving coil sets up a magnetic field about the coil. This magnetic field cuts through the aluminum former. As the moving coil swings to indicate the measured value, the magnetic field of the moving coil induces minute electrical currents in the aluminum former. These minute currents will, in turn, cause a weak opposing magnetic flux that prevents the sensitive moving coil and pointer from swinging back and forth at the current value indicated.

The core magnet mechanism shown in Fig. 11-3 is a great improvement on the U-shaped permanent magnet. It has the very important advantages of self-shielding against external magnetic fields, and light weight.

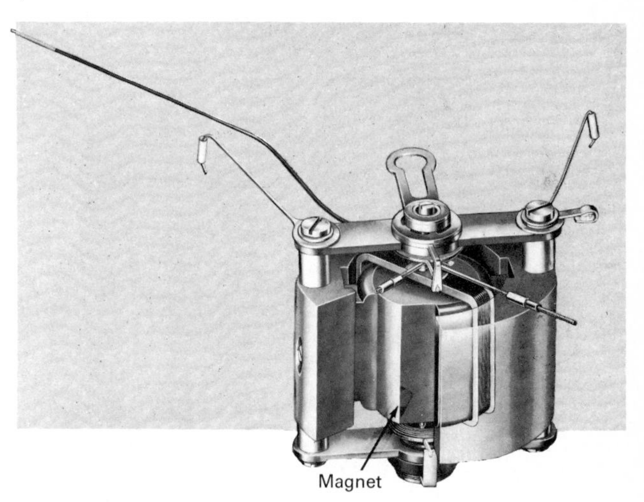

FIGURE 11-3 Core magnet mechanism. (Courtesy of Weston Instruments, Newark, NJ.)

11-2 FINDING THE INTERNAL RESISTANCE OF A METER MOVEMENT

The internal resistance of a moving-coil meter movement is the net resistance of the fine wire wound on the former. This resistance must be known in order to accurately calculate meter shunt resistances and low-range multiplier resistances. Normally, the manufacturer supplies this information with the meter. Should the information be unavailable, its internal resistance can only be measured using one of two methods. (*Note:* Using an ohmmeter will damage the meter movement beyond repair.) Only one method will be discussed here; you will be asked to devise the second method in Practice Questions 11-1.

Current-Division Method

The general steps to measure the internal resistance of a meter movement are as follows:

1. With the meter movement connected as shown in Fig. 11-4(a), the variable dc source is carefully adjusted for a full-scale (FSD) reading on the meter. Note that the limiting resisting resistor (R_{limit}) limits the current to the full-scale value of the meter movement.

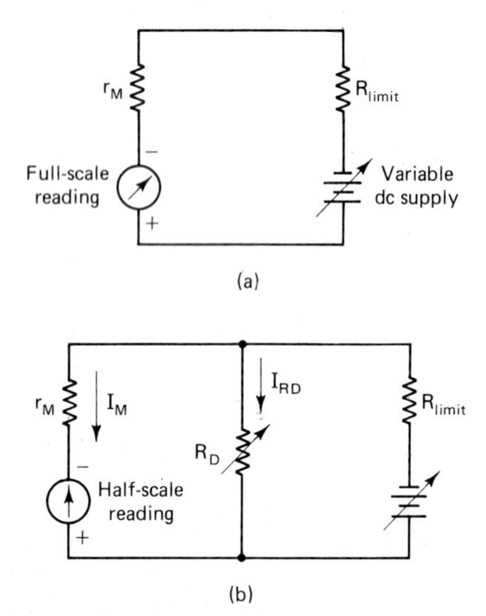

FIGURE 11-4 Measuring the internal resistance of a meter movement.

2. A decade resistance box or potentiometer, connected in parallel with the meter as shown in the schematic of Fig. 11-4(b), is adjusted so that the meter now reads half-scale.
3. The resistance read off from the decade box or measured by means of an ohmmeter from the potentiometer will be equal to the internal resistance of the meter movement. This is shown by the current-division formula, where $I_M = I_{R_D}$, or one-half of full-scale current since $R_D = r_M$.

11-3 MULTIRANGE MILLIAMMETER

Any basic d'Arsonval meter movement will be able to measure current up to its full-scale deflection capability. As an example, a 1-mA FSD movement will be capable of measuring current up to 1 mA. Having used a milliammeter, you realize that this would limit the use of this meter, and a means to extend the range was provided through the "range switch."

Extending the Range of the Milliammeter

The range of any basic meter movement can be extended by connecting a parallel or shunt resistance across the meter itself, as shown in Fig. 11-5. Since the shunt resistance is in parallel with the meter's internal resistance, their voltages will be common, and the current through each will be directly proportional to this parallel voltage. The shunt will then bypass a proportion of the circuit current governed by the ratio of r_M to R_S or by the current-division formula:

$$I_S = \frac{r_M}{r_M + R_S} \times I_T$$

From this formula we can derive R_S:

$$R_S = \left(\frac{r_M}{I_S} \times I_T\right) - r_M \quad \text{ohms} \tag{11-1}$$

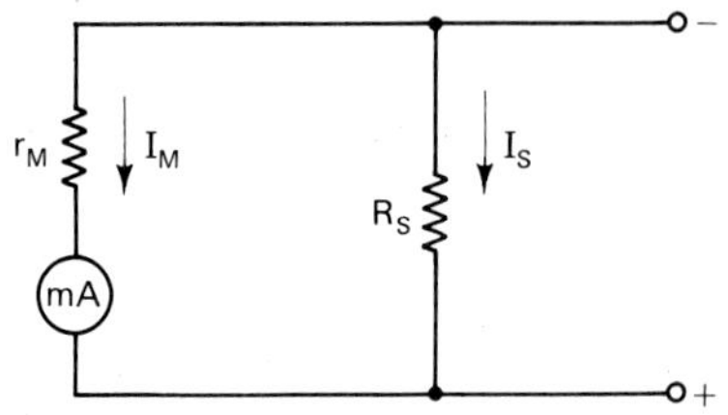

FIGURE 11-5 Milliammeter shunt resistance.

EXAMPLE 11-1

Given a meter movement of 0–1 mA with an internal resistance of 100 Ω (Fig. 11-6), calculate the shunt resistance required to extend its range to 10 mA.

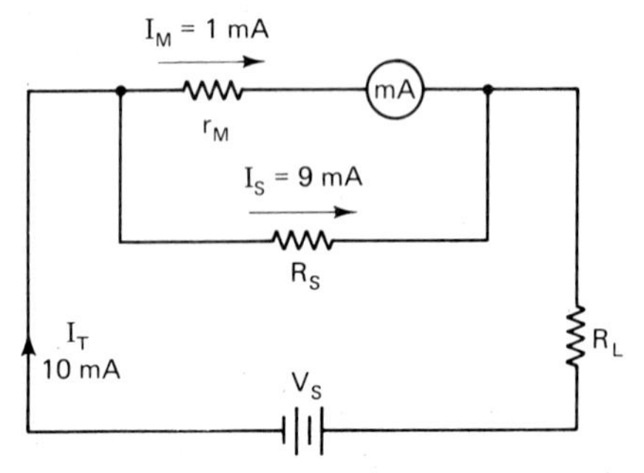

FIGURE 11-6 Milliammeter shunt current.

Solution: The full-scale deflection current of the meter movement is 1 mA and the total current or the current to be measured is 10 mA; therefore, the shunted current will be

$$I_S = I_T - I_M = 10 \text{ mA} - 1 \text{ mA} = 9 \text{ mA}$$

Use formula (11-1) to find R_S:

$$R_S = \left(\frac{r_M}{I_S} \times I_T\right) - r_M = \left(\frac{100\ \Omega}{9 \text{ mA}} \times 10 \text{ mA}\right) - 100\ \Omega = 11.11\ \Omega$$

The shunt resistance can also be calculated using Ohm's law:

$$R_S = \frac{V_S}{I_S}$$

Since the shunt resistor is in parallel with the meter movement, the voltage across both can easily be found:

$$V_S = V_M = I_M r_M$$

so that

$$R_S = \frac{V_M}{I_S} \quad \text{ohms} \qquad (11\text{-}2)$$

For the 0–1 mA meter movement where $r_M = 100\ \Omega$, $I_M = 1$ mA, and $I_S = 9$ mA; therefore,

$$R_S = \frac{V_M}{I_S} = \frac{1 \text{ mA} \times 100\ \Omega}{10 \text{ mA} - 1 \text{ mA}} = 11.11\ \Omega$$

EXAMPLE 11-2

Calculate the necessary shunt resistances to extend the range of the meter movement given in Example 11-1 to 100 mA.

Solution: For the 100-mA range using formula (11-1) gives

$$R_S = \left(\frac{r_M}{I_S} \times I_T\right) - r_M$$

$$= \left(\frac{100\ \Omega}{100 \text{ mA} - 1 \text{ mA}} \times 100 \text{ mA}\right) - 100\ \Omega$$

$$= 1.01\ \Omega$$

For the 100-mA range, using formula (11-2) gives

$$R_S = \frac{V_M}{I_S} = \frac{1 \text{ mA} \times 100\ \Omega}{100 \text{ mA} - 1 \text{ mA}} = 1.01\ \Omega$$

Shunt resistors are made from special temperature-compensated alloy wire, and are cut to length. The length will be governed by the ohms/length as well as the required calculated value for R_S.

11-4 SWITCHES AND SWITCHING THE CURRENT RANGES OF A MULTIRANGE MILLIAMMETER

For a multifunction meter it is necessary that the ranges be switched easily using one range switch. The many types of switches available necessitate that a look at these be made. The simplest and most commonly used is the single-pole single-throw (SPST) switch, for which the schematic symbol is shown in Fig. 11-7(a). Note that the "pole" in Fig. 11-7(b) is the pivot contact and the "position" is the make contact: thus "single-pole, single-throw."

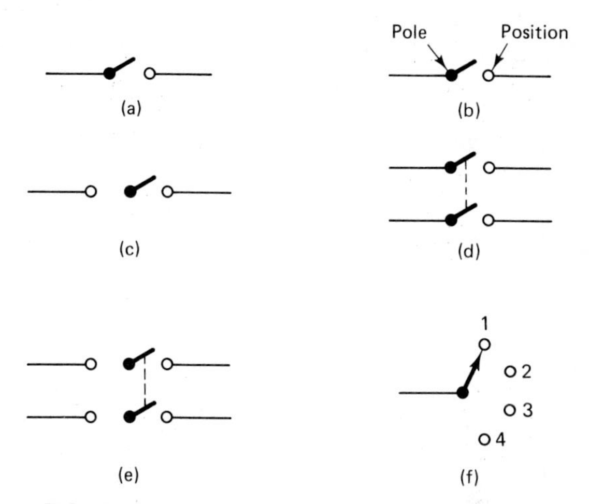

FIGURE 11-7 Various configurations of switches.
(a) Single-pole, single-throw, SPST.
(b) Pole and position designation.
(c) Single-pole, double-throw, SPDT.
(d) Double-pole, single-throw, DPST.
(e) Double-pole, double-throw, DPDT.
(f) Single-pole, 4 position, nonshorting rotary switch

Other variations of this type of switch are shown in Fig. 11-7(c) to (f). These switches come in various physical forms, such as toggle lever, pushbutton, rocker, and rotary. The dashed lines in the schematic symbol of Fig. 11-7(d) and (e) indicate that the two sets of contacts operate simultaneously but are insulated from each other. One application for a double-pole, double throw switch is shown in Fig. 11-8.

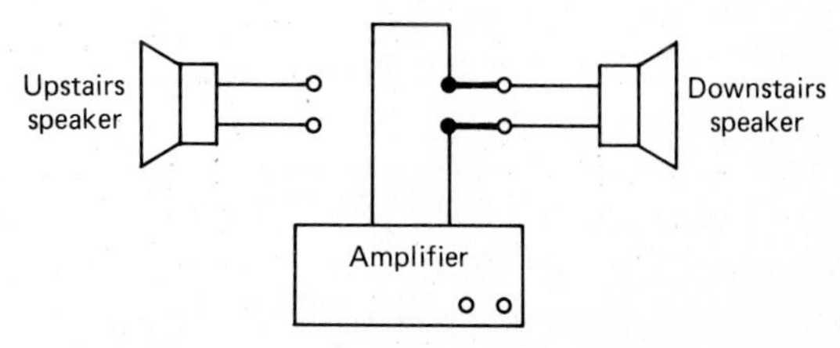

FIGURE 11-8 Application of a DPDT switch.

For the milliammeter, a rotary switch is needed to change the current ranges. With a standard type of rotary switch the contact is first opened, then closed, as the switch is rotated through its positions. This creates a brief instant of time where there is no shunt resistance connected across the meter, with resultant overloading and possible damage to the meter movement. There are two methods of switching arrangements for the shunts, with each method using a particular type of rotary switch. The first arrangement is shown in Fig. 11-9 and uses a make-before-break, nonshorting type of rotary switch.

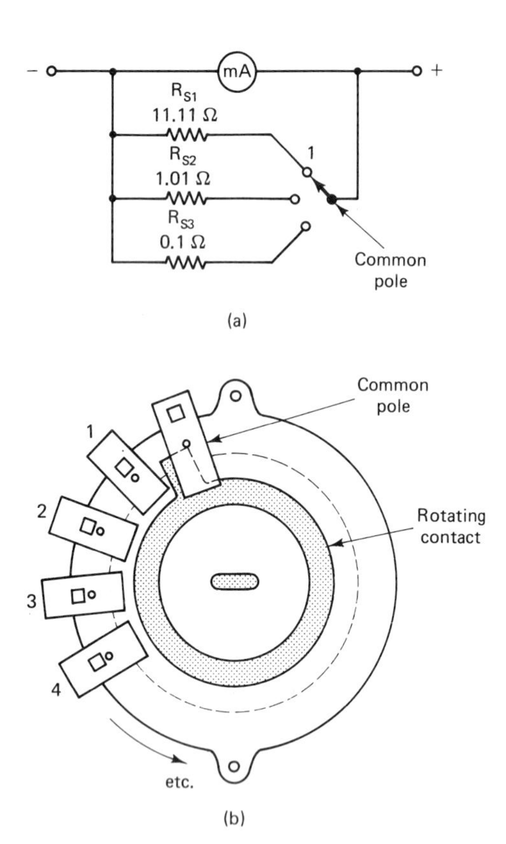

FIGURE 11-9 (a) Nonshorting make-before-break rotary, 4 position switch. (b) Make-before-break type of rotary switch wafer.

With the make-before-break type of rotary switch, the previous shunt resistance is still in the circuit as the next shunt resistance is engaged. This provides two parallel shunts in the circuit during the brief interval between switching ranges and prevents excessive currents.

The second switching arrangement is shown in Fig. 11-10 and uses a shorting-type rotary switch. This shunt resistance arrangement is called an Arytron or universal shunt. Initially, all three shunt resistors are connected across the meter, with the rotary switch contact in position 1. Note that the sum of the shunt resistances in series is equal to the required value for the 10 mA

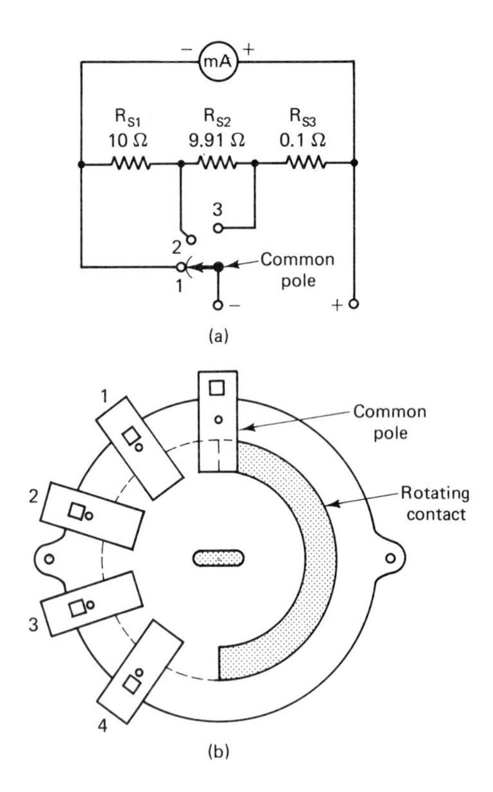

FIGURE 11-10 (a) Ayrton or universal shunt switching arrangement. (b) Shorting type of rotary switch wafer.

range, or 11.11 Ω [see Fig. 11-10(a)]. Rotating the switch to position 2 shorts out the 10-Ω shunt resistor and leaves the other two resistors connected across the meter, or 1.01 Ω, as required for the 100-mA range. Finally, in position 3 only the 0.1-Ω resistance is left across the meter, as required for the 1-A range.

PRACTICE QUESTIONS 11-1

Where applicable, choose the most suitable word or phrase to complete each statement.

1. Two principal components, ________ and ________, make up the d'Arsonval galvanometer.
2. The pivoted moving coil is deflected at a right angle to the ________ when a current flows through the coil.
3. The torque produced by the current-carrying moving coil is opposed by the ________.
4. The torque produced by the current-carrying moving coil will be proportioned to the ________ flowing through the coil.
5. The aluminum former on which the coil is wound serves to ________ the meter needle oscillations, due to its movement.

6. The ________ ________ induced in the aluminum former produce an opposing magnetic flux that prevents the meter needle from oscillating.
7. The internal resistance of d'Arsonval meter movement is due to the resistance of the ________ wound on the moving coil.
8. List the necessary steps and then draw a schematic diagram to show another method of measuring the internal resistance of a meter movement. (*Hint:* Use voltage division.)
9. Given a basic meter movement of 0 to 50 μA with an internal resistance of 2000 Ω, calculate the required shunt resistances for the following ranges.
   (a) 1 mA
   (b) 5 mA
   (c) 500 mA
10. Draw a schematic diagram to show a switching arrangement using a make-before-break, nonshorting, single-pole rotary switch for the shunt resistors and meter in Question 9.

11-5 MILLIAMMETER LOADING AND FSD ACCURACY

A number of errors derive from the operator who is making the readings and from some very small inconsistencies in manufactured meters. In addition to these "minimal" errors are two other types of errors that must be considered in order to minimize the overall difference between the actual or theoretical current value and the value indicated by the meter.

The first type of error is the loading effect of the meter. Consider the ideal circuit of Fig. 11-11(a). The ideal milliammeter has zero resistance; therefore, it will indicate the actual current, or

$$I = \frac{V}{R} = \frac{10\text{ V}}{1\text{ k}\Omega} = 10\text{ mA}$$

With a basic meter movement of 0–1 mA and 100 Ω of internal resistance, the required shunt resistance was calculated to be 11.11 Ω (see Example 11-1). The current being measured in the same circuit is now

$$I = \frac{V}{R} = \frac{10\text{ V}}{1000\ \Omega + 10\ \Omega} = 9.9\text{ mA}$$

Note: $r_M \parallel R_S = 100\ \Omega \parallel (11.11\ \Omega) \cong 10\ \Omega$

For a 10-mA reading the percentage error is

$$\%\text{ error} = \frac{10\text{ mA} - 9.9\text{ mA}}{10\text{ mA}} \times 100 = 1\%$$

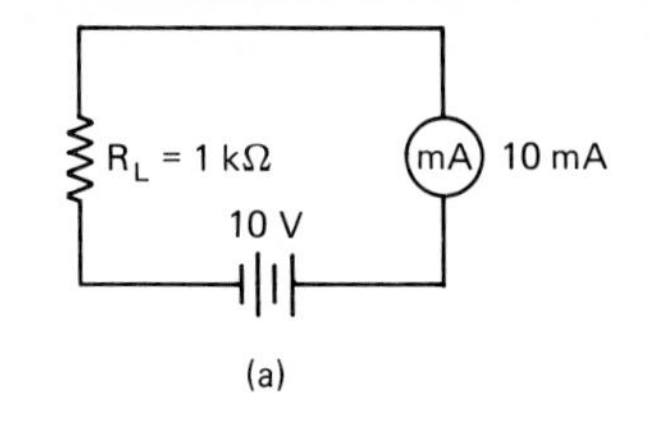

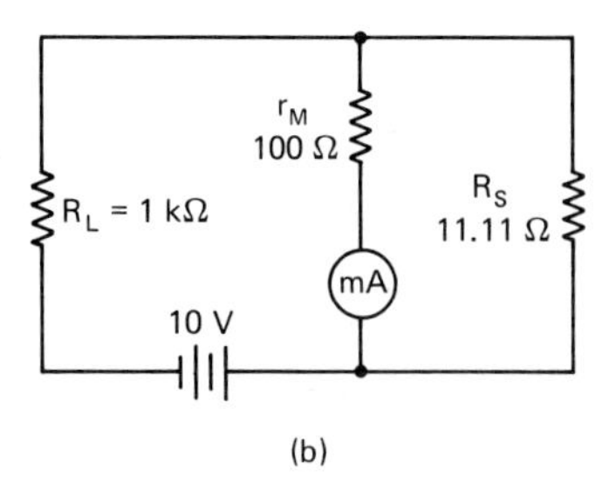

FIGURE 11-11 (a) Ideal milliammeter. (b) Shunted milliammeter.

With the milliammeter switched to the 100-mA range, the shunt resistance is now 1.01 Ω. For the same circuit of Fig. 11-11(b), the current indicated by the meter would be

$$I = \frac{V}{R} = \frac{10\text{ V}}{1000\ \Omega + 1\ \Omega} = 10\text{ mA}$$

Switching the milliammeter to a higher range decreases the loading effect. However, the accuracy of the reading along any part of the scale is subject to error due to "built-in" factors, which are:

1. Shunt resistances close but not the exact value
2. Changes in resistance and operating characteristics of the meter movement due to temperature
3. Other small discrepancies in calibration due to mass-production techniques

The overall error introduced by the built-in factors is relatively constant along the scale and is specified by the manufacturer in terms of percentage of full scale (FS). This means that a meter having specifications of ±3% FS for the milliammeter ranges would have an accuracy at any reading along the scale that would not exceed ±3% of the full-scale value for any range. An example to illustrate this follows.

EXAMPLE 11-3

Given a commercial meter with current range specifications of ±3% FS that is to measure 10 mA, find the actual current range of values and the percentage error for the actual current on (a) the 10 mA-range and (b) the 100-mA range.

Solution:

(a) For the 10-mA range the full-scale current error will be

$$\text{current error} = \frac{\pm 3\%}{100\%} \times 10 \text{ mA} = \pm 0.3 \text{ mA}$$

Since the meter is measuring 10 mA, the actual current indicated will be

$$\text{actual current range of value} = 10 \pm 0.3 \text{ mA}$$
$$= 9.7 \text{ to } 10.3 \text{ mA}$$

$$\% \text{ error} = \frac{10 \text{ mA} - 9.7 \text{ mA}}{10 \text{ mA}} \times 100 = 3\%$$

(b) On the 100-mA range the full-scale current error will be

$$\text{current error} = \frac{\pm 3\%}{100\%} \times 100 \text{ mA} = \pm 3 \text{ mA}$$

Since the meter is measuring 10 mA, the actual current indicated will be

$$\text{actual current range of value} = 10 \text{ mA} \pm 3 \text{ mA}$$
$$= 7 \text{ to } 13 \text{ mA}$$

$$\% \text{ error} = \frac{10 \text{ mA} - 7 \text{ mA}}{10 \text{ mA}} \times 100 = 30\%$$

Conclusion: Select a current range to give as close to full-scale deflection as possible in order to minimize both loading and full-scale error effect.

Readings below half-scale will normally introduce errors beyond the component tolerances that must be considered in any analysis.

11-6 THE MULTIRANGE VOLTMETER

A voltmeter is always connected in parallel or in shunt with a source, device, or component, as you are well aware by now. A voltmeter requires various ranges not only to extend its range but also to minimize loading effect and full-scale percentage error effect, the exception being an electronic and/or digital voltmeter.

The maximum voltage that a basic d'Arsonval meter movement would be able to measure is determined by its full-scale current and internal resistance. For the meter movement of Example 11-1 and by Ohm's law,

$$V = I_M r_M = 1 \text{ mA} \times 100 \ \Omega = 0.1 \text{ V}$$

Of course, this 0–0.1-V range is of very little practical use, so we must extend this into two or more ranges.

Extending the Range of the Voltmeter

Since a voltmeter is always connected across the component, voltage source, or a voltage difference, extending the range of the basic meter movement simply involves limiting the current to FSD and then calibrating the scale in volts. The current-limiting resistor, properly termed a multiplier resistor (R_M), is connected in series with the meter movement, as shown in Fig. 11-12.

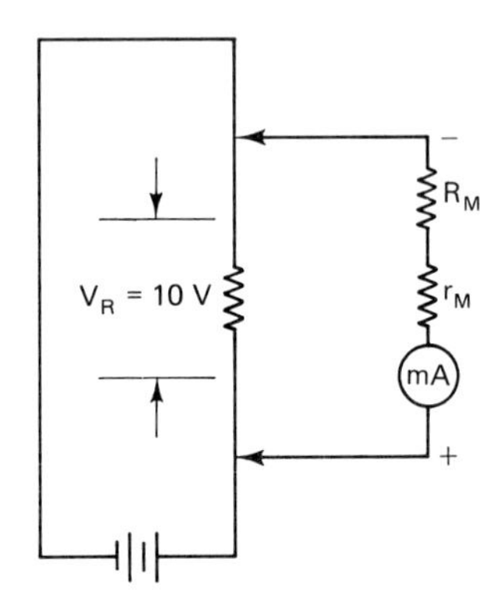

FIGURE 11-12 Voltmeter multiplier resistance is series with meter movement.

Calculating the value of the multiplier resistance is an application of Ohm's law to a series circuit (see Fig. 11-12). With 10 V applied to the series meter, the multiplier resistance is

$$R_T = \frac{\text{voltmeter (full scale)}}{\text{current (full scale)}} = \frac{V_{FS}}{I_{FS}}$$

but

$$R_T = R_M + r_M$$

and

$$R_M = R_T - r_M$$

Therefore,

$$R_M = \frac{V_{FS}}{I_{FS}} - r_M \quad \text{ohms} \qquad (11\text{-}3)$$

EXAMPLE 11-4

Given the basic meter movement of 0–1 mA with an internal resistance of 100 Ω, calculate the required multiplier resistance for the following voltage ranges: (a) 1 V, (b) 10 V, (c) 100 V, and (d) 300 V.

Solution: (a) For the 1-V range:

$$R_M = \frac{V_{FS}}{I_{FS}} - r_M = \frac{1\text{ V}}{1\text{ mA}} - 100\ \Omega = 900\ \Omega$$

(b) For the 10-V range:

$$R_M = \frac{10\text{ V}}{1\text{ mA}} - 100\ \Omega = 9900\ \Omega \quad \text{or} \quad 9.9\text{ k}\Omega$$

(c) For the 100-V range:

$$R_M = \frac{100\text{ V}}{1\text{ mA}} - 100\ \Omega = 99.9\text{ k}\Omega\text{, or } 100\text{ k}\Omega$$

(d) For the 300-V range:

$$R_M = \frac{300\text{ V}}{1\text{ mA}} - 100\ \Omega = 299.9\text{ k}\Omega\text{, or } 300\text{ k}\Omega$$

Note that in the case of the 100-V and 300-V ranges, the internal resistance of the meter is insignificant compared to the multiplier resistance value, so it can be ignored.

Switching the Voltage Ranges of a Multirange Voltmeter

Having looked at rotary switches in Section 11-4 and considering that the multiplier resistor must be series connected with the meter movement, a single-pole, four-position, nonshorting rotary switch can be selected to switch the multiplier resistors for the voltage ranges of Example 11-4. The circuit schematic is shown in Fig. 11-13.

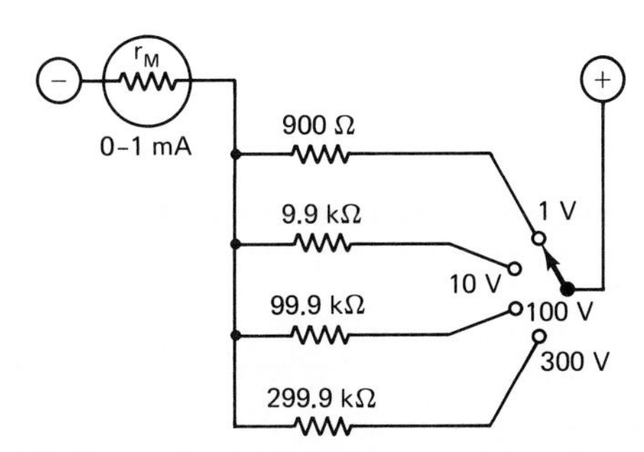

FIGURE 11-13 Switching ranges of a voltmeter using a single-pole, four-position, nonshorting rotary switch.

11-7 VOLTMETER LOADING AND FSD ACCURACY

A voltmeter is always connected across a device or resistance in order to measure the voltage difference; therefore, its multiplier resistance is always in parallel with the device or resistance, as shown in Fig. 11-14.

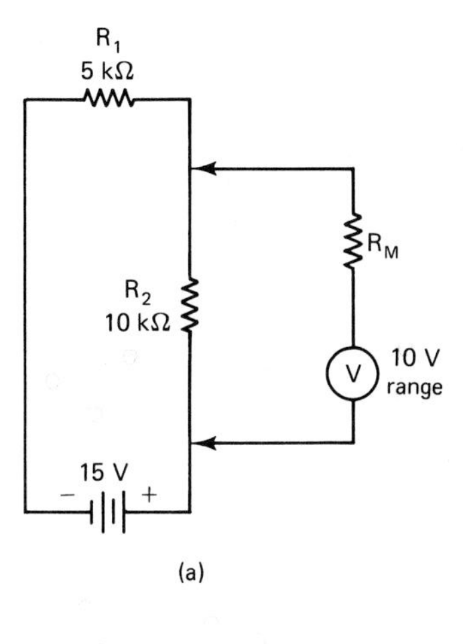

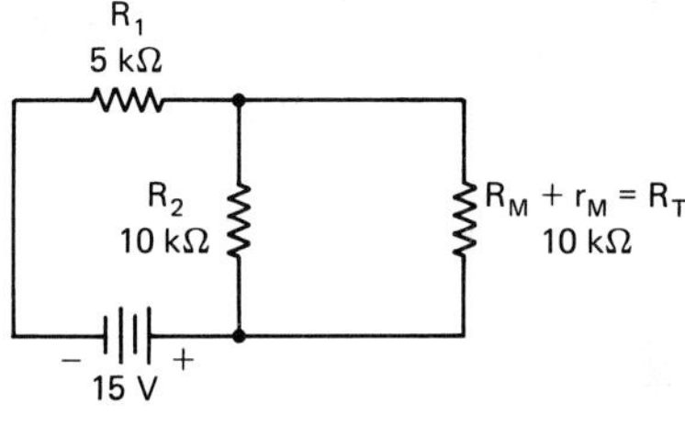

FIGURE 11-14 (a) R_M is in parallel with R_2. (b) Shunting effect of voltmeter on R_L.

The shunting effect of the voltmeter is called voltmeter loading since the voltmeter draws its own current from the circuit being measured. An ideal voltmeter with an infinite multiplier resistance would act like an open circuit and would not draw any current from the device or circuit being measured. The extent of the loading effect will be determined by the voltmeter sensitivity.

Voltmeter sensitivity (S) is the amount of series multiplier resistance required for each volt applied across the meter movement, or

$$S = \frac{\text{total resistance}}{\text{voltage range}} = \frac{R_T}{V_T} \quad \text{ohms/volt} \qquad (11\text{-}4)$$

From Example 11-4 the total resistance for the 1-V, 10-V, and 100-V ranges was 1 kΩ, 10 kΩ, and 100 kΩ, respectively. The sensitivity of this voltmeter can be shown to be 1 kΩ/V as follows:

$$\text{1-V range sensitivity} = \frac{R_T}{V_T} = \frac{1\text{ k}\Omega}{1\text{ V}} = 1\text{ k}\Omega/\text{V}$$

$$\text{10-V range sensitivity} = \frac{R_T}{V_T} = \frac{10\ \text{k}\Omega}{10\ \text{V}} = 1\ \text{k}\Omega/\text{V}$$

$$\text{100-V range sensitivity} = \frac{R_T}{V_T} = \frac{100\ \text{k}\Omega}{100\ \text{V}} = 1\ \text{k}\Omega/\text{V}$$

The sensitivity of a meter movement can also be calculated by applying Ohm's law to find the total series resistance required when 1 V is applied to give a full-scale meter current, or

$$S = R = \frac{V}{I} = \frac{1\ \text{V}}{I_{\text{FS}}} \quad \text{ohms/volt} \qquad (11\text{-}5)$$

For the 1-mA meter movement the voltmeter sensitivity is

$$S = \frac{1\ \text{V}}{1\ \text{mA}} = 1\ \text{k}\Omega/\text{V}$$

With a voltmeter sensitivity of 1 kΩ/V the total resistance of the meter shown in Fig. 11-14(a) is 10 kΩ on the 10-V range. The 10-kΩ meter resistance shunts the 10 kΩ of R_2 and the net resistance is 5 kΩ. This changes the current through the circuit and the voltage drop across R_2, which will now be

$$V_2 = \frac{R_2 \parallel R_T}{(R_2 \parallel R_2) + R_1} \times V_A = \frac{5\ \text{k}\Omega}{5\ \text{k}\Omega + 5\ \text{k}\Omega} \times 15\ \text{V}$$
$$= 7.5\ \text{V}$$

The error in the voltage reading across R_2 is 25% and is due to the loading effect of the voltmeter.

To minimize the loading effect caused by the voltmeter, we would need a meter with a greater sensitivity. Typical VOM meters have a sensitivity of 20 kΩ/V. Electronic voltmeters (FET input) and digital voltmeters have an input resistance that is constant regardless of the voltage range and is typically around 11 MΩ.

EXAMPLE 11-5

Given the circuit of Fig. 11-14(b), find the percentage error for the voltage being measured across R_2 using meters with the following sensitivity: (a) 20 kΩ/V and (b) 11 MΩ (electronic voltmeter).

Solution:

(a) For the meter with 20-kΩ/V sensitivity, the total resistance across R_2 on the 10-V range would be

$$R_T = 10\ \text{V} \times 20\ \text{k}\Omega = 200\ \text{k}\Omega$$

The voltage across R_2 is now

$$V_2 = \frac{R_T \parallel R_2}{(R_T \parallel R_2) + R_1} \times V_A$$
$$= \frac{200\ \text{k}\Omega \parallel 10\ \text{k}\Omega}{(200\ \text{k}\Omega \parallel 10\ \text{k}\Omega) + 5\ \text{k}\Omega} \times 15\ \text{V}$$
$$= 9.83\ \text{V}$$

$$\%\ \text{error} = \frac{10\ \text{V} - 9.83\ \text{V}}{10\ \text{V}} \times 100 = 1.7\%$$

which is a great improvement over the meter with the 1-kΩ/V sensitivity, whose loading effect caused a 25% error in the volts reading.

(b) For the electronic voltmeter with a constant 11-MΩ input resistance, the total resistance across R_2 will be 11 MΩ regardless of the voltage range selected. The voltage across R_2 is now

$$V_2 = \frac{R_T \parallel R_2}{(R_T \parallel R_2) + R_1} \times V_A$$
$$= \frac{11\ \text{M}\Omega \parallel 10\ \text{k}\Omega}{(11\ \text{M}\Omega \parallel 10\ \text{k}\Omega) + 5\ \text{k}\Omega} \times 15\ \text{V}$$
$$= 9.99\ \text{V}$$

$$\%\ \text{error} = \frac{10\ \text{V} - 9.99\ \text{V}}{10\ \text{V}} \times 100 = 0.1\%$$

The electronic voltmeter approaches the ideal voltmeter.

Full-Scale Deflection Accuracy of a Voltmeter

As for the case of the milliammeter, the overall error introduced by the built-in factors is relatively constant and is specified by the manufacturer in terms of percentage of full scale.

For a voltmeter specification of ±3% FS, any reading along the scale would not exceed ±3% of the full-scale value as shown for the milliammeter (see Section 11-3). Thus for a 10-V measurement on the 10-V range, the actual voltage being measured ranges from 9.7 V to 10.3 V. The same 10-V measurement on the 100-V range would give an actual voltage of ±3 V or would fall within the range 7 to 13 V.

Conclusion: Select a range to give as close to full-scale deflection as possible in order to minimize both loading and full-scale error effect. Readings below half-scale will normally introduce errors beyond the component tolerances that must be considered in any analysis.

PRACTICE QUESTIONS 11-2

Where applicable, choose the most suitable word, phrase, or value to complete each statement.

1. The maximum voltage that a 50-μA meter movement with a 2-kΩ internal resistance could measure is __________ volts.
2. To increase the voltage range of the basic meter movement, a resistance termed a __________ is connected in series with the meter movement.
3. The formula to find the resistance of a multiplier resistor is __________.
4. Given a basic meter movement of 50 μA with 2-kΩ internal resistance, calculate the required multiplier resistance for the voltage ranges 5 V, 50 V, and 150 V.
5. Draw a schematic diagram to show a switching arrangement for the voltage ranges in Question 4.
6. Shunting a voltmeter across a component or device produces a __________ effect.
7. The extent of the voltmeter loading effect will be determined by the voltmeter __________.
8. The formula for meter movement sensitivity is __________.
9. Calculate the meter sensitivity for a movement of 100 μA.
10. Given a commercial meter with voltage range specifications of $\pm3\%$ FS that is to measure 30 V, find the actual voltage range of values and the percentage error for the actual voltage on the 30-V and 300-V ranges.
11. Give two general rules to follow to minimize the loading effect and the full-scale error effect of a voltmeter.

11-8 SERIES OHMMETER

The ohmmeter is unique in that it uses its own internal battery source to function and this is one of the reasons that you must always disconnect the voltage source from the circuit before resistance measurements are made. In this section we show you why the internal battery is needed and how the ohmmeter functions.

The milliammeter movement, multiplier resistance (R_M), and the 1.5-V cell of Fig. 11-15(a) comprise the basic series ohmmeter circuit. With the leads shorted together, the full-scale meter current (1 mA for this meter movement) will be calibrated as the zero-ohms point. The multiplier resistance is calculated to give full-scale deflection with 1.5 V applied, in this case

$$R_T = \frac{V}{I_{FS}} = \frac{1.5\text{ V}}{1\text{ mA}} = 1.5\text{ k}\Omega$$

Therefore,

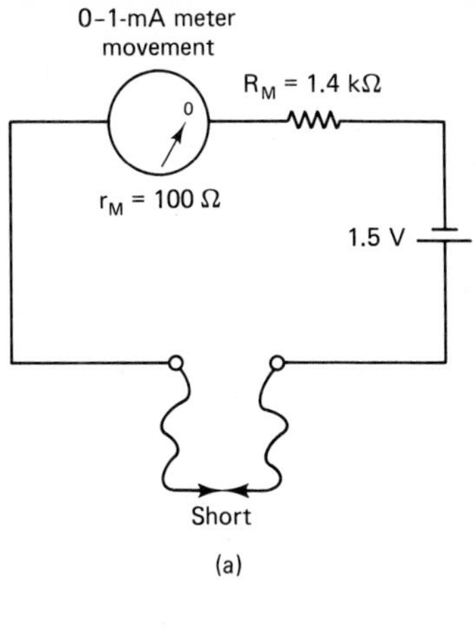

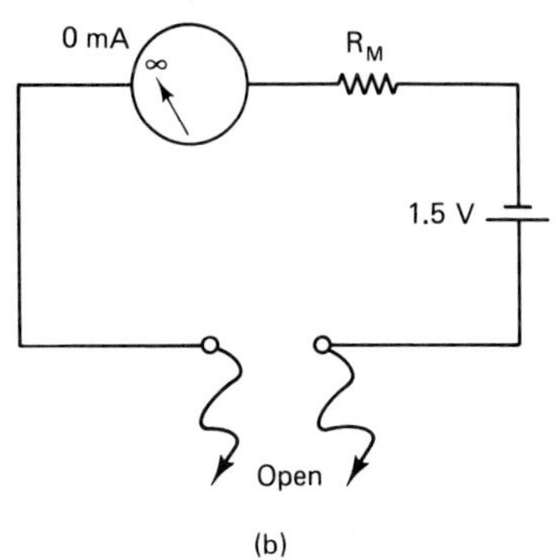

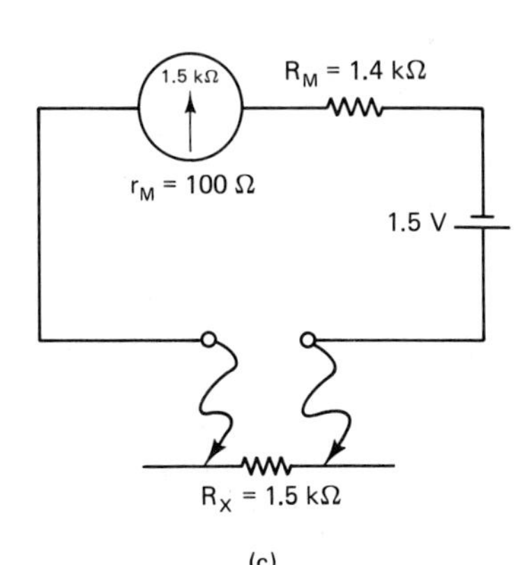

FIGURE 11-15 (a) Series ohmmeter: (a) With leads shorted together. (b) With meter leads open. (c) Half scale reading with R_X equal to 1.5 kΩ.

$$R_M = R_T - r_M = 1500\ \Omega - 100\ \Omega = 1.4\text{ k}\Omega$$

With the meter leads open [see Fig. 11-15(b)], the circuit is open and no current will flow. The meter needle will be at the zero current point, but the ohmmeter scale will be calibrated to infinite ohms (∞).

An unknown resistance that is to be measured is connected in series as shown in Fig. 11-15(c). The addition of the series resistance means that the meter current will be inversely proportional to the total resistance, since the battery voltage is constant. For the circuit of

Fig. 11-15(c), a 1.5-kΩ resistance reduces the current to half-scale (0.5 mA) since the total resistance of the circuit is now 3 kΩ. This point on the scale is calibrated to 1.5 kΩ.

In using the ohmmeter you probably noticed that the ohms scale is crowded at the high end and spread out at the low end of the scale. Two additional points on the scale can be calculated to illustrate this nonlinearity.

At the upper end of the scale the maximum readable resistance value is 15 kΩ or 10 times R_M, since the meter needle deflection at this point will be

$$I = \frac{V}{R_T} = \frac{1.5\text{ V}}{15\text{ k}\Omega + 1.5\text{ k}\Omega} = 0.09\text{ mA}$$

Any value of current lower than 0.09 mA would be barely discernible; therefore, calibrating the ohms scale beyond this point would be of no value (see Fig. 11-16). Similarly, at the lower end of the scale the minimum readable resistance value is 15 Ω or $\frac{1}{100}$ of R_M, since the meter needle deflection at this point will be

$$I = \frac{V}{R_T} = \frac{1.5\text{ V}}{1.5\text{ k}\Omega + 15\ \Omega} = 0.99\text{ mA}$$

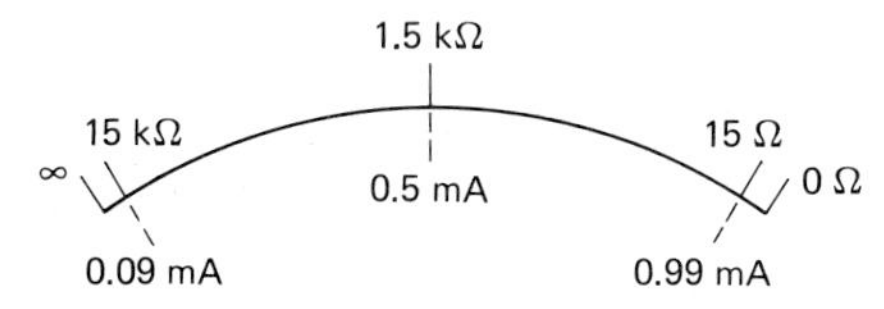

FIGURE 11-16 Series ohmmeter scale on a 0–1 mA movement.

The multiplier resistance in series with the measured resistance causes the nonlinearity along the ohmmeter scale. Additional ranges must be provided in order to take resistance readings as much as possible in the midrange or center of the scale, to ensure greater accuracy.

The zero-ohms-adjust potentiometer provides a means of adjusting the total multiplier resistance to compensate for cell voltage variations. Cell voltage changes will vary from 1.6 V when the cell is new to a lower value as the cell is used.

The series fixed multiplier resistance is calculated so that it limits the current to no more than 150% of full-scale current. For the 1-mA movement,

$$R_{M(\text{fixed})} = \frac{V_T}{I_{FS} \times 1.5} = \frac{1.5\text{ V}}{1\text{ mA} \times 1.5} = 1\text{ k}\Omega$$

Since the total resistance (see Fig. 11-17) is

$$R_M + R_{\text{zero adjust}} = 1400\ \Omega$$

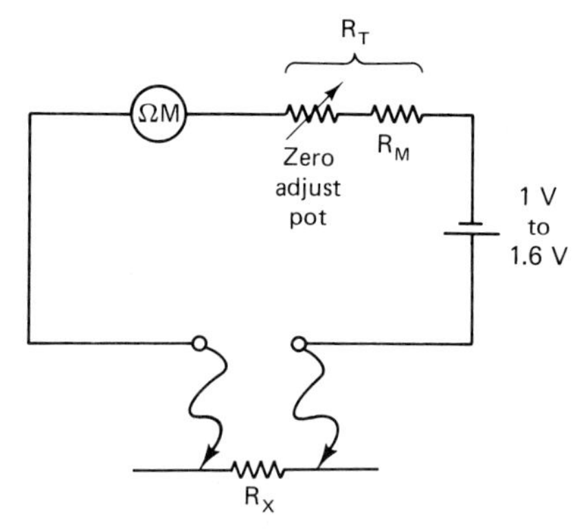

FIGURE 11-17 Series ohmmeter with zero ohms adjust pot.

then

$$R_{\text{zero adjust}} = 1400\ \Omega - 1000\ \Omega = 400\ \Omega$$

If the voltage is greater than 1.5 V, the zero-ohms adjust pot would need to be greater than 400 Ω. To provide for the voltage variation, a higher-value pot is chosen. This value will be the closest value to a commercially available pot value, in this case a 1-kΩ pot.

11-9 SHUNT OHMMETER

The advantage of the series ohmmeter was its ability to measure high resistance values. Measuring lower values than 15 Ω, however, would be a guess at the best. To obtain lower resistance ranges a shunt resistance is connected across the meter and the multiplier resistance (R_M), as shown in Fig. 11-18. The shunt resistance is always the calibrated midscale deflection value on the $R\times1$ scale. Multiples of this midscale value will give other ranges, $R\times10$, $R\times100$, $R\times1\text{k}$, and $R\times10\text{k}$. As an example, for the ohmmeter in the circuit of Fig. 11-18, it is 15 Ω.

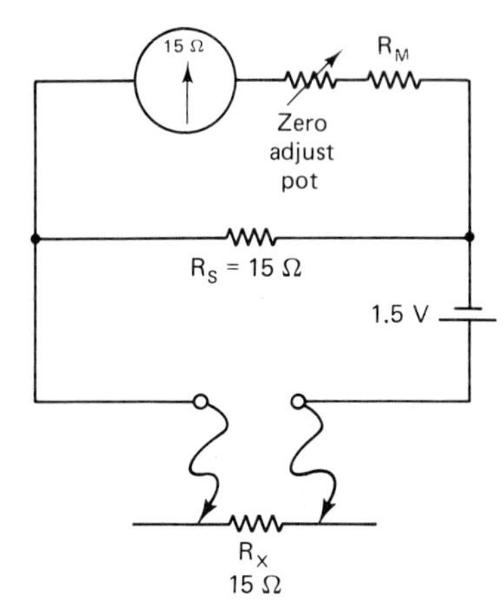

FIGURE 11-18 Shunt ohmmeter reading half scale when R_X is equal to R_S.

The voltage across the 15-Ω shunt resistor will be one-half of 1.5 V since R_S is in series with R_X and the 1.5-V dry cell, or by the voltage-division formula,

$$V_{R_S} = \frac{R_S}{R_S + R_X} \times 1.5\text{ V} = \frac{15\ \Omega}{15\ \Omega + 15\ \Omega} \times 1.5\text{ V}$$
$$= 0.75\text{ V}$$

The meter and its multiplier resistance are in parallel with R_S and therefore receive only 0.75 V, which reduces the meter current to half-scale. The ohms scale is then calibrated to 15 Ω at this midscale point.

Multirange Ohmmeter

To accommodate low and high resistance measurements, the ohmmeter has several ranges. A double-pole, five-position, nonshorting rotary switch provides a choice from typical ranges of $R\times1$, $R\times10$, $R\times100$, $R\times10$k, and $R\times1$M, as shown in Fig. 11-19.

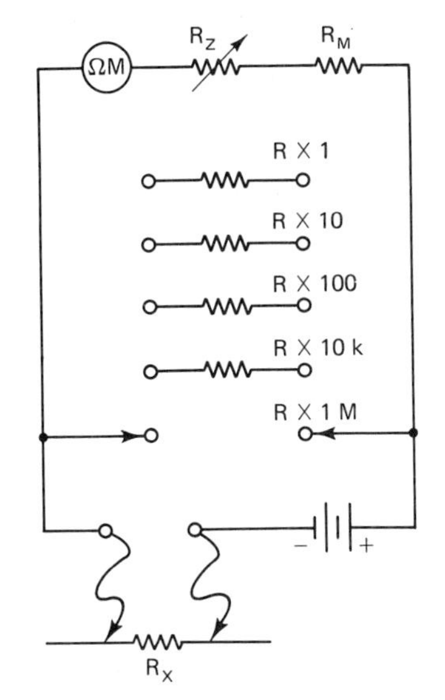

FIGURE 11-19 Multirange ohmmeter.

11-10 SPECIFICATIONS OF A VOM

The specifications of a VOM or digital multimeter will allow a purchaser to match requirements to a meter. Consider the following requirements.

Physical Size and Shape. The choice here is in the meter size, the size of the case, whether it is to be portable or bench type, and the layout of the meter functions.

Electrical Characteristics. Matching meter sensitivity to its ultimate application is most important. As an example, the minimum sensitivity for measuring electronic circuit parameters would be a meter with a sensitivity of 20 kΩ/V. An electronic or digital multimeter would be preferable since its sensitivity is constant on all ranges and is typically around 11 MΩ.

Another very important characteristic is the frequency response of the meter on the ac ranges. Again, the application of the meter will determine whether you need a frequency response of 60 Hz to 20 kHz, or radio-frequency circuits 100 kHz and higher. Overload protection is a necessity, and most of the meters on the market today provide this feature.

Multiranges and Percentage Accuracy

Voltage ranges: For most electronic circuit measurements a useful voltage range for ac or dc is provided by present electronic VOMs, mainly 200 mV, 2 V, 20 V, 200 V, and 1000 V. Full-scale accuracy ranges from ±3% for some types of VOMs to 0.03% + two digits for digital multimeters (DMMs). Refer to Fig. 11-20. DMMs have other special features, such as autozero, autopolarity, autoranging, display hold capability, and others.

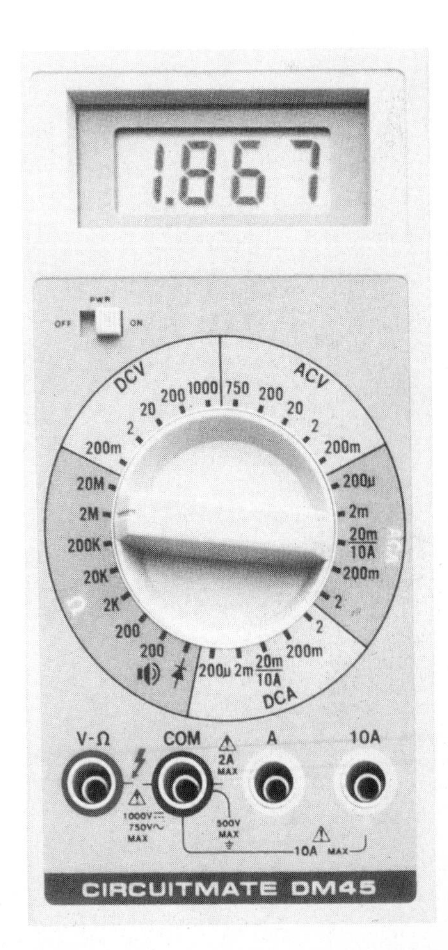

FIGURE 11-20 Digital multimeter (DMM).

Current ranges: The current ranges coincide with the voltage ranges. In the example above the current ranges would be 200 μA, 2 mA, 20 mA, 200 mA, and 2 A, on both dc and ac ranges. The full-scale accuracy would be equivalent to the voltage range accuracy.

Ohmmeter ranges: As in the case of the voltage and current ranges, the ohmmeter ranges will depend on the meter designed by a specific manufacturer. In this example the resistance ranges would be 200 Ω, 2 kΩ, 20 kΩ, 200 kΩ, 2000 kΩ, and 20 MΩ. State-of-the-art DMMs also have conductance ranges of 2 mS and 200 nS, as an example.

PRACTICE QUESTIONS 11-3

Choose the most suitable word, phrase, or value to complete each statement. *Note:* Use a 50-μA 2000-Ω meter movement for Questions 1 to 7.

1. With the meter movement above, calculate R_M for the series ohmmeter shown in Fig. 11-15(a).
2. To what resistance value will this series ohmmeter be calibrated at midscale?
3. What is the maximum readable resistance value on this ohmmeter scale?
4. What is the minimum readable resistance value on this scale?
5. Find the value of the zero-ohms-adjust potentiometer.
6. Given that $R_X = 30\ \Omega$ in Fig. 11-18, find the value of the shunt resistance. (*Note:* Use the 50-μA meter movement.)
7. The minimum sensitivity for measuring electronic circuit parameters would be a meter with a sensitivity of ________ per volt.
8. For ordinary line-voltage measurements, the frequency response of an ac voltmeter need not be greater than ________ hertz.
9. For electronic circuit voltage measurement the frequency response of VOM needs to be at least ________ kilohertz.

SUMMARY

Basic meter movement

1. The basic permanent-magnet moving-coil meter is termed a d'Arsonval galvanometer.
2. The principle of the moving coil is that a current-carrying conductor will move at a right angle to the magnetic flux lines.
3. An aluminum former helps to damp out the meter needle oscillations.
4. Springs provide an opposing torque to the moving coil.
5. General steps to follow to measure the internal resistance of a meter are:
   (a) Connect the meter movement in series with a dc source and a current-limiting resistance. Carefully adjust the source to obtain FSD.
   (b) Connect a decade resistance box in parallel with the meter movement and adjust its resistance to half-scale reading on the meter.
   (c) At half-scale reading, $R_D = r_M$.

The milliammeter: A shunt resistance extends the range of a milliammeter and is connected across the meter movement.

$$R_S = \left(\frac{r_M}{I_S} \times I_T\right) - r_M \quad \text{ohms} \qquad (11\text{-}1)$$

Also

$$R_S = \frac{V_M}{I_S} \quad \text{ohms} \qquad (11\text{-}2)$$

Figure 11-7(a) shows a single-pole, single-throw switch.
Figure 11-7(c) shows a single-pole, double-throw switch.
Figure 11-7(d) shows a double-pole, single-throw switch.
Figure 11-7(e) shows a double-pole, double-throw switch.

Figure 11-7(f) shows a single-pole, four-position, nonshorting rotary switch.

Figure 11-9(b) shows a make-before-break type of rotary switch.

Figure 11-10(b) shows a shorting type of rotary switch.

The milliammeter's shunt resistance will produce an extra series resistance in the circuit and an error in the current being measured which is termed "loading effect."

Overall "built-in" errors are stated by the manufacturer as a percentage of full scale.

A full-scale deflection percentage error of ±3% means that any reading along the scale will not be less than 3% of FS value or greater than 3% of FS value.

When measuring current: Always select a current range to give as close to FSD as possible.

A basic meter movement can be made to measure voltages by connecting a multiplier resistance (R_M) in series with the meter movement.

$$R_M = \frac{V_{FS}}{I_{FS}} - r_M \qquad \text{ohms} \tag{11-3}$$

The voltmeter: When measuring voltage: Always select a voltage range to give as close to FSD as possible.

Voltmeter sensitivity (S) is the amount of series multiplier resistance required for each volt applied across the meter movement.

$$S = \frac{\text{total resistance}}{\text{voltage range}} \qquad \text{ohms/volt} \tag{11-4}$$

or

$$S = \frac{1\text{ V}}{I_{FS}} \qquad \text{ohms/volt} \tag{11-5}$$

The multiplier resistance on any voltage range of a voltmeter will be in parallel with the resistance or device being measured, which will produce an error in the measurement. This is termed "loading effect." To minimize loading effect, use a voltmeter with greater sensitivity. For electronic circuit measurements a meter sensitivity of at least 20 kΩ/V would be desirable.

The series ohmmeter: A meter movement connected in series with a multiplier resistance and a 1.5-V cell will be able to measure resistance values along its calibrated scale.

The series ohmmeter scale is very crowded at the high-resistance end of the scale, while it is expanded at the lower end of the scale.

To obtain resistance readings with the least error, take readings along midrange, if possible.

To provide more than one resistance range, a shunt resistance across the series ohmmeter is used. The value of the shunt resistance is always the midscale ohms value of that range.

Specifications of a VOM: When choosing a VOM, some important specification considerations are as follows:

1. Physical characteristics such as meter size, portable versus bench, and general panel layout.
2. Meter sensitivity, 20 kΩ/V being desirable for most electronic circuit measurements and 11-MΩ electronic VOM preferred.
3. Frequency response on the ac ranges. For most applications a meter with a 100-kHz response is adequate.

4. Overload protection.
5. Voltage, current, and ohmmeter ranges available.

REVIEW QUESTIONS

11-1. On what principle does the d'Arsonval moving-coil galvanometer operate?

11-2. What two functions do the springs serve?

11-3. List the three steps to find the internal resistance of a meter movement using the current-division method.

11-4. Explain how the range of a milliammeter is extended by using a shunt resistance.

11-5. Give two formulas to find the value of a shunt resistance.

11-6. Explain why a make-before-break type of rotary switch is necessary to switch current shunt resistances.

11-7. What is the advantage of using the shorting type of rotary switch for switching milliammeter ranges?

11-8. Explain why a milliammeter loads the circuit being measured. What can be done to minimize the loading effect?

11-9. Explain what is meant by $\pm 3\%$ FS accuracy. At what part of the scale are readings taken to minimize percentage full-scale error?

11-10. Explain how a milliammeter movement can be used as a voltmeter.

11-11. Give the formula to find the multiplier resistance.

11-12. Explain why a voltmeter loads the circuit being measured. What can be done to minimize the loading effect?

11-13. What is meant by "voltmeter sensitivity"?

11-14. Give the formula for meter movement sensitivity.

11-15. Give a general rule to follow to minimize the full-scale error effect of a voltmeter.

11-16. What part of the scale on an ohmmeter is calibrated to measure the resistance of an open circuit? A short circuit?

11-17. Which portion of the ohmmeter scale is most linear?

11-18. What is the disadvantage of a series-type ohmmeter?

11-19. What is the advantage of a shunt-type ohmmeter?

11-20. The ohms calibration at midscale on a shunt ohmmeter is 30 Ω. What is the value of the shunt resistance for this range?

11-21. Give four important specifications of a meter needed to match a meter for electronic circuit applications.

SELF-EXAMINATION

Where applicable, choose the most suitable word or phrase to complete each statement.

11-1. A basic moving-coil meter movement is also termed a ________ galvanometer.

11-2. A current-carrying conductor will move at ________ ________ to the magnetic flux lines.

11-3. A moving coil and pointer are polarity sensitive and will be deflected from ________ to ________ when a current flows through the coil.

11-4. Reversing the coil through the meter movement will cause the pointer to be deflected from ________ to ________.

11-5. The springs of a moving-coil meter movement provide ________ which improves the linearity of the pointer deflection.

11-6. To minimize the oscillation of the moving coil and pointer, the coil former is constructed with ________, in which the induced current produces an opposing magnetic flux.

11-7. Two methods by which the internal resistance of a meter movement can be found are ________ and ________.

A basic d'Arsonval 100-μA movement with 2 kΩ internal resistance is to be used for Questions 11-8 to 11-17.

11-8. Calculate the shunt resistance value to provide a range of 10 mA.

11-9. Draw a schematic diagram to show the meter movement and its current shunt connected into a circuit to measure 10 mA. (*Note:* All circuit values are to be labeled on the components of the schematic.)

11-10. Two types of rotary switches acceptable for use in switching current ranges are ________ and ________.

11-11. Within what range of current values would a measured current of 10 mA fall given a full-scale accuracy of $\pm 1.5\%$ on the 20-mA range?

11-12. Calculate the voltmeter sensitivity for the given meter movement.

11-13. Calculate the multiplier resistance required to measure 10 V with the given meter movement.

11-14. Calculate the required multiplier resistance for a series ohmmeter using the given meter movement.

11-15. Calculate the required zero-ohms-adjust pot for the series ohmmeter in Question 11-14.

11-16. What value will a shunt resistance need to be in order to calibrate the $R \times 1\ \Omega$ midscale to 30 Ω?

11-17. What value will the shunt resistance be on the $R \times 10\ \Omega$ range of the ohmmeter in Question 11-16?

11-18. List five "main specification considerations" when choosing a VOM.

ANSWERS TO PRACTICE QUESTIONS

11-1

1. permanent magnet, moving coil
2. magnetic field
3. springs
4. current
5. dampen
6. eddy currents
7. wire
8. The voltage-division method to find the internal resistance of a meter movement is as follows:
   (a) Connect the meter movement, decade resistance box, a current-limiting resistor, and a variable dc source in series as shown in the schematic.
   (b) With the decade resistance box set to 0 Ω, carefully adjust the power supply for FSD of the meter pointer.
   (c) Adjust the decade resistance box for half-scale deflection and read off the resistance on the decade box $R_D = r_M$ since the voltage drop across R_D

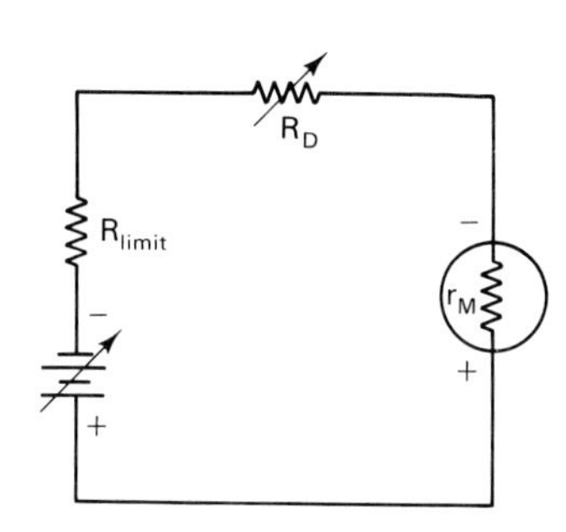

will be equal to the voltage drop across the meter movement and will be one-half of the FSD value.

9. (a) $R_S = \dfrac{V_M}{I_S} = \dfrac{50\ \mu\text{A} \times 2\ \text{k}\Omega}{1\ \text{mA} - 50\ \mu\text{A}} = \dfrac{0.1\ \text{V}}{0.95\ \text{mA}} = 105.26\ \Omega$

(b) $R_S = \dfrac{V_M}{I_S} = \dfrac{0.1\ \text{V}}{5\ \text{mA} - 50\ \mu\text{A}} = \dfrac{0.1\ \text{V}}{4.95\ \text{mA}} = 20.2\ \Omega$

(c) $R_S = \dfrac{V_M}{I_S} = \dfrac{0.1\ \text{V}}{500\ \text{mA} - 50\ \mu\text{A}} = \dfrac{0.1\ \text{V}}{499.95\ \text{mA}} = 0.20\ \Omega$

10.

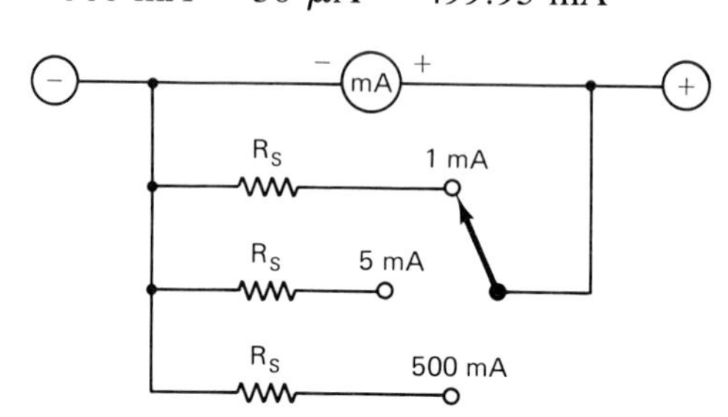

11-2

1. $V_M = I_M r_M = 50\ \mu\text{A} \times 2\ \text{k}\Omega = 0.1\ \text{V}$
2. multiplier
3. $R_M = \dfrac{V_{\text{FS}}}{I_{\text{FS}}} - r_M$
4. For the 5-V range: $R_M = \dfrac{V_{\text{FS}}}{I_{\text{FS}}} - r_M$

$$= \frac{5\ \text{V}}{50\ \mu\text{A}} - 2\ \text{k}\Omega = 100\ \text{k}\Omega - 2\ \text{k}\Omega = 98\ \text{k}\Omega$$

For the 50-V range: $R_M = \dfrac{50\ \text{V}}{50\ \mu\text{A}} - 2\ \text{k}\Omega = 998\ \text{k}\Omega$

For the 150-V range: $R_M = \dfrac{150\ \text{V}}{50\ \mu\text{A}} - 2\ \text{k}\Omega = 2.998\ \text{M}\Omega$

5.

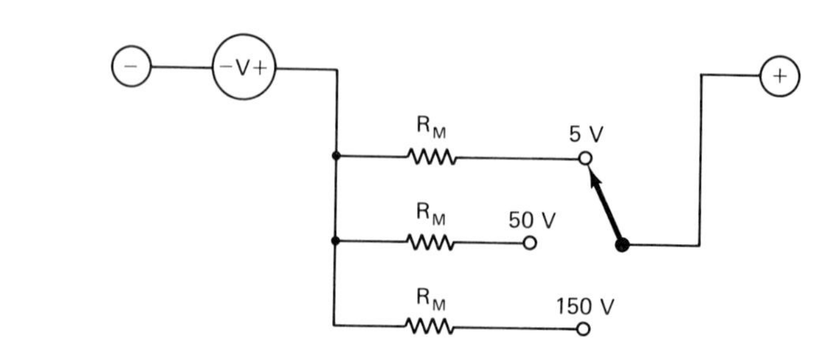

6. loading
7. sensitivity

8. $S = \frac{1\ \text{V}}{I_{\text{FS}}}$ ohms/volt or $S = \frac{R_T}{V_T}$ ohms/volt

9. $S = \frac{1\ \text{V}}{I_{\text{FS}}} = \frac{1\ \text{V}}{100\ \mu\text{A}} = 10\ \text{k}\Omega/\text{V}$

10. For the 30-V range: the full-scale error is ±3%, or

$$\text{voltage error} = \frac{\pm 3\%}{100\%} \times 30\ \text{V} = \pm 0.9\ \text{V}$$

$$\text{actual voltage range of value} = 30\ \text{V} \pm 0.9\ \text{V} = 29.1 \text{ to } 30.9\ \text{V}$$

$$\%\ \text{error} = \frac{30\ \text{V} - 29.1\ \text{V}}{30\ \text{V}} \times 100 = 3\%$$

For the 300-V range, the full-scale error is ±3%, or

$$\text{voltage error}\ \frac{\pm 3\%}{100\%} \times 300\ \text{V} = \pm 9\ \text{V}$$

$$\text{actual voltage range of values} = 30\ \text{V} \pm 9\ \text{V} = 21 \text{ to } 39\ \text{V}$$

$$\%\ \text{error} = \frac{30\ \text{V} - 21\ \text{V}}{30\ \text{V}} \times 100 = 30\%$$

11. The general rules to follow to minimize the loading and full-scale error effects of a voltmeter are:
(a) Select a range to give as close to full-scale deflection as possible.
(b) Use a voltmeter with a high ohms/volt sensitivity.

11-3

1. $R_T = \frac{V_T}{I_{\text{FS}}} = \frac{1.5\ \text{V}}{50\ \mu\text{A}} = 30\ \text{k}\Omega$

$R_M = R_T - r_M = 30\ \text{k}\Omega - 2\ \text{k}\Omega = 28\ \text{k}\Omega$

2. The ohmmeter midscale will be calibrated to 30 kΩ.
3. The maximum readable resistance value for this ohmmeter will be approximately 10 times R_M, or 300 kΩ.
4. The minimum readable resistance value of this ohmmeter will be approximately $\frac{1}{100}$ of R_M, or 300 Ω.
5. The value of the zero-ohms adjust will be

$$R_T = \frac{V_T}{I_{\text{FS}}} = \frac{1.5\ \text{V}}{50\ \mu\text{A}} = 30\ \text{k}\Omega$$

$$R'_M = R_T - r_M = 30\ \text{k}\Omega - 2\ \text{k}\Omega = 28\ \text{k}\Omega$$

With the zero-ohms adjust (R_Z):

$$R'_M + R_Z = 28\ \text{k}\Omega$$

where

$$R_{M(\text{fixed})} = \frac{V_T}{I_{\text{FS}} \times 1.5} = \frac{1.5\ \text{V}}{50\ \mu\text{A} \times 1.5} = 20\ \text{k}\Omega$$

Therefore,

$$R_Z = R'_M - R_{M(\text{fixed})} = 28\ \text{k}\Omega - 20\ \text{k}\Omega = 8\ \text{k}\Omega$$

The nearest 8-kΩ commercial value will be a 10-kΩ pot.

6. The value of the shunt resistance for the shunt ohmmeter of Fig. 11-18 will be equal to R_X, or 30 Ω.
7. 20 kΩ/V
8. 60
9. 100

12

Batteries

INTRODUCTION

The need for a source of electricity was pursued with great intensity by early eighteenth century experimenters such as Franklin. Franklin stood in a cow pasture and flew a kite during a rain storm to prove that he could draw electricity from the sky. Around the year 1800, Volta invented the first battery to provide a steady current flow with a low voltage, which he named "pila," known to us as the voltaic pile.

Demand for efficient portable power sources is just as great today. Cordless appliances, portable computers and calculators, cameras, TV sets, radios, hearing aids, heart pacemakers, wristwatches, electric cars, and many other devices rely on some type of battery to power them.

As Volta has shown, the battery is a number of cells interconnected to give a higher current and voltage capacity. Cells of a battery depend on the conversion of chemical energy into electrical energy. The type of chemical reaction will categorize the cells into two types:

1. *A primary cell,* which cannot normally be recharged since the chemical materials are used up in the process of energy conversion. The primary cells described in this chapter are the voltaic cell, carbon-zinc, alkaline-manganese, silver oxide–alkaline zinc, mercury cells, and lithium cells.
2. *A secondary cell,* which can be recharged because the chemical reaction can be reversed. Secondary cells described in this chapter include the nickel-cadmium and the lead-acid battery.

In this chapter we provide an insight into the characteristics of primary and secondary batteries.

MAIN TOPICS IN CHAPTER 12

OBJECTIVES

After studying Chapter 12, the student should be able to

1. Be familiar with the operational characteristics of various primary-type batteries.
2. Be familiar with the operational characteristics of various secondary-type batteries.

12-1 BASIC VOLTAIC CELL

Volta progressed from his famous pile and quickly discovered that a zinc and silver electrode immersed in a saltwater solution was able to produce electricity more efficiently. Volta's cell can be made using zinc and copper electrodes in a glass jar of sulfuric acid solution (see Fig. 12-1). The acid solution is called an *electrolyte*.

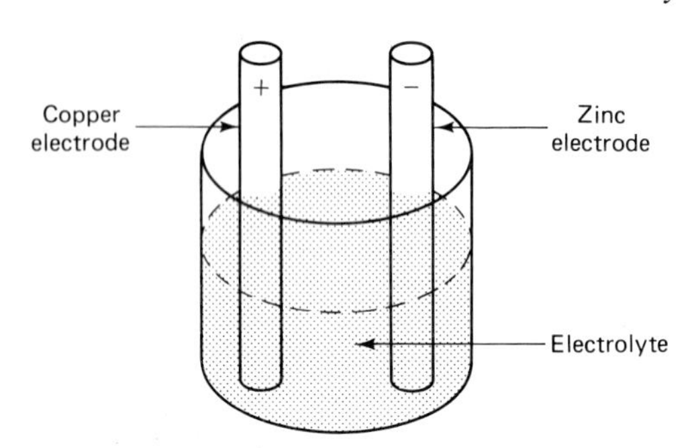

FIGURE 12-1 Basic voltaic cell.

Initially, the sulfuric acid is a combination of hydrogen and sulfate ions. Each molecule of sulfuric acid dissociates or separates into two positive hydrogen ions and one negative sulfate ion. When the two hydrogen ions separate, their valence electrons remain with the sulfate molecule, leaving it with a negative charge of two electrons.

Once the zinc electrode is immersed in the electrolyte, zinc atoms will be dissolved from the surface of the electrode. The dissolved zinc atoms combine with the sulfate ions to form zinc sulfate. Since the sulfate already has two excess electrons, the zinc atoms leave two electrons behind them on the zinc electrode. This makes the zinc electrode negatively charged.

As the zinc sulfate forms, more positive hydrogen ions are released. Immersing the copper electrode (remember that copper has a free valence electron) will attract the positive hydrogen ion (see Fig. 12-2). These ions migrate to the copper electrode, where they combine with an electron to form a neutral hydrogen atom. Each pair of hydrogen ions becomes a molecule of hydrogen gas. This gas in the form of visible bubbles, accumulates on the surface of the copper electrode, while some of the gas escapes to the surface. In giving up electrons the copper electrode becomes positively charged and a potential difference of approximately 1.1 V is established between the two electrodes.

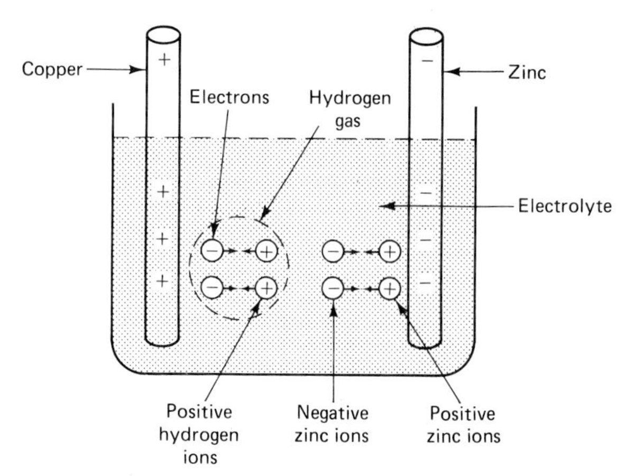

FIGURE 12-2 Chemical reaction of electrolyte.

With a load connected to the two electrodes, electrons will flow from the negative zinc terminal through the load and into the positive copper terminal. This provides more electrons to recombine with the hydrogen ions. More zinc ions are generated in the solution to replace the lost positive hydrogen ions. This chemical action is repeated until the zinc electrode is completely eaten away, or the acid solution becomes zinc sulfate.

Polarization

Hydrogen bubbles of gas that accumulate about the copper electrode impede other hydrogen ions from recombining with copper electrons and thereby reduce the cell's ability to supply current. This effect is known as *polarization*. Manufactured cells use a depolarizer to minimize the insulating effect (electrochemical resistance) of the hydrogen bubbles.

Internal Resistance

Internal resistance is a combination of the cell's electrode surface area, its depolarization efficiency or electrochemical resistance, and simple contact resistance. Thus, internal resistance is a function of cell size. The internal resistance of an unused cell is very low and is negligible for most applications. However, as the cell is used, its chemical action deteriorates and its internal resistance increases. At some point its output voltage is no longer useful. This voltage is called the *cutoff* or *end-point voltage*.

Open-Circuit Voltage versus Working Voltage

The electrodes must be two dissimilar materials in order to obtain a potential difference. The type of electrode material and electrolyte will determine the output voltage obtained from a cell. Approximately 1.1 V is typical for the copper-zinc cell. The 1.1 V is produced regardless of the electrode size or amount of electrolyte and is the *open-circuit voltage* of the cell.

The terminal voltage of a cell with a load connected is called the *working voltage* or *voltage under load*. It is the important voltage to consider. The working voltage is lower than the open-circuit voltage by an amount dependent on the current flowing through the cell and its internal resistance. The voltage at which a device powered by a cell or battery will not operate properly is going to determine the lowest value of the voltage under load.

Current Capacity

The current that a cell is capable of generating depends on the electrode area and its internal reistance. Since the internal resistance is also a function of cell size, larger cells will normally have lower internal resistance and therefore are able to provide larger current capacity.

The cell capacity is the installed rated ampere-hour capability of a cell or battery and is supplied by the manufacturer. Ampere-hour (Ah) capability is an approximation of a cell's ability to supply a certain amount of current for a length of time to a nominal cutoff voltage. A cell that has a 5-Ah rating can supply 5 A for 1 hour, or 1 ampere for 5 hours. In equation form,

$$\text{Ah rating} = I \times t \tag{12-1}$$

EXAMPLE 12-1

For the cell with a rating 5 Ah, what will be the expected life if it supplies a load current of 50 mA?

Solution: From equation (12-1),

$$t = \frac{\text{Ah}}{I} = \frac{5\ \text{Ah}}{0.05\ \text{A}} = 100\ \text{h}$$

Service Capacity

The ampere-hour rating of a cell is not indicative of its service capacity. *Service capacity* is the amount of electrical energy that can be withdrawn for a period of time under varying load and temperature conditions.

If the load current is too heavy, energy is withdrawn too rapidly from the cell, and the depolarization rate cannot be maintained. The working voltage drops off and the cell does not function efficiently. If the load current is very light, the time required to exhaust the cell is extended to its shelf deterioration period.

Between these two conditions of service, heavy and light, a point is reached where depolarization is at its optimum and shelf deterioration negligible. This is the most efficient service capacity. Figure 12-3 shows typical carbon-zinc D-cell discharge characteristics under three load conditions.

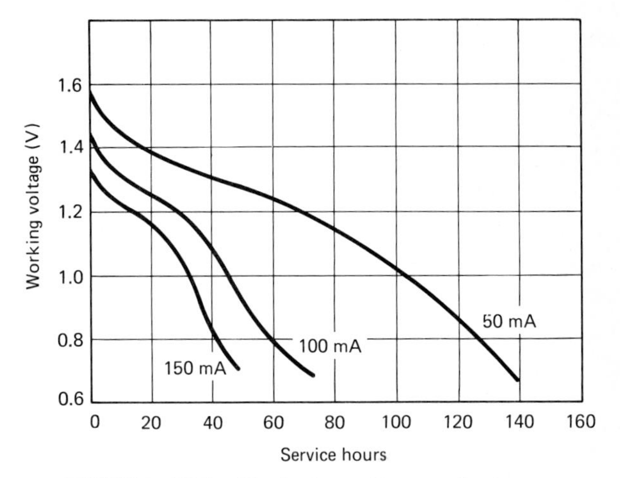

FIGURE 12-3 Typical voltage discharge characteristics of carbon-zinc D cell discharged 2 hours per day at 21 °C.

The *duty cycle* refers to the time duration and the frequency during which a cell or battery is drained. The graph of Fig. 12-3 shows a duty cycle of 2 h/day for load currents of 150 mA, 100 mA, and 50 mA.

12-2 CARBON-ZINC CELL

The carbon-zinc LeClanche-type cell is very widely used because of its low cost and reliability. Another advantage is its portability, since its electrolyte is sealed in paste rather than a liquid. These cells are referred to as *dry cells*. Various shapes and sizes of cells have been developed by industry. Round D and C cells are familiar in flashlights, games, and toys. Flat cells are stacked in series to give voltages greater than 1.5 V and are used in portable transistor radios, calculators, and other devices.

Operation of the carbon-zinc cell is similar to that of the voltaic cell described in Section 12-1. The zinc electrode that serves as a container is slowly dissolved by the electrolyte paste. This chemical process leaves an excess of electrons on the zinc to make it negatively charged. Hydrogen ions are attracted to the carbon electrode, where they combine with electrons, leaving the carbon electrode atoms deficient in electrons to make it positive. Recombination of the hydrogen ions with electrons forms hydrogen gas bubbles about the carbon electrode. A depolarizer such as manganese dioxide is used to minimize the hydrogen gas barrier.

Figure 12-4 shows the construction of a typical carbon-zinc dry cell. The electrolyte is a wet paste mixture of manganese dioxide, powdered graphite, ammomium chloride, "sal ammoniac," and zinc chloride. Manganese dioxide serves as a depolarizer. As the cell delivers current, the manganese dioxide loses oxygen, which recombines with the hydrogen gas around the carbon electrode to form water. Impurities such as carbon and iron in the zinc electrode react with the electrolyte and set up small voltaic currents that use up the zinc. This is called "local action." To reduce the local action and increase the life of the zinc electrode, mercury is coated on the interior zinc surface during manufacture.

Carbon-zinc cells cannot be hermetically sealed and must have vents for gases generated within the cell during use and while idle. Pitch, wax, rosin, waterproofed cardboard, plastic, and insulated metal covers are types of seals used to minimize air access and moisture loss by evaporation.

Working Voltage

The terminal open-circuit voltage of the carbon-zinc cell is nominally 1.5 V regardless of its physical size. However, the actual initial voltage may vary from 1.5 V to over 1.6 V depending on the type and amount of manganese.

The terminal working voltage of the cell is lower than the open-circuit voltage and depends on load current. The working voltage falls gradually as it is discharged, as shown in Fig. 12-3, to cutoff or endpoint voltage. The cutoff voltage is the voltage at which the device will no longer operate satisfactorily or below which operation is not recommended.

Current Capacity

The physical size of a dry cell determines the amount of current that it can deliver and the amount of energy stored. Small cells such as the N, AAA, AA, and C cells (American National Standards Association designations) can deliver a continuous current of a few milliamperes, while the D, F, G, and the No. 6 cells can deliver current in the range of hundreds of milliamperes.

Carbon-zinc dry cells are intended primarily for intermittent duty. Service life obtained from a cell discharged for a schedule of 2 day will be much greater than for continuous discharge. Intermittent use allows time for the diffusion of reaction products and other reactions in the recuperation process.

The LeClanche flat cell shown in Fig. 12-5 reduces waste space in assembled batteries. Individual cells are "sandwich" assembled to form batteries of different voltage and current capacities. These flat cells utilize cell materials in a laminated structure. Carbon is coated on a zinc plate to form a double electrode, a combination of the zinc of one cell and the carbon of the adjacent cell. This construction increases the amount of depolarizing mix available per unit volume and therefore the energy content. Since zinc is not used as a con-

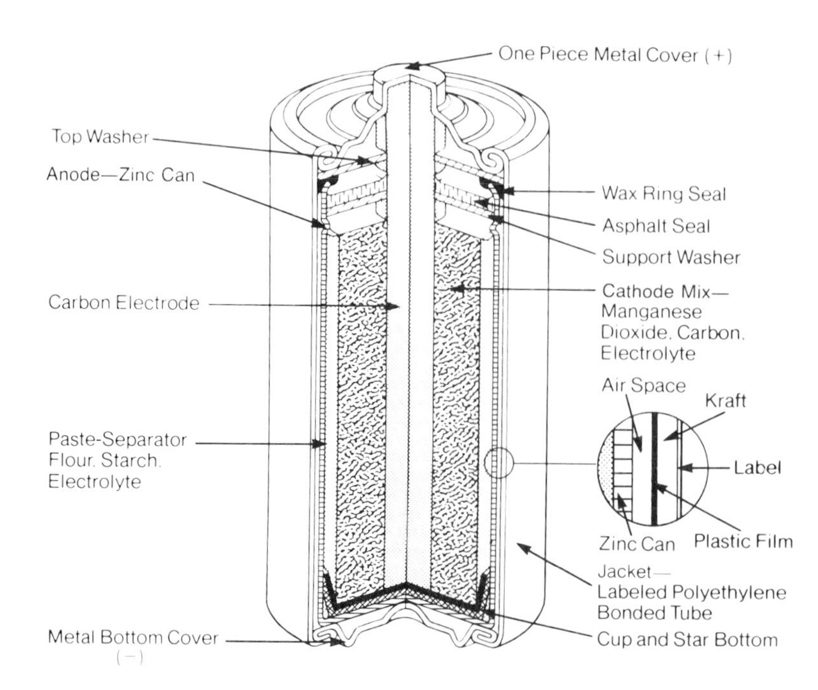

FIGURE 12-4 Cutaway of cylindrical general-purpose LeClanche cell. (Courtesy of Eveready Battery Company.)

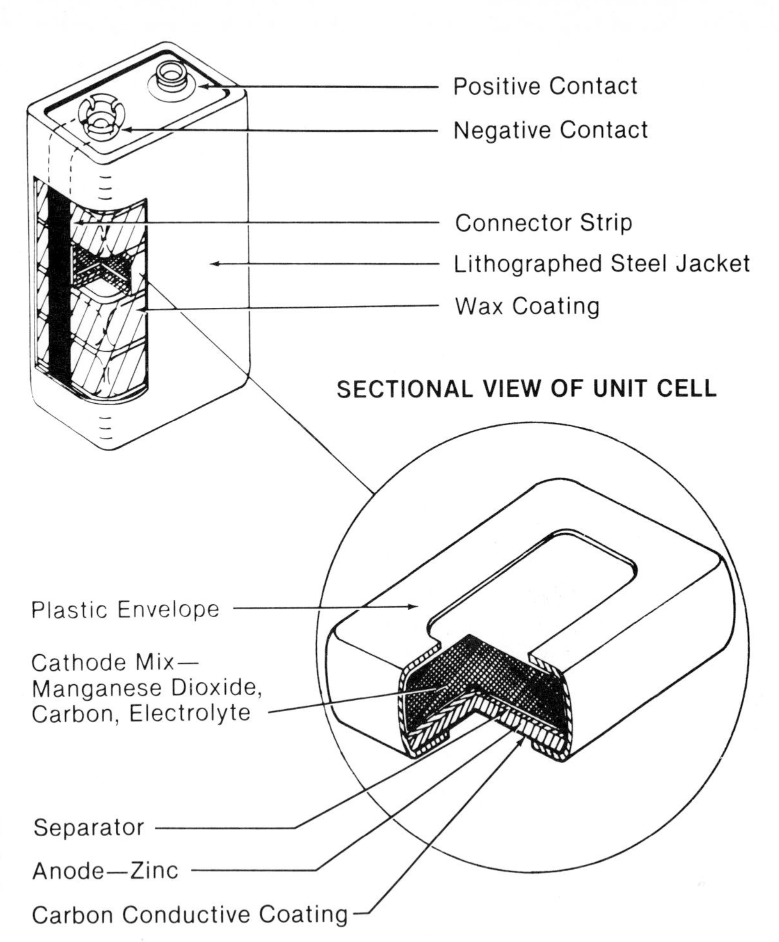

FIGURE 12-5 Cutaway of flat-cell general-purpose LeClanche battery. (Courtesy of Eveready Battery Company.)

tainer in the flat cell, leakage is almost entirely eliminated.

Rechargeability

Recharging cells that are not specifically designed for charging can be dangerous. Excessive amounts of gas may result from a high value of charging current. Buildup of gas in a tightly sealed cell may cause it to explode, resulting in personal injury and damage.

Shelf-Life

The storage of carbon-zinc cells at temperatures below 21°C will increase their shelf life. Shelf life is defined as the amount of time a cell or battery will retain a specified percent of its rated capacity. Table 12-1 shows the relative amount of cell service left of a typical Eveready carbon-zinc cell after storage at 21°C (70°F).

Batteries to be stored at low temperatures should be left in their original carton or wrapped in plastic film. When batteries are removed from low-temperature storage, they should be allowed to reach room temperature in their packaging, to avoid condensation of moisture, which may cause electrical leakage and/or destruction of the jackets. Storage at temperatures higher than 21°C for sustained periods will significantly reduce their shelf life.

TABLE 12-1
Eveready carbon-zinc cell service life

Time of Storage at 21°C (70°F)	*Typical Percent of Fresh Cell Service Retained*
1 year	90–95
2 years	85–90
3 years	75–85
4 years	65–75

Source: Courtesy Eveready Battery Company.

PRACTICE QUESTIONS 12-1

Choose the most suitable word, phrase, or value to complete each statement.

1. The acid solution in a voltaic cell is called an ________.
2. Initially, the acid solution has three ions, two ________ and one ________.

3. Sulfate ions will combine with the dissolved zinc atoms to form _________ _________.
4. The dissolved zinc atoms are _________ ions that (polarity) leave _________ electrons on the zinc electrode to (number) make it negatively charged.
5. _________ molecules migrate to the copper electrode and combine with an electron.
6. Once the hydrogen ions recombine with an electron, each pair becomes _________ _________.
7. The copper electrode becomes positively charged because it gives up an _________ to the hydrogen ions.
8. Hydrogen gas bubbles accumulate around the copper electrode to form a barrier to the hydrogen ions. This effect is known as _________.
9. The _________ _________ of a cell increases as the chemical action deteriorates.
10. When the output voltage of a cell is no longer useful, it is called the _________ or endpoint voltage.
11. Typical open-circuit voltage for a copper-zinc cell is approximately _________ volts.
12. The terminal voltage of a cell with the load connected is called the _________ voltage.
13. The current capacity of a cell will depend on _________ and its _________.
14. An approximation of a cell's ability to supply a certain amount of current for a length of time to a nominal cutoff voltage is termed _________ _________ capacity.
15. Refer to Fig. 12-3; what is the service capacity of the D cell discharged for 2 h/day at 50 mA with a cutoff voltage of 1 V?
16. The manganese dioxide serves as a _________ in a carbon-zinc dry cell.
17. The open-circuit voltage of a carbon-zinc cell is nominally _________ volts.
18. The current capacity and service capacity of a dry cell depend on its _________ _________.
19. The danger in trying to recharge a dry cell is that it might _________, due to the buildup of gases.

12-3 ALKALINE-MANGANESE CELL

One of the most important cells to the electronics industry is the alkaline-manganese system. Fundamentally, the chemical-energy-producing reaction of the alkaline-manganese cell is the same as for carbon-zinc cell. The difference exists in the electrolyte, which is alkaline. This results in different properties, as well as different construction techniques.

A cutaway view of a typical button-type alkaline-manganese cell is shown in Fig. 12-6. The cell uses manganese dioxide depolarizer (cathode) in direct contact with the cell container of nickel-plated steel. Because of the passivity of steel in alkaline electrolyte, there is no chemical reaction between the manganese dioxide and the steel. This permits the steel to be a current collector as well as a strong cell container. Porous separators saturated with potassium hydroxide form the electrolyte. Zinc-mercury amalgam is used to minimize "local action" and gas generation on an open-circuit stand, and is in contact with a steel closure to form the negative anode terminal.

Another configuration of the alkaline cell includes the cylindrical C and D cells, shown in the cutaway

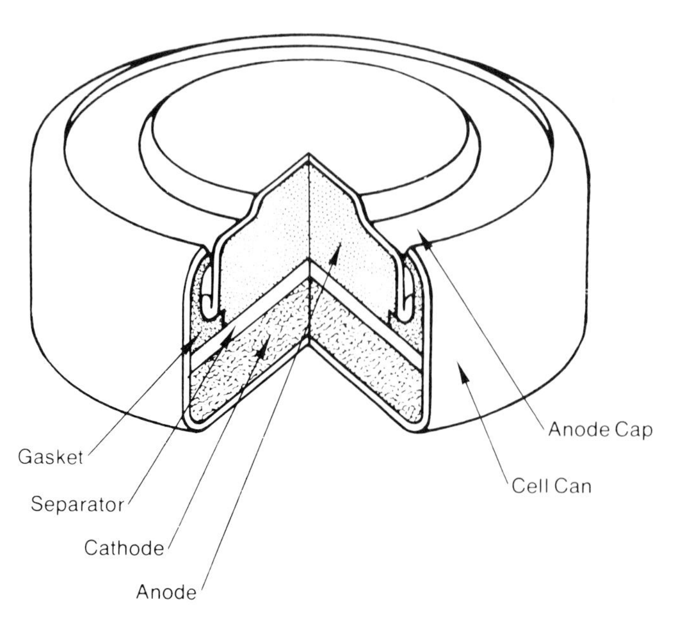

FIGURE 12-6 Cutaway view of an alkaline miniature cell. (Courtesy of Eveready Battery Company.)

view of Fig. 12-7. In this configuration the polarity is reversed from that of conventional carbon-zinc cells. Through external jacketing and terminal pieces (see the brass rivet in the cutaway view of Fig. 12-7), alkaline-manganese cylindrical cells are made to appear conventionally polarized (i.e., with the center button positive).

Working Voltage

The open-circuit voltage of the alkaline-manganese cell is approximately 1.6 V. The closed-circuit voltage declines gradually as a function of the depth of discharge; therefore, greater hours of service can be obtained as the cutoff voltage is lowered. The energy output of alkaline cells is less sensitive to variation in the discharge rate and duty cycle than in comparably sized carbon-zinc cells. Alkaline cells provide useful current capacity below 1.25 V down to 0.9 V, unlike carbon-zinc cells, which are exhausted at 1 V.

Service Capacity

Alkaline cells are capable of greater capacity and discharge rates, which can be maintained over extended periods. They will sustain a discharge rate without sacrificing capacity or the need to provide recovery periods. The alkaline-manganese system is reliable within a temperature range of approximately 0 to 50°C (−40 to 120°F).

Miniature alkaline-manganese cells typically exhibit an expansion of the cathode on discharge, which results in an overall increase in the cell's height. This increase in height is referred to as *bulge*. While miniature alkaline-manganese cells are designed to minimize bulging, they will typically bulge to a height greater than comparable silver oxide cells during discharge. Specific bulge data are given by the manufacturer.

Charging of Primary Cells

Home battery chargers that attempt rejuvenation of primary batteries are widely distributed and advertised. Also, some devices, usually radios, have ac adapters which allow current to pass through batteries in the charging direction. Advertisements for some home battery chargers claim that all types of batteries can be recharged ("rejuvenated" is a better term). These types include primary batteries, which are not designed to be rechargeable: carbon-zinc, most alkaline-manganese dioxide, mercuric oxide, and silver oxide. There are problems and hazards, however, that arise in the attempted rejuvenation of primary batteries.

It has been known for years that the LeClanche system carbon-zinc cell is rechargeable to some degree if the discharge and charge cycles are controlled with precision. On this matter the National Bureau of Standards (Letter Circular LC965) makes the following comments:

> From time to time attention has been turned to the problem of recharging dry cells. Although the dry cell is nominally considered a primary battery, it

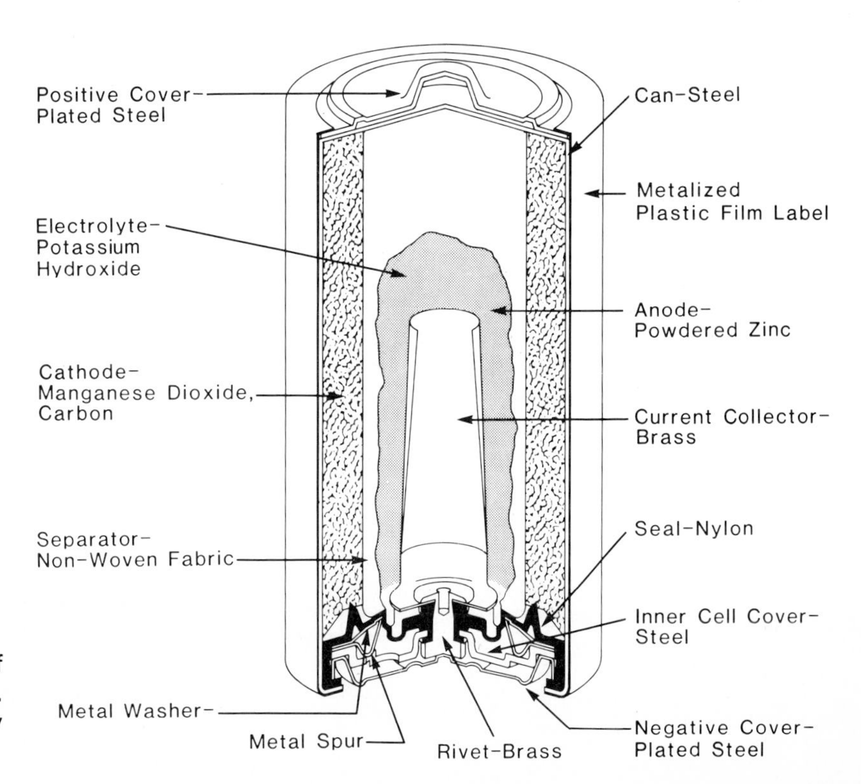

FIGURE 12-7 Cutaway view of cylindrical energizer alkaline cell. (Courtesy of Eveready Battery Company.)

may be recharged for a limited number of cycles under certain conditions.

1. The opening voltage on discharge should not be below 1.0 V per cell when the battery is removed from service for charging.
2. The battery should be placed on charge very soon after removal from service.
3. The ampere-hours of recharge should be 120%–180% of the discharge.
4. The charging rate should be low enough to distribute recharge over 12–16 h.
5. Cells must be put into service soon after charging, as the recharged cells have poor shelf life.

Recharging of dry cells may be economically feasible only when quantities of dry cells are used under controlled conditions with a system of exchange of used cells for new ones already in practice, and with equipment available to provide direct current for charging. Such a system would not be practical for home use.

Recharging cells of any chemical system which are not specifically designed for charging may be dangerous. Charging may cause leakage and may, on rare occasions, cause a cell to rupture, resulting in personal injury or damage to equipment.

12-4 SILVER OXIDE–ALKALINE ZINC CELL

Silver oxide cells contain a cathode of silver oxide with a low percentage of manganese dioxide. A highly alkaline electrolyte consisting of either sodium hydroxide or potassium hydroxide is in contact with the high surface area of the zinc anode. A cutaway view of Fig. 12-6 shows its construction.

The silver oxide cell generates energy through the reduction of silver oxides by zinc and potassium hydroxide electrolyte. Electrically, the outstanding feature of the silver oxide cell is its ability to maintain a virtually constant output voltage and current regardless of discharge time.

Working Voltage

The open-circuit voltage of silver oxide cells is 1.6 V. Typical operating voltage at a typical current drain is 1.5 V or greater. Eveready silver oxide batteries are available in voltages ranging from 1.5 to 6.0 V and in capacities ranging from 15 to 210 mAh.

Applications

Silver oxide batteries are used in watches, calculators, phototelectric devices, hearing aids, and other instrumentation.

12-5 MERCURY CELL

Mercury oxide cells are currently produced in two different designs, using either flat or cylindrical electrodes. The basic chemistry of the mercury cell system incorporates amalgamated zinc anode as a negative terminal. The anode cylinder (see Fig. 12-8) is made from highly purified compressed zinc powder. The positive cathode terminal and depolarizer form a densely pressed structure of mercuric oxide with a small percentage of conductive graphite.

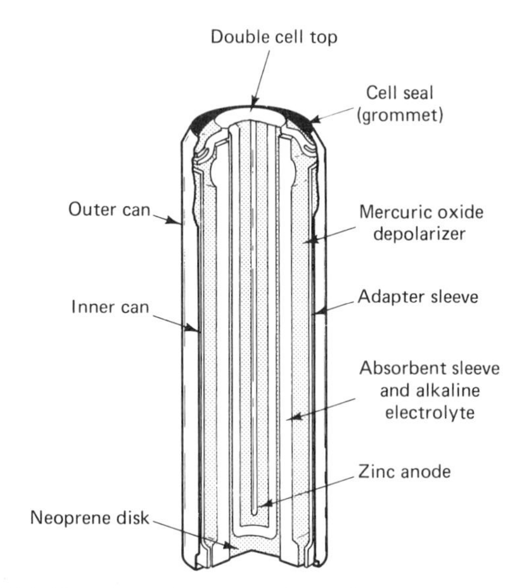

FIGURE 12-8 Cutaway view of miniature AA penlight mercuric oxide cell. (Courtesy of Eveready Battery Company.)

The type of electrolyte used with mercuric oxide cells determines the rate or current availability over a range of load current. Under heavy current drains, potassium hydroxide (KOH) electrolyte offers less resistance and allows the cell to operate at higher efficiency than does the sodium hydroxide (NaOH) electrolyte.

The cathode and the anode are separated by a microporous barrier material which is permeable to the flow of ions but will not permit the migration of structural or waste product particles between the electrodes that are employed.

Electrical energy is produced through the chemically controlled corrosion of zinc by oxygen released from the mercuric oxide. A by-product of this corrosion is the availability of free electrons at an initial open-circuit voltage of 1.35 V as well as under rated load conditions.

Service Capacity

The high energy density of the mercuric oxide system allows capacities ranging from 45 mA to 14 A in voltage ranges from 1.35 V and higher (12.6 V Eveready).

Applications

The wound-anode flat cell construction (see Fig. 12-6) provides a wide variety of voltages and current capacities by welding cells in series or parallel. Complete encapsulation in epoxy resins allows unlimited variations in form factors, packaging methods, and electrical outputs.

The low internal resistance of these cells is the main reason for their ability to deliver closely regulated voltages under useful load currents. These reference voltage values are reliable within the shelf life of the cell, which exceeds 10 years.

The extended shelf life and high-energy density of mercury batteries are employed to provide greater reliability in emergency devices such as air–sea rescue radios. Three or more years of standby service is generally obtained from battery packs built for this application.

The reference capability of the cell is widely used to provide accuracy in electrical measuring circuits. Instruments using mercury batteries can supply measurement accuracies within 1% over several years of service since both duty and drain rates are light.

A low noise factor (less than 10 μV on a larger battery) provides an excellent dc source where delicate signal detection is required. This allows filters and other regulators to be eliminated from product circuitry. Mercuric oxide batteries can be used in environmental temperatures up to 54°C (130°F), and operation at 93°C (200°F) is possible for a few hours. Mercuric oxide batteries ulitizing KOH as an electrolyte will operate with fair efficiency at low temperatures down to −28°C (−20°F), while the comparable batteries using NaOH electrolyte are less efficient at temperatures down to −10°C (14°F).

12-6 LITHIUM CELL

The lithium–sulfur dioxide battery is the latest in primary power sources. It outperforms conventional batteries, with superior energy output, high discharge-rate capability, low temperature and long shelf life. Lithium cells are used where weight is an important consideration, in such applications as instant-picture cameras, hand-held calculators, and liquid-crystal watches.

The lithium cell, Li/SO_2, uses lithium as the anode and a porous carbon cathode electrode with sulfur dioxide as the true active cathode material. Since lithium reacts with water, either a solid or a nonaqueous electrolyte is used. Sulfur dioxide and an organic solvent acetonitrile containing dissolved lithium bromide is a liquid electrolyte which is poured through a temporary port on the bottom of the cell and welded closed.

The familiar cylindrical cell is fabricated using a "jellyroll" construction technique (see Fig. 12-10). Rectangular strips of lithium foil, a microporous polypropylene separator, the cathode electrode (a Teflon-carbon mix pressed on a supporting aluminum screen), and a second separator layer are rolled into a compact cylinder as shown in Fig. 12-9(a). This design provides the high surface area and low cell resistance that produce the high-current and low-temperature performance characteristics.

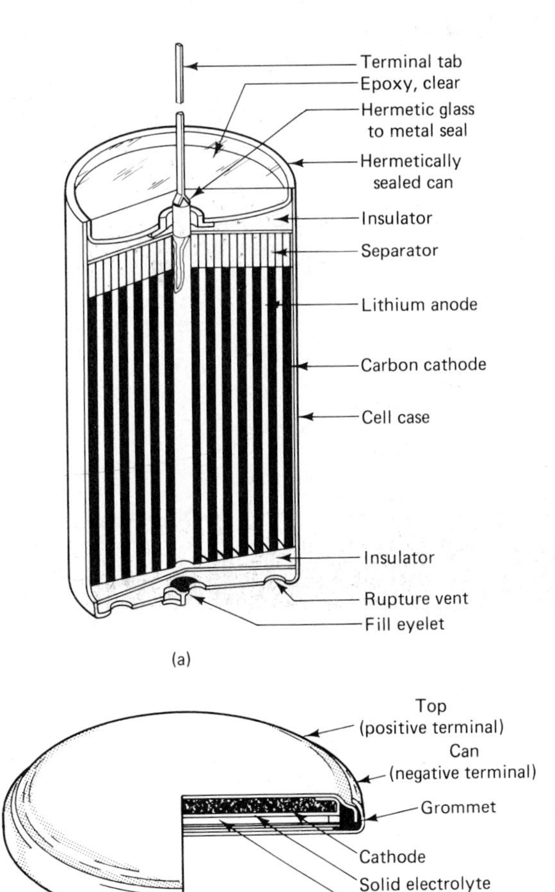

FIGURE 12-9 (a) Lithium–sulfur dioxide cell. (b) Lithium flat cell. (Courtesy of Duracell, Inc.)

The roll is inserted into a nickel-plated steel can which is connected electrically to the lithium anode. The cathode is connected to the center terminal of the glass-to-metal hermetic seal. The electrolyte/depolarizer is added through the fill port and the opening is welded closed.

Fig. 12-9(b) shows a typical flat cell configuration that uses a solid type of electrolyte. This type of electrolyte results in a highly stable cell but is only capable of supplying very low current drain rates in the range of microamperes. Although this is typical of other primary cells, the lithium cell produces a somewhat higher voltage, 1.9 V. Light weight and higher output voltage enhances applications of this cell in liquid-crystal watches and semiconductor memory and sensor circuits.

Working voltage. Typical operating voltage is 3 V compared to 1.5 V for most primary cells.

Current capacity. The "rated capacity" of the lithium cell is, of course, dependent on the cell size. For example, the Duracell LO 32S, weighing 12.0 g (0.42 oz), has a rated capacity of 0.85 Ah, while the Duracell LO-26SH, weighing 75 g (2.65 oz), is rated at 6.4 Ah.

Service capability. The lithium cell is capable of providing its rated capacity over a wide temperature range, from −18° to 70°C (−65° to 160°F). A shelf life of 5 to 10 years is projected; a storage capability of more than 1 year at a temperature of 70°C (160°F) has been demonstrated.

PRACTICE QUESTIONS 12-2

Choose the most suitable word, phrase, or value to complete each statement.

1. The chemical-energy-producing reaction of the alkaline-manganese cell is the same as for the ________ ________ cell.
2. The manganese dioxide acts as a ________ and cathode in the alkaline cell.
3. The open-circuit voltage of a alkaline-manganese cell is ________ volts.
4. Alkaline cells provide useful current capacity down to ________ volts.
5. Alkaline cells are capable of greater ________ and ________ rates which can be maintained over extended periods.
6. Home recharging of primary cells could be ________, as a cell may rupture, resulting in personal injury or damage to equipment.
7. The open-circuit voltage of silver oxide cells is ________ volts.
8. The outstanding feature of the silver oxide cell is its ability to maintain a constant ________ and ________ regardless of discharge time.
9. The open-circuit voltage of a mercury cell is ________ volts.
10. The mercuric oxide cell has the ability to deliver closely regulated voltage under useful load conditions because of its low ________ ________.
11. The high energy density of the mercuric oxide system allows capacities ranging from ________ milliamperes to ________ amperes.
12. Instruments using mercuric batteries can supply measurement accuracies within ________ percent over several years of service since both duty and drain rates are light.
13. A low noise factor (less than ________) on a larger battery provides an excellent dc source where delicate signal detection is required.
14. Lithium cells are used where ________ is an important consideration.
15. The working voltage of a lithium cell is typically ________.
16. Lithium cells are capable of working over a wide temperature range from ________°C to ________°C.
17. Lithium cells have a shelf life of ________ to ________ years.

12-7 NICKEL-CADMIUM SECONDARY CELL

One of the most important battery sources is the nickel-cadmium alkaline battery. This reliable and portable, rechargable battery is used for a wide variety of consumer, commercial, industrial, and military applications.

The nickel-cadmium secondary cell is a combination of active materials that can be electrolytically oxidized and reduced repeatedly. The oxidation of the negative electrode occurring simultaneously with the reduction of the positive electrode generates electric power.

Both electrode reactions are reversible in a secondary cell, and the cell can be recharged. The input current in the proper direction from a charging source will drive the primary or discharge reaction backwards and in effect recharge the electrodes.

In the charged state the positive electrode is nickelic hydroxide; the negative electrode is metallic cadmium. In the uncharged state the positive electrode of a nickel-cadmium cell is nickelous hydroxide; the negative is cadmium hydroxide.

Preparation of the positive or negative plates requires the sintering of a fine nickel wire screen. Sintering involves the conversion of a powdered or earthy substance into a coherent solid mass by heating without thoroughly melting the substance.

The sintered nickel screen acts as a matrix conductor that has great strength and flexibility. This results in a thin, highly porous nickel plaque, which is then impregnated with nickel salt solution for the positive plate and cadmium salt solution for the negative plate.

The separators (see Fig. 12-10) are made of an absorbent dielectric material that separates the negative electrode from the positive electrode. Being absorbent, they hold electrolyte while permitting ions or an electrical current to flow between the plates. The electrolyte used is a jellied solution of potassium hydroxide.

The cell container is usually a nickel-plated steel can (see Fig. 12-10) and cover. Assembly of the cell is done by rolling both plates separated by the dielectric into a tight roll, as shown in Fig. 12-10. This is then placed into the can with the electrolyte and sealed.

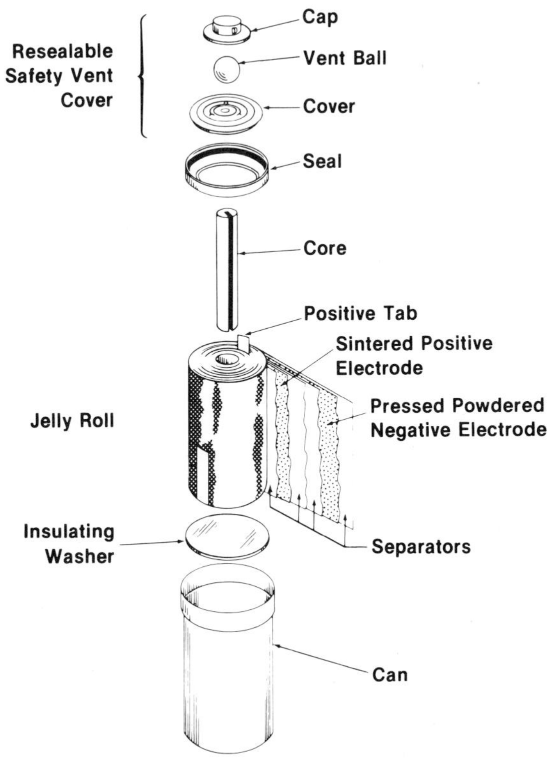

FIGURE 12-10 Cutaway of cylindrical rechargeable nickel-cadmium cell. (Courtesy of Eveready Battery Company.)

The negative electrode tab is welded to the bottom of the can, making the entire cell case the negative terminal. The entire cell asembly is then charged, discharged, and completely inspected by the manufacturer.

Cell Voltage

Open-Circuit Voltage. This is the voltage of a cell without a load. Except in the case of complete discharge, neither cell condition nor state of charge can be determined by the open-circuit voltage. At room temperature a nickel-cadmium cell has a terminal voltage of 1.33 V.

Closed-Circuit Voltage. This is the voltage of a cell connected to its rated load. During discharge the average voltage of a sealed nickel-cadmium battery is approximately 1.2 V per cell.

End-of-Discharge Voltage. This is the final voltage to which a cell is discharged. The nickel-cadmium cell provides most of its energy above 1.0 V per cell. Levels below 1.0 V should be avoided, as the cell's capacity is exhausted at voltage levels somewhere between 1.10 and 1.15 V.

End-of-Charge Voltage. This is the final voltage across the cell at the end of a charge period with a charging current still flowing through the cell. The voltage of a cell being charged soon rises to a voltage of 1.40 V and can climb to 1.47 V or more. The final voltage is an average of 1.43 V, and a cell is questionable if it does not closely approximate these values.

Ampere-Hour Capacity

Ampere-hour capacity is based on the discharge current output at an hour rate to an endpoint of 1.0 V per cell. It is the product of discharge current and the time under load.

Capacities vary with the discharge rate. If current is withdrawn at faster rates, capacity is decreased. Figure 12-11 shows the average performance characteristics of an Eveready CF4 cylindrical high rate–fast charge D-size cell. The rating of the cell is 4 Ah. The graph of Fig. 12-11 shows what voltage output might be expected if the discharge rate is changed from 4 Ah.

High-Current Capability

Sealed sintered-plate cells can deliver a high current discharge rate. As an example, the performance characteristic graph of Fig. 12-11 shows that the CF4 Eveready D cell will deliver 10 A for over 20 min. Ordinary cells would be ruined by such a discharge demand.

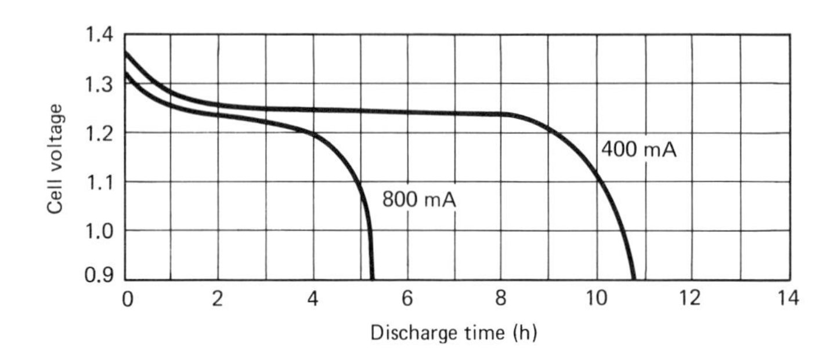

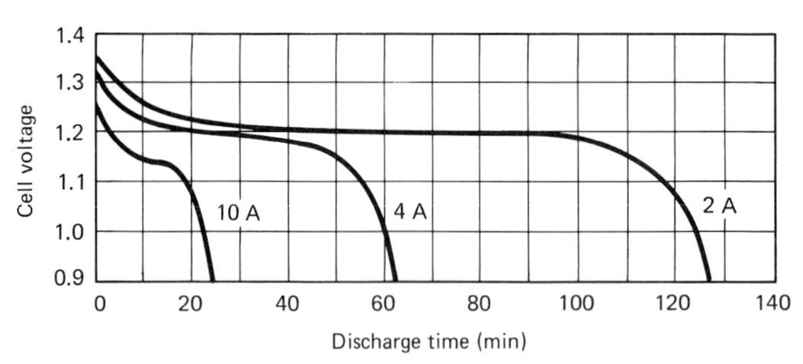

FIGURE 12.11 Typical discharge curves for Eveready CF4 and CF4T. Average performance characteristics at 70 °F (21 °C). These performance curves assume that the cells were fast-charged at 1.4 A. A slight reduction in capacity will be experienced when charged at lower rates. (Courtesy of Eveready Battery Company.)

Shelf Life

As all electrochemical storage devices, the sealed nickel-cadmium cell will lose a percentage of its charge while in storage. Generally, at room temperature the cell will retain 75% of its capacity after a 1-month stand, 60% after a 2-month stand, and 50% after a 3-month stand. This is, of course, only a temporary condition since the cell or battery can be recharged. A nickel-cadmium battery can be kept up to its full capacity when setting idle by applying a trickle charge.

Charging Techniques

Since the voltage of a nickel-cadmium system does not reflect residual capacity, there is no simple practical method of measuring the state of charge on the battery. Furthermore, the electrolyte serves mainly as an ion carrier, and the specific gravity does not change with discharge.

Practically, the best and standard procedure for maintaning a charge on either a vented or a sealed nickel-cadmium battery is either to apply a refreshing charge for 14 h at $\frac{1}{10}$ of the Ah rating, or to apply a continuous trickle charge at $\frac{1}{100}$ of the Ah rating.

Fast-charge cells are specifically designed to withstand overcharge without special control circuitry. As an example, Eveready Fast Charge cells develop the desired temperature rise and have the built-in ability to withstand short-term overcharge at rates to 1-h values without physical damage or loss in cell capacity. This means that the C4F D cell can be charged at the rate of 4 A for 1 h, since it is rated at 4 Ah.

For home use standard battery chargers can be purchased from a store outlet or battery dealer. It is not within the scope of this text to detail the design and construction of half-wave and full-wave rectifiers to be used as battery chargers. Refer to the Eveready Battery Engineering Data Book, Volume IV, for details regarding battery charging and battery-charger circuit design.

12-8 LEAD-ACID SECONDARY CELL

The lead-acid cell is capable of being recharged hundreds of times. Its use in automobile, marine, and emergency applications is universal. The cell consists of a lead peroxide (PbO_2) positive electrode, reddish brown in color, and a sponge lead negative electrode, gray in color.

With an electrolyte solution consisting of approximately 30% sulfuric acid and the balance distilled water (specific gravity 1.3), each cell produces 2.2 V. Six cells are series-connected to obtain a 12-V automobile-type battery.

Upon discharge, the chemical reaction of the electrolyte on the sponge lead negative plate slowly changes the plate to lead sulfate. This chemical reaction liberates atoms from the sponge lead electrode. These atoms combine with sulfate molecules to form lead sulfate (a whitish material) and release hydrogen. The liberated atoms leave electrons behind, making the sponge lead plate negative. The released hydrogen ions of the electrolyte combine with the oxygen atoms in the lead peroxide plate to form water. The recombining of the hydrogen and oxygen atoms makes the lead peroxide plate positive while diluting the sulfuric acid and making the electrolyte weaker.

The lead sulfate that is produced at both the negative and positive plates tends to coat the surface of the active material in layers. This reduces the efficiency of the chemical reaction of the electrolyte on the plates. The buildup of lead sulfate on both plates increases the internal resistance of the cell and reduces the voltage and current the cell produces, with a resultant loss of charge. Figure 12-12 summarizes the basic chemical change of a lead-acid cell during discharge.

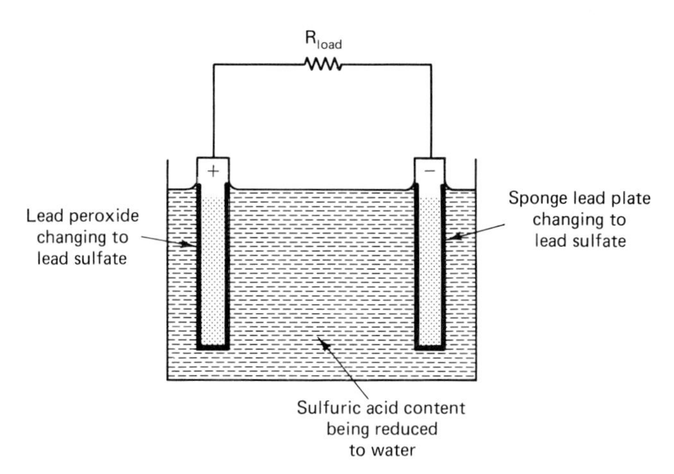

FIGURE 12-12 Summary of the basic chemical change of a lead-acid cell during discharge.

Charging Action

When the voltage of the cell drops to approximately 2 V, or the specific gravity of the electrolyte solution falls to 1.180 or approximately 1.2, the battery should normally be recharged.

Charging the battery is done by connecting a dc source to the battery terminals. It is necessary that the positive of the charging source be connected to the positive of the battery terminal. Similarly, the negative of the source is connected to the negative battery terminal (see Fig. 12-13).

The *charging voltage must be higher* than the normal battery voltage in order to force a reverse current through the battery. This reverse current causes the lead sulfate at the negative plate to recombine with hydrogen ions and reform sulfuric acid in the electrolyte. Once the lead sulfate is completely recombined, the negative plate returns to its normal sponge lead texture. At the same time, the lead sulfate on the surface of the positive plate recombines with water to increase the sulfuric acid in the electrolyte as well as to regenerate the lead peroxide positive plate.

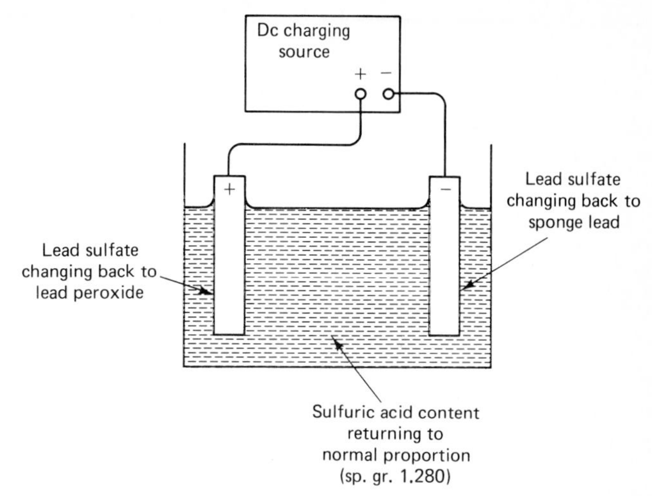

FIGURE 12-13 Summary of the basic chemical change of a lead-acid cell during a charge.

The charging action thus re-forms both the positive and negative plates as well as returning the electrolyte to its original strength. Figure 12-13 summarizes the basic chemical change of a lead-acid cell during a charge. With a proper charging technique and maintenance, the lead-acid battery is capable of being charged and discharged a great many times, to give satisfactory service over many years.

Construction

The electrodes are single plates of lead-antimony alloy with checkerboard pockets or grid cavities. Lead oxide is then pressed into the cavities, as shown in Fig. 12-14(a).

To increase the surface area with resultant current capacity, a number of positive and negative plates are alternately interweaved and separated by porous rubber separators, as shown in Fig. 12-14(b). The negative plates are then interconnected to a common electrical bus or link. The positive plates are similarly electrically linked to a common bus.

The cells are assembled into a case of hard rubber or plastic. The electrode of each cell is series interconnected so that the final terminal voltage of the battery is either 6 V or 12 V.

Vents through threaded filler caps must be provided to release the hydrogen gas produced during recharging. The filler caps also allow distilled water to be added as the water evaporates from the electrolyte.

The battery is then charged by one of the following methods.

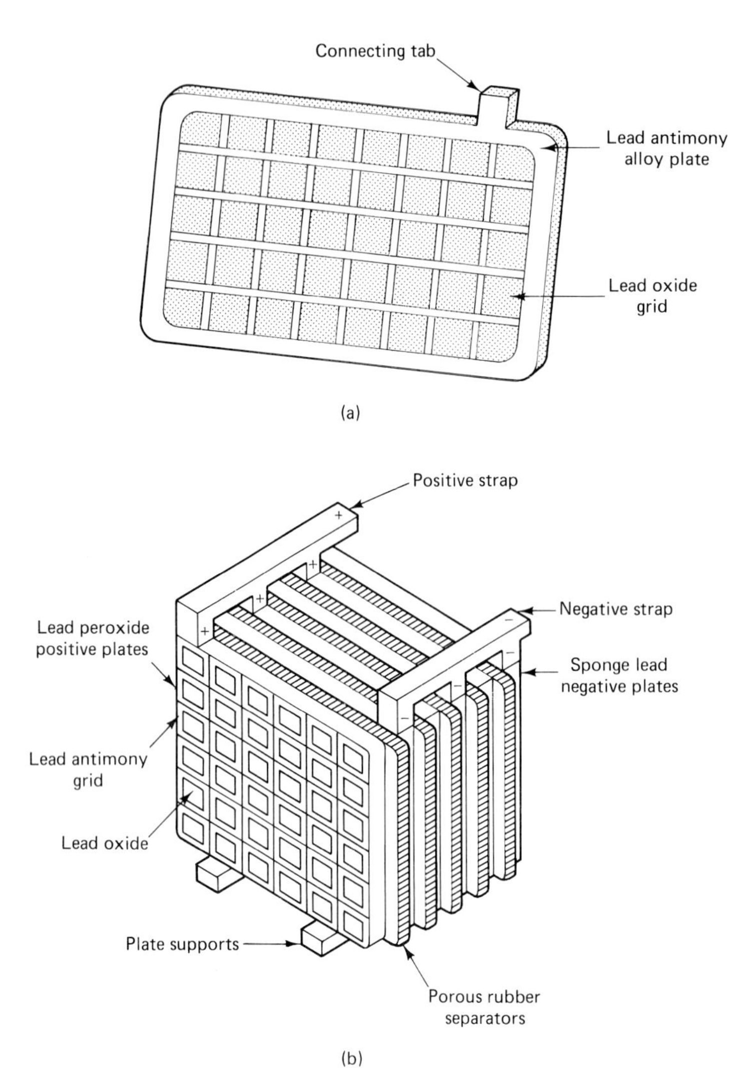

FIGURE 12-14 (a) Single plate of a lead-acid cell. (b) Plates and separator arrangement and connections of a lead-acid cell.

1. *Wet-charge method.* Once the cells are completely encased, the electrolyte is added and the battery is charged. The charging current forms the electrodes to lead peroxide and sponge lead.

2. *Dry-charge method.* In this method, lead peroxide is pressed into the grid cavities of the lead-antimony alloy positive plates and sponge lead into the grid cavities of the lead-antimony alloy negative plates. Electrolyte can be added to the battery at the time it is required for service as the battery does not need to be charged.

Maintenance-Free Lead-Acid Battery

All lead-acid automobile batteries have no filler caps and are completely sealed. This is possible because the amount of hydrogen produced by charging is eliminated.

Antimony is normally alloyed with lead to enhance the castability of the grid cavities into the lead plate. It has been found that reducing the amount of antimony in the lead reduces the amount of hydrogen gas produced.

Maintenance-free batteries use cells made of antimony-free plates, such as lead-calcium or lead-calcium-tin. Since there is no gas buildup, the battery can be completely sealed.

Service Capacity

Service capacity is the amount of electrical energy that can be withdrawn for a period of time under varying load and temperature conditions. The ampere-hour rating, Ah, of a battery is not indicative of its service capacity.

Ampere-hour rating specifies the rate of discharge in amperes at 25°C for an 8-h period. After this rate of discharge, the battery voltage falls to about 1.7 V per cell and the battery is discharged.

Typically, a 12-V lead-acid auto battery ampere-hour rating ranges from 100 to 300 Ah. This means that the battery will be able to deliver 100 A for 1 h or 12.5 A for 8 h without any cell voltage dropping below 1.7 V. From equation (12-1) the ampere-hour rating can be stated as

$$\text{Ah rating} = I \times t$$

therefore,

$$t = \frac{\text{Ah}}{I}$$

At lower temperatures the chemical reaction slows down, so that the battery is not able to deliver its ampere-hour capacity. As an example, the lead-acid cell will lose approximately 40% of its capacity when the temperature drops to −18°C (0°F). For auto applications the cold-cranking amperes are specified. Depending on the battery size, the cranking amperes range from 445 to 600 A at −18°C (0°F).

SUMMARY

A voltaic cell can be made using zinc and copper electrodes in a container of sulfuric acid solution termed electrolyte.

The insulating effect of the hydrogen gas about the positive copper electrode is termed polarization.

The internal resistance of a cell is a combination of the cell's electrode surface area, its depolarization efficiency, and contact resistance, and is a function of cell size.

The open-circuit voltage for a voltaic cell is approximately 1.1 V.

When the internal resistance of a cell causes the output voltage to fall below a useful value, this voltage value is termed the cutoff or endpoint voltage.

The terminal voltage of a cell with a load connected is called the working voltage or voltage under load.

The ampere-hour capability is an approximation of a cell's ability to supply a certain amount of current for a length of time to a nominal endpoint voltage.

The service capacity is the amount of electrical energy that can be withdrawn for a period of time under varying load and temperature conditions.

The duty cycle refers to the time duration and the frequency with which a cell or battery is drained.

A carbon-zinc cell has a positive carbon electrode and a negative zinc electrode which also acts as a container.

Manganese dioxide is used as a depolarizer.

The working voltage of a carbon-zinc cell is 1.5 V.

LeClanche cells are "sandwich"-assembled in series to reduce waste space for higher-voltage requirements.

Carbon-zinc cells are not rechargable.

Carbon-zinc cells should be stored at a low temperature to increase their shelf life.

The alkaline-manganese cell uses manganese dioxide depolarizer as the cathode in direct contact with the cell current collector and container of nickel-plated steel. The electrolyte is potassium hydroxide. Its working voltage is 1.6 V.

Silver oxide cells contain a cathode of silver oxide with a low percentage of manganese dioxide. A highly alkaline electrolyte consisting of either sodium hydroxide or potassium hydroxide is in contact with the high surface area of the zinc anode. Its working voltage is 1.5 V.

Mercury oxide cells contain an amalgamated zinc anode as the negative terminal.

The positive cathode terminal and depolarizer form a densely pressed structure of mercuric oxide with a small percentage of conductive graphite. Its working voltage is 1.35 V.

The extended shelf life, high energy density, and constant rated voltage of 1.35 V make the mercury battery a dc source of greater reliability.

Lithium cells are used where weight is important.

Lithium cells use lithium as the anode and a porous carbon cathode electrode with sulfur dioxide as the true active cathode material.

The working voltage of a lithium cell is 3 V.

The lithium cell is capable of providing its rated capacity over a wide temperature range. A shelf life of 5 to 10 years is projected.

In the charged state of a nickel-cadmium cell the positive electrode is nickelic hydroxide; the negative electrode is metallic cadmium.

In the uncharged state of a nickel-cadmium cell, the positive electrode is nickelous hydroxide and the negative electrode is cadmium hydroxide.

The working voltage of a nickel-cadmium cell is 1.2 V. It has a high current discharge rate.

Nickel-cadmium batteries should be left on a trickle charge of $\frac{1}{100}$ of the Ah rating or a refreshing charge of $\frac{1}{10}$ of the Ah rating when not in use.

The lead-acid cell consists of a lead peroxide positive electrode, reddish brown in color, and a sponge lead negative electrode, gray in color. The electrolyte solution consists of approximately 30% sulfuric acid and the balance distilled water.

The working voltage of a lead-acid cell is 2.2 V, with the A capacity ranging from 100 to 300 Ah per cell.

A discharged lead-acid cell has a layer of lead sulfate on both electrodes, which increases the internal resistance of the cell and reduces the voltage and current that the cell produces.

The lead-acid battery should be recharged when the cell voltage drops to approximately 2 V or the specific gravity falls to approximately 1.2.

Charging the battery is done by connecting a dc source having a higher voltage with the positive terminal of the battery connected to the positive terminal of the dc source, and vice versa for the negative terminal.

Maintenance-free lead-acid batteries use cells made of antimony-free plates such as lead-calcium or lead-calcium-tin. Since there is no gas buildup, the battery can be completely sealed.

REVIEW QUESTIONS

12-1. Describe the basic construction of a voltaic cell.

12-2. Explain why the zinc electrode is negatively charged in the voltaic cell.

12-3. Explain why the copper electrode is positively charged in the voltaic cell.

12-4. What is polarization?

12-5. List three factors that affect the internal resistance of any cell.

12-6. What is the cutoff or endpoint voltage of a cell?

12-7. What is the working voltage of a cell?

12-8. List two factors that determine the current capacity of a cell.

12-9. What is the ampere-hour rating of a cell?

12-10. "The ampere-hour rating of a cell is not indicative of its service capacity." Explain.

12-11. Describe the basic construction of the carbon-zinc cell.

12-12. Name and explain the purpose of the depolarizer.

12-13. What is the working voltage of a carbon-zinc cell?

12-14. What is local action? Explain how it is minimized.

12-15. List the American National Standards Association designations for small cells.

12-16. What is the shelf life for small cells?

12-17. What effect has the duty cycle on the service life of a dry cell?

12-18. What environmental conditions should be provided when storing cells and batteries for a length of time?

12-19. Describe the basic construction of the alkaline-manganese cell.

12-20. What is the working voltage of the alkaline-manganese cell?

12-21. Name three advantages of the alkaline cell over the carbon-zinc cell.

12-22. Explain why it is not advisable to recharge cells of any chemical system which are not specifically designed for recharging.

12-23. Describe the basic construction of the silver oxide–alkaline zinc cell.

12-24. What is the working voltage of the silver oxide cell?

12-25. Explain the basic construction of the mercury cell.

12-26. What is the working voltage of the mercury cell?

12-27. Give three distinct advantages of the mercury cell over other types of cells.

12-28. What type of cell is most commonly used in instant-picture cameras, hand-held calculators, and liquid-crystal watches?

12-29. What is the working voltage of the lithium cell?

12-30. What are the advantages of the lithium cell over other cells for (a) temperature and (b) shelf life?

12-31. Describe the basic construction of the nickel-cadmium cell.

12-32. Why is the nickel-cadmium cell called a secondary cell?

12-33. What is the working voltage of the nickel-cadmium cell?

12-34. What is the end-of-discharge voltage of the nickel-cadmium cell?

12-35. What is the end-of-charge voltage of the nickel-cadmium cell?

12-36. From Fig. 12-12, what is the end-of-discharge voltage given that the CF4 cell has been providing a discharge current of 400 mA for 10 h?

12-37. Compare the discharge current rate capability of the nickel-cadmium cell to that of an ordinary cell such as carbon-zinc.

12-38. What is the best and standard procedure for maintaining a charge on a nickel-cadmium cell?

12-39. Describe the basic construction of a lead-acid cell.

12-40. What is the working voltage of a lead-acid cell?

12-41. Explain why the lead-acid battery must be charged in the wet-charge method of battery construction.

12-42. Explain why the lead-acid battery does not have to be charged in the dry-charge method of construction.

12-43. Give two indications which show that a lead-acid battery needs to be recharged.

12-44. Explain why a maintenance-free lead-acid battery is completely sealed and does not need filler caps.

12-45. What are two primary requirements when charging any secondary battery?

12-46. Why does a lead-acid cell lose up to 40% of its Ah capacity at −18°C (0°F)?

SELF-EXAMINATION

Choose the most suitable word, phrase, or value to complete each statement.

12-1. The positive electrode of a basic voltaic cell is made of ________.

12-2. The negative electrode of a basic voltaic cell is made of ________.

12-3. The electrolyte is an ________ solution.

12-4. The chemical reaction producing the current gradually disolves the ________ electrode.

12-5. The open-circuit voltage of a copper-zinc cell is ________ volts.

12-6. The terminal voltage of a cell with a load connected is called the ________ voltage.

12-7. The current capacity that a cell is able to generate will depend on the ________ ________ and its ________ ________.

12-8. ________ ________ capability is an approximation of a cell's ability to supply a certain amount of current for a length of time to a nominal cutoff voltage.

12-9. What would be the expected life of a cell with an Ah rating of 2 A if it supplies a load current of 100 mA continuously?

12-10. The amount of electrical energy that can be withdrawn for a period of time under varying load and temperature conditions is termed ________ ________.

12-11. A carbon-zinc cell is discharged at 100 mA on a 2-h duty cycle. What is its life span if the endpoint voltage is 1 V? (See Fig. 12-3.)

12-12. Recharging cells that are not specifically designed for charging can be dangerous since gas buildup may cause them to ________.

12-13. Storing carbon-zinc cells at temperatures below ________°C will increase their shelf life.

12-14. Batteries or cells stored at low tempertures should be left in their original carton or wrapped in ________ ________.

12-15. In the alkaline-manganese system the electrolyte is ________ ________.

12-16. The alkaline cell uses ________ ________ depolarizer (cathode) in direct contact with the cell container of nickel-plated steel.

12-17. The working voltage of an alkaline cell is ________ volts.

12-18. Alkaline cells are capable of greater ________ and ________ rates which can be maintained over extended periods.

12-19. Silver-oxide cells contain a cathode of ________ ________ with a low percentage of manganese dioxide.

12-20. A highly alkaline electrolyte consisting of either sodium hydroxide or potassium hydroxide is in contact with the ________ anode.

12-21. Electrically, the outstanding feature of silver oxide is its ability to maintain a virtually constant ________ ________ and current regardless of discharge time.

12-22. The working voltage of a silver-oxide cell is ________ volts.

12-23. A mercury-oxide cell incorporates ________ anode as the negative terminal.

12-24. The _________ _________ terminal and depolarizer is a densely pressed structure of mercuric oxide with a small percentage of conductive graphite.

12-25. The working voltage of a mercury-oxide cell is _________ volts.

12-26. The output voltage of a mercury cell is reliable within the shelf life of the cell, which exceeds _________ years.

12-27. Instruments using mercury batteries can supply measurement accuracies within _________ percent over several years.

12-28. The working voltage of a lithium cell is _________ volts.

12-29. Outstanding features of the lithium cell are its _________ and _________ life.

12-30. The output voltage and service capacity of a lithium cell are reliable over a temperature range of _________ to _________.

12-31. A cell in which both electrode reactions are reversible and the cell can be recharged is termed a _________ cell.

12-32. In the charged state the positive electrode of a nickel-cadmium cell is _________ _________.

12-33. In the charged state the negative electrode of a nickel-cadmium cell is _________ _________.

12-34. The working voltage of a nickel-cadmium cell is _________ volts.

12-35. The nickel-cadmium cell provides most of its energy above _________ volt per cell.

12-36. A nickel-cadmium cell or battery can deliver a _________ current discharge rate.

12-37. A nickel-cadmium battery can be kept up to its full capacity when setting idle by applying a _________ _________.

12-38. The standard trickle charge rate is _________ of the ampere-hour rating of the nickel-cadmium cell.

12-39. The standard refreshing charge rate is _________ of the ampere-hour rating for a period of 14 h.

12-40. The positive electrode of a charged lead-acid cell consists of _________ _________.

12-41. The negative electrode of a charged lead-acid cell consists of _________ _________.

12-42. The electrolyte of a lead-acid cell consists of _________ percent of sulfuric acid and the balance _________ _________.

12-43. The working voltage of a lead-acid cell is _________ volts.

12-44. Upon discharge, the chemical reaction of the electrolyte on the sponge lead negative plate slowly changes the plate to _________ _________.

12-45. When the voltage of the cell drops to approximately _________ volts, the lead-acid cell should normally be recharged.

12-46. When the specific gravity of the electrolyte solution falls to approximately _________, the battery should normally be recharged.

12-47. The charging voltage must be _________ than the normal battery voltage.

12-48. It is necessary that the _________ polarity of the charging source be connected to the positive of the battery, and vice versa to the negative terminal.

12-49. Maintenance-free lead-acid batteries use cells made of _________ _________.

12-50. Since there is no gas buildup, the maintenance-free battery can be completely _________.

ANSWERS TO PRACTICE QUESTIONS

12-1

1. electrolyte
2. hydrogen, sulfate
3. zinc sulfate
4. positive, two
5. hydrogen
6. hydrogen gas
7. electron
8. polarization
9. internal resistance
10. cutoff
11. 1.1
12. working
13. electrode area, internal resistance
14. ampere-hour
15. approximately 100 h
16. depolarizer
17. 1.5
18. physical size
19. explode

12-2

1. zinc-carbon
2. depolarizer
3. 1.6
4. 0.9
5. capacity, discharge
6. hazardous
7. 1.6
8. voltage, current
9. 1.35
10. internal resistance
11. 45 to 14
12. 1
13. 10 μV
14. weight
15. 3 V
16. −18, 70
17. 5, 10

13

Conductors and Insulators

INTRODUCTION

While conductors conduct current from one point to another and an insulator insulates the conductor from its environment, the difference between an insulator and a conductor is the amount of voltage it takes to cause current to flow. Thus there is no perfect insulator since an insulation breakdown can occur with a high voltage which is termed the breakdown voltage.

In between conductors and insulators are semiconductors. Semiconductors will conduct a current provided that some external energy such as heat or a potential difference is applied. Silicon and germanium, used as a substrate for transistors and integrated-circuit chips, are two common semiconductors.

Whether a material has conducting or insulating ability depends on its atomic and interatomic structure. Good conductors are materials with free valence electrons. Materials that do not have free electrons and in which the valence electrons are tightly bonded to the atom are insulators.

In Chapter 2 we looked at the resistance properties of a conductor, which were dependent on four factors:

1. Resistivity of the material
2. Length
3. Cross-sectional area
4. Temperature

In this chapter we provide an insight into what makes a good conductor versus a poor conductor, insulator, or semiconductor. Wire sizes, wire gauges, and safe current capacity for a copper conductor are discussed. An electronics technician should be able to select the appropriate wire cable size for fixed or portable electronic equipment and have a fundamental understanding of wire size, insulation, and conductor overcurrent protection.

MAIN TOPICS IN CHAPTER 13

OBJECTIVES

After studying Chapter 13, the student should be able to

1. Define an insulator, a conductor, and a semiconductor.
2. Define wire sizes, wire gauge, and safe current capacity for copper wire.

13-1 CONDUCTORS

Recall from Section 1-3 that all matter had shells or bands of electrons. Each shell had a specific number of electrons and the outermost shell was called the valence shell or band. Electrons in the valence band of good conductors are loosely bonded to the nucleus and therefore require very little energy to break away from the valence band to become an electrical current or current carrier.

Whether a material is a conductor, an insulator, or a semiconductor depends on the number of free electrons available when the atoms bond together to form a solid, and the ease with which the current carriers drift through the material under the influence of an electric field. These two factors, plus the charge on an electron, govern the resistivity of a material. Materials with the lowest resistivity per unit volume are the best conductors. Table 2-4 shows these to be silver as the best, then copper, gold, and aluminum.

As an example, silver and copper have one valence electron (see Table 1-1, Electronic Structure of Atoms). Valence electrons are easily dislodged from their shell to become current carriers drifting about the inner atomic space. The loss of an electron produces a hole or a positive ion. Electrostatic bonding between the negative electrons and the positive ions, termed metallic bonding, determines the ease with which the current carriers drift through the material. Table 13-1 shows that silver has the greatest number of carriers per unit volume as well as the greater electron mobility, with copper next best.

13-2 INSULATORS

Unlike conductors, materials that are insulators are compounds that have two or more atoms that form molecules. Each atom shares its valence electrons with an adjacent atom to form what is called covalent bonding, as illustrated in Fig. 13-1. Covalent bonding takes place particularly in rubber and plastics, as well as in semiconductors such as germanium and silicon.

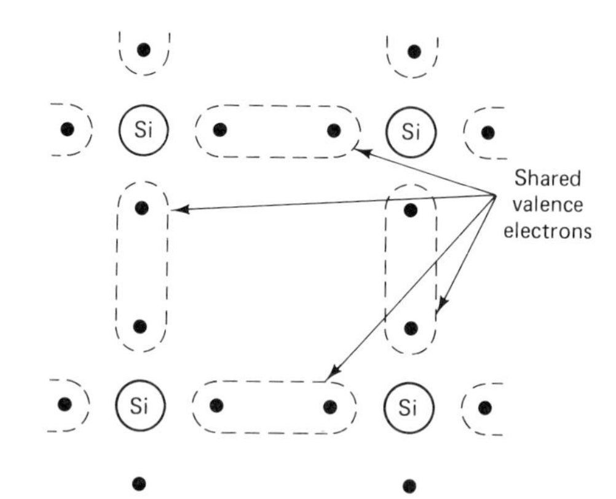

FIGURE 13-1 Covalent bonding between silicon atoms.

In other materials, atoms form ionic bonding. With ionic bonding some atoms give up electrons and become positive ions, while adjacent atoms accept these electrons to become negative ions. Electrostatic attraction between the two unlike charged ions forms ionic bonds. Materials such as glass and porcelain are examples of ionic bonding between atoms. Because free electrons are very scarce, no current can flow and the material is an insulator.

A perfect insulator would have a resistivity of infinite resistance. Table 13-2 shows the resistivity of typical insulators. Although very high, some insulators are better than others and certainly not perfect.

A sufficiently high voltage across an insulator sets up a very high electrostatic force which "pulls out" electrons from the atoms. The resultant large current flow ruptures the insulation and destroys it. This high voltage, termed *breakdown voltage,* is shown for various insulators in Table 13-3.

TABLE 13-1

Nominal values of current densities and mobility at 20°C

Material	*Carriers/cm^3*	*Electron Mobility, μ_n (cm^2/V-s)*	*Hole Mobility, μ_p (cm^2/V-s)*
Silver	9.0×10^{22}	46	Not applicable
Copper	8.5×10^{22}	43	Not applicable
Aluminum	5.1×10^{22}	44	Not applicable
Germanium	2.4×10^{13}	3900	1900
Silicon	1.5×10^{10}	1350	480

TABLE 13-2

Nominal values of resistivity for typical insulators

Material	*Resistivity, ρ (Ω-cm)*
Air	Not applicable
Glass	1.0×10^{14}
Bakelite	1.0×10^{12}
Waxed paper	1.0×10^{9}
Rubber	1.0×10^{13}
Teflon	1.0×10^{14}

TABLE 13-3

Nominal values of breakdown voltage for typical insulators

Material	*Approx. Breakdown Voltage (kV/cm)*
Air	30
Glass	1200
Bakelite	150
Waxed paper	500
Rubber	270
Teflon	600

Insulated Wire and Cables

The Canadian Standards Association in Canada and The Underwriters' Laboratories in the United States are the governing agencies that set the maximum conditions under which electrical equipment can be operated. Rules and regulations for the installation and maintenance of all electrical equipment are published in the form of the National Electrical Code® in the United States and the Canadian Electrical Code in Canada. In conjunction with the electronic and electrical industry, trade groups, electrical utilities, fire marshals, and many others, the American and Canadian national standards associations oversee, revise, and publish the various rules and standards.

Insulated wire used as hookup wire in electronic circuits, nonmetallic sheath cable for house wiring, flexible cord cable used for portable equipment, and other types of wire and wire cables are rated according to the maximum current and voltage at which the insulation is designed to be used. For example, the maximum voltage rating for hookup wire with CSA type TR-64 vinyl insulation is 300 V to a maximum temperature of 80°C. CSA type TEW vinyl-insulated wire has a rating of 600 V at 105°C. Military Type E MIL-W-16878 Teflon-insulated wire has a rating of 600 V at 200°C. Flexible power cords used under varying conditions must have a specific type, thickness, and voltage rating. Table 13-4 shows some uses, CSA designations, and voltage ratings for flexible power cords used on portable electrical equipment.

TABLE 13-4

CSA designations and voltage rating for flexible cable

Condition	*Use*	*Type*	*CSA Designation*	*Voltage Rating*	*Temp. Rating (°C)*
Damp or dry locations	For hard usage	Flexible cord	SJ	300	60
			SJO	300	60
			SJT	300	60
			SPT-3	300	60
	For extra hard usage	Flexible cord	S	600	60
			SO	600	60
			ST	600	60
		Power supply cable	SG	600	60
			SGO	600	60
			SW	600	60
			SWO	600	60
			SWT	600	60
Wet or damp or dry locations	For hard usage	Outdoor flexible cord	SJOW	300	60
			SJTW	300	60
		Outdoor flexible cord	SOW	600	60
			STW	600	60

Note: See National Electrical Code® (USA) Table 400-4.
Source: From Canadian Electrical Code, courtesy of Canadian Standards Association.

CSA type SJ or S is most commonly used for the power cord on oscilloscopes and other electronic equipment. This type of cable is rubber or thermoplastic insulated.

13-3 SEMICONDUCTORS

Recall that the valence band of a material seeks to have eight electrons to fill its outermost shell. Atoms in silicon and germanium have four valence electrons (see Table 1-1). Because of the crystalline structure of silicon and germanium atoms, each can accept four additional electrons into the holes or valence band, or share its four electrons with an adjacent atom. This sharing of valence electrons, termed *covalent bonding*, is illustrated in Fig. 13-1. Since the valence shell of each atom is now filled with eight electrons, there are no free electrons to drift about the silicon or germanium, so they behave like insulators. Applying heat or electrical energy, however, will add energy to some of the valence electrons and cause them to break away from their atoms. The resulting current, although very small, can be made to flow through the silicon or germanium. Thus silicon and germanium can be said to be semiconductors.

PRACTICE QUESTIONS 13-1

Choose the most suitable word or phrase to complete each statement.

1. A good conductor has many ________ ________ when the atoms bond themselves together to form a solid.
2. ________ is the best conductor since it has the lowest resistivity.
3. Metallic bonding between atoms of a material determines the ease with which the current carriers drift through a material and is termed ________ of carriers.
4. Atoms that share their valence electrons are said to form ________ bonding.
5. Covalent bonding takes place in materials such as germanium and silicon that are called ________.
6. Covalent bonding takes place in insulators such as ________ and ________.
7. An electrostatic attraction between two unlike charged ions forms ________ bonds.
8. Materials such as ________ and ________ are examples of ionic bonding between atoms.
9. Insulators are not perfect and will conduct electricity with a sufficiently high voltage called ________ ________.
10. The maximum voltage rating of insulated conductors as well as the maximum current-carrying capacity is set by a governing agency called ________.
11. Oscilloscopes and similar types of electronic equipment must use CSA type ________ or ________ flexible power cord.
12. Before a slight current can flow through a semiconductor such as silicon or germanium, ________ or ________ energy must be applied.

13-4 THE CIRCULAR MIL

Wire, whether it is single-conductor or multiconductor cable, must be manufactured in a predetermined size. To standardize electrical wiring, a wire gauge was developed and accepted by the industry. This wire size gauge shown in Fig. 13-2 is known as the American wire gauge (AWG). It is similar to the Brown and Sharpe (B&S) British gauge.

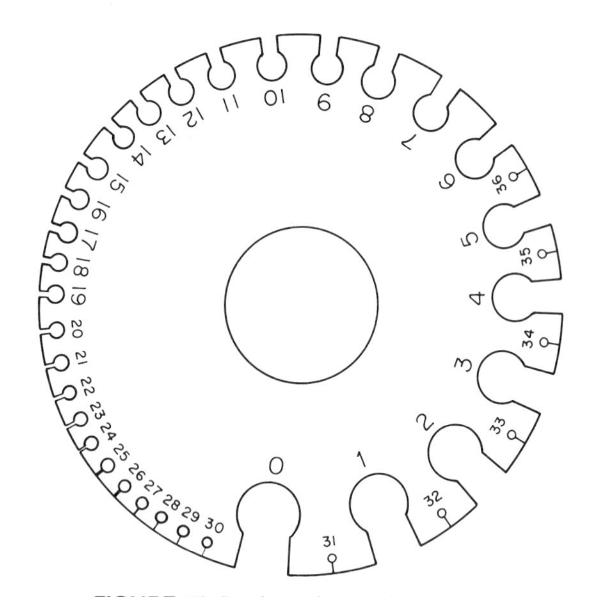

FIGURE 13-2 American wire gauge.

The American wire gauge measures the diameter of a wire in mils and its cross-sectional area in circular mils (cmil). One circular mil is defined as the area of a circle having a diameter of 1 mil, where 1 mil equals 10^{-3} in. or 1/1000 in. (Fig. 13-3).

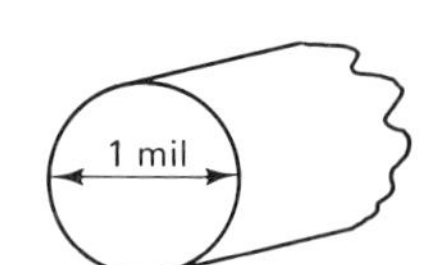

FIGURE 13-3 Circular-mil area.

TABLE 13-5
Cross-sectional area of a wire with an increase in diameter

Diameter (mils)	*Area* $= \pi \frac{d^2}{4}$ *(*10^{-6} *in.*^{2}*)*	*Area* $= d^2$ *(cmil)*	*Increase in Area*
1	0.785	1	× 1
2	3.142	4	× 4
3	7.068	9	× 9
4	12.56	16	× 16

Since the area of a circle is found using the formula $A = \pi r^2$ or $\pi d^2/4$, therefore the area of a circle *increases* as the *square* of the diameter. This is illustrated in Table 13-5.

Stating the diameter of a wire in mils and squaring the diameter gives the area of a wire in terms of circular mils:

$$\text{area} = d^2 \quad \text{circular mils} \qquad (13\text{-}1)$$

Note: Diameter is in mils. The circular mil reflects the *increase* or *decrease* in cross-sectional area of a wire from an increase (or decrease) in its diameter.

EXAMPLE 13-1

A copper conductor having a diameter of 0.0142 in. is replaced by a conductor 0.0201 in. in diameter. Calculate the increase in cross-sectional area for the replaced conductor.

Solution: The diameter of each conductor must be converted to mils; then the circular-mil area can be calculated using formula (13-1). For the 0.0201-in. conductor its diameter is equal to 20.1/1000 in. or 20.1 mils and its cross-sectional area is

$$A\text{ (cmil)} = d^2\text{ (mils)} = [20.1\text{ (mils)}]^2 = 404.0\text{ cmil}$$

For the 0.0142-in. conductor, its diameter is equal to 14.2 mils and its cross-sectional area is

$$A\text{ (cmil)} = d^2\text{ (mils)} = [14.2\text{ (mils)}]^2 = 201.6\text{ cmil}$$

The increase in cross-sectional area is

$$\frac{404.0\text{ cmil}}{201.6\text{ cmil}} = 2\times$$

13-5 WIRE GAUGE AND WIRE TABLES

North American wire manufacturers conform to the American wire gauge system of measuring wire size. This wire gauge, shown in Fig. 13-2, allows the cross-sectional area of a wire to be measured simply by snugging the wire into the appropriate-fitting hole. Note that the gauge numbers increase from 0 to 40. Lower numbers indicate a large wire size; higher numbers indicate a smaller wire size. Table 13-6 illustrates this system. From the table we can see that No. 0 wire has a cross-sectional area of 106,000 cmil, while No. 36 has a cross-sectional area of 25 cmil.

Another feature of the American wire gauge is that the conductor sizes were selected so that a decrease of one gauge number results in a 25% increase in cross-sectional area. Examining the numbers further shows that a decrease of three gauge numbers results in doubling the area. Of course, doubling the area will halve the resistance of the wire. Alternatively, an increase of three gauge numbers will halve the area and result in doubling the resistance.

The resistance of any conductor can be approximated using the principles outlined above. Consider that the resistance of 1000 ft of AWG No. 10 wire is approximately 1 Ω; then the resistance of any wire gauge size can be estimated very accurately without the use of tables.

EXAMPLE 13-2

Estimate the resistance of 250 ft of AWG No. 16 copper wire.

Solution: The resistance of 1000 ft of AWG No. 10 wire is 1 Ω or 0.001 Ω/ft. Decreasing its length to 250 ft will decrease its resistance:

$$R = 250\text{ ft} \times \frac{0.001\ \Omega}{\text{ft}} = 0.25\ \Omega$$

Increasing the cmil area of the same wire by three gauge sizes will double its resistance to 0.5 Ω. Increasing by three more gauge numbers will again double its resistance, to 1 Ω. Table 13-6 gives the value of 1000 ft of No. 16 wire, as 4.02 Ω dividing by 4 gives 1 Ω. (*Note:* The length is reduced by four times and $R \propto L$.) The resistance between

TABLE 13-6
American wire gauge AWG wire sizes and metric equivalent

Gauge No.	Diameter (mils)[a]	Area (cmil)	Resistance (Ω/1000 ft)	Diameter (mm)[b]	Resistance (Ω/km)[c]
4/0	460.0	212,000	0.049	11.68	0.161
3/0	410.0	168,000	0.062	10.41	0.203
2/0	365.0	133,000	0.078	9.271	0.256
1/0	325.0	106,000	0.098	8.255	0.321
1	289.3	83,690	0.124	7.348	0.407
2	257.6	66,370	0.156	6.543	0.512
3	229.4	52,640	0.197	5.827	0.646
4	204.3	41,741	0.248	5.189	0.814
5	181.9	33,100	0.313	4.620	1.027
6	162.0	26,250	0.395	4.115	1.30
7	144.3	20,820	0.498	3.665	1.63
8	128.5	16,510	0.628	3.264	2.06
9	114.4	13,090	0.792	2.906	2.60
10	101.9	10,380	0.999	2.588	3.28
11	90.74	8,234	1.26	2.305	4.13
12	80.81	6,530	1.59	2.053	5.22
13	71.96	5,178	2.0	1.828	6.56
14	64.08	4,107	2.52	1.628	8.27
15	57.07	3,257	3.18	1.450	10.4
16	50.82	2,583	4.02	1.291	13.2
17	45.26	2,048	5.06	1.150	16.6
18	40.30	1,624	6.39	1.024	21.0
19	35.89	1,288	8.05	0.912	26.4
20	31.96	1,022	10.1	0.814	33.1
21	28.46	810.1	12.8	0.723	42.0
22	25.35	642.4	16.1	0.644	52.8
23	22.57	509.5	20.3	0.573	66.6
24	20.10	404.0	25.7	0.511	84.0
25	17.90	320.4	32.4	0.455	106
26	15.94	254.1	41	0.405	134
27	14.20	201.5	51.4	0.361	168
28	12.64	159.8	64.9	0.321	213
29	11.26	126.7	81.4	0.286	267
30	10.03	100.5	103	0.255	338
31	8.928	79.70	130	0.227	426
32	7.950	63.21	164	0.202	538
33	7.080	50.13	206	0.180	676
34	6.305	39.75	261	0.160	856
35	5.615	31.52	329	0.143	1079
36	5.000	25.00	415	0.127	1361
37	4.453	19.83	523	0.113	1715
38	3.965	15.72	655	0.101	2148
39	3.531	12.47	832	0.090	2729
40	3.145	9.88	1044	0.080	3424

[a]Diameter in inches = diameter in mils/1000.
[b]Diameter in millimetres = 25.4 × diameter in inches.
[c]Ω/km = Ω/1000 ft/0.3048

three gauge numbers can be estimated roughly using an average value among the three values.

EXAMPLE 13-3

Estimate the resistance of 250 ft of AWG No. 18.

Solution: From Example 13-2 the resistance of AWG No. 16 was found to be 1 Ω. Increasing three gauge numbers to No. 19 will double the resistance to 2 Ω. Averaging the difference of 1 Ω for three gauge numbers gives 0.33 Ω per gauge number, so No. 18 has a resistance of 1 Ω plus 0.33 Ω or 1.33 Ω for gauge No. 17 and 1.66 Ω for 250 ft of No. 18 wire. Although not as accurate as the three-gauge estimate, it serves, nevertheless, as a reasonable estimate without the use of tables.

Metric System versus Circular Mil System

In Canada the conversion to the metric system has been made for the majority of measurements. With any system, however, conversion can be made only if the system is universally accepted. This is not the case with the wire gauge. The AWG has been the North American standard for some time, and as long as it continues to be used in the United States, it will also continue to be used in Canada. To this end, Canada has converted the diameter and length of a wire to metric, leaving the circular-mil area out of the tables but retaining the gauge numbers, as shown in Table 13-6.

Since the metric system uses the AWG gauge, the same principle of resistance doubling for every increase in three gauge numbers can also be applied. Table 13-6 shows the resistance of 1 km of AWG No. 5 conductor to be approximately 1 Ω. An increase of three gauge numbers doubles the resistance to approximately 2 Ω. Similarly, a decrease of three gauge numbers will halve the resistance to approximately 0.5 Ω.

The resistance of any length of copper conductor can be approximated by using the principles outlined above.

EXAMPLE 13-4

Estimate the resistance of 200 m of AWG No. 14 copper wire.

Solution: The resistance of 1 km of AWG No. 5 wire is 1.027 Ω, or approximately 1 Ω (i.e., 0.001 Ω/m). Decreasing its length to 200 m will decrease its resistance:

$$R = 200 \text{ m} \times \frac{0.001\ \Omega}{\text{m}} = 0.2\ \Omega$$

Increasing three gauge numbers, to No. 8, doubles its resistance to 0.4 Ω. Increasing three more gauge numbers, to No. 11, doubles the resistance to 0.8 Ω. Finally, increasing three more gauge numbers, to No. 14, doubles the resistance to 1.6 Ω. [Note that this is approximately one-fifth of the total resistance for a 1-km length of wire (i.e., 200 m).]

Thus the American wire gauge is integrated with the metric system by retaining the gauge numbers while stating the diameter in equivalent millimetres and the length in kilometres.

PRACTICE QUESTIONS 13-2

Choose the most suitable word, phrase, or value to complete each statement.

1. One mil is equal to ________ inches.
2. A wire having a diameter of 1 mil a has cross-sectional area of 1 ________.
3. Convert the diameter of a 0.0403-in. wire to mils.
4. Convert the diameter of a 101.9-mil wire to inches.
5. Calculate the cross-sectional area of the 0.0403-in. diameter wire.
6. As American wire gauge numbers increase, the cross-sectional area of a wire ________.
7. Which is the largest wire size: No. 40, No. 14, or No. 30?
8. In the AWG the cross-sectional area is halved for an increase of ________ gauge numbers.
9. In the AWG the resistance of the wire is ________ for a decrease of three gauge numbers.
10. Estimate the resistance of an AWG No. 20 copper conductor which is 1 km long, given that AWG No. 5 is 1 Ω/1 km.

13-6 MAXIMUM SAFE CURRENT CAPACITY FOR AN INSULATED CONDUCTOR

Having discovered that a conductor has a resistance that depends on its length and cross-sectional area, it follows that a current flowing through the conductor will cause the conductor to heat. If the current is sufficiently great, the insulation will reach the combustion temperature and ignite. This is what causes an electrical fire.

The maximum safe current-carrying capacity of a conductor depends on three factors:

1. Cross-sectional area
2. Permissible temperature rise
3. Ambient temperature

Cross-Sectional Area. The AWG gauge shows that for a decrease of three gauge numbers, the resistance of a conductor is halved. The power formula $P = I^2R$ tells us that reducing the resistance by one-half will reduce the power dissipation of the conductor by one-half with the same current. Conversely, increasing the wire by three AWG sizes will double its resistance;

therefore, the power dissipation will double with the same current.

Permissible Temperature Rise. In this case the insulating material will govern the permissible temperature rise. Some materials, asbestos, for example, do not ignite and will withstand extremely high temperatures. Other materials, such as vinyls, will withstand much lower temperatures and will ignite given the right conditions. Some types of insulation are slow burning and smolder but do not burst into flame.

Ambient Temperature. The ambient temperature is the environmental temperature surrounding the conductor. Asbestos-insulated wire is used in wiring toasters, electric stoves, dryers, and other appliances because the environment of the wires is hot, very hot in toasters and dryers. Normally, the environment about electronic equipment is not as severe as the toaster or stove. Nevertheless, care must be exercised, particularly with regard to the insulation characteristics, to avoid a fire in the equipment.

As in the case of all electrical equipment, the safe current-carrying capacity of copper and aluminum conductors is determined by the national standards associations. Tables 13-7 and 13-8 show examples of conductor ratings for a single conductor in free air and three conductors in a cable, or raceway, respectively. Table 13-9 shows examples of allowable current capacity for flexible cord and equipment wire.

TABLE 13-7

Allowable ampacities for single copper conductors in free air based on an ambient temperature of 30°C

	Allowable Ampacity[a]		
	60°C	*75°C*	*85–90°C*
AWG No.	*Types T, TW*	*Types RW 75, TWH*	*Types R 90 V, RW 90, THHN, A-18*
14	20	20	20
12	25	25	25
10	40	40	40
8	55	65	75

Note: See National Electrical Code® (USA) for comparable ratings.
[a]See Table 5A of the Canadian Electrical Code for the correction factors to be applied for ambient temperatures over 30°C.
Source: Excerpt from Table 1 of Canadian Electrical Code. Courtesy of Canadian Standards Association.

TABLE 13-8

Allowable ampacities for not more than three copper conductors in raceway or cable based on an ambient temperature of 30°C

	Allowable Ampacity[a]		
	60°C	*75°C*	*85–90°C*
AWG No.	*Types T, TW*	*Types RW 75, TWH*	*Types R 90 V, RW 90, THHN, A-18*
14	15	15	15
12	20	20	20
10	30	30	30
8	40	45	50

Note: National Electrical Code® (USA) for comparable ratings.
[a]See Table 5A, Canadian Electrical Code, for the corrections to be applied for ambient temperatures over 30°C.
Source: Excerpt from Table 2 of Canadian Electrical Code. Courtesy of Canadian Standards Association.

TABLE 13-9

Allowable ampacity of flexible cord and equipment wire based on an ambient temperature of 30°C

	Allowable Ampacity		
	Flexible Cord		
	Christmas-Tree Cord	*Portable Equipment Cord*	
	Types	*Types SJ, SJO, SPT-3 S, SO, SG, SGO SW, SWO, SWT, SJOW, SJTW SOW, STW*	
AWG No.	*TX, CXWT, PXT*	*Two Conductors*	*Three Conductors*
27	—	—	—
26	—	—	—
24	—	—	—
20	2	2	—
18	5	10	7
16	7	15	10
14	15	18	15
12	20	25	20

Note: National Electrical Code® (USA), Table 400-5, Ampacity of Flexible Cords and Cables.
Source: Part of Table 12 from the Canadian Electrical Code. Courtesy of Canadian Standards Association.

13-7 OVERCURRENT DEVICES

Overcurrent devices such as the fuse or circuit breaker protect electrical and electronic equipment from damage due to an excessive or short-circuit current that could cause a fire. They also protect personnel from shock hazard due to a sudden defect in the wire insulation, especially in the power cord. There are two types of overcurrent devices: the fuse and the circuit breaker.

Fuses

A fuse is always connected in series with the high-potential side of the source. Figure 13-4 shows the symbol for a fuse connected in series with a load.

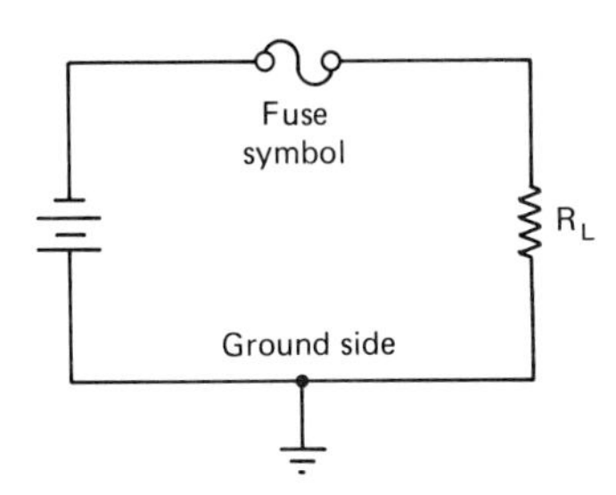

FIGURE 13-4 Fuse connected in series with a load.

Fuses are made from tinned copper, aluminum, or nickel wire that melts at a predetermined temperature. This temperature depends on the amount of current flowing through the fuse. Since the fuse is in series, the circuit current flows through it, and any overload or short circuit current will melt or "blow" the fuse open, disconnecting the high side of the source from the load. Since one side of the source voltage is grounded or at zero volts, and the other side of the source voltage is disconnected from the source by the blown fuse, shock hazard due to faulty equipment is eliminated.

Most electronic equipment has a fuse holder mounted at the rear of the equipment. Typical equipment fuses and holders are shown in Fig. 13-5.

Fuse Specifications and Selection. Fuses are selected according to their current rating, voltage rating, and blow characteristic. Instrument fuses are available from 2 mA to 30 A. Normally, an instrument fuse is replaced with the same current rating as the blown fuse. To protect the equipment from overloads, manufacturers' specifications or nameplate maximum current ratings for the fuse must never be exceeded.

The voltage rating of instrument fuses is the maximum circuit voltage at which the fuse and fuse holder are designed to interrupt. Standard voltage ratings are 32 V, 125 V, 250 V, and 600 V.

Two types of blow characteristics are the fast-acting and the slow-blow. Fast-acting fuses will blow in approximately 10 ms, with a current five times their ampere rating. Slow-blow fuses take about 2 s to blow with a current five times their ampere rating. For some equipment a 10 to 50% overload current can be tolerated for a short period, so a properly chosen slow-blow fuse avoids the nuisance of a blown fuse with every temporary overload.

Circuit Breakers

Circuit breakers serve the same function as the fuse. Although very much larger, their chief advantage is that they can be reset after each overload or short circuit. There are two types, the thermal and the magnetic. The thermal type is common in TV sets and other appliances. Its operation as a fuse depends on a bimetal strip (two dissimilar metals bonded together) that heats up with current flow through it. With a different temperature coefficient of expansion for each metal, the heat

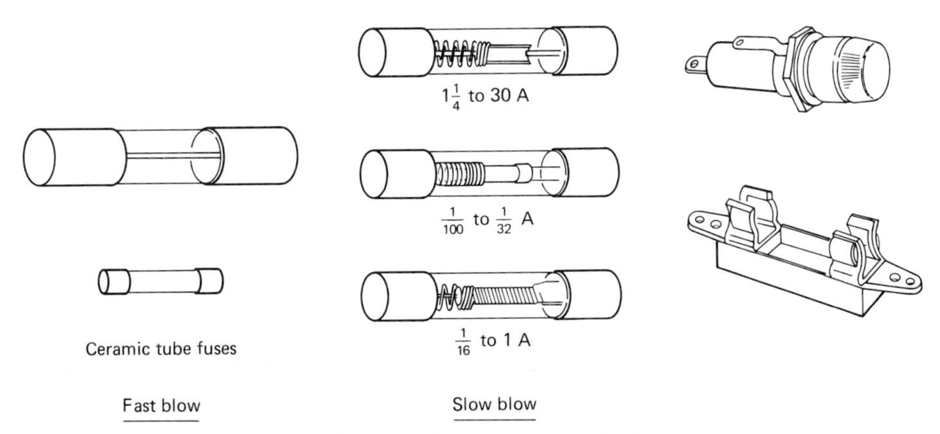

FIGURE 13-5 Fuses and fuse holders.

causes the bimetal strip to bend, tripping the circuit open. The circuit can be reset after the bimetal cools. A circuit breaker symbol is shown in Fig. 13-6.

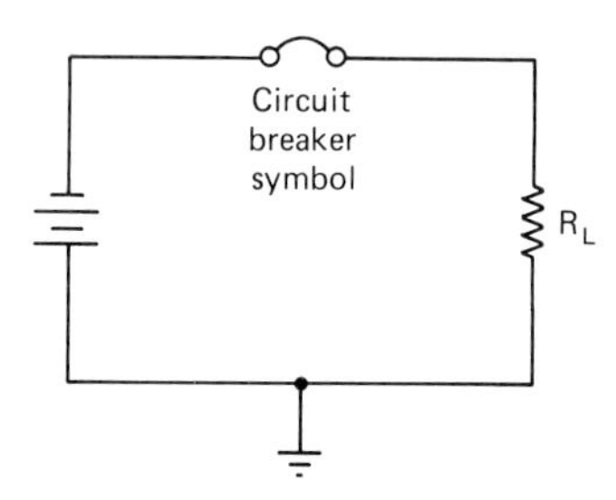

FIGURE 13-6 Circuit breaker connected in series with the load.

The magnetic breaker operates on the principle of an electromagnet. With normal current flowing through the coil of the electromagnet, the breaker contact is held closed by a spring-loaded mechanism. When the breaker current rating is exceeded, the breaker contact is tripped open by the action of the electromagnet and can be reset once the overload or short is removed. Table 13-10 shows the rating of fuses and circuit breakers for conductor current capacities up to 70 A.

TABLE 13-10

Rating or setting of overcurrent devices protecting conductors[a]

	Rating or Setting Permitted	
Ampacity of Conductor	*Fuse (A)*	*Circuit Breaker (A)*
0–15	15	15
16–20	20	20
21–25	25	30
26–30	30	30
31–35	35	40
36–40	40	40
41–45	45	50
46–50	50	50
51–60	60	60
61–70	70	70

[a]For general use where not otherwise specifically provided for.
Source: Part of Table 13 from Canadian Electrical Code. Courtesy of Canadian Standards Association.

Troubleshooting a Blown Fuse

An easily accessible fuse can be checked quickly by the use of a voltmeter. With the voltmeter leads across the fuse as shown in Fig. 13-7, an open fuse will indicate the full source voltage. A good fuse will indicate 0 V. Other types of fuses must be withdrawn from their holder and checked on an ohmmeter.

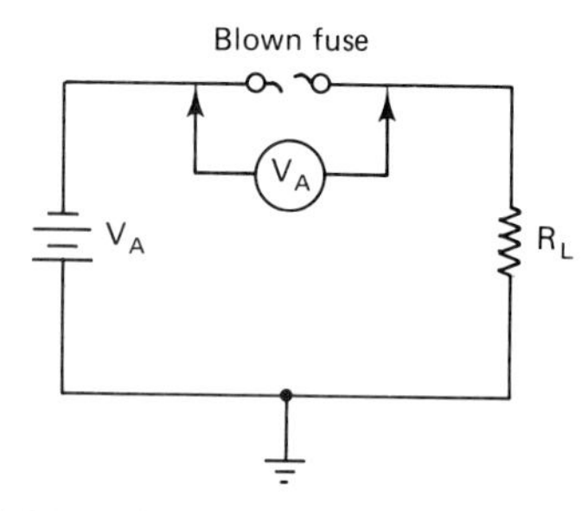

FIGURE 13-7 Voltmeter indicates V_A across open fuse.

Once it is determined that a fuse is blown, the cause of the overload or short must be determined and corrected. A convenient method to determine an overload or short-circuit condition is to connect a wattmeter in series with the equipment. A higher-rated slow-blow fuse is replaced and the equipment is connected to a variac or autotransformer. The ac input voltage is reduced to one-half (i.e., 55 V for 110 V ac input). Note the wattage drawn at this point.

The faulty component is found by disconnecting various dc supply sections, one at a time, and noting the wattage drawn each time. Excessive current and resultant wattage will drop noticeably when the defective section or component is disconnected. Once the section is isolated, an ohmmeter can normally be used to find the shorted or faulty component. *Do not forget* to replace the fuse with one of proper type and rating.

SUMMARY

Two factors that govern whether a material is a conductor or an insulator are:

1. The number of free electrons available
2. The ease with which the current carriers drift through the material

Silver, followed by copper, gold, and aluminum, are the best conductors. Silver is the best conductor since it has the greatest number of current carriers per unit volume as well as the greatest electron mobility.

In insulators the valence electrons form covalent bonds or ionic bonds between atoms.

Covalent bonding means that each atom of a material shares its valence electrons with adjacent atoms.

Covalent bonding between atoms is found in materials such as rubber and plastics, and in semiconductors such as germanuim and silicon.

Ionic bonding between atoms is found in materials such as glass and porcelain, which are good insulators.

A sufficiently high voltage, termed the breakdown voltage, will rupture an insulator and result in a high current flow through the insulator.

Underwriters' Laboratories and the Canadian Standards Association are the governing agencies that set maximum conditions under which electrical equipment can be operated.

Semiconductors such as silicon and germanium behave like insulators until external energy such as heat or electrical energy is applied; then they become conductors.

The mil is a round conductor with a diameter of 10^{-6} in.

The circular mil is the cross-sectional area of a round conductor having a diameter of 1 mil.

The circular-mil area is calculated using formula (13-1), A (cmil) $= d^2$(mils).

American wire gauge numbers increase from 0 to 40, with the lower numbers measuring a larger-diameter wire and the higher numbers measuring a smaller-diameter wire.

Increasing three AWG wire gauge sizes means that the cross-sectional area of a conductor is halved.

Decreasing three AWG wire gauge sizes means that the cross-sectional area of a conductor is doubled.

The resistance of any length of copper conductor can be approximated by starting at AWG No. 5 with a resistance of 1 Ω/km and doubling the resistance for every increase of three gauge numbers.

The resistance of any length of copper conductor can be approximated by starting at AWG No. 5 with a resistance of 1 Ω/km and halving the resistance for every decrease of three gauge numbers.

The maximum safe current capacity of a conductor depends on:

1. The cross-sectional area
2. The permissible temperature rise
3. The ambient temperature

Overcurrent devices such as a fuse or circuit breaker protect electrical and electronic equipment from damage due to overload or short-circuit current.

Fuses or circuit breakers protect personnel from shock hazard due to equipment defect by opening the circuit to disconnect the voltage source.

Fuses are always connected in series and into the high side (away from ground) of the source.

Fuses are selected according to their current rating, voltage rating, and blow characteristics.

A standard blow fuse having a current of five times its rating will blow in 10 ms.

A slow-blow fuse having a current of five times its rating will blow in 2 s.

Using slow-blow fuses on equipment that can tolerate overloads for a short period avoids the nuisance of changing fuses with every temporary overload.

A thermal circuit breaker operates on the principle of a bimetal strip. With a different temperature coefficient of expansion for each metal, the heat causes the bimetal strip to bend and trip the circuit open.

The magnetic breaker operates on the principle of an electromagnet. With normal current through the electromagnet the breaker contacts are held closed by a spring-loaded mechanism. An overload current will trip the breaker contacts open through the action of the electromagnet.

Measuring full voltage across a fuse indicates that it is blown. An ohmmeter across the fuse will indicate zero ohms if the fuse is okay and infinite ohms if it is blown.

REVIEW QUESTIONS

13-1. List two factors that determine whether a material is a conductor or an insulator.

13-2. Explain why silver is the best conductor.

13-3. Explain the difference between ionic bonding and covalent bonding, and give examples of insulators that have these types of atomic bonding.

13-4. Explain at what voltage level an insulator becomes a conductor, and the reason why.

13-5. What is the function of the Underwriters' Laboratories and the Canadian Standards Association?

13-6. List the CSA designation and the voltage rating for flexible cords under hard usage in damp or dry locations.

13-7. Under what condition does a semiconductor such as silicon or germanium become a conductor?

13-8. Define the circular mil.

13-9. Which two wire gauges are based on the circular mil?

13-10. How does the cross-sectional area of a wire increase with a decrease of three AWG numbers ?

13-11. Explain how the resistance of a wire can be approximated without the use of resistance tables.

13-12. Give three factors that govern the safe current-carrying capacity of a conductor.

13-13. List the safe current-carrying capacity for a cable having three No. 14 conductors using RW75 and TWH insulation.

13-14. List two functions of an overcurrent device such as a fuse or circuit breaker.

13-15. What is the difference between a slow-blow and a fast-acting instrument fuse?

13-16. Explain the difference between a thermal circuit breaker and a magnetic circuit breaker.

13-17. Explain how a voltmeter can be used to check a blown fuse.

13-18. Explain why the current capacity of copper conductors shown in Table 13-10 must not be exceeded.

SELF-EXAMINATION

Choose the most suitable word, phrase, or value to complete each statement.

13-1. Materials that have low resistivity per unit volume are the best (*conductors, insulators, semiconductors*).

13-2. ________ bonding between positive ions and free electrons drifting about the interatomic space of a material will determine the ease with which current carriers drift through the material.

13-3. Two materials that have the greatest number of carriers per unit volume as well as the greatest electron mobility are ________ and ________.

13-4. Two examples of covalent bonding between atoms in an insulator are ________ and ________.

13-5. Two examples of covalent bonding between atoms in semiconductors are ________ and ________.

13-6. Two examples of ionic bonding between atoms in an insulator are ________ and ________.

13-7. The installation and maintenance of electrical equipment is regulated by CSA and is known as the ________ ________ ________.

13-8. CSA designations for three types of flexible cords that are exposed to extra hard usage are ________, ________, and ________.

13-9. The American wire gauge measures the diameter of a wire in ________ and its cross-sectional area in ________.

13-10. To find the cross-sectional area of a wire, the diameter of the wire must be given in ________.

13-11. One mil is the equivalent of ________ inch.

13-12. Given that the diameter of No. 20 wire is 0.03196 in., find its cmil area.

13-13. AWG wire (No. 6, No. 14, No. 30) has the smallest cmil area.

13-14. What is the resistance of 1 km of No. 5 copper wire?

13-15. Estimate the resistance of an AWG No. 17 copper conductor which is 200 m long.

13-16. The maximum safe current-carrying capacity of a conductor depends on three factors: ________, ________, ________.

13-17. What is the safe current-carrying capacity of a three copper-conductor cable with AWG No. 12 TW insulation at 30°C?

13-18. ________ and ________ protect electrical equipment from damage due to short circuits or overloads.

13-19. Fuses or circuit breakers protect personnel from shock hazard because a blown fuse disconnects the ________ side of the source.

13-20. A fast-acting fuse blows in approximately ________ milliseconds with a current five times its ampere rating.

13-21. A slow-blow fuse blows in approximately ________ seconds with a current five times its ampere rating.

13-22. There are two types of circuit breakers: ________ and ________.

13-23. A voltmeter connected across a blown fuse will read the ________ voltage.

ANSWERS TO PRACTICE QUESTIONS

13-1

1. free electrons
2. silver
3. mobility
4. covalent
5. semiconductors
6. rubber, plastic
7. ionic
8. glass, porcelain
9. breakdown voltage
10. Underwriters' Laboratories or Canadian Standards Association
11. SJ or S
12. heat, electrical

13-2

1. 10^{-6}
2. circular mil
3. 40.3 mils
4. 0.1019 in.
5. $A = d^2 = 1624$ cmil
6. decreases
7. No. 14
8. three
9. halved
10. 32 Ω

14

Magnetism

INTRODUCTION

Lodestone, a natural magnetic material, was recognized in about 60 B.C. by Lucretius, a Roman poet, who wrote that the magnet derives from the province of Magnesia, a district of Asia Minor, where lodestone was found in plentiful supply.

There are many legends about the powers of lodestone. One of the most repeated legend tells of "mountains in the north of such powers of attraction that ships had to be built with wooden pegs because if iron nails were used they would be drawn from the timber."

More practically, it was not until some 1200 years after Lucretius that the magnetic compass appeared on Arab, Italian, Portuguese, and Norwegian ships, and possibly sooner on Finnish and Laplander ships sailing the Baltic Sea.

Peter Peregrinus, an Army engineer with Charles of Anjou, who held the city of Lucera in southern Italy at siege from 1266 to 1269, found time during this period to experiment with lodestone. He was the first to write a treatise on magnetism and the first to call the two ends of a lodestone north and south poles. He wrote of placing the lodestone on a piece of wood sufficient to float it in water and then noting which end pointed north. He described how to make a mariner's compass with a scale and a lubber's line. Peregrinus is credited with stating the basic laws of magnetism: Like poles repel each other; unlike poles attract each other.

Today, research and interest in magnetism are intense. Transformers, motors, computers, magnetic tape recorders, and all electrical and electronic equipment and devices that apply the use of magnetism require a sophisticated approach to their magnetic properties. In this chaprter we provide a basic insight and understanding of the principles and properties of magnetism.

MAIN TOPICS IN CHAPTER 14

14-1 Magnetic Field and Its Properties
14-2 Magnetic Domain Theory of Magnetism
14-3 Magnetic and Nonmagnetic Materials
14-4 Magnetic Induction
14-5 Permanent-Magnet Applications
14-6 Magnetic Shielding and Applications
14-7 Magnetic Flux Density and Units of Measurement

OBJECTIVES

After studying Chapter 14, the student should be able to

1. Define the principles and laws of magnetism with respect to permanent magnets.
2. Define magnetic materials, non-magnetic materials, and shielding against magnetism.
3. Define the SI units of flux density, the weber and the tesla.

14-1 MAGNETIC FIELD AND ITS PROPERTIES

Lodestone, known as magnetite, is an iron oxide which was probably magnetized by the earth's magnetic field while cooling from a molten state. Pieces of magnetite are called natural magnets since they exhibit magnetic characteristics.

Man-made magnets are called *permanent magnets* because they retain their magnetic properties indefinitely. Today, these properties are so greatly improved over those of the lodestone that a sensitive galvanometer using a permanent magnet can be made to measure a current in the range of picoamperes.

Permanent-Magnet Properties

Most present-day permanent magnets are made of Alnico 5, an alloy of 14% nickel, 8% aluminum, 24% cobalt, 51% iron, and 3% copper, heat-treated to produce an alignment of the magnetic domains.

To provide a rationale for magnetic laws and properties and using Peregrinus's nomenclature, one end of a bar magnet is called the north pole and is labeled N, while the opposite end is called the south pole and is labeled S, as shown in Fig. 14-1. Peregrinus broke the lodestone into a number of pieces and was able to prove that each piece retained its magnetic north and south poles. If we divide a permanent bar magnet into two or more pieces (see Fig. 14-2), each piece will retain its north and south poles.

FIGURE 14-1 Permanent bar magnet.

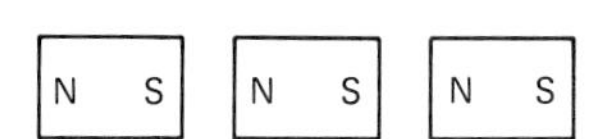

FIGURE 14-2 Permanent bar magnet divided into three pieces forms three magnets.

Sprinkling iron filings onto a sheet of paper or clear plastic over a bar magnet will visually reveal the magnetic force or field. Since Faraday we have accepted the idea that the magnetic force consists of *lines of magnetic force* or flux. Iron filings align themselves along the lines of force to show that magnetic lines repel each other and "flow" independently of each other, therefore they never cross one another.

Magnetic Laws

Magnetic lines of force about a bar magnet can be explored using miniature compasses, as shown in Fig. 14-3. The compass is a small magnet with a north pole and a south pole that is free to rotate. Each magnet in Fig. 14-3 aligns itself with *the magnetic lines of force that are said to flow from the north pole to the south pole.* We should note that it is assumed for convenience that magnetic lines flow and that the magnetic field is made up of magnetic lines of force.

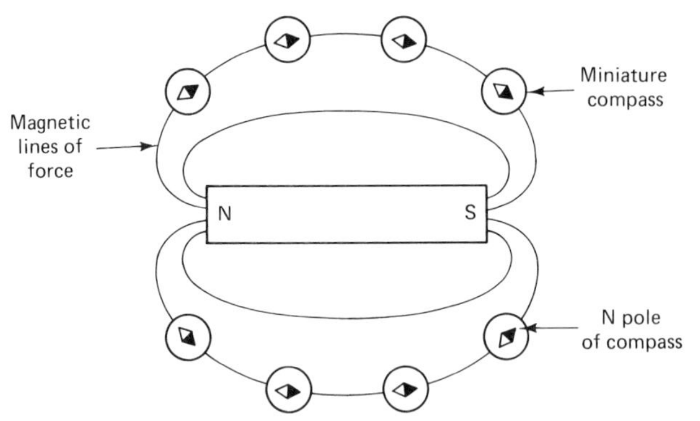

FIGURE 14-3 Exploring magnetic lines of force about a bar magnet using miniature compasses.

This leads to the *first law of magnetism:* that *magnetic lines of force form continuous loops from the north pole around the bar magnet and into the south pole,* through the magnet back to the north pole—therefore, they *never cross one another.*

Magnetic Flux. The total magnetic lines of force "flowing" from a north magnetic pole is generally termed the magnetic field. For measurement purposes the total number of magnetic lines of force flowing out of the north pole is referred to as *magnetic flux.* The symbol for this quantity is the Greek capital letter phi (Φ) (pronounced "fee" or "fi").

The compass serves as an aid in air and sea navigational or guidance systems. Since the earth itself is a huge magnet whose magnetic poles are close to the geographic north and south poles, the north end of a compass needle will align itself with the earth's magnetic lines and point to the south magnetic pole. By convention we have accepted the compass needle as pointing to the north geographic pole. A magnetic navigational compass must be compensated for its magnetic deviation from true north. The amount of deviation will vary depending on the latitude of the observed compass.

Figure 14-4(a) shows that the magnetic lines of force will "flow" from the north pole to the south pole and will contract to attract or pull the two poles together. This illustrates another basic law of magnetism: *Unlike poles attract.*

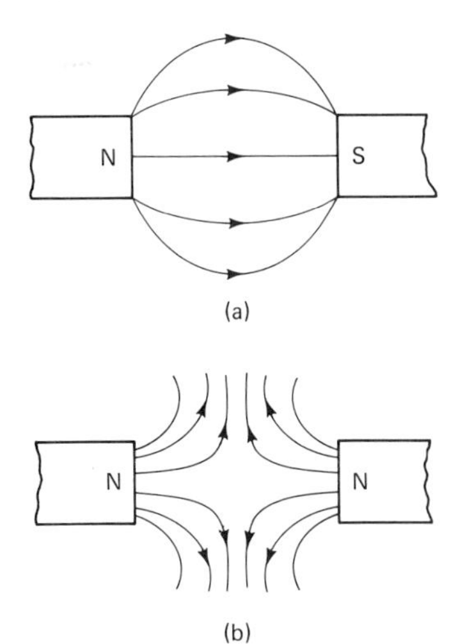

FIGURE 14-4 (a) Unlike poles attract each other. (b) Like poles repel each other.

Figure 14-4(b) shows that the magnetic lines of force will push away or repel each other. This illustrates the other basic law of magnetism: *Like poles repel.*

14-2 MAGNETIC DOMAIN THEORY OF MAGNETISM

During the eighteenth century attempts to explain the source of magnetism led to fluid theories, paralleling the fluid theory of electricity. Rutherford's discovery of the atom brought an understanding of the source of magnetism as orginating in the atomic structure of materials.

Electrons not only revolve about the nucleus of the atom but also spin, similar to the earth, on their own axis. Since the electrons are particles of electricity in motion (i.e., a tiny current flow), they set up their own magnetic field about an atom. Thus the atoms of *ferromagnetic materials* mainly iron, nickel, and cobalt, act like tiny magnets. The present-day visualization of these tiny magnets is as an infinitesimal loop of wire that carries a current. The loop of wire will have a magnetic

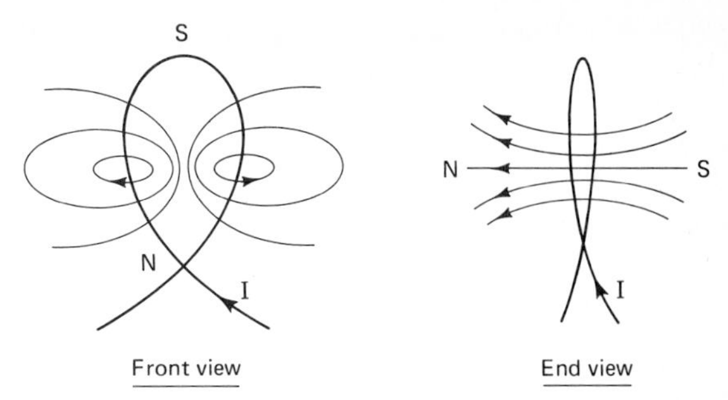

FIGURE 14-5 The magnetic dipole depicted as an infinitesimal loop of wire carrying current and having a north and a south pole.

field with north and south poles about it, as shown in Fig. 14-5. This elementary magnet is called a *magnetic dipole*—meaning that it has a north and a south pole.

All ferromagnetic materials are composed of small domains or regions in which a number of tiny magnetic dipoles are oriented in parallel with each other as depicted in Fig. 14-6. In an unmagnetized state the net

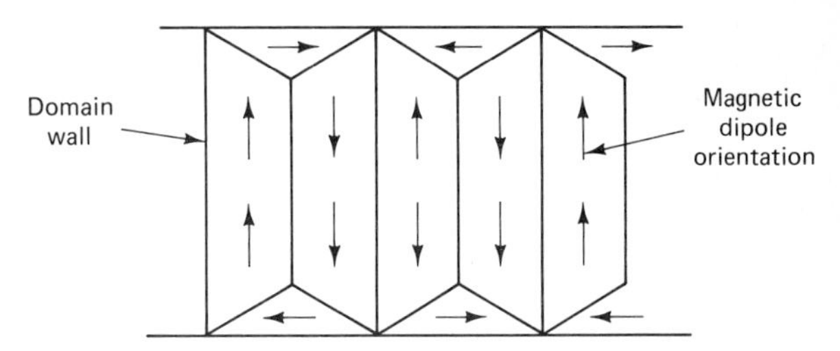

FIGURE 14-6 Domain arrangement in an unmagnetized section of iron. Copyright 1962, Bell Telephone Laboratories Incorporated. Reprinted by permission.

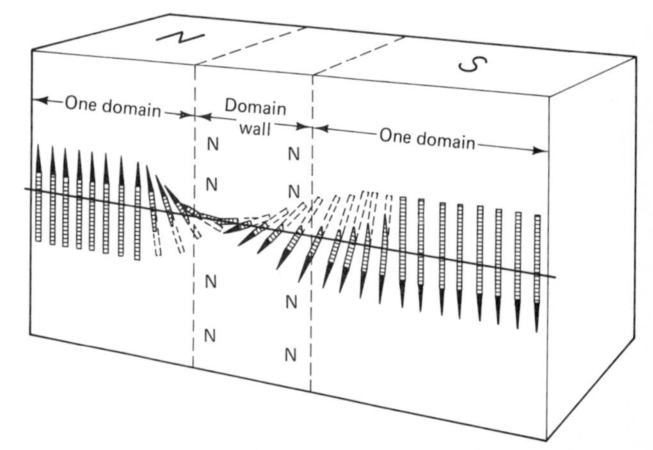

FIGURE 14-7 The change in orientation of magnetic dipoles across a 180° domain boundary. In the domain wall, magnetic poles at the surface will attract magnetic colloid. (Copyright 1962, Bell Telephone Laboratories, Incorporated. Reprinted by permission.)

magnetostatic energy is reduced to a minimum by adjacent domains with reversed dipoles. The domain wall will move under the influence of a magnetic field and cause the magnetic dipoles to reverse themselves in the direction of the magnetic field to form magnetized iron, as depicted in Fig. 14-7. The overall effect in ferromagnetic materials is that the domains in the iron or Alnico orient themselves along an easy axis of magnetization to form a magnetized bar magnet, as shown in the schematic diagram of Fig. 14-8.

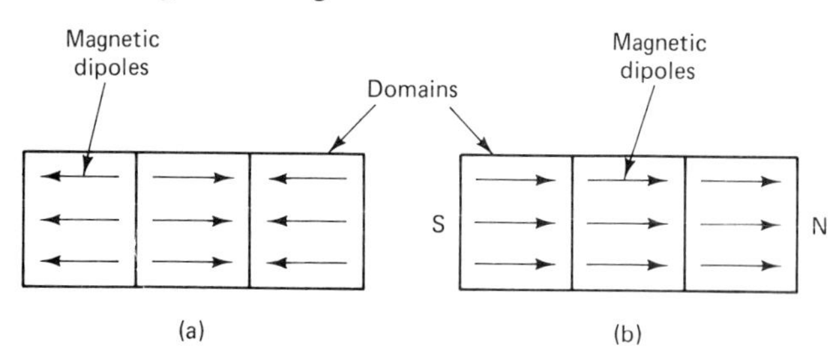

FIGURE 14-8 (a) Unmagnetized bar magnet schematic diagram. (b) Magnetized bar magnet schematic diagram. (Copyright 1962, Bell Telephone Laboratories Incorporated. Reprinted by permission.)

14-3 MAGNETIC AND NONMAGNETIC MATERIALS

We have seen that the domains in ferromagnetic materials such as iron, nickel, and cobalt align themselves easily along an axis of magnetization. The interactions between parallel magnetic dipoles of a single domain is very strong; therefore, they can be magnetized easily.

In other materials the interactions of magnetic dipoles are either zero or very weak. The dipoles behave as free and independent magnets and no domains exist. Materials such as aluminum, platinum, manganese, and chromium behave in this manner and are called *paramagnetic materials*.

Copper, zinc, bismuth, and antimony are examples of materials where the interactions between magnetic dipoles is approximately zero. These materials are called *diamagnetic materials*.

PRACTICE QUESTIONS 14-1

Choose the most suitable word or phrase to complete each statement.

1. A natural magnet is called ________ or ________.
2. A magnet that retains its magnetism indefinitely is called a ________ magnet.
3. A permanent magnet has two magnetic poles called ________ and ________.
4. When a permanent magnet is broken into two or more pieces, each piece will have a ________ and a ________ pole.
5. The magnetic field or force about a bar magnet is conventionally accepted as being composed of ________ of magnetic force.
6. Magnetic lines of force never ________ one another.
7. Magnetic lines of force are said to flow from the ________ pole to the ________ pole.
8. The north end of a compass needle aligns itself with the earth's magnetic field and points to the ________ magnetic pole.
9. Like poles ________ each other.
10. Unlike poles ________ each other.
11. The source of magnetism in materials originates in the ________ structure of the materials.
12. Materials that can easily be magnetized are called ________.
13. The three magnetic materials are ________, ________, and ________.
14. Atoms of ferromagnetic materials are called magnetic ________.
15. Ferromagnetic materials are composed of magnetic ________ in which a number of tiny dipoles are oriented in parallel with each other.
16. In ferromagnetic materials adjacent domains with reversed dipoles will cancel each other's magnetostatic energy so that the material is in an ________ state.
17. Ferromagnetic materials in a magnetized state have their domain ________ oriented along an easy axis of magnetization.
18. Materials that have very weak interactions between parallel magnetic dipoles of a single domain are called ________ materials.
19. Three examples of paramagnetic material are ________, ________, and ________.
20. Examples of materials where the interactions between dipoles is approximately zero are ________, ________, and ________.

14-4 MAGNETIC INDUCTION

Magnetic induction means to produce a magnetic state in an unmagnetized body by proximity to a magnetized body. Figure 14-9 shows a bar magnet close to but not

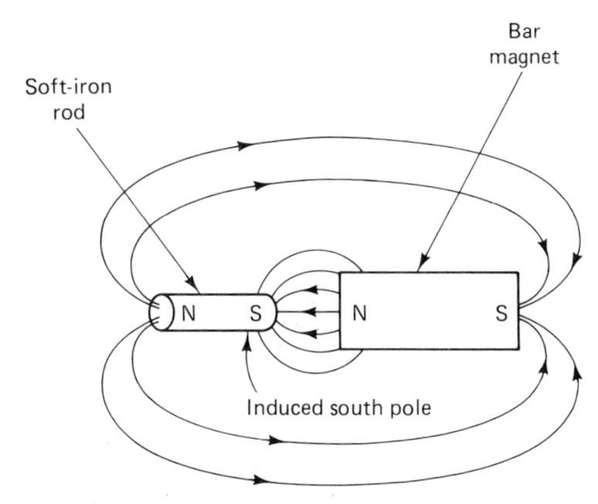

FIGURE 14-9 South pole in soft-iron rod induced by the north pole of the bar magnet.

touching (in proximity to) a soft-iron rod. The iron rod has a south pole induced (magnetic state) in the end closest to the north pole of the bar magnet. The reason for the induced pole is that the magnetic field of the bar magnet influences and moves the magnetic domain walls within the iron rod. This causes the dipoles to reverse themselves so that they align along the easy axis of magnetization, north to south.

Polarity of Induced Poles

If the bar magnet is reversed so that the south pole of the magnet is in the proximity of the soft-iron rod, a north pole will be induced in the end of the rod. Again the magnetic dipoles in the soft-iron rod will reverse and align themselves along the easy axis of magnetization (i.e., north to south). For this reason the induced pole will always be the polarity opposite to that of the magnetized body.

Permeability

Permeate means to diffuse through or penetrate. Magnetic lines of force will always diffuse or flow along the easiest axis of magnetization (i.e., through the iron rod), as shown in Fig. 14-9. As we have discovered, only iron, nickel, and cobalt have any significant magnetic ability; therefore, only these materials will influence magnetic lines of force. *Permeability (μ) is a relative measure of a material's ability to diffuse or concentrate magnetic lines within itself.*

For paramagnetic and diamagnetic materials, the relative permeability, μ_r, is close to unity (1.0). Thus the permeability for nonmagnetic materials is a constant (μ).

The permeability of ferromagnetic materials is measured relative of the permeability of nonmagnetic materials. This means that for ferromagnetic materials permeability is greater than unity and varies with the material or alloy of the materials. Typical permeability values for magnetic alloys range from 100 to 80,000. There is no unit for permeability since it is only a ratio of the total lines of force that flow through the material compared with what would flow through a nonmagnetic material or air.

Retentivity

If the bar magnet is completely removed from the vicinity of the iron rod, it can be noted that the iron rod will retain weak magnetic north and south poles. *The ability of a material to remain magnetized once the magnetizing force is removed is called retentivity.* Steel and Alnico have high retentivity values.

Reluctance

Similar to resistance in an electrical circuit, *reluctance is a measure of the opposition offered to the diffusion of magnetic flux through a material.* The symbol for reluctance is the script letter $\mathcal{R}$. Reluctance is inversely proportional to permeability. Iron has low reluctance and high permeability compared to aluminum, which has high reluctance and low permeability.

14-5 PERMANENT-MAGNET APPLICATIONS

Other than to demonstrate magnetic principles and some applications to toys or novelty objects, the bar magnet serves no other purpose. In novelty objects the bar magnet is usually a molded figure that jumps out of a coffin, which of course is another bar magnet.

In 1887, Edward Weston, the founder of Weston Electrical Instrument Corporation, was probably the first to develop a greatly improved permanent magnet for use in electrical measuring instruments. This was the horseshoe magnet, which concentrated a uniform magnetic field between the two pole pieces, shown in Fig. 11-1. Today's modern moving-coil electrical measuring instruments use a modified version of the horseshoe magnet that greatly improves the sensitivity of the meter movement.

A very important application of a permanent magnet is for loudspeakers used in radios, TV sets, tape cassettes, indoor/outdoor paging, and so on. Alnico 5 is used to produce a permanent magnetic field. A coil (called a voice coil) mounted within the air gap of the pole pieces is free to move laterally. An audio-frequency electrical current flowing into the voice coil causes a magnetic field to be set up around the voice coil which interacts with the permanent magnetic field. This moves the coil in and out, which in turn pushes air in and out to give sound. Figure 14-10 is a cross-sectional diagram of a permanent-magnet loudspeaker.

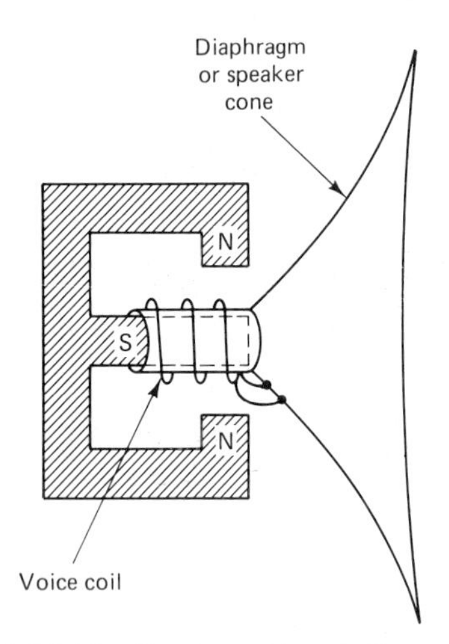

FIGURE 14-10 Cross-sectional diagram of a permanent-magnet speaker.

The bicycle generator is an application of permanent magnets. A rotor called an *armature* is wound with many coils connected to a commutator. The armature rotates by contact pressure with the bicycle wheel. Its rotation passes the coils through a permanent magnetic field surrounding the armature, and an electrical current and voltage are induced in the coils. The current is brought out from the commutator through carbon brushes, then through insulated conductors to terminal posts, and of course to the bicycle lamp.

An electrical instrument called a *megger* that is used extensively in the electrical construction field to measure insulation resistance uses the same principle of operation as the bicycle generator.

14-6 MAGNETIC SHIELDING AND APPLICATIONS

A simple application of shielding a device from stray magnetic fields is the mechanical wristwatch. Another need for magnetic shielding is the electron beam in cathode ray tubes of oscilloscopes. In both cases the principle is the same and is illustrated in Fig. 14-11. A soft-iron ring placed in the path of magnetic lines of force will show that the magnetic flux flows through the path of least reluctance (i.e., the soft-iron ring). The

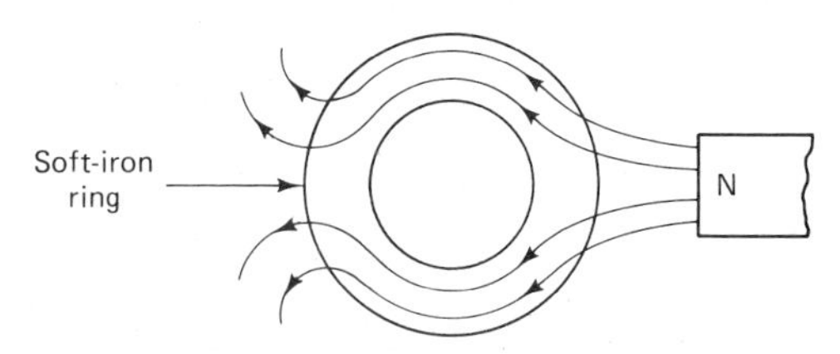

FIGURE 14-11 Magnetic field and soft-iron ring to illustrate magnetic shielding.

magnetic field then flows out of the ring and back to the south pole of the bar magnet. This leaves the inside of the ring free of magnetic flux, effectively shielding a device inside the ring of soft iron.

There is no material other than soft iron that will provide a low-reluctance path for magnetic flux lines and therefore act as a shield. Magnetic lines of force pass right through materials such as lead, copper, glass, and aluminum.

14-7 MAGNETIC FLUX DENSITY AND UNITS OF MEASUREMENT

In the electrical circuit we were able to measure quantities of voltage, current, and resistance in order to evaluate a specific circuit. A magnetic field can also be considered as a magnetic circuit, and magnetic units of measurements must be used to evaluate that circuit.

To honor one scientist who contributed greatly to our knowledge of electricity, a single line of magnetic force was named the maxwell. When studying at Cambridge University in 1898, James Clerk Maxwell was able to show the mathematical relationship between magnetic lines of force and electrostatic lines of force, thus proving that electromagnetic waves travel through the air. This enabled other scientists to pave the way to wireless transmission of telegraphy, radio, and television.

The Maxwell

The maxwell (Mx) is a small unit of flux in the centimetre-gram-second (cgs) system of measurement. One maxwell is simply one magnetic line of force or flux flowing out of the north magnetic pole.

The Weber

The weber (Wb), named after the German physicist Wilhelm Weber (1804–1890), is used in the SI system of units, and is much larger than the maxwell. One weber is equal to 1×10^8 lines of flux flowing from the north magnetic pole (Fig. 14-12). Since a typical bar

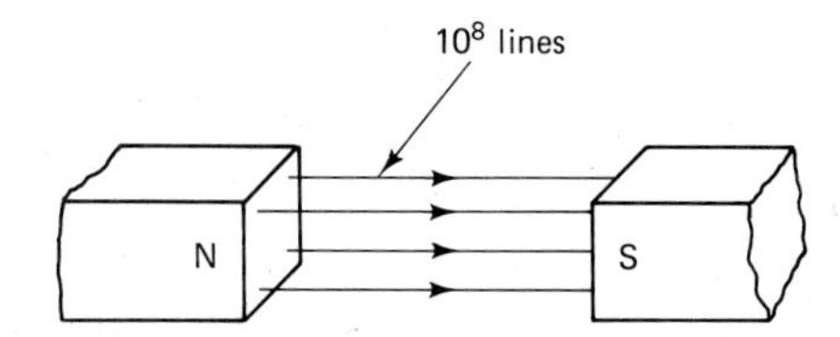

FIGURE 14-12 Representative schematic of 1 weber.

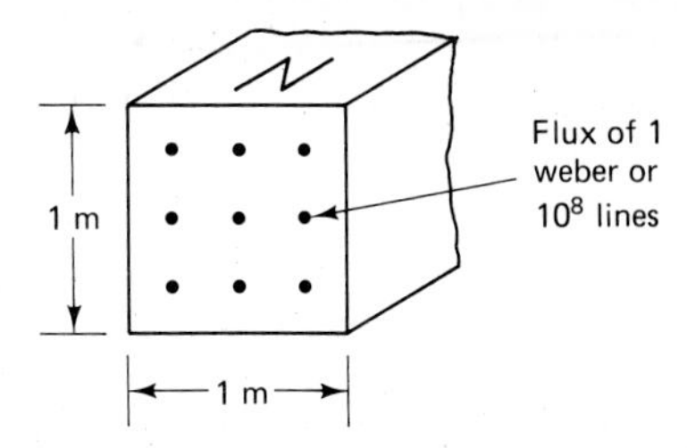

FIGURE 14-13 Flux density, *B*, of 1 tesla.

magnet can provide a magnetic flux of about 5000 Mx, converting this to webers would give

$$\Phi = \frac{5 \times 10^3}{1 \times 10^8} = 5 \times 10^{-5}\ \text{Wb}$$

As can be seen, the weber is a large unit, so the microweber is a more convenient unit.

$$\begin{aligned} 1\ \mu\text{Wb} &= 1 \times 10^{-6}\ (\text{Wb}) \\ &= 1 \times 10^{-6} \times 10^8 \text{ lines of flux} \end{aligned}$$

Therefore,

$$1\ \mu\text{Wb} = 10^2 \text{ lines of flux}$$

or

$$1\ \mu\text{Wb} = 100 \text{ lines or Mx}$$

For the example of a typical small magnet of 5000 lines, the flux measured in microwebers is

$$\Phi = \frac{5000 \text{ lines}}{100 \text{ lines}} = 50\ \mu\text{Wb}$$

To convert microwebers to lines, multiply the number of microwebers by 100 lines From the example above,

$$\begin{aligned} \text{Mx or lines} &= 50\ \mu\text{Wb} \times 100 \text{ lines} \\ &= 5000 \text{ Mx or lines} \end{aligned}$$

Flux Density

Since flux lines always form closed loops and repel each other, the greatest concentration of flux lines is at the north and south poles of a magnet. The magnetic strength of a magnet, therefore, is at its pole pieces. To reflect accurately the strength of a large magnet versus a small magnet, the number of flux lines per unit area must be compared. In the SI system of measurement a cross-sectional area of 1 square metre is used, as shown in Fig. 14-13.

One tesla is the total flux of one weber passing through a cross-sectional area of 1 square metre. Note: The contradiction here is that the flux density (B) would naturally be webers per square metre. However, the SI system has chosen to honor Nikola Tesla (1857–1943), a researcher and inventor of the alternating-current system, and flux density is stated in the unit tesla.

The flux density is 1 weber of flux passing through an area of 1 square metre, or

$$B = \frac{\Phi}{A} \tag{14-1}$$

where B is in teslas
Φ is in webers
A is in square metres

EXAMPLE 14-1

Calculate the flux density of a bar magnet 1.5 cm × 2.5 cm with a total flux of 5000 lines.

Solution: Since equation (14-1) calls for the total flux in webers,

$$\text{Wb} = \frac{5000 \text{ Mx}}{100 \text{ Mx}/\mu\text{Wb}} = 50\ \mu\text{Wb}$$

$$\begin{aligned} \text{area} &= 1.5 \text{ cm} \times 2.5 \text{ cm} = 3.75 \text{ cm}^2 \\ &= 3.75 \times 10^{-4} \text{ m}^2 \end{aligned}$$

Therefore,

$$B = \frac{\Phi}{A} = \frac{50 \times 10^{-6} \text{ Wb}}{3.75 \times 10^{-4} \text{ m}^2} = 0.13 \text{ T}$$

or

$$B = 13 \times 10^{-2} \text{ T}$$

EXAMPLE 14-2

Calculate the flux density of the magnetic flux of Example 14-1 at a point from the pole pieces that is distributed over a 3 cm × 3 cm area.

Solution: The total flux of the bar magnet is 50 μWb. Therefore,

$$B = \frac{\Phi}{A} = \frac{50 \times 10^{-6}\ \text{Wb}}{(3\ \text{cm} \times 3\ \text{cm} \times 10^{-4})\ \text{m}^2}$$
$$= 5.6 \times 10^{-2}\ \text{T}$$

EXAMPLE 14-3

Assume that the soft-iron rod of Fig. 14-9 is very close to the bar magnet pole with negligible flux leakage. Assume that the bar magnet has a total flux density of 13×10^{-2} T. Calculate the total flux in maxwells passing through the rod, whose diameter is 1 cm.

Solution: From equation (14-1), where 1 tesla = 1 Wb/m^2,

$$\text{area} = \frac{\pi d^2}{4} = 0.785\ \text{cm}^2$$
$$\Phi = BA$$
$$= 13 \times 10^{-2}\ \frac{\text{Wb}}{\text{m}^2} \times 0.785 \times 10^{-4}\ \text{m}^2$$
$$= 10.21\ \mu\text{Wb}$$
$$1\ \mu\text{Wb} = 100\ \text{M}$$

Therefore,

$$10.21\ \mu\text{Wb} = 100 \times 10.21 = 1021\ \text{Mx}$$

SUMMARY

Permanent magnets retain their magnetic properties indefinitely.

A permanent magnet has a north magnetic pole and a south magnetic pole.

Magnetic lines of force are said to flow from the north pole of a magnet to the south pole.

Magnetic lines of force form continuous loops and never cross one another.

The total magnetic lines flowing out of the north pole is termed magnetic flux, and the symbol is Φ (the Greek capital letter phi).

Like poles repel each other; unlike poles attract each other.

Materials that can easily be magnetized are called ferromagnetic.

Iron, nickel, and cobalt are ferromagnetic materials.

Only ferromagnetic materials are composed of magnetic domains.

In ferromagnetic materials the magnetic dipoles in each domain orient themselves along an easy axis of magnetization to form a magnet.

Aluminum, platinum, manganese, and chromium are a few examples of paramagnetic materials.

Paramagnetic materials do not have domains and their dipoles behave independently, so a magnetic field has very little effect on them.

Materials such as copper, zinc, bismuth, and antimony are examples of diamagnetic materials.

Diamagnetic materials are not affected by a magnetic field.

Magnetic induction means to produce a magnetic state in an unmagnetized body by proximity to a magnetized body.

The induced pole will always be of polarity opposite to that of the magnetized body producing it.

Permeability (μ) is a relative measure of a material's ability to diffuse or concentrate magnetic lines within itself.

Retentivity is the relative ability of a material to retain its magnetic characteristics once the magnetizing force is removed.

Reluctance is the relative measure of the opposition offered to the diffusion of magnetic flux through a material. Its symbol is the script letter $\mathcal{R}$.

Reluctance is inversely proportional to permeability.

Magnetic shielding means that the magnetic lines of force are bypassed or rerouted through a soft-iron shield about the object to be shielded from the magnetic flux.

A single line of magnetic force is named a maxwell.

One weber (Wb) is equal to 10^8 magnetic lines of force.

One tesla is equal to the total flux of 1 weber passing through a cross-sectional area of 1 square metre.

$$B = \frac{\Phi}{A} \qquad (14\text{-}1)$$

where B is in teslas
Φ is in webers
A is in square metres

REVIEW QUESTIONS

14-1. List two properties of magnetic lines of force about a permanent magnet.

14-2. List the three laws of magnetism.

14-3. Explain briefly how it can be shown that magnetic lines of force are said to flow from the north pole to the south pole of a magnet.

14-4. List three ferromagnetic materials.

14-5. Explain briefly the magnetic dipole theory of materials.

14-6. Explain briefly the domain theory of ferromagnetic materials.

14-7. Explain briefly why materials such as copper, zinc, bismuth, and antimony, for example, cannot be magnetized.

14-8. What pole is induced in a soft-iron bar when the south end of a bar magnet is brought close to but not touching the end of the bar? Use the domain dipole theory to explain the induced pole.

14-9. Define the following.
(a) Permeability
(b) Retentivity
(c) Reluctance

14-10. Why can only soft iron be used for magnetic shielding?

14-11. Define the following.
(a) The maxwell **(b)** The weber **(c)** The symbol Φ
(d) The microweber **(e)** Flux density **(f)** The tesla

14-12. A bar magnet has a cross-sectional area of 6 cm^2. The flux density at the poles is 2×10^{-3} T. What are the total flux lines in microwebers? In maxwells?

SELF-EXAMINATION

Choose the most suitable word, phrase, or value to complete each statement.

14-1. A permanent magnet has a ________ and a ________ pole.

14-2. The north end of a compass aligns itself along the magnetic lines of force, which are said to flow from ________ to ________.

14-3. Magnetic lines of force form continuous loops and never ________ one another.

14-4. The total number of magnetic lines of force flowing out of the north pole is referred to as ________ ________.

14-5. The symbol for magnetic flux is the Greek letter ________.

14-6. ________ magnetic poles attract.

14-7. ________ magnetic poles repel.

14-8. Three ferromagnetic materials are ________, ________, and ________.

14-9. The atoms of ferromagnetic materials behave like elementary magnets and are called magnetic ________.

14-10. All ferromagnetic materials are composed of regions called ________ in which a number of dipoles are oriented in parallel with each other.

14-11. A ferromagnetic material that has been magnetized will have the dipoles in its ________ oriented along an easy axis of magnetization.

14-12. Materials that do not contain domains and in which the magnetic dipoles behave as free and independent magnets are called ________ materials.

14-13. Materials in which the interaction between the magnetic dipoles is zero are called ________.

14-14. When brought into the proximity of a ferromagnetic material, a south magnetic pole will always induce a ________ pole in that material.

14-15. The relative measure of a material's ability to concentrate magnetic lines of force is called ________, whose symbol is ________.

14-16. The ability of a material to remain magnetized once the magnetizing force is removed is called ________.

14-17. ________ is a measure of the opposition offered to the diffusion of magnetic flux through a material, whose symbol is ________.

14-18. To shield a device from magnetic lines of force, a ________ ________ ring or enclosure must be used.

14-19. One maxwell is equal to ________ magnetic lines of force flowing out of the north magnetic pole.

14-20. One weber is equal to ________ magnetic lines of force flowing out of the north magnetic pole.

14-21. One microweber is equal to ________ lines or maxwell.

14-22. One tesla is the total flux of 1 weber passing through a cross-sectional area of ________ ________ ________.

14-23. The symbol for the flux density measured in teslas is ________.

14-24. Calculate the flux density of a horseshoe magnet whose pole piece has an area of 6.5 cm^2 with a total flux of 10,000 Mx.

14-25. Calculate the flux density of the magnetic flux at a point from the pole pieces of the horseshoe magnet that is distributed over a 4 cm × 4 cm area with a total flux of 10,000 Mx.

ANSWERS TO PRACTICE QUESTIONS

14-1

1. lodestone, magnetite
2. permanent
3. north, south
4. north, south
5. lines
6. cross
7. north, south
8. south
9. repel
10. attract
11. atomic
12. ferromagnetic
13. iron, cobalt, nickel
14. dipoles
15. domains
16. unmagnetized
17. dipoles
18. paramagnetic
19. aluminum, platinum, manganese
20. copper, zinc, bismuth

15

Electromagnetism

INTRODUCTION

Hans Christian Oersted's (Ur-sted) discovery in 1820 that electricity produces magnetism was somewhat by accident. Like many other experimenters in that period of time, Oersted spent many hours exploring the region around a charged Leyden jar with a compass, and he had often placed a compass near a current-carrying wire without any reaction by the compass needle. Up to this point he had always placed the compass needle at right angles to the direction of the current flow, expecting it to swing around parallel to the wire.

Teaching students in a class one day he said, "Let us try once, now that the battery is handy, placing the magnetic needle *parallel* to the wire." (The story was told in a letter to Faraday some years later by a Professor Hansteen.) Seeing the needle move weakly excited Oersted, but he was skeptical. After class he obtained a large battery and verified that *indeed the magnetic force acts at right angles to the current-carrying conductor*.

In this chapter we explore the characteristics of the magnetic field about a single current-carrying conductor, a coil, or solenoid carrying current, and compare Ohm's law of the magnetic circuit to that of an electrical circuit.

MAIN TOPICS IN CHAPTER 15

- 15-1 Magnetic Flux about a Current-Carrying Conductor
- 15-2 Magnetic Field about a Solenoid
- 15-3 Magnetomotive Force
- 15-4 Electromagnetic Field Intensity
- 15-5 Permeability
- 15-6 *B–H* Magnetization Curve
- 15-7 Hysteresis Loop
- 15-8 Ohm's Law for Magnetic Circuit

OBJECTIVES

After studying Chapter 15, the student should be able to

1. Define the principles and laws of electromagnetism with respect to a single current-carrying conductor and a solenoid.
2. Define and apply electromagnetic units of measurement for magnetomotive force, field intensity, and permeability.

3. Define the B–H magnetization curve and its relationship to the hysteresis loop.
4. Apply Ohm's law to the magnetic circuit.

15-1 MAGNETIC FLUX ABOUT A CURRENT-CARRYING CONDUCTOR

Oersted's discovery that a magnetic force acts at right angles to a current-carrying conductor can be demonstrated by the use of compasses placed around a current-carrying conductor as shown in Fig. 15-1.

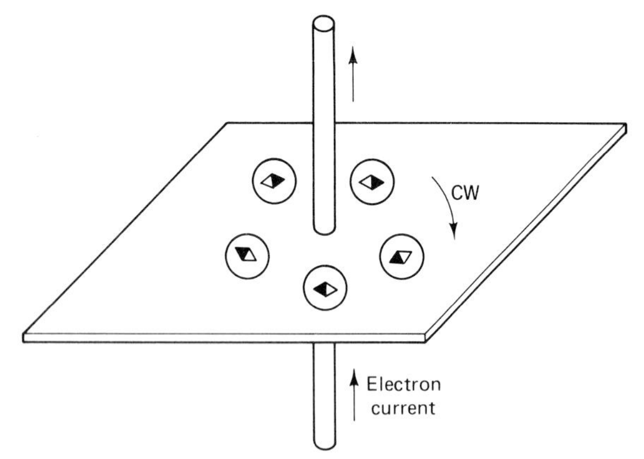

FIGURE 15-1 Magnetic lines of force flow in continuous loops at a right angle to current-carrying conductor.

Electromagnetic lines of force about a single current-carrying conductor flow in *continuous loops around the conductor* and do not exhibit any poles. As in the case of the bar magnet, *magnetic lines repel each other* and become weaker as the distance from the wire increases. This indicates that *the strength of the magnetic flux decreases as the distance from the conductor increases*. As Oersted discovered by using a larger battery, *the strength of the magnetic field increases with an increase in current and decreases with a decrease in current*. When the direction of the current through the conductor is reversed, the magnetic compasses will reverse direction. *This indicates that the flux field has a specific direction, dependent on the direction of the current flow*.

Left-Hand Rule

The direction that the flux field takes can be proven by using compasses, however, exploring a current-carrying conductor with compasses is not convenient or practical. A simple method called the left-hand rule can easily be used to determine the direction of the flux:

The left-hand rule method is illustrated in Fig. 15-2 and proceeds as follows. Grasp the conductor in the left hand with the thumb pointing in the direction of electron current flow. The fingers wrapped around the conductor will point in the direction of the flux field.

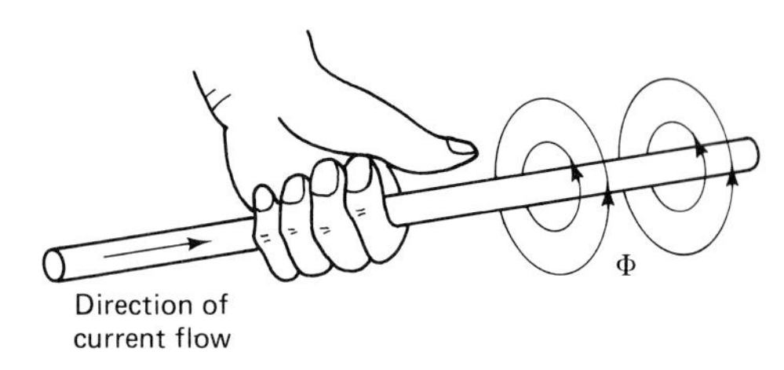

FIGURE 15-2 Left-hand-rule method to determine the direction of the flux about a current-carrying conductor.

> Note: *In all cases the left-hand rule is used for an electron direction of current flow.* The same rule can be applied using the *right hand for conventional direction of current flow.*

Reversing the left-hand rule procedure by knowing the direction of the flux field will enable us to determine the direction of current flow.

The cross-sectional view of the conductor gives a more convenient method of visualizing the magnetic flux about a conductor. The cross-and-dot method shown in Fig. 15-3 is a symbolic method that uses an ar-

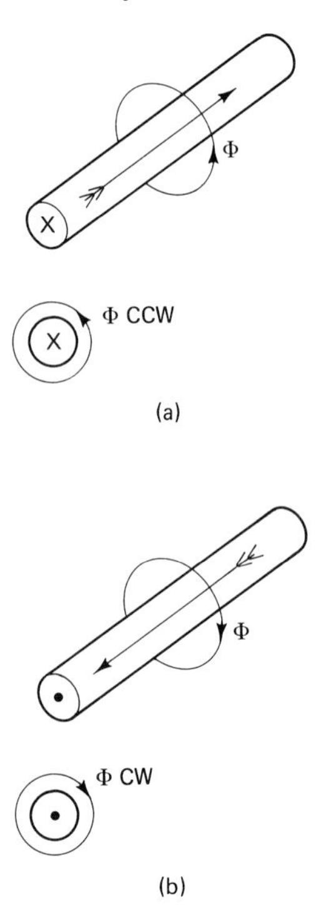

FIGURE 15-3 Cross-and-dot methods to indicate direction of current flow and direction of resultant flux.

rowhead and tail feathers of an arrow, which indicates the current direction through the conductor.

Using the left-hand rule for the conductor of Fig. 15-3(a), where the current is flowing away from the observer (tail feathers represented by a cross), will give the flux a counterclockwise direction. In Fig. 15-3(b), the current is flowing toward the observer (arrowhead represented by a dot). Applying the left-hand rule gives the flux a clockwise direction.

Magnetic Flux between Adjacent Conductors

It was André Ampère who picked up where Oersted left off and developed the principles of electrodynamics. *Two adjacent current-carrying conductors attract each other when current flows in the same direction through both conductors.* This is shown by a plot of magnetic flux in Fig. 15-4(a). Flux lines will unite and act as a single field that contracts and tends to pull the two conductors together. *Two adjacent current-carrying conductors repel each other when current flows in opposite direction, through both conductors.* This is shown by a plot of magnetic flux in Fig. 15-4(b). Flux lines will oppose each other and tend to push the conductors apart.

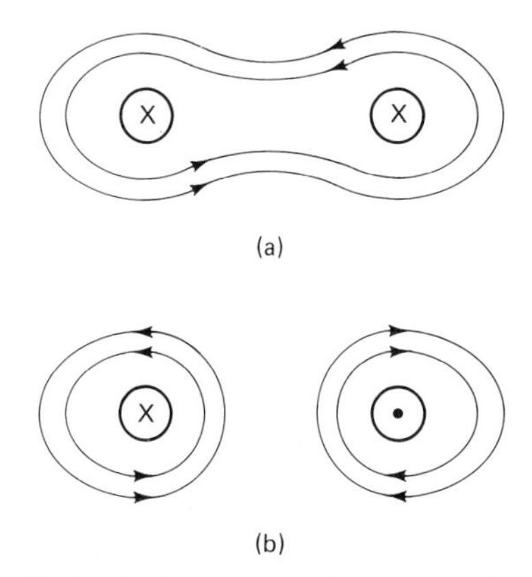

FIGURE 15-4 Magnetic flux attracting and repelling between two adjacent current-carrying conductors. (a) Attracting conductors. (b) Repelling conductors.

15-2 MAGNETIC FIELD ABOUT A SOLENOID

When a current-carrying conductor is formed into a loop, the magnetic lines of force take on a direction, as shown in Fig. 15-5(a). The flux lines leave one side of the loop and flow around and enter the other side of the loop. This creates a north pole on one side of the loop where they leave, and a south pole on the other side of the loop where they enter, as shown in Fig. 15-5(b).

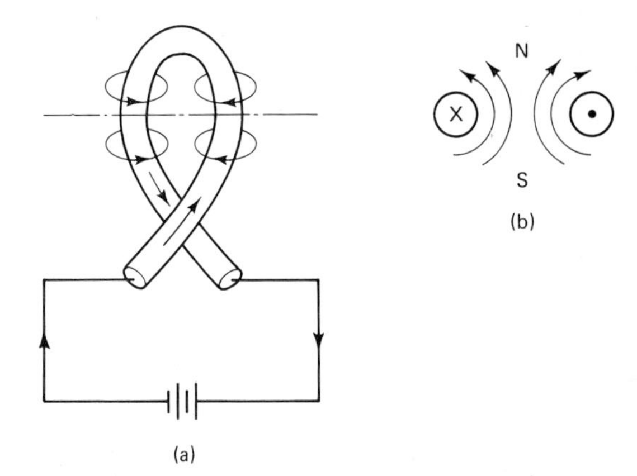

FIGURE 15-5 (a) Current-carrying loop. (b) Cross-sectional view of loop.

Coiling more loops or turns together adds more magnetic lines of force and increases the strength of the magnetic flux. The cross-sectional view of such a coil, also called a *solenoid,* is shown in Fig. 15-6(b). Current

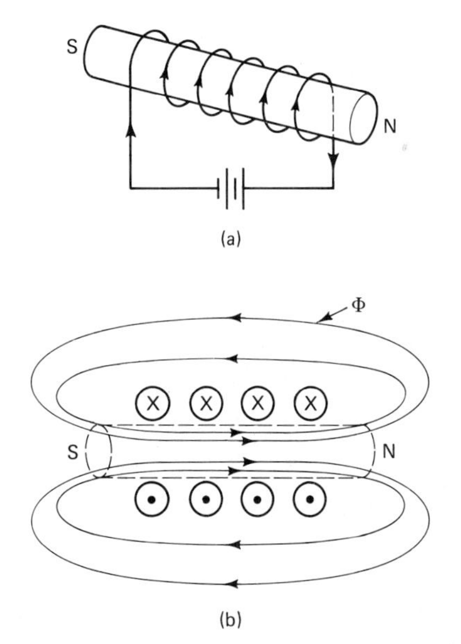

FIGURE 15-6 Solenoid and cross-sectional view of turns and resultant magnetic field.

flowing toward the observer in one side of each loop will set up a magnetic flux that couples and aids adjacent turns (see also Fig. 15-4). In the other side of the coil, current flowing away from the observer sets up a magnetic field in the same direction, so that the magnetic lines of force leave one end of the coil and enter the opposite end. The magnetic field thus created is exactly the same as that about the permanent bar magnet, with a north and a south pole. This coil or solenoid is called an *electromagnet*.

Left-Hand Rule for Solenoids

The magnetic poles or the direction of current flow through the turns can easily be determined by applying the left-hand rule for solenoids, shown in Fig. 15-7.

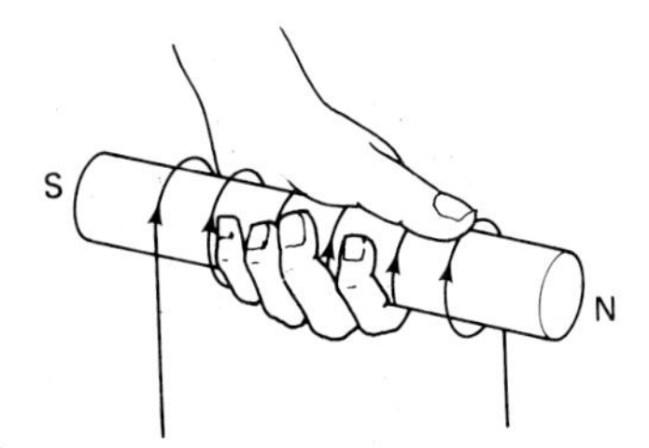

FIGURE 15-7 Applying left-hand rule for solenoids. Fingers point in direction of current flow through the turns, then extended thumb points to north magnetic pole.

With the left hand, grasp the coil or solenoid so that the fingers point in the direction of current flow through the turns. The extended thumb will then point to the north pole of the electromagnet.

Reversing Current and Turns Direction

Reversing the direction of the current or the direction of the turns will reverse the direction of the flux about each turn. This reverses the electromagnetic poles from a north pole to a south pole, and vice versa.

Air-Core versus Iron-Core Solenoid

A solenoid that has its turns of wire wound on a hollow nonmagnetic tube or cylindrical form will have an air core. With an air core the magnetic lines of force repel each other and expand out through the core, while many other flux lines expand from the turns to the external perimeter of the electromagnet. In this way the magnetic lines of force are very thinly distributed over a greater area and the electromagnetic field is weak.

Using a soft-iron core in place of the air core will improve the electromagnet tremendously. The soft iron provides a low-reluctance path for the magnetic flux since the magnetic domains in the soft iron will conduct the flux. This allows a much greater amount of flux to shrink or contract within the iron-core area. The large increase of magnetic flux per unit area multiplies the strength of the electromagnet by several thousand times compared to an air-core solenoid.

Applications for electromagnets or solenoids vary from a lifting magnet in a scrap iron yard to relays in washing machines, refrigerators, air conditioners, and many other types of electrical equipment, as well as electronic equipment. In all cases the relay has a solenoid that magnetically attracts and pulls a soft-iron armature [see Fig. 15-8(a)]. The armature engages insulated contacts or a switch to control the running and stopping of the equipment. This necessitates that the iron core of the solenoid have low *retentivity*. A soft-iron core provides low retentivity, so that once the current ceases to flow through the turns on the solenoid, the solenoid loses its magnetization ability and releases the armature. Note that the armature is restrained by a spring.

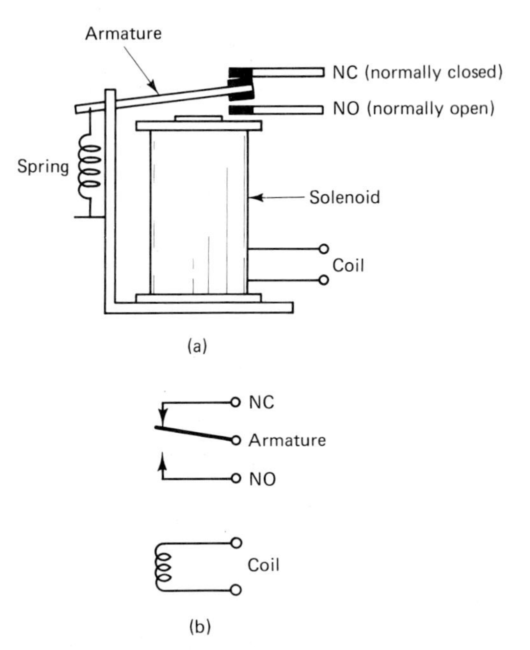

FIGURE 15-8 (a) Form C electromagnetic relay. (b) Schematic symbol for form C relay.

PRACTICE QUESTIONS 15-1

Choose the most suitable word or phrase to complete each statement.

1. Electromagnetic lines of force about a single current-carrying conductor flow in ________ loops around the conductor.
2. Magnetic lines of force (*attract, repel*) each other.
3. The strength of the magnetic flux about a current-carrying conductor decreases as the distance from the conductor ________.
4. The strength of the flux increases with an ________ in current.
5. The magnetic flux about a current-carrying conductor has a specific direction dependent on the direction of ________.
6. In the left-hand-rule method to determine the direction of magnetic flux about a single current-carrying conductor, the ________ always point in the direction of electron current flow.
7. In the left-hand-rule method to determine the direction of magnetic flux about a single current-car-

rying conductor, the __________ always points in the direction of the flux.

8. In the cross-and-dot convention method of visualizing magnetic flux direction about a current-carrying conductor, the __________ represents current flow toward the observer and the flux is shown in a __________ direction.
9. Two adjacent current-carrying conductors __________ each other when current flows in the same direction.
10. In a current-carrying conductor formed into a loop, the magnetic flux takes on a direction such that the loop has a __________ and a __________ pole.
11. Adding more __________ to a loop increases the strength of the magnetic flux.
12. In the left-hand rule for solenoids, the __________ point in the direction of current flow through the turns.
13. In the left-hand rule for solenoids, the __________ points in the direction of the __________ pole.
14. Reversing the direction of the turns or the current through the turns on a solenoid will reverse its magnetic __________.
15. A __________ __________ core improves the strength of an electromagnet by a factor of several thousand times.
16. The ability of a ferromagnetic material to retain its magnetism is called __________.

15-3 MAGNETOMOTIVE FORCE

Oersted's first experimental result with a current-carrying conductor and compass needle gave a weak deflection of the compass needle. He then increased the size of the battery and got a stronger deflection of the compass needle. This meant that the strength of the magnetic flux about the current-carrying conductor is dependent on the amount of current flow.

With a solenoid the strength of the magnetic field will depend on the amount of current as well as the number of turns. The magnetic flux that is established about a solenoid is directly proportional to the current and the number of turns. Since the current and the number of turns produce the magnetic flux, the term *magnetomotive force* (mmf), symbol $\mathcal{F}$, is applied to the product of the current and the number of turns. An analogy comparing the electrical emf that causes current flow can be made to the mmf of the magnetic circuit, where the mmf causes a magnetic flux to flow. In equation form,

$$\text{mmf} = \mathcal{F} = N \times I \quad \text{amperes} \qquad (15\text{-}1)$$

where mmf is the SI unit "ampere"
I is in amperes
N is number of turns

Note: The metric unit for mmf is ampere-turns (At).

EXAMPLE 15-1

Calculate the mmf of a coil with 600 turns and a current of 500 mA. What would the magnetizing force (mmf) be with double the turns and the same current?

Solution: From equation (15-1),

$$\text{mmf} = \mathcal{F} = N \times I = 500 \times 10^{-3} \times 600 = 300 \text{ A}$$

The magnetizing force with double the turns is

$$\text{mmf} = \mathcal{F} = N \times I = 500 \times 10^{-3} \times 1200 = 600 \text{ A}$$

15-4 ELECTROMAGNETIC FIELD INTENSITY

The mmf that sets up a magnetic flux about a coil is concentrated within the turns of the coil. In the magnetic circuit of Fig. 15-9 magnetic flux flows out of the coil

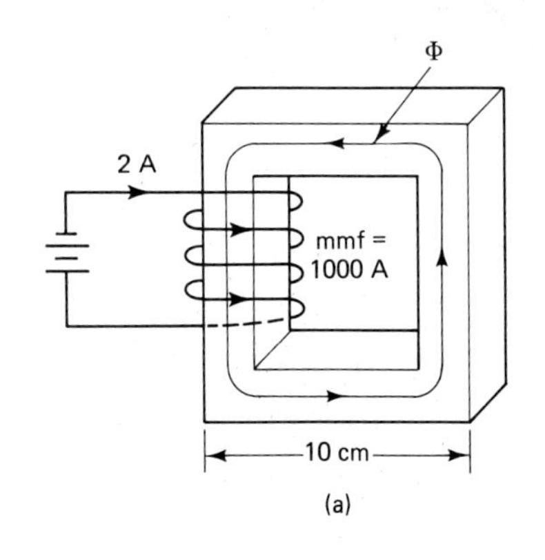

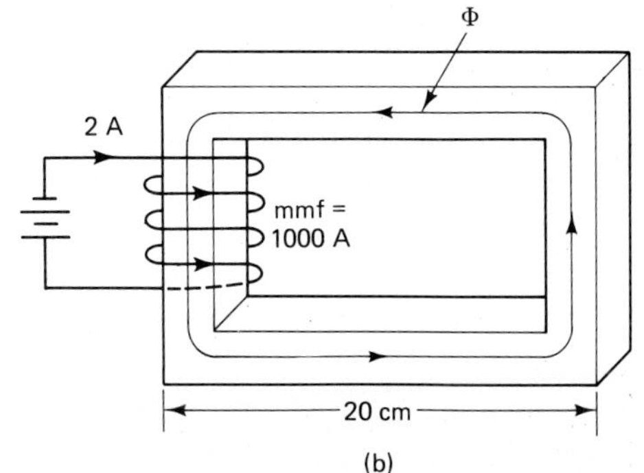

FIGURE 15-9 Magnetic field intensity produced by the same mmf through similar iron cores having the same cross-sectional area but different lengths. (a) $H = 10$ kA/m. (b) $H = 5$ kA/m.

through the path of least reluctance (i.e., through and around the iron core and back to the coil). If both iron cores have the same cross-sectional area, the circuit in Fig. 15-9(b) has the longest dimension, therefore will have the greatest reluctance, double in this case.

Thus the flux density produced by the mmf applied to similar cores having the same cross-sectional area but different length will be inversely proportional to the length. This is analogous to an electric circuit, where the current is inversely proportional to the resistance with a constant emf.

The new quantity of magnetic flux is called *magnetic field intensity* (symbol H), and is directly proportional to mmf and inversely proportional to the length of the magnetic circuit. The unit of magnetic field intensity is amperes per metre. In equation form, magnetic field intensity,

$$H = \frac{\mathcal{F}}{l}$$

or

$$H = \frac{NI}{l} \quad \text{amperes/metre} \qquad (15\text{-}2)$$

where H is the SI unit "amperes per metre"
N is the number of turns on the coil
I is in amperes
l is in metres

EXAMPLE 15-2

Calculate the magnetic field intensity for the magnetic circuits of Fig. 15-9.

Solution: For the circuit in Fig. 15-9(a),

$$H = \frac{NI}{l} = \frac{1000\text{ A}}{10 \times 10^{-2}\text{ m}}$$
$$= 10{,}000\text{ A/m} \quad \text{or} \quad 10\text{ kA/m}$$

For the circuit in Fig. 15-9(b),

$$H = \frac{NI}{l} = \frac{1000\text{ A}}{20 \times 10^{-2}\text{ m}}$$
$$= 5000\text{ A/m} \quad \text{or} \quad 5\text{ kA/m}$$

PRACTICE QUESTIONS 15-2

Choose the most suitable word, phrase, or value to complete each statement.

1. The strength of the magnetic field about a coil or solenoid is dependent on ________ and the number of ________.
2. The product of the current and the number of turns is called ________. Its symbol is ________.
3. A toroidal core and coil (circular form) has 200 turns of wire and 1.2 A when connected to a dc source. Calculate the mmf.
4. Calculate the magnetic field intensity for the toroidal core and coil in Question 3 if the average length of the magnetic circuit is 20 cm.
5. What current is required in the coil of Question 3 to produce an mmf of 720 A?
6. If the average length of the magnetic circuit of Question 5 is 40 cm, calculate the magnetic field intensity.

15-5 PERMEABILITY

In Chapter 14 we learned the definition of permeability as a relative measure of a material's ability to diffuse or concentrate magnetic lines of force within itself. It was also stated that the permeability values for magnetic alloys ranges from 100 to 80,000. With the same mmf applied to a solenoid the amount of magnetic flux density (B), will be directly proportional to its core permeability and the magnetic field intensity (H) produced by the mmf. In equation form,

$$B = \mu \times H \qquad (15\text{-}3)$$

where B is in tesla (webers/m^2)
H is in amperes/metre

Permeability of Air and Nonmagnetic Materials

The permeability of air is termed the *permeability of free space*. Permability (μ) is actually the product of two factors, the relative permeability (μ_r) of a material and the permeability of free space (μ_0). Most nonmagnetic materials have a permeability very close to unity; therefore, μ_r is nearly 1. The permeability of free space is a constant, so $\mu_0 = 4\pi \times 10^{-7} = 1.26 \times 10^{-6}$. Therefore,

$$\mu = \mu_r \times \mu_0 = 1 \times 4\pi \times 10^{-7}$$

or

$$\mu = 1.26 \times 10^{-6} \qquad (15\text{-}4)$$

for free space and most nonmagnetic materials.

EXAMPLE 15-3

In the magnetic circuit of Fig. 15-9 the 20-cm air-core coil consists of 300 turns. Calculate the flux density for a current of 2 A through the coil.

Solution: Equation (15-2) gives

$$H = \frac{NI}{l} \quad \text{amperes/metre}$$

Equation (15-3) gives

$$B = \mu \times H$$

Therefore,

$$B = 1.26 \times 10^{-6} \times \frac{300 \times 2 \text{ A/m}}{20 \times 10^{-2} \text{ m}}$$
$$= 37.8 \times 10^{-4} \text{ T}$$

EXAMPLE 15-4

If the core in the magnetic circuit of Example 15-3 was replaced with silicon having a relative permeability of 4000, calculate the flux density.

Solution

$$\mu = \mu_r \times \mu_0 = 4000 \times 1.26 \times 10^{-6}$$
$$B = \mu \times H$$
$$= 4000 \times 1.26 \times 10^{-6} \times \frac{300 \times 2 \text{ A/m}}{20 \times 10^{-2} \text{ m}}$$

Therefore,

$$B = 4000 \times 37.8 \times 10^{-4} \text{ T} = 15.1 \text{ T}$$

As a conductor of magnetic flux the silicon steel core, with a permeability of 4000, proved to have a greater flux density than the air core for the same mmf. Permeability is analogous to conductivity in an electric circuit.

15-6 *B–H* MAGNETIZATION CURVE

A ferromagnetic material is magnetized by aligning the magnetic domains in the direction of the magnetizing field. Reorientation of the magnetic domains increases as the strength of the magnetizing field increases. Fig. 15-10 is a graphical representation of a typical *B–H* magnetization curve. The value of the magnetic field intensity *H* can be varied easily by varying the current through the coil.

From the graph, an increase in the value of *H* from zero to point *a* does not produce a linear change in the flux density *B* of the material. This is because the per-

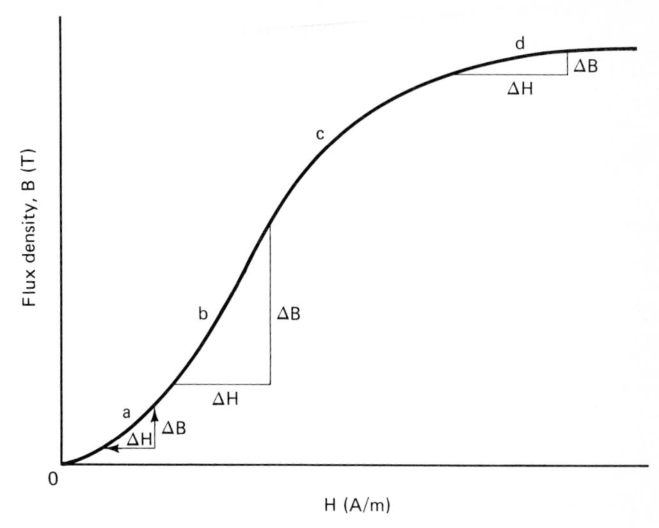

FIGURE 15-10 Magnetization curve for a ferromagnetic material.

meability of ferromagnetic materials is not constant but can vary with the applied value of magnetic field intensity *H*. The nonlinearity of the magnetization curve for ferromagnetic materials can best be explained in terms of magnetic domains.

An increase in current from zero to point *a* begins to reorient the dipoles of the magnetic domains along the easy axis of magnetization. This change in the direction of the dipoles and resultant change in orientation of the domains is very gradual.

Increasing the magnetizing field intensity *H* to point *b* on the graph gradually increases the amount of domains that have aligned themselves along the easy axis of magnetization. Note that the slope of the magnetization curve along *b*, is almost linear.

For applications where the greatest incremental change in flux density is desired for a small change in *H*, a dc biasing current sets the magnetizing field intensity at the center of the linear portion of the *B–H* curve. These applications include magnetic tape heads for tape recorders, audio transformers, and others.

A further increase in current gradually increases the flux density (see section *C* of the curve) until a point is reached where it takes a very large increase in *H* to change the flux density *B* very slightly. This point is called *saturation* and is shown as the *d* section of the curve. Although complete saturation is not practical, it is the point where most of the magnetic domains have aligned themselves along the magnetizing axis.

Devices such as transformers and chokes operate on the principle of a changing current which causes a change in the magnetic flux. Saturation of the magnetic core of these devices would make them act like a dc resistance with a resultant increase in current and burnout of the winding.

Ideally, the permeability of a material should be the largest possible value in order to have the greatest flux density with a small magnetizing field H. From equation (15-3) the permeability μ of a material depends on B and H, or

$$\mu = \frac{B}{H}$$

Since the B–H curve is nonlinear, the changing value of $\Delta B/\Delta H$ will change the permeability for different values of H. The curve of Fig. 15-10 shows that the greatest possible permeability and the greatest resultant flux, with the smallest magnetizing field intensity H, are achieved at the center of the linear portion of the B–H curve.

PRACTICE QUESTIONS 15-3

Choose the most suitable word, phrase, or value to complete each statement.

1. ________ is a relative measure of a material's ability to concentrate magnetic lines of force within itself. Its symbol is ________.
2. The amount ________ ________ will be directly proportional to the permeability of the core of a coil or solenoid.
3. The amount of flux density will be directly proportional to the ________ produced by the coil.
4. Flux density B is measured in webers/m^2 or ________.
5. Magnetic field intensity H or mmf is measured in ________.
6. The permeability of free space or nonmagnetic material is a constant and is ________.
7. Calculate the length of a magnetic air core if a magnetic flux density of 6.7×10^{-3} T produces a field intensity of 320 A/m.
8. A silicon-steel core with a μ of 8000 has been replaced for the core in the magnetic circuit of Question 7. Calculate the flux density of the circuit.
9. The permeability of a ferromagnetic core is not constant but will vary with the applied value of ________.
10. A ________ magnetization curve is a graphical plot of magnetic flux density with an increase in magnetic field intensity.
11. The ________ portion of the B–H curve is used in applications where there is a need for the greatest incremental change in flux density with a small change in H.
12. The point on the B–H curve where a further increase in magnetic field intensity no longer causes much of a change in magnetic flux density is called ________.

15-7 HYSTERESIS LOOP

The hysteresis loop is of more practical importance than the magnetization curve, especially in motors, transformers, and magnetic memory devices. In motor and transformer applications the alternating current magnetizes the iron core first in one direction, then reverses and magnetizes the iron core in the opposite direction. This cycle is repeated 60 times a second for the line voltage frequency of 60 Hz.

To better understand the effects of this constant reversal of magnetic flux density B with a reversal of magnetic field intensity H, a hysteresis loop can be obtained through measurements, either plotting the curve or from an oscilloscope. Fig. 15-11 shows such a typical hysteresis loop of a ferromagnetic material.

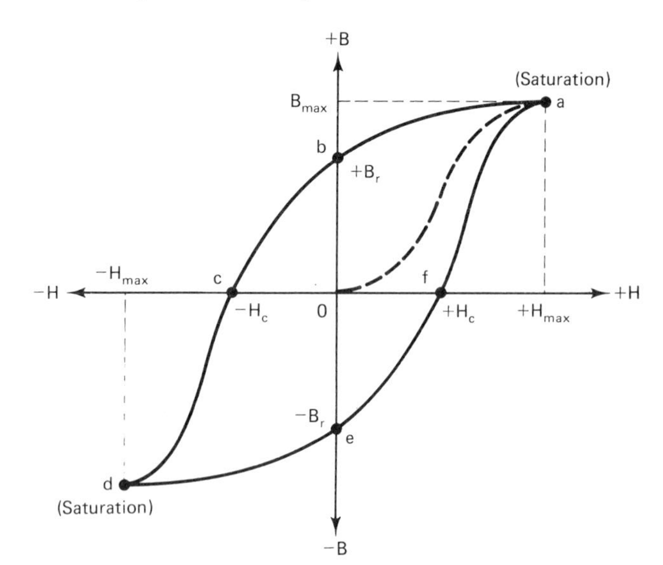

FIGURE 15-11 Typical hysteresis loop of a ferromagnetic material.

Magnetization from Zero to Point *a*

With a current applied to a coil, the initial magnetization follows the dashed-line curve from zero to a. At a the flux density is maximum with applied maximum magnetizing field intensity, H. When point a is reached on the curve, the domains in the ferromagnetic material are oriented mainly in the direction of the magnetizing field, H. Point a is called the *saturation point,* since a further increase in the magnetization intensity H does not produce any appreciable increase in flux B.

Demagnetization from Point *a* to Point *b*

Decreasing the magnetizing field intensity H to zero does not reduce the flux density to zero, but to a value shown as b on the graph. The value of the flux density at this point is known as the *remanent* (remaining or residual) *flux density* B. At this point not all of the magnetic domains have returned to their original unmagnetized state. The demagnetization curve a to b does *not* follow the original magnetization curve zero to a but *lags behind*. This phenomenon was named *hysteresis* by Sir James Ewing.

Reverse Magnetization from Point *b* to Point *c*

Reversing the direction of current flow through the coil reverses the direction of the applied magnetizing field intensity H. The value of the magnetization intensity H to bring the magnetic domains back to their initial or unmagnetized state (zero point on the curve) is termed the *coercive force*, $-H_c$.

Reverse Magnetization from Point *c* to Point *d*

Once the magnetic domains have returned to their unmagnetized state at point c, increasing the reverse magnetizing field gradually aligns the domains in the opposite direction. At point d most of the magnetic domains are aligned in this manner. This saturation point corresponds to point a on the positive half of the curve.

Demagnetization from Point *d* to Point *e*

Decreasing the magnetizing field intensity reduces the amount of domains that are oriented in the direction of the applied reverse magnetizing field. Point e is symmetrically opposite point b and is representative of the remanent flux density $-B_R$ with zero magnetizing field H.

Remagnetization from Point *e* to Point *f*, Then from Point *f* to Point *a*

With an increase in magnetizing field intensity $+H$ the magnetic domains gradually return to their unmagnetized state as shown by points e to f on the graph. Point f is symmetrically opposite point c and is termed the coercive force $+H_c$. The coercive force $+H_c$ is representative of the field intensity required to overcome the coerciveness of the domains. A further increase in the magnetizing field remagnetizes the material as shown by the curve f to a.

The *B–H* loop shows that the magnetization of the material always lags the applied magnetizing field intensity. This lag is due to the energy required to reorient the magnetic domains along the magnetizing field. With a silicon-iron alloy core such as that found in power transformers, the 60 current reversals every second cause the magnetic field to cycle through the *B–H* loop 60 times per second. The energy that is needed to reorient and overcome the coerciveness of the magnetic domains is absorbed by the iron alloy core. This energy heats the core and is dissipated into the surrounding air. The amount of energy loss in one complete magnetization cycle is proportional to the area enclosed by the hysteresis loop. To minimize the power loss due to the hysteresis loop the symmetrically opposite coercive points $-H_c$ and $+H_c$ must be as narrow as possible. With a narrow loop the area within the loop will be smaller; therefore, the power loss measured by the area within the loop will be less.

For applications where the magnetic field reversals are much greater than 60 Hz (cycles per second), the core must be made of soft iron, or a molded powder iron core. Figure 15-12 shows typical hysteresis loops for a soft-iron and silicon-steel core.

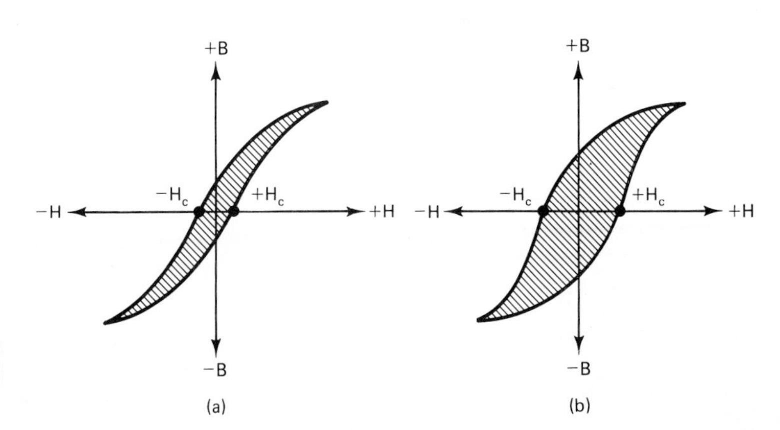

FIGURE 15-12 Typical hysteresis loop. (a) Soft iron or molded powdered iron. (b) Silicon steel.

15-8 OHM'S LAW FOR MAGNETIC CIRCUIT

Examining the magnetic circuit of Fig. 15-13 shows it to be analogous to the electric circuit. An emf is required before current can flow through the conductors of the electric circuit. In the magnetic circuit we need a mmf to set up a magnetic flux which flows around the circuit either through an air core or through a ferromagnetic core. The amount of current that will flow in the electric circuit will depend upon the circuit resistance *and* will be inversely proportional to the resistance. For the magnetic circuit the magnetic flux that will flow will depend on the reluctance of the magnetic circuit. Table 15-1 shows the comparative circuit terms, their units, and their symbols.

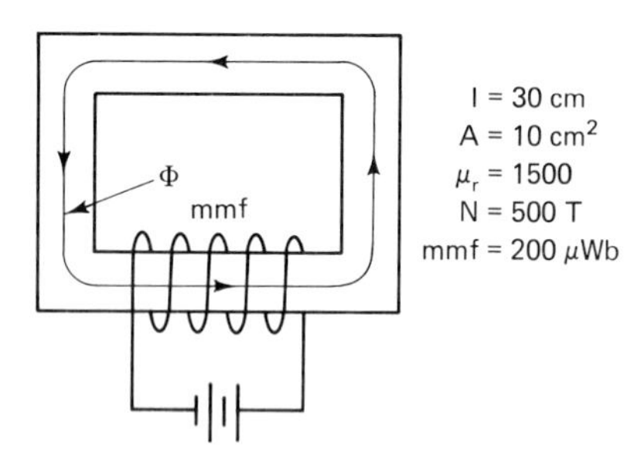

FIGURE 15-13 Magnetic circuit.

Ohm's law for a resistive circuit states that the current will be directly proportional to the emf and inversely proportional to the resistance. In the magnetic circuit the flux is *directly proportional to the mmf and inversely proportional to the reluctance.* In equation form,

$$\Phi = \frac{\mathcal{F}}{\mathcal{R}} \qquad (15\text{-}5)$$

where $\mathcal{F}$ = mmf = NI in amperes
$\mathcal{R}$ = reluctance, in amperes/weber
Φ = flux, in webers

EXAMPLE 15-5

A 500-turn coil wound on an iron core has a 200-mA current flowing through it. The reluctance of the iron core is 5×10^6 A/Wb; calculate the total flux in webers.

Solution: Equation (15-5) gives

$$\Phi = \frac{\mathcal{F}}{\mathcal{R}} = \frac{500\ \text{T} \times 200 \times 10^{-3}\ \text{A}}{5 \times 10^6\ \text{A/Wb}}$$

$$= 20 \times 10^{-6}\ \text{Wb}, \quad \text{or} \quad 20\ \mu\text{Wb}$$

Reluctance

Reluctance is the opposition offered by a magnetic circuit to the setting up of magnetic flux. It is analogous to the resistance of a conductor in an electric circuit, where the resistance is directly proportional to length and inversely proportional to the area. As in the case of the electric circuit, reluctance is directly proportional to the length of the magnetic "path" and inversely proportional to its cross-sectional area. Since various grades of ferromagnetic alloys have varying permeability, the permeability will also affect the reluctance. A larger μ will result in lower reluctance for a given circuit. Reluctance is inversely proportional to permeability, μ. The relationship of length, cross-sectional area, and permeability to reluctance can be stated in equation form as

$$\mathcal{R} = \frac{1}{\mu} \times \frac{l}{A} \quad \text{amperes/weber} \qquad (15\text{-}6)$$

Note that the reciprocal of μ corresponds to the resistivity ρ of the conductor in an electric circuit. Just as ρ is the resistance of a cubic metre of a conductor in the electric circuit, the inverse of μ is termed the reluctance of a cubic metre of magnetic material.

EXAMPLE 15-6

Given the sheet steel core and its parameters as shown in Fig. 15-13, calculate its reluctance.

Solution: Since μ_r is the permeability of the material, the absolute permeability μ must be calculated first using equation (15-4):

$$\mu = \mu_r \times \mu_0 = 1500 \times 1.26 \times 10^{-6}$$

$$= 18.9 \times 10^{-4}\ \frac{\text{T}}{\text{A/m}}$$

TABLE 15-1

Electric Circuit			*Magnetic Circuit*		
Term	*Unit*	*Symbol*	*Term*	*Unit*	*Symbol*
emf, voltage	Volt	V	Magnetomotive force (mmf)	Ampere	$\mathcal{F}$
Current flow	Ampere	I	Flux	Weber (Wb)	Φ
Resistance	Ohm (Ω)	R	Reluctance	A/Wb	$\mathcal{R}$
Conductance	Siemens	G	Permeance	Wb/A	ρ

Having calculated the absolute μ of the core, its reluctance can now be calculated using equation (15-6).

$$\mathcal{R} = \frac{l}{\mu A} = \frac{30 \times 10^{-2}\ \text{m}}{18.9 \times 10^{-4}\ \text{T/(A/m)} \times 10 \times 10^{-4}\ \text{m}^2}$$

$$= 15.9 \times 10^4\ \text{A/Wb}$$

EXAMPLE 15-7

Calculate the required current through the coil of Fig. 15-13 to produce a flux of 200 μWb.

Solution: Transposing equation (15-5) to find mmf gives

$$\text{mmf} = \mathcal{F} = \Phi \times \mathcal{R}$$

$$\mathcal{F} = 200 \times 10^{-6}\ \text{Wb} \times 15.9 \times 10^4\ \text{A/Wb}$$

$$= 31.8\ \text{A}$$

From equation (15-1),

$$\text{mmf} = N \times I$$

$$31.8\ \text{A} = 500\ \text{T} \times I$$

Therefore,

$$I = \frac{31.8\ \text{A}}{500\ \text{T}} \cong 64\ \text{mA}$$

Effect of an Air Gap on Reluctance

In magnetic circuits such as motor control relays, other types of relays, motors, speakers, permanent-magnet moving-coil meters, tape recording heads, and others, the magnetic flux lines must flow across an air gap. The air gap provides a space for the moving parts of such devices.

A deliberate air gap is introduced in some magnetic circuits to prevent undesirable saturation of the core. Saturation of the core is reached when a further increase in current (mmf) does not appreciably increase the total flux.

The air gap introduces a path of higher reluctance into the ferromagnetic circuit as shown in Fig. 15-14.

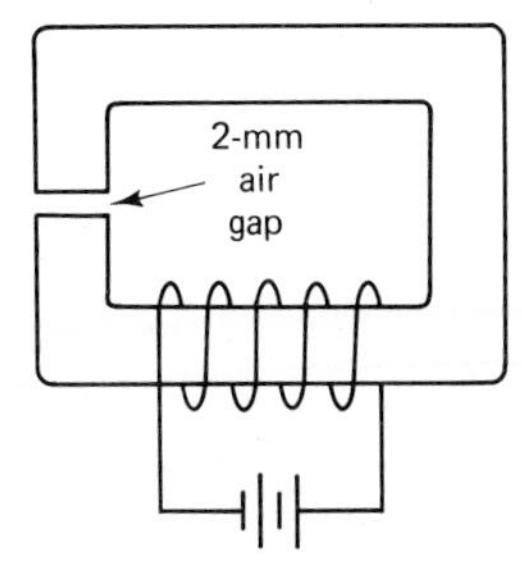

FIGURE 15-14 Magnetic circuit with an air gap.

Since the flux must flow through the gap, the magnetic circuit can be treated as having the higher-reluctance air gap in series with the lower-reluctance ferromagnetic path. Thus the total reluctance in the circuit is the simple sum of the air-gap reluctance plus the ferromagnetic path reluctance. This is analogous to the series electric circuit where the sum of the resistances is equal to the total resistance. For the magnetic circuit the total reluctance can be stated as

$$\mathcal{R}_T = \mathcal{R}_{\text{core}} + \mathcal{R}_{\text{gap}}$$

Rewriting equation (15-5) to incorporate the total reluctance gives

$$\Phi = \frac{\mathcal{F}}{\mathcal{R}_T} = \frac{\mathcal{F}}{\mathcal{R}_{\text{core}} + \mathcal{R}_{\text{gap}}} \qquad (15\text{-}7)$$

EXAMPLE 15-8

Calculate the required current through the coil of Fig. 15-14 to produce a flux of 200 μWb.

Solution: First the air-gap reluctance of the magnetic circuit must be found using equation (15-6).

$$\mathcal{R}_{\text{gap}} = \frac{l}{\mu_0 A}$$

where $l = 2\ \text{mm} = 2 \times 10^{-3}\ \text{m}$
$\mu_0 = 1.26 \times 10^{-6}$
$A = 10\ \text{cm}^2 = 10 \times 10^{-4}\ \text{m}^2$

$$\mathcal{R}_{\text{gap}} = \frac{2 \times 10^{-3}\ \text{m}}{1.26 \times 10^{-6} \times 10 \times 10^{-4}\ \text{m}^2}$$

$$= 158.7 \times 10^4\ \text{A/Wb}$$

Transposing equation (15-7) to find mmf gives

$$\text{mmf} = \mathcal{F} = \Phi \times (\mathcal{R}_{\text{core}} + \mathcal{R}_{\text{gap}})$$

$$= 200 \times 10^{-6}\ \text{Wb} \times (15.9 \times 10^4 + 158.7 \times 10^4)\ \text{A/Wb}$$

$$= 349\ \text{A}$$

The required current through the coil to produce an mmf of 349 A can be calculated by transposing equation (15-1).

$$I = \frac{\text{mmf}}{\text{T}} = \frac{349\ \text{A}}{500\ \text{T}} = 698\ \text{mA} \quad \text{or} \quad 0.7\ \text{A}$$

As shown by the magnetic circuit example, an air gap increases the reluctance significantly. In this case the current must be approximately 10 times greater to produce the same flux as the circuit without the gap.

SUMMARY

Magnetic lines of force flow at right angles to a current-carrying conductor.

Electromagnetic lines of force about a single current-carrying conductor flow in continuous loops around the conductor.

The strength of the magnetic field increases with an increase in current and decreases with a decrease in current.

The flux field about a current-carrying conductor has a specific direction which is dependent on the direction of the current flow. This can be determined using the left-hand-rule method (see Fig. 15-2).

The cross-and-dot method to show the direction of flux about a current-carrying conductor is shown in Fig. 15-3.

Two adjacent current-carrying conductors attract each other when current flows in the same direction and repel each other when current flows in opposite directions.

A loop or a number of loops of insulated wire having a current flowing through will have north and south magnetic poles at its ends.

The left-hand rule to determine the magnetic poles or the direction of current flow through the turns is shown in Fig. 15-7.

Reversing the direction of current flow through the coil will reverse its magnetic poles.

Reversing the direction of turns on the coil will reverse its magnetic poles.

The strength of the magnetic field about a coil or solenoid can be increased a large number of times with a soft-iron core instead of an air core.

Solenoids with a soft-iron core are used in relays.

The magnetic flux that is established about a solenoid is directly proportional to the current and the number of turns.

The product of the current and the number of turns is termed magnetomotive force (mmf) with the symbol $\mathcal{F}$, or

$$\mathcal{F} = N \times I \qquad \text{amperes} \tag{15-1}$$

Magnetic field intensity is a quantity of flux and is directly proportional to mmf and inversely proportional to the length of the magnetic circuit, or

$$H = \frac{NI}{l} \qquad \text{amperes/metre} \tag{15-2}$$

The amount of flux per square metre is termed flux density and is directly proportional to the core permeability and the magnetic field intensity, or

$$B = \mu \times H \tag{15-3}$$

The permeability of free space and most nonmagnetic materials is

$$\mu = \mu_0 = 1.26 \times 10^{-6} \tag{15-4}$$

An increase in magnetizing intensity H does not produce a linear change in flux density B of a material. This is known as the B–H magnetization curve (Fig. 15-10).

A material will reach its saturation point shown on its B–H curve, when a further increase in mmf changes the flux density very slightly.

A hysteresis loop is a *B–H* plot of magnetization, demagnetization of a material in one direction and a *B–H* plot in the reverse magnetization, demagnetization direction.

The energy that is needed to reorient or overcome the coerciveness of the magnetic domains in a material is termed hysteresis loss.

Hysteresis losses are proportional to the area enclosed by the loop. This energy is absorbed by the core material, which heats up and is dissipated into the air.

Ohm's law for a magnetic circuit states that the flux is directly proportional to the mmf and inversely proportional to the reluctance, or

$$\Phi = \frac{\mathcal{F}}{\mathcal{R}} \quad \text{webers} \tag{15-5}$$

The relationship of reluctance to the magnetic circuit length, cross-sectional area, and permeability can be stated in equation form as

$$\mathcal{R} = \frac{1}{\mu} \times \frac{l}{A} \quad \text{amperes/weber} \tag{15-6}$$

An air gap is considered to be in series, and its reluctance must be summed with the reluctance of the ferromagnetic material to obtain the total reluctance of a magnetic circuit, or

$$\mathcal{R}_T = \mathcal{R}_{\text{core}} + \mathcal{R}_{\text{gap}}$$

REVIEW QUESTIONS

15-1. Describe three characteristics of magnetic lines of force about a current-carrying conductor.

15-2. How can the strength of the magnetic lines of force about a current-carrying conductor be increased or decreased?

15-3. State the left-hand rule for a current-carrying conductor.

15-4. What effect does the electromagnetic flux have on two conductors:
(**a**) When current flows in the same direction in both conductors?
(**b**) When current flows in the opposite direction?

15-5. Explain briefly the effect on the magnetic flux when a current-carrying conductor is formed into a loop.

15-6. State the left-hand rule for a solenoid.

15-7. What is the effect on the electromagnetic flux about a solenoid of the following?
(**a**) Reversing the direction of current.
(**b**) Reversing the direction of turns.
(**c**) Increasing the number of turns.
(**d**) Increasing the current through the turns.
(**e**) Decreasing the current through the turns.
(**f**) Using a soft-iron core in place of an air core.

15-8. What is the function of a relay? Describe briefly the components of a relay.

15-9. What two factors govern the mmf of a coil or solenoid?

15-10. Calculate the mmf of a coil with 400 turns and 300 mA of current flowing through it.

15-11. Explain the difference between flux density and magnetic field intensity H.

15-12. State how the length of a magnetic circuit and the mmf affect the magnetic field intensity H.

15-13. Calculate the magnetic field intensity for the magnetic circuit of Fig. 15-9 given that the new mmf is 600 A.

15-14. State how the permeability, μ, and the magnetic field intensity, H, affect the flux density, B, of a magnetic circuit.

15-15. Explain the difference between the permeability of free space and the permeability of a ferromagnetic material.

15-16. Explain briefly in terms of magnetic domains why the B–H curve is nonlinear.

15-17. Explain why a ferromagnetic material heats up when the magnetization force is constantly reversing direction.

15-18. What type of material and shape of hysteresis loop should a ferromagnetic core have in order to minimize power losses?

15-19. Explain saturation briefly, including why it is undesirable in transformers, chokes, motors, and other devices.

15-20. List the magnetic circuit quantities and their units of measurement that are comparative to the electric circuit.

15-21. State in equation form the relationship of length, cross-sectional area, and permeability to the reluctance of a magnetic circuit.

15-22. What affect does an air gap have on the reluctance of a magnetic circuit? How is the total reluctance found?

15-23. List five devices that contain an air gap in their magnetic circuit.

SELF-EXAMINATION

Choose the most suitable word or phrase to complete each statement.

15-1. The magnetic flux that is established about a current-carrying conductor is proportional to the ________.

15-2. Electromagnetic lines of force about a single current-carrying conductor flow in ________ loops.

15-3. A current-carrying conductor has a magnetic flux that flows clockwise about the wire. Reversing the ________ will cause the flux to flow counterclockwise about the wire.

15-4. In using the left-hand rule for a single current-carrying conductor, the ________ points to the direction of the current and the ________ points to the direction of the magnetic flux.

15-5. In using the cross-and-dot method to show the direction of current flow, the ________ represents current flowing away from the observer and the ________ represents current flowing toward the observer.

15-6. Two adjacent current-carrying conductors ________ each other when current flows in the same direction through both conductors, and ________ when current flows in opposite directions.

15-7. Current flowing through the turns on a solenoid establish a ________ and ________ magnetic ________.

15-8. The magnetic flux that is established about a solenoid is directly proportional to the ________ and the ________.

15-9. In using the left-hand rule for solenoids, the ________ points to the north magnetic pole of the solenoid and the ________ point in the direction of current flow through the turns on the solenoid.

15-10. If the current flow is reversed through the turns on a solenoid, the ________ of the magnetic field will also reverse.

15-11. Reversing the direction of the turns on a solenoid will ________ the north and south magnetic poles.

15-12. Using a soft-iron core in place of an air core will greatly reduce the ________ of the magnetic circuit and greatly increase the strength of the electromagnet.

15-13. Retentivity is the ability of a ferromagnetic material to ________ its magnetism.

15-14. The emf of an electric circuit can be compared to the ________ symbol ________ of the magnetic circuit.

15-15. The mmf of a magnetic circuit is directly proportional to ________ and ________.

15-16. A certain coil has a current flowing through it of 1.3 A. Find the number of turns on the coil if it produces a mmf of 260 A.

15-17. Find the current through a coil if 500 turns set up a mmf of 300 A.

15-18. Magnetic field intensity H is directly proportional to ________ and inversely proportional to ________ of the magnetic circuit.

15-19. Find the magnetic field intensity for the magnetic circuit of Question 15-16 if the length of the magnetic path is 16 cm.

15-20. Find the flux density for the magnetic circuit of Question 15-17 given that the air core has a length of 28 cm.

15-21. Find the flux density for the same magnetic circuit of Question 15-17 with a length of 28 cm where the air core has been replaced with an iron core having a permeability factor of 1500.

15-22. For any given B–H curve a point is reached called ________, where it takes a very large increase in magnetic field intensity to change the flux density B very slightly.

15-23. Refer to the B–H curve of Fig. 15-10. The curve shows that the greatest possible permeability and the greatest possible resultant flux with the smallest magnetizing field intensity H is achieved at the center of the ________ portion of the B–H curve.

15-24. Refer to the hysteresis loop of Fig. 15-11. The value of flux density that remains in the magnetic material when the magnetizing field H is reduced to zero is termed ________ flux density and is labeled ________ on the curve.

15-25. Refer to the hysteresis loop of Fig. 15-11. The value of the magnetization intensity H to bring the magnetic domains back to their initial unmagnetized state is termed the ________ ________ and is labeled ________ on the curve.

15-26. For any given hysteresis loop the amount of energy loss in one complete magnetization cycle is proportional to the ________ of the hysteresis loop.

15-27. Refer to Fig. 15-12. For applications where the magnetic field reversals are greater than 60 Hz, the material with the hysteresis loop of Fig. 15-12 ________ must be used.

For Questions 15-28 to 15-31, assume a silicon-iron core with a μ of 2500 has a length of 35 cm and a cross-sectional area of 6.5 cm^2. A coil with 2400 turns carries a current of 0.6 A.

15-28. Calculate the mmf of the circuit.

15-29. Calculate the core's reluctance.

15-30. Calculate the flux in the core.

15-31. Calculate the flux density in the core.

For Questions 15-32 and 15-33, use the magnetic circuit of Questions 15-28 to 15-31 which has been altered to include an air gap of 0.2 cm.

15-32. Calculate the new core's reluctance.

15-33. Calculate the current required to maintain a flux of 84.2×10^{-4} Wb with the air gap.

ANSWERS TO PRACTICE QUESTIONS

15-1

1. continuous
2. repel
3. increases
4. increase
5. current
6. thumb
7. fingers
8. dot, clockwise
9. attract
10. north, south
11. turns
12. fingers
13. thumb, north
14. poles
15. soft iron
16. retentivity

15-2

1. current, turns
2. mmf, $\mathcal{F}$
3. 240 A
4. 1200 A/m
5. 3.6 A
6. 1800 A/m

15-3

1. permeability, μ
2. flux density
3. mmf
4. teslas
5. amperes/metre
6. 1.26×10^{-6}
7. 6 cm
8. 53.6 T
9. field intensity, H
10. B–H
11. linear
12. saturation

16

Generating Ac and Dc Voltages

INTRODUCTION

Volta's discovery paved the way to producing electricity through chemical action. Such batteries were used mainly for experimentation and were not produced commercially to any extent.

Oersted's discovery that a current-carrying conductor had a magnetic field surrounding it led to another discovery by Michael Faraday in 1831—that magnetism could produce electricity. Faraday wound two coils on opposite sides of an iron ring. With one coil connected to a battery, he shorted the second coil together. A compass meter placed parallel to the second coil moved whenever Faraday interrupted the current flow. Faraday was disappointed because he was looking for a steady current flow. Further experiments by Faraday proved that *magnetism induced a current* into the secondary coil whenever there *was a change* (i.e., rising or collapsing of the magnetic field).

Other scientists quickly grasped the principle that a changing magnetic field induced a current into a secondary coil and the induction coil was rapidly developed. By 1838, most scientists were generating electricity by means of the induction coil.

Today, the induction coil powers the spark plugs of every automobile on the road. The principle of an induced current through the action of a changing magnetic field applies to many electronic and electrical devices, such as motors, generators, transformers, receiving antennas, and many others. This chapter will provide the student with the principles of how voltage and current are generated within a single conductor as well as how ac and dc current is generated.

MAIN TOPICS IN CHAPTER 16

OBJECTIVES

After studying Chapter 16, the student should be able to

1. Master the concepts of an induced current and voltage.
2. Master the concepts of the factors that govern the polarity, magnitude, and frequency of an induced voltage.
3. Apply induction principles to the generation of an ac voltage.
4. Apply induction principles to the generation of dc voltage and operation of a dc motor.

16-1 ELECTROMAGNETIC INDUCTION

Faraday's first experiment with two separate coils wound on an iron ring with the primary connected to a battery produced a current in the second coil, called the secondary, whenever he interrupted the current to the primary. He knew that the current in the secondary was produced by the magnetic field set up by the primary, but *only when he closed or opened the circuit to the primary*.

Needing further proof that the current in the secondary winding of his induction coil was produced by the proximity of the magnetic field (by induction), he set up another experiment, shown in Fig. 16-1. He proved conclusively that the magnetic flux must cut through the turns to induce a current and voltage in the coil. Thrusting the bar magnet into the coil caused the galvanometer to deflect in one direction. Pulling the bar magnet out of the coil made the galvanometer needle deflect in the opposite direction, as illustrated in Fig. 16-1(b).

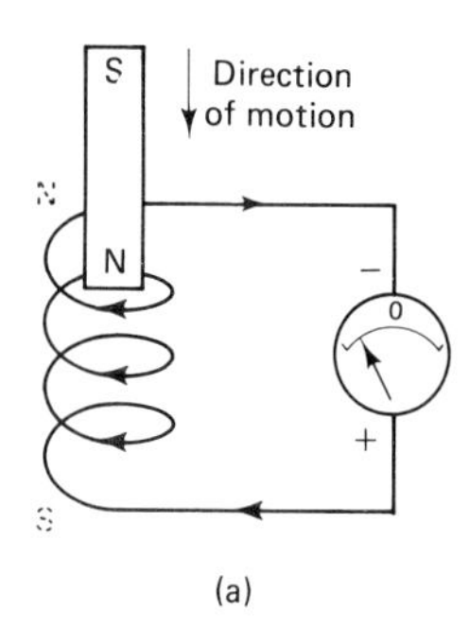

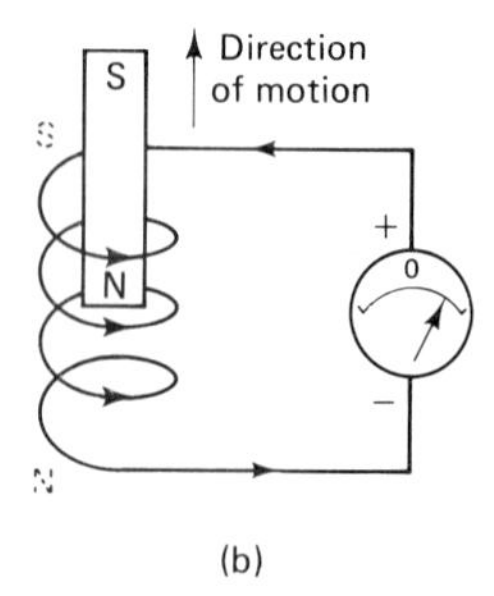

FIGURE 16-1 (a) Induced current flows in one direction. (b) Induced current flows in opposite direction.

Current Generation in a Conductor Cutting through a Magnetic Field

Reversing the process used by Faraday, we can examine the induced current in a conductor cutting through the magnetic lines of force at right angles, as shown in Fig. 16-2(a). The upward motion of the conductor moving through the magnetic field cuts through the flux lines. The reaction of the magnetic flux cutting through the conductor induces free electrons to flow from the lower left end of the conductor through the galvanometer and back to the conductor, completing a circuit. This current direction is shown by the zero-center galvanometer being deflected from zero to the right of the scale, as

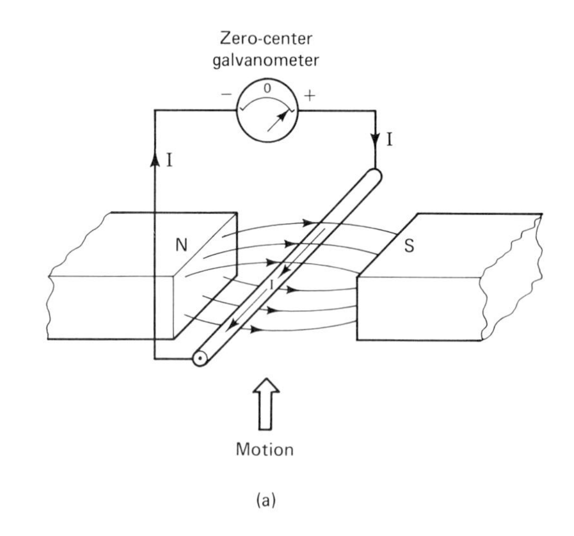

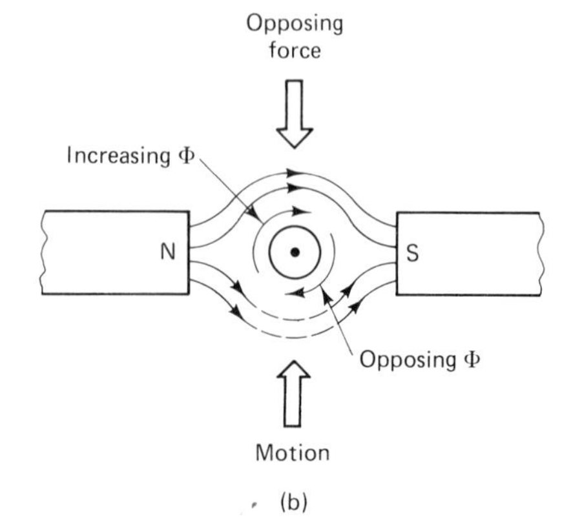

FIGURE 16-2 (a) Conductor moving up through a magnetic field. (b) Direction of flux about a moving conductor (electron flow).

shown in Fig. 16-2(a). Note that the electron current is flowing toward the observer; therefore, the current-carrying conductor will have its own magnetic flux in a clockwise direction (left-hand rule), as shown in Fig. 16-2(b). The magnetic flux above the conductor strengthens the permanent magnet flux while it opposes and weakens the flux below the conductor. *Thus the flux produced by the current through the conductor opposes the direction of motion of the conductor.*

With a downward motion of the conductor, shown in Fig. 16-3(a), the induced current in the conductor now flows in the reverse direction. This is shown by the galvanometer needle being deflected to the left of zero-center. The induced electron current is now flowing away from the observer (left-hand rule), as shown in Fig. 16-3(b). The magnetic flux below the conductor strengthens the permanent-magnet flux, while it opposes and weakens the flux above the conductor. Again, the flux produced by the current through the conductor opposes the direction of motion of the conductor.

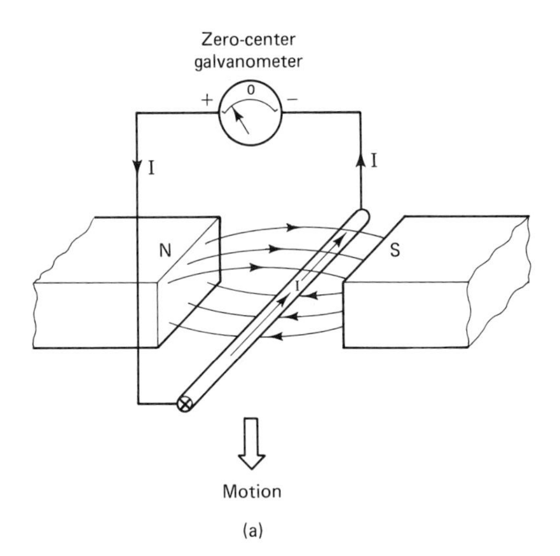

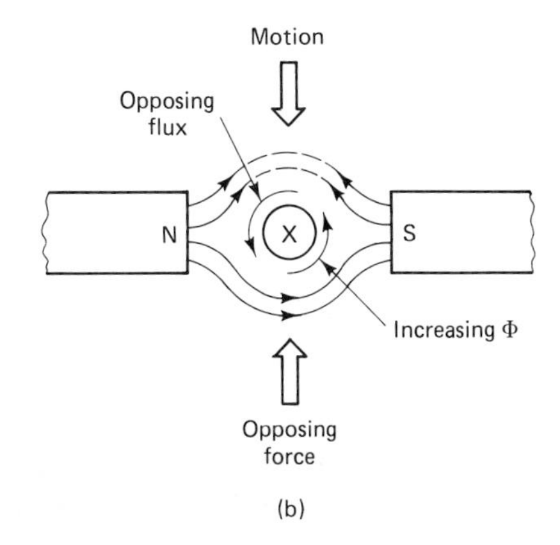

FIGURE 16-3 (a) Conductor moving down through a magnetic field. (b) Direction of flux about a moving conductor (electron flow).

Induced Current versus Induced Emf

In Figs. 16-2 and 16-3, the action of the conductor cutting through the magnetic flux lines induced electrons to flow to one end of the conductor, which will make that end negative. This means that the opposite end of the conductor is left with a deficiency of electrons, making it positive. This instantaneous action of either a downward or upward motion of the conductor produces an induced emf or voltage in the conductor. The polarity of this voltage will always be such as to produce a current once the circuit is completed, to oppose the action producing the induced voltage and current.

16-2 LENZ'S LAW OF ELECTROMAGNETIC INDUCTION

By now you may have noticed that the direction of the induced current and the polarity of the induced voltage is such that it opposes the action producing it. This can be shown through Figs. 16-1, 16-2(b), and 16-3(b).

In Fig. 16-1(a) the induced voltage will produce a current that sets up a north and south magnetic pole to oppose the downward motion of the bar magnets north pole. Similarly, the induced voltage in Fig. 16-1(b) will produce a current so that it sets up a south and north magnetic pole to attract or oppose the upward motion of the bar magnet's north pole. Figures 16-2(b) and 16-3(b) illustrate the direction of current and magnetic flux that opposes the upward or downward motion of the conductor.

Whenever there is *a change* either *in* the magnetic flux or a motion of a conductor through the flux, the induced voltage and current will always oppose the direction of this change. This law of electromagnetic circuits was first formulated by Heinrich Lenz, a German scientist working in Russia, and is known as *Lenz's law,* stated simply as:

> The induced voltage and current will always set up a magnetic flux that opposes the change producing the voltage and current.

A good "close to home" example of Lenz's law is a magneto or bicycle generator (besides all the electric motors). It takes a force to move the armature (moving coils) through the magnetic flux because the magnetic field produced by current in the armature coils opposes the turning motion, producing the induced current.

16-3 FARADAY'S LAW OF ELECTROMAGNETIC INDUCTION

Faraday's experiments with electromagnetic induction showed that the induced voltage in a conductor cutting through a magnetic flux was dependent on three factors:

1. The density of magnetic flux lines
2. The rate at which the conductor cuts through the magnetic flux
3. The number of turns on a coil

Increasing the number of magnetic flux lines or the strength of the magnetic field, will increase the number of free electrons that are induced to flow in one direction. Similarly, increasing the rate at which the conductor cuts through the magnetic flux will increase the rate at which induced free electrons flow in one direction. Forming the conductor into a coil with a number of turns means a greater length of conductor cutting through the magnetic flux. This in turn means that more free electrons within the conductor will be induced to flow in one direction.

Faraday's Law of Induced Voltage

The first two factors apply to a conductor moving through a magnetic flux or conversely, a moving magnetic flux cutting through a conductor. Faraday's law states that the induced voltage in a conductor is directly proportional to the rate at which the conductor moves or cuts through the magnetic flux. In equation form,

$$v_{\text{ind}} = \frac{\Delta\Phi\ (\text{Wb})}{\Delta t\ (\text{s})} \quad \text{volts} \qquad (16\text{-}1)$$

where v_{ind} is the induced emf, in volts

$\Delta\Phi$ is a change (Δ) in the amount of magnetic flux cut by the conductor, in webers

Δt is a change (Δ) in time required for the conductor to move, in seconds

Δ is the Greek capital letter delta, symbol for a change

$\dfrac{\Delta\Phi}{\Delta t}$ is the rate of change at which the conductor cuts through the flux

Note that the lowercase v is used to denote the induced voltage. This is because the induced voltage is not a constant. v_{ind} is dependent on the *rate of change* through the magnetic flux and also the *angle at which the conductor cuts* through the flux; therefore, it is an instantaneous value.

Induced Voltage in a Coil

A current-carrying coil that is wound around an iron core will have all its turns linked by the magnetic flux through the iron core. Since 100% of the flux cuts through each turn, the total induced voltage will be the sum of all the individual induced voltages in each turn. For 100% flux linkage through a coil, the induced voltage will be directly proportional to the number of turns, N. Modifying equation (16-1) gives

$$v_{\text{ind}} = N\frac{\Delta\Phi\ (\text{Wb})}{\Delta t\ (\text{s})} \quad \text{volts} \qquad (16\text{-}2)$$

EXAMPLE 16-1

Assume that the total flux field in Fig. 16-2 is 1 Wb (10^8 lines). The conductor cuts through the magnetic flux in one second. What is the induced voltage in the conductor?

Solution: From equation (16-1),

$$v_{\text{ind}} = \frac{\Delta\Phi}{\Delta t} = \frac{1\ \text{Wb}}{1\ \text{s}} = 1\ \text{V}$$

If the rate at which the conductor cuts through the magnetic flux is speeded up to a half-second, the induced voltage would be

$$v_{\text{ind}} = \frac{\Delta\Phi}{\Delta t} = \frac{1\ \text{Wb}}{0.5\ \text{s}} = 2\ \text{V}$$

EXAMPLE 16-2

Consider two coils N_p and N_s wound on an iron core as shown in Fig.16-4. The cross-sectional area of the core is 4 cm^2 and its length is 24 cm. The relative permeability of the iron is 500. With 200 turns on the primary N_p and 400 turns on the secondary N_s, calculate the induced emf in the secondary coil if at the instant the switch is closed the primary current rises from zero to 3 A in 4 ms.

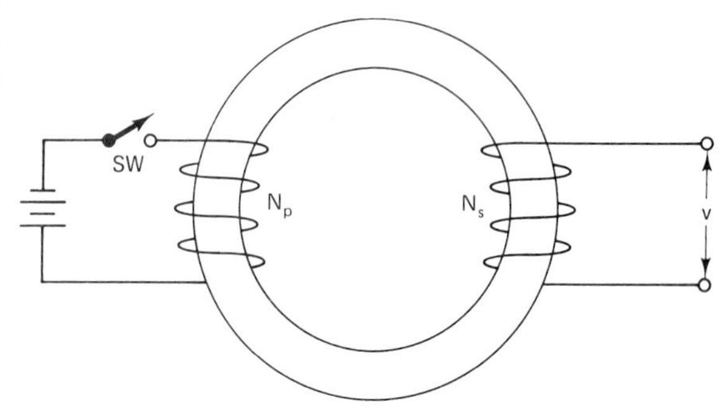

FIGURE 16-4 Primary winding and secondary winding on an iron core.

Solution: Find the magnetic field intensity H using equation (15-2):

$$H = \frac{N_p I}{l} = \frac{200 \times 3 \text{ A}}{24 \times 10^{-2} \text{ m}} = 2500 \text{ A/m}$$

Find the permeability μ using equation (15-4):

$$\mu = \mu_r \times \mu_0 = 500 \times 1.26 \times 10^{-6}$$
$$= 630 \times 10^{-6}$$

Find the flux density B using equation (15-3):

$$B = \mu \times H = 630 \times 10^{-6} \times 2500 \text{ A/m}$$
$$= 1.58 \text{ T}$$

Find the total flux Φ by transposing equation (14-1):

$$\Phi = B \times A = 1.58 \text{ T} \times 4 \times 10^{-4} \text{ m}$$
$$= 632\ \mu\text{Wb}$$

Find the induced emf using equation (16-2):

$$v_{\text{ind}} = N_s \frac{\Delta\Phi \text{ (Wb)}}{\Delta t \text{ (s)}} = 400 \times \frac{632 \times 10^{-6} \text{ Wb}}{4 \times 10^{-3} \text{ (s)}}$$
$$= 63 \text{ V}$$

PRACTICE QUESTIONS

Choose the most suitable word or phrase to complete each statement.

1. A voltage will be induced in a conductor moving through a magnetic flux at ________ angle.
2. The induced current in the conductor produces its own magnetic flux that (*opposes, aids, does not affect*) the direction of motion of the conductor.
3. The polarity of the induced voltage or current will always be such as to (*oppose, aid, not affect*) the action producing it.
4. For a conductor cutting through a magnetic flux, the induced voltage will depend on the ________ of the magnetic flux.
5. For a conductor cutting through a magnetic flux, the induced voltage will depend on the ________ at which the conductor cuts through the magnetic flux.
6. For 100% flux linkage through a coil, the induced voltage will be ________ proportional to the number of turns.
7. Consider the conductor of Fig. 16-2 cutting through a magnetic flux of 4000 μWb in 17 ms. Calculate the induced voltage.
8. Calculate the induced voltage if the conductor in Question 7 was replaced by a coil having 400 turns.

Refer to the coil and bar magnet in Fig. 16-1(a) when answering the following questions.

9. If the coil of 200 turns is cut by a magnetic flux of 0.4 Wb/s, calculate the induced voltage.
10. What polarity of induced voltage is at the top of the coil with respect to the bottom if the bar magnet's south pole is thrust into the coil?

16-4 SIMPLE AC GENERATOR

An electrical generator converts mechanical energy into electrical energy. An ac generator converts mechanical energy into alternating current, which is a flow of electrons first in one direction, then in the opposite direction. This conversion of energy is based on the fundamental process of a wire cutting through a magnetic flux to produce a voltage and current by induction.

Generating a current and voltage through induction is easily done by rotating coils of wire mounted on an *armature* or *rotor* within a magnetic field produced by the mmf of a coil wound on each pole piece of the *stator*, shown in Fig. 16-5.

A single turn or loop of wire rotating through the magnetic flux set up by the stator will be used to simplify the explanation and illustration of how a current and voltage is produced by a generator, as shown in Fig. 16-6(a).

Rotating the loop of Fig. 16-6(a) counterclockwise (CCW) through 360° or one revolution will cause the loop to cut through the magnetic flux at various angles. The induced voltage will then vary in intensity from zero to a maximum value depending on the angular position of the loop.

Rotating Coil Moving through 0°

With some means of mechanically rotating the loop through one revolution, four "stop-action" positions, at 0°, 45°, 90°, and 180°, are shown to represent the in-

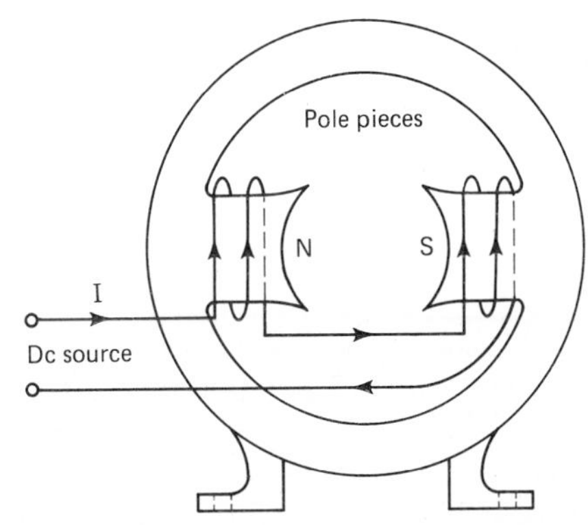

FIGURE 16-5 Cross-sectional view of a basic electrical generator stator.

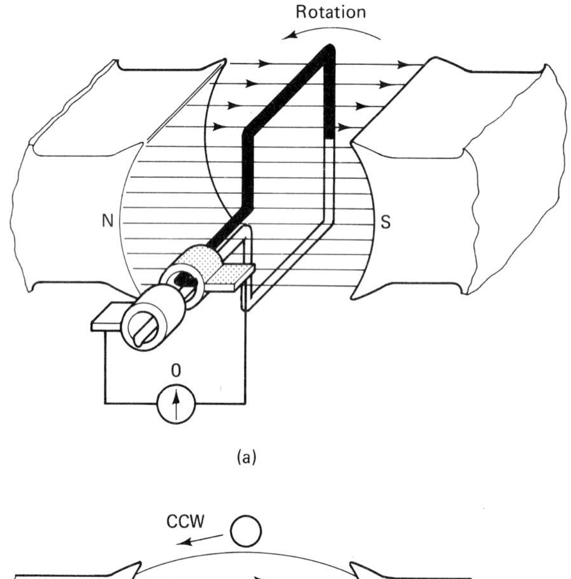

(a)

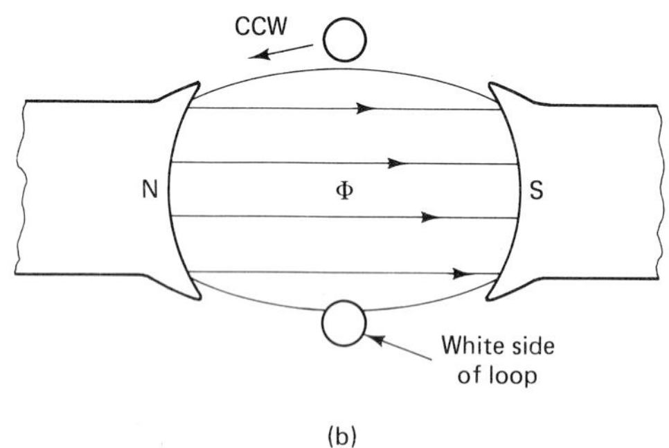

(b)

FIGURE 16-6 (a) One turn coil rotating through a magnetic flux from its initial position of zero degrees. (b) Conductor begins to rotate counterclockwise and cut through the magnetic flux.

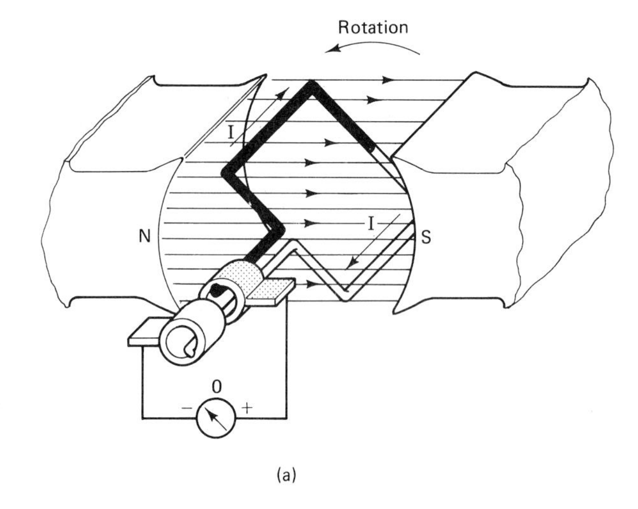

(a)

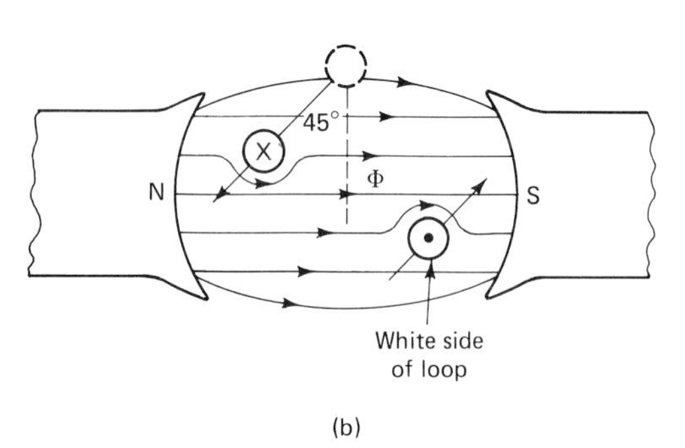

(b)

FIGURE 16-7 (a) One turn coil rotating through a magnetic flux from 0° to 45°. (b) Conductor moves at 45° through the flux; small voltage is induced.

duced voltage. In the initial position at 0° the conductor just begins to move, but does not cut through any magnetic flux lines, as shown in Fig. 16-6(b); therefore, the induced voltage will be zero.

Note that one side of the loop is identified black while the other side is identified white. Each end of the loop is connected to a *slip ring,* which is a bronze ring attached to, but insulated from, the shaft. Both rings are insultated from each other. *Carbon brushes* are held by spring tension against the slip rings. Leads are brought out from the carbon brushes to the external circuit, in this case to a galvanometer.

Rotating Coil Moving through 45°

As the rotation of the loop is continued counterclockwise from 0°, the black side of the loop cuts through the magnetic flux to our next stop-action point of 45° [Fig. 16-7(a)]. The induced voltage was gradually increasing and was proportional to the angle at which the conductor cut the magnetic flux.

Polarity of Induced Voltage. As shown in Fig. 16-7(b), the black side of the loop now has a magnetic flux that is warped in a counterclockwise direction about it. This induces a current that is flowing away from the black slip ring and into the white side of the loop.

At the same instant the white side of the loop has a magnetic flux that is warped in a clockwise direction about it. This induced current aids the current flowing from the black side of the loop so that current flows into the white slip ring. Since current is flowing into the white slip ring, then to the external circuit, the polarity of the white slip ring will be negative while the black ring is positive.

Rotating Coil Moving through 90°

Continuing the rotation of the loop from 45° to 90° shows that an ever-increasing angle at which the loop cuts the magnetic flux induces an ever-increasing voltage. At 90° the loop is cutting the flux at a right angle, as shown in Fig. 16-8, and the induced voltage will be maximum. The polarity of the voltage has not changed. Only the magnitude of the induced voltage has built up in a smooth upward sweep from zero value at 0° to maximum value at 90°.

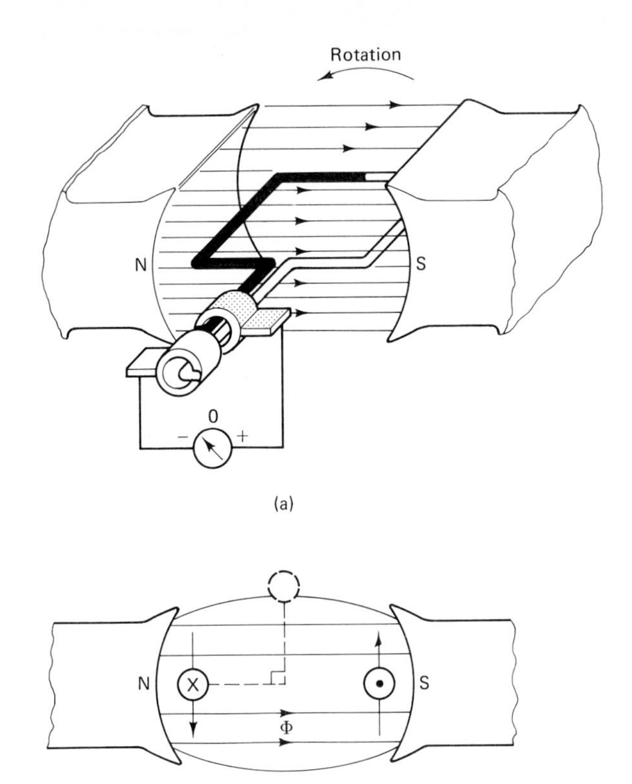

FIGURE 16-8 (a) One turn coil rotating through a magnetic field from 0° to 90°. (b) Conductor moves at 90° through the flux; max voltage is induced.

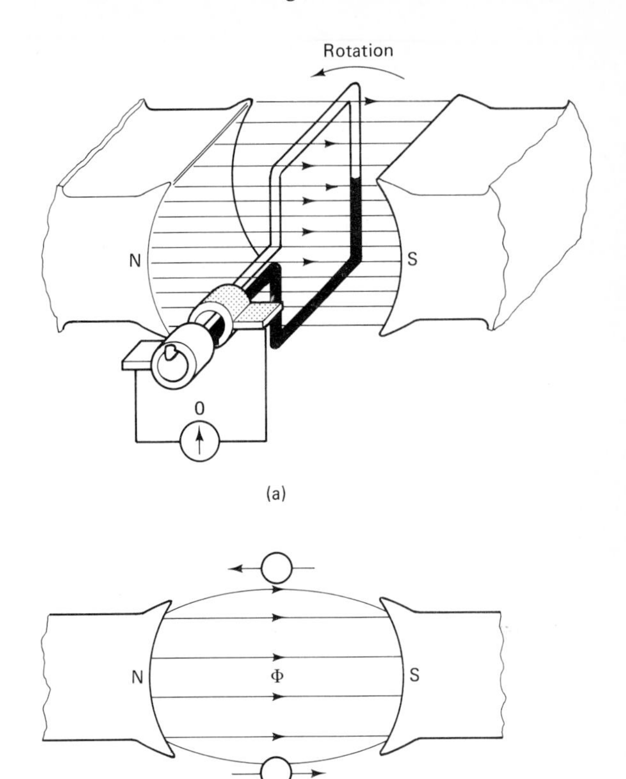

FIGURE 16-9 (a) One turn coil rotating through a magnetic flux from 0° to 180°. (b) Conductor moves at 180° parallel to flux, therefore zero volts will be induced.

Rotating Coil Moving from 90° to 180°

From the 90° position the induced voltage in the loop decreases in a smooth downward sweep to zero volts at 180°. This gradual decrease in voltage is the result of the loop cutting the flux lines at an ever-decreasing angle until at 180°, the coil has reached its initial position. The rotor has now completed one-half of a revolution and the induced voltage has passed through *one alteration* or one-half of a cycle.

Rotating Coil Moving from 180° to 360°

At 180° the black side of the loop has reversed its position with respect to the white side of the loop as shown in Fig. 16-9.

As the loop continues its rotation through 180° to 360°, the induced current will flow in the reverse direction. The white side of the loop is cutting the flux so that the current flows away from the white slip ring. In the black side of the loop the induced current is flowing toward the black slip ring. Thus the polarity of the black slip ring is now negative while the white slip ring polarity is positive. At 360° the rotor has reached its initial position. This position is called the *neutral* position or plane.

The second alternation is a repeat of the first half alternation. Only the current has changed direction, with a resultant polarity change of voltage. The coil has now completed one revolution or cycle. *The induced voltage has passed through two alternations or one cycle.*

16-5 AC VOLTAGE/CURRENT WAVEFORM

The complete ac voltage/current cycle is shown in Fig. 16-10. From the loop's neutral position, 1 on the diagram, the voltage rises in a positive direction.

It builds up in a smooth upward sweep to its maximum value at one-quarter, (position 2) of the coil revolution, or 90°, of the voltage cycle. The voltage then falls from its maximum value in a smooth downward sweep to its zero value at one-half of the coil revolution (position 3), or 180° of the voltage cycle. The graph shows the polarity of the ac voltage waveform to be positive going for 180° of the cycle. This portion of the ac voltage is termed one *alternation*.

During the next half of the coil revolution the polarity of the voltage reverses and is shown to be nega-

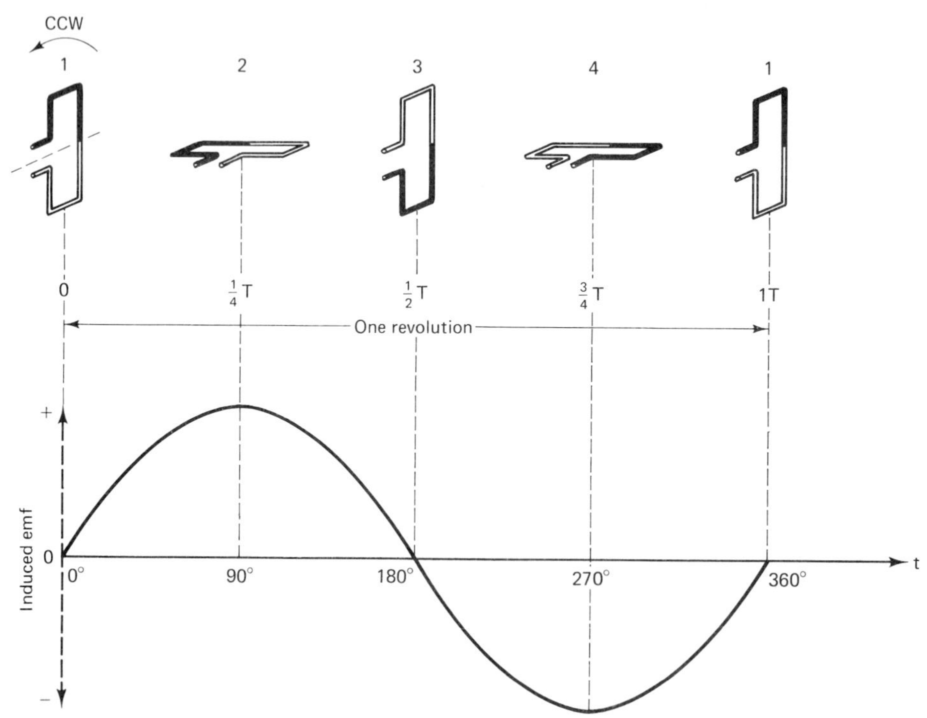

FIGURE 16-10 Complete ac voltage cycle.

tive-going in the graph of Fig. 16-10. As the coil continues its rotation from position 3, the voltage follows the same upward sweep in the negative direction. It reaches its maximum value at 270° (coil position 4), then falls in a smooth downward sweep back to zero volts (coil position 1). The second portion of the ac cycle is also termed one alternation. Since the induced current in the coil flows in one direction during one-half of the cycle and then flows in the opposite direction during the next half of the cycle, it is termed an *alternating current* or simply *ac current.*

16-6 AMPLITUDE OF AN AC VOLTAGE

The simple one-turn loop served to illustrate how the ac voltage was generated. If a conductor cuts through 1 weber (10^8 magnetic flux lines) in 1 second, 1 volt is induced or generated by the conductor.

Practical ac alternators have coils wound on a rotor. The stator also has two or more pole pieces with coils wound around each. Dc current is supplied to the stator winding to provide a constant magnetic flux.

Faraday's laws of electromagnetic induction can be applied to a practical generator since the number of turns in each coil on the rotor and mode of connection (i.e., series aiding) will govern the amplitude of the induced voltage.

Increasing or decreasing the magnetic field strength can be accomplished by varying the dc current supplied to the stator windings. Reducing the magnetic field strength reduces the voltage. This is sometimes done by the electric utilities. The voltage is deliberately reduced at peak power periods (suppertime or extremely hot days) because they are unable to meet the power demand with their generating equipment.

Increasing or decreasing the speed of the rotor will increase or decrease the rate at which the coils cut the magnetic flux. This will vary the amplitude and the frequency of the output voltage of an alternator. However, the speed of an alternator is always kept constant since the frequency, that is, the number of ac voltage cycles for every second, must be kept constant.

16-7 FREQUENCY OF AN AC VOLTAGE

With one pair of magnetic poles, one mechanical revolution of the single coil produced one cycle of voltage. If the coil rotation is increased, say to 60 revolutions every second, the number of voltage cycles would also increase to 60 every second. The number of voltage or current cycles produced every second is called the *frequency* (f). Frequency is defined simply as cycles per (symbol Hz) in honor of German physicist Heinrich Hertz. In the example above, the frequency of the ac voltage has been increased from 1 Hz to 60 Hz.

Adding another pair of poles to the stator, as shown in Fig. 16-11, will increase the frequency of the generated voltage. Assuming that the rotor speed is the

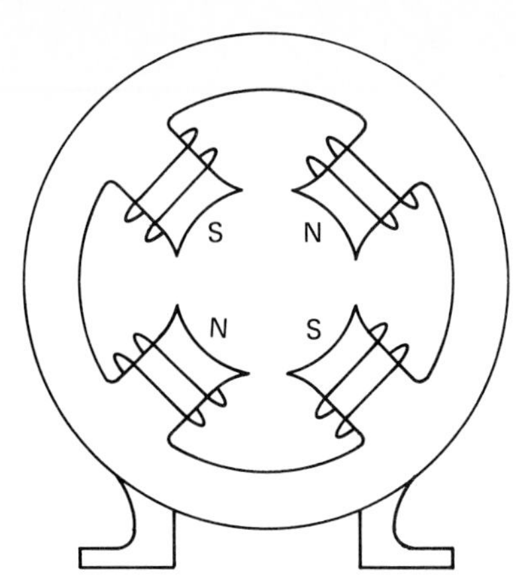

FIGURE 16-11 Four-pole generator. Pole-piece windings are series or compound connected.

same, the coil will generate a complete cycle of voltage in just one half-revolution, and two cycles with one revolution of the coil.

PRACTICE QUESTIONS 16-2

Choose the most suitable word, phrase, or value to complete each statement.

1. Rotating coils mounted on a ________ within a magnetic flux will induce a current and voltage in the coils.
2. Magnetic flux is set up by the mmf produced by a coil wound on each pole piece of the ________.
3. To induce a ________ or ________ in the coil, the loop or coil must cut through the magnetic flux.
4. The induced voltage is proportional to the ________ function of the angle at which the conductor cuts the magnetic flux.
5. The direction of the induced current and resultant polarity of the voltage depends on the direction of the magnetic flux about the ________.
6. Zero induced voltage in the coil takes place at ________ degrees and at ________ degrees of its rotation.
7. Maximum induced voltage in the coil takes place at ________ degrees and at ________ degrees of its rotation.
8. When the induced voltage in a coil is zero volts, the mechanical position at that point is termed the ________ position.
9. A voltage or current has gone through one ________ when it rises from zero to maximum, then back to zero in 180°.
10. When the voltage or current goes through two alternations, it has completed one ________.
11. ________ current is a flow of current in one direction for one-half of a cycle, and then a flow in the opposite direction for the next half of a cycle.
12. Increasing the number of turns in a coil will (*increase, decrease*) the amplitude of the induced voltage.
13. Two other factors that will increase or decrease the amplitude of the induced voltage are ________ and ________.
14. The number of voltage or current cycles that take place every second is called ________.
15. The present-day term ________, symbol ________, is used to designate cycles per second.

16-8 SIMPLE DC GENERATOR

Any current that is generated by a coil rotating in a magnetic field is an alternating current. It must go through two alternations as the coil cuts through the magnetic flux. Since dc is a steady-state or unchanging value of current, some means of rectifying the ac to dc must be provided. The ac alternator is changed to a dc generator by means of a *commutator,* as shown in Fig. 16-12.

The commutator provides the rectifying action to change the ac to dc. A simple form of a commutator shown in Fig. 16-12 is a split ring of two segments each insulated from one another as well as from the shaft. Spring-loaded carbon brushes press against opposite segments and provide connection to the external circuit. The brushes are set one on each side of the insulated gaps between the segments when the coil is in the neutral position.

As in the case of the ac alternator, rotating the coil one-half of a revolution (i.e., 180°) will produce one alternation of voltage. Note that the black side of the coil has cut downward across the face of the north pole while the white side has moved upward across the face

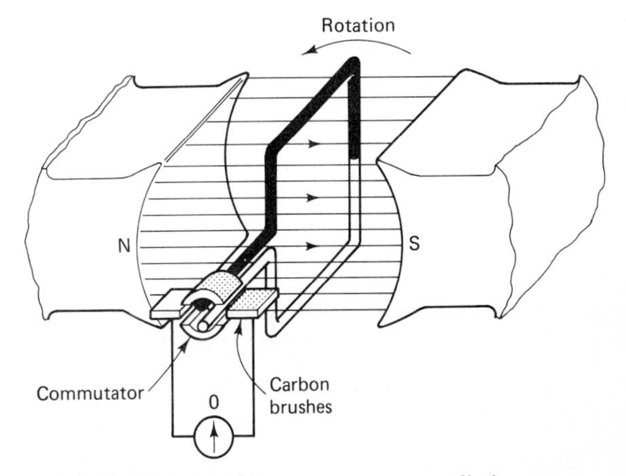

FIGURE 16-12 Simple one-turn coil dc generator. The coil is in its initial position of 0°.

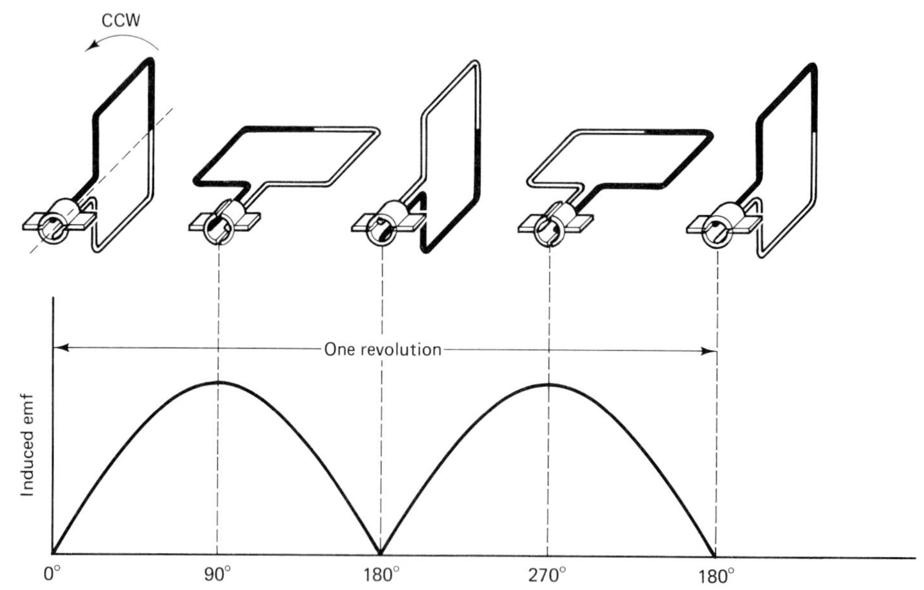

FIGURE 16-13 Pulsating dc voltage produced by a one-turn coil, two-pole dc generator.

of the south pole. This means that the induced current flows out of the white commutator segment, making it negative, and returns to the black segment through the black brush, making it positive. The voltage that is produced will be one positive alternation, as shown in Fig. 16-13.

For the next one-half revolution from 180° to 360°, the black side of the loop is now cutting the magnetic flux upward across the face of the south pole, while the white side of the loop is moving downward across the face of the north pole. The direction of the current has reversed in the loop; however, the segments have also been displayed by 180° so that the terminal polarity of the voltage that is produced will be another positive alternation, as shown in Fig. 16-13.

Figure 16-13 shows that the voltage produced during one revolution of a single loop is a *pulsating type of dc*. If additional coils are added at right angle to each other, the voltage produced by the generator will very nearly approach pure dc, as shown in Fig. 16-14. Practical dc generators have various configurations of armature windings (e.g., lap winding and wave winding), so that the output voltage for practical applications is dc.

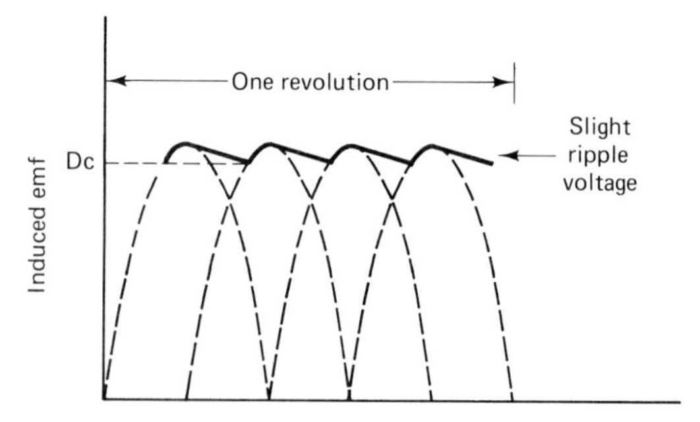

FIGURE 16-14 Voltage output of a four-pole dc generator.

SUMMARY

A current and voltage will be induced in a conductor that is moving or cutting through a magnetic flux at a right angle.

The direction of the induced current in a conductor is dependent on the direction of the magnetic flux being cut and the direction of the conductor cutting through the flux.

Lenz's law states that the induced voltage and current will always set up a magnetic flux that opposes the change producing the voltage and current.

Faraday's law of electromagnetic induction shows that the induced voltage in a conductor cutting through a magnetic flux is dependent on three factors,

1. The amount or strength of the magnetic flux

2. The rate at which the conductor cuts through the magnetic flux
3. The number of turns on a coil

The induced voltage in a conductor is directly proportional to the rate at which the conductor cuts through the magnetic flux. In equation form,

$$v_{\text{ind}} = \frac{\Delta\Phi}{\Delta t}\,\frac{\text{Wb}}{s} \quad \text{volts} \tag{16-1}$$

For 100% flux linkage through a coil, the induced voltage will be directly proportional to the number of turns. In equation form,

$$v_{\text{ind}} = N\,\frac{\Delta\Phi\ \text{Wb}}{\Delta t\ s} \quad \text{volts} \tag{16-2}$$

An ac generator converts mechanical energy into electrical energy.

A current and voltage are generated through magnetic induction by rotating coils of wire mounted on a rotor or armature.

The magnetic flux for the generator is produced by dc current-carrying coils wound on poles pieces of the generator stator.

Rotating a coil mounted on a shaft between north and south pole pieces for one revolution will produce two alternations of voltage, a positive-going alternation and a negative-going alternation.

Two alternations of voltage is termed on alternating current.

An ac voltage has gone through one cycle for every two alternations of voltage or 360°.

The number of cycles that the voltage goes through every second is termed frequency and is in hertz (Hz).

The amplitude of an ac voltage is determined by the number of turns on the rotating coil and the number of flux lines, or strength of the magnetic flux.

An ac alternator is changed to a dc generator by means of a commutator.

A dc commutator is a split ring of two segments each insulated from one another as well as from the shaft.

A dc generator produces positive alternations of voltage termed pulsating dc.

With additional coils on the armature the output voltage for practical purposes will be dc.

REVIEW QUESTIONS

16-1. What basic principle of electromagnetic induction did Faraday discover?

16-2. A conductor or coil is moved through the magnetic field set up by a north and a south pole piece. What determines the direction of the induced current in the conductor?

16-3. State Lenz's law in your own words.

16-4. What three factors determine the induced voltage of a conductor cutting through a magnetic flux?

16-5. Explain flux linkage.

16-6. What is the induced voltage in a conductor cutting through a magnetic flux of 500 μWb in 10 ms?

16-7. Explain briefly how a current and voltage can be produced by rotary equipment.

16-8. Explain briefly the voltage waveform that is obtained by a single loop rotating one revolution through a magnetic flux.

16-9. What two factors determine the amplitude of an ac voltage produced by an alternator?

16-10. What is the effect on the ac output voltage if the speed of the rotor is increased?

16-11. Why must the rotor speed be kept constant?

16-12. Explain the frequency of an ac voltage. Give the symbol that denotes cycles per second.

16-13. How is the ac alternator changed to a dc generator?

16-14. Sketch and label the axis of a dc single-coil-generator output voltage.

16-15. Explain how the dc voltage output is improved to correspond to pure dc in a practical generator.

SELF-EXAMINATION

Choose the most suitable word or phrase to complete each statement.

16-1. A current and voltage will be induced into a conductor by ________ the conductor through a magnetic field at a right angle.

16-2. For a single conductor the amplitude of the induced voltage depends on the ________ at which the flux is cut.

16-3. The magnetic flux produced by the induced current (*aids, opposes, does not affect*) the motion of the conductor.

16-4. ________ law states that the induced voltage and current will always set up a magnetic flux that opposes the change producing the voltage and current.

16-5. The magnitude of an induced voltage in a conductor cutting through a magnetic flux is dependent on the amount of ________ ________ ________.

16-6. The magnitude of an induced voltage in a conductor cutting through a magnetic flux is dependent on the ________ at which the conductor cuts through the magnetic flux.

16-7. The induced voltage in a conductor cutting through a magnetic flux is dependent on the number of ________ on a coil.

16-8. Calculate the induced voltage in a conductor cutting through a magnetic flux of 800 μWb at the rate of 1 ms.

16-9. What is the induced voltage of Question 16-8 if the strength of the magnetic flux is doubled to 1600 μWb?

16-10. What would the induced voltage be in the coil of Question 16-9 if the number of turns were doubled?

16-11. Calculate the induced voltage in the secondary coil if the primary current rises to 1 A in 100 ms (Fig. 16-15). Consider the following parameters for the magnetic circuit.

$N_p = 1600$ T   area of the core $= 6.5$ cm^2

$N_s = 400$ T $l = 24$ cm

$\mu_r = 1500$

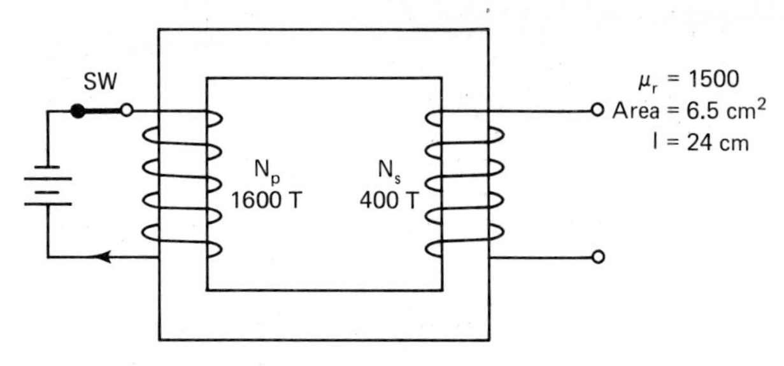

FIGURE 16-15 Self-Examination Question 16-11.

16-12. An ac voltage that rises from zero, reaches a maximum value in a positive direction, then falls back to zero is called a positive ________.

16-13. An ac voltage that completes two alternations has gone through one ________.

16-14. There are ________ degrees in one ac cycle.

16-15. An ac voltage completes one alternation in ________ degrees of a cycle.

16-16. The maximum value of an ac voltage alternation occurs at ________ degrees and at ________ degrees.

16-17. The term ________ of a voltage is used to indicate the number of cycles that a voltage has gone through in 1 s.

16-18. The universally accepted term ________, symbol ________, is used to designate cycles per second.

16-19. An ac generator can be made to produce dc by modifying the slip rings into ________.

16-20. The dc generator produces an irregular dc voltage termed ________ dc.

ANSWERS TO PRACTICE QUESTIONS

16-1

1. right
2. opposes
3. oppose
4. magnitude
5. rate
6. directly
7. 0.24 V
8. 94 V
9. 80 V
10. negative

16-2

1. rotor
2. stator
3. current, voltage
4. sine
5. loop or turn
6. 0, 360
7. 90, 270
8. neutral
9. alternation
10. cycle
11. ac
12. increase
13. strength of flux, speed of rotor
14. frequency
15. hertz, Hz

17

Alternating Current and Voltage

INTRODUCTION

The world's first dc generator, put together by Faraday, was a working model of a copper disk mounted between two magnetic poles. He was able to draw dc current from the disk through two sliding contacts or brushes. Soon other experimenters improved this crude generator by substituting electromagnets for the permanent magnets. Coils were wound lengthwise in slots on a soft-iron core called an armature. These machines were known as dynamos. All early generators of electric power were of the commutator-armature type and produced dc.

It was Nikola Tesla who introduced alternating current to the world. Tesla was born in Yugoslavia. After he graduated from Prague University and a postgraduate course in Budapest, he worked for a Paris telephone company. Tesla immigrated to the United States in 1884, where in 1887–1888 he was granted a range of patents to cover single-phase and polyphase ac systems. George Westinghouse bought out Tesla's ac patents for $1 million.

At the same time, Edison and an associate, Brown, already had been generating and selling dc power from their first power station. They quickly realized that the Westinghouse ac system was going to provide competition. Both Edison and Brown lectured, argued, and used their considerable influence to discredit the ac system. One of the best examples of the extremes that Edison and Brown were willing to go can be demonstrated by a challenge Brown issued to Westinghouse in a number of newspapers on December 18, 1888.

"I challenge Mr. Westinghouse to meet me in the presence of competent electrical experts and take through his body the alternating current while I take through mine a continuous current. We will begin with 100 volts and will gradually increase the pressure 50 volts at a time . . . until either one or the other has cried enough, and publicly admits his error." In spite of all the "hullabaloo" surrounding the dc versus ac debate, the economics of transmitting power over a distance won out and the ac system was adopted. Westinghouse won the bid to build the generators for the International Niagara Commission at Niagara Falls. Power was transmitted at 22,000 V to Buffalo 26 miles away without any great power loss. Today, ac power is generated not only by water power but also by nuclear power.

Having discovered in Chapter 16 how alternating current and direct current are generated, we now investigate the characteristics of ac as well as ac measuring instruments such as the oscilloscope and the voltmeter.

MAIN TOPICS IN CHAPTER 17

OBJECTIVES

After studying Chapter 17, the student should be able to

1. Master the characteristic concepts of an ac voltage and current.
2. Master the concepts of representing ac values by means of a phasor diagram.
3. Master the basic functions of the controls on an oscilloscope and be able to use the scope to measure the amplitude and frequency of an ac voltage.
4. Interchange values of ac voltage and current by calculations.

17-1 AC COMPARED TO A SINE WAVE

The ac waveform generated by an alternator resembles a sine-function wave. In fact, the design of alternators must be such as to follow this sine function as closely as possible with each rotor rotation. An irregular ac waveform produces large losses in motors, transformers, line conductors, and insulation, since current variations are not uniform.

Sine-wave or sinusoidal-wave voltage can also be generated easily by special electronic circuits. Modern electronic sine-wave generators or oscillators can produce a sine-wave voltage with a frequency range of up to gigahertz (10^9 Hz).

Generated Ac Waveform Resemblance to Sine Function

Recall the simple coil conductor of Chapter 16, rotating through one revolution or 360° in a stationary magnetic field. The angle at which the conductor cuts through the magnetic flux determines the *instantaneous* induced voltage for each degree of rotation. Figure 17-1 shows a plot of sample instantaneous values for one cycle of output voltage from a two-pole alternator.

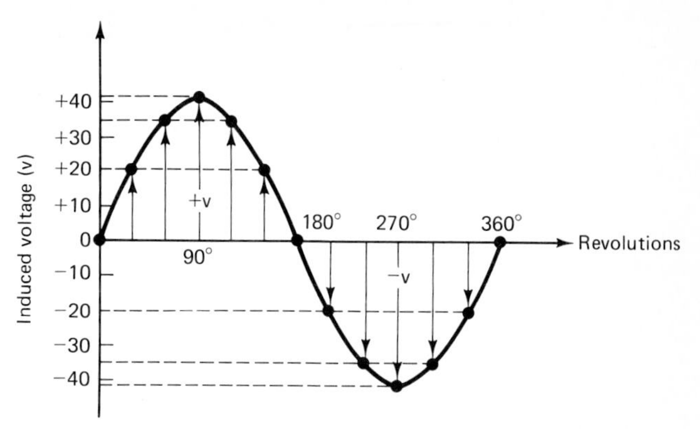

FIGURE 17-1 Instantaneous ac voltage values plotted for one cycle or one revolution of a two-pole single coil alternator.

Whether the ac voltage and current are produced by an alternator or electronically by an oscillator, the waveform will resemble a sine wave fairly accurately. To plot a sine wave a rotating vector is used since the instantaneous ac values are changing continuously through 360° or a cycle of voltage.

> NOTE: A vector shows magnitude and direction.
> A rotating vector is termed a phasor.

Figure 17-2 shows a rotating vector or phasor. The direction of the sine-wave value is shown by the arrowhead, which rotates *counterclockwise* (CCW) by convention through 360°. The magnitude or maximum value is shown by the length of the phasor arm drawn to scale. Since the phasor is to represent a voltage or current, the length of the phasor will be stated in terms of the maximum voltage or current. For the rotating phasor of Fig. 17-2, the length represents the maximum voltage and is labeled V_{max}. The instantaneous voltage value is shown in the lowercase letter v or i for current, to differentiate them from other values.

A plot of the sine angle (θ) voltage or current values can be made by rotating the phasor arm counterclockwise from $\theta = 0°$ to $\theta = 360°$. (θ is the Greek lowercase letter theta.) Assuming 1 V as the maximum unit value for the length of the phasor arm, the instantaneous value shown as the vertical projection for the angle θ is projected to the right side of the appropriate angle to obtain a plotting point(s) of the sine wave. From trigonometric functions for a right-angle triangle, the sine function of an angle is

$$\sin \theta = \frac{\text{opposite side}}{\text{hypotenuse}}$$

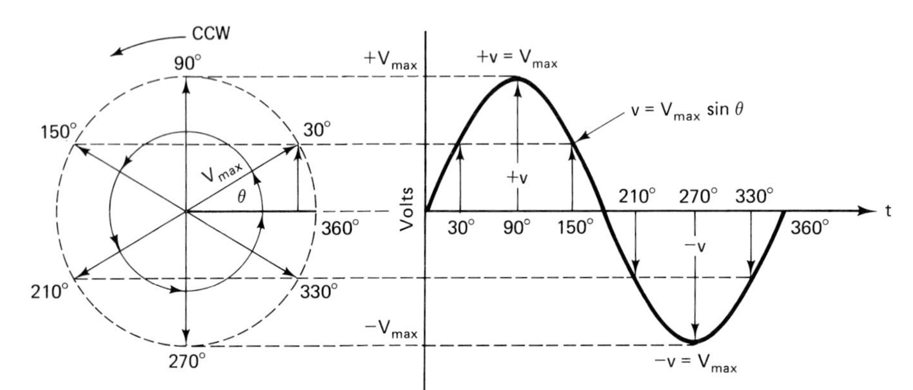

FIGURE 17-2 Rotating phasor and resultant sine wave produced by instantaneous values from 0 through 360°.

Since the arm of the phasor represents the maximum value and is the hypotenuse, as shown in Fig. 17-3, the instantaneous value forms the right angle; then

$$\sin \theta = \frac{\text{opposite side}}{\text{hypotenuse}} = \frac{a}{b} = \frac{v}{V_{\text{max}}}$$

or

$$\sin \theta = \frac{v}{V_{\text{max}}}$$

Therefore,

$$v = V_{\text{max}} \sin \theta \tag{17-1}$$

As an example, at 30° with a maximum value of 1 V, the instantaneous value will be

$$V = V_{\text{max}} \sin \theta = 1 \text{ V} \times \sin 30° = 0.5 \text{ V}$$

Similarly at 45°, $\sin \theta = 0.707$; therefore, the instantaneous value will be 0.707 V. At 90° the instantaneous value will be equal to 1 V or the maximum value since $\sin 90° = 1$. For all other angles in the second, third, and fourth quadrants, the instantaneous values can be found by applying equation (17-1). For example, at 150° $\sin \theta = \sin (180° - 150°) = 0.5$; therefore,

$$v = V_{\text{max}} \times \sin \theta = 1 \text{ V} \times 0.5 = 0.5 \text{ V}$$

At 210°, $\sin \theta = \sin (180° - 210°) = -0.5$ and $v = 1 \text{ V} \times (-0.5) = -0.5$ V. The negative sign shows that the instantaneous voltage has changed polarity so that its vertical projection is below the horizontal axis, as shown in Fig. 17-2.

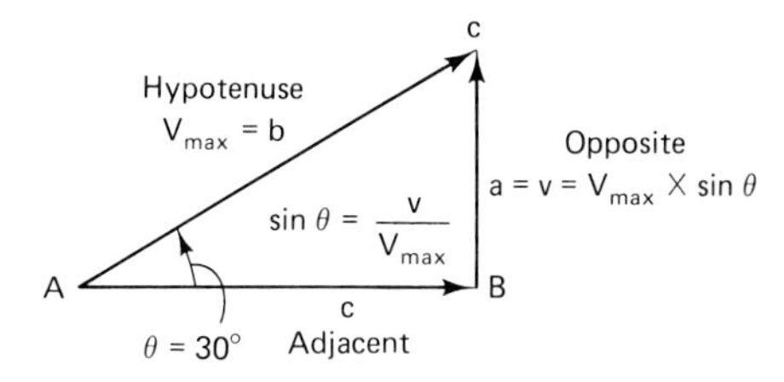

FIGURE 17-3 Trigonometric function for a right-angle triangle.

17-2 VALUES OF AC VOLTAGE AND CURRENT

Instantaneous values are particularly useful to explain the resultant voltage of two or more voltages that are out of phase with each other. Out-of-phase voltages and currents are discussed later in the chapter.

Of more immediate interest to us is being able to read ac values using an oscilloscope or voltmeter. The sine-wave voltage you see in Fig. 17-4 can be seen on an oscilloscope screen. The zero-axis line shown in Fig. 17-4 is not seen on an oscilloscope. It is a horizontal line drawn through the zero-crossing ac values and is equidistant from both the positive and negative peak values.

Peak and Peak-to-Peak Value

The amplitude of the sine-wave voltage from the zero-crossing voltage value to its maximum value is termed the *peak value*, V_p. Measuring the voltage from the positive peak value to the negative peak value gives the peak-to-peak value, $V_{p\text{-}p}$. Both peak and peak-to-peak values can be read easily from an oscilloscope screen. An analog multimeter voltmeter is capable of reading a peak-to-peak value. Note that the peak-to-peak value is twice the peak value since the amplitude of each peak value is equal.

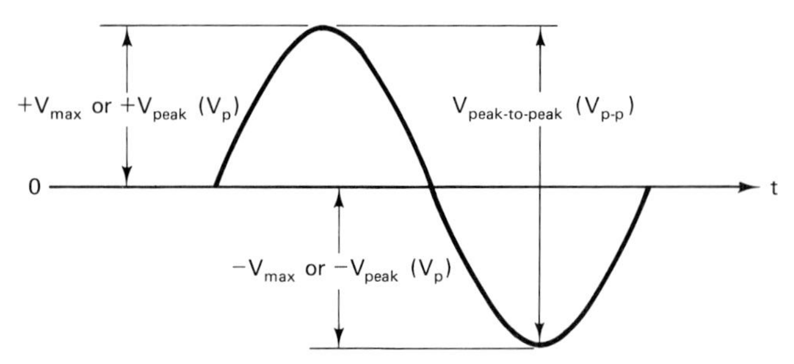

FIGURE 17-4 Amplitude values for a sine-wave voltage readable on an oscilloscope.

Any peak or peak-to-peak voltage or current value that is specified must be labeled with its appropriate subscript. The subscript for the peak value is the lowercase p, with pp for peak-to-peak. It is understood that the peak value can also be termed the maximum value. The peak-to-peak value is twice the peak value; in equation form;

$$V_{pp} = 2V_p \tag{17-2}$$

$$I_{p\text{-}p} = 2I_{p\text{-}p} \tag{17-3}$$

Average Value

The instantaneous values for a sine-wave voltage shown in Fig. 17-2 can be determined using equation (17-1). The average of these values is found by summing the values and dividing the sum by the number of samples. Of course, the greater the number of samples, the greater will be the accuracy of the average value. An average value for one cycle of ac will be zero since one alternation is positive and the next alternation is negative, thus canceling each other. An average value taken over one and two alternations, however, is useful for half-wave and full-wave ac-to-dc rectifier circuits. The dc output voltage will be positive or negative pulses, which will give an average dc value over one or two alterations of the ac voltage.

Given a unit value of 1 A, Table 17-1 shows 18 sample instantaneous values. Summing the values gives

TABLE 17-1
Instantaneous values of ac for one unit of current or voltage used to find the average and rms values

Angle (deg)	*Current I (A)*	*Current I^2 (A)*
0	0.00000	0.00000
10	0.17365	0.03015
20	0.34202	0.11698
30	0.50000	0.25000
40	0.64279	0.41318
50	0.76604	0.58682
60	0.86603	0.75000
70	0.93969	0.88302
80	0.98481	0.96985
90	1.00000	1.00000
100	0.98481	0.96985
110	0.93969	0.88302
120	0.86603	0.75000
130	0.76604	0.58682
140	0.64279	0.41318
150	0.50000	0.25000
160	0.34202	0.11698
170	0.17365	0.03015
180	0.00000	0.00000
Total	11.43006	9.00000

11.43006. Dividing the sum by 18 gives 0.635, the approximate average of the peak value. To obtain an exact average value, we must use a much greater sampling of instantaneous values or calculate the value using integral calculas. The exact value is

$$V_{avg} = 0.637 \times V_p \tag{17-4}$$

Equation (17-4) can easily be manipulated on a calculator by using $2/\pi$ for 0.637, or

or $$V_{avg} = \frac{2}{\pi} \times V_p \tag{17-5}$$

A particular application of average values is in rectifier circuits. A rectifier circuit changes or rectifies the alternating current and voltage into pulses of dc current and voltage, as shown in Fig. 17-5. In Fig. 17-5(a) the rectifier conducts on both ac alternations, so that two dc pulses are produced for every cycle of ac voltage. This rectifier is termed a *full-wave rectifier* since it rectifies the full ac waveform of one cycle. The average dc value that would be measured by a dc voltmeter would then be 0.637 of the peak voltage taken for both alternations.

In Fig. 17-5(b) the rectifier conducts only on one ac alternation, so that one dc pulse is produced for every

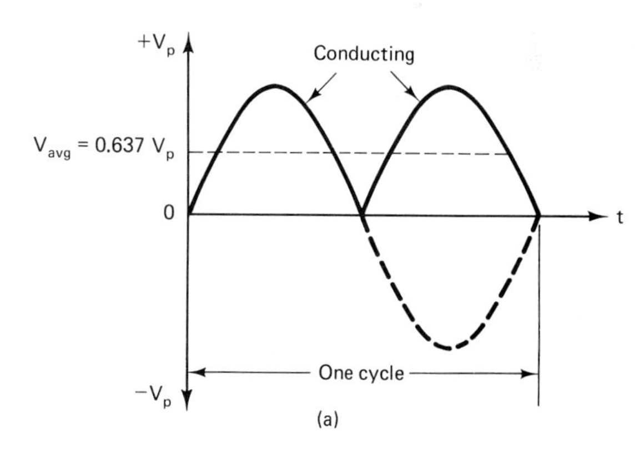

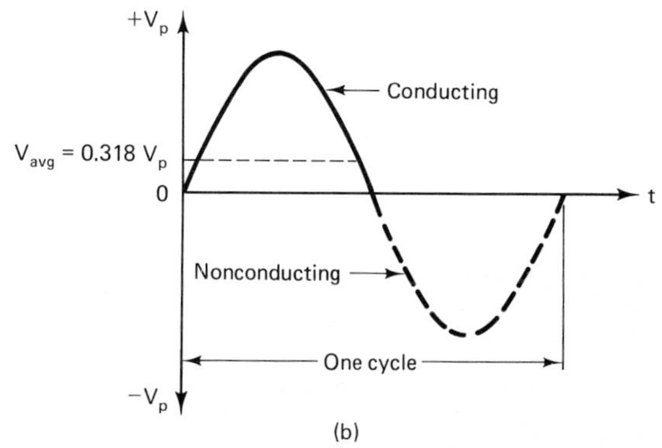

FIGURE 17-5 (a) Voltage output of a full-wave rectifier. (b) Voltage output of a half-wave rectifier.

cycle of ac voltage. This rectifier is termed a *half-wave rectifier* since it rectifies only one-half of the ac waveform cycle. The average dc value that would be measured by a dc voltmeter would then be 0.318 of the peak voltage.

PRACTICE QUESTIONS 17-1

Choose the most suitable word, phrase, or value to complete each statement.

1. The basic ac waveform is called a sine wave because it closely resembles the ________ function of a rotating vector.
2. An ac voltage that rises from zero and reaches its maximum value in one direction is called the ________ value of ac.
3. An ac voltage that is measured from its positive maximum value to its negative maximum value is termed the ________ value of ac.
4. There are ________ degrees in one cycle of ac.
5. The maximum value of the positive alternation of an ac voltage occurs at ________ degrees.
6. The maximum value of the negative alternation of an ac voltage occurs at ________ degrees.
7. The instantaneous value of an ac voltage is ________ percent of its maximum value at 150°.
8. The instantaneous value of an ac voltage is ________ percent of its maximum value at 30°.
9. The instantaneous value of an ac voltage is ________ percent of its maximum value at 270°.
10. Find the instantaneous value of an ac voltage given that $\theta = 45°$ and the peak value is 156 V.
11. Find the peak value of an ac voltage given that the instantaneous value $v = 135$ V at 60°.
12. Find the average value of the voltage in Question 11.

17-3 EFFECTIVE OR Rms VALUE OF AC

In order that an ac voltage or current can be equated to a dc value of voltage or current, some value other than peak, peak-to-peak, or the average value is used. The term *root-mean-square* (rms), also termed the *effective value,* is the value equivalent to dc in a pure resistive circuit. Since the ac voltage and current go through two alternations in one cycle, the power in a circuit will also go through two alternations. The alternate instantaneous peak power, zero power, and so on, will give an overall heating effect which will be the average of the peak power in the circuit. In a dc circuit the heat dissipated by a resistor (i.e., heating effect) was calculated using any one of the three power formulas.

From $P = I^2R$, squaring the instantaneous values of current shown in Table 17-1, then summing the square, gives 9.0. Dividing this by the number of values (18 in this case) to obtain the average gives 0.5. Finally, taking the square root of this mean average value, we obtain 0.707, which also represents a value of $\sqrt{2}/2$. This means that 0.707 of the ac peak voltage or current value will produce the same heating effect as the equivalent value of dc voltage and current, or rms value = 0.707 × peak value, and

$$V_{\text{rms}} = 0.707 \times V_{\text{p}} \tag{17-6}$$

Similarly,

$$I_{\text{rms}} = 0.707 \times I_{\text{p}} \tag{17-7}$$

Or using calculator values, equations (17-6) and (17-7) can be manipulated easily by using $\sqrt{2}$:

$$V_{\text{rms}} = \frac{\sqrt{2}}{2} \times V_{\text{p}} \tag{17-8}$$

Similarly,

$$I_{\text{rms}} = \frac{\sqrt{2}}{2} \times I_{\text{p}} \tag{17-9}$$

EXAMPLE 17-1

Calculate the power dissipation in the dc circuit of Fig. 17-6(a) and in the ac circuit of Fig. 17-6(b).

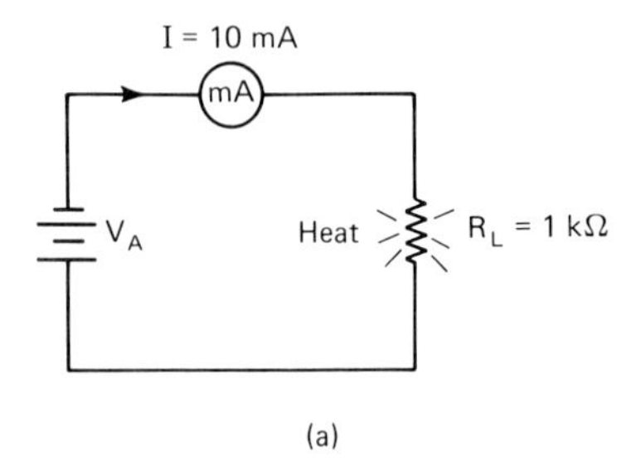

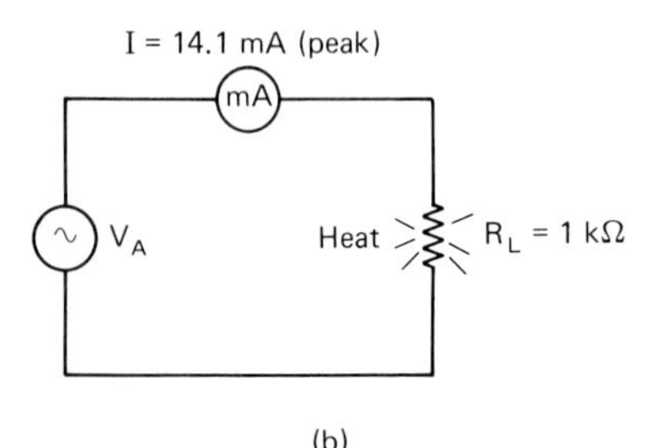

FIGURE 17-6 Dc and ac circuits for Example 17-1.

Solution: The power dissipation in the dc circuit of Fig. 17-6(a) can be found by

$$P = I^2R = (10 \text{ mA})^2 \times 1 \text{ k}\Omega = 100 \text{ mW}$$

The power dissipation in the ac circuit will be calculated using the rms value for current. Using equation (17-9) will give the rms current:

$$I_{rms} = \frac{\sqrt{2}}{2} \times 14.1 \text{ mA peak} = 10 \text{ mA}$$

Therefore,

$$P = I^2R = (10 \times 10^{-3} \text{ A})^2 \times 1 \times 10^3 \ \Omega$$
$$= 0.1 \text{ W} \quad \text{or} \quad 100 \text{ mW}$$

The rms value of 10 mA produces the same heating effect and is equal to the dc current value.

The relationship of ac-to-dc power equivalency can be shown to be the average of the ac peak power. From the circuit schematic of Fig. 17-6(b) the peak power is

$$P_p = (I_p)^2 \times R = (14.1 \text{ mA})^2 \times 1 \text{ k}\Omega = 200 \text{ mW}$$

Since there are two peak power pulses for every cycle of ac, the average of the peak power will be equivalent to the dc power and the ac power dissipation value.

$$P_{avg} = P_{dc} = P_{ac} = \frac{200 \text{ mW}}{2} = 100 \text{ mW}$$

EXAMPLE 17-2

Calculate the applied voltage V_A in the dc circuit and the ac circuit of Fig. 17-6.

Solution: The applied voltage in the dc circuit can easily be calculated using Ohm's law:

$$V = IR = 10 \text{ mA} \times 1 \text{ k}\Omega = 10 \text{ V}$$

In the ac circuit the equivalent or rms current was 10 mA, so that the applied rms voltage has to be 10 V by Ohm's law. The peak voltage, however, is

$$V_p = IR = 14.1 \text{ mA peak} \times 1 \text{ k}\Omega = 14.1 \ V_p$$

A circuit diagram or other data specifies the kind of voltage or current you are measuring, such as peak or peak-to-peak.

By convention, unless the type of value for voltage and current is specified it is assumed to be an rms value.

As an example, the current shown in the ac circuit of Fig. 17-6(b) was specified as 14.1 mA *peak*. All Ohm's law and power calculations could have been done using this peak value. The results, however, must be specified as "peak," knowing that a peak value is not equivalent to dc.

17-4 RELATIONSHIP OF AC VALUES

There are many instances where an rms current or voltage must be converted to a peak or peak-to-peak value, and vice versa. Voltage peak-to-peak readings taken from a calibrated oscilloscope screen are one example of the possible requirement to convert peak-to-peak values to rms values. Converting rms values to peak or peak-to-peak values can be done easily by transposing the basic equations for the average and rms value.

Finding Peak Value Given Average Value

From equation (17-5),

$$V_{avg} = \frac{2}{\pi} \times V_p$$

Transposing equation (17-5) gives

$$V_p = \frac{V_{avg}}{2/\pi}$$

By the simple arithmetic rule for fractions, we can invert and multiply:

$$V_p = V_{avg} \times \frac{\pi}{2} \qquad (17\text{-}10)$$

Therefore,

$$V_p = V_{avg} \times 1.57 \qquad (17\text{-}11)$$

EXAMPLE 17-3

Given that the average value of the pulsating dc of Fig. 17-5(a) is 24 V, find the peak value.

Solution: Using equation (17-11), we have

$$V_p = V_{avg} \times 1.57 = 24 \text{ V} \times \frac{\pi}{2} = 37.7V_p$$

Finding Peak Value Given Rms Value

A voltage waveform value on an oscilloscope screen is easily determined by counting the number of divisions from peak to peak and then multiplying the div obtained by the volts/div setting of the vertical gain control. This, of course, will give the peak-to-peak value. To use this value in relationship to voltage calculations, we must convert the peak-to-peak value to rms value. From equation (17-6) or (17-7), $V_{rms} = 0.707 \times V_p$. Transposing equation (17-6) gives

$$V_p = \frac{V_{rms}}{0.707} = V_{rms} \times \frac{1}{0.707}$$

Therefore,

$$V_p = V_{rms} \times 1.414 \qquad (17\text{-}12)$$

Equation (17-12) can easily be manipulated on a calculator by using $\sqrt{2}$ for 1.414:

$$V_p = V_{rms} \times \sqrt{2} \qquad (17\text{-}13)$$

The peak-to-peak value must be twice the peak value:

$$V_{p\text{-}p} = V_{rms} \times 2 \times 1.414$$

or

$$V_{p\text{-}p} = V_{rms} \times 2\sqrt{2} \qquad (17\text{-}14)$$

EXAMPLE 17-4

The domestic and commercial power-line voltage is given as 120 V. The peak value of this voltage would be found using equation (17-13).

$$V_p = V_{rms} \times \sqrt{2} = 120 \text{ V} \times 1.414 = 169.7 \text{ V}$$

The peak-to-peak value of this voltage is found using equation (17-14).

$$V_{p\text{-}p} = V_{rms} \times 2\sqrt{2} = 120 \text{ V} \times 2.828 = 339.4 \text{ V}$$

Form Factor

Converting an average value to rms value can be done using the form factor 1.11. It is the ratio of rms to average values or 0.707/0.637 = 1.11. Therefore,

$$\text{rms value} = 1.11 \times \text{average value} \qquad (17\text{-}15)$$

17-5 PERIOD AND FREQUENCY OF AC

The time that it takes an ac voltage or current to go through one alternation or 180° is one-half of the total time period for the two alternations in one cycle. Since each cycle of a sine wave contains 360° the time it takes the ac to complete 360° for one cycle is termed the *time period T,* as shown in Fig. 17-7. Of course, the ac voltage or current repeats its cycle over and over again. At the end of the second cycle the time period is twice that for the first cycle or $2T$, and so on. *The number of cycles that an ac voltage or current goes through in 1 second is termed the frequency* (f). Frequency, then, is simply cycles per second, and the unit hertz (Hz) designates cycles per second.

If the ac voltage of Fig. 17-7 goes through two cycles in 1 second, the frequency is 2 Hz (i.e., two cycles per second). The time period for two cycles, then, is 1 s, so that the time period for one cycle would be $\frac{1}{2}$ s. The frequency of an ac voltage can be related to the time period for one cycle. Dividing the time period for one cycle into 1 s will give the frequency:

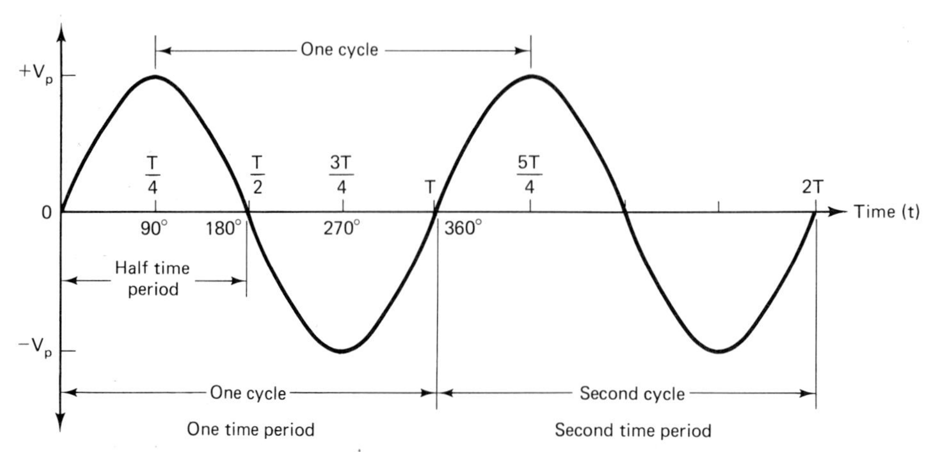

FIGURE 17-7 Time period and frequency of an ac voltage or current.

$$f = \frac{1}{T} \quad \text{Hz} \qquad (17\text{-}16)$$

For the waveform of Fig. 17-7 the frequency is

$$f = \frac{1}{T} = \frac{1}{0.5 \text{ s}} = 2 \text{ Hz}$$

The time period can similarly be found by dividing the number of cycles per second (or frequency) into 1 second:

$$T = \frac{1}{f} \quad \text{s} \qquad (17\text{-}17)$$

A practical example to illustrate the time period and frequency for an ac voltage is the domestic and commercial power line, which operates at 60 Hz. The time period for one cycle of this ac using equation (17-17) is

$$T = \frac{1}{f} = \frac{1}{60} \text{ s} \quad \text{or} \quad 16.7 \text{ ms}$$

Measuring the Time Period and Frequency on an Oscilloscope

Outside a digital frequency counter the oscilloscope is the only method to measure the frequency of voltage waveform. With the horizontal sweep gain and time/division controls in the calibrate position, the time period is measured by counting the number of divisions for one cycle. This is easily done as shown in Fig. 17-7 by taking one cycle from one ac peak to the next ac peak. The time/division oscilloscope sweep setting is then multiplied by the total divisions to obtain the time period. Once the time period is found, equation (17-17) ($f = 1/T$) can be applied to find the frequency. Further examples are given in Section 17-10, "The Oscilloscope."

Units of Time and Frequency

Frequencies greater than 1000 Hz are stated in the multiple units shown in Table 17-2. The time period for one cycle for frequencies greater than 1 Hz is stated in submultiple units of the second which are also shown in Table 17-2.

Angular Frequency

In later chapters we will see such formulas as $X_L = 2\pi fL$, $X_C = 1/2\pi fC$, and $f_r = 1/2\pi\sqrt{LC}$. Notice that the constant 2π is used in all three cases having to do with frequency. This section will show the derivation of 2π.

The angular measurement of a sine wave can be related to the angular rotation of an ac generator rotor producing the voltage. When the rotor of a two-pole generator goes through one revolution of 360°, the resulting voltage produced is one full cycle of a sine wave.

The angular distance that one cycle of ac has gone through is measured in radians. When the length of the arc along the circumference is equal to the length of the radius, the sine wave has gone through 57.3°, as shown in Fig. 17-8. For the two-pole generator the angular distance that the ac cycle goes through in one revolution would be $360°/57.3° = 6.28$ rad or 2π radians, as shown in Fig. 17-9.

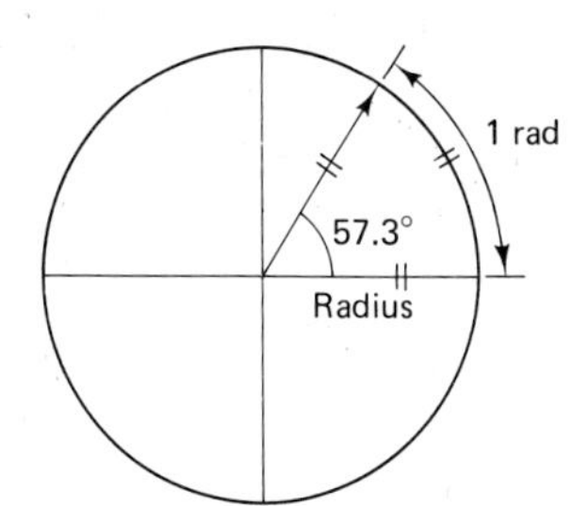

FIGURE 17-8 Angular distance of 1 radian.

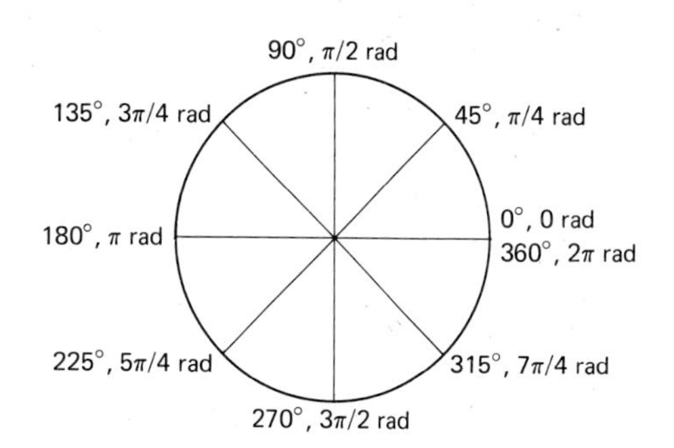

FIGURE 17-9 $360° = 2\pi$ radians.

TABLE 17-2

Prefixes for frequencies and time period

Frequency	*Prefix*	*Time period*	*Prefix*
1 kilohertz, 1×10^3 Hz	kHz	1 millisecond, 1×10^{-3} s	ms
1 megahertz, 1×10^6 Hz	MHz	1 microsecond, 1×10^{-6} s	μs
1 gigahertz, 1×10^9 Hz	GHz	1 nanosecond, 1×10^{-9} s	ns

Increasing the pairs of poles, or increasing the speed of the rotor, increases the frequency of the output voltage for the same rotation of the rotor. This means that the angular distance 2π radians covered by the sine wave has gained velocity.

As an example, the coil of a simple two-pole generator that rotates a full 360° in 1 s to produce a voltage with a frequency of 1 cycle has an angular velocity of 2π rad/s. If an extra pair of poles is added or the speed of the rotor is doubled and thus the frequency of the output voltage is doubled, the angular velocity will double to 4π rad/s.

This relationship can be shown by

$$\text{angular velocity} = 2\pi \text{ rad/s} = \frac{2\pi \text{ rad}}{T}$$

Substituting $1/f$ for T from equation (17-17) gives

$$\text{angular velocity} = \frac{2\pi}{1/f} \quad \text{or} \quad 2\pi f \text{ rad/s} \qquad (17\text{-}18)$$

We do not need to concern ourselves with the angular velocity of a sine-wave current or voltage since $2\pi f$ incorporates this angular velocity as shown in equation (17-19).

The term *angular frequency* is used for $2\pi f$. The Greek lower case letter omega (ω) is used to abbreviate the $2\pi f$ term, or

$$\omega = 2\pi f \qquad \text{radians/second} \qquad (17\text{-}19)$$

EXAMPLE 17-4

A sine-wave voltage has a frequency of 1 kHz. Determine the angular frequency and the time period for one cycle.

Solution: From equation (17-19),

$$\omega = 2\pi f = 6.28 \times 1 \times 10^3 \text{ rad/s}$$
$$= 6.28 \times 10^3 \text{ rad/s}$$

The time period for one cycle can be calculated using equation (17-17):

$$T = \frac{1}{f} \text{ s} = \frac{1}{10^3} \text{ s} \quad \text{or} \quad 1 \text{ ms}$$

PRACTICE QUESTIONS 17-2

Choose the most suitable word, phrase, or value to complete each statement.

1. The rms value of an ac voltage is the same as its instantaneous value at ________ degrees of its cycle.
2. What is the peak value of an ac voltage on a 220-V power line?
3. What peak voltage rating should be chosen for a capacitor which is to be connected across a 250-V ac source?
4. The average value of an ac voltage is 137 V. Find the rms value.
5. The value of voltage or current is always understood to be an ________ value unless otherwise stated.
6. The time period of one alternation of 60 Hz ac is ________ milliseconds.
7. Express 100 cycles per second as kilohertz.
8. Express 4.5 megacycles per second as megahertz.
9. There are ________ degrees in 1 rad.
10. There are ________ radians in one ac cycle.
11. The ratio of a circle's circumference to its radius gives a mathematical constant termed ________.
12. The value 6.28 appears commonly in ac formulas. It is equal to ________.
13. The angular frequency is the product of ________ and ________.
14. The symbol for angular frequency is the Greek letter ________.
15. The angular frequency of 60 Hz is ________ radians per second.

17-6 PHASE RELATIONSHIP OF VOLTAGES

In ac circuits the instantaneous value of voltages and currents are relative to one another. For example, two ac voltages connected series-opposing will cancel. Each instantaneous value of V_A is opposite going to that of V_B. They are said to be 180° out of phase, as shown by the waveforms of Fig. 17-10.

The output voltage of two similar series-connected generators will be zero when they are connected series bucking. Their instantaneous voltage are 180° out of phase; therefore, the net voltage will be zero.

Leading and Lagging Waveform

Figure 17-11(a) illustrates a phase shift of a sine wave. Waveform V_B is shifted by 90° from V_A. Thus the instantaneous value of V_B is zero while V_A is maximum. Since time increases along the x axis, it can be said that V_A occurs first. If we use *voltage V_A as a reference,* it can be said that *V_A leads V_B by 90°.*

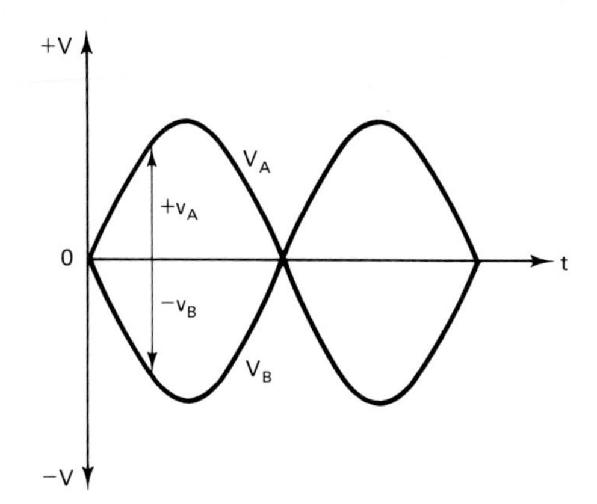

FIGURE 17-10 Two voltages 180° out of phase.

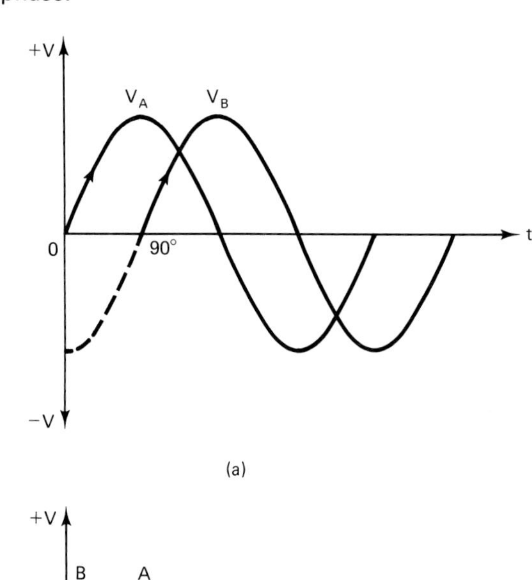

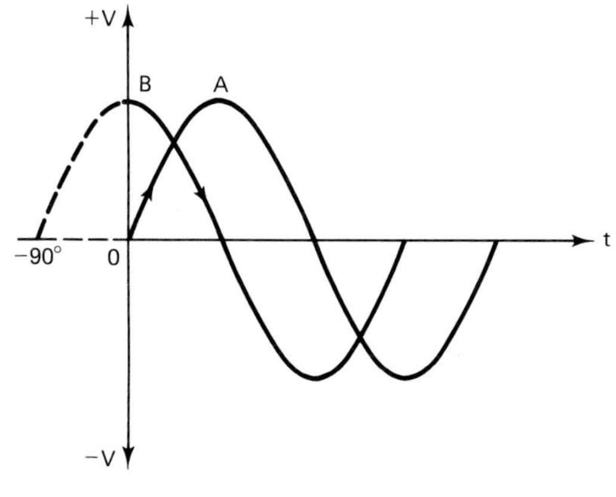

FIGURE 17-11 (a) Voltage *B* is shifted by 90° from voltage *A*; voltage *A leads* voltage *B* by 90°. (b) Voltage *B* is shifted by 90° from voltage *A*; voltage *A* lags voltage *B* by 90°.

Waveform *B* of Fig. 17-11(b) is shifted by 90° to the left. The instantaneous peak value of *B* occurs sooner than that of *A*. Since *B* voltage started earlier in time, it can be said that V_B now leads V_A by 90°. If voltage *A* is referenced to *B*, it can be said that *voltage A lags voltage B by 90°.*

Resultant Waveform with In-Phase and Out-of-Phase Voltages

The instantaneous values of two out-of-phase voltages will sum algebraically. Figure 17-12 shows the resultant voltage of two in-phase voltages. At every instantaneous value along the sine wave, both V_A and V_B instantaneous values rise in the same direction so that the resultant will be the sum of both instantaneous values.

Figure 17-13 shows the resultant voltage with two 90° out-of-phase voltages. The instantaneous values of both voltages will sum algebraically, so that all values rising in the same direction will add to each other. For the waveform of Fig. 17-13 the instantaneous value at 135° sum so that the resultant maximum or peak value will be the sum of the instantaneous values V_A and V_B. Instantaneous values that rise in opposite directions cancel each other. For the waveform of Fig. 17-13 the instantaneous values cancel each other at 45°/225° and the resultant is zero.

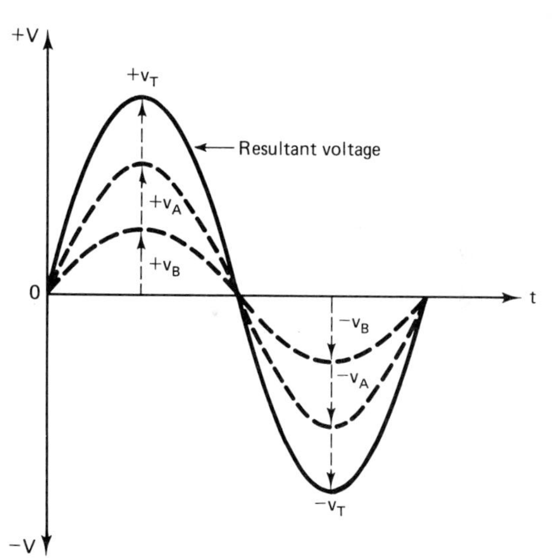

FIGURE 17-12 Resultant voltage of two in-phase voltages.

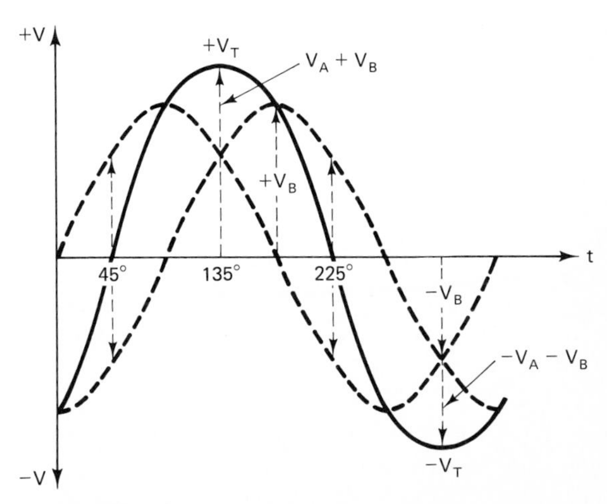

FIGURE 17-13 Resultant voltage of two voltages out-of-phase by 90°.

17-7 PHASOR DIAGRAMS

Drawing a graphical sine-wave representation of out-of-phase voltages, currents, and their resultant values can become tedious. A much easier and simple method of showing the phase relationship and resultant of out-of-phase ac values is by the rotating vector or phasor. But first, we look at the difference between a scalar quantity, a vector quantity, and a phasor quantity.

A scalar quantity has magnitude only and is represented by units such as temperature, linear measure, volume, and resistance, as a few examples.

A quantity such as force and velocity must be shown to have a direction as well as magnitude. An example of this is a prevailing wind that blows an aircraft off its course. Figure 17-14 shows the resultant direction of flight of an aircraft with a wind velocity at right angles to the flight path. The length of each line of the vector diagram must be scaled to represent units of velocity of both wind and aircraft. The arrowhead on each line shows the direction of the aircraft and the wind. Note that the wind direction is drawn with *reference* to the aircraft direction.

An ac alternator produces instantaneous values of alternating current that are continuously changing in time and value. The instantaneous values of alternating voltage and current must be represented by a rotating vector, as shown in Fig. 17-2. A vector that is rotating at a constant angular velocity is termed a phasor. The phasor, then, represents both magnitude and a time-varying quantity.

Two out-of-phase sine-wave voltages or currents having different amplitudes but the same frequency can be represented by a phasor diagram as shown in Fig. 17-15. The length of the phasors can represent peak, rms, or average value, but both values must be consistent. The arrowhead identifies the varying voltage or current, while the opposite end is the axis of rotation. *By convention, the direction of phasor rotation is always taken to be counterclockwise*.

Once a phasor is drawn, it represents the phase angle between two or more sine waves since the frequency of both must be the same. The phase angle must be referenced to one of the phasors. It is not sufficient simply to state the phase angle; it must be stated with reference to either sine wave. The phasor diagram of Fig. 17-15(a) shows that voltage V_A leads voltage V_B by 90°. Voltage V_A is the *reference phasor*. If the reference phasor is V_A in Fig. 17-15(b), the phase angle will remain the same and voltage V_A will lag voltage V_B.

In Fig. 17-16(a), sine wave V_A occurs first since time increases along the axis, so that V_A leads sine wave V_B by 90°. The phasor representation is shown with sine wave V_A as the reference phasor. Figure 17-16(b) shows sine wave V_A as the reference phasor along the axis. Since time increases along the axis, sine wave V_A will lag V_B by 90°.

A phasor representation of two in-phase voltage is shown in Fig. 17-17(a). Both *A* and *B* phasors are drawn along the 0° axis. Figure 17-17(b) shows two 180° out-

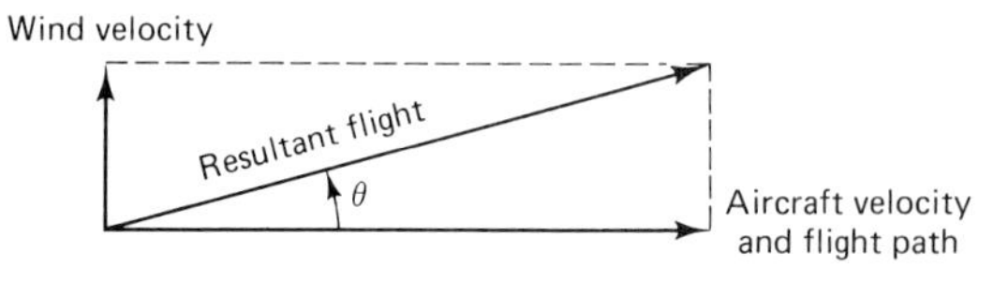

FIGURE 17-14 Vector representation of wind and aircraft velocity.

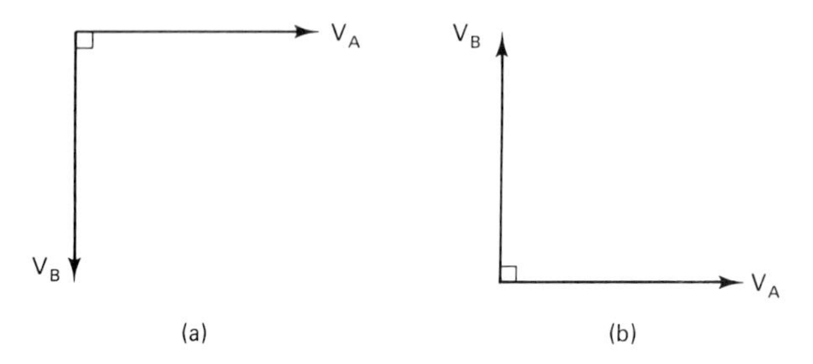

FIGURE 17-15 (a) Voltage *A* is the reference phasor and V_A leads V_B by 90°. (b) Voltage *A* is the reference phasor and V_A lags V_B by 90°.

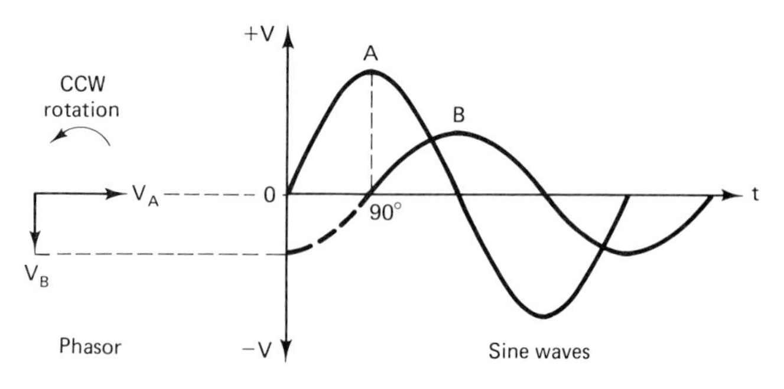

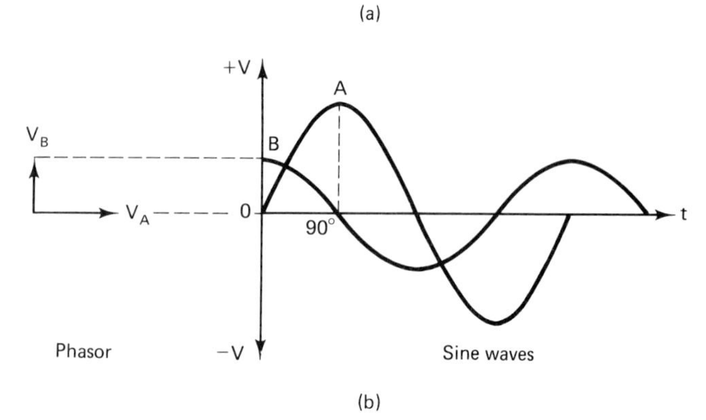

FIGURE 17-16 (a) Phasor and sin-wave representation of voltages *A* and *B*. Note that voltage *A* is the reference phasor, and V_A leads V_B by 90°. (b) Phasor and sine-wave representation of voltages *A* and *B*. Note that voltage *A* is the reference phasor, and V_A lags V_B by 90°.

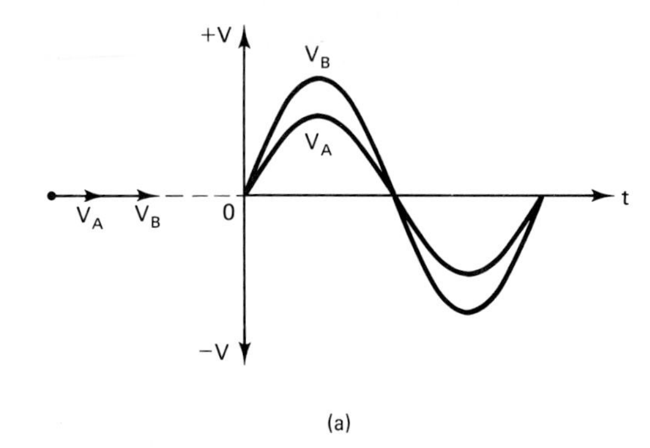

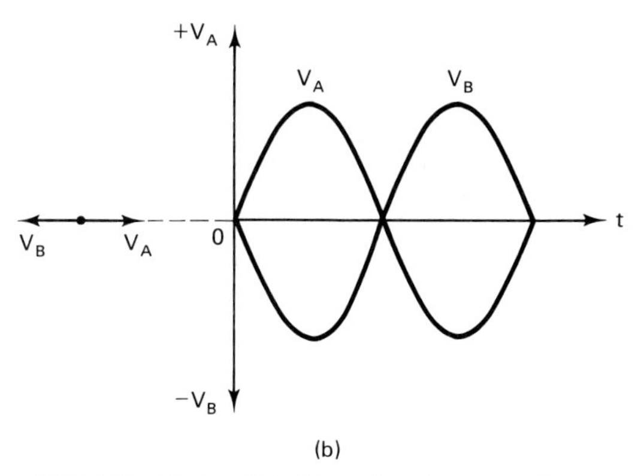

FIGURE 17-17 (a) Two in-phase voltages. (b) Two 180° out-of-phase voltages.

of-phase voltages and their phasor diagram. The phasor arrowheads are drawn opposite each other to indicate the 180° displacement. Voltages that are 180° out of phase cancel each other.

Finding the Resultant of a Phasor

The instantaneous values for out-of-phase sine-wave voltages (or currents) must be summed algebraically to obtain the resultant voltage. A simple example of this is two 180° out-of-phase voltages. The resultant voltage will be the algebraic sum, or $V_R = V_A - V_B$, since one voltage will be positive going at the same instant that the other voltage is negative going.

For voltages and currents less than 180° out of phase, a phasor diagram and its resultant can be found using the parallelogram method, as shown in Fig. 17-18. The length of the resultant is converted to voltage units.

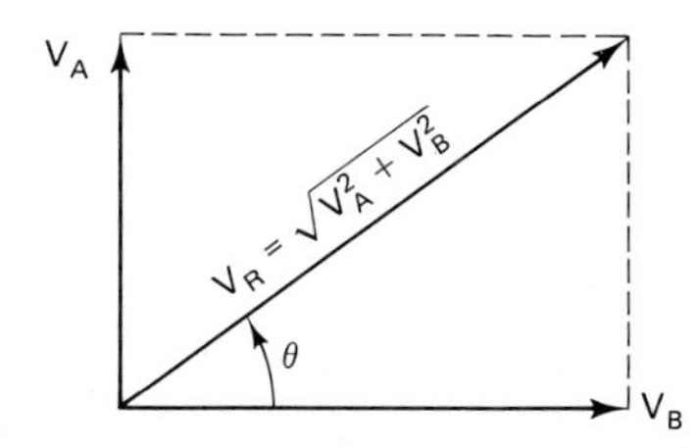

FIGURE 17-18 Phasor resultant voltage.

A much quicker solution is to use trigonometric relations. For two voltages or currents 90° out of phase, the Pythagorean theorem for a right-angle triangle is used:

$$V_R = \sqrt{V_A^2 + V_B^2} \qquad (17\text{-}20)$$

EXAMPLE 17-5

Calculate the resultant voltage of Fig. 17-18 if both V_A and V_B = 100 V.

Solution: Using equation (17-20) yields

$$V_R = \sqrt{V_A^2 + V_B^2} = \sqrt{100^2 + 100^2} = 141 \text{ V}$$

Voltages that are leading or lagging by an angle less or more than 90° will not concern us, therefore will not be discussed here. Refer to trigonometric relations for these cases.

Phase Angle between V_B and V_R

The phase angle θ can be solved using the tan function for θ:

$$\tan \theta = \frac{\text{opposite side}}{\text{adjacent side}}$$

$$\tan \theta = \frac{V_A}{V_B} \qquad (17\text{-}21)$$

EXAMPLE 17-6

Calculate the phase angle between the resultant and V_B of Fig. 17-18.

Solution: Using the trig function of (17-21), we obtain

$$\tan \theta = \frac{V_A}{V_B} = \frac{100}{100} = 1$$

Therefore,

$$\theta = 45°$$

17-8 NONSINUSOIDAL VOLTAGE WAVEFORMS

An ac voltage produced by a rotary generator will always be a sine wave because the induced voltage is proportional to the angle at which the coil cuts through the magnetic field. Two important types of waveforms used in electronic circuits are the square wave and the triangular wave. Waveforms that are not a sine wave are termed *nonsinusoidal* waveforms.

Square waves and similar pulse types of waveform voltages or currents are essentially switching voltages that have different characteristics from a sinusoidal waveform. Figure 17-19(a) shows a square-wave type of switching voltage. Ideally, the voltage rises to its maximum value, +8 V in this case, and remains ON for a period (0.5 ms), then reverses polarity to −8 V and remains ON for another period (0.5 ms).

One cycle of a switching waveform is defined as the period between any two leading or trailing edges of the waveform. In the waveform of Fig. 17-19(a) the period for one cycle is 1 ms; therefore, $f = 1/T = 1/1 \text{ ms} = 1 \text{ kHz}$

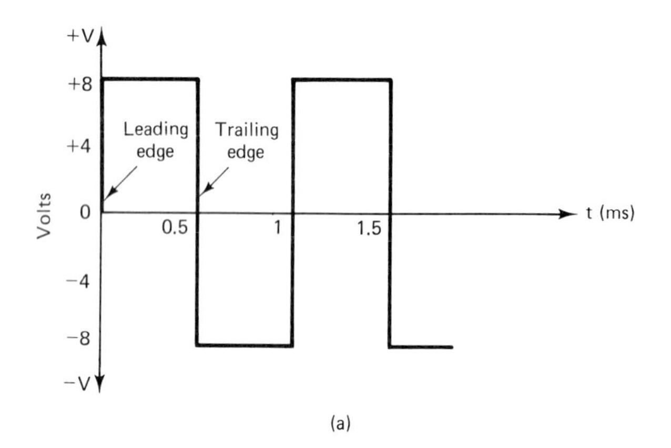

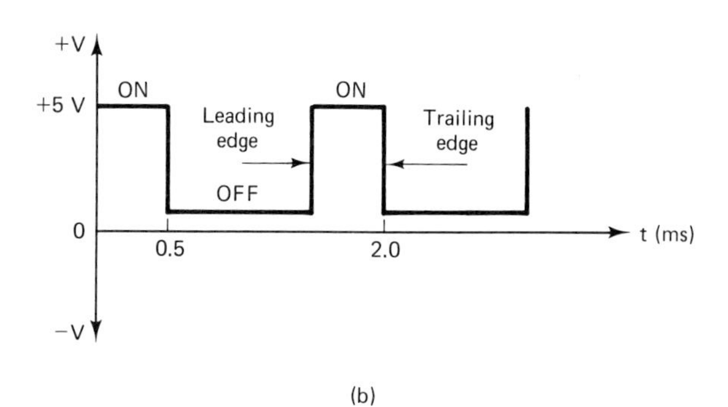

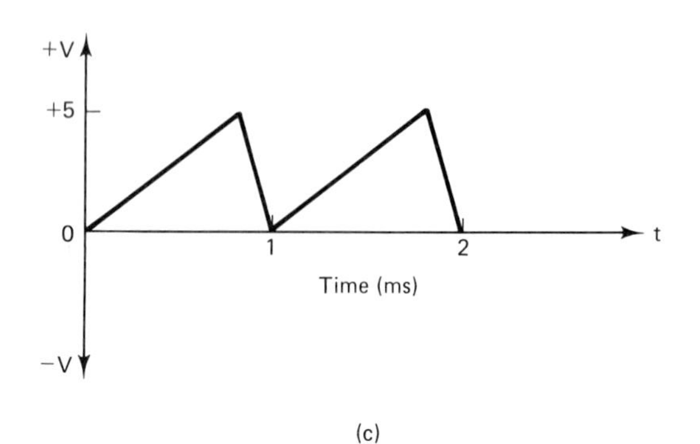

FIGURE 17-19 (a) Square-wave switching voltage. (b) Logic circuit switching voltage. (c) Sawtooth or ramp waveform.

Rms values do not apply to switching types of voltages and currents. They are basically a dc pulse of positive and negative values, and as such, only the peak value or the peak-to-peak value is inferred. The peak value of the voltage waveform of Fig. 17-19(a) is 8 V, while the peak-to-peak value is 16 V.

When the period for the ON part of a switching voltage is equal to the OFF time, it is said to be a square wave. Depending on the application, the switching voltage could be a rectangular wave. That is, the ON and OFF periods would not be equal. In both cases the ON and OFF period is equal to one cycle.

Figure 17-19(b) shows a typical logic circuit switching voltage or pulse. In this case the voltage pulse is in one direction only, +5 V. The duration of the ON time, or pulse width, is 0.5 ms, while the OFF time is 1 ms. The period for one cycle is from one leading edge to another leading edge (or from one trailing edge to another trailing edge) and is 1.5 ms. From $f = 1/T$, the frequency of these pulses is 667 Hz approximately. Note that the zero voltage is slightly *offset* from zero and is typically around 0.4 V. Zero voltage is referenced to chassis ground.

The sawtooth or ramp voltage waveform shown in Fig. 17-19(c) is used to deflect or sweep the electron beam horizontally across a cathode-ray tube (CRT) of an oscilloscope. The rate at which the beam is swept across the CRT screen is determined by the frequency of the sawtooth wave. In Fig. 17-19(c) the frequency of the waveform is 1 kHz since the period is 1 ms.

PRACTICE QUESTIONS 17-3

Where applicable, choose the most suitable word, phrase, or value to complete each statement.

1. Two 60-Hz alternators, each having an output of 200 V peak, are wired in series. If they are exactly out of phase condition, what is the resultant voltage ________ rms?
2. If the two alternators of Question 1 are in perfect phase-aiding condition, the net voltage will be ________ rms.
3. What is the net voltage produced when a 6.3-V secondary of a power transformer is connected series phase-opposing to the 24-V secondary?
4. Sketch a sine-wave diagram to show the condition in Question 1.
5. A scalar quantity indicates only ________.
6. A vector quantity indicates both ________ and ________.
7. Voltages and currents that are out of phase with one another can be represented by a ________ diagram.

8. Show by a representative phasor diagram 110 V leading 55 V by 45°.
9. Show by a representative phasor diagram a 3-A current lagging a 1-A current by 90°.
10. Calculate the resultant current of Question 9.
11. Draw a representative phasor diagram showing a 2-A current 180° out of phase with a 3-A current. Indicate the value and position of the resultant.
12. For any nonsinusoidal waveform the ________ or ________ voltage is inferred and rms values do not apply.
13. The time period for one cycle is from one ________ edge to another ________ edge of the square waveform.
14. Calculate the period in microseconds for one cycle of a 5-kHz square wave.
15. If the square wave has a peak-to-peak value of 10 V, what is its equivalent value to an ac rms value?

17-9 THE AC VOLTMETER

Having used a dc meter, you have probably noticed that there is a separate calibrated scale for ac. All permanent-magnet moving-coil meters are dc meters, so that with an ac voltage of say 60 Hz, the pointer would vibrate at 60 Hz and read zero volts. Given sufficient voltage or current, the movement would be permanently damaged by the vibrations.

To make the meter useful for ac measurements, a rectifier circuit must be used. The rectifier circuit is normally a full-wave bridge rectifier, as shown in Fig. 17-20. The diodes D_1 to D_4 rectify the ac into pulsating dc.

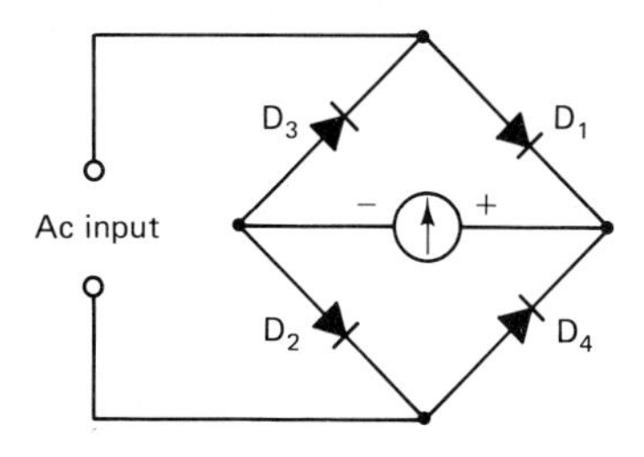

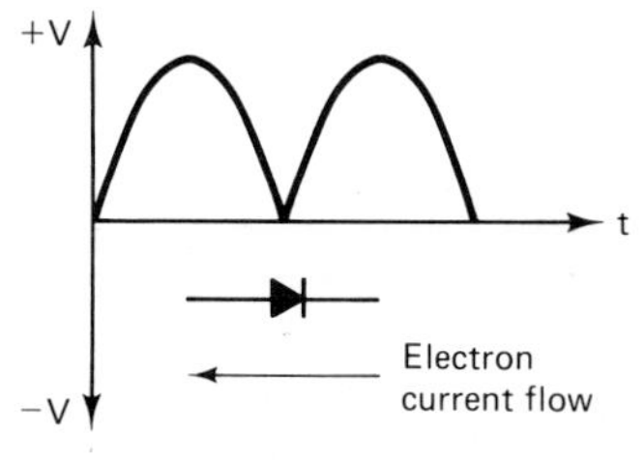

FIGURE 17-20 Full-wave bridge rectifier and pulsating dc output.

Diodes D_1 to D_4 are semiconductor diodes that conduct current in one direction only. The ac voltage applied to the bridge circuit will be conducted alternately by D_1 and D_2 on one alternation of the ac sine wave, and D_4, D_3 by the next alternation of the sine wave. This will give two dc pulses for every cycle of ac, as shown in Fig. 17-20.

The average value for this type of pulsating waveform is 0.637 of the peak value. Since the meter movement responds to this average dc value and not to the rms value for ac, the scale must be calibrated in terms of rms values. That is why there is a separate scale for ac. In most meters there is also a peak-to-peak scale, which is identified by a separate colored scale, usually red.

The ac meter sensitivity of permanent-magnet moving-coil meters is much lower on the ac ranges than on the dc ranges because of the need to rectify the ac. Loading effect is therefore more severe on the ac ranges. Typically, a 20-kΩ/V meter movement has a sensitivity of 5 kΩ/V on the ac ranges. Electronic field-effect transistor (FET) input and multimeters that have a 11-MΩ input resistance on the dc ranges will drop to an input resistance of 1 MΩ on the ac ranges.

The frequency response is also another factor to be considered. For audio voltage measurements the meter must respond uniformally over the range of frequencies from 30 Hz to 20 kHz. Manufacturers' specifications should be checked for all these factors in order to select a meter for a wide range of uses.

Other types of ac meters such as the moving-vane meter movement and the electrodynamometer movement, are only suitable to measure low-frequency voltages, particularly the 60-Hz line voltage, and they are for this reason discussed in a later chapter. The electrodynamoter movement is of particular use as a wattmeter.

17-10 THE OSCILLOSCOPE

Present-day electronics technology, especially computer technology, demands viewing voltage/current pulses for correct timing, shape, and amplitude. The oscilloscope is the only device that allows us to view any type of waveform as well as measure its parameters.

When you view the screen of an oscilloscope, you are looking at the screen of the cathode-ray tube (CRT). The CRT of Fig. 17-21 is similar to the television CRT. It is a glass tube from which the air has been highly evacuated. The screen of the CRT is coated with a fine layer of phosphor. This phosphor screen provides the "picture" of the voltage waveform.

An electron gun is mounted at the rear or neck of the CRT. Its function is to provide a source of electrons by means of a heater and cathode. The electrons are

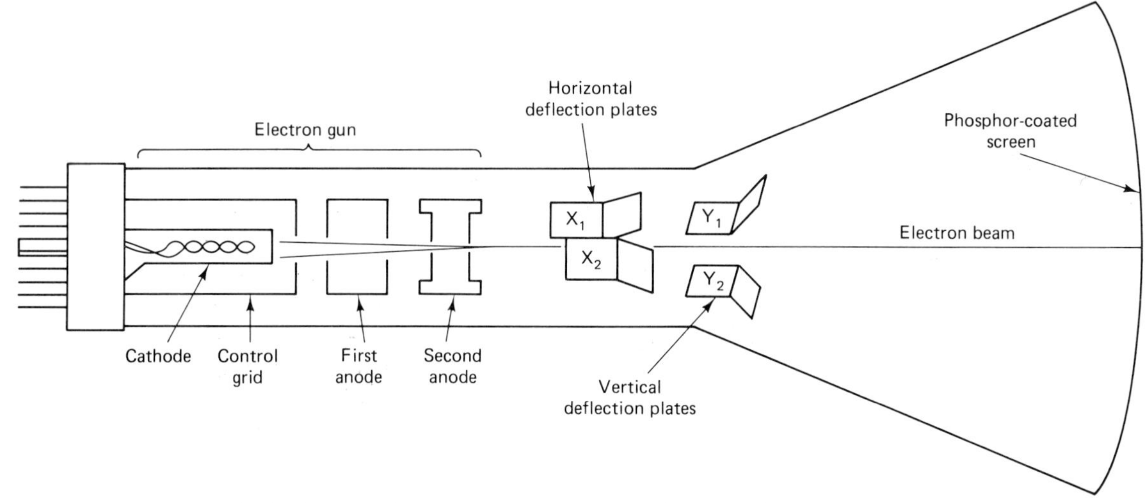

FIGURE 17-21 Schematic cross-sectional view of an electrostatic deflection cathode-ray tube.

then accelerated and "shot" at the CRT phosphor screen. Gathering velocity, the electron beam hits the phosphor, which makes it glow.

The color of the phosphor glow is governed by the type of phosphor used for the screen. The phosphors of CRT oscilloscopes are designated as follows: P_1, green medium; P_2, blue-green medium; P_5, blue very short; and P_{11}, blue short. The medium, short, and very short designation refers to the persistence of phosphor. *Persistence* is the time it takes the phosphor glow to decay since the decay of the light emitted is gradual rather than abrupt. The designations are combined in the CRT type number. Thus 5 GPI is a 5-in. tube with a medium-persistence green trace.

There are many phosphors. Different phosphors can be made to glow with all the colors of the rainbow, a fact not overlooked by color-television inventors. Zinc orthosilicate gives off a green light, cadmium tungstate a blue light, and cadmium borate a red light. Zinc sulfide combined with silver provides the white light for black-and-white (monochrome) TV.

Focus and Accelerating Anodes

Before the electron beam hits the phosphor screen it must pass through both the focus and accelerating anodes. After passing the control grid, the electrons would disperse were it not for the electrostatic field that exists between the first and second anodes. The strength of this field is varied by means of the FOCUS control, which varies the first anode's voltage. This causes the electron beam to converge, similar to a lens' action, and gives a small sharp spot of light on the screen. The second anode connected to a positive voltage in the range 2 to 4 kV accelerates the electron beam to the CRT screen.

Deflecting the Electron Beam

Once the beam passes the second accelerating anode, it must pass between pairs of deflection plates as shown in Fig. 17-21. Viewing the face of the CRT, these would appear as shown in Fig. 17-22. X_1 and X_2 are the horizontal deflection plates. If both are connected to a voltage source and X_1 is made positive with respect to X_2, the negative electron beam will be attracted to X_1 plate and be deflected to the left of center. Similarly, if the X_2 plate were made positive with respect to X_1, the beam would strike the CRT screen to the right of center.

X Shift. This control provides a controllable difference of potential on the horizontal plates so that the beam can be positioned horizontally to the left or right of center.

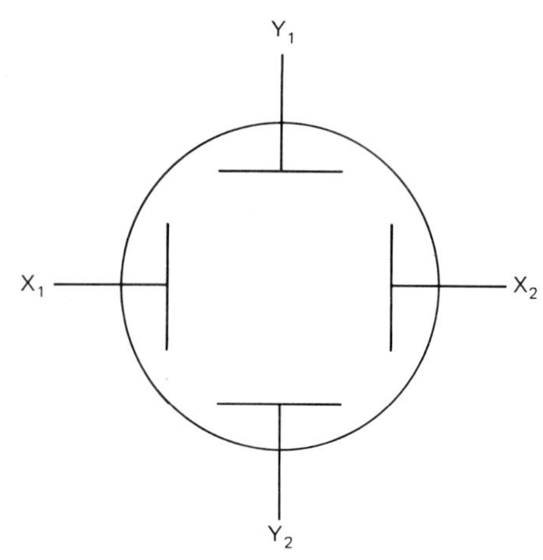

FIGURE 17-22 Vertical and horizontal deflection plates. View from face of the CRT.

Y Shift. This control provides a conrollable difference of potential on the vertical plates so that the beam can be positioned vertically up or down from center. The vertical deflection plates Y_1 and Y_2 shown in Fig. 17-21 control the electron beam in exactly the same manner as the horizontal deflection plates.

Obtaining a Waveform on the Screen

Suppose that we were able to vary the X-SHIFT control back and forth at a very fast rate. The beam would travel back and forth across the CRT screen so fast that the eye would be unable to see the moving spot. Only a horizontal line would be seen across the screen because of the persistence of the phosphor and of vision. The same action could be taken with the vertical deflection or Y SHIFT, to obtain a vertical line. But first, back to the horizontal deflection system.

More practically, the beam of the CRT is swept from the left of the screen to the right, by what is known as the sawtooth sweep oscillator or the sweep generator. This is a built-in electronic circuit that generates a sawtooth waveform as shown in Fig. 17-23 and is coupled to the horizontal deflection plates X_1 and X_2.

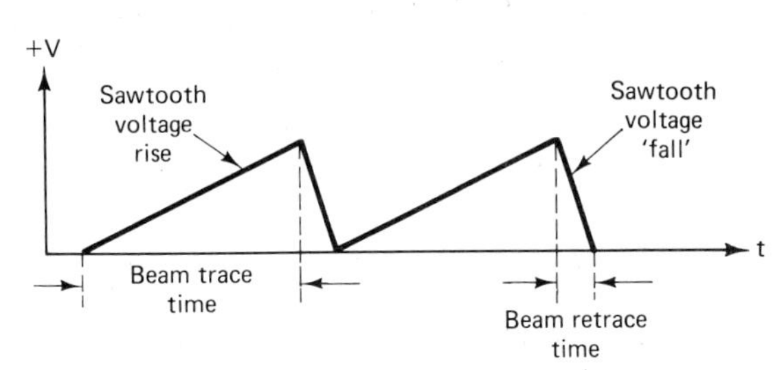

FIGURE 17-23 Sawtooth waveform generated by the sweep generator.

Time/cm Control

The sawtooth generator frequency can be varied by the Time/cm (sweep) selector switch and the frequency vernier. The vernier is a potentiometer that fine-tunes the frequency. By varying the frequency of the sawtooth sweep voltage, we change the time base for the electron beam trace period, thus changing the number of cycles that will be displayed on the screen during this period (see Fig. 17-23).

Now we are ready to obtain a waveform on the CRT screen. The input voltage to be sampled is coupled through a coaxial cable and BNC connector on the front panel of the scope. This input voltage, after the appropriate Volts/cm setting, is coupled through an amplifier to the vertical deflection plates Y_1 and Y_2.

As the sweep generator sweeps the trace horizontally the instantaneous value of the input voltage moves the beam up or down depending on the polarity of the instantaneous value. An input voltage waveform is the simultaneous result of the trace being swept across the screen by the sweep generator and being deflected vertically by the input voltage waveform. Figure 17-24 shows a sweep trace period which is equal to the input waveform period so that one cycle of the input voltage will be traced out on the scope CRT screen.

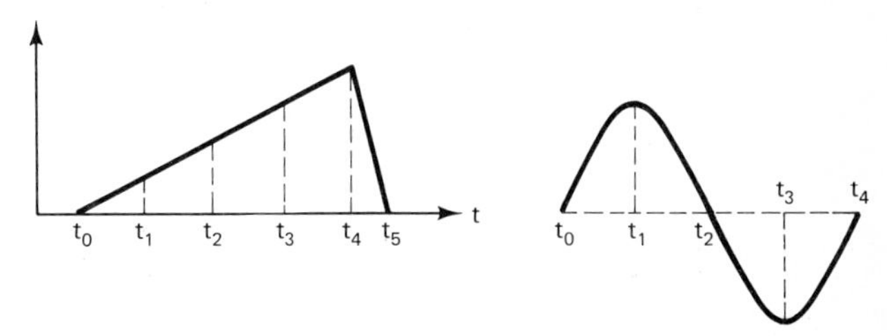

FIGURE 17-24 Sweep-trace time period is equal to one cycle of input waveform time period.

For ease of waveform viewing, the most desirable situation is to have either two or three cycles of the waveform displayed by the trace. This means that the input waveform's frequency must be respectively two or three times higher than the sweep oscillator frequency. The *Time/cm* control should therefore *be set* to about *one-half* or *one-third* of *the input voltage frequency.*

If you set the sweep frequency too low, so that it is, say, $\frac{1}{10}$ of the waveform's frequency, 10 cycles of the waveform will be displayed by the trace and detail will be difficult to observe. Conversely, if you set the sweep frequency too high, so that it is, say, twice the waveform frequency, one-half cycle of the waveform will be displayed.

Synchronization of the Waveform

Normally, the waveform has a tendency to drift or jitter horizontally. This is due to frequency drift of the sawtooth oscillator as well as the input voltage. To "lock in" the waveform and prevent drift or jitter, the sweep generator voltage is synchronized with either the input voltage (internal sync) or the external voltage (external sync). The sweep generator can be synchronized from either the positive or negative half-cycle of the input or external source voltage. This selection is made by a pushbutton switch on the front panel of a scope and is labeled (−) and (+). Adjusting the TRIG LEVEL control will stabilize and lock the waveform into a steady pattern. For most scopes the TRIG LEVEL control can be set in the AUTO position, usually CCW, where you will hear a click, which is the action of a switch.

Having gone through the fundamental function sections of an oscilloscope, the remaining functions can be picked up "on the job" (i.e., in your lab). Of more immediate importance is the need to use the scope as a measuring instrument for frequency and amplitude of the viewed waveform.

Measuring the Input Waveform Frequency

Accurate frequency or voltage measurements using the oscilloscope can be made only if the appropriate controls are set in the calibrate mode. For most oscilloscopes, a potentiometer termed a *vernier* allows the Time/Div setting of the sweep sawtooth to be slightly altered by rotating this control knob. If the vernier is set to CAL then the step calibrations on the Time/Div dial can be used to calculate the frequency of the viewed waveform as follows.

Having set the vernier in the CAL mode, count the number of centimetres or divisions for one cycle. For the example shown in Fig. 17-25 the number of divisions for one cycle can be taken from one peak to the next waveform peak and is 3.0 Div. Note that the peak value of the waveform has been set along the graduated horizontal graticule. This allows better accuracy when interpreting the number of centimetres or divisions per cycle. From this reading the time period T can be calculated by noting the setting of the Time/Div calibration dial. In the example of Fig. 17-25, the setting was 1 ms, so that

$$T = \text{number of Div for one cycle} \times \text{Time/Div} \quad (17\text{-}22)$$

or

$$T = 3.0 \text{ Div} \times 1 \text{ ms} = 3.0 \text{ ms}$$

Using equation (17-16) to find the frequency gives

$$f = \frac{1}{T} = \frac{1}{3.0 \text{ ms}} = \frac{1}{3.0 \times 10^{-3} \text{ s}} = 333 \text{ Hz}$$

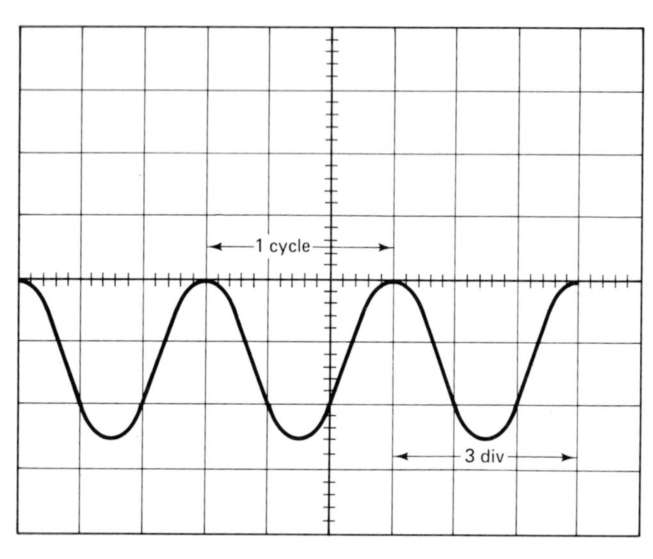

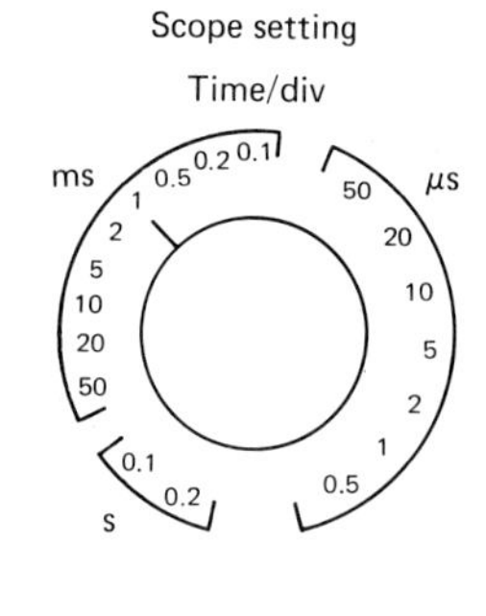

FIGURE 17-25 Frequency measurement, 3 cm for one cycle with time/cm set at 1 ms.

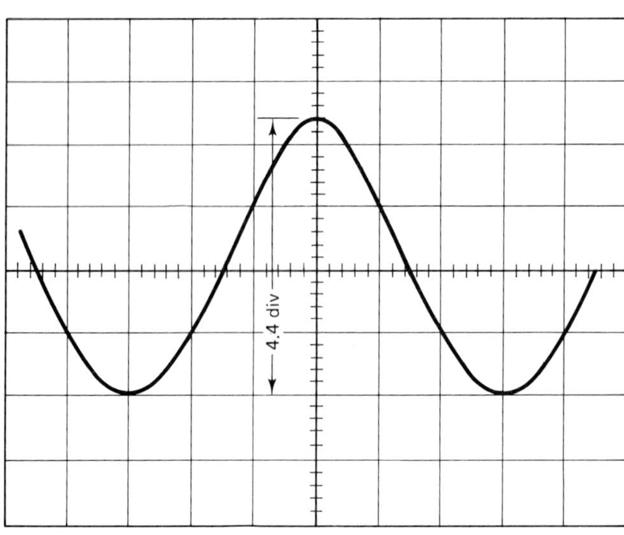

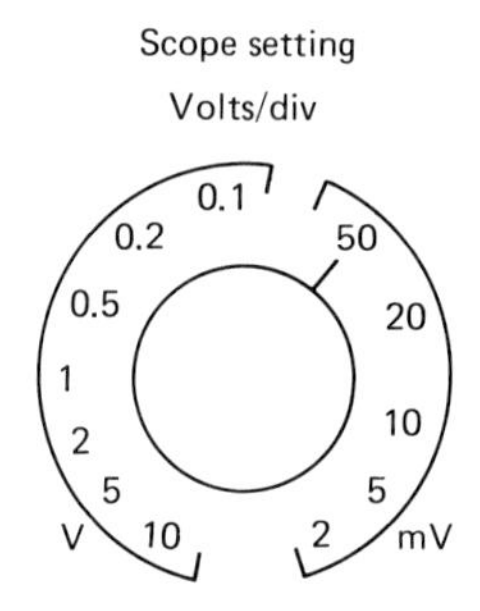

FIGURE 17-26 Peak-to-peak voltage measurement, 4.4 cm from peak to peak with volts/cm set at 50 mV.

Measuring the Input Waveform Amplitude

As was the case for measuring frequency, the Volts/Div vernier must be set to the CAL position before proceeding to measure the amplitude of a waveform. Having set the vernier in the CAL mode, count the number of centimetres or divisions vertically from one peak to the next peak. For the example shown in Fig. 17-26, the number of divisions from one peak to the other peak is 4.4.

Note that the bottom peak is referenced along a horizontal graticule while the top peak is referenced along the graduated vertical graticule. This allows better accuracy when interpreting the number of centimetres or divisions from peak to peak. From this reading the peak-to-peak value can be calculated by noting the setting of the Volts/Div calibration dial. In the example of Fig. 17-26 the setting was 50 mV, so that

$$V_{\text{p-p}} = \text{number of Div} \times \text{volts/cm} \qquad (17\text{-}23)$$

Therefore,

$$V_{\text{p-p}} = 4.4 \text{ Div} \times 50 \text{ mV} = 120 \text{ mV}$$

SUMMARY

A generated alternating current resembles the mathematical sine function.

The instantaneous value of voltage or current can be determined using equation (17-1),

$$v = V_{\text{max}} \times \sin \theta$$

The amplitude of the sine-wave voltage from the zero-crossing voltage value to its maximum value is termed the peak value.

Measuring the voltage from the positive peak value to the negative peak value gives the peak-to-peak value.

$$V_{\text{pp}} = 2V_{\text{p}} \qquad (17\text{-}2)$$

$$I_{\text{pp}} = 2I_{\text{p}} \qquad (17\text{-}3)$$

If the instantaneous values for one-half of a sine wave are summed, and the sum is divided by the number of samples taken, an average value is obtained, which is

$$V_{\text{avg}} = 0.637 \times V_{\text{p}} \qquad (17\text{-}4)$$

Also,

$$V_{\text{avg}} = \frac{2}{\pi} \times V_{\text{p}} \qquad (17\text{-}5)$$

The rms value, also termed the effective value of ac, is the value equivalent to dc.

$$V_{\text{rms}} = 0.707 \times V_{\text{p}} \qquad (17\text{-}6)$$

$$I_{\text{rms}} = 0.707 \times I_{\text{p}} \qquad (17\text{-}7)$$

Using $\sqrt{2}$ yields

$$V_{\text{rms}} = \frac{\sqrt{2}}{2} \times V_{\text{p}} \qquad (17\text{-}8)$$

$$I_{\text{rms}} = \frac{\sqrt{2}}{2} \times I_{\text{p}} \qquad (17\text{-}9)$$

By convention, unless otherwise specified, an ac value is assumed to be an rms value.

Given the average value, the peak value can be found using equation (17-10) or (17-11).

$$V_p = V_{avg} \times \frac{\pi}{2} \tag{17-10}$$

or

$$V_p = V_{avg} \times 1.57 \tag{17-11}$$

Given the rms value, the peak value can be found using equation (17-12) or (17-13).

$$V_p = V_{rms} \times 1.414 \tag{17-12}$$

or

$$V_p = V_{rms} \times \sqrt{2} \tag{17-13}$$

The peak-to-peak value is twice the peak value, or

$$V_{p\text{-}p} = V_{rms} \times 2\sqrt{2} \tag{17-14}$$

A form factor converts an average value to an rms value and is found using equation (17-15), or

$$\text{rms value} = 1.11 \times \text{average value} \tag{17-15}$$

The number of cycles that an ac voltage or current goes through in 1 second is termed the frequency (f).

The unit for cycles per second is hertz (Hz).

Dividing the time period for one cycle into 1 second will give the frequency, or

$$f = \frac{1}{T} \quad \text{hertz} \tag{17-16}$$

The time period for one cycle can be found from equation (17-16) and is

$$T = \frac{1}{f} \quad \text{seconds} \tag{17-17}$$

The angular distance that one cycle of ac has gone through is measured in radians.

$$1 \text{ radian} = 57.3^\circ$$

There are 2π radians in 360°.

Angular velocity is given by

$$\text{angular velocity} = 2\pi f \text{ rad/s} \tag{17-18}$$

Angular frequency (ω) is given by

$$\omega = 2\pi f \tag{17-19}$$

Out-of-phase instantaneous values of ac voltages either aid or cancel each other.

Two 180° out-of-phase voltages cancel each other.

Two in-phase voltages aid each other.

A phasor diagram can represent peak, rms, or average values of ac.

By convention the direction of phasor rotation is always taken to be counterclockwise.

If the reference phasor is along the x axis, a 90° leading voltage or current will be drawn in the first quadrant, and a 90° lagging voltage or current in the fourth quadrant.

The resultant of two 90° out-of-phase voltages or currents can be found using the Pythagorean theorem, or

$$V_R = \sqrt{V_A^2 + V_B^2} \qquad (17\text{-}20)$$

The phase angle of two voltages or currents can be found using the tan trig function for θ, or

$$\tan\theta = \frac{V_A}{V_B} \qquad (17\text{-}21)$$

Rms values do not apply to nonsinusoidal waves. Their values are stated as peak or peak-to-peak.

A moving-coil permanent-magnet meter is calibrated to read rms as well as peak-to-peak values of ac.

An oscilloscope is an ideal instrument to measure ac since the waveform can be viewed as well as measured.

To measure the frequency of an ac waveform from the screen of a scope, the Time/Div vernier is set to CAL and then the number of centimetres or divisions is counted horizontally for one cycle. The period T is found by multiplying the number of divisions by the setting of the Time/Div calibration:

$$T = \text{number of Div for one cycle} \times \text{Time/Div} \qquad (17\text{-}22)$$

The frequency is given by

$$f = \frac{1}{T} \quad \text{Hz} \qquad (17\text{-}16)$$

To measure the input waveform amplitude from the screen of a scope, the Volts/cm vernier is set to CAL and then the number of centimetres or divisions is counted vertically from one peak to the next peak. The peak-to-peak value is found by multiplying the number of centimetres by the setting of the Volts/cm calibration, or

$$V_{\text{p-p}} = \text{number of cm} \times \text{volts/cm} \qquad (17\text{-}23)$$

REVIEW QUESTIONS

17-1. An ac voltage and current must resemble a sine wave as closely as possible. Explain why.

17-2. Instantaneous values of an ac can be plotted using a rotating vector or phasor. Explain how an ac value is represented by a phasor and its conventional direction of rotation.

17-3. Given an ac peak value of 300 V, find the instantaneous value at 60°.

17-4. Find the peak-to-peak value of the voltage given in Question 17-3.

17-5. A full-wave rectifier's pulsating peak output voltage is 34 V. Calculate the average dc value that a voltmeter would read.

17-6. A 56-kΩ resistor is connected to 250 V. Determine its power dissipation in terms of dc and ac power. What will be the peak power in the ac circuit?

17-7. How can the average ac power dissipation be found? Explain why it is equivalent to the dc power dissipation.

17-8. Given that the rms value of an ac voltage is 110 V, find its peak and peak-to-peak values.

17-9. Define frequency and its unit.

17-10. Explain how the time period for one cycle changes with frequency.

17-11. Explain how the frequency of a voltage can be found from the sinusoidal waveform on an oscilloscope screen.

17-12. Give the symbol and equation for angular frequency.

17-13. Explain how a phasor diagram represents two out-of-phase voltages or currents.

17-14. Voltage V_B of 100 V lags voltage V_A, also 100 V, by 90°. Draw a phasor diagram to show the resultant voltage. Find the value of the resultant voltage and the phase angle between the resultant and V_A.

17-15. Explain why rms values do not apply to nonsinusoidal voltage or current waveforms.

17-16. What two periods constitute one cycle of a square or rectangular voltage waveform?

17-17. Explain the importance of setting the vernier control to CAL when measuring the amplitude or frequency of a waveform on an oscilloscope.

17-18. Briefly, list the steps required to measure the amplitude of an input voltage waveform on an oscilloscope.

17-19. Briefly, list the steps required to measure and find the frequency of an input voltage waveform on an oscilloscope.

SELF-EXAMINATION

Where applicable, choose the most suitable word or phrase to complete each statement.

17-1. An irregular ac voltage waveform will cause __________ in transformers, motors, conductors, and insulators.

17-2. An ideal ac waveform closely resembles a __________ wave function.

17-3. A rotating vector is termed a __________ and its conventional direction of rotation is __________.

17-4. Draw a phasor diagram of two 100-V in-phase generators.

17-5. What is the resultant voltage of the two generators in Question 17-4?

17-6. Draw a phasor diagram to represent two 100-V out-of-phase generators.

17-7. What is the resultant voltage of the two generators in Question 17-6?

17-8. Draw a phasor diagram to represent V_A lagging V_B by 90°. Each has a peak value of 100 V.

17-9. What is the resultant voltage of the two generators in Question 17-8?

17-10. Given an ac peak value of 200 V, find the instantaneous value at 60°.

17-11. A full-wave rectifier has an ac input voltage of 50 V peak-to-peak. Calculate the average dc output value of the rectifier.

17-12. A 12-kΩ resistor is connected to a voltage with a peak value of 200 V. Calculate its average power dissipation.

17-13. A 2.7 kΩ resistor is connected to 50 V dc. Calculate its equivalent ac power dissipation.

17-14. What must be the peak voltage rating of a device if it is to be connected across an ac voltage of 300 V?

17-15. The peak value of a sine-wave voltage is 250 V. Calculate the rms value.

17-16. The peak value of a square-wave voltage is 250 V. What is its equivalent value to ac rms?

17-17. The number of cycles in 1 second is defined as ________, whose unit is the ________.

17-18. The time period for one cycle can be found by dividing the ________ into 1 second.

17-19. Given the time period for one cycle, the ________ can be found by dividing the time period into 1 second.

17-20. List four steps that are required to determine the frequency of a waveform from oscilloscope readings.

17-21. List three steps that are required to determine the peak-to-peak value of a waveform from oscilloscope readings.

ANSWERS TO PRACTICE QUESTIONS

17-1

1. sine
2. peak
3. peak-to-peak
4. 360
5. 90
6. 270
7. 50
8. 50
9. 100
10. 110 V
11. 156 V
12. 99 V

17-2

1. 45
2. 311 *V* peak
3. 354 V
4. 152 V
5. rms
6. 16.7
7. 0.1 kHz
8. 4.5 MHz
9. 57.3
10. 2π
11. 2π radians
12. 2π radians
13. $2\pi, f$
14. ω
15. 376.8

17-3

1. zero
2. 282.8 V
3. 17.7 V
4. 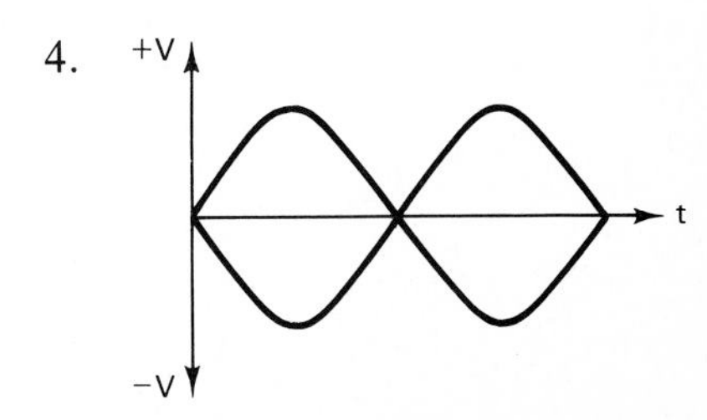

5. magnitude
6. magnitude, direction
7. phasor
8.

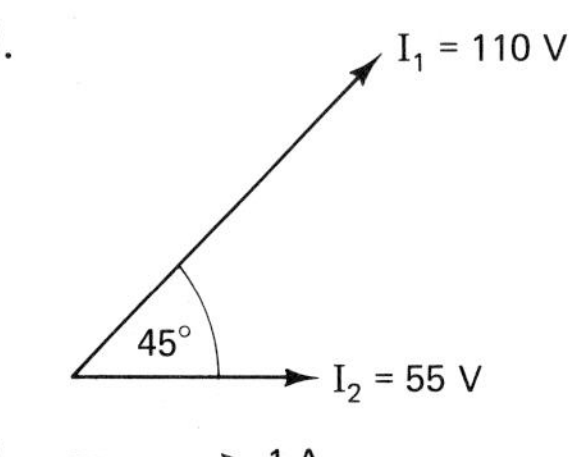

9.

10. 3.16 A
11. 2 A ← • → 3 A

→ 1 A resultant

12. peak, peak-to-peak
13. leading or trailing
    leading or trailing
14. 200 μs
15. nonsinusoidal waveforms do not have rms values; 10 V is equivalent to dc, which is equivalent to rms

18

The Ac Power Line

INTRODUCTION

The electrical power outlet that we use for home, commercial, and industrial uses is pretty well taken for granted. Having seen how ac was generated in Chapter 16, and having looked at its characteristics in Chapter 17, we should become familiar with its transmission and how to use it safely.

The National Electrical Code® or Canadian Electrical Code specifies extensively how the electric wiring and equipment must be installed to provide safety from electric shock and from electrical fire. At a minimum we must be aware of such regulations and must certainly be aware of the safety aspects of using electricity. To cite an example from a newspaper article, a man was found dead under his car. He died from electric shock due to a defective electric drill that was not properly grounded because he was using a two wire extension cord.

In this chapter we provide the reader with basic knowledge of the ac power line and the built-in protection to minimize and prevent electric shock as well as fire hazard.

MAIN TOPICS IN CHAPTER 18

18-1 Ac Power Transmission
18-2 The Standard 120/240-V Power Line
18-3 Ground and Shock Hazard
18-4 Industrial Three-Phase Power
18-5 Delta and Wye (Star) Circuits
18-6 Universal Motor
18-7 Induction Motor

OBJECTIVES

After studying Chaper 18, the student should be able to

1. Master residential, commercial, and industrial power-line basic concepts.
2. Apply the requirements for grounding to prevent a shock hazard.
3. Apply precautions when working on higher voltages to prevent shock.

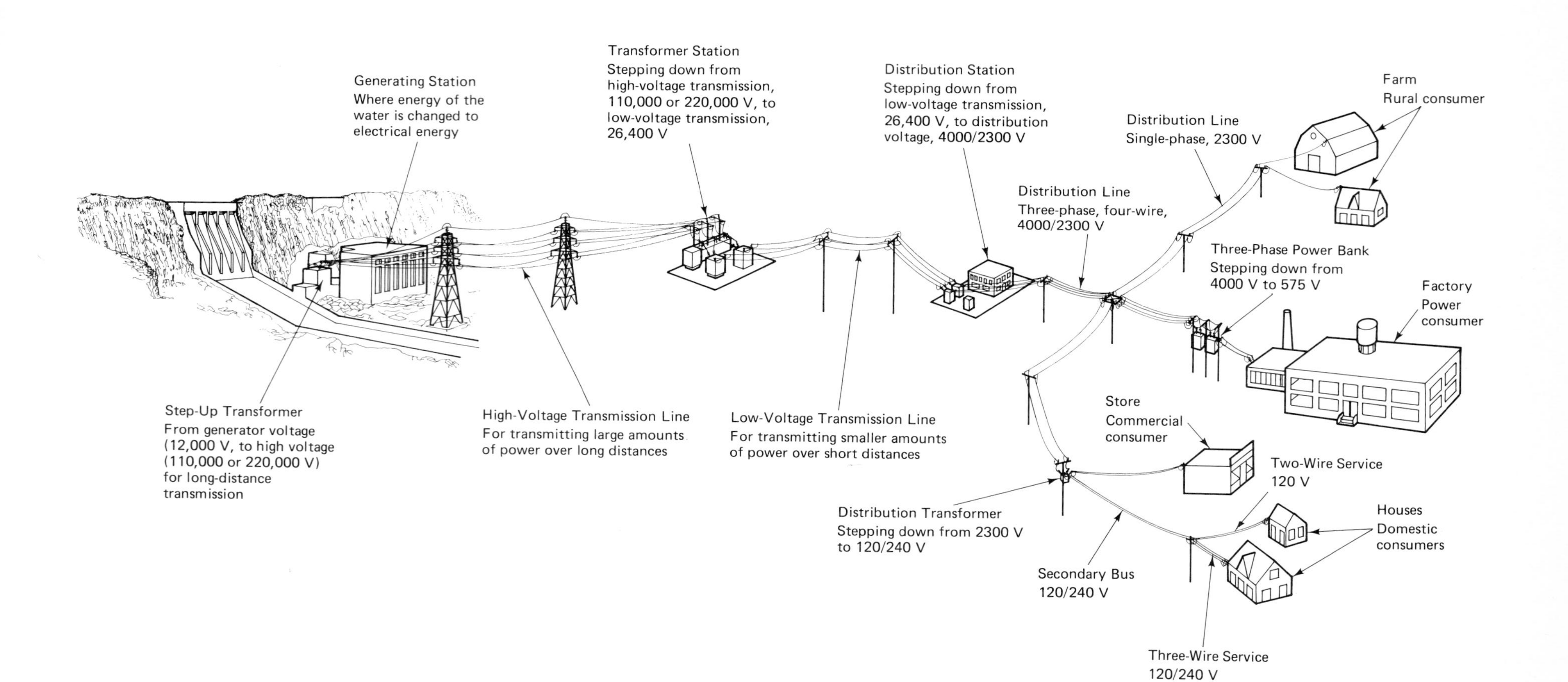

FIGURE 18-1 Ac power transmission and distribution.

4. Acquire an insight into the operation of the universal and induction motors.

18-1 AC POWER TRANSMISSION

Transmitting ac power over a great distance requires that the ac voltage is stepped up to hundreds of thousandths of volts. The main reason for the stepped-up voltage is the I^2R power loss due to the resistance of the conductor. To reduce the power loss, the resistance of the conductor can be decreased. However, resistance is directly proportional to the length and inversely proportional to the area of the conductor. With physical and economic limitations the conductor resistance is fixed and dependent on these factors. The remaining factor to reduce the power losses is to decrease the current through the conductor. This can be done only if the voltage is increased, since $P = VI$.

Modern high-voltage transmission of ac power varies from 110 to 300 kV, with experiments being conducted to transmit power at 500 kV. Transmitted at this high voltage, the current travels over the high-tension lines to an electric utility substation, as shown in the diagram of Fig. 18-1.

At the substation the voltage is stepped down to typically 4000/2300 V and transmitted to residential, commercial, and industrial locations. There, further step-down transformers reduce this voltage to 120/240 V single phase for residential and commercial use and to 575 V three-phase for factory or industrial use.

18-2 THE STANDARD 120/240-V POWER LINE

The frequency of North American ac power is 60 Hz. Early engineers discovered that 60 Hz was the best frequency for transmission of ac power, in terms of iron-core losses for transformers, generators, and motors as well as transmission-line losses.

Voltages for residential use were also optimized at 120 V and 240 V. These voltages were derived from a center tap and either one of the outside terminals, as shown in Fig. 18-2. Double the voltage, or 240 V can be obtained between the two outside terminals.

Lighting and small appliances are "plugged" into the 120-V outlet, while dryers, electric stoves, and other power equipment are connected to the 240-V side of the line.

The line voltage that is "delivered" by the electric utility is not always 120 V. All utilities try to maintain 110 V as the minimum voltage and not to exceed 120 V as the maximum voltage. In most cases at peak electric usage hours, the voltage does in fact drop to the bare minimum of 110 V. In "brownouts" it drops even lower! That is why the lights are dimmer during these periods.

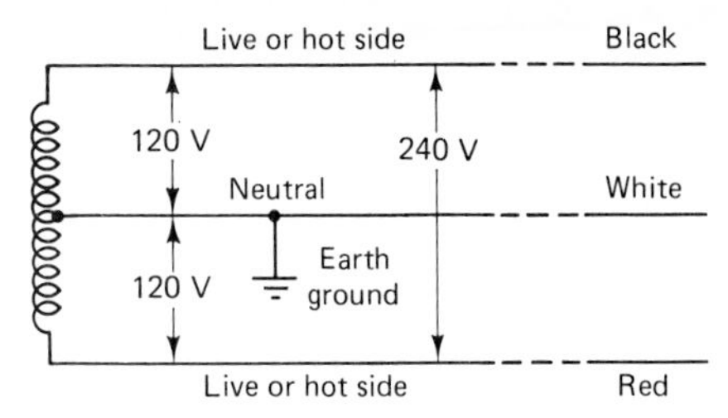

FIGURE 18-2 Three-wire 120/240-V single-phase system used for residential and commercial power.

Line-Wire Color Coding

The center tap of the three-wire system shown in Fig. 18-2 is called the *neutral*. The neutral conductor is always identified by having a *white* or *gray* covering or insulation color. The two remaining conductors are termed the *live, hot, or high side* of the system. These are always identified by having *any color insulation except white and green*. Black and red are the customary insulation colors used to identify the hot or live conductors.

Electric Service Entrance and Distribution

Ac power is brought into residential and commercial buildings by means of the electric service entrance. This is a conduit with three large gauge conductors or buried underground that are connected to a fused disconnect or circuit breaker. From the circuit breaker the electric meter is connected in series and then the conductors are brought into the distribution panel as shown in Fig. 18-3. The distribution panel is a number of individual circuit breakers or fuses that service the convenience outlets and lighting outlets. The purpose of fuses or circuit breakers is to protect the conductors and the electrical device from overloads and short circuits (see Section 13-7). In all cases the fuse or circuit breaker should never exceed the safe current-carrying capacity of the installed conductors. Typically, size AWG No. 14 is used in residential wiring and is rated at 15 A.

18-3 GROUNDING AND SHOCK HAZARD

The neutral of the 120/240-V residential system is connected to the city water pipe system at the service entrance switch. If a water system is unavailable, such as in a rural farmhouse, two 3-m rods are driven into the ground and the neutral wire is connected to the rods

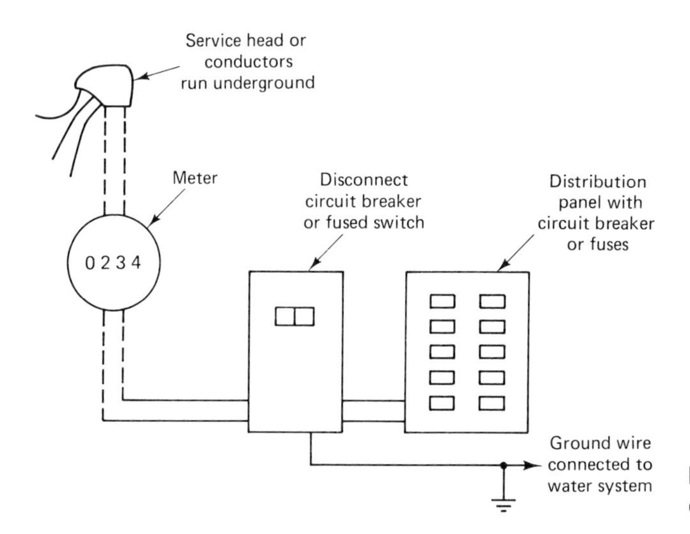

FIGURE 18-3 Typical residential electrical service entrance.

through a heavy copper wire. The symbol for the ground is shown in Fig. 18-3.

Grounding the neutral puts that conductor at zero potential with respect to ground. Thus only two conductors of the three-wire system are live with respect to ground. By this means the maximum voltage that a person handling or accidentally contacting one of the other two conductors will be 120 V. Of course, contacting the two outside conductors at the same time would expose the person to a 240-V shock, which could prove lethal. This type of shock would be due to carelessness, and with carelessness you might not get another chance.

U-Ground Duplex Outlet

To minimize shock hazard all wall electrical boxes must be grounded either through a separate bare copper wire enclosed along with two or three other insulated conductors in a fire-resisting, moisture-proofed nonmetallic sheath (NMSC).

Flexible armored cable also provides a ground through its outer metallic sheath. Instead of nonmetallic sheathing, the insulated conductors are covered with an insulating paper spirally wrapped, then covered with a spiral interlocking galvanized metal or aluminum strip. The metal covering gives good mechanical protection while providing a continuous grounding conductor.

Electrical outlet boxes must be grounded by means of the bare copper conductor in the nonmetallic sheath cable, as shown in Fig. 18-4(a). In the case of duplex receptacle outlets commonly known as wall plugs, the receptacles must be of a U-ground type, shown in Fig. 18-4(b). The U-ground terminal is connected to the bare copper ground wire. Note that one slot of the receptacle is longer than the other. The neutral wire connects to the wider-blade side and the hot wire connects to the narrow-blade side of the receptacle.

Grounding Appliance

The whole purpose of grounding the neutral and then providing an extra ground wire is to protect people from a shock hazard. A shock hazard can develop in any of

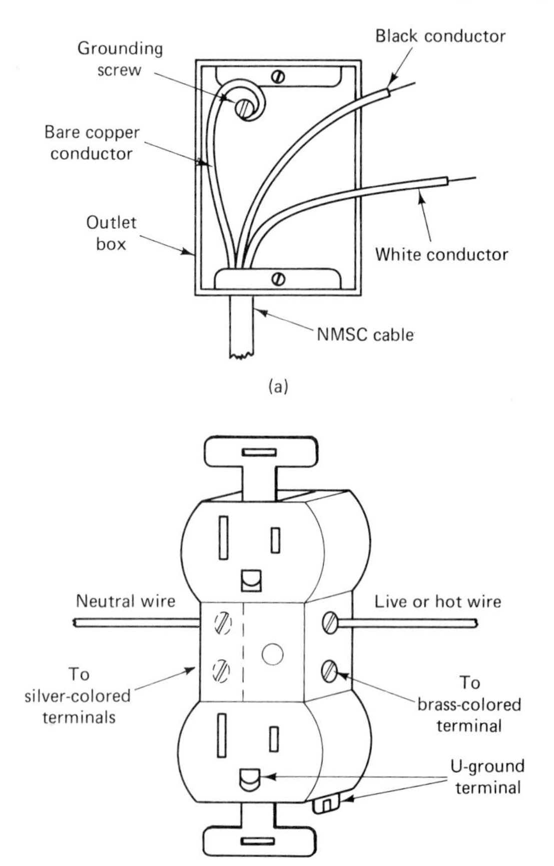

FIGURE 18-4 (a) Bare copper conductor grounds outlet box. (b) U-ground duplex outlet. *Note:* Narrow slot connects to hot wire, wide slot to neutral.

many appliances, such as toaster, frying pans, electric dryers, stoves, mixers, drills, and others. With constant usage and flexing, the insulation on the hot conductor could break open and expose the bare live wire. The user is then exposed to a 120-V shock should the bare live wire touch the metal frame of the appliance.

Grounding the metal frame of the device to the U-ground provided in an outlet receptacle prevents a shock hazard from happening. This is provided by a third, *green insulated* conductor in the flexible cable or cord that serves as the ground connection between the metal frame of the device and the U-ground of the 120-V outlet, as shown in Fig. 18-5.

When the bare wire touches the metal frame there is an immediate short circuit, which blows the fuse or trips the circuit breaker open. This disconnects the 120-V main from the device, eliminating any shock hazard. Of course, the device will trip the breaker continually until the fault is repaired.

The green conductor in a flexible portable extension cord is the ground conductor. The *green color is reserved exclusively to identify this grounding conductor*. It must always be properly connected to the terminal of a U-ground cap or plug and never be bypassed by a two-wire cap.

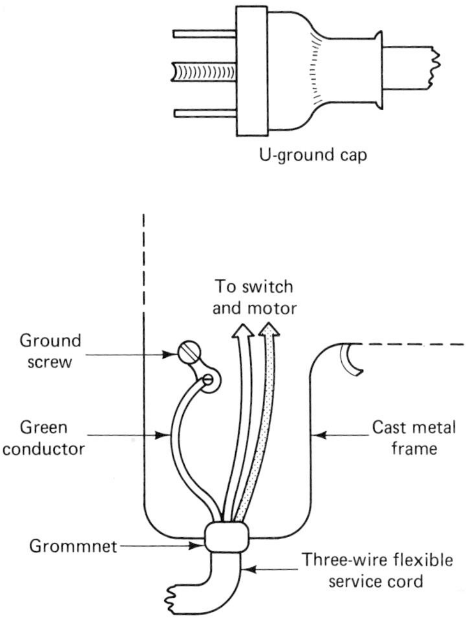

FIGURE 18-5 Grounding the metal frame of a portable electric drill or device.

Shock Hazard

Shock due to an electrical contact and resultant current flow through a human body is very unpleasant, to say the least. Under the best of conditions the human body's resistance between the fingertips of one hand and the fingertips of the other hand will vary between 10 to 40 kΩ. This means that it would take a high voltage to have 150 mA flow through the body and cause death, as shown in Fig. 18-6.

Conditions in most cases are not ideal, so that voltages greater than 30 V can be considered as potentially capable of producing a shock hazard beyond the painful level. Electrical installations above 30 V must conform to the National Electrical Code® or the Canadian Electrical code.

Working on live circuits poses a shock hazard. You probably have to make a voltage or current measurement so that the power has to be turned ON. In this case a few rules should be followed which will minimize the chance of a shock. One is always to use one hand to measure the voltages, with the other hand behind your back. Another rule is to ensure that you stand on a nonconducting floor or mat. In all cases you must be alert and very cautious.

PRACTICE QUESTIONS 18-1

Choose the most suitable word, phrase, or value to complete each statement.

1. To reduce power loss ac power is transmitted at very high (*current, resistance, voltage, conductance*).

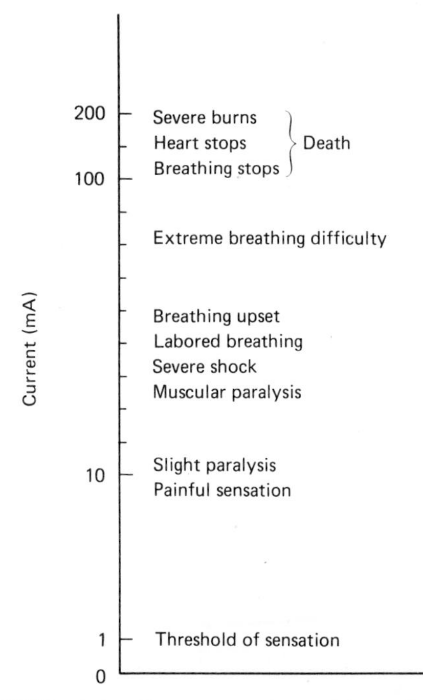

FIGURE 18-6 Chart to show various levels of current and their effect on the human body.

2. The power loss of a conductor is termed ________ loss.
3. Increasing the length of a conductor will (*decrease, increase*) its resistance.
4. Doubling the area of a conductor will (*double, halve, increase, triple*) its resistance.
5. The North American frequency of ac is ________.
6. The North American residential electric utility voltage is ________ volts.
7. The (*hot, live, neutral*) wire is grounded to the water system.
8. The hot or live wires are color coded any color other than ________ and ________.
9. The neutral is color coded ________ exclusively.
10. The purpose of fuses or circuit breakers is to protect the conductors and devices from ________ and ________.
11. Grounding the neutral and electrical device helps to protect personnel from ________ hazard.
12. The hardware of an electrical device is grounded through the ________ terminal of a duplex receptacle or wall outlet.
13. The ground wire in a flexible service cord is identified by its ________ color insulation.
14. Grounding eliminates the danger to shock hazard since the fuse blows, or the circuit breaker trips the circuit open, should the ________ wire insulation become defective and the wire touch the metal frame of the device.
15. Working with one hand with the other behind your back minimizes the danger to ________.

18-4 INDUSTRIAL THREE-PHASE POWER

A three-phase ac voltage is used extensively in industrial operations. The requirement in factories is for heavy-duty induction-type motors that start up easily under heavy load conditions. Closer to home, the automobile has a three-phase alternator to charge the battery. The three-phase ac is easily rectified and filtered into dc since there are three dc pulses every second.

A simple alternator with three coils spaced physically 120° apart would produce a three-phase sine-wave voltage, as shown in Fig. 18-7. Voltage V_B leads V_A by 120° and voltage V_C lags V_A by 120°, as shown by the phasor diagram of Fig. 18-7.

The three windings of the alternator are connected in either a wye or a delta configuration. Connecting the windings in a wye configuration [see Fig. 18-8(a)] will give three *line-to-line* voltages V_{ab}, V_{bc}, and V_{ac}. The line-to-line voltage can be found in terms of the phase voltage by the use of the phasor diagram of Fig. 18-7, and can be shown to be greater by a factor of 1.73 times. For a phase voltage of 120 V the line-to-line voltage will be $1.73 \times 120 = 208$ V. Note that $1.73 = \sqrt{3}$, so that the line voltage can be determined from the equation

$$V_{\text{line}} = \sqrt{3} \times V_{\text{phase}} \tag{18-1}$$

In the delta connection of Fig. 18-8(b), two windings are in series, with the third in parallel for any line-to-line voltage. The resultant voltage is still 208 V, but the current will be increased by a factor of $\sqrt{3}$.

Four-Wire Three-Phase System

Grounding the common connection point of a wye configuration gives a unique ac voltage source. This type of ac system is widely used in commerce and industry. Its advantage is that the 120 V, single phase, can be used for lighting while the three-phase voltage can be used to supply the motors that run the equipment. The line-to-line, three-phase voltage remains at 208 V.

Grounding the common connection point of the wye configuration will give 120 V single phase from

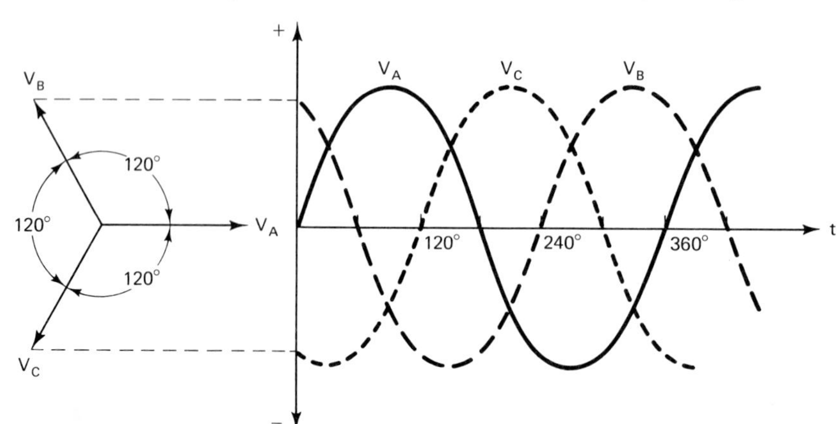

FIGURE 18-7 Three-phase sine-wave voltage.

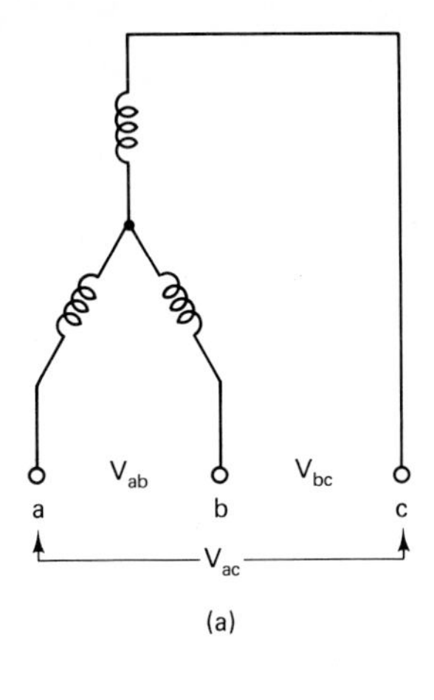

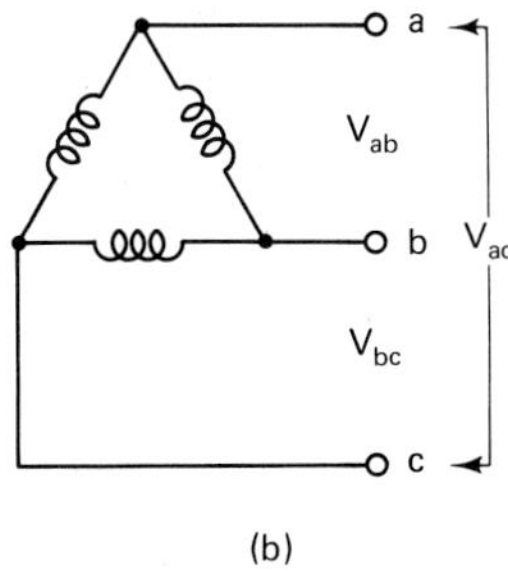

FIGURE 18-8 Wye and delta connections of alternator or transformer windings. (a) Wye connection. (b) Delta connection.

each of the three windings, as shown in Fig. 18-9. The added advantage is that the ac system now has a grounded conductor for safety. The three-phase line-to-line voltage is used to run the motors of the installed air-conditioning or other heavy-duty equipment in commercial establishments or industry. Of course, the 120 V loads should be balanced. Distributing the total load current equally between the three windings ensures that no one winding has less current flowing through it, thereby affecting its voltage.

18-5 DELTA AND WYE (STAR) CIRCUITS

A delta (Δ) or wye (Y) circuit (also termed a star connection), shown in Fig. 18-10, is used extensively for connecting electrical machinery such as transformer, motor, and generator windings to a three-phase ac power source.

A delta circuit can be redrawn in the form of a π (pi) circuit as shown in Fig. 18-11(a). The wye circuit [Fig. 18-11(b)] can be converted to a T circuit. Both the π and T circuits are common configurations in electronic circuits such as filter circuits.

Delta-to-Wye Conversion

The analysis of a delta connection could be done using other forms of network theorems, such as loop equations or nodal analysis. The analysis can be greatly simplified, however, if the Δ circuit is converted to a Y circuit. Converting the Δ circuit to a Y circuit means that the resistance as measured between A and B, A and C, and B and C in Fig. 18-12(a) must be equivalent in both circuits.

To find each of the resistances in the Y circuit requires a set of conversion formulas that will give R_a, R_b, and R_c in terms of the delta circuit resistance R_A, R_B, and R_C. The derivation of the three formulas will not be gone into since a conversion rule can easily be followed:

> To find any equivalent Y resistance, take the product of the adjacent two Δ resistances [$R_A R_C$ for R_a Fig. 18-12(b)] and divide it by the sum of the Δ resistances.

From the circuit of Fig. 18-12(b), R_A and R_C are adjacent to R_a; therefore,

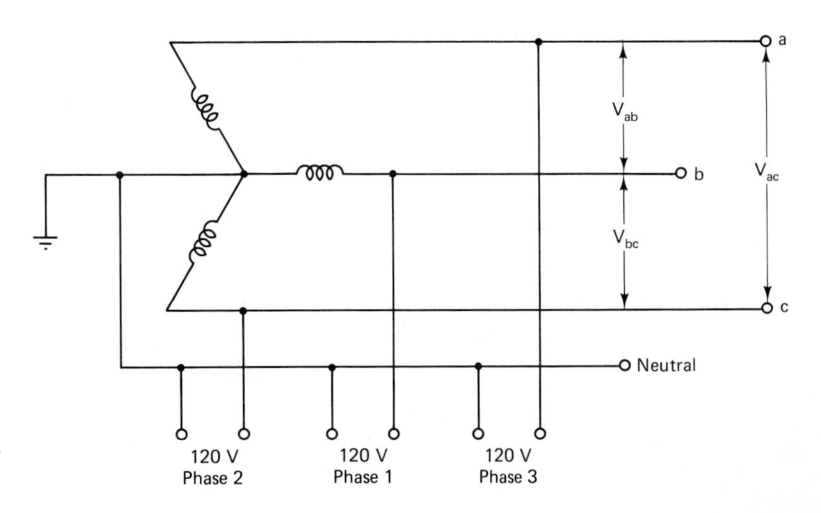

FIGURE 18-9 Four-wire three-phase and single-phase system.

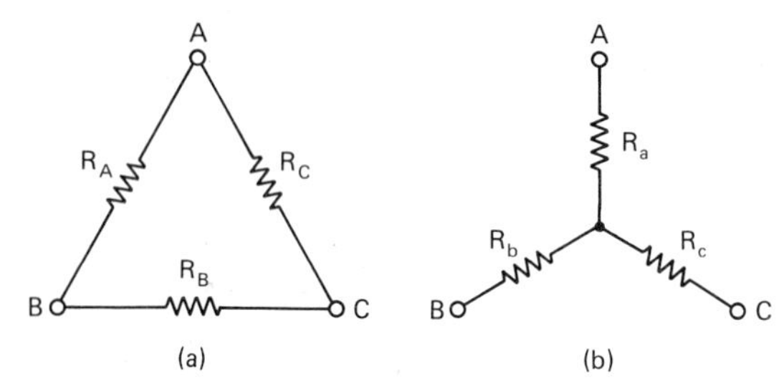

FIGURE 18-10 (a) Delta (Δ) network. (b) Wye (y) network.

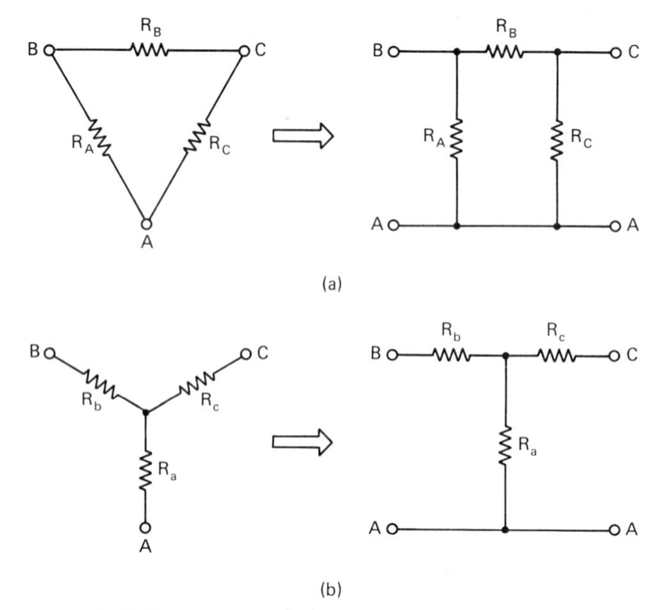

FIGURE 18-11 (a) π-network configuration. (b) T-network configuration.

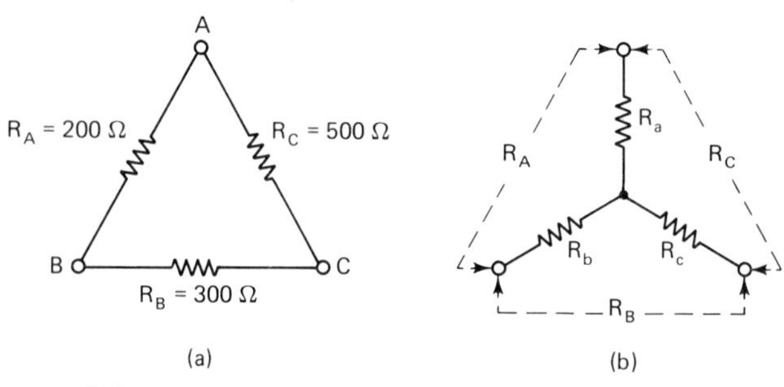

FIGURE 18-12 Delta-to-wye conversion. (a) Delta. (b) Wye.

$$R_a = \frac{R_A R_C}{R_A + R_B + R_C} \quad \text{ohms} \qquad (18\text{-}2)$$

R_A and R_B are adjacent to R_b; therefore,

$$R_b = \frac{R_A R_B}{R_A + R_B + R_C} \quad \text{ohms} \qquad (18\text{-}3)$$

R_B and R_C are adjacent to R_c; therefore,

$$R_C = \frac{R_B R_C}{R_A + R_B + R_C} \quad \text{ohms} \qquad (18\text{-}4)$$

EXAMPLE 18-1

Convert the Δ circuit to Y in Fig. 18-12 given $R_A = 200\ \Omega$, $R_B = 300\ \Omega$, and $R_C = 500\ \Omega$.

Solution: Using equations (18-2) to (18-4) yields

$$R_a = \frac{R_A R_C}{R_A + R_B + R_C} = \frac{200\ \Omega \times 500\ \Omega}{200\ \Omega + 300\ \Omega + 500\ \Omega} = 100\ \Omega$$

$$R_b = \frac{R_A R_B}{R_A + R_B + R_C} = \frac{200\ \Omega \times 300\ \Omega}{1000\ \Omega} = 60\ \Omega$$

$$R_c = \frac{R_B R_C}{R_A + R_B + R_C} = \frac{300\ \Omega \times 500\ \Omega}{1000\ \Omega} = 150\ \Omega$$

Wye-to-Delta Conversion

The Y-to-Δ conversion requires a set of formulas that will give R_A, R_B, and R_C in terms of the Y-circuit resistance R_a, R_b, and R_c. The conversion rule is as follows:

> To find any equivalent Δ resistance [R_A, R_B, or R_C of Fig. 18-12(a)], take the sum of all possible combination product pairs of the Y resistance [$R_aR_b + R_aR_c + R_bR_c$ of Fig. 18-12(b)] and divide the sum by the Y resistance connected diagonally opposite from the Δ resistance.

From the circuit of Fig. 18-12(b), R_c is diagonally opposite R_A; therefore,

$$R_A = \frac{R_aR_b + R_aR_c + R_bR_c}{R_c} \quad \text{ohms} \qquad (18\text{-}5)$$

R_a is diagonally opposite R_B; therefore,

$$R_B = \frac{R_aR_b + R_aR_c + R_bR_c}{R_a} \quad \text{ohms} \qquad (18\text{-}6)$$

R_b is diagonally opposite R_C; therefore,

$$R_C = \frac{R_aR_b + R_aR_c + R_bR_c}{R_b} \quad \text{ohms} \qquad (18\text{-}7)$$

EXAMPLE 18-2

Convert the Y circuit back to a Δ circuit in Fig. 18-12 given the solution values of $R_a = 100\ \Omega$, $R_b = 60\ \Omega$, and $R_c = 150\ \Omega$ in Example 18-1.

Solution: Using equations (18-5) to (18-6) gives us

$$R_A = \frac{R_aR_b + R_aR_c + R_bR_c}{R_c}$$

where, $R_aR_b + R_aR_c + R_bR_c = 30{,}000\ \Omega$; therefore,

$$R_A = \frac{30{,}000\ \Omega}{R_c} = \frac{30{,}000\ \Omega}{150\ \Omega} = 200\ \Omega$$

$$R_B = \frac{30{,}000\ \Omega}{R_a} = \frac{30{,}000\ \Omega}{100\ \Omega} = 300\ \Omega$$

$$R_C = \frac{30{,}000\ \Omega}{R_b} = \frac{30{,}000\ \Omega}{60\ \Omega} = 500\ \Omega$$

18-6 UNIVERSAL MOTOR

A great many of the household appliances, such as mixers, vacuum cleaners, electric lawn mowers, and drills, use the universal motor. Its construction is similar to a dc generator and it is capable of operating on either ac or dc.

As shown in the schematic diagram of Fig. 18-13, the armature coils are brought out to the segments of the commutator. The brushes contacting the segments connect the armature in series with the field coil windings. This type of motor is always connected to its load since its speed tends to run away. In electric drills it is coupled to a gear train, while in vacuum cleaners it is directly coupled to one or more fans. An overload will increase the current through the armature coils and field winding and quickly cause overheating of the motor.

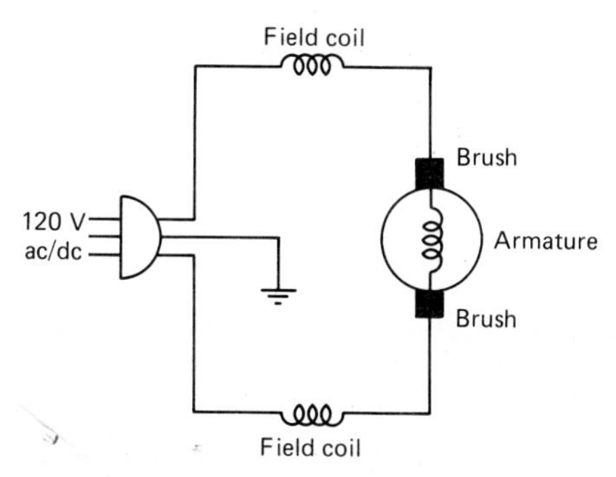

FIGURE 18-13 Schematic diagram of a universal motor.

18-7 INDUCTION MOTOR

The induction motor is a common type of three-phase motor used in industry. With a modified start winding it is also used on single phase for such appliances as washers and driers.

The usual pair of field pole windings are connected across the 120-V ac line. A rotor is used instead of an armature and brushes. The rotor has a number of large-gauge copper wire or bus brazed-to-copper end pieces, as shown in Fig. 18-14.

This type of rotor is termed a *squirrel-cage rotor.* Each copper bus is a shorted turn. A very large current is induced as each shorted turn cuts through the ac magnetic field produced by the field pieces (stator). Lenz's law tells us that the magnetic field produced by the induced current in each of the shorted turns will be opposite to the magnetic field producing it. Thus the rotor produces a repulsion flux which causes the squirrel cage rotor to rotate.

Induction motors, however, require some method of starting them. The field poles and rotor field are opposing, but it is necessary that the magnetic field in the motor be rotating for the rotor to start turning. For a three-phase motor each phase is 120° apart. The three pairs of field poles, then, will have a true rotating magnetic field capable of pulling the squirrel-cage rotor into rotation.

For a single-phase squirrel cage motor the starting torque is produced by what is termed as a *split-phase winding*. If a capacitor is connected in series with a second or split-phase winding (see Fig. 18-15), the current

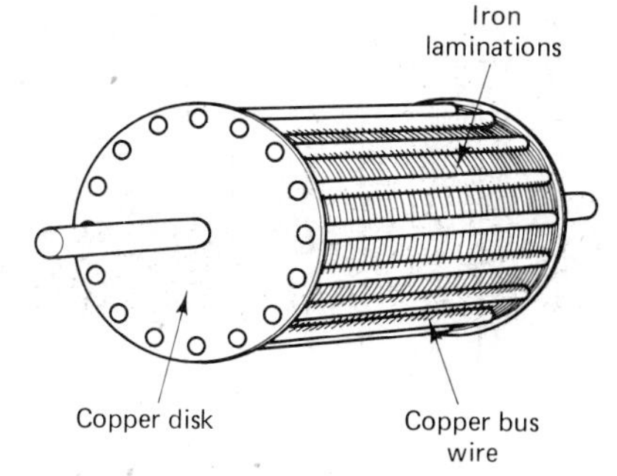

FIGURE 18-14 Squirrel-cage rotor.

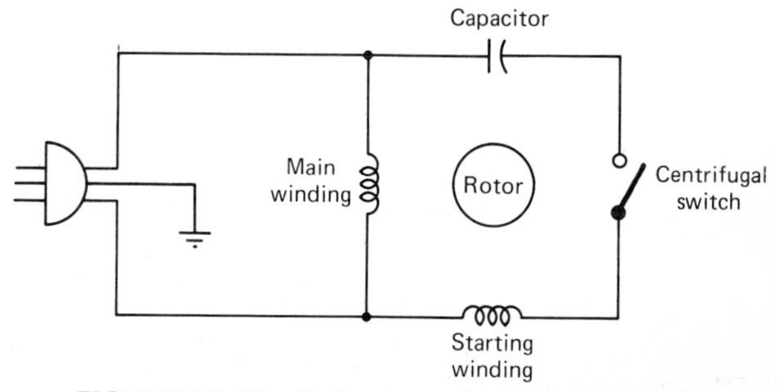

FIGURE 18-15 Split-phase induction motor.

flowing through this winding will be out of phase with the main field windings. The result is that the main magnetic field produced is out of phase with the start or phase-winding magnetic field. This produces the rotating magnetic field necessary to make the motor self-starting. Once the rotor reaches its running speed, the phase winding is disconnected by a centrifugal switch mounted on the rotor shaft. The rotor then continues to follow the alternating field set up by the main winding.

Reversing Motor Rotor Direction

The rotation direction, clockwise (CW) or counterclockwise (CCW), of a rotor can be changed simply by reversing the connection of the field coil windings. For the three-phase motor, phase sequence will govern the direction of rotation. Simply interchanging one of the three-phase line wires to the field winding will change the rotor direction.

In the case of the split-phase motor, the main windings are connected in series; therefore, their connections must be reversed. This can be done by interchanging one of the ends of the field coil connected to the line with one of the series connection ends.

SUMMARY

Electricity is transmitted at a high voltage to minimize I^2R losses.

Power loss due to the resistance of the conductors is termed I^2R loss.

The ac high voltage is stepped down to 120/240 V for domestic and commercial use.

The frequency of the North American ac power is 60 Hz.

The center tap of the three-wire ac system is called the neutral.

The two outside conductors of a three-wire ac system are termed the line or hot wires.

The neutral is grounded to the water system and is color coded white or gray.

The live or hot wires are color coded any color except white and green. Black and red are the usual colors used.

Grounding the neutral minimizes the shock hazard from three wires to two wires.

Electrical outlet boxes are grounded by means of a ground conductor to eliminate shock hazard.

The U-ground of receptacles or wall plugs is connected to the ground conductor.

Appliances are connected to the U-ground through a third conductor color-coded green.

Grounding appliances and devices protects personnel from a shock hazard.

Voltages greater than 30 V are considered to be potentially capable of producing a shock hazard.

When working on a live circuit, keep one hand behind your back and stand on an insulated floor mat.

A three-phase system is generally used in industry.

Connecting the three 120-V windings of an alternator or transformer in a wye configuration will give a line-to-line voltage of 208 V.

Grounding the common connection point of a wye connection gives a neutral to the system and allows 120 V single phase to be tapped off the system.

The universal motor is commonly found in small household appliances.

The armature windings are connected in series with the field windings in a universal motor.

An induction motor is used for heavy-duty applications, particularly in industry.

A rotor that has shorted turns is used in an induction motor.

To start an induction motor, some means of rotating the magnetic field must be provided.

The magnetic field rotates in a three-phase motor since each phase is 120° apart, so that the magnetic field of the three pairs of field coils will follow the current and start the motor.

An induction motor that is used on single-phase has a capacitor in series with a separate start winding which is termed a split-phase winding.

The split-phase winding is disconnected by a centrifugal switch once the motor reaches its run speed.

The rotor direction of a motor can be reversed by reversing the connection of the field windings.

***Δ-to-Y Conversion* (*Fig. 18-12*):** To find any equivalent Y resistance, take the product of the adjacent two Δ resistances and divide it by the sum of all three Δ resistances.

***Y-to-Δ Conversion* (*Fig. 18-12*):** To find any equivalent Δ resistance [R_A, R_B, or R_C of Fig. 18-12(a)] take the sum of all possible combination product pairs of the Y resistance [$R_aR_b + R_aR_c + R_bR_c$ of Fig. 18-12(b)] and divide the sum by the Y resistance connected diagonally opposite from the Δ resistance.

REVIEW QUESTIONS

18-1. Explain why ac power is transmitted at high voltages of 110 to 220 kV.

18-2. What is the difference between single-phase and three-phase ac voltage?

18-3. Of what voltage and frequency is the line voltage delivered by an electric utility?

18-4. What is the color code for the hot wires and the neutral of a three-wire residential and commercial power line?

18-5. Why is the neutral grounded?

18-6. What function does the U-ground of a wall receptacle provide?

18-7. Why does grounding a device render it safe against a shock hazard?

18-8. What is the color code for the ground conductor of three-wire flexible cable or cord?

18-9. Approximately what value of current causes muscular paralysis?

18-10. Why do voltages beyond 30 V present a potential for a shock hazard?

18-11. What two precautions can be used when working on a live circuit with higher voltages?

18-12. Explain the difference between a single-phase, three-wire system and a three-phase, three-wire system.

18-13. What two advantages are gained if the common connection point of a three-phase wye configuration is grounded?

18-14. What is the difference between an armature of a universal motor and the rotor of an induction motor?

18-15. What must be provided in order to start the rotor of an induction-type motor rotating?

18-16. What is the purpose of a split-phase winding in a single-phase induction motor?

18-17. How can the rotational direction of a rotor be reversed?

18-18. From Fig. 18-12, list the three formulas to convert the delta circuit to a wye circuit.

18-19. From Fig. 18-12, list the three formulas, to convert the wye circuit to a delta circuit.

SELF-EXAMINATION

Choose the most suitable word, phrase, or value to complete each statement.

18-1. Heating of a conductor due to its resistance and current flow is termed ________ loss.

18-2. To minimize ________ loss ac power is transmitted at voltages from 110 to 300 kV.

18-3. Single-phase ac voltage used for residential purposes is ________ volts and ________ volts.

18-4. The frequency of the North American ac power is ________ Hz.

18-5. The center tap of a three-wire ac system is called the ________.

18-6. The neutral conductor is always identified by having a ________ or ________ covering or insulation color.

18-7. ________ and ________ are the customary insulation colors used to identify the hot or live conductors.

18-8. The ________ conductor of a three-wire, single-phase ac system is always grounded.

18-9. The ________ terminal of a wall receptacle is connected to the bare copper ground wire.

18-10. Grounding the metal frame of a device to the U-ground provided in an outlet receptacle prevents a ________ hazard from happening.

18-11. The ________ colored conductor in a flexible portable extension cord is the ground conductor.

18-12. Electrical installations above ________ volts must conform to the National Electrical Code® or the Canadian Electrical Code.

18-13. Any voltage great enough to cause a current of ________ mA to flow through a person can cause paralysis of both lungs and heart and death.

18-14. In a three-phase ac system the line-to-line voltages are ________ degrees out of phase with each other.

18-15. For a phase voltage of 120 V the line-to-line voltage will be larger by a factor of ________.

18-16. Grounding the common connection point of a wye, three-phase configuration will give ________ volts single-phase for lighting and small tools use.

18-17. Household appliances such as mixers and vacuum cleaners use a ________ motor.

18-18. The ________ motor is a common type of three-phase motor used in industry.

18-19. A ________ ________ rotor has a number of large-gauge copper wires or buses brazed to copper end pieces.

18-20. For a single-phase induction motor the starting torque is produced by what is termed a ________ winding.

18-21. Once the rotor reaches its running speed, the phase winding of a single-phase induction motor is disconnected by a ________ switch mounted on the rotor shaft.

18-22. Reversing the connections to the ________ ________ windings will reverse the rotor direction.

18-23. For a three-phase motor, interchanging one of the ________ wires will change the rotor direction.

18-24. Interchanging one of the ends of the field coil leads connected to the ________ with one of the series connection ends of a split-phase motor will reverse its rotor direction.

18-25. Convert the delta circuit of Fig. 18-16 to a wye circuit.

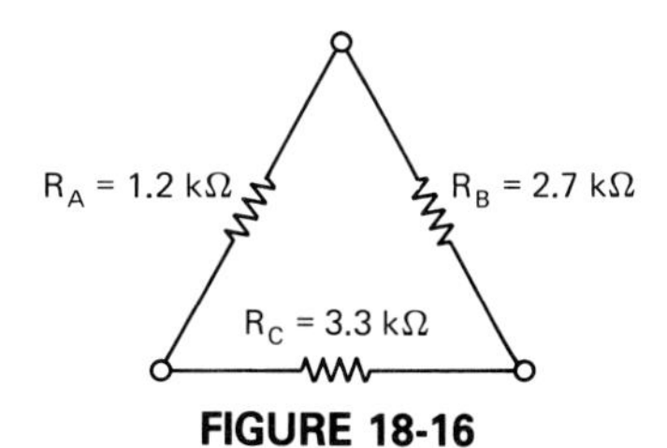

FIGURE 18-16

18-26. Convert the wye circuit of Fig. 18-17 to a delta circuit.

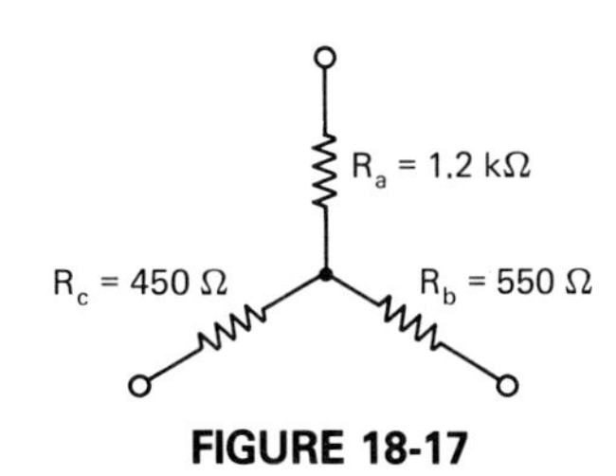

FIGURE 18-17

ANSWERS TO PRACTICE QUESTIONS

18-1

1. voltage
2. I^2R
3. increase
4. halve
5. 60 Hz
6. 120/240
7. neutral
8. white, green
9. white
10. shorts, overloads
11. shock
12. U-ground
13. green
14. line or hot
15. shock

19

Inductance

INTRODUCTION

Faraday's discovery that magnetism induced a current into a secondary coil whenever there was a *change* in the magnetic flux led to his third discovery of electromagnetism, self-inductance.

This discovery of self-inductance came about through a "problem" that a young man named William Jenkins posed to him. Jenkins was puzzled by a strange occurrence in the simple circuit of Fig. 19-1. A spark appeared between the main switch contacts whenever he opened the switch without the coil, but why does a much larger spark appear when a coil is connected into the circuit?

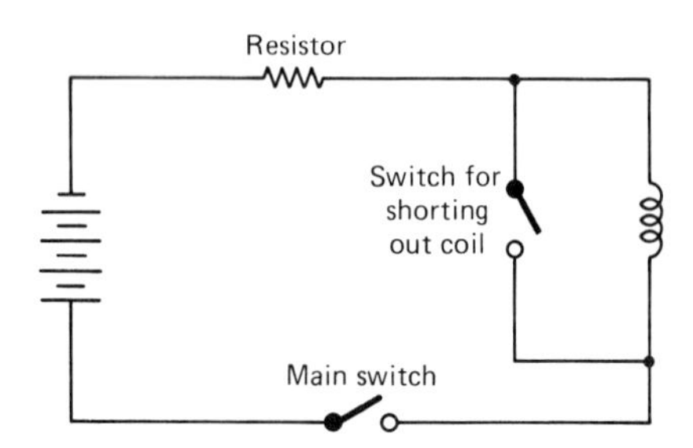

FIGURE 19-1 A much larger spark occurs between the switch contacts if the coil is in the circuit upon opening the main switch.

Faraday soon had the answer. In his original primary and secondary coil experiment the magnetic flux also cuts across the primary itself. After all, the primary coil is much closer than the secondary coil to its own magnetic field. This would create *self-inductance*. The falling magnetic field cuts through the turns on the coil and induces a much larger voltage than the source voltage to cause the extra large spark across the main switch contacts.

> The self-induced voltage always opposes any change producing it.

If the applied voltage is rising, the self-induced voltage opposes its rise. Similarly, if the applied voltage is falling, the self-induced voltage opposes its fall. This opposition to the rise and fall of the applied voltage is the "choke" action of an inductance or coil.

Unknown to Faraday, Joseph Henry, an American physicist, had already developed the concepts of inductance. For his work the unit of inductance was named the henry.

In this chapter we examine the characteristics of the inductor or choke (i.e., self-inductance as well as mutual inductance) and their applications to electronics.

MAIN TOPICS IN CHAPTER 19

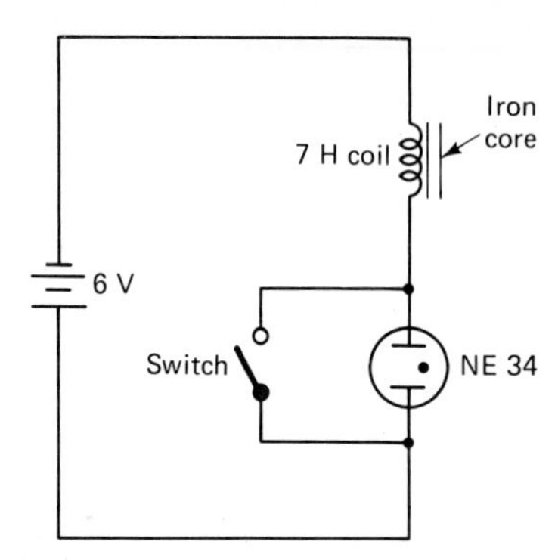

FIGURE 19-2 Circuit to demonstrate self-inductance.

OBJECTIVES

After studying Chapter 19, the student should be able to

1. Master the concepts of self-inductance and mutual inductance.
2. Solve for total inductance of inductors connected in series and parallel.
3. Have an understanding of stray inductance and shielding electromagnetic fields.

19-1 SELF-INDUCTANCE

Faraday's explanation of self-inductance can be demonstrated with the circuit shown in Fig. 19-2. The schematic diagram of Fig. 19-2 shows a neon lamp connected in series with a 6-V dc source and inductance. The NE 34 is a neon-gas-filled bulb with two electrodes. A voltage of at least 67 V must be connected across the two electrodes before the lamp ignites. The higher voltage is needed to "strip" the neon gas atoms of electrons. The electrons dislodge other electrons. Energy is given up by the atoms and the gas glows. Notice that the battery voltage is 6 V, which immediately suggests that the neon lamp will never glimmer. But wait! Close the switch and then open it. The lamp flashes! Why?

The neon lamp acts like an open circuit; therefore, we must close the switch to have a current and magnetic field build up in the coil. When the switch is opened, the collapsing magnetic field induces a voltage within the coil which is sufficiently great (67 V) to ignite the neon gas and make it glow for an instant.

Every automobile's ignition system uses the principle of a coil's self-inductance to fire the spark plugs. Although the battery voltage is only 12 V, the secondary coil gives a voltage of up to 25 kV. Figure 19-3 shows a schematic diagram of a basic auto ignition system.

The ignition coil is actually a step-up autotransformer. The primary has a few hundred turns of heavy wire capable of carrying several amperes. The secondary has a large number of turns of fine wire.

With the ignition switch ON and the engine being cranked by the starting motor, the cam rotates. This alternately opens and closes the primary circuit. Current flows into the primary coil when the switch is closed and builds up a magnetic field. Then the cam opens the switch and the magnetic field collapses. The collapsing magnetic field induces a voltage into the primary. By

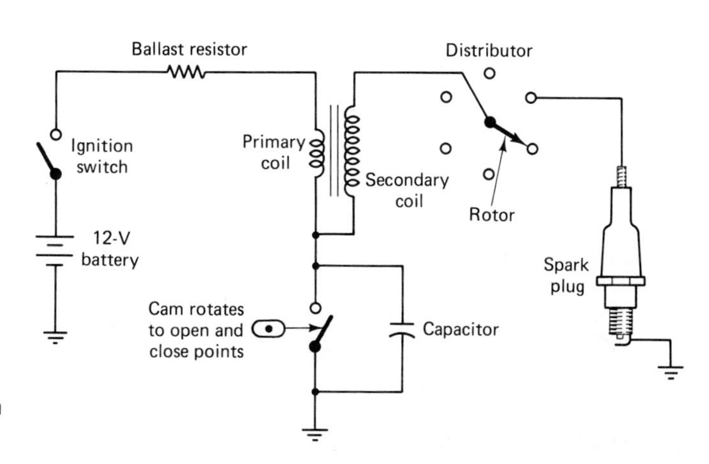

FIGURE 19-3 Basic auto ignition system.

mutual inductance or coupling the secondary self-induced voltage is stepped up by the very large number of turns on the secondary coil. The high voltage from the secondary is coupled through the distributer to each spark plug in turn by the rotor switch.

> Self-inductance is the ability of a conductor or coil to induce a voltage within itself with a current change.

In both example circuits the current was interrupted and fell to zero. When the current dropped to zero the magnetic field collapsed and induced a voltage within the coil. The induced voltage and current will be in such a direction as to oppose the action producing it. In these cases the induced voltage and current will try to maintain the magnetic field. This is known as Lenz's law. Because this induced emf is in opposition to the applied voltage, it is termed *counter emf.*

In electronic systems the inductor or choke (*schematic* symbols shown in Fig. 19-4) is used in ac and pulsating dc circuits. An ac current produces four changes in the magnetic flux for every cycle, as shown in Fig. 19-5. These magnetic flux changes produce a counter emf that opposes the applied voltage. The amplitude of the counter emf induced in a choke is determined by the rate of current change and the choke's inductance.

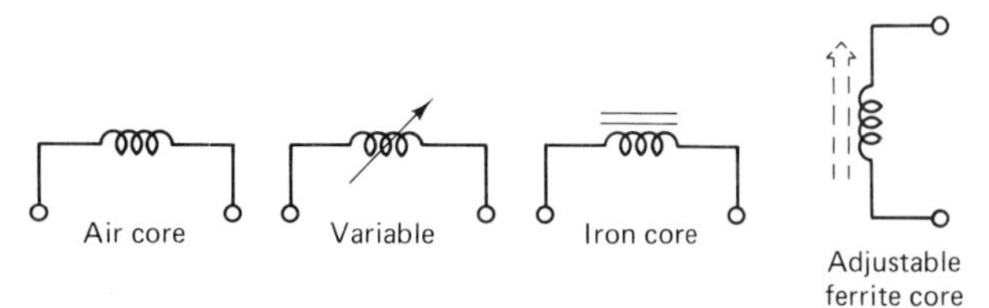

FIGURE 19-4 Schematic symbols for inductance.

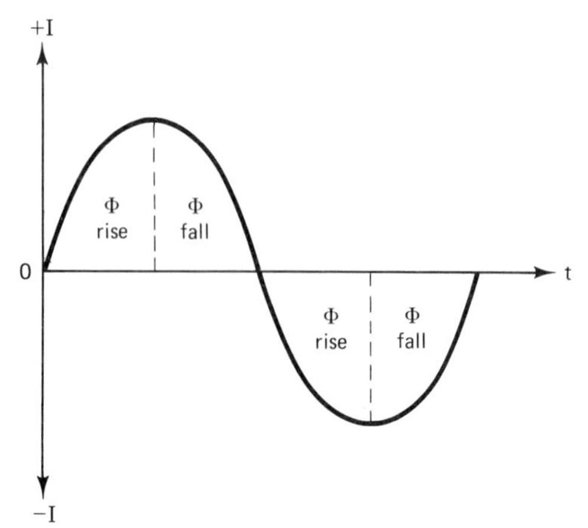

FIGURE 19-5 Ac current produces four changes in the magnetic flux for every cycle.

19-2 UNITS OF INDUCTANCE

The unit of inductance is the henry (H), named in honor or the American physicist Joseph Henry. A henry is defined as follows:

> A coil has a self-inductance of 1 henry when it generates 1 volt induced by a current change of 1 ampere per second.

In equation form

$$L = \frac{v_L}{\Delta i/\Delta t} \quad \text{henrys} \tag{19-1}$$

where L is in henrys
v_L is the induced voltage, in volts
Δi is the change in current, in amperes
Δt is the change in time, in seconds

Submultiples of the Henry

As in the case of current and voltage, the prefix for the henry is identical:

$$\text{millihenry} = \text{mH} = 10^{-3}\ \text{H}$$

$$\text{microhenry} = \mu\text{H} = 10^{-6}\ \text{H}$$

Although not commonly encountered, the nanohenry (10^{-9} H) and the picohenry (10^{-12} H) are applicable.

EXAMPLE 19-1

What is the inductance of a coil if a current change at the rate of 35 mA per millisecond produces a counter emf of 70 V?

Solution: Using equation (19-1) yields

$$L = \frac{v_L}{\Delta i/\Delta t} = v_L \times \frac{\Delta t}{\Delta i} = 70\ \text{V} \times \frac{1 \times 10^{-3}\ \text{s}}{35 \times 10^{-3}\ \text{A}}$$
$$= 2\text{H}$$

EXAMPLE 19-2

A coil has an inductance of 1 H. At what rate must the current change in order to induce a counter emf of 1 V?

Solution: Dividing both sides of equation (19-1) by v_L and rearranging the equation gives

$$\frac{\Delta i}{\Delta t} = \frac{v_L}{L} = \frac{1.0\ \text{V}}{1\ \text{H}} = 1.0\ \text{A/s}$$

EXAMPLE 19-3

A current of 75 ma flows through a 7-H coil connected to a 60-Hz ac source. Find the counter emf produced by the coil.

Solution: The current in the 60 Hz ac circuit will change four times every cycle (see Fig. 19-5). The time period, Δt, for the current will be

$$\Delta t = \frac{1/F}{4} = \frac{1/60}{4} = 4.17 \text{ ms}$$

Cross-multiplying equation (19-1) gives

$$v_L = L \times \frac{\Delta i}{\Delta t} = 7 \text{ H} \times \frac{75 \times 10^{-3} \text{ A}}{4.17 \times 10^{-3} \text{ s}} = 126 \text{ V}$$

Note that the 126 V is the counter emf voltage that opposes the applied voltage.

19-3 FACTORS AFFECTING INDUCTANCE

Any conductor can have some inductance as long as there is a changing current through it. All inductors are constructed by winding a length of insulated conductor around a core. This core is usually of some magnetic material such as powdered iron core, laminated iron core, or simply a plastic coil former on which the wire is wound. In this case the inductor has an air core. The physical construction of both the core and the windings about the core determines the inductance. These physical factors are as follows.

Number of Turns

For a given change in current, increasing the turns on a coil increases the induced emf since each turn adds to the overall induced emf. Thus the inductance of a coil increases directly with an increase in induced voltage, as equation (19-1) shows.

Increasing the number of turns, however, affects the magnetic circuit so that inductance will increase in proportion to the turns squared (N^2). Doubling the number of turns will increase inductance by four times.

Reluctance of the Magnetic Circuit

The *cross-sectional area,* the length, and the *core permeability* govern the amount of flux linking each turn and therefore the amount of induced emf in the turns. With a lower reluctance the inductance will be greater. Conversely, a higher reluctance will lower the inductance. Inductance is inversely proportional to the reluctance of the magnetic circuit. In equation form,

$$L = \frac{N^2}{\mathcal{R}} \quad \text{henrys} \qquad (19\text{-}2)$$

By substituting formula (15-6) for reluctance $\mathcal{R} = l/\mu A$ as given in Chapter 15, we obtain

$$L = \frac{N^2}{l/\mu A} = \frac{N^2 \mu A}{l} \quad \text{henrys} \qquad (19\text{-}3)$$

where N^2 is the number of turns squared
μ is the absolute permeability
A is the area in square metres
l is in metres

Since μ is the absolute permeability of the core material, we must use equation (15-4) or $\mu = \mu_r \times 1.26 \times 10^{-6}$. Substituting equation (15-4) for μ in equation (19-3) gives

$$L = \frac{N^2 A}{l} \times \mu_r \times 1.26 \times 10^{-6} \quad \text{henrys} \quad (19\text{-}4)$$

Equation (19-4) shows that the inductance will be increased directly by the μ factor of the core and the cross-sectional area of the coil. For an air core, μ_r is 1. With a ferrite or iron core the magnetic field is concentrated within the core; therefore, more flux lines cut through the turns to induce a greater voltage and increase inductance.

EXAMPLE 19-4

A 100-turn coil with an air core is 10 cm long. Its cross-sectional area is 7 cm^2. Calculate its inductance.

Solution: For the air core the relative permeability, $\mu_r = 1$, $A = 7 \times 10^{-4}$ m^2, and $l = 0.1$ m. Using equation (19-4) gives

$$L = \frac{N^2 A}{l} \times \mu_r \times 1.26 \times 10^{-6} \text{ H} \qquad (19\text{-}4)$$

$$= \frac{100^2 \times 7 \times 10^{-4} \text{ m}^2}{0.1 \text{ m}} \times 1 \times 1.26 \times 10^{-6}$$

$$= 88.2\ \mu\text{H}$$

EXAMPLE 19-5

The coil in Example 19-4 has a soft-iron core with a relative permeability of 100. Find its inductance.

Solution: Since L is directly proportional to μ_r, the inductance will be 100 times greater, or

$$L = 100 \times 88.2\ \mu\text{H} = 8.8 \text{ mH}$$

For the same turns and diameter, the inductance will decrease with the length of the coil, as shown in Example 19-6.

EXAMPLE 19-6

Find the inductance of the coil in Example 19-4, if the length of the coil is increased to 20 cm.

Solution: Using equation (19-4) gives us

$$L = \frac{100^2 \times 7 \times 10^{-4}}{0.2 \text{ m}} \times 1 \times 1.26 \times 10^{-6}$$

$$= 44.1 \ \mu\text{H}$$

PRACTICE QUESTIONS 19-1

Choose the most suitable word, phrase, or value to complete each statement.

1. A single conductor can have inductance provided that the current value is ________.
2. The inductance of a single conductor can be increased by forming the wire into a ________.
3. Self-inductance is the ability of a coil to ________ a voltage within itself with a changing current.
4. The self-induced voltage produced by inductance opposes the applied voltage and is termed ________.
5. A coil has a self-inductance of 1 ________ when it generates 1 V induced by a current change of 1 A/s.
6. A coil has an inductance of 300 mH. At what rate must the current change in order to induce a voltage of 70 V?
7. For tightly wound coils the inductance will be directly proportional to the ________.
8. Reducing the turns on a coil by one-half will (*increase, decrease*) the inductance by ________.
9. A certain air-core coil has an inductance of 15 μH. If the cross-sectional area is doubled with the same number of turns, what is its inductance?
10. The coil of Question 9, now has a soft-iron core that has a μ_r factor of 100. What is the inductance of the coil?
11. The length of the same coil of Question 10 has now been doubled. What is its new inductance?

19-4 INDUCTANCE IN SERIES

With no flux linkage between the two 1-H coils of Fig. 19-6 and a current change of 1 A/s, each coil will have a counter emf of 1 V. The total counter emf will be the simple sum of the counter emfs or

$$v_{LT} = v_{L1} + v_{L2} \quad \text{etc.}$$

$$v_{LT} = 1 \text{ V} + 1 \text{ V} = 2 \text{ V} = \text{V}_A$$

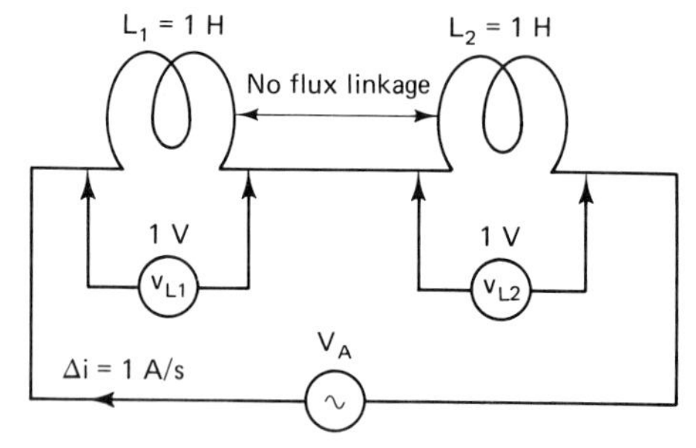

FIGURE 19-6 Two inductors connected in series v_{L1} and v_{L2} are additive; therefore, inductance is additive.

Using equation (19-1) shows that the total inductance of the circuit will equal the simple sum of the inductances.

$$L_T = \frac{v_{LT}}{\Delta i/\Delta t} = \frac{v_{L1} + v_{L2}}{1 \text{ A/s}} = \frac{2 \text{ V}}{1 \text{ A/s}} = 2 \text{ H}$$

For two or more inductors connected in series the total inductance can be found as follows:

$$L_T = L_1 + L_2 + L_3 \quad \text{etc.} \qquad \text{henrys} \qquad (19\text{-}5)$$

EXAMPLE 19-7

Three inductors, 300 mH, 600 mH, and 2 H, are connected in series. Find the total inductance of the circuit.

Solution: Using equation (19-5) yields

$$L_T = L_1 + L_2 + L_3$$

$$= (300 \times 10^{-3} \text{ H}) + (600 \times 10^{-3} \text{ H}) + 2 \text{ H}$$

$$= 2.9 \text{ H}$$

19-5 INDUCTANCE IN PARALLEL

With no flux linkage between the two 1-H coils of Fig. 19-7 and a current change of 1 A/s, each coil will have a counter emf of 1 V. Since the coils are connected in parallel, the total counter emf will be equal to $v_{L1} = v_{L2}$ (i.e., the counter emfs are common). The total current change, however, will be 2 A/s. Using equation (19-1), the total inductance is found to be

$$L_T = \frac{v_L}{\Delta i/\Delta t} = \frac{1 \text{ V}}{2 \text{ A/s}} = \frac{1}{2} \text{H}$$

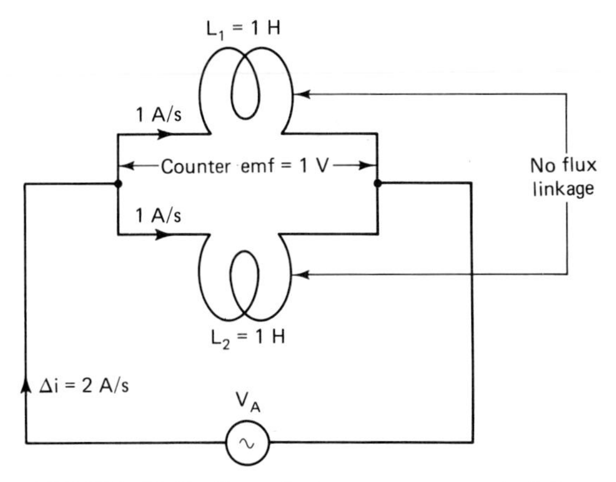

FIGURE 19-7 Two inductors connected in parallel v_{L1} and v_{L2} are common.

Or rearranging equation (19-1) in terms of the current gives

$$\frac{\Delta i}{\Delta t} = \frac{v_L}{L_T}$$

Since current is the simple sum in a parallel circuit, we can substitute v_L/L to obtain,

$$\frac{v_{LT}}{L_T} = \frac{v_{L1}}{L_1} + \frac{v_{L2}}{L_2} \quad \text{etc.}$$

The counter emf is common in a parallel circuit; therefore,

$$\frac{1}{L_T} = \frac{1}{L_1} + \frac{1}{L_2} + \frac{1}{L_3} \quad \text{etc.} \qquad \text{henrys} \qquad (19\text{-}6)$$

Note that the total inductance of two or more parallel coils will be less than the smallest inductance. Or using the product-over-sum equation for not more than two inductances at one time,

$$L_T = \frac{L_1 \times L_2}{L_1 + L_2} \qquad \text{henrys} \qquad (19\text{-}7)$$

EXAMPLE 19-8

Three inductors, 300 mH, 600 mH, and 2 H, are connected in parallel. Find the total inductance of the circuit.

Solution: Convert values of millihenrys to henrys and use equation (19-6) to solve for total inductance:

$$\frac{1}{L_T} = \frac{1}{L_1} + \frac{1}{L_2} + \frac{1}{L_3} = \frac{1}{0.3 \text{ H}} + \frac{1}{0.6 \text{ H}} + \frac{1}{2\text{H}}$$

$$= \frac{1}{5.5}$$

Therefore,

$$L_T = 181.8 \text{ mH} \quad \text{or} \quad \text{approx. } 182 \text{ mH}$$

Using equation (19-7) gives us

$$L_T' = \frac{L_1 \times L_2}{L_1 + L_2}$$

$$= \frac{0.3 \text{ H} \times 0.6 \text{ H}}{0.3 \text{ H} + 0.6 \text{ H}} = \frac{0.18}{0.9} = 0.2 \text{ H}$$

$$L_T = \frac{L_T' \times L_3}{L_T' + L_3} = \frac{0.2 \text{ H} \times 2 \text{ H}}{0.2 \text{ H} + 2 \text{ H}} = \frac{0.4}{2.2} \cong 182 \text{ mH}$$

19-6 STRAY INDUCTANCE AND SHIELDING

As long as there is a changing current flowing through a conductor, the conductor will have some inductance. The amount of inductance is particularily dependent on the frequency of the current. Point-to-point wiring or even printed-circuit wiring carrying radio-frequency voltages and currents will have some *stray inductance*.

Unshielded test leads or other connecting leads are another source of stray inductance. An ordinary wire-wound resistor cannot be used in RF circuits and must be *noninductively wound*. The resistance wire is first wound in one direction, then doubled back and wound in the opposite direction, as shown in Fig. 19-8. The magnetic fields generated by the adjacent turns cancel each other; therefore, no counter emf is induced and the resistive coil is noninductive. This type of resistive winding is used in precision wire-wound resistors.

In all these cases the stray inductance is in the range of a few microhenrys, which is very small. However, at high frequencies, and up, stray inductance becomes significant. The symbol for stray inductance is shown in Fig. 19-9.

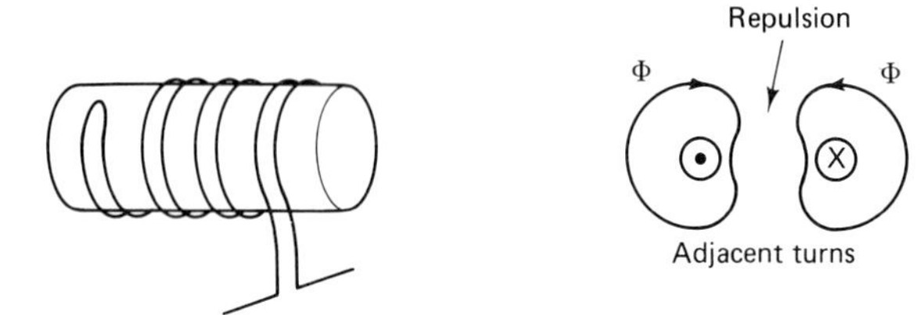

FIGURE 19-8 Noninductive wound coil. Adjacent turns carrying current in opposite direction will have their magnetic fields repelling each other.

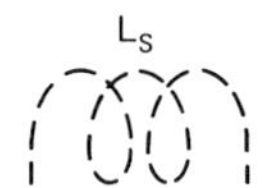

FIGURE 19-9 Symbol for stray inductance.

Electromagnetic Shielding

Some 60-Hz transformers have a copper metal cover or shield about the iron core, as well as a soft-steel enclosure about the windings. Stray magnetic lines from the core induce eddy currents in the copper shield. The energy of the stray magnetic flux is then dissipated in the form of heat produced by the eddy currents.

Inductors and RF transformers used in high-frequency circuits are shielded in a separate totally enclosed compartment. The compartments are either aluminum or copper. As in the case of the 60-Hz transformer, eddy currents induced in the aluminum or copper walls reduce the stray electromagnetic flux to zero and prevent coupling to adjacent circuit components. The distance between the walls and the inductor must be at least equal to its diameter to minimize loading effect.

19-7 TYPES OF CORES AND WINDINGS

As discussed in Section 19-3, the inductor is a length of insulated conductor wound on a core. There are three types of cores to consider, the air core, powdered iron or ferrite material, and the laminated iron core.

Air Core

A phenolic or plastic coil form on which the wire is wound is considered to be an air core, since the plastic is a nonmagnetic material. The relative permeability of air is 1; therefore, we were able to compare the increase in inductance with iron, powered iron, or a ferrite core.

Low-inductance chokes are used in radio-frequency (RF) circuits such as AM, FM, TV, and communication receiver and transmitter circuits. Generally, chokes below 10 μH are wound on a nonmagnetic bobbin and are "solenoid" wound, as shown in Fig. 19-10, either single layer or multilayer.

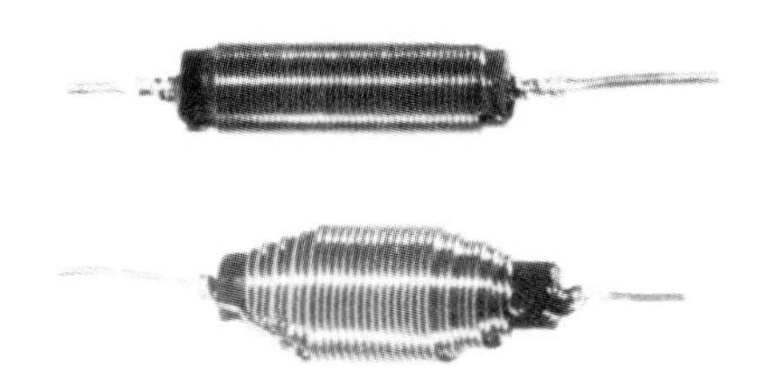

FIGURE 19-10 (a) Single-layer and (b) multilayer windings on a plastic coil former. (Courtesy of Hammond Manufacturing Co. Ltd.)

Powdered Iron and Ferrite Core

Powdered iron or ferrite material is mixed with a binding agent and molded into a threaded core which screws into the insulated air core. Inductance of the coil is increased or decreased by adjusting the threaded core by means of an insulated screwdriver or hexdriver. The inductance of a choke can be increased to a great extent by winding the coil on a ferrite or powdered iron bobbin. Powdered iron bobbins or cores are usually used for chokes between the range 5 to 100 μH. Particularly ferrite, but also powdered iron, is used for chokes higher than 200 μH.

The powdered iron and ferrite core will have some eddy current and hysteresis losses. These losses dissipate the inductors energy in the form of heat; therefore, less energy is available for circuit operation. The higher the frequency, the greater will be these losses.

RF chokes are often "pie-wound," as shown in Fig. 19-11. This is a type of winding where the wire is zigzagged around the circumference of the bobbin and built up in many layers. The purpose of this type of winding is to reduce the total stray capacitance between the turns. This method of winding is also referred to as "universal" winding. Pie-wound RF chokes may have one, two, or more pies connected in series to make up an inductance.

FIGURE 19-11 Pie-wound choke. (Courtesy of Hammond Manufacturing Co. Ltd.)

Laminated Iron Core

This type of core is made of iron or silicon-steel alloy strips. After a varnish process, the laminated strips of the steel are stacked and clamped to form the core on which the wire is wound, as shown in Fig. 19-12.

FIGURE 19-12 Laminated iron core. (Courtesy of Hammond Manufacturing Co. Ltd.)

The purpose of the laminations is to reduce the losses due to eddy currents and hysteresis. Eddy currents are tiny electrical currents set up in the iron core due to the changing magnetic flux field cutting through the iron core. As outlined in Section 15-7, hysteresis loss is due to the constant magnetization and demagnetization of the magnetic domains in the iron core. In both cases heat is generated within the iron core. This heat energy must be made up by the source.

Silicon steel or iron core chokes are used for high-inductance and low-frequency applications such as 120-Hz full-wave rectifier filtering.

Series Resistance of a Choke

Since the choke is a length of conductor wound around a core, the wire will have some dc resistance. This dc resistance can be measured using an ohmmeter. The dc resistance of the choke may need to be taken into account in a circuit, particularly where a high-inductance choke is used. High-value inductors can have a dc resistance anywhere from several hundred ohms to several thousand ohms.

The equivalent series resistance of a choke is made up of the actual dc resistance of the winding plus the RF resistance of the wire due to the "skin effect" of RF current flow. This topic is beyond the scope of this book.

PRACTICE QUESTIONS 19-2

Choose the most suitable word, phrase, or value to complete each statement.

1. Given that two coils are connected in series as shown in Fig. 19-13, L_1 produces a counter emf of 10 V and L_2 produces a counter emf of 20 V. Which coil has the greatest inductance, and by how much?

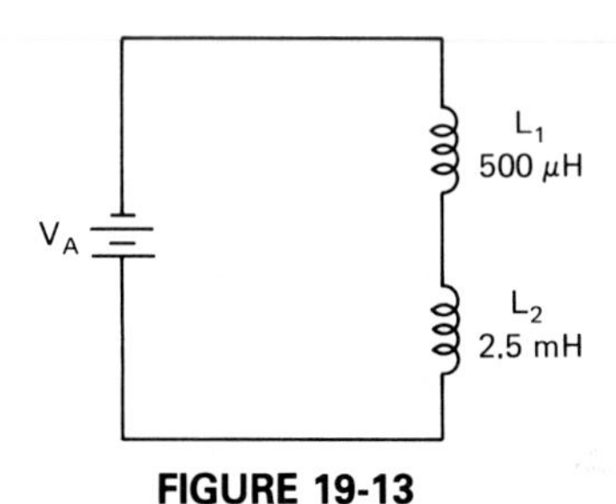

FIGURE 19-13

2. Find the total inductance in the circuit of Fig. 19-13.
3. What is the total inductance of the circuit if the inductors are connected to 30 V ac?
4. The two inductors of Question 3 are connected in parallel. Find the total inductance of the circuit.
5. Any conductor that has an ac current flowing through it will have ________ inductance.
6. Precision wire-wound resistors are ________ wound to minimize stray inductance.
7. Inductors and RF transformers used in high-frequency circuits are shielded in separate totally enclosed ________ compartments.
8. Shielding a choke or RF transformer prevents the electromagnetic flux from ________ to adjacent circuit components.
9. A phenolic or plastic coil form used for an inductor is considered to be an ________ core inductor.
10. Much greater inductance values can be obtained by using a ________ or ________ core.
11. A zigzag type of winding built up in many layers is called a ________ choke.
12. The purpose of laminated strips of steel used as a core for inductors or transformers is to reduce losses due to ________ and ________.
13. The dc resistance of a choke can be found by using an ________.
14. High-value inductors can have a dc resistance in the range ________ to ________ ohms.

19-8 MUTUAL INDUCTANCE

Faraday's two coils wound on a cardboard tube had one thing in common, a changing electromagnetic field. Since each coil was wound on either end of the cardboard tube, the changing primary current through L_1 produced a common changing flux that cut through both coils, as shown in Fig. 19-14. The two coils are said to be *magnetically linked* or *coupled*. Flux linkage or coupling takes place when the proximity of one coil links the magnetic flux of another coil.

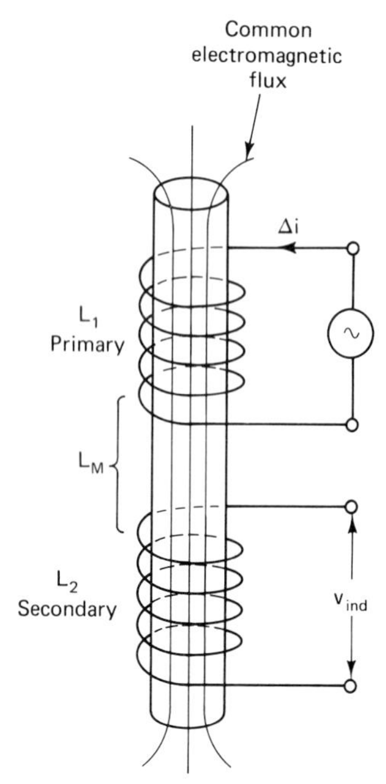

FIGURE 19-14 Mutual inductance between two coils that are electromagnetically coupled.

As a result of the changing electromagnetic field, a voltage is induced into the second coil L_2 by mutual inductance. Mutual inductance (L_M) is a measure of how much voltage is induced in the secondary coil L_2 due to the changing current in the primary coil L_1. *Two coils have an L_M of 1 henry when a current change of 1 ampere per second in one coil induces 1 volt in the second coil.* In equation form,

$$L_M = \frac{v_L \text{ secondary}}{\Delta i/\Delta t \text{ primary}} \quad \text{henrys} \qquad (19\text{-}8)$$

There are three factors that govern the mutual inductance of the two coils: the coefficient of coupling, the inductance of L_1, and the inductance of L_2.

Coefficient of Coupling

The amount of magnetic flux lines that couple the secondary coil vary depending on the core material and the proximity of the secondary coil to the primary coil, particularly for air-core coils.

If L_1 and L_2 are wound around a ferromagnetic core, then effectively 100% of the magnetic flux will couple the secondary. Since magnetic lines repel, an air core will cause the flux lines to disperse, so that magnetic coupling is drastically reduced. The coils may be classified as *lightly coupled* or *loosely coupled*, depending on how much of the primary flux links the secondary coil.

The ratio of the flux linking both coils to the total flux produced by the primary coil is defined as the *coefficient of coupling, k.*

$$k = \frac{\text{flux linking primary and secondary coil}}{\text{total flux produced by primary coil}}$$

The coefficient k will depend on the core material and the physical proximity of the coils. With a ferromagnetic core, k is 1 or unity. Practically all of the magnetic flux lines produced by the primary coil will couple the secondary coil.

Moving the coils closer together on an air core as shown in Fig. 19-15(a), or winding the secondary coil over the primary coil, termed *bifilar winding* [see Fig. 19-15(b)], will increase the flux linkage and therefore increase k.

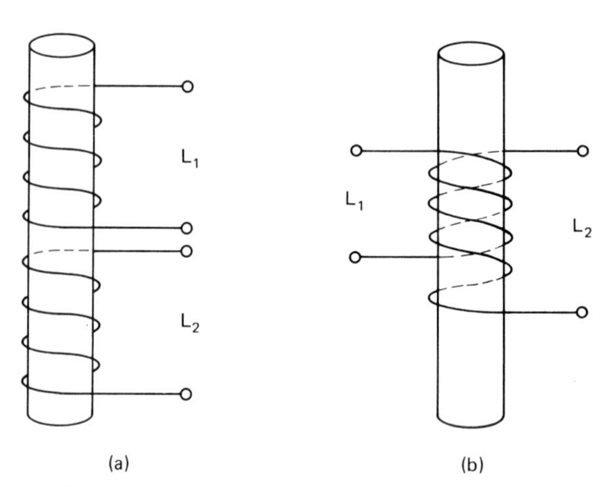

FIGURE 19-15 Physical placement of secondary coil to the primary coil will increase or decrease k. (a) Proximity winding for loose coupling. (b) Bifilar winding for tight coupling.

Inductance of L_1 and L_2

Since both coils share the same magnetic flux, the induced volts per turn in the primary coil L_1 will be the equal to the induced volts per turn in the secondary coil L_2.

With higher values of L_1 and L_2 the induced emf will be greater. In equation form the mutual inductance of two coils on a common core is

$$L_M = k\sqrt{L_1L_2} \quad \text{henrys} \qquad (19\text{-}9)$$

EXAMPLE 19-9

Two coils are wound on a common air core. The primary coil L_1 produces a flux field of 60 μWb. The secondary coil L_2 is linked by 9 μWb. What is its coefficient of coupling, k?

Solution

$$k = \frac{\text{flux linking primary to secondary}}{\text{total flux produced by primary}}$$

$$= \frac{9\ \mu\text{Wb}}{60\ \mu\text{Wb}} = 0.15$$

Since k is a ratio, it has no unit.

EXAMPLE 19-10

Calculate the mutual inductance of the two coils in Example 19-9 if L_1 has an inductance of 12 μH and L_2 has 15 μH.

Solution: Using equation (19-9), we have

$$L_m = k\sqrt{L_1 \times L_2}$$

$$= 0.15\sqrt{(12\ \mu\text{H}) \times (15\ \mu\text{H})} = 2.0\ \mu\text{H}$$

EXAMPLE 19-11

The core of the two coils of Examples 19-9 and 19-10 is changed to ferrite. What is the mutual inductance of the two coils?

Solution: With a ferromagnetic core the coefficient of coupling is approximately 1. Using equation (19-9) yields

$$L_m = k\sqrt{L_1 \times L_2} = 1\sqrt{(12\ \mu\text{H}) \times (15\ \mu\text{H})}$$

$$= 13.4\ \mu\text{H}$$

19-9 CHECKING AN INDUCTOR

An inductor is one of the components that rarely causes a problem in a circuit. It can be checked easily and quickly for continuity by using an ohmmeter. Inductor resistance values range from a fraction of an ohm to hundreds of ohms.

If the resistance of a choke is given on a schematic diagram, it becomes a simple step to set the ohmmeter on the appropriate range to check this value. For unknown values of an inductor's resistance, the ohmmeter range must be stepped up or down in order to measure its resistance. With some experience a technician is able to judge the credibility of the ohmmeter reading by the appearance of the wire size and coil size. In all cases a coil winding will give a continuity reading with some finite resistance value on the ohmmeter.

A coil that has been carrying an overload current can easily be seen. It will be discolored and/or charred. Of course, the source of the overload current must be corrected before a new coil is replaced.

In the case of a terminal lead breaking off, an open circuit results and the ohmmeter reading will be infinite (∞) ohms. This type of fault is rare and probably due to "poking" or excessive vibration. The actual inductance of a choke must be determined using a commercial Wheatstone bridge.

SUMMARY

Self-inductance is the ability of a conductor or coil to induce a voltage within itself with a current change.

The induced voltage is in apposition to the applied voltage and is termed counter emf.

A coil has a self-inductance of 1 henry when it generates 1 volt induced by a current change of 1 ampere per second. In equation form

$$L = \frac{v_L}{\Delta i/\Delta t} \quad \text{henrys} \qquad (19\text{-}1)$$

Factors that affect inductance are:

1. The number of turns
2. The reluctance of the magnetic circuit

Given the number of turns on a coil and the reluctance of the magnetic circuit, the inductance can be found using equation (19-3).

$$L = \frac{N^2 \mu A}{l} \quad \text{henrys} \qquad (19\text{-}3)$$

If the core of the magnetic circuit is other than air that has a relative permeability greater than 1, the inductance of a coil is found using equation (19-4).

$$L = \frac{N^2 A}{l} \times \mu_r \times 1.26 \times 10^{-6} \quad \text{henrys} \qquad (19\text{-}4)$$

For inductors connected in series, the total inductance is found using equation (19-5).

$$L_T = L_1 + L_2 + L_3 \quad \text{etc.} \quad \text{henrys} \qquad (19\text{-}5)$$

For inductors connected in parallel, the total inductance is found using equation (19-6).

$$\frac{1}{L_T} = \frac{1}{L_1} + \frac{1}{L_2} + \frac{1}{L_3} \quad \text{etc.} \quad \text{henrys} \qquad (19\text{-}6)$$

Or using the product-over-sum equation (19-7) for not more than two inductors connected in parallel, the total inductance is found:

$$L_T = \frac{L_1 \times L_2}{L_1 + L_2} \quad \text{henrys} \qquad (19\text{-}7)$$

Stray inductance is an undesired inductance between conductors that have a changing current flowing through them.

Stray inductance of precision wire-wound resistors is minimized by winding the resistive wire in one direction, then winding it in the opposite direction.

Inductors and RF transformers used in high-frequency circuits are shielded in a separate totally enclosed aluminum compartment to protect them from stray inductance due to their EM field.

A powdered iron or ferrite core increases the inductance of a choke many times.

A laminated core is made of iron or silicon-steel alloy strips that are insulated with varnish and then stacked to form the core for an inductance.

The purpose of laminations is to reduce the losses due to eddy currents and hysteresis.

The conductor wound on the core of a choke will have some dc resistance that can be measured using an ohmmeter.

Two coils have a mutual inductance, L_M, of 1 henry when a current change of 1 ampere per second in one coil induces 1 volt in the second coil.

The ratio of the flux linking both coils to the total flux produced by the primary coil is defined as the coefficient of coupling k.

The mutual inductance of two coils with a common flux is found using equation (19-9).

$$L_M = k\sqrt{L_1 L_2} \qquad \text{henry} \qquad (19\text{-}9)$$

An inductor can be checked for continuity or an open circuit by using an ohmmeter. Inductor resistance values range from a fraction of an ohm to hundreds of ohms.

REVIEW QUESTIONS

19-1. Explain briefly why the secondary of an ignition coil gives a voltage up to 25 kV when the battery voltage is only 12 V.

19-2. Define self-inductance.

19-3. What does Lenz's law tell you about the self-induced voltage of an inductor?

19-4. Define the henry.

19-5. How much voltage is induced across an inductor of 7 H when the current changes at the rate of 2 A/s?

19-6. Explain briefly how the number of turns on a coil affects its inductance.

19-7. Explain briefly how the reluctance of the magnetic circuit affects the inductance of a choke.

19-8. A 50-turn coil with an air core is 10 cm long. Its cross-sectional area is 7 cm^2. Calculate its inductance.

19-9. Calculate the inductance of the same coil in Question 19-8 with an iron core that has a permeability of 200.

19-10. Three inductors, 30 μH, 300 mH, and 150 μH, are connected in series. Find the total inductance of the circuit.

19-11. The three inductors of Question 19-10 are connected in parallel. Find the total inductance of the circuit.

19-12. Explain briefly what is meant by stray inductance. List three examples of stray inductance.

19-13. Why does an aluminum or copper metal compartment shield adjacent components from the stray electromagnetic field of an RF transformer?

19-14. Besides an air core, what other types of cores do chokes have?

19-15. What will determine the dc resistance of a choke?

19-16. Define mutual inductance.

19-17. Give two methods of increasing the k of two coils.

19-18. Calculate the mutual inductance of two coils wound on a ferrite core having 3 mH and 6 mH inductance, respectively.

19-19. Explain how an inductor can be checked using an ohmmeter. What resistance range could be expected depending on the choke?

SELF-EXAMINATION

Choose the most suitable word, phrase, or value to complete each statement.

19-1. The ability of a conductor or coil to induce a voltage within itself with a current change is termed ________.

19-2. The induced voltage always opposes the applied voltage and is termed __________.

19-3. A coil has a self-inductance of 1 __________, when it generates 1 V induced by a current change of 1 A/s.

19-4. What is the inductance of a coil if a current change at the rate of 50 mA per millisecond produces a counter emf of 200 V?

19-5. Any conductor can have some inductance as long as there is a __________ current through it.

19-6. The inductance of a coil will increase in proportion to the turns __________.

19-7. Inductance will be inversely proportional to the __________ of the magnetic circuit.

19-8. Calculate the number of turns of a 100 μH coil wound on a plastic core with a cross-sectional area of 10×10^{-5} m^2 and a length of 5 cm.

19-9. A certain air-core coil has an inductance of 9 mH. If the cross-sectional area is halved with the same amount of turns, what is its inductance?

19-10. The coil of Question 19-9 now has a soft-iron core that has a μ_r factor of 50. What is the inductance of the coil?

19-11. The length of the same coil of Question 19-9 has now been halved. What is its new inductance?

19-12. Calculate the total inductance of the circuit schematic shown in Fig. 19-16.

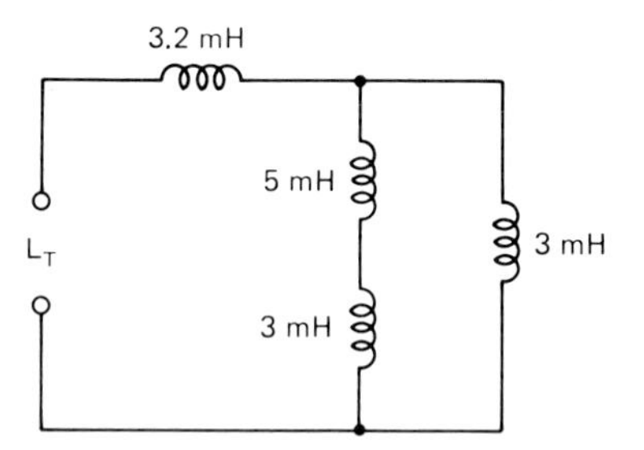

FIGURE 19-16 Schematic for Self-Examination Question 19-12.

19-13. Precision wire-wound resistors are __________ wound to minimize stray inductance.

19-14. Any conductor will have significant stray inductance at __________ frequencies and up.

19-15. Shielding an RF coil in an __________ compartment eliminates stray electromagnetic flux from coupling to adjacent circuit components.

19-16. Generally chokes above 200 μH have a __________ core.

19-17. Silicon-steel core chokes are used for (*high, low*) inductance and (*high, low*)-frequency applications.

19-18. An ohmmeter measures the __________ resistance of the winding on a choke.

19-19. To reduce losses due to hysteresis and eddy currents, a choke's steel core is __________.

19-20. Flux __________ or __________ takes place when the proximity of one coil links the magnetic flux of another coil.

19-21. Mutual inductance is a measure of how much voltage is induced in the __________ coil due to the changing current in the primary coil.

19-22. The ratio of the flux linking both coils to the total flux produced by the primary coil is defined as the ________ of ________.

19-23. For an iron and ferrite core, k will be ________.

19-24. Two coils, 20 μH and 30 μH, are wound on a ferrite core. Calculate their mutual inductance.

ANSWERS TO PRACTICE QUESTIONS

19-1

1. changing
2. coil
3. induce
4. counter emf
5. henry
6. 233 A/s
7. N^2
8. decrease, $\frac{1}{4}$
9. 30 μH
10. 1.5 mH
11. 0.75 mH

19-2

1. L_2, twice
2. zero
3. 3 mH
4. 0.42 mH
5. stray
6. noninductive
7. aluminum or copper
8. coupling
9. air
10. powdered iron, ferrite
11. pie-wound
12. eddy currents, hysteresis
13. ohmmeter
14. 200 Ω, 2 kΩ

20

Transformers

INTRODUCTION

Faraday's two coils wound on a cardboard tube constituted an air-core transformer. This device, discussed in Chapter 18, is able to transfer energy from the primary to the secondary. The changing current in the primary produces a changing flux that induces a voltage into the secondary by mutual inductance. Tesla designed the alternators and the transformers for the first ac power generation and transmission system installed by Westinghouse at Niagara Falls.

Transformers can be grouped into three main applications. There are radio-frequency (RF) transformers, audio-frequency (AF) transformers, and power transformers. In the power transformer applications there are auto, isolation, step-up, step-down, tapped, and current transformers. Radio and audio-frequency transformers are mainly impedance-matching devices that transfer energy efficiently from one stage to another stage of the electronic system.

This chapter is concerned mainly with the ac power transformer and its applications. Later studies in radio- and audio-frequency circuits should involve you with those types of transformers.

MAIN TOPICS IN CHAPTER 20

OBJECTIVES

After studying Chapter 20, the student should be able to

1. Master the concept of transformer action and construction.

2. Apply voltage and turns-ratio principles to master the concept of isolation, step-up, step-down, auto, and current transformers.
3. Apply power calculations to show transformer loading and losses.

20-1 IRON-CORE TRANSFORMER

The power transformer consists of a primary and a secondary coil wound on a laminated silicon-iron alloy core as shown in the schematic of Fig. 20-1.

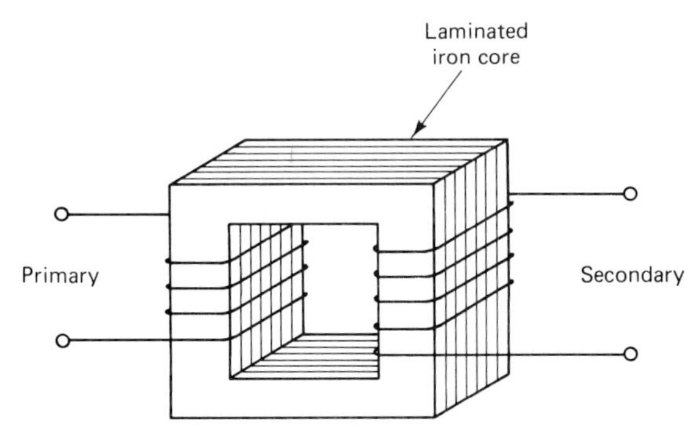

FIGURE 20-1 Laminated silicon-iron alloy core transformer.

Laminated iron cores are constructed in several types of configurations. The most common is the E and I, shown in Fig. 20-2(a). The E and I iron strips are stacked alternately, with the primary and secondary coils having been prewound on a coil former and placed over the center leg of the E, as shown in Fig. 20-2(b).

The L laminations shown in Fig. 20-3 are also alternately stacked to build the core. This type of core is often used in high-voltage transformers since this core shape is easily insulated.

The laminated silicon-iron alloy core provides a closed magnetic circuit so that the flux linkage between the primary and the secondary is, for practical purposes, unity or 100% ($k = 1$).

Connecting the primary to an ac power source causes a small no-load primary current called "excitation" current to flow and sets up a magnetic flux in the iron core. The alternating magnetic flux set up by the primary current links the secondary turns and induces a voltage in the secondary coil. The schematic symbol for the transformer is shown in Fig. 20-4.

Polarity of Secondary-to-Primary Voltage

The instantaneous polarity of the secondary voltage with respect to the primary instantaneous voltage will be de-

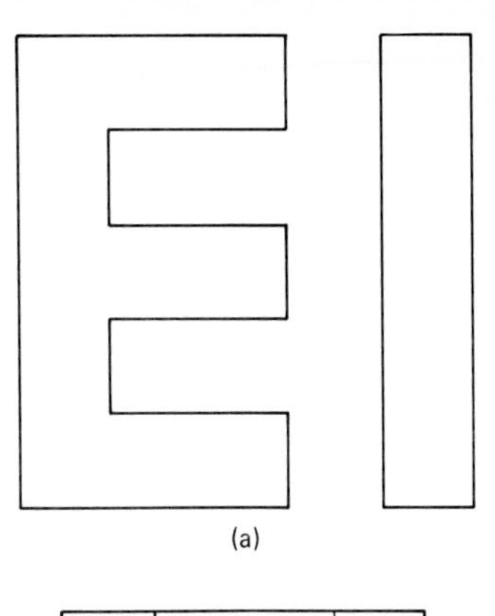

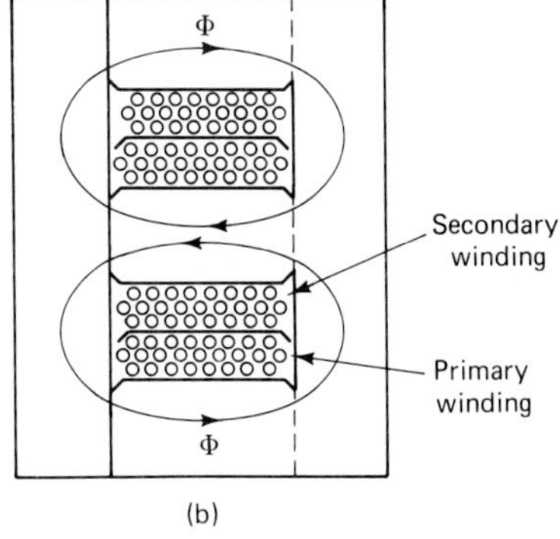

FIGURE 20-2 E and I laminations with primary and secondary coils mounted over center leg of E laminations. (a) Steel laminations. (b) Layered wound primary and secondary coils.

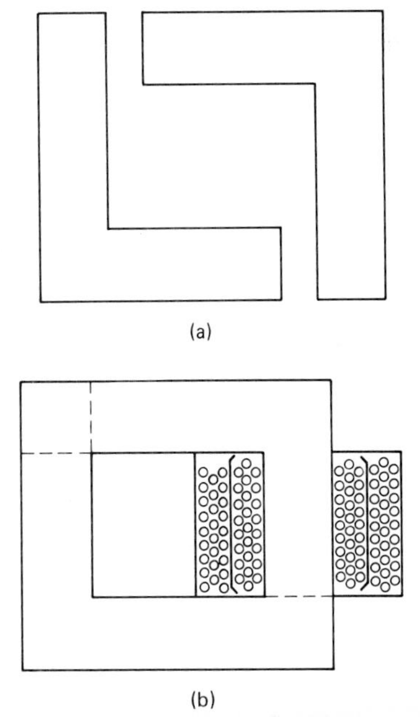

FIGURE 20-3 L laminations with primary and secondary coils mounted on one side of the core. (a) L laminations. (b) Layered wound primary and secondary coils.

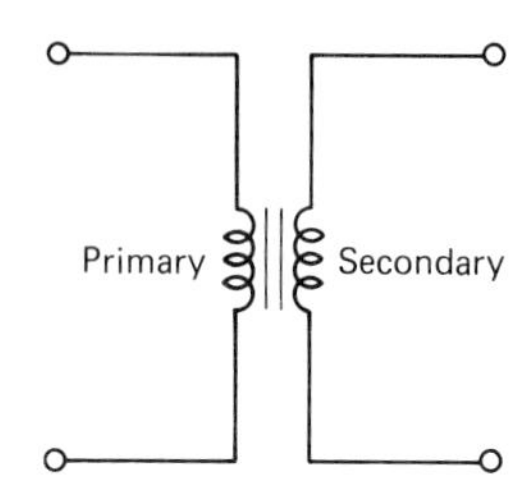

FIGURE 20-4 Schematic symbol for an iron-core transformer.

termined by the direction in which the secondary turns are wound. With the E and I types of core laminations, the primary and secondary are layer wound on a coil former that fits over the center leg of the E laminations as shown in Fig. 20-2(b). The magnetic flux therefore cuts through the turns of both coils in the same direction.

Given that the turns of both windings are in the same direction, the instantaneous voltage polarity of the secondary to primary will be in phase, as shown by the waveforms in Fig. 20-5(a).

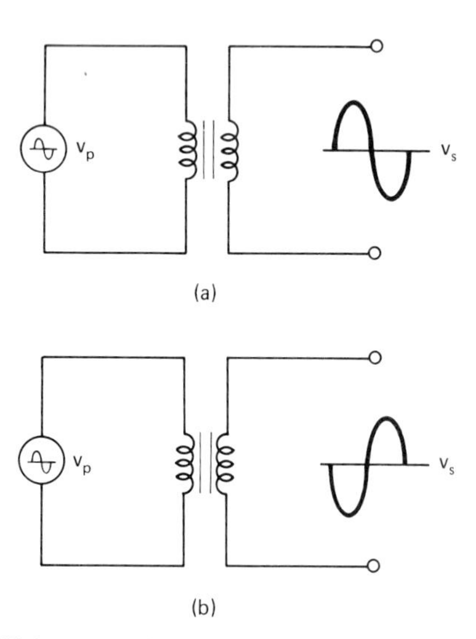

FIGURE 20-5 Voltage polarity of primary to secondary voltage is determined by the direction in which the secondary turns are wound. (a) Secondary instantaneous v_s is in phase with primary voltage v_p. (b) Secondary instantaneous voltage v_s is 180° out of phase with primary voltage v_p.

If the secondary turns are wound in reverse to the primary shown in Fig. 20-5(b), the secondary voltage will be 180° out of phase with the primary. Although not significant in the case of power transformers, this phase inversion is important to audio circuit applications.

Dot Convention. To aid theoretical explanation of ac-to-dc rectifier circuits, audio circuits, and others, a dot is often used to indicate the *instantaneous polarity* of the windings. The dot simply tells us that the polarity of the lead where the *dot appears will be positive,* as shown in Fig. 20-6.

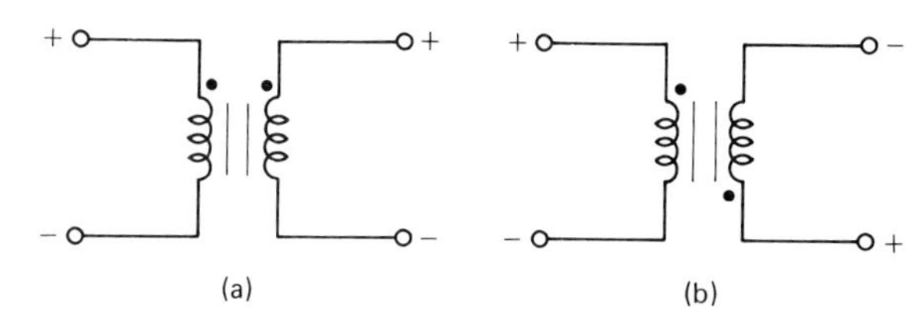

FIGURE 20-6 Dot convention to show polarity of primary and secondary instantaneous voltages. (a) In phase. (b) 180° out of phase.

20-2 TURNS RATIO AND VOLTAGE RATIO

Turns Ratio

A transformer may be used to either increase or decrease the applied voltage. This means that the number of turns on the secondary increases to increase the voltage and decreases to reduce the voltage. The ratio of the number of turns on the primary, N_p, to the number of turns on the secondary, N_s, is termed the turns ratio:

$$\text{turns ratio} = \frac{N_p}{N_s} \tag{20-1}$$

Voltage Ratio

With 100% flux coupling between the primary and secondary, the induced volts per turn in the primary will be equal to the induced volts per turn in the secondary. (*Note:* Ignore voltage loss due to the resistance of the copper wire.) Thus the voltage ratio will be in the same proportion as the turns ratio. This can be shown by using Faraday's law [equation (16-2)] for both primary and secondary induced voltages.

$$V_p = N_p \frac{\Delta\Phi}{\Delta t} \quad \text{volts}$$

$$V_s = N_s \frac{\Delta\Phi}{\Delta t} \quad \text{volts}$$

The two equations can be combined in ratio format using equation (20-1) or dividing the equations by each other:

$$\frac{V_P}{V_s} = \frac{N_P}{N_s} \times \frac{\Delta\Phi/\Delta t}{\Delta\Phi/\Delta t}$$

The change in flux over a change in time cancels:

$$\frac{V_P}{V_s} = \frac{N_P}{N_s} \qquad (20\text{-}2)$$

EXAMPLE 20-1

A power transformer has a 30-V secondary. There are 600 turns on the primary. Calculate the turns on the secondary, if the primary is 120 V.

Solution: Using equation (20-2), we know that the primary will be connected to the 120-V line; therefore,

$$\frac{V_P}{V_s} = \frac{N_P}{N_s} \qquad \frac{120\text{ V}}{30\text{ V}} = \frac{600\text{ turns}}{N_s}$$

Cross-multiply and solve for N_s:

$$N_s = \frac{600\text{ turns} \times 30\text{ V}}{120\text{ V}} = 150\text{ turns}$$

EXAMPLE 20-2

A power transformer has 600 turns on the primary and 300 turns on the secondary. Calculate the secondary voltage, if the primary is 120 V.

Solution: Using equation (20-2), we know that the primary voltage will be 120 V; therefore,

$$\frac{V_P}{V_s} = \frac{N_P}{N_s} \qquad \frac{120\text{ V}}{V_s} = \frac{600\text{ turns}}{300\text{ turns}}$$

Cross-multiply and solve for V_s:

$$V_s = \frac{120\text{ V} \times 300\text{ turns}}{600\text{ turns}} = 60\text{ V}$$

PRACTICE QUESTIONS 20-1

Choose the most suitable word, phrase, or value to complete each statement.

1. The power transformer consists of a ________ and a ________ coil wound on a laminated iron core.
2. Two common types of laminations are the ________ and the ________-shape core.
3. With a laminated iron core the primary-to-secondary flux coupling will be practically ________.
4. An ________ current flows through the primary to set up a magnetic flux in the core even though there is no load connected to the secondary.
5. The instantaneous voltage polarity of the secondary with respect to the primary voltage will be determined by the ________ of the secondary winding.
6. Usually, the secondary voltage is ________ degrees out of phase with the primary voltage.
7. The dot on a transformer schematic tells us that the instantaneous polarity of the lead where the dot is located will be ________ with respect to the primary lead.
8. A power transformer has a 240-V 2400-turn primary. The secondary voltage is 24 V. Calculate the turns on the secondary. Assume a 120-V input to the primary.
9. A power transformer has a secondary voltage of 12 V with 120 turns. Calculate the turns on the primary. Assume a 120-V input to the primary.
10. The turns ratio of a power transformer is 2:1. Calculate the secondary voltage. Assume a 120-V input to the primary.

20-3 ISOLATION TRANSFORMER

Any electronic equipment that plugs into the 120-V ac line must be provided with a ground-line cord. The chassis of the equipment is then properly grounded through the ground, as required by the National Electrical Code®. This eliminates shock hazard, discussed in Section 18-3.

Most electronic equipment has a power transformer that is the integral part of an ac-to-dc power rectifier. Since the primary is electrically insulated from the secondary, the ac power line is electrically isolated from the dc voltage and chassis. The transformer is termed an isolation transformer.

Eliminating the isolation transformer reduces the cost of the equipment. Isolation transformers are not installed in some lower-cost TV sets, but the sets are still shock-hazard-free. If the TV or equipment chassis is connected directly to the 120-V ac line, provision must be made by the manufacturer to insulate the chassis completely from possible shock hazard. This is done by enclosing the chassis in a cabinet made of plastic or other insulating material. No shock hazard exists as long as the chassis is in the cabinet.

Servicing the TV set or that particular equipment means that the test equipment, such as an oscilloscope ground, must be connected to the chassis ground of the equipment as shown in Fig. 20-7. No shock hazard exists as long as the line-cord cap connects the neutral wire of the ac line to the chassis.

Fig. 20-8 shows that the line-cord cap has been plugged into the wall receptacle in the reverse polarity. The hot side of the line wires is now connected to the chassis. Immediately, a shock hazard exists between the chassis and *any* test equipment. If the oscilloscope ground is connected to the TV or equipment chassis, a short circuit occurs and serious damage to the TV transistor circuits could be done, as well as possible eye injury from flying sparks.

To eliminate this shock hazard, the special external isolation transformer shown in Fig. 20-9(a) is used. Equipment to be isolated plugs into the receptacle on the transformer. The receptacle is connected to the secondary winding while the primary is connected to the line cord. Since the primary and secondary windings of the transformer are electrically insulated from each other, the ac power line is electrically isolated from the equipment, as shown in Fig. 20-9(b).

Turns and Voltage Ratio of an Isolation Transformer

The special isolation transformer serves only its one function; therefore, the secondary voltage is equal to the primary voltage. This means that the turns ratio will be 1:1. This is shown by equation (20-2):

$$\frac{V_p}{V_s} = \frac{N_p}{N_s} \qquad \frac{120\text{ V}}{120\text{ V}} = \frac{1}{1}$$

Note: The secondary current rating of the isolation transformer must not be exceeded or it will overheat and burn out.

20-4 STEP-UP TRANSFORMER

When the requirement is for a voltage greater than the ac line voltage, a step-up transformer is used. Oscilloscopes, x-ray equipment, and vacuum-tube RF power amplifiers are some examples when a step-up trans-

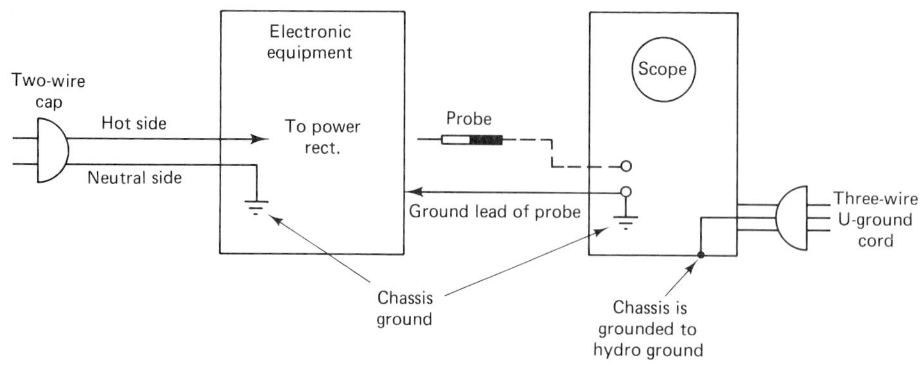

FIGURE 20-7 Equipment chassis ground must be connected to test equipment ground for proper servicing.

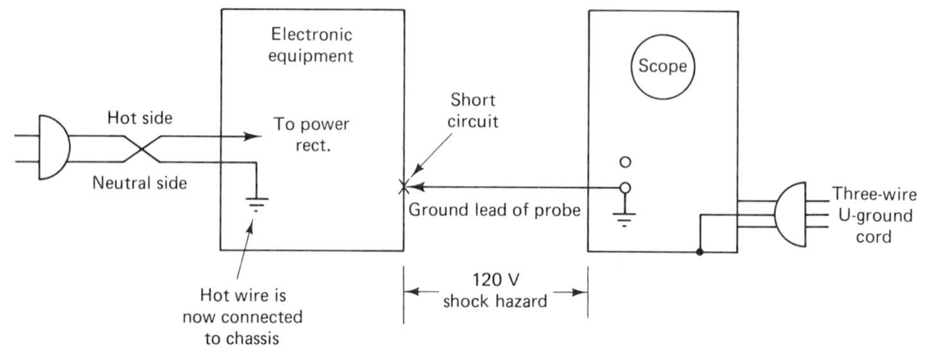

FIGURE 20-8 Equipment chassis is connected to the hot side of the ac power line because of the transposed two-wire cap. This presents both a shock hazard and a short-circuit hazard.

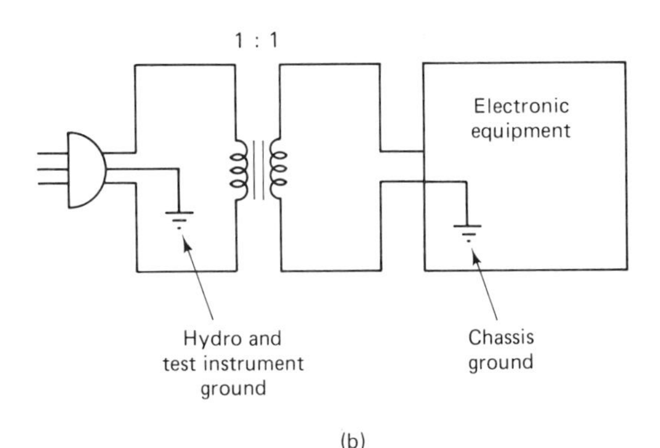

FIGURE 20-9 (a) Typical isolation transformer. (Courtesy of Hammond Manufacturing Co. Ltd.) (b) Isolation transformer electrically isolates the equipment chassis ground from the hydro and test instrument ground. Note that the turns ratio is shown on the schematic as 1:1.

former would be used. The turns ratio will determine the amount by which the voltage is stepped up and rectified to dc.

EXAMPLE 20-3

A transformer has 600 turns on the primary and 2400 turns on the secondary. What is the secondary voltage?

Solution: Using equation (20-2) gives us

$$\frac{N_p}{N_s} = \frac{V_p}{V_s} \qquad \frac{600 \text{ turns}}{2400 \text{ turns}} = \frac{120 \text{ V}}{V_s}$$

$$V_s = 480 \text{ V}$$

Note that the turns ratio is greater than 1, primary to secondary; therefore, this is a step-up transformer.

$$\frac{N_p}{N_s} = \frac{600 \text{ turns}}{2400 \text{ turns}} = 1{:}4$$

20-5 STEP-DOWN TRANSFORMER

Transistor and integrated circuits (ICs) require dc voltages lower than the line voltage. A step-down transformer is used to provide the lower ac voltage, which is then rectified to dc.

EXAMPLE 20-4

A transformer has a 600-turn primary winding and a 200-turn secondary winding. What is the secondary voltage?

Solution: Using equation (20-2) yields

$$\frac{N_p}{N_s} = \frac{V_p}{V_s} \qquad \frac{600 \text{ turns}}{200 \text{ turns}} = \frac{120 \text{ V}}{V_s}$$

$$V_s = 40 \text{ V}$$

Note that the turns ratio is 3:1 primary to secondary; therefore, this is a step-down transformer.

$$\frac{N_p}{N_s} = \frac{600 \text{ turns}}{200 \text{ turns}} = 3{:}1$$

20-6 AUTOTRANSFORMER/VARIAC

An autotransformer has one winding that serves as both a primary and a secondary. Its function is a variable ac voltage source. In the schematic of Fig. 20-10(a), the turns are tapped and leads brought out from the taps. The variac is an application of the autotransformer. The variac has a brush contact that is varied or slides along the turns. This arrangement allows small incremental voltage changes with rotation of a control knob. The schematic symbol for a variac is shown in Fig. 20-10(b). Note that the autotransformer and variac do not provide ac line isolation.

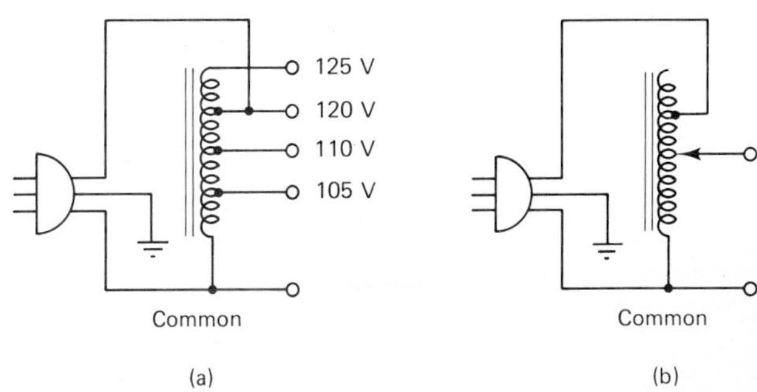

FIGURE 20-10 (a) Autotransformer. (b) Variac.

20-7 TAPPED PRIMARY/SECONDARY

Electronic equipment, especially instrumentation, has an isolation transformer that can be used on a 120/240-V line voltage. The primary is tapped at 120 V, as shown in the schematic of Fig. 20-11.

To operate the transformer on 240 V, the jumper is removed and reconnected so that the full primary winding is across the 240-V line. Tapped primary is also used to adjust the primary voltage. A lower line voltage can be connected to a lower tap on the primary.

Secondary windings are center tapped where there is a requirement for a full-wave rectifier voltage. The voltage from the center tap to each outside lead will be the same as shown by S_1 in the schematic of Fig. 20-11. The center tap gives two instantaneous ac voltages that are 180° out of phase but are equal in amplitude for rectifier purpose, as shown in Fig. 20-12. Another secondary winding will provide taps at higher and lower voltages for other requirements, as shown in the schematic of Fig. 20-11.

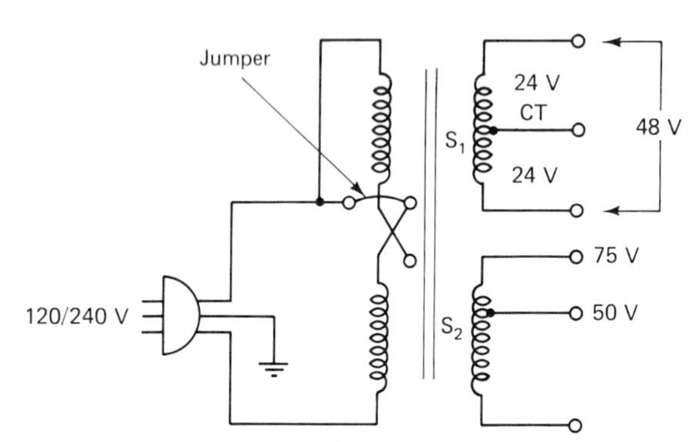

FIGURE 20-11. Tapped primary and secondary.

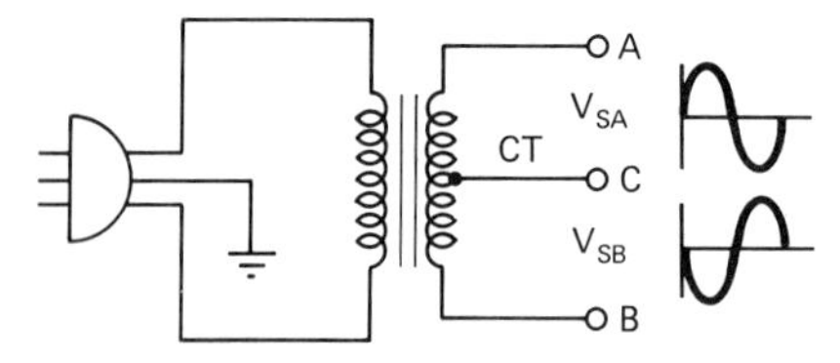

FIGURE 20-12 Instantaneous voltage V_{SA} and V_{SB} are 180° out of phase with respect to center tap.

20-8 CURRENT TRANSFORMER

Current transformers can be mentioned briefly as their application is limited to ac current measurement, especially in high-voltage power circuits. A large ac load current is connected in series with the primary of a current transformer, as shown in the schematic of Fig. 20-13. The ac ammeter is then connected to the secondary. This isolates the high-voltage from the ac ammeter. Various taps can also be provided on the secondary to give different ac ampere or milliampere ranges. This allows a large current to be measured by a low-current ammeter.

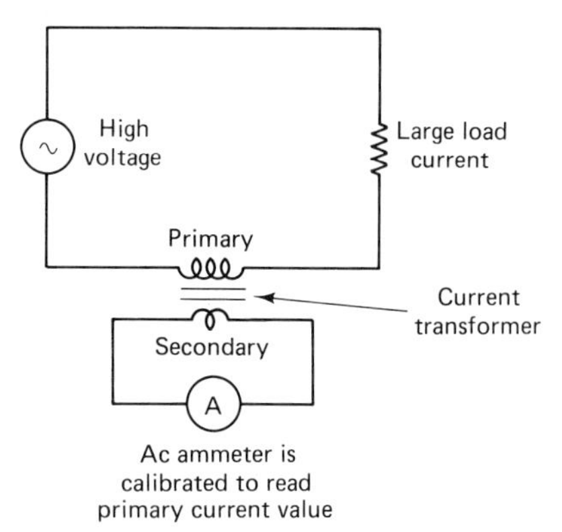

FIGURE 20-13 A current transformer used to measure a large current by using a small-current range ammeter. It also provides high-voltage isolation to the ammeter.

20-9 CURRENT RATIO

We have seen that the primary voltage can be stepped up or stepped down, depending on the turns ratio since the turns ratio will determine the voltage ratio. The current in the secondary is also stepped up or stepped down. With a voltage step down the current will be stepped up and vice versa. The current ratio is the inverse of the voltage ratio and is in terms of the secondary current. Since the transformer is an energy transfer device, and assuming no losses, the secondary current and power can only be provided by the primary, or

$$\text{power in primary} = \text{power in secondary}$$

Substituting VI for the power in both the primary and secondary gives

$$V_pI_p = V_sI_s \tag{20-3}$$

Rearranging equation (20-3) gives the current ratio:

$$\frac{V_p}{V_s} = \frac{I_s}{I_p} \tag{20-4}$$

Equation (20-4) shows that the current ratio is the inverse of the voltage ratio.

EXAMPLE 20-5

A transformer with a voltage ratio of 10:1 has its secondary connected across a 6-Ω load. What are its secondary and primary currents?

Solution: Using equation (20-2) will give the secondary voltage:

$$\frac{V_P}{V_s} = \frac{N_P}{N_s} \qquad \frac{120\text{ V}}{V_s} = \frac{10}{1}$$

Cross-multiplying to solve for V_s gives

$$10V_s = 120\text{ V} \qquad \text{therefore, } V_s = 12\text{ V}$$

The secondary current is found using Ohm's law:

$$I_s = \frac{V_s}{R_L} = \frac{12\text{ V}}{6\ \Omega} = 2\text{ A}$$

The primary current can be found using equation (20-4):

$$\frac{V_P}{V_s} = \frac{I_s}{I_P} \qquad \frac{120\text{ V}}{12\text{ V}} = \frac{2\text{ A}}{I_P}$$

Solving for I_P gives 0.2 A. Assuming no losses and using equation (20-3), the power in the primary will equal the power in the secondary:

$$V_P I_P = V_s I_s \qquad 120\text{ V} \times 0.2\text{ A} = 12\text{ V} \times 2$$

or 24 primary watts = 24 secondary watts.

PRACTICE QUESTIONS 20-2

Choose the most suitable word, phrase, or value to complete each statement.

1. The term "_________ transformer" can be applied to a power transformer since the primary and secondary are electrically insulated from each other.
2. An external isolation transformer must be used when servicing equipment without an isolation transformer to prevent _________ and _________ hazard.
3. When using an isolation transformer the _________ current rating of the transformer must not be exceeded.
4. A transformer has 800 turns on the primary and 1600 turns on the secondary. What is the turns ratio?
5. What is the secondary voltage of the transformer in Question 4?
6. A transformer has 800 turns on the primary and 200 turns on the secondary. What is the turns ratio?
7. What is the secondary voltage of the transformer in Question 6?
8. A transformer with only one winding that serves as both a primary and a secondary is termed an _________.
9. A type of autotransformer that has a continuously variable output is termed a _________.
10. A tapped primary transformer is used to adjust the primary to the _________ voltage.
11. The center-tapped secondary gives two instantaneous ac voltages that are _________ out of phase but are _________ in amplitude to each other.
12. A current transformer is used primarily in conjuction with a low-range _________.
13. The current ratio is the _________ of the voltage ratio.
14. A transformer with a voltage ratio of 4:1 has a secondary current rating of 1 A. What is its primary current?
15. Calculate the value of a resistor and its wattage rating to simulate a resistive load for the secondary of the transformer in Question 14.

20-10 TRANSFORMER POWER LOSSES AND EFFICIENCY

There are two types of losses in an iron-core transformer: *core losses* and *copper losses*. Core losses include hysteresis and eddy current losses. Copper losses are due to the resistance of the windings.

Core Losses

Hysteresis loss is the heat generated by the magnetic domains. With a silicon-iron alloy core, the 60 current reversals every second causes the magnetic field to cycle through the *B–H* loop 60 times per second (see Chapter 15). The energy that is needed to reorient and overcome the coerciveness of the magnetic domains is absorbed by the iron core. This energy heats the core and is dissipated into the surrounding air. The amount of energy loss in one complete magnetization cycle is proportional to the area enclosed by the hysteresis loop.

The eddy current loss is due to the heat generated in the core due to the induced emf within the core laminations. Since iron is an electrical conductor and the laminations are shorted turns, the changing magnetic flux in the core will induce an emf within the laminations. The induced emf produces a circulating eddy current which dissipates heat since iron has some resistance. Iron-core laminations increase the resistance to the circulating eddy current perpendicular to the laminated sheets, greatly reducing the power loss.

Copper Loss (I^2R Loss)

Both the primary and the secondary windings have resistance. As a result of this resistance, current flowing through the windings will dissipate heat into the core and into the air. By measuring the dc resistance of the winding with an ohmmeter, and the current through the windings *under load,* the copper loss can be calculated using the equation $P = I^2R$. Thus a copper loss is known as I^2R loss.

Transformer Efficiency

The efficiency of an ideal transformer is 100% since the power input will equal the power output. Efficiency is defined as the ratio of power output to the power input:

$$\text{Efficiency} = \frac{\text{Power Output}}{\text{Power Input}}$$

If this equation is multiplied by 100, we can obtain efficiency in terms of percentage:

$$\%\ \text{Efficiency} = \frac{P_{out}}{P_{in}} \times 100 \qquad (20\text{-}5)$$

In practical transformers, core and I^2R losses are in the form of heat. Heat is dissipated from the core and the windings and is lost to the surrounding air. The energy thus lost must be made up by an increase in the primary power. This means that the primary power (input) is greater than the secondary power (output) and the percentage efficiency is less than 100%.

Power transformers in the range 50 to 300 W have an efficiency of 80 to 90%. For higher wattage transformers the primary and secondary windings use larger-diameter wire; therefore, the I^2R losses are greatly reduced and efficiency increased.

EXAMPLE 20-6

Using the schematic shown in Fig. 20-14, find:

(a) The secondary power.
(b) The primary power.
(c) The wattage loss due to core and I^2R losses.

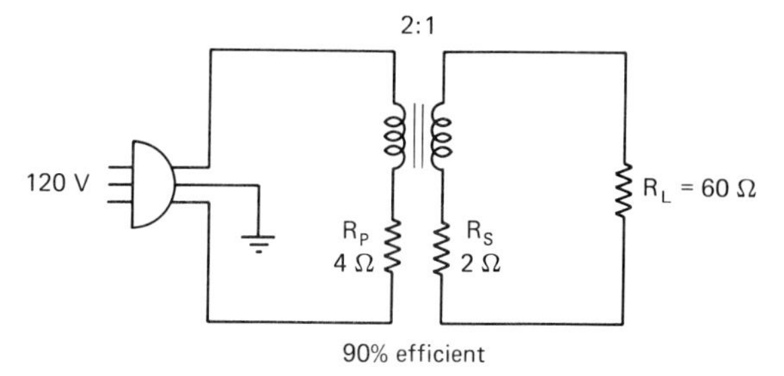

FIGURE 20-14 Circuit schematic.

(d) The wattage loss due to I^2R losses (both primary and secondary).
(e) The wattage loss due to core losses.

Solution: The secondary voltage can be found using equation (20-2).

$$\frac{N_p}{N_s} = \frac{V_p}{V_s} \qquad \frac{2}{1} = \frac{120\text{ V}}{V_s} \qquad V_s = 60\text{ V}$$

(a) The secondary power can now be found.

$$P_s = \frac{V^2}{R} = \frac{(60\text{ V})^2}{60\ \Omega} = 60\text{ W}$$

(b) The primary power can be found using equation (20-5).

$$\%\ \text{efficiency} = \frac{\text{power out}}{\text{power in}} \times 100$$

$$90\% = \frac{P_s}{P_p} \times 100$$

$$P_p = 60\text{ W} \times \frac{100}{90} = 66.7\text{ W}$$

(c) The difference between the input power and the output power is due to the losses:

$$P_{losses} = 66.7\text{ W} - 60\text{ W} = 6.7\text{ W}$$

(d) To find the I^2R losses, we must know the current in the primary and secondary. The current in the secondary can be found using Ohm's law:

$$I = \frac{V_s}{R_L} = \frac{60\text{ V}}{60\ \Omega} = 1\text{ A}$$

The current in the primary can be found using equation (20-4).

$$\frac{V_p}{V} = \frac{I_s}{I_p} \qquad \frac{120\text{ V}}{60\text{ V}} = \frac{1\text{ A}}{I_p}$$

Therefore,

$$I_p = 0.5\text{ A}$$

I^2R loss primary and secondary are

$$P_p = I^2R_p = (0.5\ \text{A})^2 \times 4\ \Omega = 1\ \text{W}$$

$$P_s = I^2R_s = (1\ \text{A})^2 \times 2\ \Omega = 2\ \text{W}$$

$$\text{total } I^2R \text{ losses} = 1\ \text{W} + 2\ \text{W} = 3\ \text{W}$$

(e) Core losses will be the difference between the total losses and the I^2R losses:

$$\text{core losses} = 6.7\ \text{W} - 3\ \text{W} = 3.7\ \text{W}$$

In a well-designed transformer the core losses will be approximately equal to the I^2R losses.

20-11 UNLOADED AND LOADED SECONDARY VOLTAGE

Voltage and turns ratio up to this point have been calculated without considering a load connected across the secondary. A practical transformer must be designed to have a step-up or a step-down voltage ratio greater than would be computed for the no-load condition. The increase in voltage ratio and therefore the turns ratio makes up for the voltage loss due to the resistance of the windings.

Increasing the turns on the primary means that the no-load voltage across the secondary will be higher than the full-load voltage, as illustrated by the following simplified example.

EXAMPLE 20-7

A transformer secondary has a rating of 12.6 V at a rated current of 2 A. The secondary winding has a resistance of 0.6 Ω as measured by an ohmmeter. Calculate the approximate no-load secondary voltage.

Solution: The full-load voltage is 12.6 V with a maximum load current of 2 A. With 2 A flowing in the secondary, the internal voltage loss due to the resistance of the winding can be calculated using Ohm's law:

$$V_{\text{loss}} = I \times R_s = 2\ \text{A} \times 0.6 = 1.2\ \text{V}$$

The no-load voltage will, therefore, be the full-load voltage plus the internal voltage loss:

$$V_s(\text{no-load}) = 12.6\ \text{V} + 1.2\ \text{V} = 13.8\ \text{V}$$

The turns ratio will give 13.8 V across the secondary with a no-load condition.

With a lighter load of, say, 1 A, the secondary voltage will be lower than 13.8 V but higher than 12.6 V because the internal voltage loss is now only 0.6 V, so that

$$V_s = 13.8\ \text{V} - 0.6\ \text{V} = 13.2\ \text{V}$$

This example serves to show the effect on the secondary voltage with no load and full load due to the internal resistance of the winding.

20-12 TRANSFORMER RATING

Power transformers used in electronic systems are rated in volt-amperes (VA), or if they are very large in kilovolt-amperes (kVA). The VA rating refers to the wattage that can safely be handled by the primary. From the equation $P = VI$, and since the primary is always connected to either 120 V or 240 V, the current through the primary determines its power capability.

The power and, as a result, the current are always dependent on what type of load is connected across the secondary. If the load is resistive, the wattage handled by the transformer and dissipated by the resistive load will be almost the number of volt-amperes in the primary circuit. To transfer 100 W from a 120-V source, it is necessary to have a primary current of $I = P/V =$ 100 W/ 120 V, or 0.83 A.

With an inductive load V and I are out of phase. As an example, if V and I are 45° out of phase, the current required by the primary to transfer 100 W to the load would be

$$I = \frac{P}{V} \cos\theta$$

or

$$I = \frac{100\ \text{W}}{120\ \text{V}} \cos 45° = 1.18\ \text{A}$$

The extra current goes into setting up a magnetic flux in the inductive load. The primary current has increased to 1.18 A. This increases the VA requirement of the transformer primary to 142 W ($P = VI$). If the primary was wound to handle a VA rating of 100 W, it would be seriously overloaded (by 42%) and the transformer would overheat.

20-13 POWER TRANSFORMER EIA LEAD COLOR CODE

The schematic diagram of Fig. 20-15 shows the Electronic Industries Association (EIA) standard color code for power transformer leads.

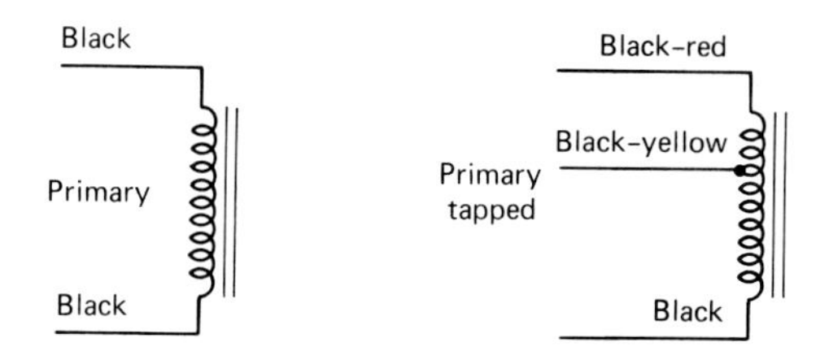

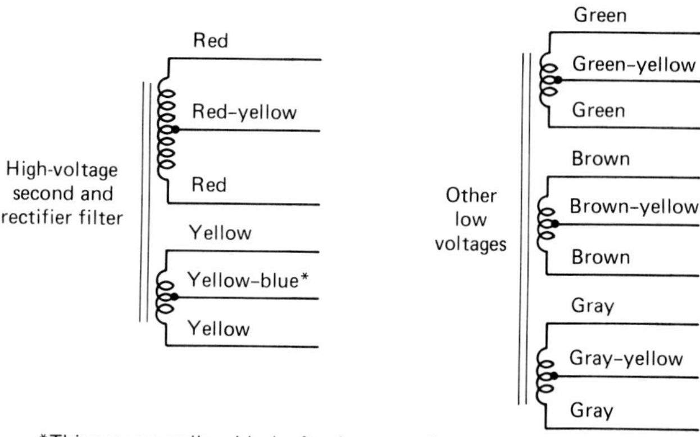

*This appears yellow-black after impregnation.

FIGURE 20-15 EIA standard power transformer lead color code. (Courtesy of Hammond Manufacturing Co. Ltd.)

SUMMARY

A power transformer consists of a primary and a secondary coil wound on a laminated silicon-iron alloy core.

Two common types of lamination configurations are the E and I, and the L.

Reversing the direction of turns on an EI type of core will cause the secondary voltage to be 180° out of phase with the primary voltage.

The dot on a transformer schematic indicates a positive instantaneous polarity to that respective lead.

The ratio of the number of turns on the primary, N_p, to the number of turns on the secondary, N_s, is termed the turns ratio:

$$\text{turns ratio} = \frac{N_p}{N_s} \tag{20-1}$$

The turns ratio is equal to the voltage ratio.

$$\frac{N_p}{N_s} = \frac{V_p}{V_s} \tag{20-2}$$

An isolation transformer isolates the ac power line from the dc voltage and chassis of electronic equipment.

Isolating the ac power line from electronic equipment and grounding the chassis through a ground plug eliminates a shock hazard.

An external isolation transformer is used between equipment that does not have a built-in isolation transformer.

A step-down transformer has a secondary voltage that is less than the line voltage.

An autotransformer has one winding that serves as both a primary and a secondary.

The primary of a transformer is tapped to allow it to be used on a line voltage other than 120 V.

Center-tapped secondary is used particularly for full-wave rectifier circuits.

Current transformers step-down high values of ac current to be measured while isolating the high voltage of the line.

The current ratio is the inverse of the voltage ratio:

$$\frac{V_p}{V_s} = \frac{I_s}{I_p} \tag{20-4}$$

For an ideal transformer the power input to the primary will equal the power output of the secondary. From $P = VI$,

$$V_pI_p = V_sI_s \tag{20-3}$$

Two types of transformer losses are core losses and copper losses. Core losses include hysteresis and eddy current losses.

Copper losses are due to the resistance of the winding and are also termed I^2R losses.

Because of the losses, the efficiency of a power transformer is less than 100%.

$$\% \text{ efficiency} = \frac{P_{\text{out}}}{P_{\text{in}}} \times 100 \tag{20-5}$$

The no-load voltage across the secondary will be higher than the full-load voltage.

Transformers are rated in volt-amperes (VA).

The Electronic Industries Association (EIA) standard color code for power transformer leads is shown in Fig. 20-15.

REVIEW QUESTIONS

20-1. What is the function of the primary in a transformer?

20-2. What is the function of the secondary in a transformer?

20-3. What function does the laminated iron core serve?

20-4. What is the effect on secondary voltage polarity with respect to the primary when the direction of the secondary turns is reversed?

20-5. Explain the purpose of the dot convention for a transformer schematic diagram.

20-6. Explain why the voltage ratio is in the same proportion as the turns ratio.

20-7. A power transformer has 1200 turns on the primary and 400 turns on the secondary. Calculate the turns ratio.

20-8. Is the transformer in Question 20-7 a step-up or step-down transformer? Prove your answer.

20-9. A transformer has 900 turns on the primary with a secondary voltage of 30 V. Calculate the number of turns on the secondary and the turns ratio.

20-10. The turns ratio of a power transformer is 1:10. Calculate the secondary voltage.

20-11. Explain what is meant by "isolating the ac power line."

20-12. How does the power transformer isolate the ac power line in electronic equipment?

20-13. Give two reasons why an external isolation transformer must be used in cases where an electronic device does not have an internal isolation transformer.

20-14. Explain why a variac does not isolate the ac line voltage.

20-15. A transformer with a voltage ratio of 1:10 has a primary current rating of 1 A; what is its secondary current?

20-16. Explain the difference between transformer core losses and copper losses.

20-17. Given the information shown on the schematic of Fig. 20-16, find the following:

(**a**) The 60-Ω load current.
(**b**) The 15-Ω load current.
(**c**) The primary power considering the transformer to be 90% efficient.
(**d**) The power loss due to core and copper losses.
(**e**) The primary current considering the transformer to be 90% efficient.

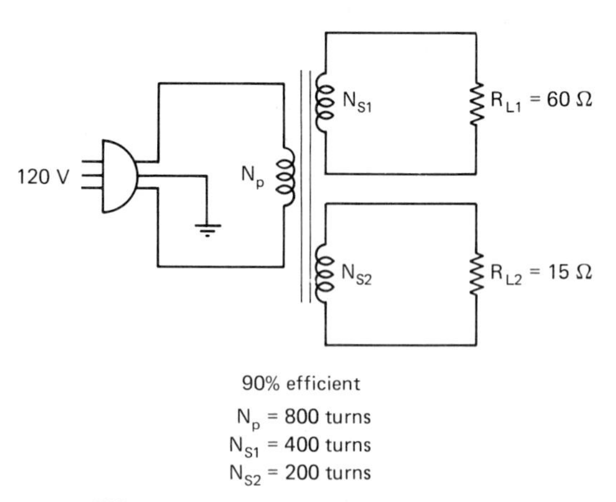

FIGURE 20-16 Circuit schematic.

20-18. What is the volt-ampere rating of the transformer in Question 20-17?

SELF-EXAMINATION

Choose the most suitable word, phrase, or value to complete each statement.

20-1. The ac voltage input to a power transformer is connected across the ________ windings.

20-2. The output of a power transformer is taken across the ________ windings.

20-3. The silicon-iron alloy core provides magnetic flux ________ between the primary and the secondary.

20-4. With the secondary wound in reverse to that of the primary, the secondary voltage will be ________ out of phase with the primary.

20-5. A transformer schematic diagram having a dot at one lead if its winding indicates that lead's instantaneous value is ________ (polarity) with respect to the other winding lead.

20-6. Since the flux coupling of an iron-core transformer is 100%, the induced voltage per turn in the primary winding will be the same as the induced voltage per turn in the ________.

20-7. The turns ratio is equal to the ________ ratio.

20-8. A power transformer has 200 turns on the primary and 800 turns on the secondary. Its turns ratio is ________.

20-9. The transformer in Question 20-8 is a (*step-down, step-up*) transformer.

20-10. A tranformer has 400 turns on the primary with a secondary voltage of 350 V. The number of turns on the secondary is ________.

20-11. The turns ratio of a power tranformer is 5:1. What is its secondary voltage?

20-12. The ac power line is isolated from electronic equipment that have a internal power transformer because the primary is ________ ________ from the secondary.

20-13. An external isolation transformer prevents ________ and ________ hazard between another device that does not have an internal isolation transformer.

20-14. A variable-voltage transformer with one winding is called a ________.

20-15. A power loss due to hysteresis and eddy currents is termed a ________ loss.

20-16. A power loss due to the resistance of the winding is termed ________ loss or also ________ loss.

Refer to the schematic of Fig. 20-17 for Questions 20-17 to 20-24.

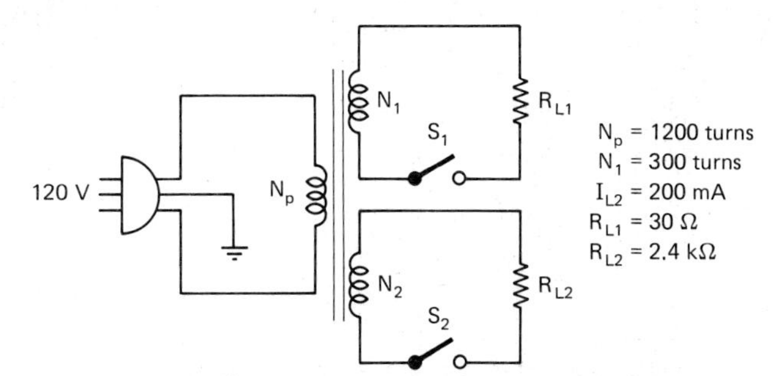

FIGURE 20-17 Circuit schematic.

20-17. What is the voltage scross R_{L1}?

20-18. What is the current through R_{L1}?

20-19. Find the voltage across R_{L2}.

20-20. Find the number of turns N_2.

20-21. Find the primary current with S_2 open and S_1 closed. (Ignore losses.)

20-22. Find the primary current with S_1 and S_2 closed. (Ignore losses.)

20-23. Find the power loss if the transformer efficiency is 90%.

20-24. What is the volt-ampere rating of the transformer in Question 20-23?

ANSWERS TO PRACTICE QUESTIONS

20-1

1. primary, secondary
2. E, I
3. 100% or unity
4. excitation
5. direction
6. 180
7. positive
8. 240 turns
9. 1200 turns
10. 60 V

20-2

1. isolation
2. shock, short circuit
3. secondary
4. 1:2
5. 240 V
6. 4:1
7. 30 V
8. autotransformer
9. variac
10. line
11. 180°, equal
12. ammeter
13. inverse
14. 0.25 A
15. 30 Ω, 30 W

21

Inductive Reactance

INTRODUCTION

With a pure resistive dc or ac circuit, the opposition to current flow is resistance. We were able to calculate this resistance using Ohm's law. An inductor in an ac circuit produces a cemf due to the changing current and resultant changing magnetic field. The effect of this cemf is to oppose the current flow. Because the cemf is frequency and inductance dependent, its opposition to the current flow is termed *reactance*. Reactance can be calculated using the formula $2\pi fL$ (ohms). Reactance is also an Ohm's law value, and therefore can be calculated using the Ohm's law formula, $X_L = V/I$. The symbol for reactance is X, with the capital letter L as a subscript to indicate that it is inductive reactance.

Any ac circuit that has inductance will have reactance as a prime source of opposition to current flow. Some examples are motors, transformers, and of course, inductors or chokes. In this chapter we look at the reactance of a choke in a simple series and parallel circuit.

MAIN TOPICS IN CHAPTER 21

21-1 Inductance in the Dc versus Ac Circuit
21-2 Ohm's Law and Inductive Reactance
21-3 Effect of Frequency on X_L
21-4 Effect of Inductance on X_L
21-5 The Equation $X_L = 2\pi fL$ and Graphical Representation of X_L versus f
21-6 Series and Parallel X_L's
21-7 Phase Relationship of Inductive Current, Counter Emf, and Applied Voltage

OBJECTIVES

After studying Chapter 21, the student should be able to

1. Master the fundamental concept of inductive reactance.
2. Master the fundamental concept of why and how X_L changes with inductance and frequency.
3. Solve X_L problems with a change in frequency and inductance.
4. Solve series and parallel X_L circuit problems using Ohm's law.
5. Master the concept of a lagging current in a pure inductive circuit.

21-1 INDUCTANCE IN THE DC VERSUS AC CIRCUIT

Inductance in the Dc Circuit

In Chapter 19 we discovered that in order to have an inductance, we must have a change in current which causes a change in magnetic flux to self-induce a counter emf within the inductor or choke. The counter emf thus induced opposes the change in current and the applied voltage.

Looking at the schematic diagram of Fig. 21-1 shows that the choke and lamp are initially (switch position 1) series-connected to a 120-V dc source. Considering that we need a changing current to produce a counter emf, how will the current and the brilliance of the lamp be affected by the choke?

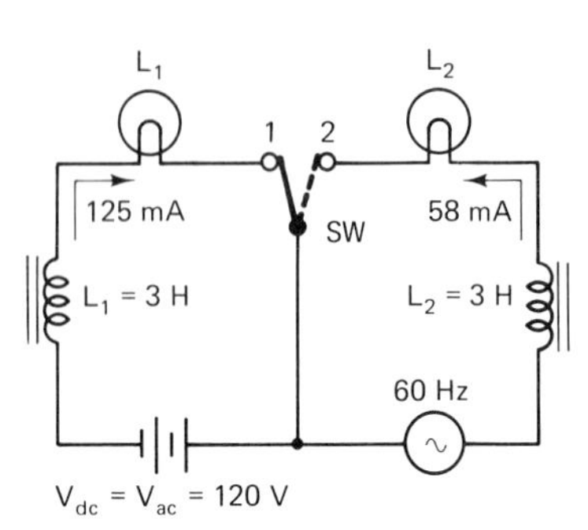

FIGURE 21-1 3-H choke connected in series with a 15-W lamp to a 120-V ac/dc source.

Of course, the lamp will light to almost full brilliance since the only current limiting or opposition in this circuit will be the series resistance of the choke and the lamp. In a dc circuit there is no such thing as inductance.

If we ignore the series resistance of the choke, the current flowing in the circuit can be calculated using $I = P/V$ or 15 W/120 V $\cong$ 125 mA. Practically, the current will be lower than this value and will depend on the internal resistance of the choke.

Inductance in the Ac Circuit

Toggling the switch to position 2 places the series lamp and choke across a 120-V 60-Hz source. This time an alternating current induces a counter emf within the coil that opposes the applied voltage and reduces the current flowing through the lamp to 58 mA. The brilliance of the lamp is reduced substantially and visually illustrates the effect of the inductance on the current in the ac circuit.

21-2 OHM'S LAW AND INDUCTIVE REACTANCE

The 15-W lamp in the circuit schematic of Fig. 21-2 needs 125 mA for full brilliance. With a source of 120 V, 60 Hz, the lamp has approximately 58 mA flowing through it. This is because the inductance produces a counter emf which opposes the change in current and therefore reduces the current. The counter emf is the reactive component that produces the opposition to the ac current flow. This opposition is termed *inductive reactance*, X_L. Inductive reactance is an Ohm's law value and can be calculated for any given voltage across, and current flowing through, the inductor.

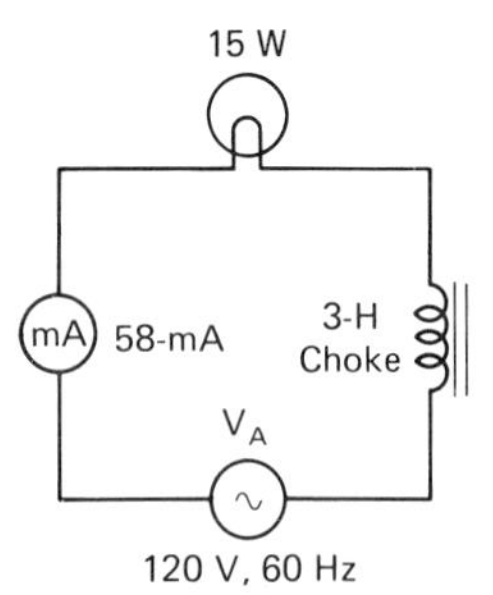

FIGURE 21-2 Schematic. Lamp needs 125 mA for full brilliance.

EXAMPLE 21-1

Calculate the reactance of the 7-H choke in the schematic of Fig. 21-3.

Solution: Using Ohm's law, we have

$$X_L = \frac{V_L}{I_L} = \frac{120\ \text{V}}{45.5\ \text{mA}} = 2.64\ \text{k}\Omega$$

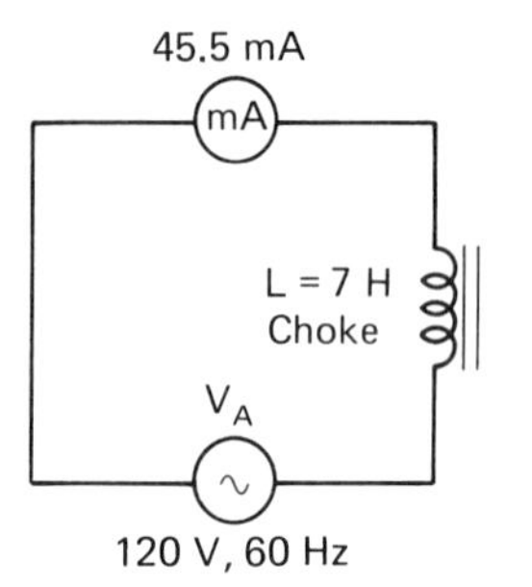

FIGURE 21-3 Circuit schematic.

21-3 EFFECT OF FREQUENCY ON X_L

Inductive reactance is entirely due to the counter emf produced by an inductor. This induced counter emf is directly proportional to the inductance and the rate of current change [see equation (19-1)]. If the rate of current change is twice as fast, the induced counter emf will be twice as great and reactance will double.

Consider the 3-H inductance in the schematic of Fig. 21-4(a) connected to a 120-V 60-Hz source. There are four current changes in $\frac{1}{60}$ of a second: that is, two current rises, and two current falls [see Fig. 21-4(b)] for one cycle. If the peak current is approximately 106 mA $\times \sqrt{2}$ = 150 mA, the average rate of change of the inductor current will be the time period t divided by 4:

$$\frac{\Delta i}{\Delta t}(\text{avg}) = \frac{i_p}{t/4} = \frac{4i_p}{t}$$

$$= \frac{4 \times 150 \times 10^{-3} \text{ A (peak)}}{1/60 \text{ s}} = 36 \text{ A/s}$$

and

$$V_L(\text{avg}) = L \times \frac{\Delta i}{\Delta t}(\text{avg}) = 3 \text{ H} \times \frac{36 \text{ A}}{1 \text{ s}} = 108 \text{ V (avg)}$$

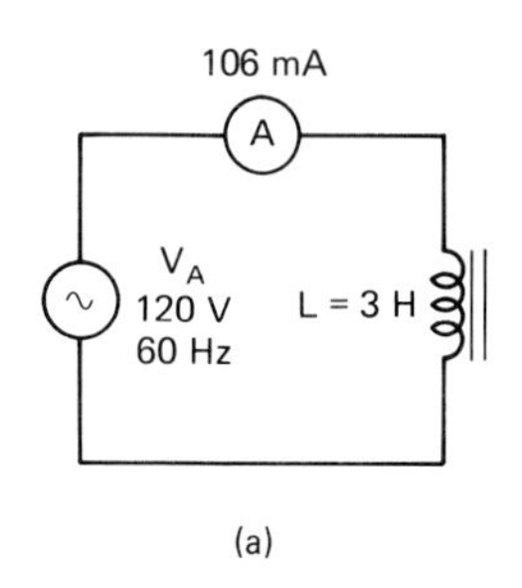

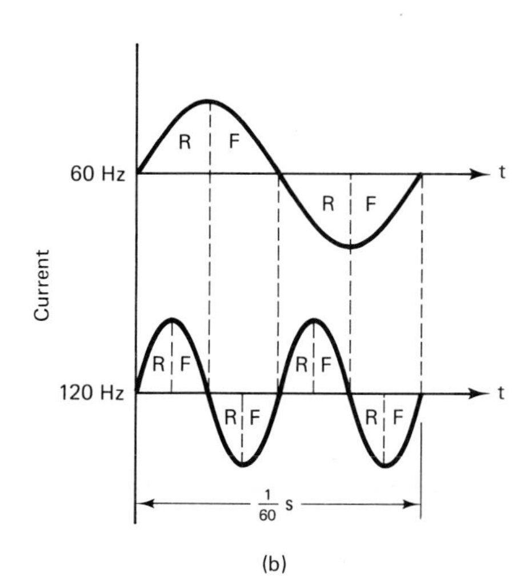

FIGURE 21-4 With double the frequency the rate of current change doubles for the same time period.

The rate of change for double the frequency, or 120 Hz, is twice as great as for 60 Hz. There are eight rise and fall changes of current for the same period of 1/60 of a second, as shown in Fig. 21-4.

If the current is kept constant at 0.15 A (peak) as a reference, we can see the effect on the counter emf with double the frequency:

$$\frac{\Delta i}{\Delta t}(\text{avg}) = \frac{i_p}{t/8} = \frac{8i_p}{t} = \frac{8 \times 15 \text{ A (peak)}}{1/60 \text{ s}} = 72 \text{ A/s}$$

and

$$V_L(\text{avg}) = L \times \frac{\Delta i}{\Delta t}(\text{avg}) = 3 \text{ H} \times \frac{72 \text{ A}}{1 \text{ s}} = 216 \text{ V(avg)}$$

With double the frequency the counter emf is doubled. As the counter emf is the reactive component that is directly proportional to reactance, we can conclude that the reactance will be directly proportional to the frequency:

$$X_L \propto f \tag{21-1}$$

> **EXAMPLE 21-2**
>
> Calculate the current for the schematic of Fig. 21-3 with double the frequency.
>
> ***Solution:*** The reactance of the 3.0-H choke with the 60-Hz source was found to be 1130 Ω. With double the frequency, the reactance of the choke will double, to 2260 Ω. The current, then, is found using Ohm's law, or
>
> $$I_L = \frac{V_L}{X_L} = \frac{120 \text{ V}}{2260 \ \Omega} = 53 \text{ mA}$$
>
> *Note:* We could have deduced that the current would have been halved from the original value of 106 mA with the 60-Hz source, since the reactance would double with the 120-Hz source.

21-4 EFFECT OF INDUCTANCE ON X_L

Since the induced counter emf is directly proportional to inductance, reactance will be directly proportional to inductance. This can again be shown by calculating the counter emf for the 3.0-H choke and then doubling the inductance while keeping the frequency and rate of current change constant.

Having calculated the average counter emf for the 3-H choke across the 120-V 60-Hz source as being

108 V, we can double the inductance while holding the rate of current change constant to find the average counter emf:

$$V_L(\text{avg}) = L \times \frac{\Delta i}{\Delta t}(\text{avg}) = 6.0 \text{ H} \times \frac{36 \text{ A}}{1 \text{ s}}$$
$$= 216 \text{ V(avg)}$$

With double the inductance the average induced counter emf has doubled; therefore, the emf is directly proportional to inductance. As the counter emf is not an Ohm's law reactance but is the reactive component, we can conclude that the reactance will also be directly proportional to the inductance:

$$X_L \propto L \qquad (21\text{-}2)$$

EXAMPLE 21-3

Calculate the current for the schematic of Fig. 21-4 if the inductance is doubled.

Solution: The reactance of the 3.0-H choke with the 60-Hz source was found to be 1130 Ω. With double the inductance, the reactance of the choke will double to 2260 Ω. The current is found using Ohm's law:

$$I_L = \frac{V_L}{L} = \frac{120 \text{ V}}{2260 \ \Omega} = 53 \text{ mA}$$

Note: We could have deduced that the current would have been halved from the original value of 106 mA with a 3.0-H choke since reactance would double with the 6.0-H choke.

21-5 EQUATION $X_L = 2\pi fL$ AND GRAPHICAL REPRESENTATION OF X_L VERSUS f

As long as the current through an inductor, and the voltage across the inductor, are known, its reactance can be calculated using Ohm's law. An equation that is able to predict the reactance for any given frequency and inductance is far more useful. This equation can be derived from the two previous statements, (21-1) and (21-2)—that reactance is directly proportional to frequency and inductance:

$$X_L \propto fL$$

The frequency component in the equation is an ac sinusoidal waveform; therefore, the angular velocity $2\pi f$ for a sine wave must be considered. With the constant 2π, the equation for inductive reactance with frequency and inductance becomes

$$X_L = 2\pi fL \quad \text{ohms} \qquad (21\text{-}3)$$

where f is in hertz
L is in henrys

EXAMPLE 21-4

What is the reactance of the choke in Fig. 21-4 if the frequency is increased to 240 Hz?

Solution: Using equation (21-3) gives

$$X_L = 2\pi fL = 6.28 \times 240 \text{ Hz} \times 3 \text{ H} = 4522\Omega$$

EXAMPLE 21-5

A certain choke has a reactance of 188 Ω at 1 kHz. What is its inductance?

Solution: Rearranging equation (21-3) gives

$$L = \frac{X_L}{2\pi f} = \frac{188 \ \Omega}{6.28 \times 1 \times 10^3 \text{ Hz}} = 30 \text{ mH}$$

Graphical Representation of X_L versus f

The inductance of some types of inductors is fixed. Other types of inductors have an adjustable ferrite core, which slightly alters their inductance. Once the inductance is adjusted it remains fixed for a particular frequency.

Since frequency is the variable factor in an electronic circuit, it would be advantageous to have a visual graphical representation of inductive reactance with a change in frequency. Equation (21-3) shows that reactance (X_L) is directly proportional to frequency and inductance. Keeping inductance fixed, a plot of X_L versus f, as graphed in Fig. 21-5, shows that X_L is a linear function of frequency. *Note:* By convention, X_L is always considered to be a positive quantity and therefore is plotted in the positive quadrant.

PRACTICE QUESTIONS 21-1

Choose the most suitable word or phrase to complete each statement.

1. A 6-H choke has 1 A of current flowing through it

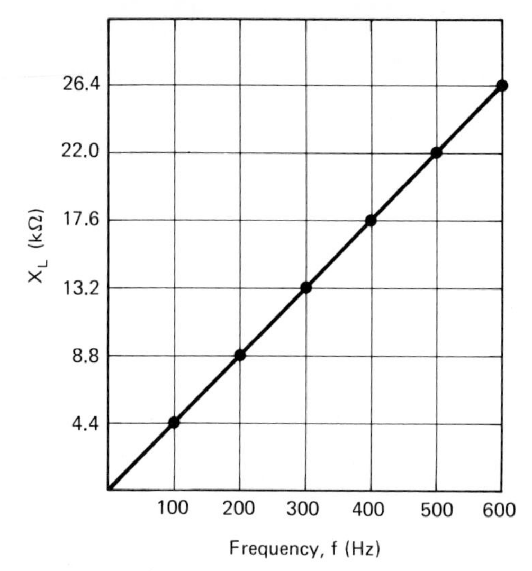

FIGURE 21-5 X_L increases linearly with frequency, L is 7 H.

after it is connected to 6 V dc. Calculate its counter emf.

2. The same 6-H choke has 53 mA of current flowing through it when it is connected to a 60-Hz source. Calculate its counter emf.
3. Calculate the reactance of the choke in Question 2 if the source voltage is 120 V.
4. Calculate the reactance of the choke in Question 2 if the frequency of the source is 400 Hz.
5. Calculate the inductance of a choke when the current through it is 250 mA at 30 V, 120 Hz.
6. Inductive reactance is directly proportional to ________ and ________.
7. The reactance of 3-H choke is 1132 Ω. What is its reactance if the frequency is tripled?
8. What is the reactance of the choke in Question 7 if the frequency is halved?
9. Two similar 3-H chokes with a reactance of 1132 Ω each are connected in series. Calculate their net inductance.
10. What is the net reactance of the two chokes in Question 9?

21-6 SERIES AND PARALLEL X_L'S

Series Circuit

Since X_L is an Ohm's law opposition to ac current, the total reactance of series connected inductors can be found by the simple sum of the individual reactances:

$$X_{LT} = X_{L1} + X_{L2} + X_{L3} \text{ etc.} \quad \text{ohms} \qquad (21\text{-}4)$$

EXAMPLE 21-6

Determine the total reactance of the inductors in the circuit schematic of Fig. 21-6. What is the total inductance of the circuit?

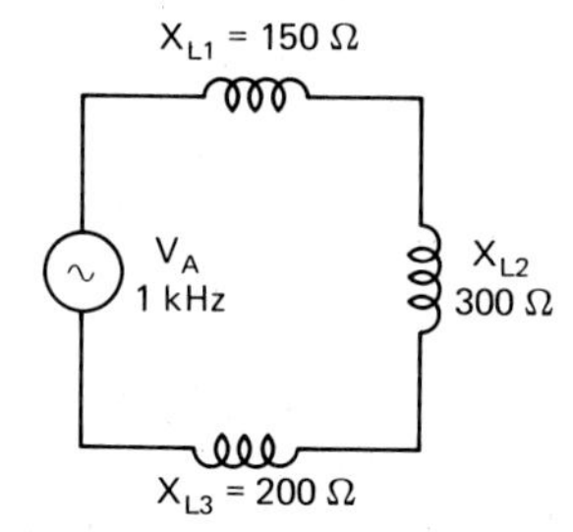

FIGURE 21-6 Circuit schematic.

Solution: The total reactance can be found by summing the individual reactances. Using equation (21-4) gives

$$X_{LT} = X_{L1} + X_{L2} + X_{L3}$$
$$= 150\ \Omega + 300\ \Omega + 200\ \Omega = 650\ \Omega$$

The total inductance of the circuit can be found by rearranging equation (21-3), which gives

$$L_T = \frac{X_{LT}}{2\pi f} = \frac{650\ \Omega}{6.28 \times 1000\ \text{Hz}} = 103.5\ \text{mH}$$

EXAMPLE 21-7

From the circuit schematic of Fig. 21-7 determine:

(a) The total reactance of the circuit.
(b) The source voltage.

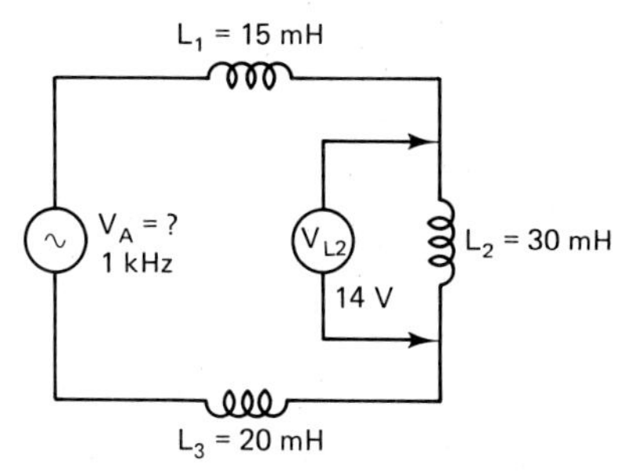

FIGURE 21-7 Circuit schematic.

Solution:

(a) The total reactance of the circuit can be found by summing the individual inductors to find the total inductance. Then using equation (21-3) to find the total reactance will give

$$L_T = L_1 + L_2 + L_3$$
$$= 15 \text{ mH} + 30 \text{ mH} + 20 \text{ mH}$$
$$= 65 \text{ mH}$$

$$X_{LT} = 2\pi f L_T$$
$$= 6.28 \times 1 \times 10^3 \text{ Hz} \times 65 \times 10^{-3} \text{ H}$$
$$= 408\ \Omega$$

(b) The applied voltage V_A can be determined by finding the current through L_2, then using Ohm's law to find V_A as follows. The current through L_2 can be found using Ohm's law, but first we must find X_{L2} using equation (21-3).

$$X_{L2} = 2\pi f L_2$$
$$= 6.28 \times 1 \times 10^3 \text{ Hz} \times 30 \times 10^{-3} \text{ H}$$
$$= 188\ \Omega$$

Then

$$I_{L2} = I_T = \frac{V_{L2}}{X_{L2}} = \frac{14 \text{ V}}{188\ \Omega} = 74.5 \text{ mA}$$

Therefore,

$$V_A = I_T \times X_{LT}$$
$$= 74.5 \times 10^{-3} \text{ A} \times 408\ \Omega$$
$$= 30.4 \text{ V}$$

Parallel Circuit

Similar to parallel resistance, the total reactance of a parallel circuit can be found using the reciprocal sum equation:

$$\frac{1}{X_{LT}} = \frac{1}{X_{L1}} + \frac{1}{X_{L2}} + \frac{1}{X_{L3}} \text{ etc.} \quad \text{ohms} \quad (21\text{-}5)$$

If two parallel inductors are taken at a time, their total reactance can be found using the product-over-sum equation:

$$X_{LT} = \frac{X_{L1} \times X_{L2}}{X_{L1} + X_{L2}} \quad \text{ohms} \quad (21\text{-}6)$$

EXAMPLE 21-8

Determine the total reactance of the inductors in the circuit schematic of Fig. 21-8.

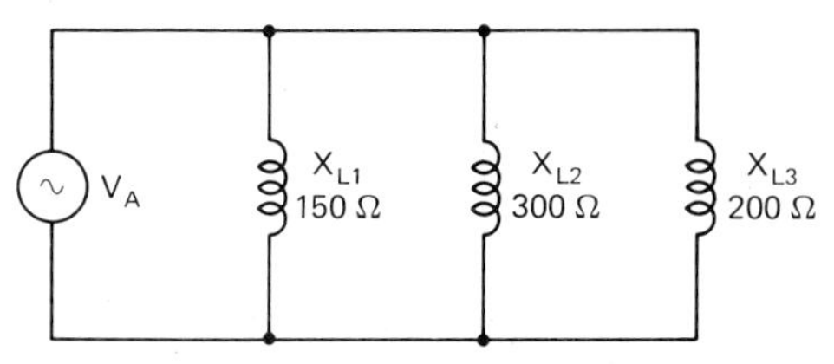

FIGURE 21-8 Circuit schematic.

Solution: The total reactance can be found by using the reciprocal equation (21-5) or product-over-sum equation (21-6) with two reactances at a time. Using equation (21-5) gives

$$\frac{1}{X_{LT}} = \frac{1}{X_{L1}} + \frac{1}{X_{L2}} + \frac{1}{X_{L3}}$$
$$= \frac{1}{150\ \Omega} + \frac{1}{300\ \Omega} + \frac{1}{200\ \Omega} = 66.7\ \Omega$$

Note: The calculation above is easily done on a calculator by summing the inverse of each reactance, then taking the inverse of the total.

Using equation (21-6) gives

$$X_{LT} = \frac{X_{L1} \times X_{L2}}{X_{L1} + X_{L2}} = \frac{150\ \Omega \times 300\ \Omega}{150\ \Omega + 300\ \Omega} = 100\ \Omega$$
$$= \frac{X_{L3} \times 100\ \Omega}{X_{L3} + 100\ \Omega} = \frac{200\ \Omega \times 100\ \Omega}{200\ \Omega + 100\ \Omega} = 66.7\Omega$$

EXAMPLE 21-9

From the circuit schematic of Fig. 21-9, determine;

(a) The total reactance of the circuit.
(b) The source voltage.
(c) The total source current.

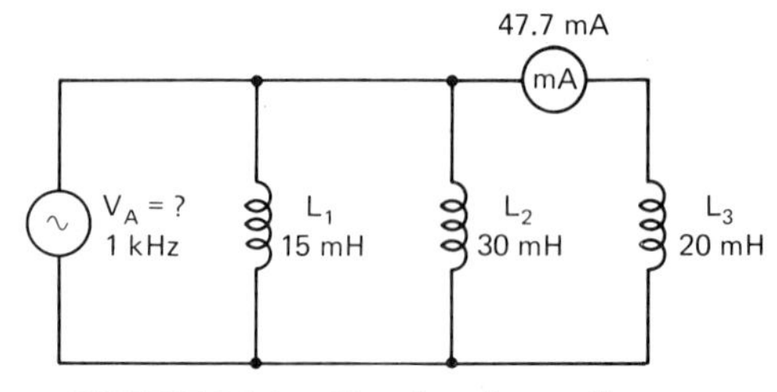

FIGURE 21-9 Circuit schematic.

Solution:

(a) If we find the individual reactance of the inductors, we can then find the total reactance. Using equation (21-3) gives us

$$X_{L1} = 2\pi f L_1 = 6.28 \times 10^3 \text{ Hz} \times 15 \times 10^{-3} \text{ H}$$
$$= 94.2\ \Omega$$

$$X_{L2} = 2\pi f L_2 = 6.28 \times 10^3 \text{ Hz} \times 30 \times 10^{-3} \text{ H} = 188.4\ \Omega$$

$$X_{L3} = 2\pi f L_3 = 6.28 \times 10^3 \text{ Hz} \times 20 \times 10^{-3} \text{ H} = 125.6\ \Omega$$

The total reactance can be calculated using the product-over-sum equation (21-6).

$$X'_{LT} = \frac{X_{L1} \times X_{L2}}{X_{L1} + X_{L2}} = \frac{94.2\ \Omega \times 188.4\ \Omega}{94.2\ \Omega + 188.4\ \Omega} = 62.8\ \Omega$$

$$X_{LT} = \frac{62.8\ \Omega \times X_{L3}}{62.8\ \Omega + X_{L3}} = \frac{62.8\ \Omega \times 125.6\ \Omega}{62.8\ \Omega + 125.6\ \Omega} = 41.9\ \Omega$$

(b) Since L_1, L_2, and L_3 are in parallel with the source, finding the voltage across X_{L3} will give this value.

$$V_A = V_{L3} = I_{L3} \times X_{L3} = 47.7 \times 10^{-3} \text{ A} \times 125.6\ \Omega = 6 \text{ V}$$

(c) The total source current can be found using Ohm's law since we know the source voltage and the total reactance.

$$I_T = \frac{V_A}{X_{LT}} = \frac{6 \text{ V}}{41.9\ \Omega} = 143.2 \text{ mA}$$

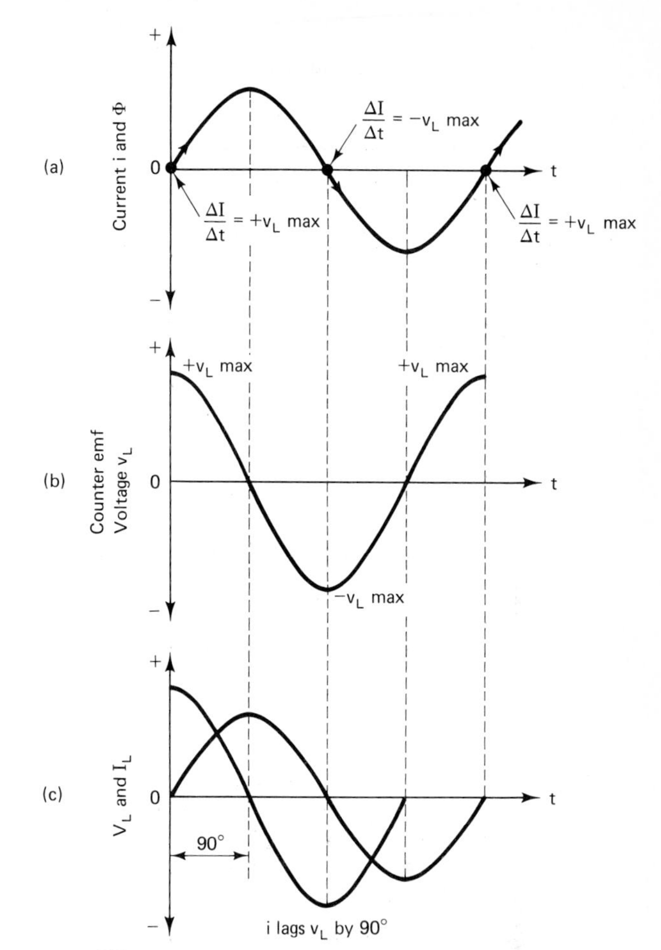

FIGURE 21-10 Waveform diagrams to represent the phase relationship of inductive current I_L and voltage V_L.

21-7 PHASE RELATIONSHIP OF INDUCTIVE CURRENT AND COUNTER EMF

A sinusoidal current waveform flows through an inductor, as shown in Fig. 21-10(a). The maximum rate of change is at each zero crossing (i.e., at 0°, 180°, and 360°). Therefore, the induced counter emf (v_L) will be maximum at these points. At the current peaks, the rate of change is zero; therefore, v_L will be zero.

Plotting v_L at the 0° current rate of change point will give a maximum value that will be positive since the current change is in the positive direction, as shown in Fig. 21-10(a). Similarly, at the 180° current rate of change point the current change is in the negative direction, so v_L will be maximum in the negative direction at this point. At the 360° point, the current change is in the positive direction, so v_L will have a positive maximum peak value.

The inductive voltage v_L can be superimposed over the inductive current i_L as shown in Fig. 21-10(c). Then the *inductive current can be shown as lagging* v_L by 90°.

Phasor Diagram

The waveforms in Fig. 21-10 are the method by which we can graphically show that the current lags the applied voltage by 90°. Once this concept is grasped, a phasor diagram is a much simpler method to show this relationship and is shown in Fig. 21-11. The current waveform of Fig. 21-11 is used as the reference waveform so that the current phasor is drawn along the x axis and is the reference phasor. The inductor voltage waveform is shown leading the current by 90°; the current is zero, while the inductive voltage is maximum, as shown in Fig. 21-11. Since the phasor is considered to be rotating counterclockwise, the voltage is shown leading the current by 90°.

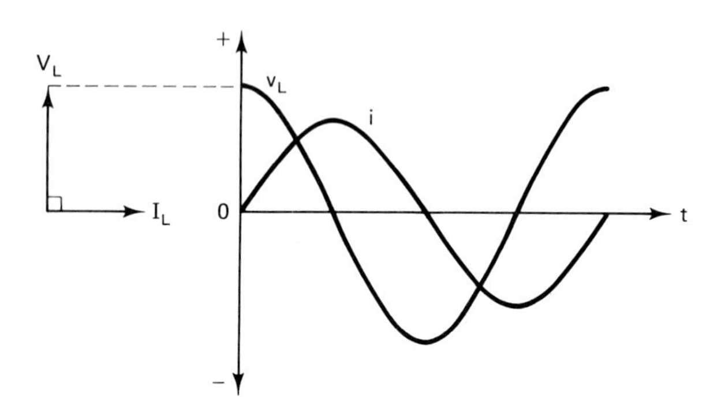

FIGURE 21-11 Waveform and phasor diagram to represent inductive current I_L lagging V_L by 90°.

SUMMARY

The ac opposition to current flow through an inductor is termed inductive reactance (X_L).

Inductive reactance is an Ohm's law value.

Inductive reactance is directly proportional to inductance and frequency.

Inductive reactance can be found using equation (21-3):

$$X_L = 2\pi fL$$

where f is in hertz
L is in henrys

The total reactance of series-connected inductors can be found by the simple sum of the individual reactances:

$$X_{LT} = X_{L1} + X_{L2} + X_{L3} \quad \text{etc.} \qquad \text{ohms} \tag{21-4}$$

The total reactance of parallel-connected inductors can be found using the reciprocal equation:

$$\frac{1}{X_{LT}} = \frac{1}{X_{L1}} + \frac{1}{X_{L2}} + \frac{1}{X_{L3}} \quad \text{etc.} \qquad \text{ohms} \tag{21-5}$$

If two parallel inductors are taken at a time, their total reactance can be found using the product-over-sum equation:

$$X_{LT} = \frac{X_{L1} \times X_{L2}}{X_{L1} + X_{L2}} \qquad \text{ohms} \tag{21-6}$$

The inductive current lags the applied voltage by 90° in a pure inductive circuit.

REVIEW QUESTIONS

21-1. Why is inductive reactance applicable only to an ac circuit and not to a dc circuit?

21-2. What reactive opposition establishes the inductive reactance of a circuit?

21-3. A current of 85 mA flows through an inductor connected across 60 V ac. What is the reactance of the inductor?

21-4. If the frequency of the source is halved, what will be the reactance of the inductor in Question 21-3?

21-5. If the frequency of the source is doubled, what will be the reactance of the inductor in Question 21-3?

21-6. What will be the reactance of the inductor in Question 21-3 if the inductance is doubled? If halved?

21-7. A certain choke has a reactance of 110 kΩ at 5 kHz. Calculate its inductance.

21-8. Three inductors, 120 mH, 56 mH, and 400 mH, are connected in series. Find their total reactance if the source frequency is 10 kHz.

21-9. The three inductors of Question 21-8 are connected in parallel. Find their total reactance.

21-10. From the schematic of Fig. 21-12, calculate the voltage drop across each inductor and the source voltage.

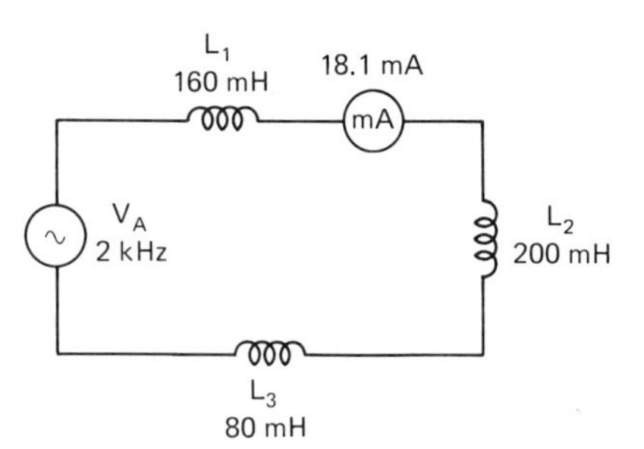

FIGURE 21-12 Circuit schematic.

21-11. From the schematic of Fig. 21-13, calculate the source voltage and the total current in the circuit.

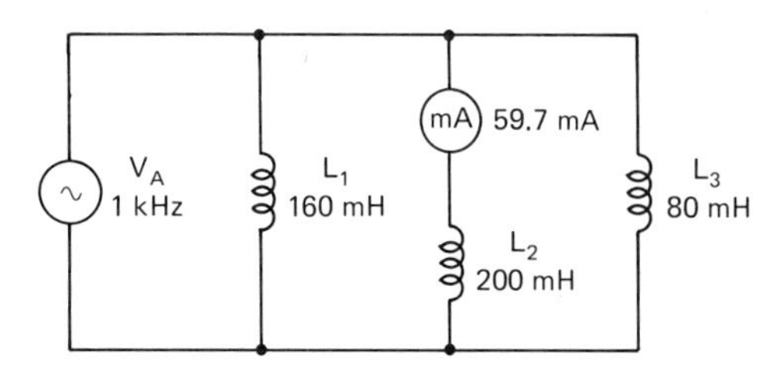

FIGURE 21-13 Circuit schematic.

21-12. What is the phase relationship of the applied voltage to the inductive current? Which one leads? Draw a representative phasor diagram to show this.

SELF-EXAMINATION

Choose the most suitable word, phrase, or value to complete each statement.

21-1. What is the inductance of the coil in the schematic of Fig. 21-14?

21-2. Find the reactance of the coil in the schematic of Fig. 21-14 if the source is now 30 V ac with a frequency of 1 kHz.

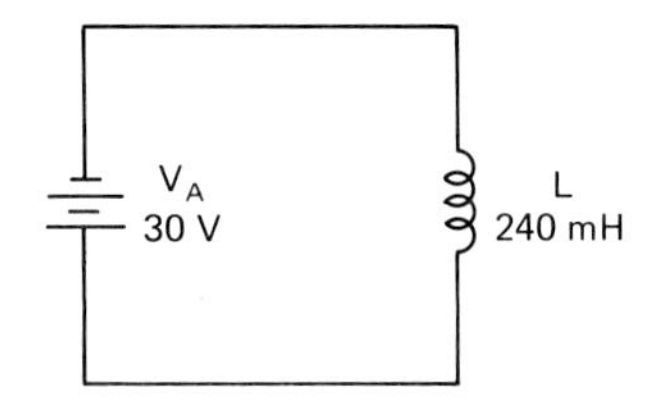

FIGURE 21-14 Circuit schematic.

21-3. What is the circuit current for the inductor of Question 21-2?

21-4. An inductance L_1 produces a counter emf of 65 V, while L_2 produces a counter emf of 130 V in a series circuit. Which coil has the greatest reactance?

21-5. What is the total reactance of two series-connected inductors with X_{L1} of 300 Ω and X_{L2} of 600 Ω?

21-6. Find the total current in the circuit schematic of Fig. 21-15.

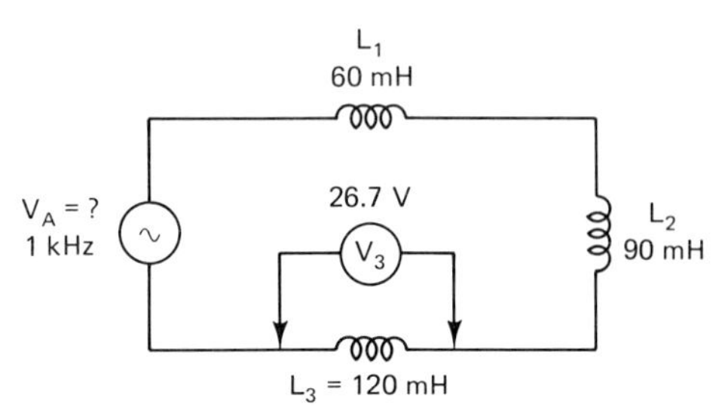

FIGURE 21-15 Circuit schematic.

21-7. Find the applied voltage in the circuit schematic of Fig. 21-15.

21-8. What is the total reactance of the circuit in Fig. 21-15?

21-9. What is the total inductance of the circuit in Fig. 21-15?

21-10. Find the inductance of L_3 in the circuit schematic of Fig. 21-16.

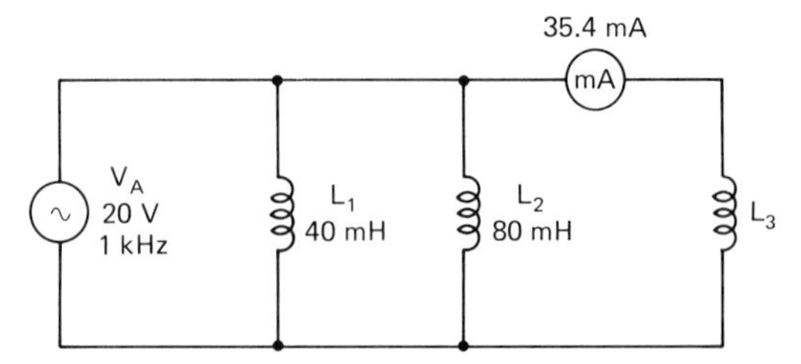

FIGURE 21-16 Circuit schematic.

21-11. Find the total inductance of the circuit in Fig. 21-16.

21-12. Find the total reactance of the circuit in Fig. 21-16.

21-13. Find the total current in the circuit schematic of Fig. 21-16.

21-14. The voltage across a pure inductive circuit (*lags*, *leads*) the current through the inductor by ________ degrees.

ANSWERS TO PRACTICE QUESTIONS

21.1

1. 0 V
2. 76 V (avg)
3. 2264 Ω
4. 15,072 Ω
5. 159 mH
6. frequency, inductance
7. 3396 Ω
8. 566 Ω
9. 6 H
10. 2264 Ω

22

Impedance in the Series and Parallel *RL* Circuit

INTRODUCTION

In Chapter 21 we looked at an ideal inductor connected to a sinusoidal voltage source whose internal resistance was ignored. Practical inductors do have some internal resistance, ranging from a fraction of an ohm to hundreds of ohms. This internal resistance alters the phase relationship of the ac voltages in the series resistance–inductance (*RL*) circuit and the phase relationship of the ac currents in the parallel *RL* circuit.

A resistor in some circuits is connected in series or in parallel with an inductor. In this case the phase relationship of the ac voltages and currents is altered further. Resistance, whether it is the resistance of the winding or an external resistor, does not react to the ac current. Its opposition remains passive even though the frequency of the current changes. Inductance, on the other hand, is reactive. Its opposition, X_L, changes with a change in frequency.

Combining the two types of oppositions, X_L and R, must be done by considering the phase relationship of the voltages and current in both the series and parallel *RL* circuit. This combined opposition to ac current flow is called *impedance*.

In this chapter we examine impedance in a series and parallel *RL* circuit and the effect on impedance with a change in frequency and inductance.

MAIN TOPICS IN CHAPTER 22

OBJECTIVES

After studying Chapter 22, the student should be able to

1. Master the current and voltage phase relationship concept for a series *RL* circuit.
2. Master the current and voltage phase relationship concept for a parallel *RL* circuit.

3. Apply phasor diagrams to series *RL* circuits and calculate voltage, current, and impedance.
4. Apply phasor diagrams to parallel *RL* circuits and calculate voltage, current, and impedance.

22-1 VOLTAGE AND CURRENT PHASE RELATIONSHIP FOR A PURE RESISTANCE VERSUS PURE INDUCTANCE

Ac current flowing through a resistor or resistance in an ac circuit does not produce a counter emf to react against the changing current. As in the case of dc, its opposition to ac current flow is due to the acceleration of conduction electrons, which continually collide with the resistive materials atoms. The voltage required to produce a certain level of current flow through a resistor is its voltage drop or loss, or $V = IR$. Since we are dealing with a sinusoidal ac voltage and current, the phase difference between the current flowing through the resistance and the voltage drop across the resistance will be zero degrees. The voltage and current are said to be in phase as shown by the waveform diagram of Fig. 22-1(a). The phasor diagram for the current and voltage across the resistance is drawn with the voltage phasor superimposed or "in line" with the current phasor, indicating a zero-degree phase angle [see Fig. 22-1(a)].

Ac current flowing through an inductor in an ac circuit produces a changing magnetic flux that induces a counter emf within the inductor. This counter emf not only opposes the current flow but causes it to lag 90° behind the voltage. The voltage and current are said to be 90° out of phase with each other, as shown by the waveform of Fig. 22-1(b).

The phasor diagram for the 90° lagging current or leading voltage is shown in Fig. 22-1(b). Note that the voltage and current phasors are drawn to an arbitrary length. The lengths are not important except where a graphical solution is required. Each phasor is labeled with a capital letter, as shown.

22-2 VOLTAGE AND CURRENT PHASE RELATIONSHIP IN THE SERIES *RL* CIRCUIT

In analyzing the series *RL* circuit, we start with the basic series circuit characteristic—that current is common in all parts of the circuit. With this in mind, the current waveform shown at the top of the waveform diagram of Fig. 22-2 is used as our reference for all the other waveforms.

The voltage waveform across the series resistor, V_R, is in phase with the current through the resistor, as shown by the waveform diagram of Fig. 22-2. Since the

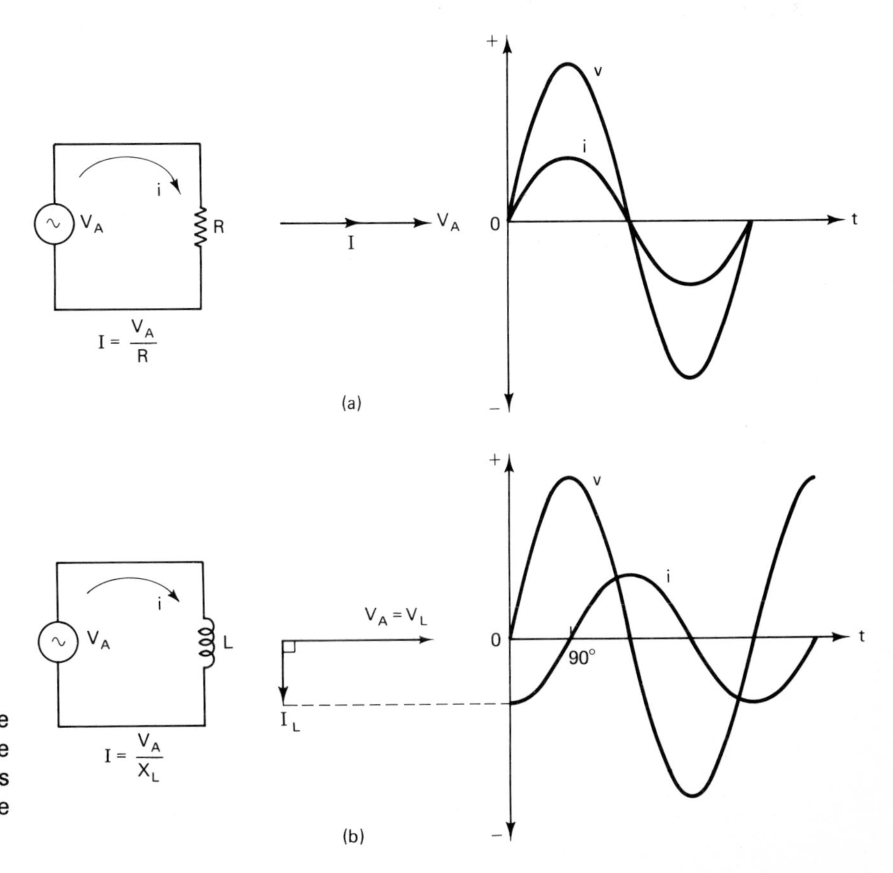

FIGURE 22-1 Current is in phase with the applied voltage in the pure resistive circuit. Current lags the applied voltage by 90° in the pure inductive circuit.

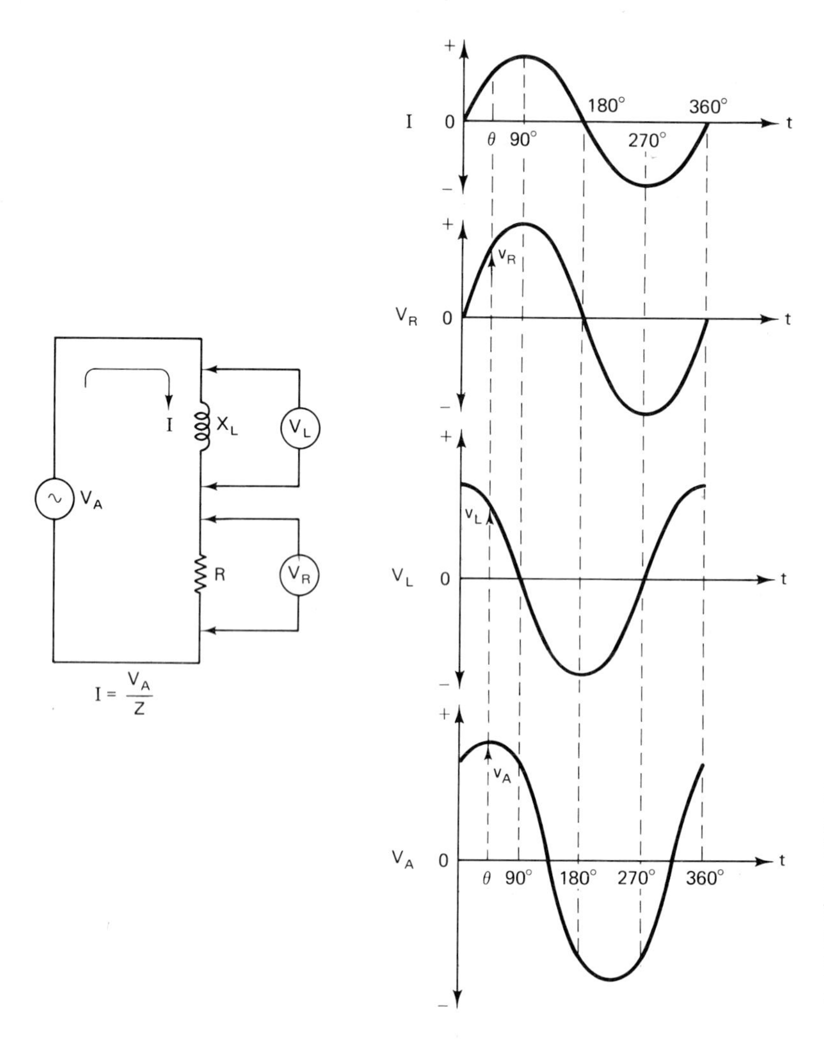

FIGURE 22-2 Series *RL* circuit and waveforms showing that V_L is 90° out of phase with V_R. The applied voltage, V_A, is the resultant of summing the instantaneous values of V_R and V_L algebraically at each instant of time, or is the phasor sum of V_R and V_L.

voltage across the inductor will lead the current through the inductor by 90°, V_L will be 90° out of phase with V_R, as shown by the waveform diagram of Fig. 22-2.

The applied voltage, V_A, is the *resultant of summing the instantaneous values* of v_R and v_L algebraically at each instant of time. The voltage drop V_R and V_L can be added as phasor quantities, shown in the phasor diagrams of Fig. 22-3. V_L and V_R form the two sides of a right-angle triangle with V_A the hypotenuse.

The rms, peak, or peak-to-peak values (usually rms) for V_L and V_R are representated by the lengths of the two sides of the right-angle triangle. The length of the hypotenuse represents the rms, peak, or peak-to-peak value of the applied voltage. *Note:* In all cases the values must be consistent.

Voltages in a Series *RL* Circuit

V_R and V_L can be found by using Ohm's law, $V_R = IR$, and $V_L = IX_L$. The applied voltage, V_A, however, can be found by using the phasor sum, that is, by applying the Pythagorean theorem for a right-angle triangle, shown in Fig. 22-3(c), to give

$$V_A = \sqrt{V_R^2 + V_L^2} \quad \text{volts} \qquad (22\text{-}1)$$

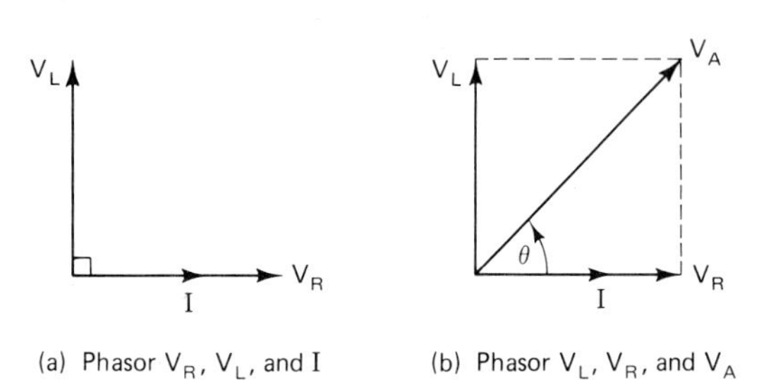

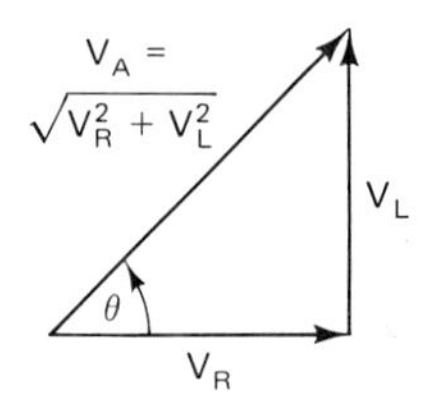

FIGURE 22-3 Voltage phasor diagrams for the series *RL* circuit.

V_A can also be found using Ohm's law provided that we know the current and the total opposition, called impedance (Z), which is the combined opposition of X_L and R with due regard to the phase relationship of the voltage and current. Therefore,

$$V_A = I_T Z \quad \text{volts} \tag{22-2}$$

Phase Angle between V_A and I

Since the current is in phase with V_R, we can find the phase angle by which the applied voltage leads the current using a trigonometric function. In this case the tangent (tan) lends itself as the most convenient, since V_R and V_L are adjacent and opposite sides to the angle θ [see Fig. 22-3(c)].

$$\tan\theta = \frac{V_L}{V_R} \tag{22-3}$$

or

$$\theta = \tan^{-1}\frac{V_L}{V_R} \quad \text{or} \quad \arctan\frac{V_L}{V_R} \tag{22-3a}$$

Note: The symbols $\tan^{-1}$ or arctan V_L/V_R mean that θ is the angle whose tangent is V_L/V_R. Using a scientific calculator, this phase angle can be obtained by dividing V_L by V_R on the calculator, then pressing the INV and TAN buttons in that sequence.

EXAMPLE 22-1

Find (a) V_R, (b) V_L, (c) V_A, and (d) θ for the series *RL* circuit schematic of Fig. 22-4.

Solution: Using Ohm's law for both V_R and V_L since current is common gives

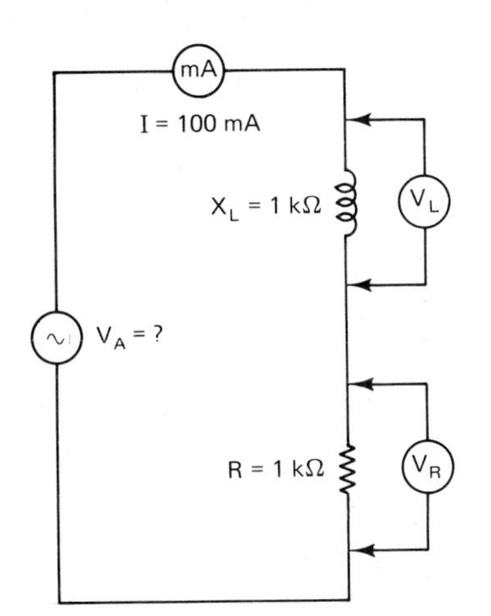

FIGURE 22-4 Series *RL* circuit schematic.

(a) $V_R = IR = 100\text{ mA} \times 1\text{ k}\Omega = 100\text{ V}$.

(b) $V_L = IX_L = 100\text{ mA} \times 1\text{ k}\Omega = 100\text{ V}$.

(c) Applying equation (22-1) to find V_A gives

$$V_A = \sqrt{V_R^2 + V_L^2} = \sqrt{(100\text{ V})^2 + (100\text{ V})^2}$$
$$= 141\text{ V}$$

(d) Applying equation (22-3a) and using a calculator gives

$$\theta = \tan^{-1}\frac{V_L}{V_R} = \tan^{-1}\frac{100\text{ V}}{100\text{ V}} = \tan^{-1}1 = 45°$$

Having obtained 1 on the calculator ($100/100 = 1$), press the INV button and then press the TAN button to obtain 45°; therefore, $\theta = 45°$. This means that the angle θ represents the phase angle by which V_A leads V_R or 45°. Since V_R is in phase with the circuit current, the phase angle θ also represents the phase angle by which the applied voltage leads the current.

EXAMPLE 22-2

From the series *RL* schematic of Fig. 22-5, find:

(a) The circuit current.

(b) The voltage drop V_L across the inductor.

(c) The phase angle between the applied voltage and the current.

(d) The inductance of the coil.

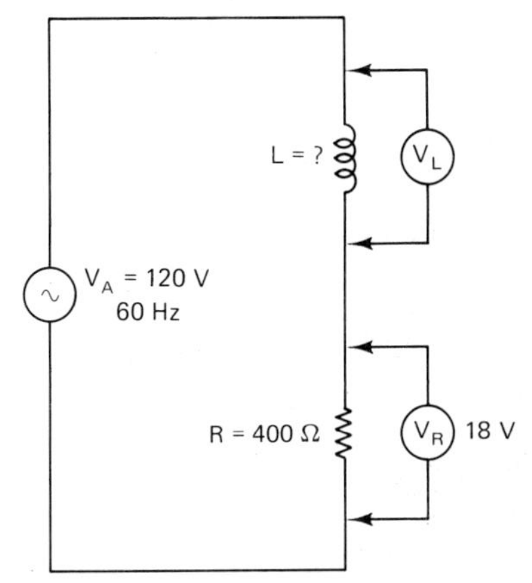

FIGURE 22-5 Series *RL* circuit.

Solution:

(a) The current is common in a series circuit. Finding the current through the resistor will give the circuit current.

$$I_R = I_T = \frac{V_R}{R} = \frac{18\ \text{V}}{400\ \Omega} = 45\ \text{mA}$$

(b) Rearranging equation (22-1) will give V_L.

$$V_A = \sqrt{V_R^2 + V_L^2}$$

Therefore,

$$V_L = \sqrt{V_A^2 - V_R^2} = \sqrt{(120\ \text{V})^2 - (18\ \text{V})^2}$$
$$= 118.6\ \text{V}$$

(c) The phase angle between V_A and the current can be found using equation (22-3) or (22-3a).

$$\theta = \tan^{-1}\frac{V_L}{V_R} = \tan^{-1}\frac{118.6\ \text{V}}{18\ \text{V}}$$
$$= \tan^{-1} 6.58 = 81.4°$$

This shows that V_A leads I by approximately 81°.

(d) The inductance of the coil can be found by rearranging equation (21-3), $X_L = 2\pi fL$. But first we must find X_L using Ohm's law.

$$X_L = \frac{V_L}{I} = \frac{118.6\ \text{V}}{45\ \text{mA}} = 2.64\ \text{k}\Omega$$

and

$$L = \frac{X_L}{2\pi f} = \frac{2.6 \times 10^3\ \Omega}{6.28 \times 60} = 6.9\ \text{H}$$

22-3 IMPEDANCE IN THE SERIES *RL* CIRCUIT

The total opposition to ac current flow in a series *RL* circuit is called *impedance*, *Z*. Impedance can be found by using Ohm's law provided that we know the total circuit current. Therefore,

$$Z = \frac{V_A}{I_T} \quad \text{ohms} \qquad (22\text{-}4)$$

Inductive reactance is a reactive component, while resistance is a passive component. Their combined ac opposition in a series *RL* circuit is determined with due regard to the phase relationship of the current and voltages in the circuit.

The impedance phasor is developed from the voltage phasor of Fig. 22-6(a). With V_L transferred to form a closed triangle, as shown in Fig. 22-6(b), IR, IX_L, and IZ are substituted for each of the phasor voltages. Since current is the same or common in all parts of a series circuit, the current of each voltage phasor can be dropped, leaving the impedance triangle, as shown in Fig. 22-6(c). This can be shown by developing equation (22-1) as follows:

$$V_A = \sqrt{V_R^2 + V_L^2} \quad \text{volts} \qquad (22\text{-}1)$$

and

$$V_R = IR \qquad V_L = IX_L \qquad V_A = IZ$$

Substituting for each voltage gives

$$V_A = IZ = \sqrt{(IR)^2 + (IX_L)^2}$$
$$= \sqrt{I^2(R^2 + X_L^2)}$$
$$= \sqrt{I^2(R^2 + X_L^2)}$$
$$= I\sqrt{R^2 + X_L^2}$$

Therefore,

$$Z = \sqrt{R^2 + X_L^2} \quad \text{ohms} \qquad (22\text{-}5)$$

where Z is the impedance, in ohms
R is the resistance, in ohms
X_L is the inductive reactance, in ohms

Phase Angle between V_A and I

Since the impedance triangle corresponds to the voltage triangle, the phase angle between V_A and I_T will correspond to the phase angle between Z and R. Using the tangent (tan) trigonometric function for the impedance triangle of Fig. 22-6(c) gives

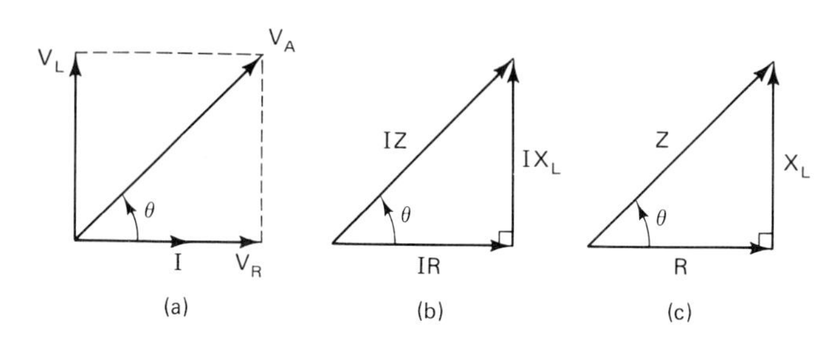

FIGURE 22-6 Voltage and impedance triangle for a series *RL* circuit. (a) Voltage phasor. (b) Equivalent triangle. (c) Impedance triangle.

$$\tan \theta = \frac{X_L}{R} \qquad (22\text{-}6)$$

or

$$\theta = \tan^{-1} \frac{X_L}{R} = \arctan \frac{X_L}{R} \qquad (22\text{-}6a)$$

EXAMPLE 22-3

Use Fig. 22-4 series *RL* schematic to find:

(a) The impedance of the circuit.
(b) The phase angle between V_A and the current.

Solution:

(a) The impedance of the circuit can be found using equation (22-5).

$$Z = \sqrt{R^2 + X_L^2} = \sqrt{(1\text{ k}\Omega)^2 + (1\text{ k}\Omega)^2}$$
$$= 1.41\text{ k}\Omega$$

(b) The phase angle between V_A and I can be found using equation (22-6a).

$$\theta = \tan^{-1} \frac{X_L}{R} = \tan^{-1} \frac{1\text{ k}\Omega}{1\text{ k}\Omega} = \tan^{-1} 1 = 45^\circ$$

Note that whenever X_L equals R, the phase angle between V_A and I_T will be 45°.

EXAMPLE 22-4

Use Example 22-2 solutions to find the following in the series *RL* schematic of Fig. 22-5.

(a) The impedance of the circuit.
(b) The phase angle between V_A and I.
(c) The total current in the circuit by using the Ohm's law equation.

Solution:

(a) The impedance of the circuit can be found by using equation (22-4) or (22-5).

$$Z = \frac{V_A}{I_T} = \frac{120\text{ V}}{45\text{ mA}} = 2667\ \Omega$$

or

$$Z = \sqrt{R^2 + X_L^2} = \sqrt{(400\ \Omega)^2 + (2640\ \Omega)^2}$$
$$= 2670\ \Omega$$

(b) The phase angle between V_A and I can be found using equation (22-6a).

$$\theta = \tan^{-1} \frac{X_L}{R} = \tan^{-1} \frac{2640\ \Omega}{400\ \Omega}$$
$$= 81.4^\circ$$

This shows that V_A leads I by approximately 81.4°

(c) The total current in the circuit can be found using Ohm's law.

$$I_T = \frac{V_A}{Z} = \frac{120\text{ V}}{2667\ \Omega} = 45\text{ mA}$$

EXAMPLE 22-5

The resistance in the circuit of Fig. 22-7 represents the internal resistance of the coil. Find (a) Z, (b) V_L, (c) V_R, and (d) the phase angle θ by which the applied voltage leads the current.

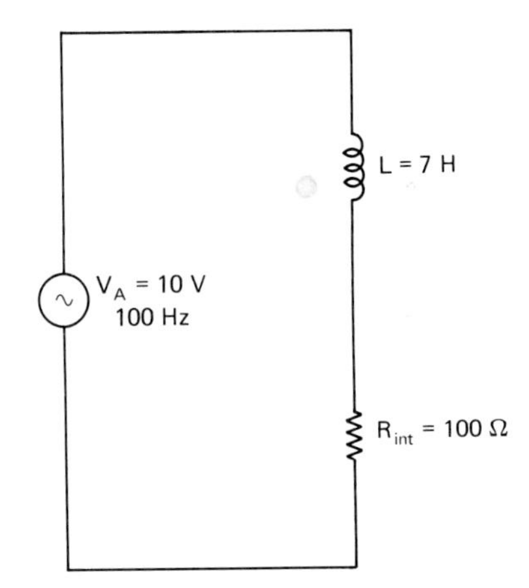

FIGURE 22-7 Series *RL* circuit.

Solution: The reactance of the inductor must be found first. Using equation (21-3) yields

$$X_L = 2\pi f L = 6.28 \times 100 \times 7\text{ H} = 4396\ \Omega$$

(a) Then the impedance of the circuit is found using equation (22-5).

$$Z = \sqrt{R^2 + X_L^2} = \sqrt{(100\ \Omega)^2 + (4396\ \Omega)^2}$$
$$= 4397\ \Omega$$

Note that the impedance is almost equal to the reactance of the coil. The current can be found using Ohm's law.

$$I = \frac{V}{Z} = \frac{10\text{ V}}{4397\ \Omega} = 2.27\text{ mA}$$

(b) The voltage drop across the inductor is then

$$V_L = IR = 2.27 \times 10^{-3}\text{ A} \times 4397\ \Omega = 9.98\text{ V}$$

(c) The voltage drop due to the internal resistance of the inductor is

$$V_R = IR = 2.27 \times 10^{-3}\text{ A} \times 100\ \Omega = 0.227\text{ V}$$

(d) The phase angle θ by which the applied voltage leads the current is

$$\theta = \tan^{-1}\frac{X_L}{R} = \tan^{-1}\frac{4396}{100}$$

$$= 88.7°$$

22-4 EFFECT OF FREQUENCY, INDUCTANCE, AND RESISTANCE ON IMPEDANCE IN THE SERIES *RL* CIRCUIT

Equation (21-3), $X_L = 2\pi fL$, shows us that X_L is directly proportional to frequency and inductance. This means that X_L doubles if the frequency or inductance is doubled and is halved if the frequency or inductance is halved, and so on.

The impedance triangles of Fig. 22-8 show that as f or L increases, X_L and Z will increase. If X_{L1} is much smaller than R, the phase angle, θ between the resistive voltage and leading applied voltage approaches 0°. Most of the voltage drop will be across the resistance; therefore, Z_1 will be mostly resistive with a low f or L.

When X_{L2} is equal to R, the phase angle θ_2 between the resistive voltage and applied voltage will be 45°. The voltage drops across X_L and R will be equal to each other; therefore, Z_2 will be a combination of X_L and R.

When X_{L3} is much greater than R, the phase angle θ_3 between the resistive voltage and the leading applied voltage approaches 90°. Most of the voltage drop will be across the inductance; therefore, Z_3 will be mostly inductive with a higher f or L.

The impedance triangles of Fig. 22-9 exemplify the effect of resistance on impedance. If R_1 is at least 1/10 smaller than X_L, the phase angle, θ, approaches 90° and the impedance, Z_1, will be mostly X_L.

When R_L is equal to X_L, the phase angle θ_2 is 45°. The impedance Z_2 is a combination of both R and X_c.

When R_3 is at least 10 times greater than X_L, the phase angle θ_3 approaches 0° and the impedance Z_3 will be mostly resistive.

PRACTICE QUESTIONS 22-1

Choose the most suitable word, phrase, or value to complete each statement.

1. An ac current flowing through a pure resistance is (*in phase, out of phase*) with its voltage drop by (*0°, 90°*).
2. An ac current flowing through a pure inductance is (*in phase, out of phase*) with its voltage drop by (*0°, 90°, 45°*).
3. The voltage across the inductor will (*lag, lead*) the current through the inductor by ________ degrees.

Refer to Fig. 22-10, the series *RL* schematic, for Questions 4 to 9.

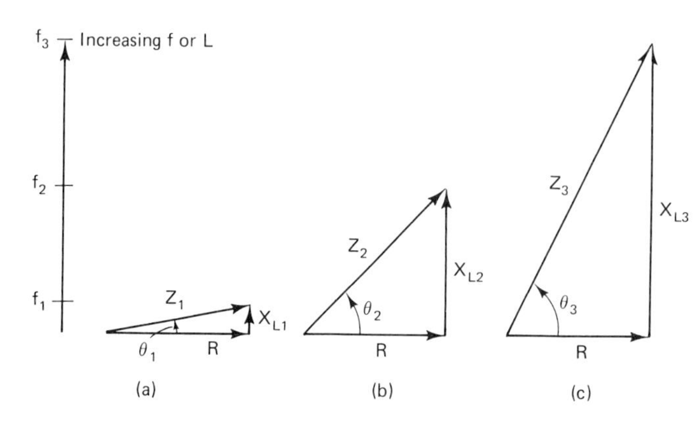

FIGURE 22-8 Effect of frequency and inductance on impedance in the series *RL* circuit. (a) $\frac{1}{10}X_{L1} \ll R$. (b) $X_{L2} = R$. (c) $X_{L3} \gg R$.

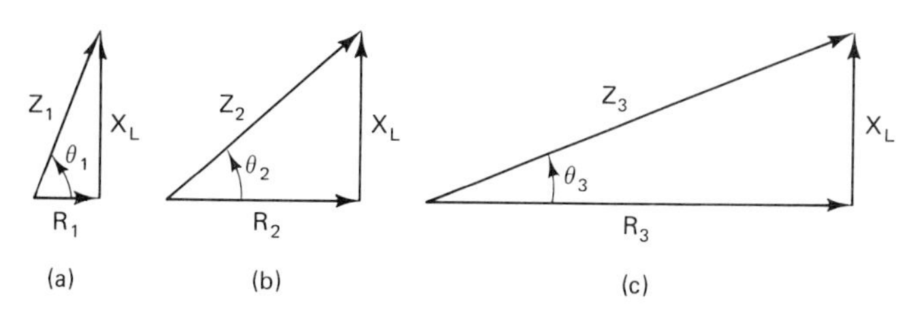

FIGURE 22-9 Effect of resistance on impedance in the series *RL* circuit. (a) $\frac{1}{10}R_1 \ll X_L$. (b) $R_2 = X_L$. (c) $10\ R_3 \gg X_L$.

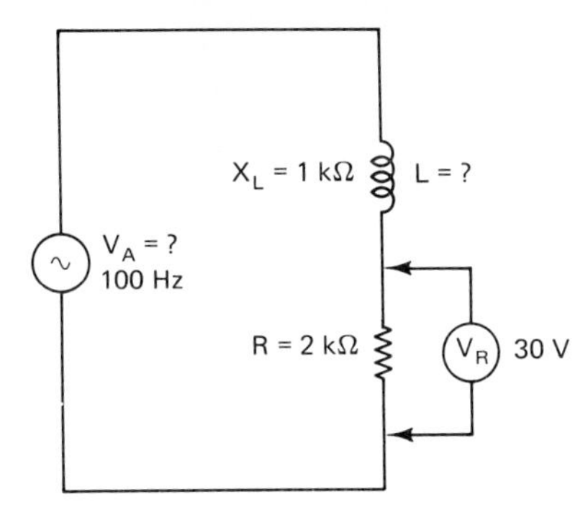

FIGURE 22-10 Series *RL* circuit schematic.

4. Find the total circuit current.
5. Find the voltage drop across the inductor.
6. Find the applied voltage V_A.
7. Find the inductance of the coil.
8. Find the impedance of the circuit.
9. Find the phase angle between the current and the applied voltage.

Refer to Fig. 22-11, the series *RL* schematic, for Questions 10 to 17.

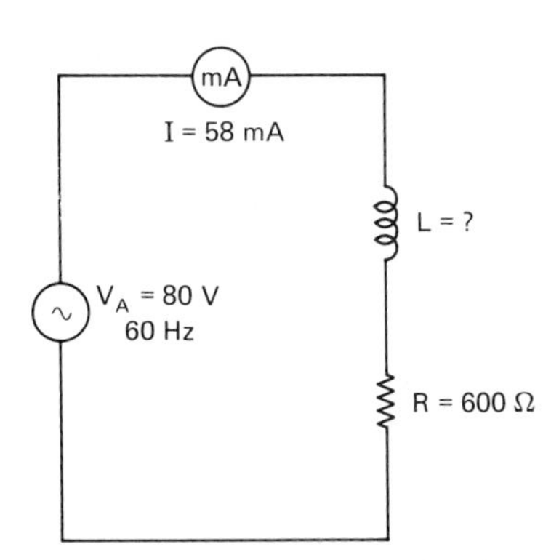

FIGURE 22-11 Series *RL* circuit schematic.

10. Find the voltage drop across the inductor.
11. Find the voltage drop across the resistance.
12. Find the phase angle between the current and the applied voltage.
13. Find the inductance of the coil.
14. What is the reactance of the coil if the frequency of the source is increased to 300 Hz?
15. What is the impedance of the circuit if the frequency of the source is increased to 300 Hz?
16. What is the phase angle between the current and the applied voltage of Question 15?
17. What type of impedance is mostly present in the circuit of Question 15?

22-5 CURRENT AND VOLTAGE PHASE RELATIONSHIP IN THE PARALLEL *RL* CIRCUIT

In analyzing the parallel *RL* circuit, we start with the basic parallel circuit characteristic that voltage is common to all parts of the circuit. With this in mind, the applied voltage waveform shown at the top of the waveform diagram of Fig. 22-12 is used as our reference for all the other waveforms.

The current through the parallel resistor is in phase with the applied voltage, as shown by the waveform diagram of Fig. 22-12. Since the current through the inductor will lag the applied voltage by 90°, I_L will be 90° out of phase with I_R, as shown by the waveform diagram of Fig. 22-12.

The total circuit current, I_T, is the resultant of summing the *instantaneous values* of i_R and i_L algebraically at each instant of time.

The currents I_R and I_L can be added as phasor quantities, shown in the phasor diagrams of Fig. 22-13. I_R and I_L form the two sides of a right-angle triangle and I_T the hypotenuse, as shown in the current triangle of Fig. 22-13(c).

The rms, peak, or peak-to-peak values (usually rms) for I_R and I_L are represented by the length of the two sides of the right-angle triangle. The length of the hypotenuse represents the rms, peak, or peak-to-peak value of the total circuit current. Note that in all cases the values must be consistent.

Currents in a Parallel *RL* Circuit

I_R and I_L can be found by using Ohm's law, $I_R = V_A/R$ and $I_L = V_A/X_L$. The total circuit current, however, must be found by using the phasor sum or Pythagorean theorem for a right-angle triangle,

$$I_T = \sqrt{I_R^2 + I_L^2} \qquad \text{amperes} \qquad (22\text{-}7)$$

I_T can also be found by using Ohm's law provided that we know the current and the impedance, Z, for the circuit:

$$I_T = \frac{V_A}{Z} \qquad \text{amperes} \qquad (22\text{-}8)$$

Phase Angle between I_T and V_A

Since the applied voltage is in phase with I_R, we can find the phase angle by which the total current lags the applied voltage using the tan function for the angle θ (see Fig. 22-13). The negative sign indicates that the to-

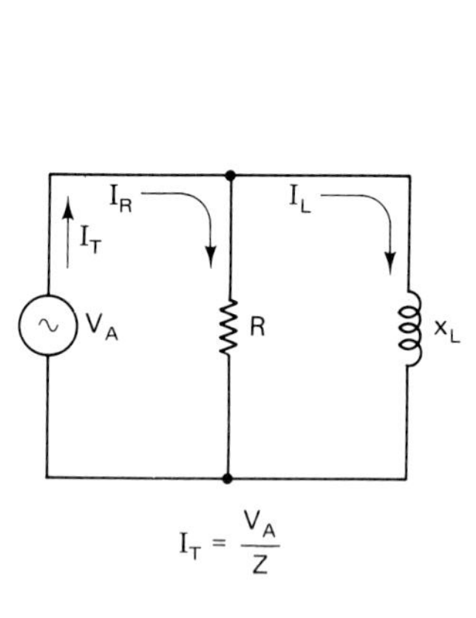

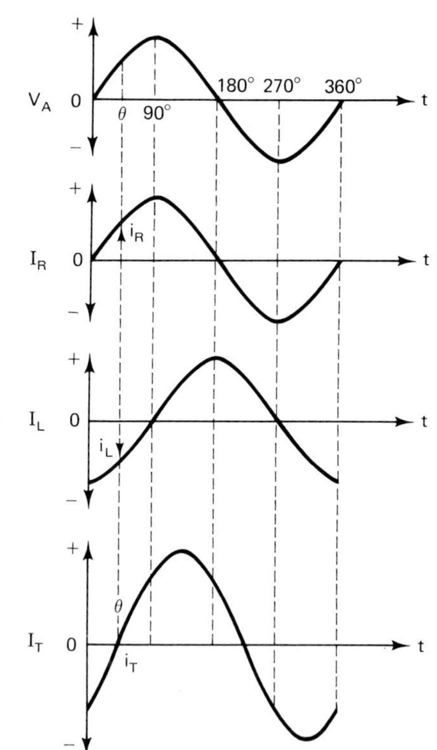

FIGURE 22-12 Parallel *RL* circuit and sine waves showing that I_L is 90° out of phase with I_R. The total circuit current I_T is the resultant of summing the instantaneous values of i_R and i_L algebraically at each instant of time, or is the phasor sum of I_R and I_L.

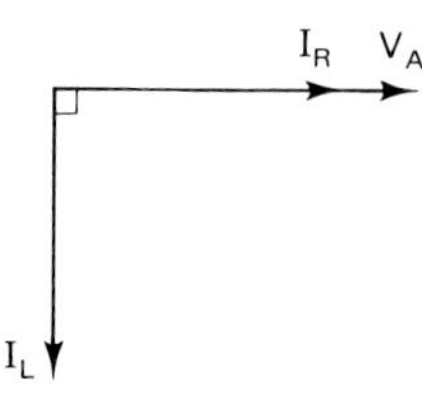

(a) Phasor I_R, I_L, and V_A

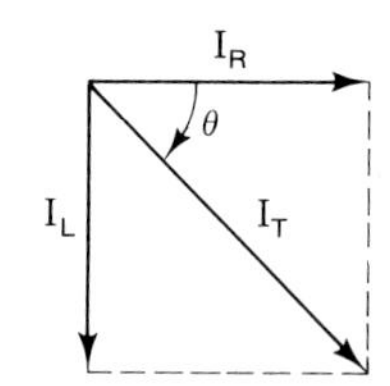

(b) Phasor I_R, I_L, and I_T

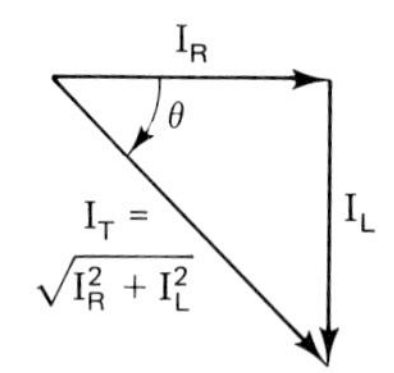

(c) Current triangle

FIGURE 22-13 Current phasor diagrams for the parallel *RL* circuit.

tal circuit lags the applied voltage similar to the series *RL* circuit.

$$\tan \theta_I = \left(-\frac{I_L}{I_R}\right) \qquad (22\text{-}9)$$

or

$$\theta_I = \tan^{-1}\left(-\frac{I_L}{I_R}\right) \qquad (22\text{-}9a)$$

EXAMPLE 22-6

Find (a) I_R, (b) V_A, and (c) L in the schematic of Fig. 22-14.

Solution:

(a) To find I_R, rearrange equation (22-7).

$$I_R = \sqrt{I_T^2 - I_L^2} = \sqrt{(100 \text{ mA})^2 - (70.7 \text{ mA})^2}$$
$$= 70.7 \text{ mA}$$

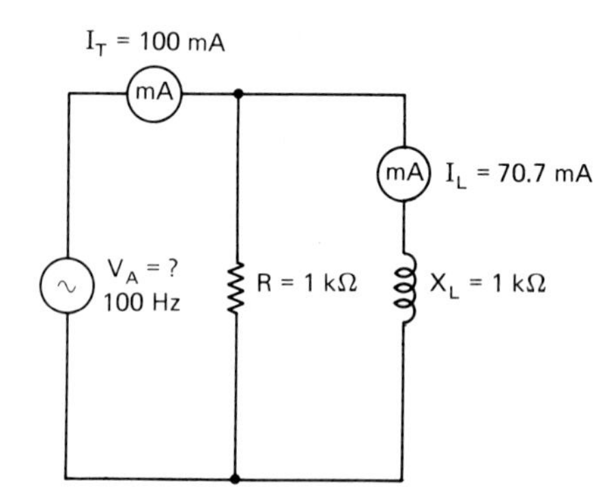

FIGURE 22-14 Parallel *RL* circuit schematic.

(b) Finding the voltage across the resistor or the inductor will give V_A since they are connected in parallel.

$$V_A = I_L \times X_L = 70.7 \text{ mA} \times 1 \text{ k}\Omega = 70.7 \text{ V}$$

(c) Since we know X_L, rearranging equation (21-3) will give the inductance L.

$$L = \frac{X_L}{2\pi f} = \frac{1000\ \Omega}{6.28 \times 100\ Hz} \cong 1.6\ H$$

EXAMPLE 22-7

Find (a) I_R, (b) I_T, (c) θ_I, and (d) L in the parallel RL schematic diagram of Fig. 22-15.

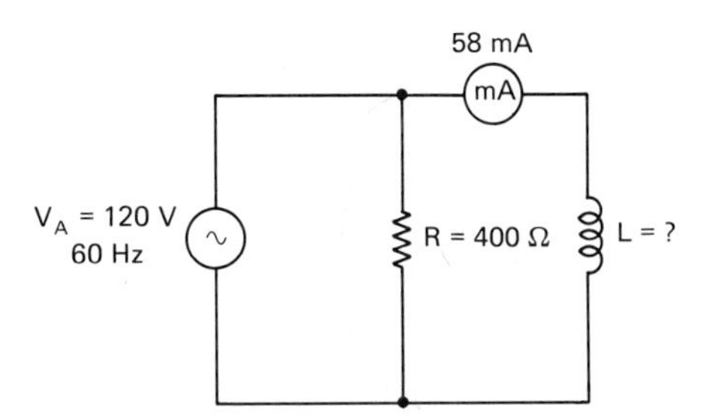

FIGURE 22-15 Parallel *RL* circuit schematic.

Solution:

(a) The current through the resistance can be found using Ohm's law.

$$I_R = \frac{V_A}{R} = \frac{120\ \text{V}}{400\ \Omega} = 300\ \text{mA}$$

(b) The total current can be found using equation (22-7).

$$I_T = \sqrt{I_R^2 + I_L^2} = \sqrt{(300\ \text{mA})^2 + (58\ \text{mA})^2}$$
$$= 306\ \text{mA}$$

(c) The phase angle between the lagging current and the applied voltage can be found using equation (22-9) or (22-9a).

$$\theta_I = \tan^{-1}\left(-\frac{I_L}{I_R}\right) = \frac{58\ \text{mA}}{300\ \text{mA}} = -11°$$

The small phase angle of $-11°$ between V_A and the circuit current indicates that the current is mostly resistive. This is due to the much lower resistance value of R shunting X_L.

(d) To calculate the coil inductance we must first find the reactance. Reactance can be solved by using Ohm's law.

$$X_L = \frac{V_A}{I_L} = \frac{120\ \text{V}}{58\ \text{mA}} = 2.07\ \text{k}\Omega$$

We can calculate the inductance of the coil by rearranging equation (21-3).

$$L = \frac{X_L}{2\ \pi f} = \frac{2070\ \Omega}{6.28 \times 60\ \text{Hz}} = 5.5\ \text{H}$$

22-6 IMPEDANCE IN THE PARALLEL *RL* CIRCUIT

The total opposition to ac current flow in a parallel RL circuit is called impedance, Z. Impedance can be found by using Ohm's law, $Z = V_A/I_T$, provided that we know the total circuit current.

Since the inductive current is out of phase with the resistive current, the total current in the circuit can be found using equation (22-7), or

$$I_T = \sqrt{I_R^2 + I_L^2}$$

Removing the square-root sign and substituting Ohm's law for each current gives

$$\left(\frac{V_A}{Z}\right)^2 = \left(\frac{V_A}{R}\right)^2 + \left(\frac{V_A}{X_L}\right)^2$$

Voltage is common in a parallel circuit, therefore we can drop V_A to give

$$\frac{1}{Z^2} = \frac{1}{R^2} + \frac{1}{X_L^2} \quad \text{ohms} \qquad (22\text{-}10a)$$

From equation (22-10a) we can derive the product-over-the-phasor-sum equation:

$$Z = \frac{\text{product}}{\sqrt{\text{sum}}} = \frac{R \times X_L}{\sqrt{R^2 + X_L^2}} \quad \text{ohms} \qquad (22\text{-}10)$$

EXAMPLE 22-8

Find the impedance in the circuit schematic of Fig. 22-14.

Solution: We can find Z using Ohm's law since the total circuit current is given.

$$Z = \frac{V_A}{I_T} = \frac{70.7\ \text{V}}{100\ \text{mA}} = 0.707\ \text{k}\Omega$$

Using the product-over-the-phasor-sum equation (22-10) gives

$$Z = \frac{R \times X_L}{\sqrt{R^2 + X_L^2}} = \frac{1\ \text{k}\Omega \times 1\ \text{k}\Omega}{\sqrt{(1\ \text{k}\Omega)^2 + (1\ \text{k}\Omega)^2}}$$
$$= 0.707\ \text{k}\Omega$$

Or using equation (22-10a) gives,

$$\frac{1}{Z^2} = \frac{1}{R_2} + \frac{1}{X_L^2} = \frac{1}{(1\ \text{k}\Omega)^2} + \frac{1}{(1\ \text{k}\Omega)^2}$$
$$Z = \sqrt{1/2\ \text{k}\Omega} = 0.707\ \text{k}\Omega$$

EXAMPLE 22-9

Find the impedance of the circuit in the schematic of Fig. 22-15.

Solution: Since the total current was found to be 306 mA from Example 22-7, the impedance can be found using Ohm's law,

$$Z = \frac{V_A}{I_T} = \frac{120\text{ V}}{306\text{ mA}} = 392\ \Omega$$

Using the product-over-phasor sum will also give Z. We must find X_L first since it is not given.

$$X_L = \frac{V_A}{I_L} = \frac{120\text{ V}}{58\text{ mA}} = 2.07\text{ k}\Omega$$

Therefore,

$$Z = \frac{R \times X_L}{\sqrt{R^2 + X_L^2}} = \frac{400\ \Omega \times 2070\ \Omega}{\sqrt{(400\ \Omega)^2 + (2070\ \Omega)^2}}$$
$$= 393\ \Omega$$

Or using equation (22-10a) gives

$$\frac{1}{Z^2} = \frac{1}{R^2} + \frac{1}{X_L^2} = \frac{1}{(400\ \Omega)^2} + \frac{1}{(2070\ \Omega)^2}$$

$$Z = \sqrt{1/65 \times 10^{-7}}\ \Omega \cong 393\ \Omega$$

22-7 EFFECT OF FREQUENCY, INDUCTANCE, AND RESISTANCE ON IMPEDANCE IN THE PARALLEL *RL* CIRCUIT

The resistance connected in parallel remains fixed for any frequency changes, so that we need only consider X_L. Inductive reactance is directly proportional to frequency and will vary with frequency. If X_L increases, the current through L will decrease, and vice versa. The increase or decrease in I_L will affect I_T, which in turn affects Z.

Consider the current triangles of Fig. 22-16. Given a parallel RL circuit where X_L is at least 10 times smaller than R, the current through X_L will be 10 times greater than I_R. As shown by the current triangle of Fig. 22-16(a), the total current or circuit current is mostly inductive, I_L. The phase angle between the lagging total current and the applied voltage will approach 90°; therefore, Z will be approximately equal to X_L.

Conversely, if the frequency of the source is changed so that X_L is 10 times greater than R, the current through the inductor will be 10 times less than I_R, as shown by the current triangle of Fig. 22-16(b). The phase angle between the lagging total current and the applied voltage will approach 0°; therefore, Z will be approximately equal to R.

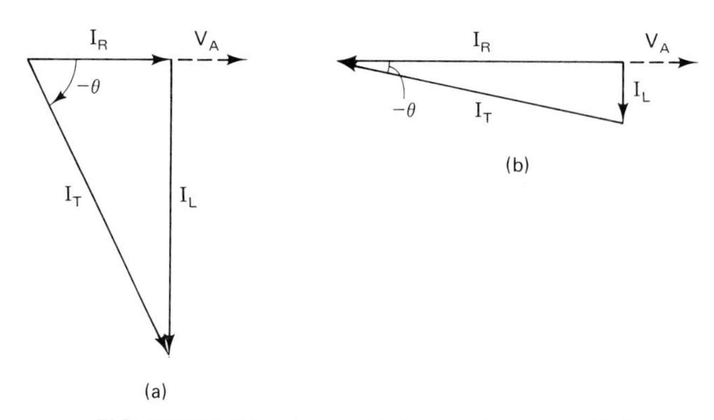

FIGURE 22-16 Current triangle for a parallel *RL* circuit where (a) X_L is 10 times smaller than R and (b) X_L is 10 times greater than R.

PRACTICE QUESTIONS 22-2

Choose the most suitable word, phrase, or value to complete each statement.

1. In a parallel *RL* circuit the current through the resistor is in phase with ________.
2. In a parallel *RL* circuit the current through the inductor (*leads, lags*) the applied voltage by ________ degrees.

Refer to Fig. 22-17, the parallel *RL* schematic, for Questions 3 to 6.

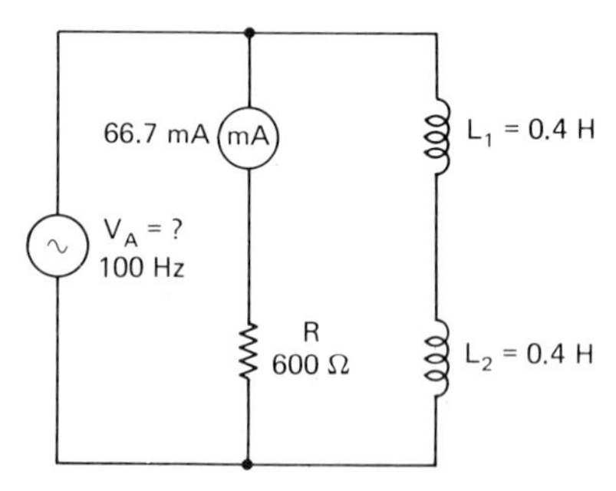

FIGURE 22-17 Parallel *RL* circuit schematic.

3. Find the total circuit current.
4. Find the applied voltage.
5. Find the phase angle between the total current and the applied voltage.
6. Find the impedance of the circuit.

Refer to Fig. 22-18, the parallel *RL* schematic, for Questions 7 to 14.

7. Find the current through the resistor.

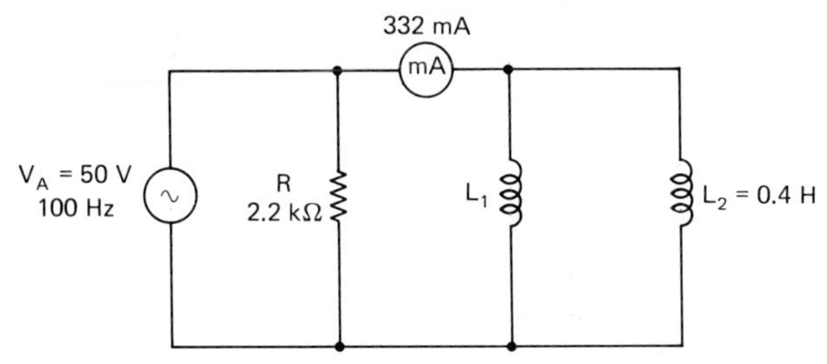

FIGURE 22-18 Parallel *RL* circuit schematic.

8. Find the current through inductor L_2.
9. Find the current through L_1.
10. Find the total circuit current.
11. Find the phase angle between the current and the applied voltage.
12. Find the impedance of the circuit.
13. Find the inductance of L_1.
14. What is the total inductance in the circuit?

SUMMARY

An ac current flowing through a resistor will be in phase with the voltage drop it produces across the resistor.

An ac current flowing through an inductor will lag the applied voltage by 90°.

In a series *RL* circuit V_R lags V_L by 90°.

In a series R_L circuit the phasor sum of the voltage drops is equal to the applied voltage:

$$V_A = \sqrt{V_R^2 + V_L^2} \quad \text{volts} \tag{22-1}$$

V_A can also be found by using Ohm's law provided that we know the impedance:

$$V_A = I_T Z \quad \text{volts} \tag{22-2}$$

The phase angle between the applied voltage and the lagging current in a series *RL* circuit can be found from the voltage triangle:

$$\tan \theta = \frac{V_L}{V_R} \tag{22-3}$$

In a series *RL* circuit the total opposition to current flow is termed impedance, *Z*, and can be found by using Ohm's law

$$Z = \frac{V_A}{I_T} \quad \text{ohms} \tag{22-4}$$

Impedance in a series *RL* circuit is also found by using the phasor sum of *R* and X_L:

$$Z = \sqrt{R^2 + X_L^2} \quad \text{ohms} \tag{22-5}$$

The phase angle between the applied voltage and the lagging current can be found from the impedance triangle:

$$\tan \theta = \frac{X_L}{R} \tag{22-6}$$

Impedance is proportional to frequency since X_L is directly proportional to frequency.

In a parallel *RL* circuit I_L lags I_R by 90°.

In a parallel *RL* circuit the phasor sum of the currents is equal to the total current:

$$I_T = \sqrt{I_R^2 + I_L^2} \quad \text{amperes} \tag{22-7}$$

I_T can also be found by using Ohm's law provided that we know the impedance:

$$I_T = \frac{V_A}{Z} \quad \text{amperes} \tag{22-8}$$

The phase angle between the lagging current through the inductor and the applied voltage can be found from the current triangle:

$$\tan \theta_I = \left(-\frac{I_L}{I_R}\right) \tag{22-9}$$

$$\theta_I = \tan^{-1}\left(-\frac{I_L}{I_R}\right) \tag{22-9a}$$

In a parallel *RL* circuit the total opposition to current flow is termed impedance, *Z*, and can be found by using Ohm's law:

$$Z = \frac{V_A}{I_T}$$

The product-over-the-phasor-sum equation for *Z* is

$$Z = \frac{R \times X_L}{\sqrt{R^2 + X_L^2}} \quad \text{ohms} \tag{22-10}$$

Since X_L is directly proportional to frequency, impedance will also be proportional to frequency in a parallel *RL* circuit.

Impedance can also be found by the reciprocal-phasor-sum:

$$\frac{1}{Z^2} = \frac{1}{R^2} + \frac{1}{X_L^2} \quad \text{ohms} \tag{22-10a}$$

REVIEW QUESTIONS

22-1. Explain why there is a phase difference between the current and voltage of an inductor, while there is no phase difference of a resistance.

22-2. Draw a phasor diagram to show the phase relationship of the applied voltage and current in a series *RL* circuit.

22-3. Draw a voltage triangle for a series *RL* circuit and give the equation to solve for V_A.

22-4. From the voltage triangle give the equation to solve for the phase angle between the applied voltage and the current in a series *RL* circuit.

22-5. Draw an impedance triangle for a series *RL* circuit and give the equation to solve for *Z*.

22-6. From the impedance triangle give the equation to solve for the phase angle between the applied voltage and the current in a series *RL* circuit.

22-7. Explain why impedance increases in a series *RL* current with an increase in frequency.

22-8. At what approximate ratio of X_L to R is the phase angle between V_A and I nearly zero?

22-9. At what approximate ratio of X_L to R is the phase angle between V_A and I nearly 90°?

22-10. Draw a phasor diagram to show the phase relationship of the total circuit current and the branch currents of a parallel *RL* circuit.

22-11. Draw a current triangle for a parallel *RL* circuit and give the equation to solve for I_T.

22-12. From the current triangle give the equation to solve for the phase angle between the total current and the applied voltage of a parallel *RL* circuit.

22-13. Give two equations to solve for Z in a parallel *RL* circuit.

22-14. What is the approximate phase relationship of V_A and I_T if the current through the inductor is 10 times greater than through the resistor in a parallel *RL* circuit?

SELF-EXAMINATION

Choose the most suitable word, phrase, or value to complete each statement.

22-1. In a sinusoidal ac circuit having a pure resistance, the phase angle between the applied voltage and the current is ________ degrees.

22-2. In a sinusoidal ac circuit having a pure inductance, the phase angle between the applied voltage and the current is ________ with the ________ leading.

Refer to Fig. 22-19, the series *RL* schematic, for Questions 22-3 to 22-8.

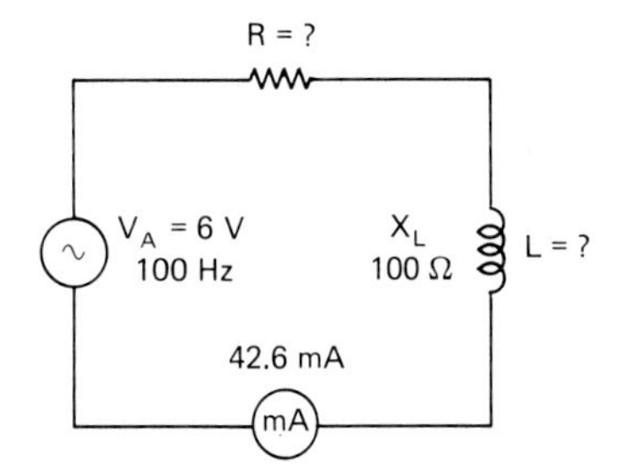

FIGURE 22-19 Series *RL* circuit schematic.

22-3. Find the impedance of the circuit.

22-4. Find the inductance of the coil.

22-5. Find the resistance of the resistor.

22-6. Find the voltage drop across R.

22-7. Find the voltage drop across X_L.

22-8. Find the phase angle between the applied voltage and the current.

Refer to Fig. 22-20, the series *RL* schematic, for Questions 22-9 to 22-13.

22-9. Find the reactance of the coil.

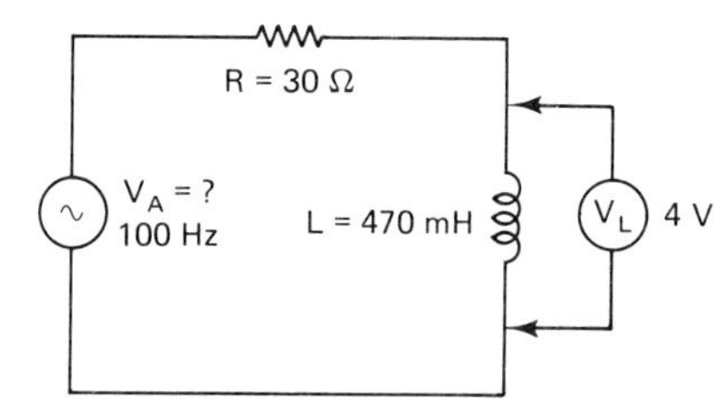

FIGURE 22-20 Series *RL* circuit schematic.

22-10. Find the impedance of the circuit.
22-11. Find the voltage drop across *R*.
22-12. Find the applied voltage.
22-13. Find the phase angle between the applied voltage and the current.

Refer to Fig. 22-21, the parallel R_L schematic, for Questions 22-14 to 22-20.

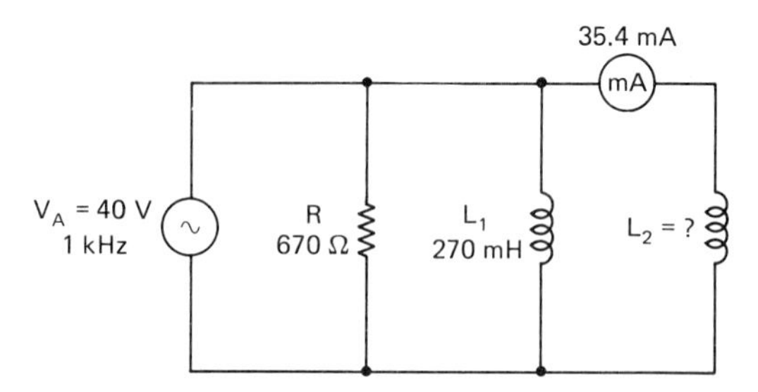

FIGURE 22-21 Parallel *RL* circuit schematic.

22-14. Find the current through *R*.
22-15. Find the current through the L_1 inductance.
22-16. Find the total circuit current.
22-17. Find the impedance of the circuit.
22-18. Find the phase angle between the applied voltage and the circuit current.
22-19. Find the inductance of L_2.
22-20. Find the total inductance of the circuit.

ANSWERS TO PRACTICE QUESTIONS

22-1

1. in phase, 0°
2. out of phase, 90°
3. lead, 90
4. $I_T = 15$ mA
5. 15 V
6. $V_A = 33.5$ V
7. $L = 1.59$ H
8. $Z = 2.24$ kΩ
9. $\theta = 26.6°$
10. $V_L = 72$ V
11. $V_R = 34.8$ V
12. $\theta = 64°$
13. $L = 3.3$ mH
14. $X_L \cong 6$ kΩ
15. $Z = 6.03$ kΩ
16. $\theta = 84.3°$
17. $Z \approx X_L$

22-2

1. V_A
2. lags, 90
3. $I_T = 104$ mA
4. $V_A = 40$ V
5. $\theta = 50°$
6. $Z = 385\ \Omega$
7. $I_R = 22.7$ mA
8. $I_{L2} \cong 200$ mA
9. $I_{L1} \cong 132$ mA
10. $I_T = 333$ mA
11. $\theta = 86°$
12. $Z = 150\ \Omega$
13. $L_1 \cong 602$ mH
14. $L_T \cong 240$ mH

23

Capacitance

INTRODUCTION

The forerunner of the modern capacitor was the Leyden jar. All that the eighteenth-century scientists were getting out of their friction machines was a bigger and better spark. Then in 1745, Von Kleist, at Kammin, Germany, discovered what is called the Leyden jar capacitor. Since Von Kleist was not a scientist, he did not document a proper description. As a result, Professor Musschenbroek of Leyden, Holland, who independently made the same discovery a year later, was given the credit.

This discovery by the professor came about accidentally. As he was changing the water in a glass jar held in his hand, he reached over and touched the conductor between the water and a friction machine with his other hand. The shock he received from this contact was immense. He later wrote that it was of such violence that his whole body was shaken as if by a lightning stroke and he thought he was done for. The early experimenters believed that an electrical "fluid" was condensed in the Leyden jar; thus it was termed a condenser.

Capacitors in modern electronic circuits operate on the same principle as the Leyden jar. This principle of capacitance, charge, discharge, construction, and types of dielectrics is discussed in this chapter.

MAIN TOPICS IN CHAPTER 23

OBJECTIVES

After studying Chapter 23, the student should be able to

1. Master the concepts of electrostatic charge and apply those concepts to capacitance.
2. Apply the concept of capacitance to various types of dielectrics.

3. Apply the concept of capacitance connected in series and parallel.
4. Use the EIA color code to determine the capacitance value of a capacitor.

23-1 ELECTROSTATIC CHARGES AND FIELDS

In Chapter 1 we discussed electric charges and how they behave. A capacitor stores these charges, so a review will aid in a better understanding of just how a capacitor behaves.

An electron, having been assigned (by convention) a negative electrical charge, is also said to have electrostatic (ES) lines of force that flow out from the electron. Similarly, a proton having a positive electrical charge is said to have ES lines of force that flow into the proton.

With these two concepts it can be shown that two electrical charges will have an electric field between them, as shown in Fig. 23-1.

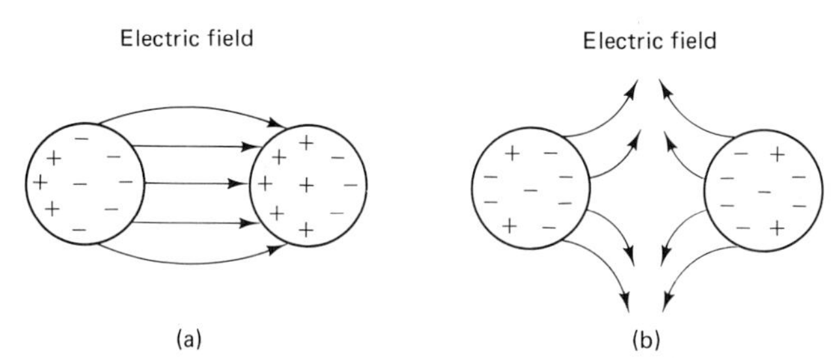

FIGURE 23-1 Attraction and repulsion between two electrical charges. (a) Unlike charges attract. (b) Like charges repel.

> Like charges repel each other. Unlike charges attract each other.

If a negatively charged body such as a charged pith ball is placed close to but not touching a neutral body, the negative charge will induce a positive charge in the neutral body by laws of electric charges (i.e., unlike charges attract each other), as shown in Fig. 23-2(a). The positive charged body shown in Fig. 23-2(b) will induce a negative charge in a neutral body. Of course, if the charged body is removed from the vicinity of the second body, it will revert to its neutral state because the electrons will revert to their normal state in the atomic structure of that body.

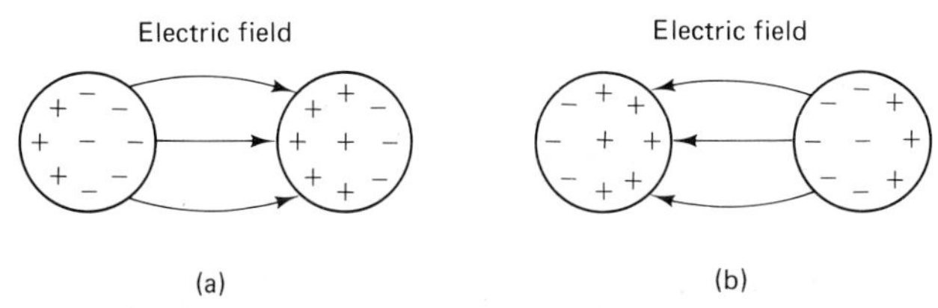

FIGURE 23-2 (a) A negative charge will induce a positive charge. (b) A positive charge will induce a negative charge.

23-2 COULOMB'S LAW OF ELECTROSTATIC CHARGES

A coulomb is defined as 6.25×10^{18} electrons. Charging two spheres or bodies and then placing them close together but not touching each other will produce a force of attraction or repulsion. This law of force between charges was discovered by the eighteenth century French scientist Charles Coulomb and is called Coulomb's law in his honor. Using Q as the symbol for the coulomb charge, Coulomb's law for the magnitude of the force between two charged bodies states that the force of attraction or repulsion varies directly as the product of the two charges and inversely as the square of the distance, or

$$F = K\frac{Q_1 Q_2}{D^2} \quad \text{newtons} \tag{23-1}$$

where F is in newtons
Q is in coulombs
D is distance, in metres
K is a constant $= 9 \times 10^9$

The significance of the force between two charged bodies can be appreciated by a simple calculation. Assume that two bodies having a charge of 1 C are separated by a distance of 1 m. Using equation (23-1) will give a force of 9×10^9 newtons (N). Recall that 1 newton is a unit of force which when applied to a 1-kg mass gives it an acceleration of 1 m/s^2. This is an astounding force (some 2 billion pounds in the old system). Electrostatic charges in electronic circuits are in microcoulombs. The force of attraction and repulsion, although greatly reduced, is quite significant.

23-3 THE CAPACITOR: CHARGE AND DISCHARGE

The Leyden jar had no purpose other than to be charged by an electrostatic machine and then discharged through a short circuit. Its ability to store an electrical charge is the basic principle of any modern capacitor. Technological materials such as Teflon, ceramic, polystyrene, and others allow modern capacitors to be made in various shapes and sizes.

A capacitor has two aluminum-foil plates which are insulated from each other as shown in Fig. 23-3.

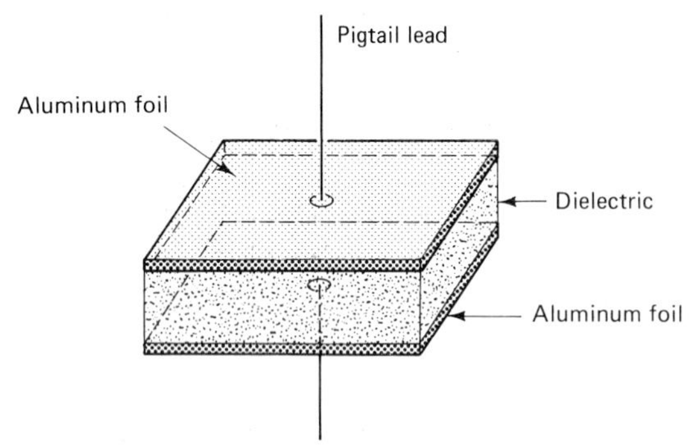

FIGURE 23-3 Construction of a basic capacitor.

The insulating material is called a dielectric. Dielectric materials and their characteristics are discussed in Sections 23-5 and 23-6. For now, let us look at the charge and discharge action of a capacitor.

Charging Action

Figure 23-4(a) shows the two plates of a capacitor connected to a dc voltage source. At the instant the switch is thrown ON electrons will flow into plate *A*. The neutral atoms in plate *B* will give up electrons, which are attracted to the positive charge of the source and flow into the source.

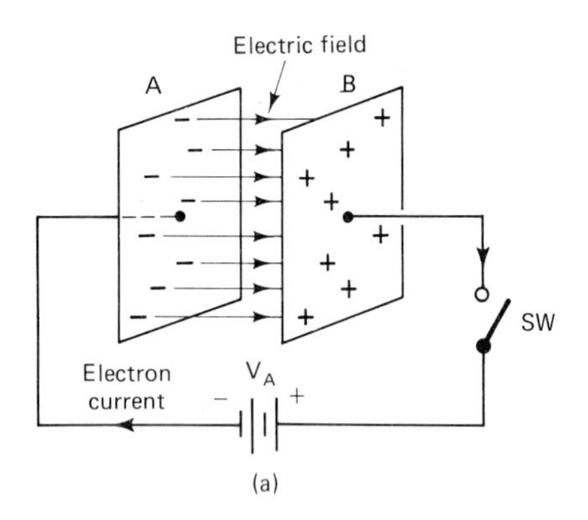

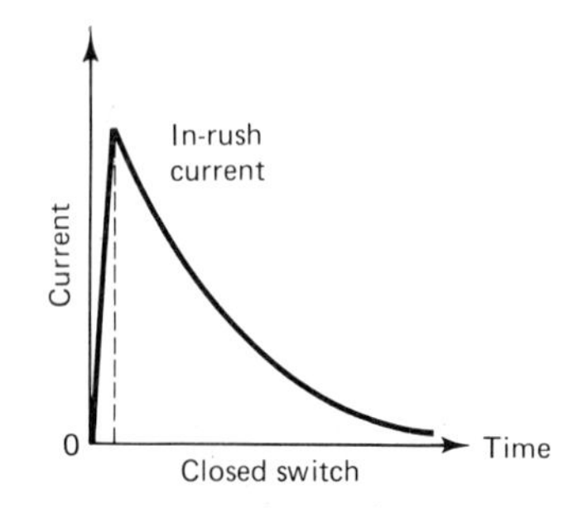

FIGURE 23-4 Charging action of a capacitor. At the instant the switch is closed the capacitor behaves like a short circuit.

Initially, the amount of electrons or current is very large, since both plates were neutral, so that the capacitor behaves like a short circuit. This "in-rush" charging current is very large and is depicted in Fig. 23-4(b). The buildup of the charge within the capacitor causes a potential difference (pd) to be built up between the two plates of the capacitor. This pd is in effect a counter emf that opposes the applied voltage, V_A. As the charge on the capacitor builds up, the counter emf builds up and reduces the current into the capacitor so that the current diminishes to zero. This is shown by the graphical representation of the charging current in Fig. 23-4(b).

The counter emf, which we now label V_C, will be equal to the applied voltage once the capacitor is fully charged, as shown by the graph of Fig. 23-5. Note that the voltage across the capacitor due to the stored charge will be equal to the applied voltage. The circuit current or charging current will be zero since $V_C = V_A$ and nullifies V_A.

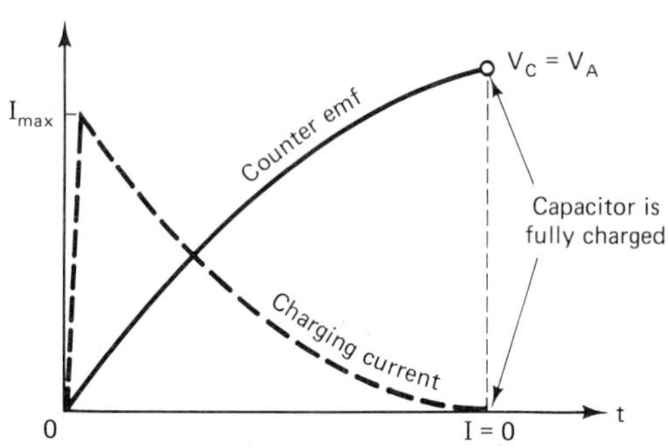

FIGURE 23-5 Graph showing charging current and counter emf or V_C built up across the capacitor.

Stored Charge. Plates *A* and *B* [Fig. 23-4(a)] now have opposite charges; therefore, the electric field exerts a force of attraction between the two charges. This electric field has a constant density and represents the new energy form in which power taken from the battery during the charging action, is now stored in the capacitor.

The stored energy is a function of capacitance, voltage, and charge. In equation form,

$$W = \frac{1}{2}CV^2 = \frac{1}{2}QV \quad \text{joules}$$

where W is the energy, in joules
C is the capacitance, in farads
V is the voltage, in volts
Q is the charge, in coulombs

In practice, this energy is expressed in watt-seconds; 1 watt-second is equal to 1 joule.

Energy-Form Contrast to Inductance. The difference between the electrostatic (ES) energy stored by a capacitor and the electromagnetic (EM) energy stored by an inductor (see Fig. 23-6) is that once the capacitor is charged, it does not consume any further power. Dielectric losses are negligible and the electric field will be maintained without any additional current flow into the capacitor. The electromagnetic field is sustained by a small but constant-current flow since the resistance in the turns of the coil consumes power (I^2R loss), which must be made up by the source.

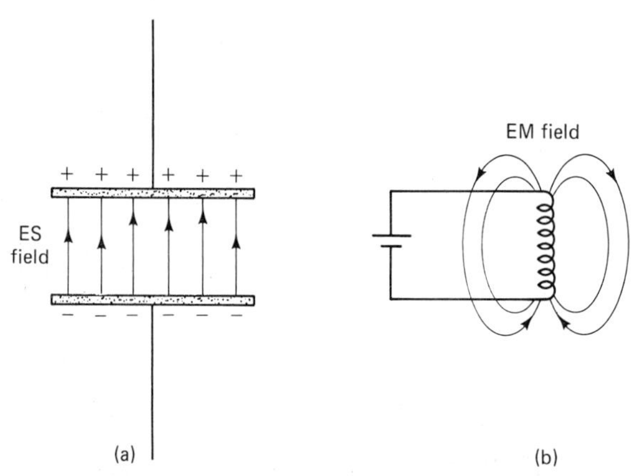

FIGURE 23-6 Contrasting capacitor and inductor stored energy. (a) Electric field is energy stored by a capacitor. (b) Electromagnetic field is energy stored by an inductor.

Discharging Action

Once the capacitor has been charged, it behaves as a battery with a voltage of V_C and a stored coulomb charge Q, as shown in Fig. 23-7(a). Discharging the capacitor through a load resistor will cause the coulomb charge to flow through the resistance and neutralize the ions on each plate. At this point the voltage and charge on the capacitor will reach zero, as shown by the graph of Fig. 23-7(b). The capacitor is said to be discharged.

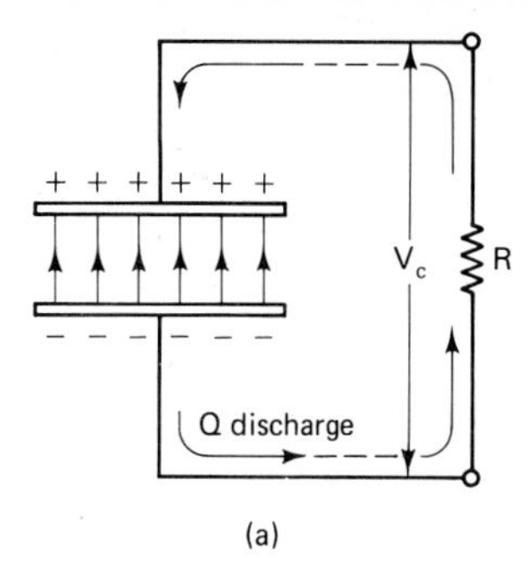

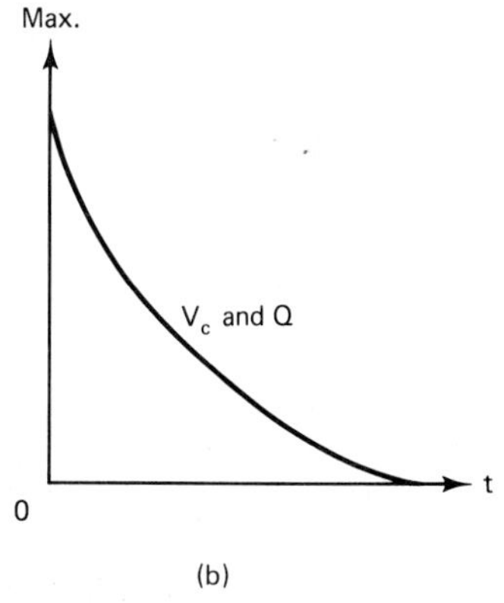

FIGURE 23-7 Discharge action of a capacitor.

23-4 UNIT OF CAPACITANCE, THE FARAD

A capacitor's ability to store a charge is called *capacitance*. The unit of capacitance is the farad (F), named after Michael Faraday. The insulation between the two plates of a capacitor was named a *dielectric* by Michael Faraday.

> A capacitor is said to have a capacitance of 1 farad (F) if it can store 1 coulomb in 1 second when the applied voltage is 1 volt.

The farad is far too large a unit for use in electronics. As an example, a bank of 2000 parallel capacitors were used at a certain aircraft company to obtain 390,000 μF (0.39 F) for a 5,000,000-A arc in order to generate shock waves in a testing air tunnel.

Capacitance is usually in tens of microfarads, hundreds of nanofarads, or thousands of picofarads. It may be tens, hundreds, or thousands of microfarads for use in power rectifier filtering or photoflash units and other applications. The common units are:

$$1 \text{ microfarad} = 1\ \mu\text{F} = 1 \times 10^{-6}\ \text{F}$$

$$1 \text{ nanofarad} = 1\ \text{nF} = 1 \times 10^{-9}\ \text{F}$$

$$1 \text{ picofarad} = 1\ \text{pF} = 1 \times 10^{-12}\ \text{F}$$

Note: The picofarad has replaced the micro-microfarad.

PRACTICE QUESTIONS 23-1

Choose the most suitable word, phrase, or value to complete each statement.

1. Two similar electrostatic charges ________ each other.
2. Two unlike electrostatic charges ________ each other.
3. A negative charge on one body will induce a ________ charge on to another neutral body.
4. One coulomb is defined as ________ electrons.

5. The symbol for the coulomb charge is ________.
6. A capacitor has ________ conducting plates which are insulated by a material called a ________
7. A capacitor that is being charged will have a ________ across itself that opposes the current flowing into it.
8. Once charged, the capacitor's voltage is equal to the ________ and no further current will flow from the source.
9. The stored energy of a capacitor is a function of ________, ________, and ________
10. Once a capacitor is fully charged it does not consume additional power because the ________ losses are negligible.
11. An inductor requires a small but constant current to sustain its electromagnetic field because of the ________ losses.
12. The coulomb charge and the voltage across the capacitor will fall to zero when the capacitor is fully ________.
13. The unit of capacitance is called the ________.
14. The ability of a capacitor to store a charge is called ________.
15. A capacitor is said to have a capacitance of 1 farad if it can store ________ in 1 second when the applied voltage is ________ volt.
16. 1 microfarad is equal to ________ farads.
17. 1 nanofarad is equal to ________ farads.
18. 1 picofarad is equal to ________ farads.

23-5 FACTORS THAT DETERMINE CAPACITANCE

The quantity of electrons or the coulomb charge that a capacitor is able to store is a direct function of its capacitance and applied voltage. Given a fixed capacitance, more charge will accumulate on the plates with an increase in voltage. Since the potential difference between the plates is greater, the electric field is stronger, which allows more electrons to be "captured" on one plate and more positive ions on the other plate.

The same charge action will take place with an increase in capacitance. A decrease in voltage or capacitance will decrease the coulomb charge. In equation form,

$$Q = CV \quad \text{coulombs} \qquad (23\text{-}2)$$

where C is in farads
V is in volts

EXAMPLE 23-1

(a) Calculate the coulomb charge on a 10-μF capacitor with 150 V applied.
(b) What happens if the voltage is decreased to 50 V?
(c) What happens if the capacitance is increased to 20 μF?

Solution: Using equation (23-2) gives:

(a) $Q = CV = 10 \times 10^{-6}\text{ F} \times 150\text{ V} = 1500\ \mu\text{C}$
(b) $Q = CV = 10 \times 10^{-6}\text{ F} \times 50\text{ V} = 500\ \mu\text{C}$
(c) $Q = CV = 20 \times 10^{-6}\text{ F} \times 150\text{ V} = 3000\ \mu\text{C}$

EXAMPLE 23-2

Assume that the coulomb charge of the capacitor in Example 23-1, part (a), was discharged instantaneously. What current would a milliammeter read?

Solution: Converting the coulomb charge to milliamperes gives

$$1500\ \mu\text{C} = 1500 \times 10^{-6}\text{ A} = 1.5 \times 10^{-3}\text{ A} = 1.5\text{ mA}$$

Plate Area

Capacitance is determined by the physical parameters of a capacitor as well as the electrical characteristics of the dielectric. Coulomb's law, equation (23-1), shows that the force of attraction is directly proportional to the coulomb charge.

Increasing the plate area will allow more electrons to be "captured" by this force of attraction on one plate and more positive ions on the other plate. If the plate area is doubled, the coulomb charge of the capacitor will double. If the plate area is halved, the coulomb charge of the capacitor will be halved. Thus, capacitance is directly proportional to the area of the plates:

$$C \propto \text{area of the plates}$$

Distance between Plates

The force of attraction (or repulsion) between two charges varies inversely as the distance between the two charges. This is shown by the Coulomb's law equation (23-1).

The charge of a capacitor is on its plates. In turn, its capacitance will depend on the force of attraction between the charges on the plates. Coulomb's law shows that capacitance will be inversely proportional to the distance between two plates. Reducing the distance by one half will double the capacitance. Doubling the distance will halve the capacitance. Since the plates are separated by a dielectric, the electric field will be stronger or weaker depending on the thickness of the dielectric. Thus, capacitance is inversely proportional to the distance between the plates:

$$C \propto \frac{1}{d}$$

Dielectric Constant

An electrical characteristic of a dielectric that affects capacitance is its dielectric constant (K). The dielectric constant (relative permittivity, ϵ_r) is a comparative measure or ratio to vacuum or air of the ability of a dielectric to concentrate electrostatic flux within itself (ϵ is the Greek lowercase letter epsilon). The dielectric constant of vacuum or air is 1 since it is the comparative reference.

Table 23-1 shows the relative permittivity or dielectric constant for various dielectrics. A material with a high dielectric constant will establish a more intense electric field, therefore a greater capacitance, than a material with a lower dielectric constant. Capacitance is directly proportional to the dielectric constant. A large value of dielectric constant produces a large capacitance.

$$C \propto K$$

TABLE 23-1

Material	*Dielectric Constant, K*	*Average Dielectric Strength (kV/mm)*
Vacuum or air	1	3
Paper	5	20
Polycarbonate	3.0	<3
Polystyrene	2.55	<3
Polypropylene	2.2	<3
Mylar[a] (PETP)	3.0	60
Mica	6.5–8.7	120
Ceramic (low-K)	5–600	40
Ceramic (high-K)	600–10,000	40
Glass	5–100	28

[a]DuPont registered trade name; PEPT, polyethylene terephthalate.

Putting the three factors, plate area, distance between the plates, and the dielectric constant, together in equation format will give the capacitance:

$$C = K\frac{A}{d}\epsilon_0 \quad \text{farads} \tag{23-3}$$

where C is the capacitance, in farads
A is the area of one plate, in square metres
d is the distance between the plates, in metres
K is the dielectric constant given in Table 23-1
ϵ_0 is equal to 8.85×10^{-12} F/m

Note: ϵ_0 is the absolute permittivity of vacuum and is dimensionless. The dielectric constant K (relative permittivity, ϵ_r) is a ratio of the absolute permittivity, ϵ, of a material to the absolute permittivity, ϵ_0, of a vacuum. The absolute permittivity of any material can therefore be calculated as $\epsilon = \epsilon_r \times \epsilon_0$ or $\epsilon = K \times \epsilon_0$, where ϵ_0 is 8.85×10^{-12} F/m.

EXAMPLE 23-3

Calculate the capacitance of a capacitor with a plate area of 60 cm² and a mica dielectric that is 0.25 mm thick separating the two plates.

Solution: Use equation (23-3) and dielectric constant for mica from Table 23-1.

$$C = K\frac{A}{d}\epsilon_0$$

$$= \frac{8.7 \times 60 \times 10^{-4}\ \text{m}^2 \times 8.85 \times 10^{-12}\ \text{F/m}}{0.25 \times 10^{-3}\ \text{m}}$$

$$= 1.8\ \text{nF}$$

23-6 DIELECTRIC STRENGTH AND DC VOLTAGE RATING

Since a capacitor is connected across a voltage, its dielectric must be able to withstand that voltage without breaking down. A dielectric that breaks down means that the current has arced through the dielectric damaging it. Once damaged in this manner, the capacitor has an internal leakage current ranging from a small value to an outright short-circuit current. Table 23-1 shows the average dielectric strength for materials.

Capacitor Dc Voltage Rating

The voltage rating of a capacitor is the maximum voltage that can be applied across a capacitor without causing arcover through the dielectric. This rating is given in V_{dc}, sometimes as DCWV (dc working volts) and is either printed on the outside case of the capacitor or given by means of a color code. Table 23-2 shows a

TABLE 23-2
Typical dc working voltage range for types of capacitors shown

Capacitor Type	V_{dc} *Range*
Paper	100–1600 V
Ceramic	25–6 kV
Mica	200–5 kV
Plastic film	100–630 V
Oxide electrolytic	10–550 V
Tantalum electrolytic	5–300 V

typical dc working voltage range for various types of capacitors.

A capacitor that is to be connected to an ac voltage must have a dc voltage rating equal to the *ac peak value*. This is because the instantaneous ac peak value would rupture the dielectric should it exceed the capacitor's dc voltage rating.

EXAMPLE 23-4

(a) Calculate the dc voltage rating of a capacitor that is to operate across an ac voltage of 200 V.

(b) Choose an appropriate voltage rating for a flat film capacitor from Fig. 23-8.

Solution:

(a) The peak ac voltage value will be

$$V_p = 200 \text{ V} \times \sqrt{2} = 283 \text{ V}$$

(b) From Fig. 23-8 the appropriate voltage rating for a flat film capacitor would be 283 V plus a safety factor of 50%, or

$$V_{dc} = 283 \text{ V} \times 1.5 = 425 \text{ V}$$

The closest rating is 400 V, which would be adequate. A 630-V rating would be more desirable.

Interpreting Printed Data on a Capacitor

Electrolytic, tantalum, and cylindrical types of paper capacitors and ceramic capacitors have their capacitance and dc working volts printed on the outer casing. Special types of ceramic capacitors also have their temperature coefficient printed on the case.

A general rule for interpreting these values can be applied as follows. If the value shown is in decimal format (e.g., 0.01, 10, etc.), the unit of capacitance is microfarads. This is particularly applicable to paper, electrolytic, and tantalum capacitors.

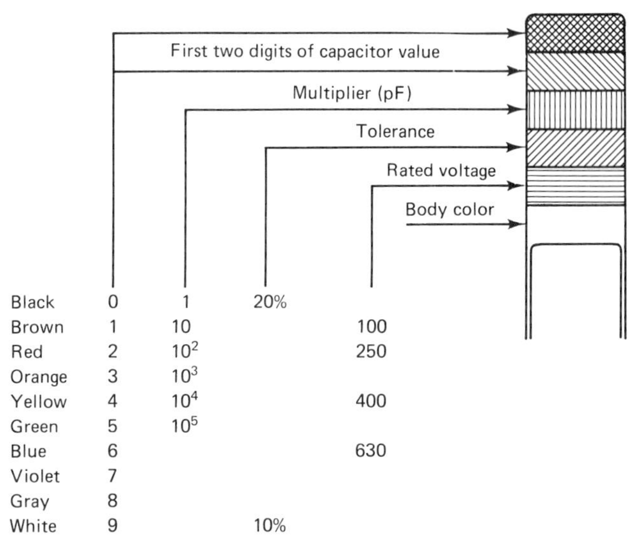

FIGURE 23-8 Flat film capacitor color code.

Electrolytic capacitors have the μF alongside the value printed on the case. Some manufacturers are still using the capital letters MF or MFD for microfarads and MMFD for picofarads. In each case this implies microfarads.

When a capacitor is physically very small, such as miniature ceramic types, a nomenclature similar to precision resistors is used. This shows the value and the capacitance unit for the capacitor. As an example, a ceramic capacitor has the value 2 n 2 printed on its case. The n tells us that the unit of capacitance is the nanofarad. The n also designates the decimal place, so the value of this capacitor is 2.2 nF.

When the decimal place is designated by the letter k (lkO) the unit of capacitance is picofarads and the k serves as a decimal marker as well as the multiplier, in this case 1.0 kpF, or 1000 picofarads.

Capacitor Color Code

Flat film capacitors have their capacitance value shown by means of a color-code system similar to that for the resistor value. Fig. 23-8 shows such a system and is self-explanatory.

23-7 TYPES OF CAPACITORS

Having discussed the characteristics of a dielectric, we now turn our attention to types of dielectrics. A capacitor is known by its dielectric; thus we have the air, paper capacitor, mica, ceramic, and the glass capacitor. The general term "flat film" is used for capacitors with dielectrics of Teflon, polystyrene, Mylar, and other plastic film. Aluminum oxide electrolytic and tantalum oxide capacitors are capable of having very large capacitance within a small physical size.

Air Capacitor

An air dielectric capacitor is desirable where the capacitance must be varied to tune in radio signals, such as tuners in FM and communication receivers. They are used especially in transmitter circuits where the voltages are very high. Figure 23-9 shows a typical air dielectric tuning capacitor. The stator plates are fixed while the rotor plates are mounted on a common shaft which can be rotated. The stator plates are insulated from the main frame of the capacitor. Rotating the shaft meshes the two plates together so that the capacitance is increased or decreased. Other miniature variable capacitors used in portable AM and FM receivers use plastic film dielectric.

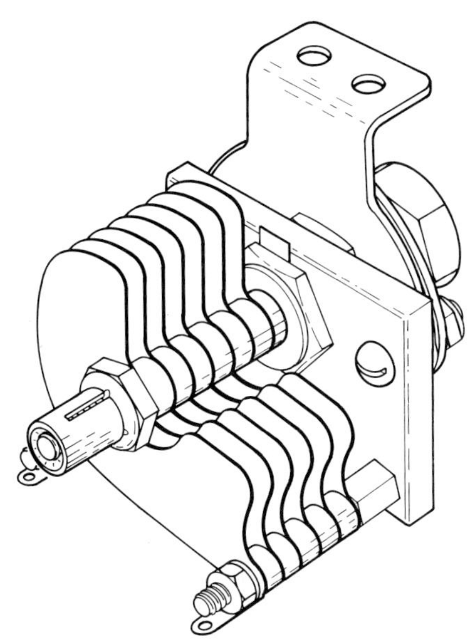

FIGURE 23-9 Variable capacitor air dielectric.

Plate spacing, plate area, and the number of plates determine the capacitance value. The voltage rating is governed by the plate spacing and, as usual, is specified by the manufacturer.

Paper Capacitor

The dielectric material of a paper capacitor is a kraft paper. There are many types and grades of kraft paper. It is manufactured from specially selected unbleached sulfate pulp made from certain species of soft woods. Use of impregnates such as liquid wax or resin improves the dielectric strength and increases the dielectric constant, thus reducing the physical size of the capacitor.

A paper capacitor is made by coiling two aluminum-foil sheets separated and insulated from each other by the kraft paper, as shown in Fig. 23-10(a). Tinned copper leads are pressed over either end of the protruding foil. Assembly in nonmetallic molded, tubular, or dipped enclosures follows the operation of attaching the lead wires.

Metalized paper capacitors are constructed by vapor-depositing a thin film of metal on the kraft paper dielectric. This enables the capacitor to be made physically smaller. Metalized capacitors should not be used where high insulation resistance is a critical requirement.

A band printed on one end of the outer case of a coiled capacitor [see Fig. 23-10(b)] indicates the lead that is connected to the outside foil. Connecting this lead to the ground or to the lower voltage point reduces stray capacitance and possible pickup of unwanted ac signals.

Plastic Film

Among some of the plastic-film dielectrics used at present are polycarbonate, polyester, polypropylene, and polystyrene. The basic construction of a capacitor using these materials is similar to that of the paper capacitor. Film foil capacitors are used for applications where high currents and step pulses occur, such as in deflection circuits of TV sets. One type of film foil construction is shown in Fig. 23-11.

Metalized film capacitors are used in general-purpose applications. The plastic film is vapor-deposited directly onto the metal electrode as a pure uniform polymer in thickness less than one-third that of free film dielectric. This enables the capacitor to be substantially reduced in physical size for an equivalent rating.

Dual-Dielectric Capacitors

Two different dielectrics, such as PETP and polystyrene, can be used together in a capacitor winding, either in series or in parallel within the same coil. The resultant electrical characteristics are substantially improved, giving zero capacitance change from 0 to 80°C to compete with NPO ceramic capacitors.

Mica Capacitors

Mica, being a natural mineral and adapted to use without physical or chemical alteration, is completely inert both dimensionally and electrically. As a dielectric, it will not exhibit aging or deterioration nor subtle variants in electrical properties. A particular variety of mica hav-

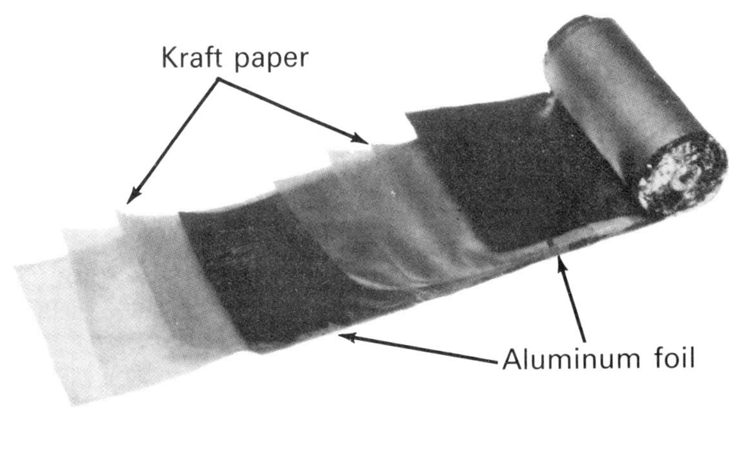

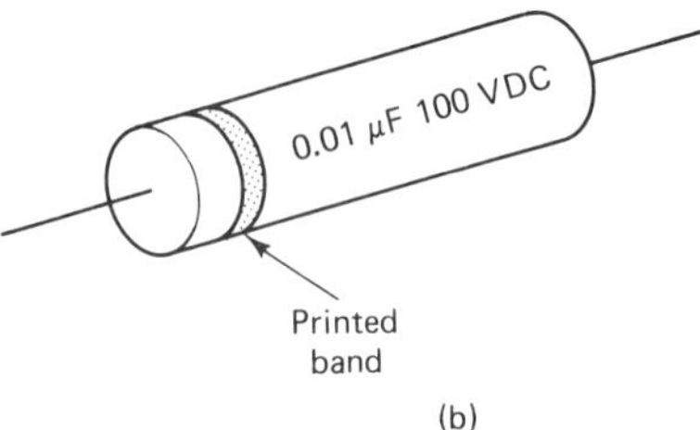

FIGURE 23-10 (a) Construction of a paper capacitor. (b) Printed band indicates outside foil lead.

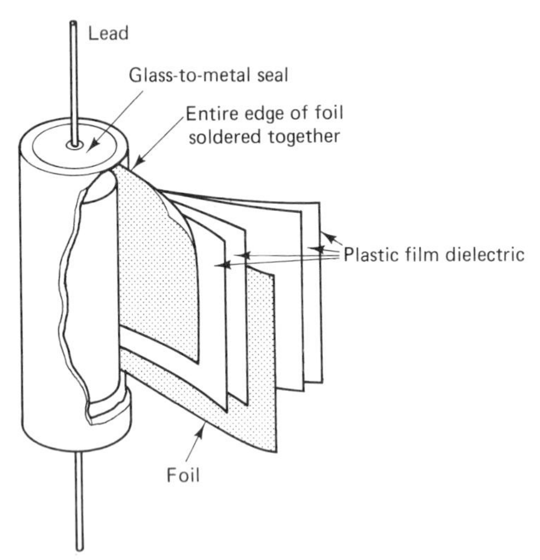

FIGURE 23-11 Construction of a metal-encased film capacitor.

ing the best electrical properties is muscovite, found primarily in South America and the Orient.

Figure 23-12 shows the "sandwich" type of construction used to make a mica capacitor. Each mica "sandwich" between two tin-lead foil conductors forms the capacitor, with capacitance being a function of area, spacing, and dielectric constant. The tin-lead plates extend alternately from opposite edges of the mica stack. All foils from each side, respectively, are shorted together and connected to one capacitor terminal. The capacitor is then encapsulated in a plastic mold or dipped several times into a combination of resins to build up a series of coats.

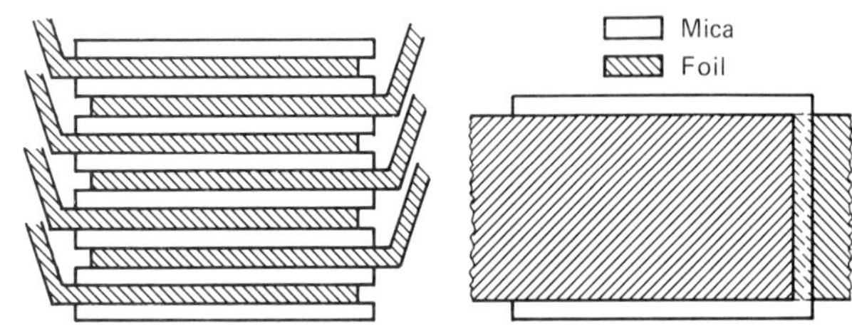

FIGURE 23-12 Construction of a mica capacitor.

The silvered mica capacitor is made by vapor-depositing a thin layer of silver directly on the surface of the mica. It has close tolerance and retains its electrical characteristics over a wide range of temperature and environmental conditions. Silvered mica capacitors are especially used in high-Q tuned circuits and timing circuits.

The transmitting mica capacitor is used extensively in low-and high-power transmitter circuits. The mica thickness is increased to give working voltages of up to 30 kV ac and currents in excess of 100 A at radio frequencies.

Ceramic Capacitor

Ceramic dielectric capacitors are used extensively in the electronics industry. Their wide range of dielectric constant (K), ability to be functional over large temperature excursions, and predictable capacitance change versus temperature (temperature coefficient TC) gives them superiority over other dielectrics.

Virtually all of the ceramics used for the dielectric in these capacitors can be divided into two broad categories: temperature compensating (low-K) and high dielectric constant (high-K).

Temperature-Compensating (Low-K) Capacitor. Temperature compensation for commercial applications is specified from 25 to 85°C. Military (MIL) specifications require measurements from −55 to 125°C.

The temperature coefficent (TC) of capacitance is expressed in parts per million per degree Celsius (ppm/°C). A capacitor can have either a positive TC (e.g., P120), a negative TC (e.g., N120), or zero temperature coefficient, designated NPO. The numerical value indicates the part-per-million change per degree Celsius. This can be related in terms of a percentage change and resultant capacitance change.

EXAMPLE 23-5

A 47-pF ceramic capacitor having a coefficient of N750 is connected across a tuned circuit. Calculate its capacitance if the temperature increases from 25°C to 45°C.

Solution: At 25°C, its capacitance = 47 pF. The capacitor has a negative temperature coefficient of 750 ppm/°C.

$$\% \text{ capacitance change}/°\text{C} = \frac{-750}{1{,}000{,}000} \times 100 = -0.075\%$$

$$\% \text{ capacitance change} = (-0.075) \times (45°\text{C} - 25°\text{C}) = -1.5\%$$

$$\text{capacitance} = 47 \text{ pF} - (47 \text{ pF} \times 0.015) = 46.295 \text{ pF}$$

The temperature coefficient of ceramic dielectric capacitors is essentially linear so they are ideal for use in tuned circuits to compensate for frequency drift due to temperature. In addition to temperature compensation, the stability, high Q, and high insulation resistance (low leakage) makes the ceramic capacitor an ideal general-purpose, high-quality capacitor. Some uses particularly are in tuned circuits, oscillators, high-frequency filters, radio-frequency transmitter circuits, or any other electronic circuit requiring high-Q stability.

Feedthrough ceramic capacitors are a combination of bypass and conductor feedthrough. This enables the designer to use a conductor lead through a chassis and simultaneously bypass the lead to RF voltages.

There are two basic physical configurations of ceramic capacitors, the disk and the tubular, as shown in Fig. 23-13. The disk type usually has its value and working volts printed on the face of the disk. For the tubular type, the industry uses a color code system, as shown in Fig. 23-13.

Glass Dielectric

The glass capacitor was first introduced by the Corning Glass Works as a substitute for the mica capacitor. A process was developed to form glass into extremely thin, flat ribbon. The capacitor is made by alternately stacking layers of aluminum foil and glass ribbon, similar to mica capacitors. The entire assembly is then fused into a monolithic block, as shown in Fig. 23-14.

Glass capacitors are very sturdy, moisture resistant, and radiation resistant. Their temperature coefficient over the entire temperature range −55 to 125°C is absolutely predictable and reliable. The temperature coefficient between other glass capacitors, regardless of capacitance value or size, will track accurately and predictable during temperature variations.

The glass capacitor is being used successfully in the aerospace electronics applications. It is used in high-quality equipment such as measurement instrumentation, process controls, high-Q filters, reliable bypassing in RF circuits and peripheral computer applications where circuit parameters cannot be allowed to wander. Their relative small size and low weight make them suitable for airborne equipment and mobile transmitters/receivers.

Polarized Electrolytic Capacitors

Electrolytic capacitors are among the most widely used where requirements call for large capacitance with small size. Aluminum foil is etched to increase its surface area. It is then treated electrochemically to form aluminum oxide on its surface. The oxide, having a high electrical resistance, acts as a dielectric. Its thickness will determine the voltage rating of the capacitor. Since the oxide is extremely thin, the surface area of the electrodes can be made very large. This is because a greater amount of aluminum foil can be rolled into the same capacitor volume, which in turn produces a very large capacitance value. Typically, electrolytic capacitors are manufactured with capacitance values in the range 0.33 to 220,000 μF. The schematic of Fig. 23-15 shows that the aluminum foil with the layer of oxide is the anode foil. Paper spacers are applied next to the aluminum oxide surface to prevent direct contact between the anode and cathode foil. For higher voltages thicker paper spac-

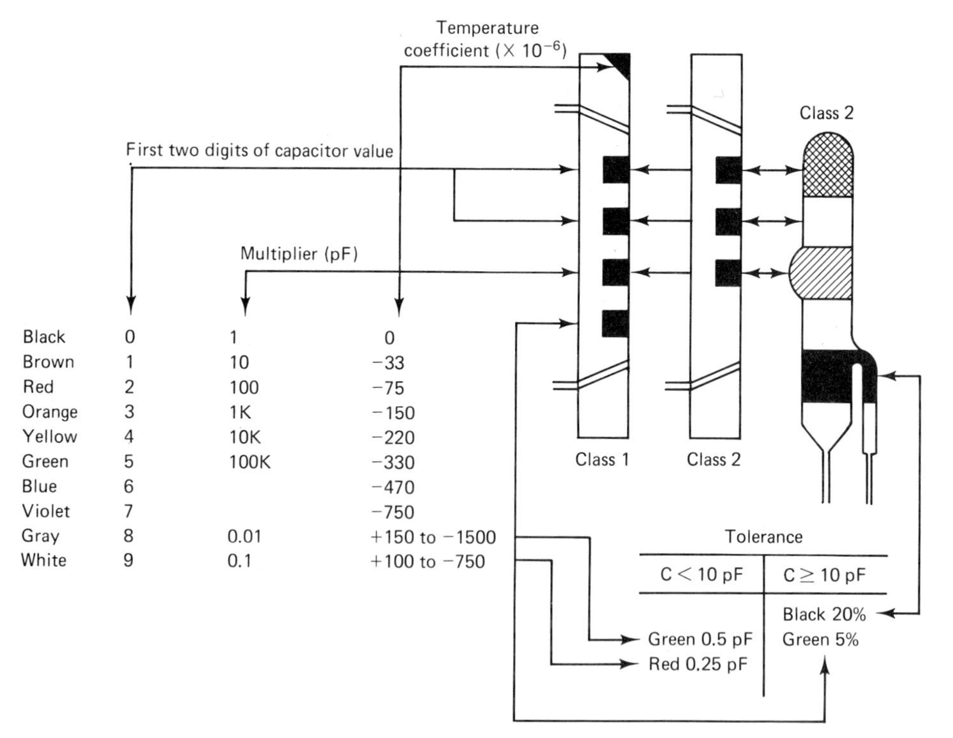

FIGURE 23-13 Tubular ceramic capacitor color code.

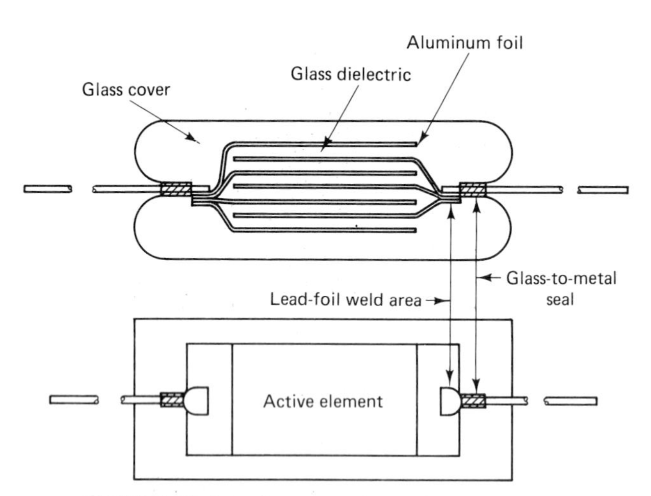

FIGURE 23-14 Construction of glass dielectric capacitor.

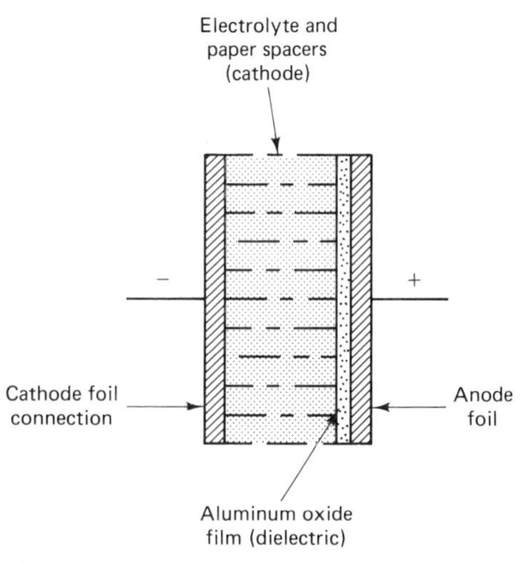

FIGURE 23-15 Construction of a polarized electrolytic capacitor.

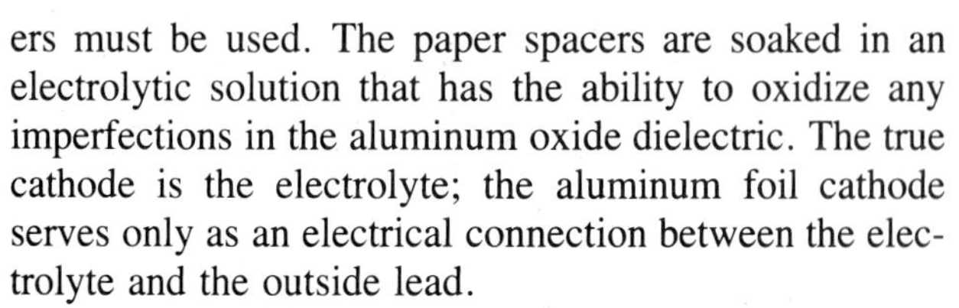

ers must be used. The paper spacers are soaked in an electrolytic solution that has the ability to oxidize any imperfections in the aluminum oxide dielectric. The true cathode is the electrolyte; the aluminum foil cathode serves only as an electrical connection between the electrolyte and the outside lead.

Precaution: When a polarized electrolytic capacitor is properly connected to a dc source (i.e., cathode to negative and anode to positive), the leakage current maintains the oxide layer on the anode. Should the capacitor polarity be reversed, the oxide is depleted, with a resultant partial short circuit and large current flow. This causes the electrolyte to be heated and release a gas, which builds up and usually results in rupturing the container in an explosion.

Two important parameters for electrolytics are leakage current and maximum rms ripple current. The maximum leakage current for an electrolytic is given in manufacturers' data sheets. As an example, a 22-μF electrolytic rated at 10 V has a leakage current of 11 μA, compared to a 3300-μF, which has a leakage current of 202 μA for the same voltage rating.

The maximum rms ripple current of an electrolytic must not be exceeded because of the heating effect due to the ESR (equivalent series resistance) of an electrolytic capacitor. ESR increases as the capacitor deteriorates and is dependent on the temperature, capacitance, and frequency of the ripple voltage. As an example, the same 10-V 22-μF capacitor has a maximum rms ripple current rating of 35 mA; the 3300-μF has a maximum rms ripple current rating of 950 mA.

Identifying Lead Polarity. Figure 23-16 shows a drawing that will aid you in identifying the correct lead polarity of a polarized electrolytic capacitor. In Fig. 23-16(a), the axial lead type, the negative lead is the one that is welded to the aluminum case (the cathode). It is identified by a black band or negative symbol, as shown. The radial lead type shown in Fig. 23-16(b) has its leads similarly labeled.

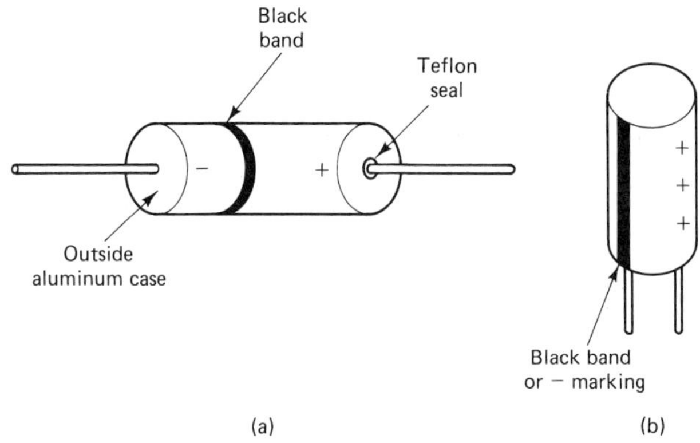

FIGURE 23-16 Two types of electrolytic capacitors. (a) Axial lead type. (b) Radial lead type.

Nonpolarized Electrolytics

Nonpolarized capacitors have two anodes, as shown in the drawing of Fig. 23-17. Because of this, they have one-half the capacitance of an equal-sized polarized unit of the same voltage rating. Applications for nonpolarized electrolytics include ac split-phase motor, audio crossover networks, and applications where large pulse signals are employed.

Electrolytic Capacitance Tolerance. Tolerance on nominal capacitance for electrolytic capacitors can be typically −10 + 50%. Specifying the minium value for an electrolytic is more important than the upper limit, for reasons of bypass, filtering, and coupling circuit design.

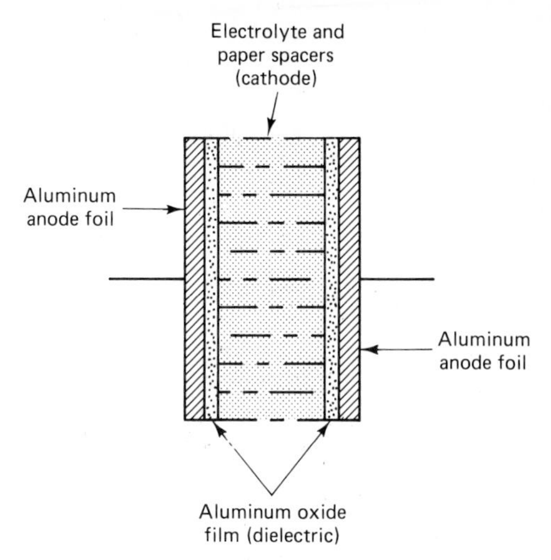

FIGURE 23-17 Construction of a nonpolarized electrolytic capacitor.

Tantalum Electrolytics

Tantalum has the same property as aluminum—of being a base-oxidizable metal. It is very inert to chemical attack, which allows the use of highly ionized acid electrolytes not possible with aluminum. Tantalum oxide has almost twice the dielectric constant of aluminum oxide and is exceptionally stable with temperature. Since it is available in a high-purity form, both as a foil and a powder, more diversified physical configurations are possible than with the aluminum electrolytic.

These properties give the tantalum electrolytic many advantages over the aluminum electrolytic:

1. Higher capacitance per volt in a given unit volume
2. Wider operating temperature range
3. Better temperature stability characteristics
4. Longer life
5. More rugged construction features
6. Improved electrical parameters and excellent shelf life

The main disadvantages are the increased cost and the lower operating voltages of the tantalum types. The most outstanding feature is the stable shelf life (storage) at temperatures of 30 to 40°C for periods of up to 10 years without any adverse effect of the electrical parameters. There are three categories of tantalum capacitors: the solid electrolytic type, the foil type, and the wet anode type.

Solid Tantalum Electrolytic. These units are made by using a sintered (baked) anode that has an oxide formed on it by electrochemical treatment. The porous anode is then impregnated with a liquid solution of manganous nitrate which is converted to manganese dioxide after being heat treated in an oven. The manganese dioxide becomes a solid electrolyte which serves as the true cathode of the capacitor. Two types of encapsulation are used: resin dipped and a tinned metal container with the positive lead brought out through a glass-to-metal seal. This type of encapsulation gives a rugged hermetically sealed package.

Foil-Type Tantalum Electrolytic. The tantalum foil types are similar in construction to the aluminum electrolytics (see Fig. 23-15). The oxide on the anode foil is the dielectric and the electrolyte is the true cathode. Electrical contact between the electrolyte is made by the aluminum case. The positive lead is brought out through a Teflon or rubber seal.

Wet-Anode Tantalum Electrolytic. These are made by pressing tantalum powder mixed with a binder into a mold or die to a given shape, usually cylindrical (see Fig. 23-18). These pellets are then sintered under high vacuum and temperature to remove the binder and impurities, leaving a rugged, porous metal pellet. The tantalum pellet is then electrochemically treated to form a layer of tantalum oxide on its surface. It is then assembled in a silver-plated outer case, filled with an electrolyte and sealed. The positive lead is brought out through this seal, which is Teflon or rubber or a combination of both, to give a hermetically sealed container.

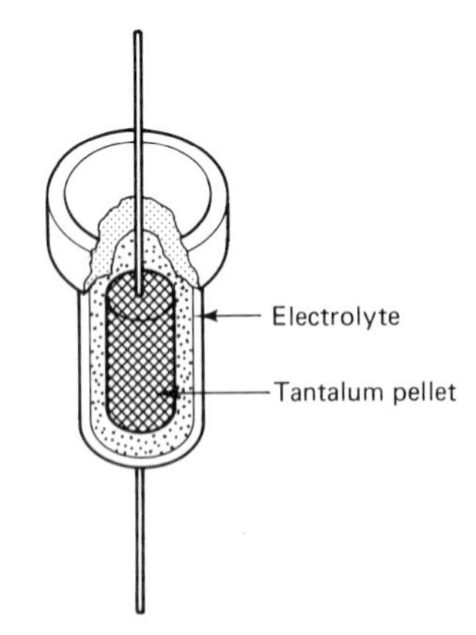

FIGURE 23-18 Construction of a tantalum electrolytic capacitor.

Solid Tantalum Color Code. The value and working volts are printed on the aluminum case for the foil and wet-anode tantalum capacitors. A color code to show the capacitance value, working volts, and tolerance is used on the greatly reduced physical size of the solid tantalum, as shown in Fig. 23-19.

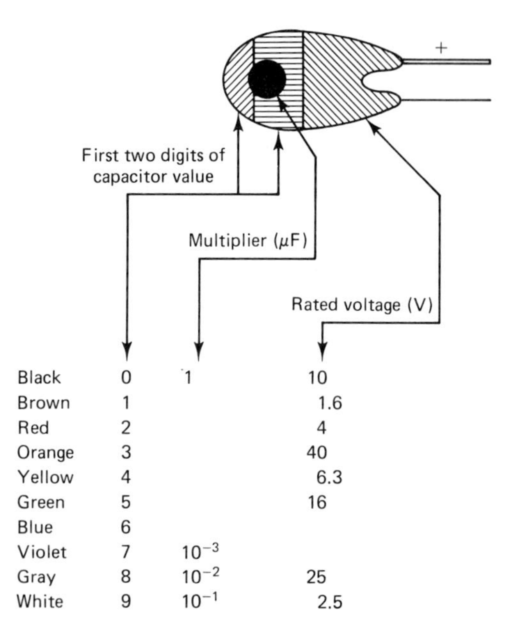

Color	First two digits of capacitor value	Multiplier (μF)	Rated voltage (V)
Black	0	1	10
Brown	1		1.6
Red	2		4
Orange	3		40
Yellow	4		6.3
Green	5		16
Blue	6		
Violet	7	10^{-3}	
Gray	8	10^{-2}	25
White	9	10^{-1}	2.5

FIGURE 23-19 Solid tantalum electrolytic color code.

Voltage-Variable Capacitors

Capacitance is one of the parameters of a semiconductor junction diode and is voltage dependent. The diode junction capacitance can be set up or varied by means of a reverse dc bias supply. This means that the anode is connected to the negative of the dc supply and the cathode to the positive of the supply. Capacitance values as high as 1800 pF at −4 V are provided by commercial silicon diodes.

Semiconductor theory would have to be studied before an explanation of this device operation could be given, which is not within the scope of this text.

PRACTICE QUESTIONS 23-2

Choose the most suitable word, phrase, or value to complete each statement.

1. A coulomb charge on a capacitor can be doubled if the ________ or ________ is doubled.
2. Capacitance is ________ proportional to the plate area. Reducing the plate area by one-half will reduce the capacitance by one-half.
3. Capacitance is ________ proportional to the distance between the plates. Doubling the distance between the plates will halve the capacitance.
4. The ability of a dielectric material to concentrate electrostatic lines of force is called the ________ ________ of the material.
5. A large value of dielectric constant will produce a ________ capacitance.

6. Calculate the capacitance of a capacitor with a plate area of 30 cm^2 and a ceramic (high K maximum value) dielectric that is 0.20 mm thick separating the two plates.
7. Calculate the ac (rms) working voltage of a capacitor that is rated as 250 V dc.
8. What is the value of a capacitor that has 22 p printed on its case?
9. Usually, where it is desirable to vary the capacitance, such as in AM and FM tuners, an ________ dielectric variable capacitor is used.
10. The type of paper used as a dielectric in a paper capacitor is called ________ paper.
11. Paper capacitors that are constructed by vapor-depositing a thin film of metal on kraft paper dielectric are termed ________ paper capacitors.
12. ________ ________ capacitors are used for applications where high currents and step pulses occur, such as in deflection circuits of TV sets.
13. Mica dielectric capacitors can be made with voltage ratings as high as ________ volts ac and currents as high as ________ amperes.
14. ________ mica capacitors are used in high-Q tuned circuits and timing circuits.
15. The greatest advantage of ceramic capacitor is its predictable ________ change with a temperature change.
16. Calculate the capacitance of a 68-pF ceramic capacitor with a temperature coefficient of −470 ppm/°C if the temperature increases from 25°C to 40°C.
17. The dielectric of an electrolytic capacitor is an ________ ________ formed on the aluminum foil surface by electrochemical treatment.
18. The ________ is the true cathode of an electrolytic capacitor.
19. The cathode of a ________ electrolytic capacitor must be connected to the negative side of a dc source.
20. The lead that is welded to the aluminum outside case and is also identified by a black band at one end of the electrolytic capacitor will be the ________ lead.
    (polarity)
21. Electrolytic capacitors that are used on ac applications such as motors and audio crossover networks are of the ________ type.
22. There are many advantages of the tantalum capacitor over the electrolytic capacitor. Its two main disadvantages are the higher ________ and the lower operating ________.

23-8 SCHEMATIC SYMBOLS FOR CAPACITANCE

The common symbol for capacitance is the one shown in Fig. 23-20(a). There are two methods of showing a variable capacitor: an arrowhead on the curved part of the symbol (not commonly used) or an arrow through the capacitor symbol, as shown in Fig. 23-20(b) and (c).

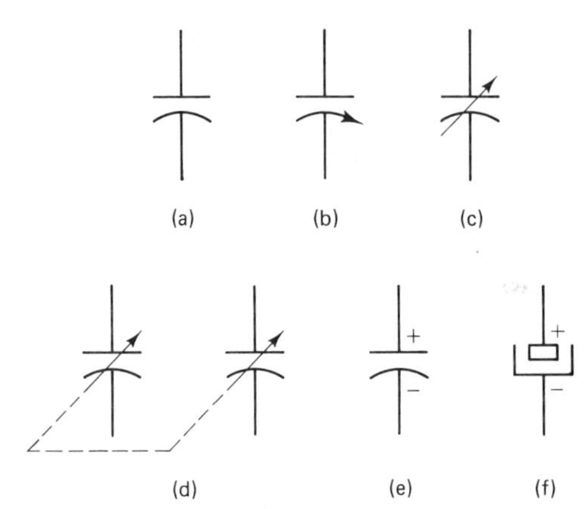

FIGURE 23-20 Schematic symbols for capacitance.

Variable tuning capacitors are coupled on a common shaft so that both rotors move or are varied simultaneously. The capacitors are said to be "ganged" and the joined dashed lines in Fig. 23-20(d) indicate that the two capacitors have one common shaft.

The polarity of electrolytics must be shown on the schematic symbol [see Fig. 23-20(e)]. A symbol for an electrolytic capacitor that is not commonly used is shown in Fig. 23-20(f).

23-9 CAPACITANCE IN SERIES

Charging Current and Coulomb Charge

Capacitors, especially electrolytics, are sometimes connected in series to increase their working voltage rating or as filters in voltage-doubler and voltage-tripler circuits. In this case their capacitance must be equal in order to have an equal division of working voltage.

Since the charging current in series-connected capacitors must be the same in both capacitors, the net coulomb charge in each capacitor will be the same regardless of its capacitance size. The net coulomb charge that each capacitor will charge to is given by equation (23-3), or $Q = CV$.

Voltage Division

The voltage that each capacitor will charge to can be found by rearranging equation (23-3):

$$V = \frac{Q}{C} \quad \text{volts} \qquad (23\text{-}4)$$

EXAMPLE 23-6

The schematic of Fig. 23-21 shows the coulomb charge to be 1000 μC. The voltage across each capacitor can be found using equation (23-4).

$$V_1 = \frac{Q}{C_1} = \frac{1000 \times 10^{-6}\ \text{C}}{5 \times 10^{-6}\ \text{F}} = 200\ \text{V}$$

$$V_2 = \frac{Q}{C_2} = \frac{1000 \times 10^{-6}\ \text{C}}{10 \times 10^{-6}\ \text{F}} = 100\ \text{V}$$

Note that the smallest capacitor has the greater voltage. The 5-μF capacitor would need to have a working voltage rating of at least twice that of the 10-μF capacitor. If both capacitors were of equal value, the voltage across each capacitor would be equal and their working voltage rating would be the same.

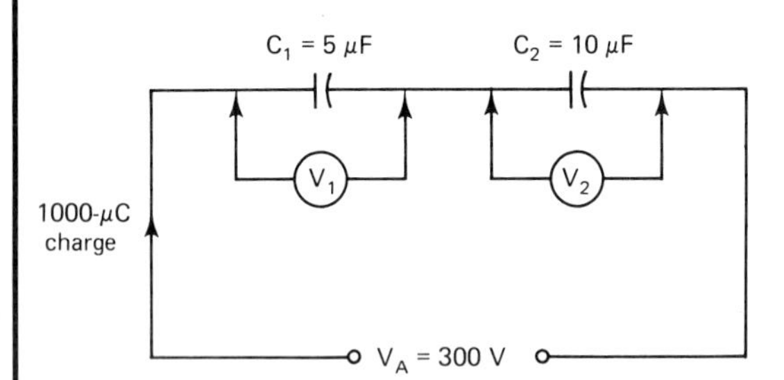

FIGURE 23-21 Series-connected capacitors.

Selecting the Working Voltage Rating of a Capacitor

All capacitors are made available by manufacturers in dc working voltage ranges. As an example, a certain manufacturer's electrolytic capacitors typically are made with a dc working voltage range from 6.3 to 100 V. However, this working voltage range is available (from this particular manufacturer) only in steps of 6.3 V, 10 V, 16 V, 25 V, 40 V, 63 V, and 100 V. Not all capacitance values are available in all voltage ratings.

Selecting a capacitor with an adequate working voltage means that at least a 50% voltage derating should be applied. A manufacturer's catalog selection closest to the derated working voltage and minimum capacitance value can then be made.

EXAMPLE 23-7

A 100-μF electrolytic is to be used as a decoupling capacitor across a 20-V dc source. Select an appropriate working voltage electrolytic from the previously mentioned voltage range steps available.

Solution: Applying a derating factor of 50% to the 20 V across which the capacitor will be connected gives

$$1.5 \times 20\ \text{V} = 30\ \text{V}$$

The closest working volt rating that is available would be 40 V. Note that although a 25-V rated capacitor might serve the purpose, increasing the derating would be preferable.

For capacitors other than electrolytics, the dc working voltage must be selected from a manufacturer's catalog. If the capacitor is to operate on ac, the peak voltage must be used as the working voltage plus a selected derating factor.

Net Capacitance in a Series Connection

The net or total capacitance of series-connected capacitors can be derived from the basic voltage equation for a series circuit:

$$V_T = V_1 + V_2 + V_3$$

Since $V = Q/C$ [equation (23-4)], we can substitute Q/C for each voltage:

$$\frac{Q_T}{C_T} = \frac{Q_1}{C_1} + \frac{Q_2}{C_2} + \frac{Q_3}{C_3}$$

The coulomb charge is common in all parts of the circuit. Dividing through by the common factor Q gives

$$\frac{1}{C_T} = \frac{1}{C_1} + \frac{1}{C_2} + \frac{1}{C_3} \quad \text{etc.} \quad \text{farads} \qquad (23\text{-}5)$$

As in the case of parallel resistors from equation (23-5) the product-over-sum equation is derived for two capacitors at a time:

$$C_T = \frac{C_1 \times C_2}{C_1 + C_2} \quad \text{farads} \qquad (23\text{-}6)$$

EXAMPLE 23-8

Find the net capacitance of the series-connected capacitors in the schematic of Fig. 23-21.

Solution: Using equation (23-6) gives

$$C_T = \frac{C_1 \times C_2}{C_1 + C_2} = \frac{5\ \mu\text{F} \times 10\ \mu\text{F}}{5\ \mu\text{F} + 10\ \mu\text{F}} = 3.3\ \mu\text{F}$$

Find the Voltage across a Capacitor with the Method of Net Capacitance to Selected Capacitance Ratio of Applied Voltage

The voltage across any one of the series capacitors can be found by using the ratio of net capacitance to the selected capacitance. This voltage-divider method is derived as follows.

From equation (23-3),

$$Q = VC$$

In the series circuit connection,

$$Q_T = Q_1 = Q_2 = Q_3$$

$$Q_T = C_1V_1 = C_2V_2 = C_3V_3$$

Substitute C_TV_A for Q_T; therefore,

$$C_TV_A = C_1V_1$$

Divide through by C_1 and rearrange the equation to give

$$V_1 = \frac{C_T}{C_1} \times V_A \qquad \text{volts} \qquad (23\text{-}7)$$

Similarly,

$$V_2 = \frac{C_T}{C_2} \times V_A$$

EXAMPLE 23-9

Find the voltage drop across C_1 and C_2 in the schematic of Fig. 23-21 using the method of net capacitance to selected capacitance ratio of applied voltage.

Solution: Using equation (23-7), the total capacitance was found to be 3.3 μF from Example 23-8; therefore,

$$V_1 = \frac{C_T}{C_1} \times V_A = \frac{3.3\ \mu\text{F}}{5\ \mu\text{F}} \times 300\ \text{V} \cong 200\ \text{V}$$

$$V_2 = \frac{C_T}{C_2} \times V_A = \frac{3.3\ \mu\text{F}}{10\ \mu\text{F}} \times 300\ \text{V} \cong 100\ \text{V}$$

23-10 CAPACITANCE IN PARALLEL

Shunting or connecting capacitors in parallel first requires that the working voltage be equal to or greater than the source voltage. Paralleling capacitors will increase the total capacitance since the coulomb charge of each capacitor will add to the total charge in the circuit. From this the equation for net capacitance in a parallel circuit is derived as follows.

The total coulomb charge of each capacitor is equal to the sum of the individual charges:

$$Q_T = Q_1 + Q_2 + Q_3$$

Substituting VC for Q [equation (23-3)] gives

$$V_AC_T = V_1C_1 + V_2C_2 + V_3C_3$$

and V_A is a common factor; therefore, it can be dropped from the equation to give

$$C_T = C_1 + C_2 + C_3 \qquad \text{farads} \qquad (23\text{-}8)$$

EXAMPLE 23-10

(a) Find the net capacitance of the parallel-connected capacitors in the schematic of Fig. 23-22.

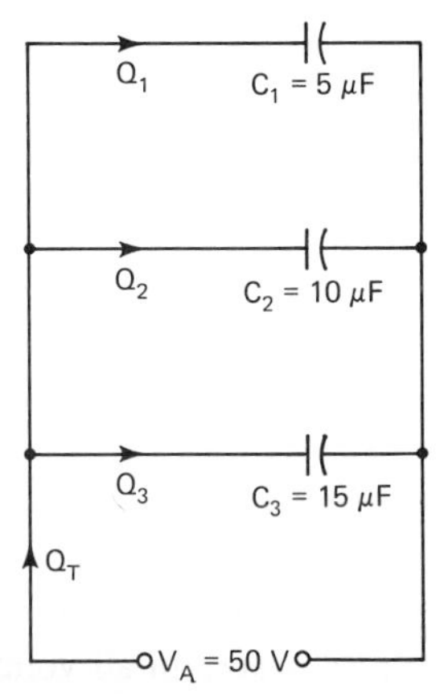

FIGURE 23-22 Parallel-connected capacitors.

(b) What is the minimum working voltage of each capacitor?
(c) What is the total charging current in the circuit?

Solution:

(a) Using equation (23-8) will find the net capacitance:

$$C_T = C_1 + C_2 + C_3$$
$$= 5\ \mu\text{F} + 10\ \mu\text{F} + 15\ \mu\text{F} = 30\ \mu\text{F}$$

(b) The minimum working voltage of each capacitor must be equal to the applied voltage V_A, or 50 V.
(c) The total charging current is

$$Q_T = C_T V_A = 30 \times 10^{-6}\ \text{C} \times 50\ \text{V}$$
$$= 1500\ \mu\text{C or } 1.5\ \text{mA}$$

23-11 DIELECTRIC ABSORPTION LOSSES

Figure 23-23(a) shows the normal condition of electron orbits about the atoms of a capacitor's dielectric that has not been charged. Once a voltage is applied to the plates, the electron orbits of the atoms are distorted toward the positive plate and the positive nucleus toward the negative plate, as shown in Fig. 23-23(b). Since the dielectric is an insulator, no current flows through it. However, the momentary electron shift constitutes a small momentary charge motion or displacement current within the dielectric.

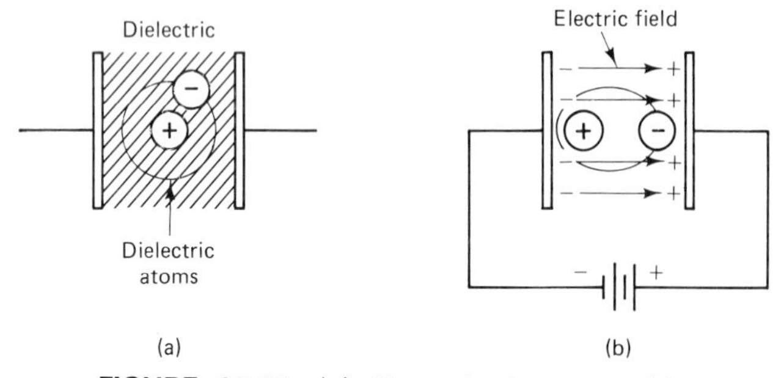

FIGURE 23-23 (a) Normal electron orbits about dielectric atoms. (b) Distorted electron orbits due to electric field.

A capacitor that is connected to an ac voltage will undergo a constant reversal of polarity. These reversals of polarity will cause a continuous displacement current within the dielectric as the insulator atom orbits are distorted alternately toward the plates. This is a form of dielectric hysteresis and heat energy is released in the dielectric. The heat energy represents a power loss, especially at radio frequencies.

Dielectric absorption is present only in capacitors that have a very high leakage resistance, such as mica, ceramic, and oil. These types of capacitors should be discharged through a resistor so that the initial surge current is not too large.

23-12 TESTING CAPACITORS

An ohmmeter test, although not the most desirable test, will indicate weather a capacitor is open, partially shorted, or has a low-resistance short. To test a capacitor, set the ohmmeter on the highest possible range (e.g., $R \times 1\ \text{M}\Omega$). Connect the test leads across the capacitor. (*Note:* Disconnect one end of an in-circuit capacitor to eliminate any parallel resistance, inductance, or other capacitance.) Once test lead clip contact is made, the ohmmeter pointer will swing toward the zero-ohms end of the scale, then gradually recede toward the infinity-ohms end. The length of time it takes for it to recede to infinity ohms depends on the capacitance. The smaller the capacitance, the faster the charging action and the shorter the time for the pointer to return to the infinity end of the ohms scale. The pointer eventually stops at infinite ohms and is measuring the insulation resistance of the dielectric. For ceramic, mica, plastic film, and other types of dielectrics, the insulation resistance of the dielectric ranges from 10,000 Ω to greater than 1000 MΩ, so the ohmmeter is incapable of measuring such a high resistance, therefore measures infinite ohms.

Electrolytics allow a small amount of dc leakage current to pass through the dielectric; therefore, their insulation resistance is much lower. Generally, an ohmmeter reading can be anywhere from about 10 kΩ and much higher, depending on the capacitance and condition of the electrolyte in the capacitor.

Voltmeter Method

Testing a capacitor with an ohmmeter does not provide a realistic voltage across the dielectric since the ohmmeter battery is normally only 1.5 to 3 V. A better method is to apply a much higher voltage, preferably the rated votage of the capacitor to see if it is leaky.

A capacitor can be tested by connecting a voltmeter in series with the capacitor, as shown in Fig. 23-24. For in-circuit capacitors it is safer to disconnect both leads of the capacitor from the circuit. (*Note*: Do not exceed the voltage rating of the capacitor.) Once the volt-

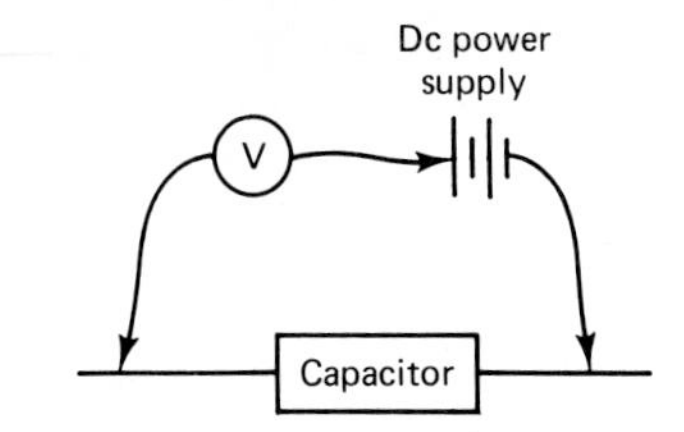

FIGURE 23-24 Testing a capacitor with a voltmeter.

meter connections are made, the power can be turned on. The voltmeter pointer will then swing to indicate the full dc supply voltage, then gradually recede to zero volts. If the capacitor is leaky or shorted, the voltmeter voltage will indicate some voltage value. If it is open, the voltmeter will not react.

Commercial Capacitor–Inductor Analyzers

The ideal method of testing a capacitor is with commercially built capacitors–inductor analyzers. These are available with various test modes. Capacitors can be tested at their full rated voltage as well as at a test frequency. With current models the capacitance and its unit are digitally displayed.

SUMMARY

Like electrical charges repel.

Unlike electrical charges attract.

Coulomb's law states that the force of attraction or repulsion varies directly as the product of the two charges and inversely as the square of the distance:

$$F = C\frac{Q_1 Q_2}{D^2} \quad \text{newtons} \qquad (23\text{-}1)$$

A charged capacitor will have an opposing voltage (counter emf or V_C) equal to the applied voltage.

The electric field of a capacitor is a form of stored energy, which is shown by the equation

$$W = \frac{1}{2}CV^2 \quad \text{joules}$$

The electromagnetic field of an inductor is a form of stored energy as compared to the electric field of a capacitor.

Once the capacitor is charged, it behaves as a battery with a voltage V_C.

The unit of capacitance is the farad (F). The common units of capacitance are:

$$\text{microfarad, } \mu\text{F} = 10^{-6}\ \text{F}$$

$$\text{nanofarad, nF} = 10^{-9}\ \text{F}$$

$$\text{picofarad, pF} = 10^{-12}\ \text{F}$$

The coulomb charge on a capacitor is given by the equation

$$Q = VC \quad \text{coulombs} \qquad (23\text{-}2)$$

Capacitance is directly proportional to the plate area and the dielectric constant, and inversely proportional to the distance between the plates:

$$C = K \times \frac{A}{d} \times 8.85 \times 10^{-12} \quad \text{farads} \qquad (23\text{-}3)$$

The voltage rating of a capacitor is the maximum voltage that can be applied across a capacitor without causing arcover through the dielectric.

A capacitor that is to be connected to an ac voltage must have a dc rating equal to the ac peak value.

Types of capacitors are air, paper, plastic film, mica, ceramic, glass, polarized electrolytic, nonpolarized electrolytic, and tantalum.

Temperature coefficient of capacitance means that the capacitance increases or decreases with a temperature change and is given in ppm/°C.

The cathode lead of a polarized electrolytic capacitor, indentified by a black band or negative symbol, must be connected to the negative polarity of a dc voltage source.

The total capacitance of capacitors connected in series can be found by

$$\frac{1}{C_T} = \frac{1}{C_1} + \frac{1}{C_2} + \frac{1}{C_3} \quad \text{farads} \tag{23-5}$$

or

$$C_T = \frac{\text{product}}{\text{sum}} = \frac{C_1 \times C_2}{C_1 + C_2} \quad \text{farads} \tag{23-6}$$

The voltage drop of a capacitor connected in series can be found using the net capacitance to the selected capacitance:

$$V_1 = \frac{C_T}{C_1} \times V_A \quad \text{volts} \tag{23-7}$$

The net capacitance of capacitors connected in parallel can be found by

$$C_T = C_1 + C_2 + C_3 \quad \text{farads} \tag{23-8}$$

The energy lost in the form of heat by a dielectric, due to displacement current, is termed absorption loss.

An ohmmeter set on its highest range will check a capacitor for an open circuit, partial short, and short circuit, as well as its charging action.

REVIEW QUESTIONS

23-1. State the two laws of two electrical charges.

23-2. What is the function of a capacitor?

23-3. Explain why the initial charging current is very large when charging a capacitor.

23-4. Give three factors that are a function of the stored energy of a capacitor.

23-5. Explain the difference between the stored energy of a capacitor and that of an inductor.

23-6. Give the unit, the symbol, and the value, expressed in powers of 10, for the three practical units of capacitance.

23-7. Explain why an increase in voltage causes a greater coulomb charge to be built up on a capacitor.

23-8. What three physical parameters determine the capacitance of a capacitor?

23-9. Why is the dc voltage rating of a capacitor important?

23-10. Why must the dc voltage rating of a capacitor be derated when it is connected across an ac voltage?

23-11. A paper capacitor has the values of 0.05, 150 V dc, printed on the case. What is its value and unit of capacitance? Its working voltage?

23-12. A ceramic capacitor has the value 20 n printed on its surface. What are its value and unit of capacitance?

23-13. A flat film capacitor has the following color-code stripes: orange, orange, orange, white, and red. Give (a) its value and unit of capacitance, (b) its tolerance, (c) its dc voltage rating, and (d) its value in microfarads.

23-14. Explain why and where air capacitors are used.

23-15. Explain the difference between a film capacitor and a metallized film capacitor.

23-16. For what particular application is the mica capacitor especially suitable? Explain why.

23-17. What is the temperature coefficient of capacitance? How is it expressed? What types of capacitors have a linear temperature coefficient, therefore are suitable for temperature compensation?

23-18. Give two differences between an electrolytic capacitor and other types of capacitors.

23-19. What type of electrolytic capacitor must be used on ac voltages? Explain why.

23-20. How is the negative lead of a polarized electrolytic indentified?

23-21. List six advantages of a tantalum electrolytic over the aluminum electrolytic.

23-22. Regarding the smallest capacitor value, what precaution must be taken when capacitors are connected in series?

23-23. Find the net capacitance of three series capacitors with values of 82 nF, 0.22 μF, and 370 nF.

23-24. What is the net capacitance of the three capacitors in Question 23-23 if they are connected in parallel?

23-25. Why are dielectric absorption losses applicable only to an ac voltage?

23-26. Explain why the initial reading on an ohmmeter is zero ohms when an uncharged capacitor's resistance is being measured.

23-27. Explain why the final reading on an ohmmeter is infinity ohms when the capacitor's resistance is being measured.

23-28. Explain why the final reading on an ohmmeter is less than infinity ohms when measuring the resistance of an electrolytic capacitor.

SELF-EXAMINATION

Choose the most suitable word or phrase to complete each statement.

23-1. The ________ on a capacitor produces a voltage or counter emf which opposes the applied voltage.

23-2. The stored energy of a charged capacitor is in the form of an ________ ________.

23-3. The ability to store a charge is called ________.

23-4. The unit of capacitance is called the ________ with symbol ________.

23-5. Convert 10,000 pF to microfarads.

23-6. Convert 10,000 pF to nanofarads.

23-7. Convert 0.05 μF to nanofarads.

23-8. Which dielectric material will produce the greatest capacitance given the same physical parameters: paper, plastic film, or ceramic?

23-9. Which dielectric—paper, plastic film, or ceramic—will produce a physically smaller capacitor for the same capacitance value?

23-10. A capacitor has a dc voltage rating of 250 V. What is its maximum rms working voltage?

23-11. A 150-pF capacitor has a temperature coefficient designated NPO. By what amount will a 20°C rise in temperature change its capacitance?

23-12. A 22-pF capacitor has a temperature coefficient designated N150. What is its new value with a change in temperature to −20°C?

23-13. A voltage divider supplying 340 V is filtered by two series-connected capacitors. Assume that the required net capacitance is to be 250 μF. What is the correct value of each capacitor?

23-14. Allowing a derating factor of 50%, what dc voltage rating would each capacitor of Question 23-13 have?

23-15. What types of capacitor would be used in this circuit?

23-16. The net capacitance of three capacitors connected in series is 6000 pF. Two capacitors have a value of 0.01 μF and 0.05 μF, respectively. What is the value of the third capacitor in microfarads?

23-17. Allowing a derating factor of 50%, what voltage rating must each capacitor of Question 23-16 have if the series string was connected to 150 V ac?

23-18. Three capacitors of 1000 pF, 0.01 μF, and 120 nF are connected in parallel. Find their net capacitance in nanofarads.

23-19. A ceramic capacitor has the value 2 n 2 printed on its surface. What is its value and unit of capacitance?

23-20. A flat film capacitor has the following color code stripes: brown, black, red, white, and brown. Give its value and unit of capacitance.

23-21. What dc voltage rating does the capacitor of Question 23-20 have?

23-22. What capacitance tolerance does the capacitor of Question 23-20 have?

23-23. What is the initial ohms reading on an ohmmeter when an uncharged capacitor's resistance is being measured?

23-24. What is the final ohms reading on an ohmmeter when a capacitor's resistance is being measured?

23-25. A power loss of a capacitor, due to displacement current in the dieletric, is applicable to radio-frequency voltages and is termed _________ _________ loss.

ANSWERS TO PRACTICE QUESTIONS

23-1

1. repel
2. attract
3. positive
4. 6.25×10^{18}
5. Q
6. two, dielectric
7. counter emf
8. applied voltage
9. capacitance, voltage, charge
10. dielectric
11. I^2R
12. discharged
13. farad
14. capacitance
15. 1 C, 1
16. 10^{-6}
17. 10^{-9}
18. 10^{-12}

23-2

1. voltage, capacitance
2. directly
3. inversely
4. dielectric constant
5. larger
6. 1.33 μF
7. 177 V
8. 22 pF
9. air
10. kraft
11. metallized
12. film foil
13. 30 kV, 100
14. silvered
15. capacitance
16. 67.5 pF
17. aluminum oxide
18. electrolyte
19. polarized
20. negative
21. nonpolarized
22. cost, voltage

24

CR and *L/R* Time-Constant Circuit

INTRODUCTION

A series capacitance–resistance (*CR*) time constant circuit is the basis for very accurate timing and wave-shaping circuits. Among the many applications for timing circuits is a photo flash unit, oscilloscope sawtooth waveform generator, logic circuit pulse waveform generator, and automatic gain control in radio and TV receivers.

The series inductance–resistance (*L/R*) time-constant circuit is the basis for filter networks. Filter networks are designed to accept or reject a frequency component of a voltage, especially audio and radio frequencies. There are many types of filter circuits using various combinations of resistors, capacitors, and inductors.

In this chapter we examine the *CR* time-constant circuit as well as the *LR* time-constant circuit.

MAIN TOPICS IN CHAPTER 24

OBJECTIVES

After studying Chapter 24, the student should be able to

1. Master the fundamental concepts of *CR* and *L/R* time constants.

2. Solve basic time-constant applications.
3. Master the fundamental concepts of a short and long time constant and its significance to a square-wave input voltage pulse width.
4. Recognize the difference between an integrator and a differentiator circuit.

24-1 CURRENT VERSUS TIME IN A PURE RESISTIVE AND CAPACITIVE DC CIRCUIT

In the pure resistive circuit of Fig. 24-1(a), the dc current reaches its maximum Ohm's law value of 10 mA at the instant the switch is closed, as shown graphically in Fig. 24-1(b). A pure resistance is nonreactive. It does not produce a counter emf, due to an electric field like the capacitor or a magnetic field of the inductor.

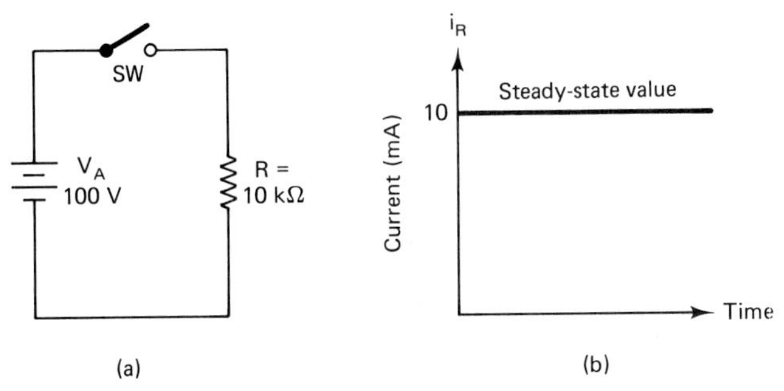

FIGURE 24-1 A pure resistive circuit produces an Ohm's law current at the instant the switch is closed.

In the pure capacitive circuit of Fig. 24-2(a), the charging current reaches a peak value almost instantly. The charging current is limited only by the applied voltage and the capacitance and is given by the equation

$$i = \frac{Q}{t} = C\frac{\Delta V}{\Delta t}$$

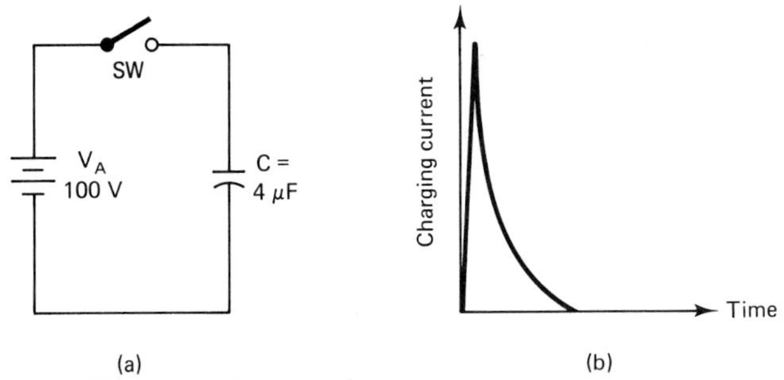

FIGURE 24-2 A pure capacitive circuit produces a "spike" of charging current at the instant the switch is closed.

At the instant the switch is closed, the capacitor appears as a short circuit; therefore, the initial charging current is quite large. This initial current is called the *surge current*. Because a counter emf is quickly built up by the coulomb charge on the capacitor, the surge current diminishes as the counter emf increases. Thus the surge current forms a spike or pulse of current that lasts for a very short period, as shown in Fig. 24-2(b).

24-2 *CR*; v_C, i_R, AND v_R CHARGING TRANSIENT RESPONSE CURVES

The schematic of Fig 24-3(a) shows that the current must flow through the nonreactive resistor before it can charge the capacitor. Initially, the capacitor is not charged, so it appears as a short circuit at the instant the switch is closed, as shown by the equivalent circuit schematic of Fig. 24-3(b).

The series resistor limits the surge current or the charging current of the capacitor. Since the instantaneous applied voltage is across the resistor, the charging current will be limited to its Ohm's law value, in this case $i_C = V/R = 100\text{ V}/10\text{ k}\Omega = 10\text{ mA}$. Initially, then, the voltage across the capacitor will be zero volts while the voltage across the resistor will be equal to the applied voltage, as shown in the transient response curves of Fig. 24-3(c). The graphical representation of v_C with time in Fig. 24-3(c) is termed the transient response of the *CR* circuit. *Transient response* is a temporary condition existing only until the steady-state condition is reached.

As the capacitor builds up a coulomb charge, the voltage v_C across the capacitor increases. This means that the voltage v_R across the resistor decreases since the sum of the series voltage drops cannot exceed the applied voltage. The decreasing voltage drop across R causes a similar decrease in charging current, as shown by the transient response curves of Fig. 24-3(c). After a time interval the coulomb charge on the capacitor reaches its maximum value given by $Q = VC$ and the charging current falls to zero. Since no further current flows in the circuit, the capacitor appears as an open circuit to the source voltage, and the voltage drop across the resistor falls to zero, as shown in Fig. 24-3(c).

24-3 *CR*; v_C, v_R, AND i_R DISCHARGING TRANSIENT RESPONSE CURVES

The charged capacitor can now be discharged by removing the source and short circuiting the terminals as shown in the schematic diagram of Fig. 24-4(a). Closing the switch will discharge the capacitor through the resistor. The equivalent-circuit schematic of Fig. 24-

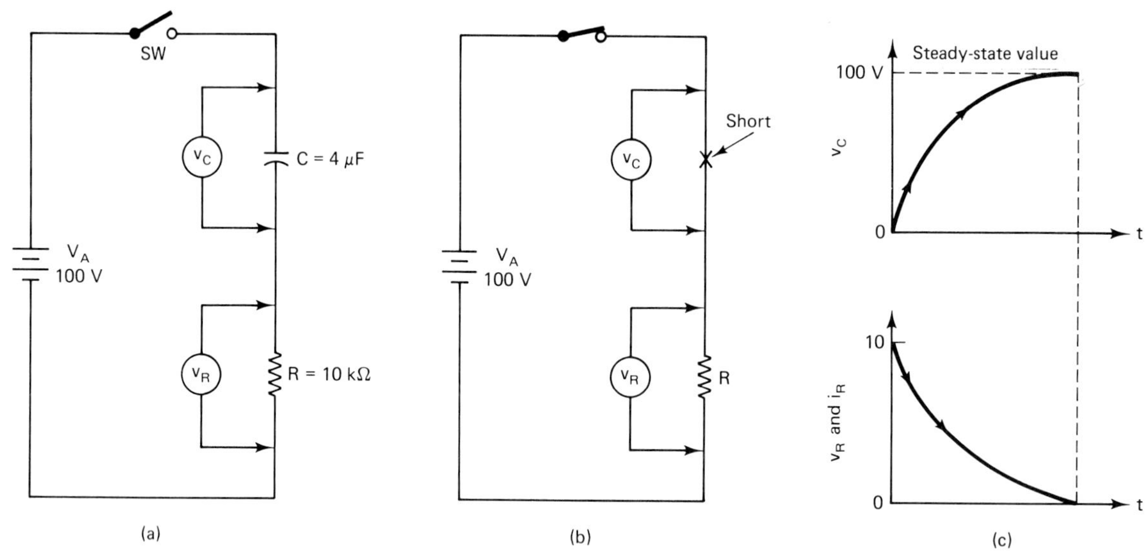

FIGURE 24-3 (a) Series *CR* circuit. (b) Equivalent circuit at the instant the switch is closed. (c) Graphical transient response of charging v_C, v_R, and i_R.

4(b) shows that the fully charged capacitor appears as a source of voltage and current connected across the resistor. The current now flows from the negatively charged plate of the capacitor through the resistor and into the positive plate of the capacitor. This discharge action converts the potential or stored energy into heat energy which is released by the resistor.

Initially, the full v_C appears across the resistor but it decays with time as the coulomb charge on the capacitor decays. This is shown by the representative graph of v_C, v_R, and i_R of Fig. 24-4(c). Note that since v_C is effectively across the resistor, the resistor voltage will follow the decaying discharge current and voltage of the capacitor. After a time interval the charge on the capacitor is completely depleted and v_C, v_R, and i_R fall to zero.

24-4 *CR* CHARGING TIME CONSTANT

The current-limiting action of the series resistor will affect the rate at which the capacitor charges or discharges. The rate at which the capacitor charges or discharges is determined by the time constant of the series *CR* circuit and is the product of *CR*:

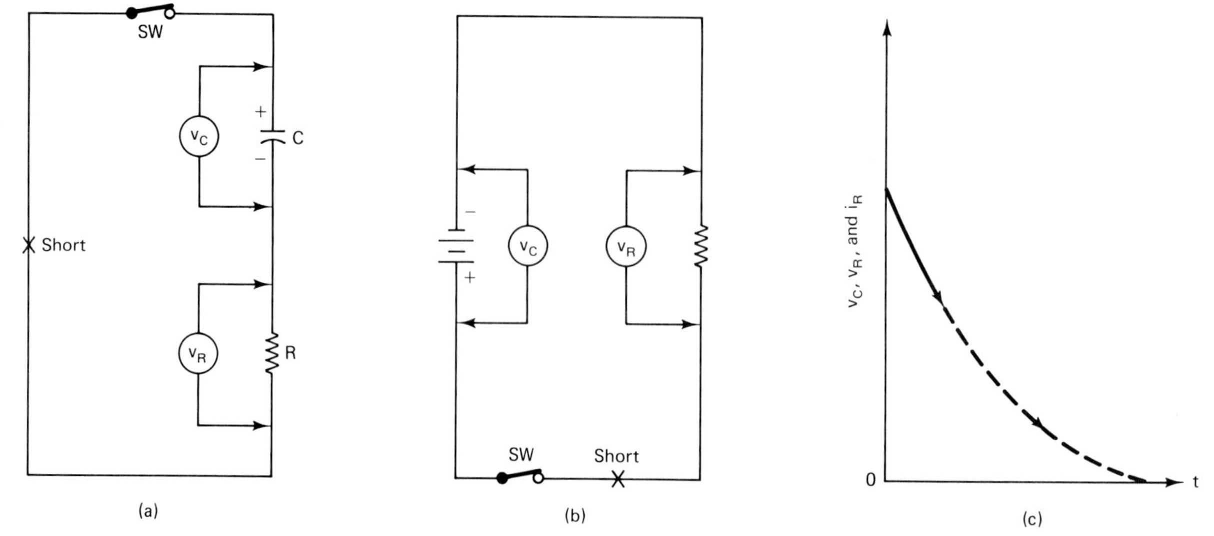

FIGURE 24-4 (a) Discharging a capacitor. (b) Equivalent circuit v_C appears as a source voltage. (c) Graphical transient response of decaying v_C, v_R, and i_R.

$$T = CR \qquad \text{seconds} \qquad (24\text{-}1)$$

when C is in microfarads
R is in megohms
T is in seconds

Using the units of measurement for C and R, we can show that

$$R = \frac{V}{I} = \frac{\text{volts}}{\text{amperes}} = \frac{\text{volts}}{\text{coulomb/second}}$$

and

$$C = \frac{Q}{V} = \frac{\text{coulombs}}{\text{volt}}$$

Therefore,

$$CR = \frac{\text{coulombs}}{\text{volt}} \times \frac{\text{volts}}{\text{coulomb/second}} = \text{seconds}$$

or

$$T = CR$$

> The time interval that it takes a capacitor in a series *CR* circuit to charge to 63% of its steady-state voltage is termed one time constant.

A discharged capacitor that is charged by a *CR* circuit will gain 63% of its voltage during each time-constant interval. Its voltage at the end of each time-constant interval will then be an additional 63% of its voltage at the beginning of the next time-constant interval. The symbol for a time constant (*CR*) interval is the Greek lowercase letter tau (τ).

Value of v_C after One Charging Time Constant (τ)

In the series *CR* circuit of Fig. 24-5(a), at the instant the switch is closed, the voltage across the capacitor will be zero (uncharged capacitor), while the instantaneous voltage across the resistor will be 100 V. For the first and subsequent time-constant intervals, the capacitor voltage will increase by 63% of the voltage across the resistor.

Figure 24-5(a) illustrates the resultant voltages of a series *CR* circuit at the end of one time-constant intervals. The time constant of the circuit is $T = CR = 10\ \mu\text{F} \times 0.01\ \text{M}\Omega = 0.1$ s. After 0.1 s the capacitor

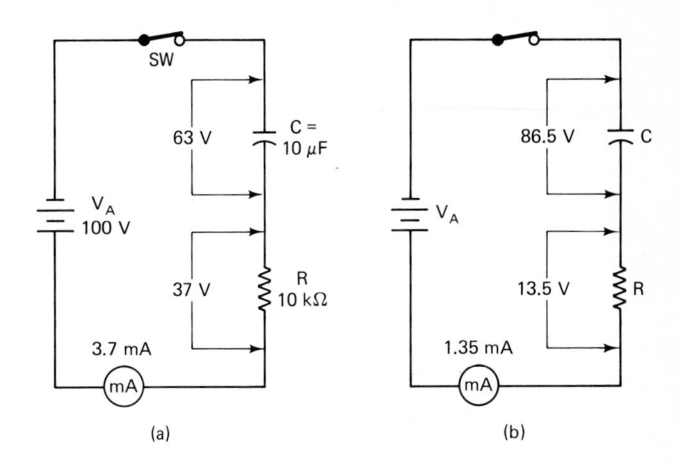

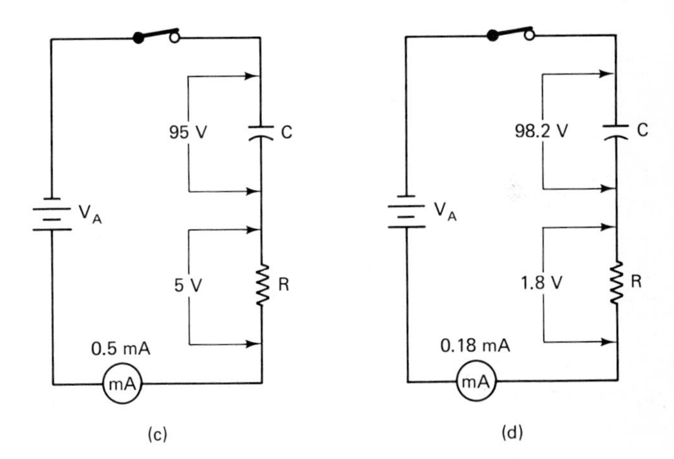

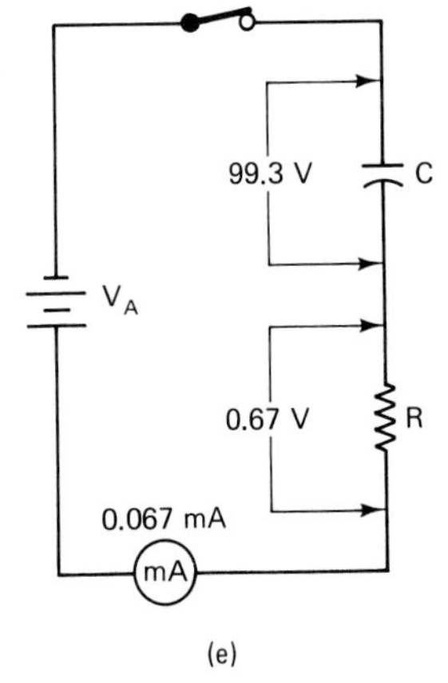

FIGURE 24-5 Voltages across *C* and *R* after each time constant. After 5 TC the capacitor is considered as being fully charged. (a) After first TC (0.1 s). (b) After second TC (0.2 s). (c) After third TC (0.3 s). (d) After fourth TC (0.4 s). (e) After fifth TC (0.5 s).

will be charged to 63 V (63% of 100 V), while the remaining voltage across the resistor drops to 37 V (100 V − 63 V = 37 V).

Determining v_C after One Charging Time Constant by Using a Calculator. The voltage v_C across the capacitor for any charging time-constant interval can easily be determined using a calculator and the general equation

$$v_C = V_A(1 - e^{-t/CR}) = V_A(1 - e^{-t/\tau}) \quad \text{volts} \quad (24\text{-}2a)$$

$$v_C = V_A\left(1 - \frac{1}{e^x}\right) \quad \text{volts} \quad (24\text{-}2)$$

where e is the base of natural logarithm and has a constant value of 2.718. x is the exponent of the natural log base, in this case the desired time constant (τ).

EXAMPLE 24-1: For the First Charging Time Constant

1. Enter 1; the display reads 1.
2. Press the e^x key (or INV then INx); the display reads 2.718.
3. Press the $1/x$ key; the display reads 0.3678.
4. Press the STO key; the display reads 0.3678.
5. Enter 1; the display reads 1.
6. Press the − key; the display reads 1.
7. Press the RCL key; the display reads 0.3678.
8. Press the = key; the display reads 0.6321.
9. Press the X key; the display reads 0.6321.
10. Enter the value of V_A; the display reads 100.
11. Press the = key; the display reads 63.21.

63 volts or 63% is the general value for the first charge time constant.

Value of v_C and v_R after Second Charging Time Constant

In Fig. 24-5(a) at the end of the first time-constant interval of 0.1 s, the resistor voltage is 37 V. The voltage across the capacitor at the end of the two charge time-constant intervals (0.2 s) will increase by 63% of the remaining voltage across the resistor, or 0.63 × 37 V = 23.3 V. This means that at the end of two time constant intervals the voltage across the capacitor has increased to 86.5 V, or 63 V + 23.3 V = 86.5 V. The remaining voltage across the resistor will be 100 V − 86.5 V = 13.5 V, as shown in Fig. 24-5(b).

Determining v_C after Two Charging Time Constants by Using a Calculator. The voltage v_C across the capacitor for any charge time-contant interval can be determined on a calculator using the general equation (24-2).

EXAMPLE 24-2: For the Second Charging Time Constant

1. Enter 2; the display reads 2.
2. Press the e^x key (or INV then INx); the display reads 7.389.
3. Press the $1/x$ key; the display reads 0.135.
4. Press the STO key; the display reads 0.135.
5. Enter 1; the display reads 1.
6. Press the − key; the display reads 1.
7. Press the RCL key; the display reads 0.135.
8. Press the = key; the display reads 0.8646.
9. Press the X key; the display reads 0.8646.
10. Enter the value of V_A; the display reads 100.
11. Press the = key; the display reads 86.46.

The value of v_C at the end of the second time-constant interval is 86.5 V rounded off to three significant figures.

Value of v_C and v_R after Three Charging Time Constants.

In Fig. 24-5(b), at the end of two charge time-constant intervals (0.2 s) the resistor voltage is 13.5 V. The voltage across the capacitor at the end of three charge time-constant intervals (0.3 s) will increase by another 63% of the remaining voltage across the resistor, or 0.63 × 13.5 V = 8.5 V. This means that the end of three time constants the voltage across the capacitor has increased to 95 V, or 86.5 + 8.5 V = 95 V. The remaining voltage across the resistor will be 100 V − 95 V = 5 V, as shown in Fig. 24-5(c).

Determining v_C after Three Charging Time Constants by Using a Calculator. The voltage v_C across the capacitor for any charge time-constant interval can be determined on a calculator using the general equation (24-2). For the third TC,

$$v_C = 1 - \frac{1}{e^3} \times V_A$$

Value of v_C and v_R after Four Charging Time Constants

In Fig. 24-5(c), at the end of three charge time-constant intervals (0.3 s) the resistor voltage is 5 V. The voltage across the capacitor at the end of four time-constant intervals (0.4 s) will increase by another 63% of the remaining voltage across the resistor or $0.63 \times 5\text{ V} = 3.2\text{ V}$. This means that after four time constants the voltage across the capacitor has increased to 98.2 V, or $95\text{ V} + 3.2\text{ V} = 98.2\text{ V}$. The remaining voltage across the resistor will be $100 - 98.2 = 1.8\text{ V}$, as shown in Fig. 24-5(d).

Determining v_C after Four Charging Time Constants by Using a Calculator. The voltage v_C across the capacitor for any charge time-constant interval can be determined on a calculator using the general equation (24-2). For the fourth TC,

$$v_C = 1 - \frac{1}{e^4} \times V_A$$

Value of v_C and v_R after Five Charging Time Constants

In Fig. 24-5(d), at the end of four charge time-constant intervals (0.4 s), the resistor voltage is 1.8 V. The voltage across the capacitor at the end of five charge time-constant intervals (0.5 s) will increase by another 63% of the remaining voltage across the resistor, or $0.63 \times 1.8\text{ V} = 1.1\text{ V}$. The voltage across the capacitor will therefore be increased to 99.3 V, or $98.2\text{ V} + 1.1\text{ V} = 99.3\text{ V}$, as shown in Fig. 24-5(e).

The voltage across the capacitor at the end of five time constants is so close to the full charge voltage that for practical purposes it is considered fully charged. A five-time-contant interval (5T) is termed the *transient time*. Table 24-1 summarizes the percentage capacitor and resistor voltages for five charge time constants.

TABLE 24-1

Percentage capacitor and resistor voltage for each charge time constant

Time Constant, τ	*Capacitor v_C (% volts)*	*Resistor v_R (% volts)*
1	63	37
2	86	14
3	95	5
4	98	2
5	99	1

24-5 *CR* DISCHARGING TIME CONSTANT

A fully charged capacitor that is now discharged into a *CR* circuit will lose 63% of its voltage during each time-constant interval. Its voltage at the end of each time-constant interval will then be 37% of its initial voltage at the beginning of each time-constant interval.

Value of v_C and v_R after One Discharge Time Constant

In Fig. 24-6(a), at the instant the switch is closed, the initial voltage across the capacitor will be equal to its full charged state voltage V_A, or 100 V. Since the resistor is effectively connected across the capacitor then $v_R = v_C$, which will equal 100 volts. After a one time-constant interval, the voltage across the capacitor will decay to 37% of its initial value, or $0.37 \times 100\text{ V} = 37\text{ V}$.

Figure 24-6(a) illustrates the resultant voltages of a series *CR* circuit after one time-constant interval. The time constant of the circuit was previously found to be 0.1 s. After 0.1 s of capacitor discharge, its voltage will decay to 37 V (37% of 100 V), which is also the remaining voltage across the resistor.

Determining v_C after One Discharge Time Constant by Using a Calculator. The voltage v_C across the capacitor for any discharge time constant can be determined on a calculator and with the general equation

$$v_C = V_A e^{-t/RC} \quad \text{volts} \qquad (24\text{-}3a)$$

or

$$v_C = \frac{1}{e^x} \times V_A \quad \text{volts} \qquad (24\text{-}3)$$

EXAMPLE 24-3: For the First Discharge Time Constant

1. Enter 1; the display reads 1.
2. Press the e^x key (or INV then INx); the display reads 2.718.
3. Press the $1/x$ key; the display reads 0.3678.
4. Press the X key; the display reads 0.3678.
5. Enter the value of V_A; the display reads 100.
6. Press the = key; the display reads 36.78.

37 volts or 37% is the general value for the first discharge time constant.

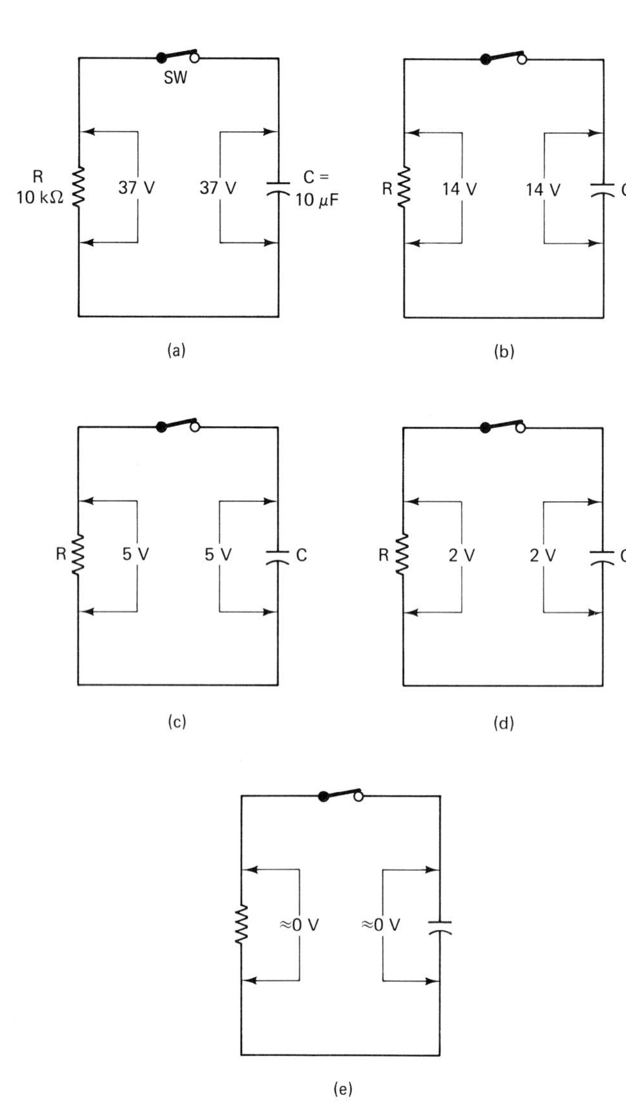

FIGURE 24-6 Voltages across C and R after each time constant. After 5 TC the capacitor is considered as being fully discharged. (a) After first TC (0.1 s). (b) After second TC (0.2 s). (c) After third TC (0.3 s). (d) After fourth TC (0.4 s). (e) After fifth TC (0.5 s).

Value of v_C and v_R after Two Discharge Time Constants

In Fig. 24-6(a) at the end of one discharge time-constant interval (0.1 s) the voltage left across the capacitor and resistor was 37 V. At the end of the second time-constant interval (0.2 s) the voltage across the capacitor and resistor will decay to 37% of its initial value, or 14 V (37% of 37 V = 13.69 V), as shown in Fig. 25-6(b).

Determining v_C after Two Discharge Time Constants by Using a Calculator. The voltage v_C and v_R for any discharge time constant can be determined on a calculator using the general equation (24-3):

$$v_C = \frac{1}{e^2} \times V_A \qquad \text{volts}$$

EXAMPLE 24-4: For the Second Discharge Time Constant

1. Enter 2; the display reads 2.
2. Press the e^x key (or INV then INx); the display reads 7.389.
3. Press the $1/x$ key; the display reads 0.1353.
4. Press the X key; the display reads 0.1353.
5. Enter the value of V_A; the display reads 100.
6. Press the = key; the display reads 13.53.

Value of v_C and v_R after Three Discharge Time Constants

In Fig. 24-6(b), at the end of two discharge time-constant intervals (0.2 s) the voltage left across the capacitor and resistor was approximately 14 V. At the end of the third time-constant interval (0.3 s) the voltage across the capacitor and resistor will decay to 37% of its initial value, or 5 V (37% of 14 V = 5.18 V), as shown in Fig. 25-6(c).

Determining v_C after Three Discharge Time Constants by Using a Calculator. The voltages v_C and v_R for any discharge time-constant can be determined on a calculator using the general equation (24-3).

$$v_C = \frac{1}{e^3} \times V_A \qquad \text{volts}$$

Value of v_C and v_R after Four Discharge Time Constants

In Fig. 24-6(c), at the end of the third discharge time-constant interval (0.3 s) the voltage left across the capacitor and resistor was approximately 5 V. At the end of the fourth time-constant interval (0.4 s) the voltage across the capacitor and resistor will decay to 37% of its initial value, or approximately 2 V (37% of 5 V = 1.85 V), as shown in Fig. 25-6(d).

Determinig v_C after Four Discharge Time Constants by Using a Calculator. The voltages v_C and v_R for any discharge time constant can be determined on a calculator using the general equation (24-3).

$$v_C = \frac{1}{e^4} \times V_A \qquad \text{volts}$$

Value of v_C and v_R after Five Discharge Time Constants

In Fig. 24-6(d), at the end of the fourth discharge time-constant interval (0.4 s) the voltage left across the capacitor and resistor was approximately 2 V. After the fifth time constant interval (0.5 s) the voltage across the capacitor and resistor will decay to 37% of its initial value, or approximately 0 V (37% of 2 V = 0.7 V), as shown in Fig. 25-6(e).

The capacitor at the end of the fifth time-constant interval is so close to being completely discharged that for practical purposes its voltage is considered to be zero. Table 24-2 summarizes the percentage capacitor and resistor voltage for five discharge time constants.

TABLE 24-2
Percentage capacitor and resistor voltage for each discharge time constant

Time Constant, τ	*Capacitor v_C (% volts)*	*Resistor v_R (% volts)*
1	37	37
2	14	14
3	5	5
4	2	2
5	0	0

24-6 UNIVERSAL TIME-CONSTANT CURVES

The universal time-constant curves of Fig. 24-7 give the instantaneous value of v_C or i as a percentage of the initial or steady-state value, with time given in time constants tau (τ). These curves can be used to determine graphically capacitor voltage or current values other than those at each specific time constant.

EXAMPLE 24-5

(a) Use the universal time-constant curves to determine the value of the capacitor voltage in Fig. 24-5(a) after 0.24 s.

(b) Determine the time it takes for the capacitor to charge to 75 V.

Solution:

(a) From the universal time-constant curves the 100% level is 100 V. One time constant is equal to 0.1 s, which is 0.24 s/0.1 s = 2.4τ after 0.24 s. At 2.4τ the percentage change is 90%; therefore, the capacitor reaches 90% of 100 V, or 90 V.

(b) At 75 V the change on the capacitor is 75 V/100 V × 100% = 75%. From the universal time-constant curves, at 75% the time constant is 1.4τ. The time in seconds is then 1.4τ × 0.1 s = 0.14 s or 140 ms.

PRACTICE QUESTIONS 24-1

Choose the most suitable word, phrase, or value to complete each statement.

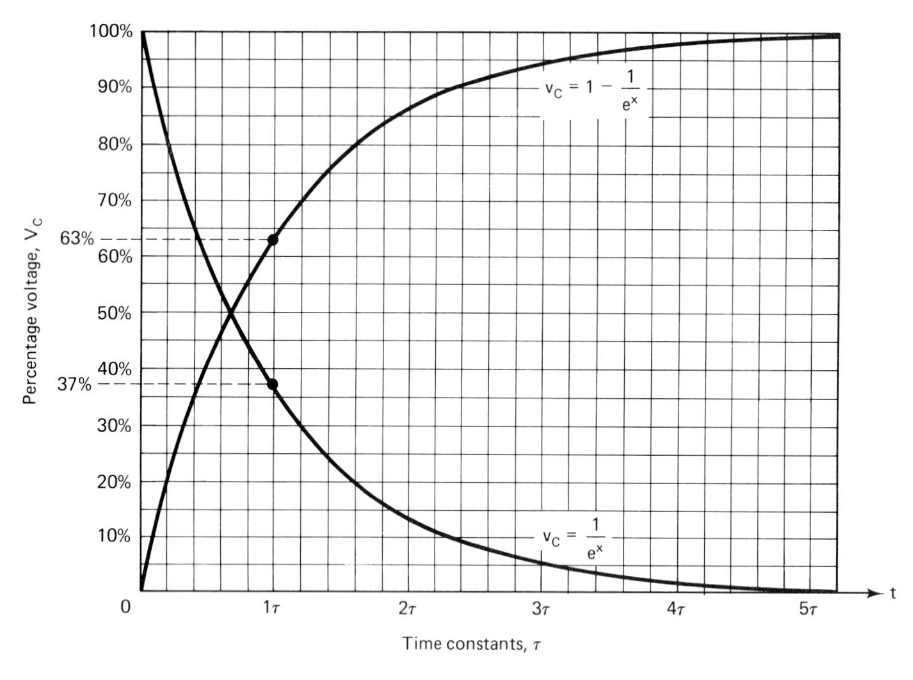

FIGURE 24-7 Universal time constant curves.

1. A pure resistance is ________ because it does not produce a counter emf like the capacitor or inductor.
2. In a pure resistive circuit the current reaches its ________ ________ value instantly and remains at this value as long as the circuit is complete.
3. In a pure capacitive circuit the current reaches its peak value almost instantly, then falls to ________.
4. In a capacitive circuit the initial charging current is termed a ________ current.
5. In a series *CR* circuit the series ________ limits the charging or surge current of the capacitor.
6. The initial voltage across the resistor of a series *CR* circuit will be equal to ________ voltage because of the instantaneous value of the charging current.
7. Once charged, the voltage across the capacitor due to its charge will be equal to the ________ voltage.
8. Once the capacitor is charged, the voltage across the resistor will be equal to ________.
9. In a series *CR* circuit the capacitor discharges through the resistor so that the resistor voltage is equal to ________ voltage.
10. The stored energy of a capacitor is released in the form of ________ by the resistor when the capacitor is discharged.
11. The slow discharge of a capacitor's coulomb charge is termed the ________ of the capacitor's charge or voltage.
12. The time it takes for a series *CR* capacitor to charge to 63% of the applied voltage is termed one ________ ________.
13. The time constant of a series *CR* circuit is determined by the ________ of *CR* and is in seconds when *C* is in ________ and *R* is in ________.
14. Calculate the time constant of a series *CR* circuit when $C = 0.05\ \mu\text{F}$ and $R = 1\ \text{M}\Omega$.
15. If the applied voltage in the series *CR* circuit of Question 14 is 150 V, what is the voltage across *C* after one time constant? Assume a discharged capacitor.
16. For the same capacitor as in Question 15; what is the voltage across *C* after 1 s?
17. For the capacitor in Questions 14 and 15; what is the voltage across the resistor after 0.1 s?
18. For the capacitor in Questions 14 and 15, that is, being discharged from a fully charged condition, what is the voltage across the capacitor after 0.1 s? What is the voltage across the resistor?
19. From the universal time-constant curves, in what time does it take the capacitor of Questions 14 and 15 to reach a charge of 75 V?
20. From the universal time-constant curves, in what time does it take the capacitor of Questions 14 and 15 to reach a discharge voltage of 30 V.

24-7 *CR* TIME-CONSTANT APPLICATIONS

The basic circuit of Fig. 24-8(a) is used to trigger a device/circuit at a rate determined by the *CR* time con-

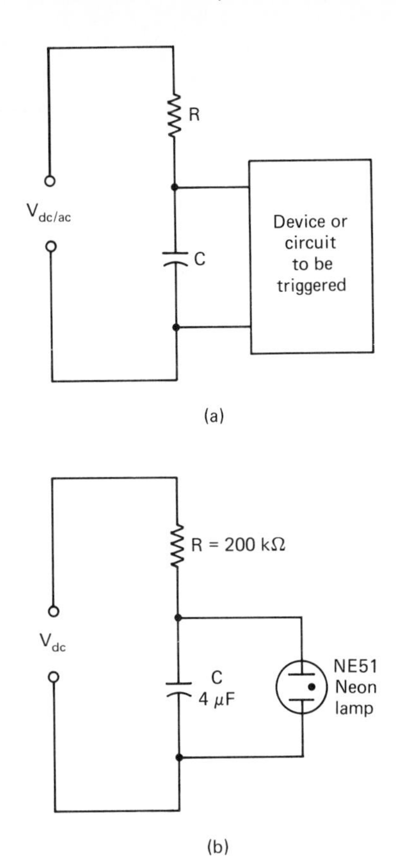

FIGURE 24-8 Basic *CR* time constant circuit used to trigger other device or circuits. (a) *CR* circuit used to trigger a device or circuit. (b) A basic neon flasher.

stant. Some of these applications include triggering relays, zener diodes, silicon-controlled rectifiers, and unijunction transistors, which in turn control circuits to operate devices such as photoflash units, phototimers automatic spot welding, strobe lights, and others.

A demonstration of a triggered timing circuit can be made with the neon flasher shown in the schematic of Fig. 24-7(b). The neon lamp is triggered into glowing at approximately 67 V. Once it glows, the voltage must be reduced to turn it off. This is accomplished by the *CR* time-constant circuit.

In this example circuit the capacitor will take *CR* seconds (or $4\ \mu F \times 0.2\ M\Omega = 0.8$ s) to charge to 63% of the dc source voltage. With a 105-V source the capacitor will charge to 67 V after 0.8 s $[1 - (1/e^x) \times V_A]$. This will trigger the neon lamp into glowing. The capacitor now discharges through the neon lamp. Once the capacitor's voltage falls below 60 V, the lamp goes out, and the cycle is repeated. The lamp in this particular circuit will flash approximately ($f = 1/T = 1/0.8$ s) once every second.

Applications such as oscillators, pulse generators, ramp or square-wave generators, one-shot multivibrators, burglar alarms, windshield wiper timers, and voltage monitors all require a circuit capable of producing timing intervals.

The most popular integrated-circuit (IC) timer is the 555 IC. Connected as a monostable multivibrator, it uses an external resistor and capacitor to vary the time duration of a square-wave output voltage (see the schematic of Fig. 24-9).

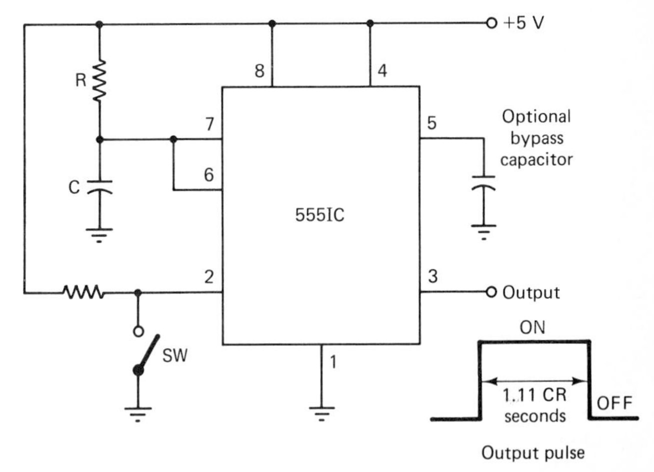

FIGURE 24-9 555 IC timer connected as a monostable multivibrator or timed pulse generator. Numbers shown are PIN numbers of IC chip.

Figure 24-9 shows the 555 timer connected as a monostable multivibrator. Once the switch is closed, the ON time of the square wave or pulse will be determined by the length of time it takes *C* to charge from zero to $\frac{2}{3}$ V and is given by the equation $t = 1.11\ CR$. The input is normally triggered by narrow 5-V trigger pulses. The 555 timer is capable of producing timed pulses from microseconds to hours in duration.

24-8 CAPACITOR AND RESISTOR VOLTAGE WAVEFORMS WITH AN APPLIED SQUARE-WAVE VOLTAGE

The circuit schematic of Fig. 24-10(a) is analogous to a generator similar to a 555 timer, that is, capable of producing a square-wave voltage. Toggling the switch to the ON position for a time interval will cause an Ohm's law current of 1 mA to flow, with a resultant voltage of 10 V across the resistor.

The switch is then quickly toggled to the OFF position and the current falls to zero immediately for a time interval. If the toggling cycle is repeated, a square wave is generated as shown in Fig. 24-9(b). The voltage waveform will be identical to the current waveform

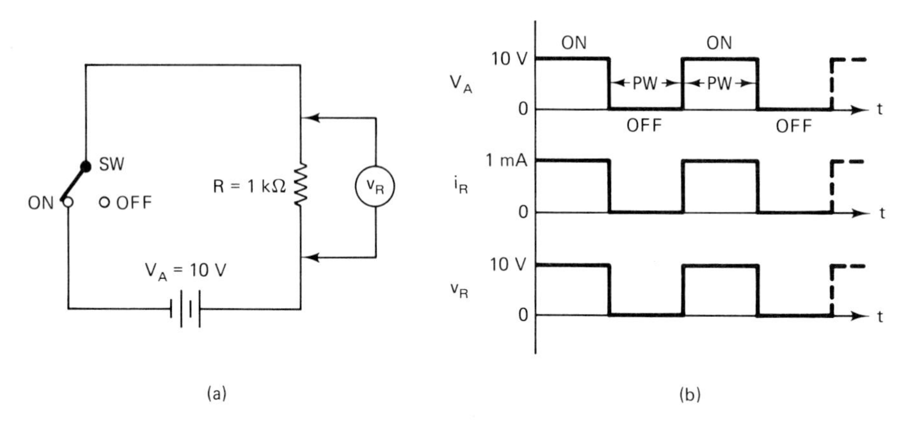

FIGURE 24-10 (a) On–off action of switch sets up a square-wave voltage. (b) Resultant current and voltage waveforms of resistor.

since the resistor is a nonreactive component (i.e., it does not produce a counter emf). The time interval that the voltage is ON or OFF is considered to be the pulse width (PW) of the square wave. Note that the waveform could be a rectangular shape; however, it is referred to as a square wave.

Capacitor Fully Charging and Discharging with a Square-Wave Voltage Input

In the *CR* circuit of Fig. 24-11(a), the rate at which the capacitor charges and discharges will depend on the time constant of the circuit. This, in turn, will determine the current flowing through the resistor and its resultant voltage waveform.

With a fixed time constant the voltage that the capacitor charges to will be determined by the length of time that the square-wave pulse in ON [i.e., its pulse width (PW)]. Similarly, the OFF time of the pulse will determine the discharge extent of the capacitor, with resultant fall in capacitor voltage.

If the pulse width of the square wave is equal to or greater than five time constants (PW $\geq 5\tau$), the capacitor will *fully charge* on the first pulse. At the end of the OFF portion of the pulse, the capacitor will fully discharge back into the source, as shown in the representative waveforms of Fig. 24-11(b).

Figure 24-11(b) shows the resultant waveforms for both capacitor and resistor voltages with a long time constant and a fixed pulse width that is equal to five time constants (5τ).

At the instant of the input pulse, the charging current will be a 1-mA spike. The current spike then decays to zero, as the capacitor becomes fully charged during the time period of five time constants of the pulse width.

When the capacitor discharges back into the source, the current through the resistor is reversed; therefore, the voltage polarity will be reversed, as shown in the representative waveforms of Fig. 24-11(b) and (c).

Since the charging and discharging current flows through the resistor, the resistor voltage waveform will be identical to the current waveform shown in the representative waveforms of Fig. 24-11. Note that the waveform shape of the capacitor voltage approaches that of the input voltage waveform.

Capacitor Partially Charging and Discharging with a Square-Wave Voltage Input

If the frequency of the square-wave source voltage is increased, the pulse width and the time between pulses become shorter. When the pulse width and the time between pulses are shorter than five time constants, the capacitor will never fully charge or completely discharge.

Consider the case where the input voltage pulse width into the circuit of Fig. 24-12 is equal to one time constant.

For each of the following explanations, refer to Fig. 24-13 and the corresponding voltage waveform.

Beginning of the First Pulse Interval. Assuming that the capacitor is not charged, its voltage reaches 63% of the applied voltage or 6.3 V, as shown in the representative v_C waveform of Figure 24-13. At the start of the pulse, the instantaneous voltage across the resistor falls to 3.7 V, the difference between v_C and v_R (10 V − 6.3 V) = 3.7 V.

Interval between the First and Second Pulses. The voltage across the capacitor is now 6.3 V as it starts to discharge through the resistor and into the source. Since the discharge current flows in the opposite direction to the charge current, the resistor voltage polarity is reversed and is equal to −6.3 V.

At the end of the time interval between the first and second pulses the capacitor voltage falls to 37% of 6.3 V, or 0.37 × 6.3 V = 2.33 V, which is also equal to the resistor voltage since the capacitor is effectively in parallel with the resistor on discharge.

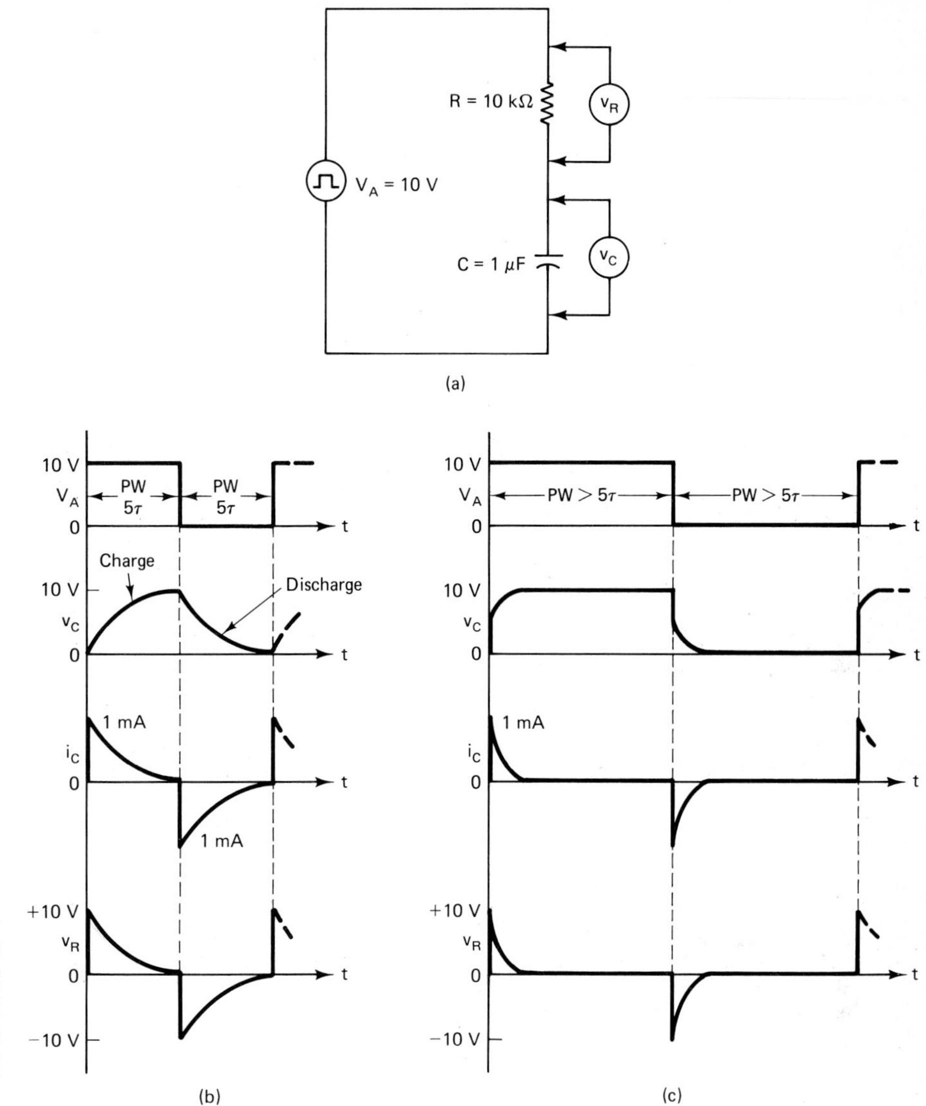

FIGURE 24-11 (a) Series *CR* circuit with an applied square-wave voltage. (b) Waveforms with PW = 5 time constants and short *CR* time constant. (c) Waveforms with PW > 5 time constants and short *CR* time constant.

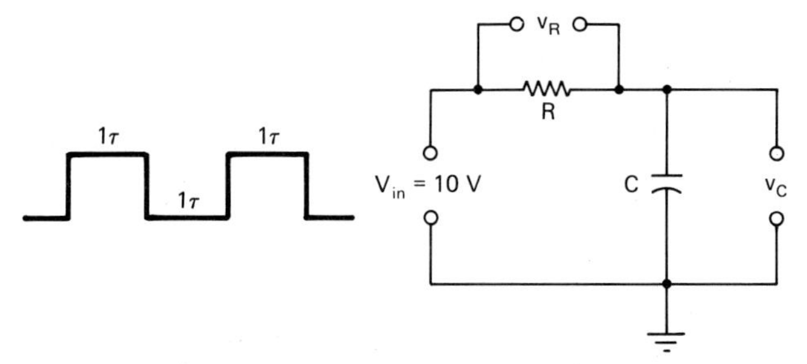

FIGURE 24-12 Pulse width of the input voltage is equal to one time constant (T).

Beginning of the Second Pulse Interval. On the second 10-V input pulse the voltage that is going to charge the capacitor will be the difference between the final capacitor voltage and the source voltage, or 10 V − 2.33 V = 7.67 V. Since the capacitor voltage begins at 2.33 V and charges an additional 63% of 7.67 V, or 0.63 × 7.67 V = 4.83 V, the net voltage across the capacitor will be 2.33 V + 4.83 V = 7.16 V at the end of the second pulse interval. The resistor voltage at the end of the second pulse interval will be the difference between the source voltage and the capacitor voltage, or 10 V − 7.16 V = 2.84 V.

Interval between the Second and Third Pulses. The voltage across the capacitor is now 7.16 V as it starts to discharge through the resistor and into the source. Since the discharge current flows in the opposite direction to the charge current, the resistor polarity is reversed and is equal to −7.16 V.

At the end of the time interval between the second and third pulses the capacitor voltage falls to 37% of 7.16 V, or 0.37 × 7.16 V = 2.65 V, which is also equal to the resistor voltage since the capacitor is effectively in parallel with the resistor on discharge.

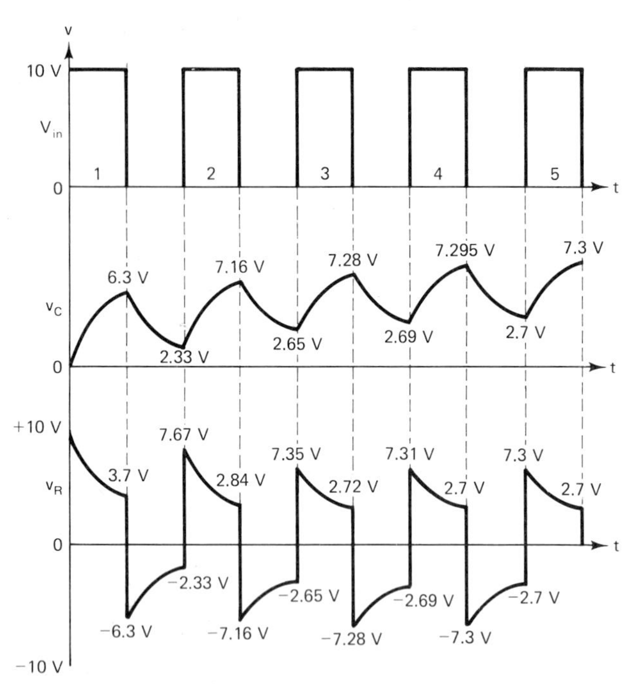

FIGURE 24-13 Capacitor and resistor voltage waveforms after each input voltage pulse width of one time constant (T).

Beginning of the Third Pulse Interval. On the third 10 V input pulse the voltage that is going to charge the capacitor will be the difference between the final capacitor voltage and the source voltage, or 10 V − 2.65 V = 7.35 V. Since the capacitor voltage begins at 2.65 V and charges an additional 63% of 7.35 V, or 0.63 × 7.35 V = 4.63 V, the net voltage across the capacitor will be 2.65 V × 4.63 V = 7.28 V at the end of the third pulse interval.

Interval between the Third and Fourth Pulses. The voltage across the capacitor is now 7.28 V as it starts to discharge through the resistor and into the source. Since the discharge current flows in the opposite direction to the charge current, the resistor polarity is reversed and is equal to −7.28 V.

At the end of the time interval between the third and fourth pulses the capacitor voltage falls to 37% of 7.28 V, or 0.37 × 7.28 V = 2.69 V, which is also equal to the resistor voltage since the capacitor is effectively in parallel with the resistor on discharge.

After Five Pulse Intervals. Continuing this analysis for the fourth and fifth pulse intervals will show that after five time constants the capacitor and resistor voltage reach a steady-state value as shown in Fig. 24-14.

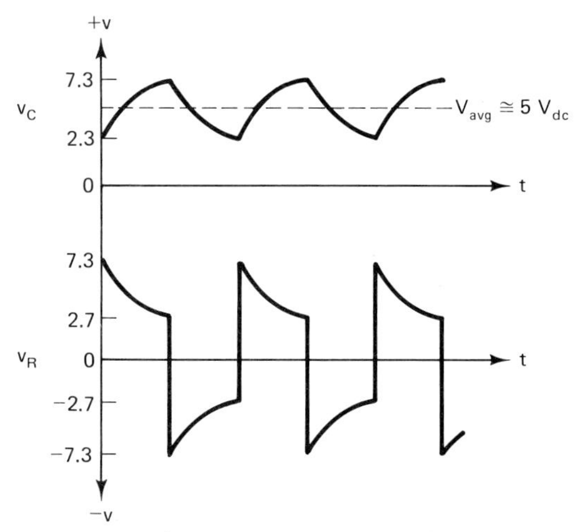

FIGURE 24-14 Steady-state value of v_C and v_R after five pulses.

Long *CR* Time Constant

Increasing the frequency of the source voltage will decrease the pulse width. When the pulse width is much smaller than the time constant, the capacitor never fully charges or discharges. As a result, the capacitor output voltage changes very little, as shown in the representative waveform v_C of Fig. 24-15. Note that the shape of the resistor's voltage waveform approaches the input voltage waveform shape.

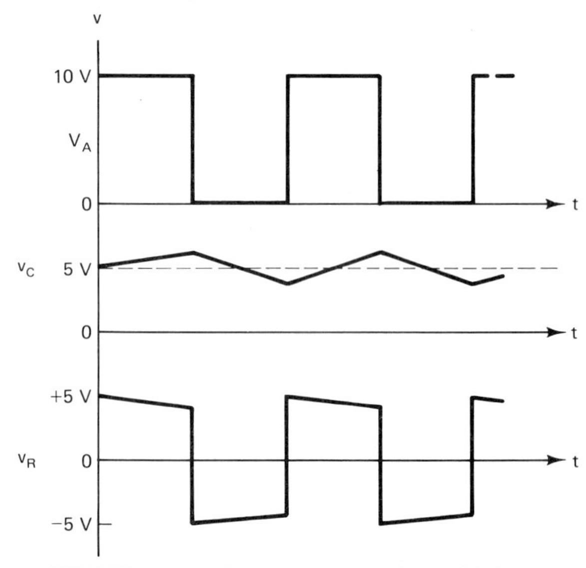

FIGURE 24-15 Resultant v_C and v_R with long *CR* time constant and narrow pulse width input voltage.

24-9 *CR* INTEGRATOR AND DIFFERENTIATOR CIRCUIT

CR Integrator

When the output voltage of a series *CR* circuit is taken across the capacitor, the circuit is called an *integrator* in terms of its time response, as shown in the schematic of Fig. 24-16. The term "integrator" derives from a calculus function that this circuit simulates under certain conditions. The shape of the capacitor voltage waveform is determined by the time constant and the input voltage pulse width, as shown by the representative waveforms of Figs. 24-14 and 24-15. The width of the input pulse voltage is determined by the frequency of the source.

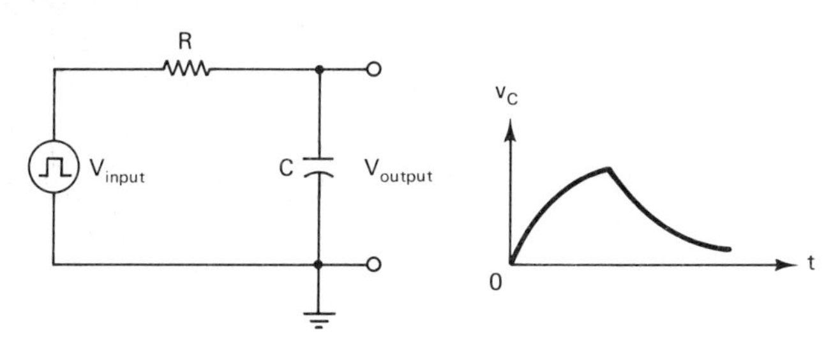

FIGURE 24-16 *CR* integrator.

CR Differentiator

When the output voltage of a series *CR* circuit is taken across the resistor, the circuit is called a *differentiator* in terms of its time response, as shown in Fig. 24-17. The shape of the resistor voltage waveform is determined by the time constant of the circuit and the input voltage pulse width, as shown by the representative waveforms of Figs. 24-14 and 24-15.

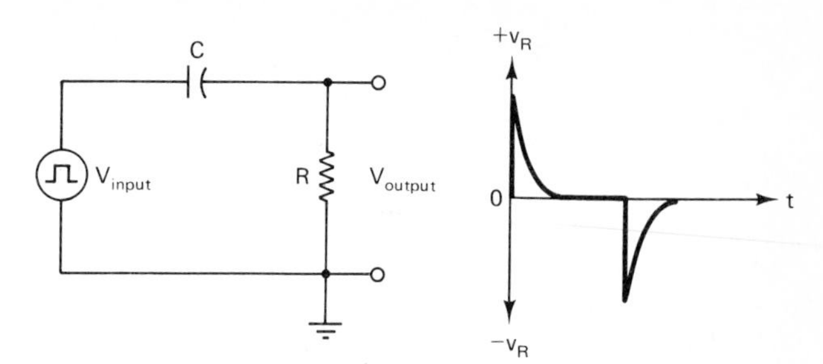

FIGURE 24-17 *CR* differentiator.

24-10 CURRENT VERSUS TIME IN A PURE INDUCTIVE DC CIRCUIT

In the pure inductive circuit of Fig. 24-18(a), the self-induced voltage (v_L) reaches a maximum value at the instant the switch is closed and is given by the equation

$$v_L = V_A = L \times \frac{\Delta i}{\Delta t}$$

This initial voltage opposes the change in current which flows into the inductor to build up a magnetic flux. As the rate of change in current decreases, the counter emf diminishes and allows more current to flow. This diminishing counter emf and increasing current, as shown by the graphical representation of Fig. 24-18(b), continue until the counter emf reaches zero. At this point the current is no longer being limited by the counter emf. The inductor now appears as a short circuit to the source, and the resultant short-circuit current is limited only by the current capacity of the source.

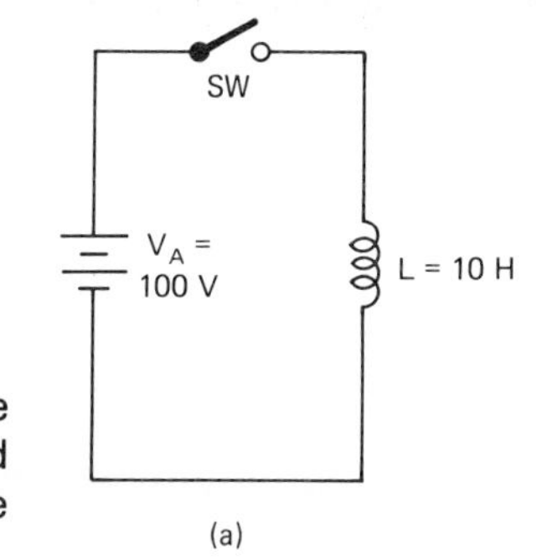

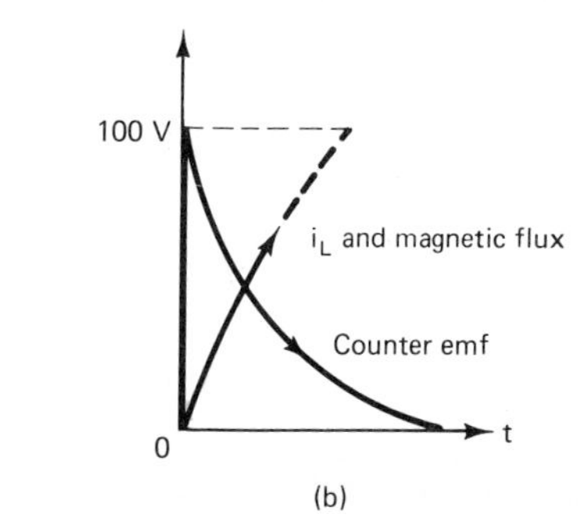

FIGURE 24-18 A pure inductive circuit produces a self-induced voltage equal to the source voltage.

24-11 *LR*; v_L, i_L, AND v_R CHARGING TRANSIENT RESPONSE CURVES

A practical inductor has an internal resistance due to the resistance of the copper wire which makes up the turns on the coil. In the schematic of Fig. 24-19(a) the series resistor represents this internal resistance and is 400 Ω in this case. Note that the internal resistance of an inductor varies and is dependent on the current capacity of the inductor as well as the number of turns, which in turn determines its inductance.

The circuit of Fig. 24-19(a) behaves similarly to that of the pure inductance. At the instant the switch is

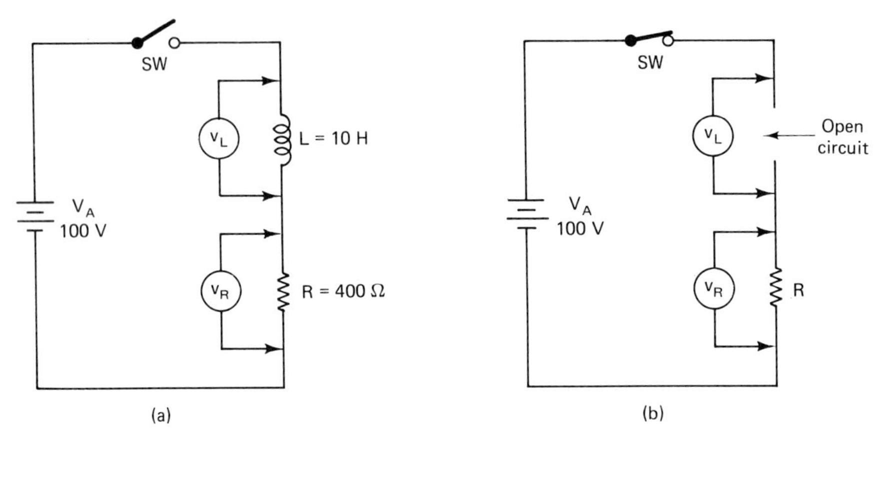

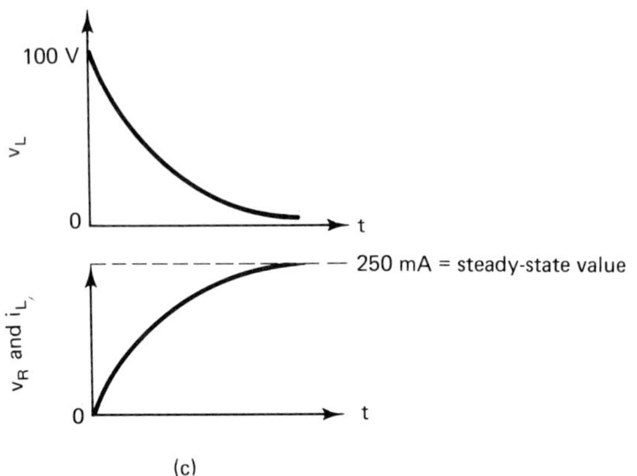

FIGURE 24-19 (a) Series *LR* circuit. (b) Equivalent circuit at instant the switch is closed. (c) v_L, v_R, and i_L transient response curves with rising flux.

closed, the self-induced voltage across the inductor is 100 V and opposes the applied voltage. This causes the conductor to appear instantaneously as an open circuit to the source, as shown in the equivalent-circuit schematic of Fig. 24-19(b). Initially, then, the current will be zero but will increase as the induced voltage decreases. This is shown by the i_L, v_R, and v_L transient response curves of Fig. 24-19(c).

After a time interval the induced voltage reaches zero, while the current reaches a steady-state Ohm's law value. For the circuit in the schematic of Fig. 24-19(a), the steady-state current will be $I = 100\ \text{V}/400\ \Omega = 250\ \text{mA}$.

24-12 *LR*; v_L, i_L, AND v_R DISCHARGING TRANSIENT RESPONSE CURVES

Removing the source and short circuiting its terminals, as shown in Fig. 24-20(a), will cause the inductor's stored energy to be discharged. Closing the switch will cause the magnetic flux to collapse. The collapsing flux induces a voltage and current that oppose the falling flux. The equivalent-circuit schematic of Fig. 24-20(b) shows that the inductor appears as a source of voltage and current connected across the resistor. The current now flows from the negative side of the inductor through the resistor and back to the positive side. This flow of induced current converts the stored energy of the magnetic flux into heat energy, which is released by the resistor.

Initially, the steady-state current flows through the resistor but decays with time as the flux falls to zero. This is shown by the transient response curve of v_L, v_R, and i_R of Fig. 24-20(c). Note that since v_L is effectively across the resistor, the resistor voltage will follow the decaying current and voltage of the inductor. After a time interval the flux field falls to zero and v_C, v_R, and i_R fall to zero.

24-13 RISING CURRENT *L/R* TIME CONSTANT

The current-limiting action of the series resistance will affect the rate at which the current rises and decays. The rate at which the current rises or decays is determined by the time constant of the series *LR* circuit and is the ratio of L/R:

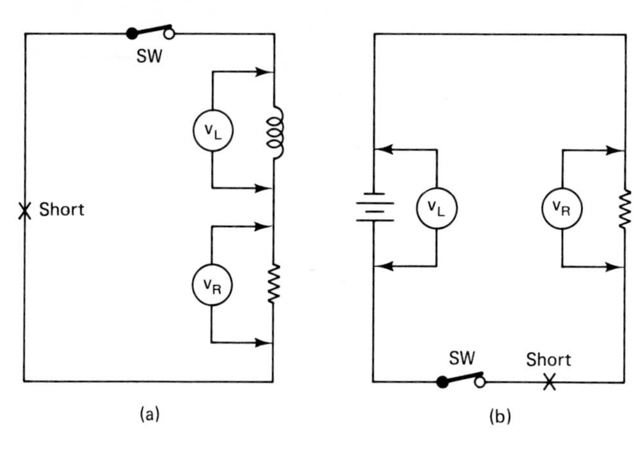

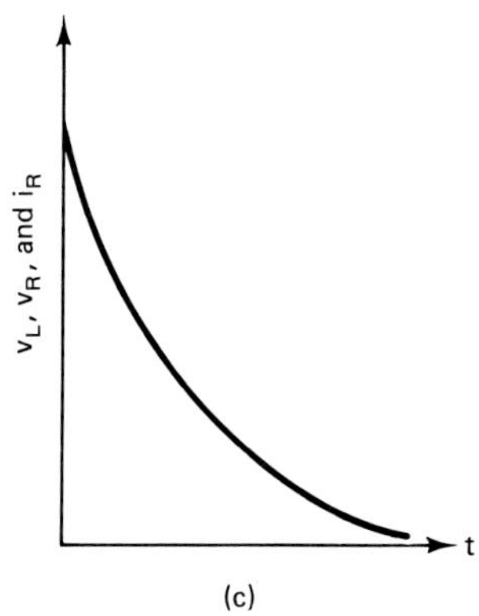

FIGURE 24-20 (a) Discharging an inductor. (b) Equivalent circuit. (c) Transient response curve of v_L, v_R, and i_R with falling flux.

$$T = \frac{L}{R} \quad \text{seconds} \qquad (24\text{-}4)$$

when L is in henrys
R is in ohms
T is in seconds

Using equation (19-1) for the inductance we can show that

$$L = v_L \times \frac{\Delta T}{\Delta i}$$

Dividing through by R gives

$$\frac{L}{R} = \frac{v_L \times T}{iR}$$

iR and v factor cancel to give

$$T = \frac{L}{R}$$

One L/R time constant is the time interval that it takes the current to change 63% of its steady-state value in a series LR circuit.

EXAMPLE 24-6

(a) Find the time constant of the series LR circuit in the schematic of Fig. 24-21.
(b) What is the current after one time constant?

Solution:

(a) Using equation (24-4) gives

$$T = \frac{L}{R} = \frac{10 \text{ H}}{10 \times 10^3 \ \Omega} = 1 \text{ ms}$$

(b) The steady-state current value is found using Ohm's law:

$$I = \frac{V}{R} = \frac{100 \text{ V}}{10 \text{ k}\Omega} = 10 \text{ mA}$$

After one time constant (1 ms), the current rises from zero to 63% of its steady-state value:

$$i_L = 0.63 \times 10 \text{ mA} = 6.3 \text{ mA}$$

The current will rise to 6.3 mA after one time constant or after 1 ms.

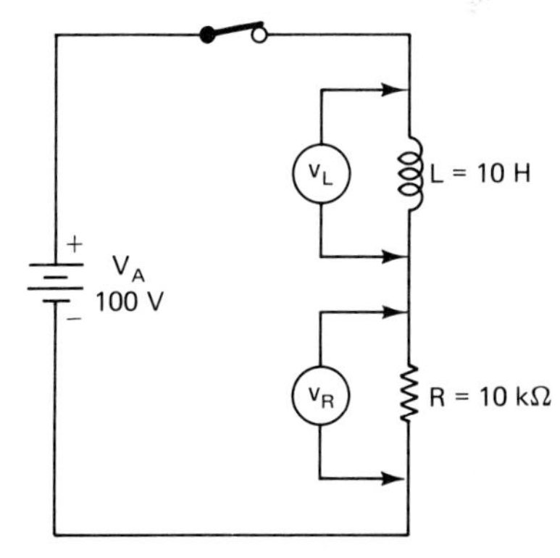

FIGURE 24-21 *LR* time-constant circuit.

Value of i_L, v_L, and v_R after One Current Rise Time Constant

In the series LR circuit of Fig. 24-22(a), at the instant the switch is closed the inductor develops a self-induced voltage drop of 100 V and the current is zero. For the first and subsequent time constants the current will increase by 63% of its steady-state value. The steady-state

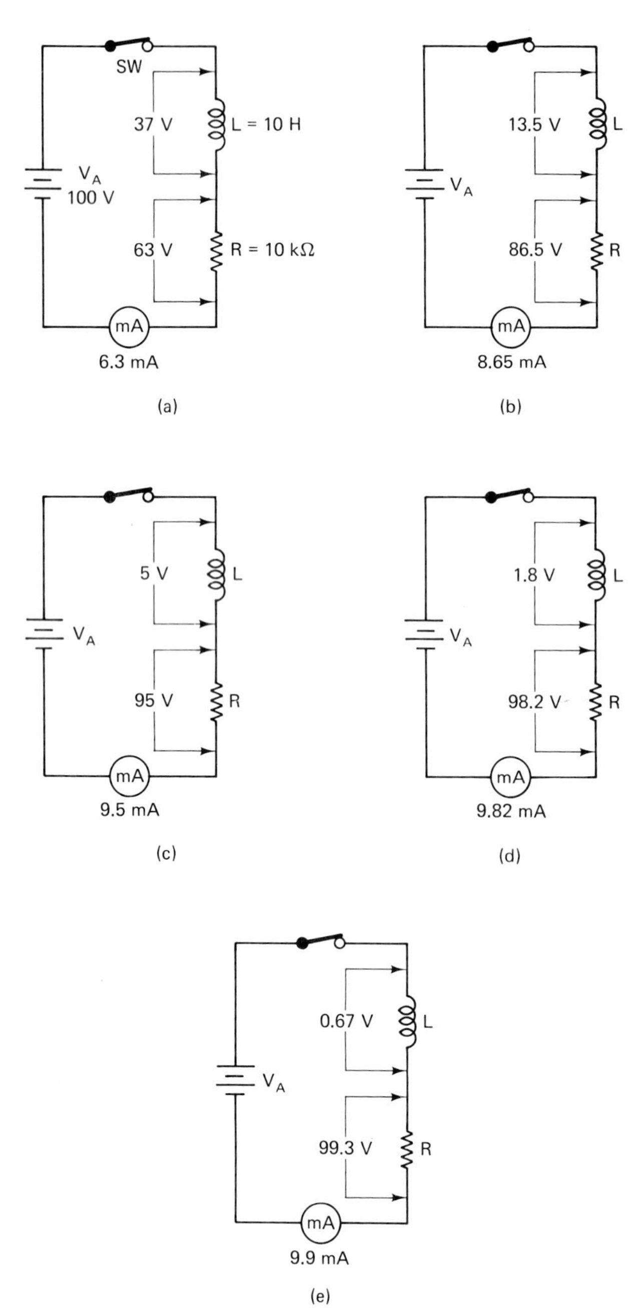

FIGURE 24-22 Current and voltages across L and R after each time constant. After 5 TC the current reaches a steady-state value. (a) After first TC (1 ms). (b) After second TC (2 ms). (c) After third TC (3 ms). (d) After fourth TC (4 ms). (e) After fifth TC (6 ms).

value of the current in this circuit is calculated using Ohm's law, or $I = 100\text{ V}/10\text{ k}\Omega = 10\text{ mA}$. (*Note:* Ignore the internal resistance of the coil.)

Figure 24-22(a) illustrates the resultant current and voltage drops about a series *LR* circuit after one time-constant interval. The time constant of the circuit is $T = L/R = 10\text{ H}/10\text{ k}\Omega = 1\text{ ms}$. After 1 ms the current rises from zero to $0.63 \times 10\text{ mA} = 6.3\text{ mA}$. Note that now 63% of the voltage drop is across the resistor while the remaining 37% is across the inductor after one time constant.

Determining the Current after One Current Rise Time Constant by Using a Calculator. The current in the *RL* circuit after any rising current time constant can be calculated using equation (24-2) and substituting the steady-state current I_T.

$$i_L = I_T\left(1 - \frac{1}{e^x}\right) \quad \text{amperes} \qquad (24\text{-}5)$$

Value of i_L, v_L, and v_R after Two Current Rise Time Constants

In Fig. 24-22(a) after one current rise time constant interval of 1 ms, the current is 6.3 mA. The remaining amount of current required to reach the steady-state value is now $10\text{ mA} - 6.3\text{ mA} = 3.7\text{ mA}$. After the second time-constant interval of 2 ms, the current will rise a further 63% of this remaining steady-state current of 3.7 mA, or $0.63 \times 3.7\text{ mA} = 2.33\text{ mA}$. Since there already is 6.3 mA flowing in the circuit, the total current at the end of the second time-constant interval will be $6.3\text{ mA} + 2.33\text{ mA} = 8.65\text{ mA}$.

With this value of current flowing in the circuit, the voltage drop across the resistor is determined by Ohm's law and is shown in Fig. 24-22(b). The self-induced voltage drop across the inductor will be the difference between the applied voltage and the resistor voltage drop, or 13.5 V, as shown in Fig. 24-22(b).

Determining the Current after Two Current Rise Time Constants by Using a Calculator. The current value after two time constants can easily be determined using a calculator and the general equation (24-5).

$$i_L = I_T\left(1 - \frac{1}{e^2}\right) \quad \text{amperes}$$

Value of i_L, v_L, and v_R after Three Current Rise Time Constants

In Fig. 24-22(b) after the second current rise time-constant interval of 2 ms, the current is 8.65 mA. The remaining amount of current required to reach the steady-state value is now $10\text{ mA} - 8.65\text{ mA} = 1.35\text{ mA}$. After the third time-constant interval of 3 ms, the current will rise a further 63% of this remaining steady-state current of 1.35 mA, or $0.63 \times 1.35\text{ mA} = 0.85\text{ mA}$. Since there already is 8.65 mA flowing in the circuit, the total current at the end of the third time-constant interval will be 8.65 mA × 0.85 mA = 9.5 mA.

With this value of current flowing in the circuit, the voltage drop across the resistor is determined by Ohm's law and is shown in Fig. 24-22(c). The induced voltage drop across the inductor will be the difference between the applied voltage and the resistor voltage drop, or 5 V, as shown in Fig. 24-22(c).

Determining the Current after Three Current Rise Time Constants by Using a Calculator. The current value after three time constants can easily be determined by using a calculator and the general equation (24-5).

$$i_L = I_T\left(1 - \frac{1}{e^3}\right) \qquad \text{amperes}$$

Value of i_L, v_L, and v_R after Four Current Rise Time Constants

In Fig. 24-22(c), after the third current rise time-constant interval of 3 ms, the current is 9.5 mA. The remaining amount of current required to reach the steady-state value is now 10 mA − 9.5 mA = 0.5 mA. After the fourth time-constant interval of 4 ms, the current will rise a further 63% of this remaining steady-state current of 0.5 mA, or 0.63 × 0.5 mA = 0.32 mA. Since there already is 9.5 mA flowing in the circuit, the total current at the end of the fourth time-constant interval will be 9.5 mA + 0.32 mA = 9.82 mA.

With this value of current flowing in the circuit, the voltage across the resistor is determined by Ohm's law and is shown in Fig. 24-22(d). The induced voltage drop across the inductor will be the difference between the applied voltage and the resistor voltage drop, or 1.8 V, as shown in Fig. 24-22(d).

Determining the Current after Four Current Rise Time Constants by Using a Calculator. The current value after four time constants can easily be determined by using a calculator and the general equation (24-5).

Value of i_L, v_L, and v_R after Five Current Rise Time Constants

In Fig. 24-22(d), after the fourth current rise time-constant interval of 4 ms, the current is 9.82 mA. The remaining amount of current required to reach the steady-state value is now 10 mA − 9.82 mA = 0.18 mA. After the fifth time-constant interval of 10 ms, the current will rise a further 63% of this remaining steady-state current of 0.18 mA, or 0.63 × 0.18 mA = 0.11 mA. Since there already is 9.82 mA flowing in the circuit, the total current at the end of the fifth time-constant interval will be approximately 10 mA. For practical purposes the current has reached its steady-state value after five time-constant intervals, as shown in Fig. 24-22(e).

With this value of current flowing in the circuit, the voltage across the resistor is determined by Ohm's law and is shown in Fig. 24-22(e). The induced voltage drop will now be zero since the changing current is zero.

24-14 DECAYING CURRENT *L/R* TIME CONSTANT

The steady-state current of a series *LR* circuit will decay by 63% of its value during each time-constant interval. The current value at the end of each time-constant interval will then be 37% of its initial value at the beginning of that time-constant interval.

Value of Decaying Current, v_L, and v_R after One Time Constant

In Fig. 24-23(a), at the instant the switch is closed, v_L was 100 V, v_R was 100 V, and the steady-state current was 10 mA. After one time-constant interval the current decays to 37% of the steady-state value, or 0.37 × 10 mA = 3.7 mA. Since the resistor is effectively connected across the inductor, $V_R = v_L$. Fig. 24-23(a) illustrates the resultant voltages of a series *LR* circuit at the end of one time-constant interval. The time constant of the circuit was previously found to be 1 ms. After 1 ms of decaying current the self-induced voltage drop across the inductor will be 37 V which is also the remaining voltage across the resistor.

Determining the Value of the Decaying Current after One Time Constant by Using a Calculator. The decaying current at the end of one time constant can be determined using a calculator and the general equation

$$i_L = \frac{1}{e^x} \times I_T \qquad \text{amperes} \qquad (24\text{-}6)$$

This method was illustrated using the capacitor discharge time-constant example. Refer to Example 24-3. (*Note:* I_T is the steady-state value for current and is calculated using Ohm's law. Ignore the internal resistance of the coil.)

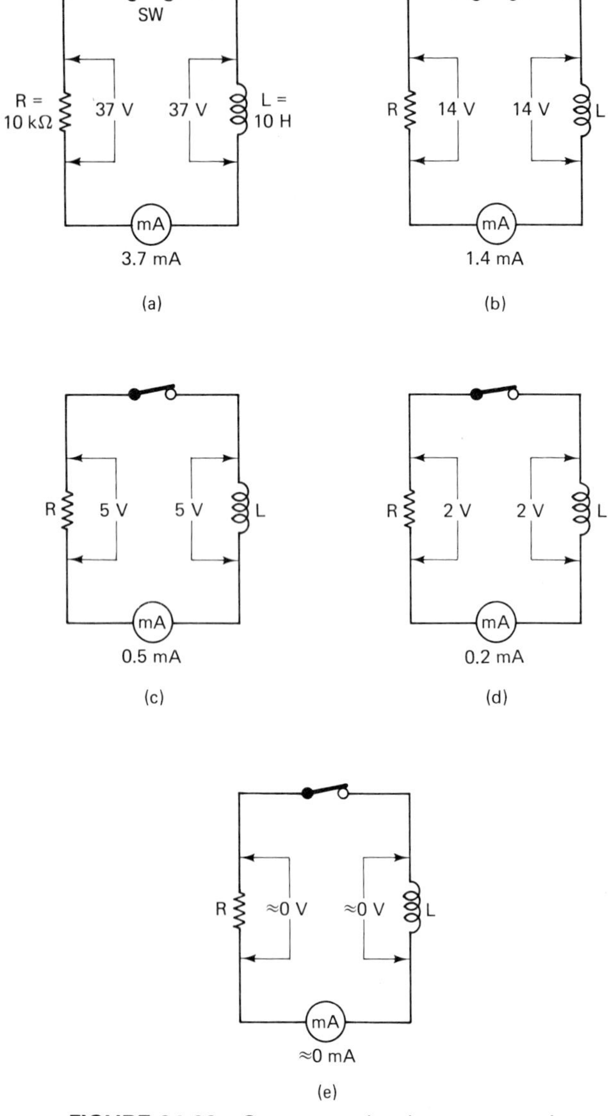

FIGURE 24-23 Current and voltage across L and R after each time constant. After 5 TC the inductor stored energy is depleted and all values are zero. (a) After first TC (1 ms). (b) After second TC (2 ms). (c) After third TC (3 ms). (d) After fourth TC (4 ms). (e) After fifth TC (5 ms).

Value of Decaying Current, v_L, and v_R after the Second Time Constant

At the end of one time-constant interval (1 ms), the current decayed to 3.7 mA while the self-induced voltage of the inductor decayed to 37 V, as shown in Fig. 24-23(a). Beginning with an initial value of 3.7 mA, the current decays to 37% of this value at the end of the second time-constant interval (2 ms), or 0.37×3.7 mA = 1.4 mA, as shown in Fig. 24-23(b). After 2 ms of decaying current the self-induced voltage drop across the inductor will be 14 V, which is also the remaining voltage across the resistor.

A similar analysis can be made for each succeeding time constant. The resultant current and voltages are shown in the schematics of Fig. 24-23(c), (d), and (e). After five time-constant intervals the decaying current is almost zero; therefore, the inductor's stored energy is considered to be "discharged."

24-15 *LR* DIFFERENTIATOR AND INTEGRATOR CIRCUIT

LR Differentiator

When the output voltage of a series *LR* circuit is taken across the inductor, the circuit is called a differentiator in terms of its time response [see Fig. 24-24(a)]. The shape of the inductor voltage waveform is determined by the time constant of the circuit and the input voltage pulse width, as shown in the representative waveform, v_L of Fig. 24-24(c),

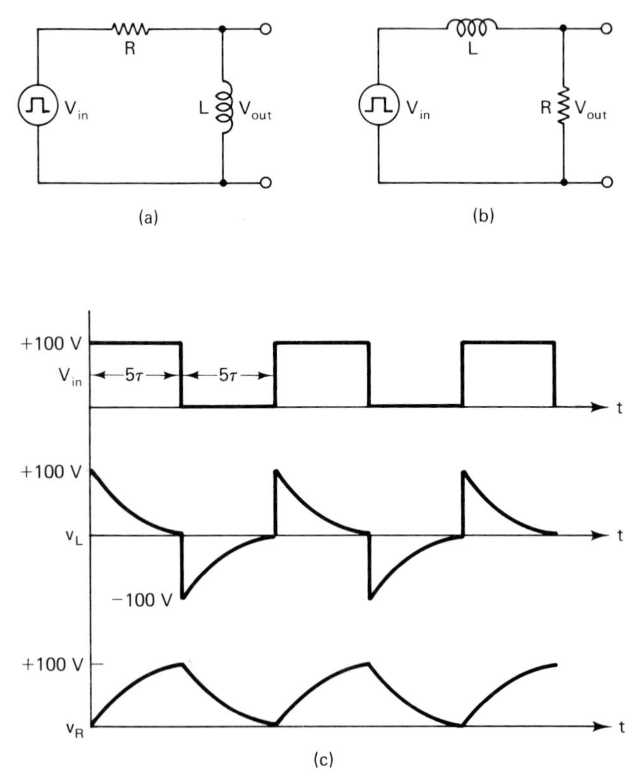

FIGURE 24-24 (a) *LR* differentiator with v_L output voltage waveform. (b) *RL* integrator with v_R output voltage waveform.

Compare the output of the *CR* differentiator in Fig. 24-17 and note its similarity. The pulse waveform of the *LR* differentiator will have the same characteristic wave shape for a given time constant and input pulse width. Varying the pulse width of the input voltage results in similar wave shapes for v_L, as shown in the waveform diagrams for v_R in Figs. 24-11(b) and (c), 24-13, and 24-14.

LR Integrator

When the output voltage of a series *LR* circuit is taken across the resistor, the circuit is called an integrator in terms of its time response [see Fig. 24-24(b)]. The shape of the resistor voltage waveform is determined by the time constant of the circuit and the input voltage pulse width, as shown in the representative waveform [Figure 24-24(c)].

Compare the output of the *CR* integrator in Figure 24-16 and note its similarity. The pulse waveform of the *LR* integrator will have the same characteristic wave shape for a given time constant and input pulse width. Varying the pulse width of the input voltage results in similar wave shapes for v_R as shown in the waveform diagrams for v_C in Figs. 24-11(b) and (c), 24-14, and 24-15.

SUMMARY

The current response with time of a resistor connected to a dc source is constant since the resistor is a nonreactive component (see Fig. 24-1).

The current response with time of an uncharged capacitor is a pulse or "spike" because the capacitor develops a counter emf as it charges (see Fig. 24-2).

In a series *CR* circuit, the length of time that it takes the capacitor to charge will be determined by the resistor value.

The initial instantaneous voltage across a discharged capacitor connected in a series *CR* charging circuit will be zero volts, while the voltage across the resistor will be equal to the applied voltage [see Fig. 24-3(c)].

After a time interval the final voltage across the charged capacitor will be equal to the applied voltage, while the voltage across the resistor will be zero, and the capacitor is said to have reached a steady-state condition.

The initial instantaneous voltage across a charged capacitor connected in a series *CR* discharging circuit will be equal to the applied voltage, which will also be across the resistor [see Fig. 24-4(b)].

The rate at which the capacitor charges or discharges is determined by the time constant of the series *CR* circuit and is the product of *CR*:

$$T = CR \qquad \text{seconds} \tag{24-1}$$

One charging time constant is defined as the time interval that it takes a series *CR* capacitor to charge to 63% of the steady-state voltage.

The symbol for a time constant (*CR*) interval is the Greek lowercase letter tau (τ).

The capacitor voltage for any charging time-constant interval can easily be determined using a calculator and the general equation (see Examples 24-1 and 24-2)

$$v_C = V_A\left(1 - \frac{1}{e^x}\right) \qquad \text{volts} \tag{24-2}$$

Once v_C is determined, the voltage across the resistor will be the difference between the source voltage and the capacitor's voltage at each respective time-constant interval.

One discharge time constant is defined as the time interval that it takes a series *CR* fully charged capacitor to discharge to 37% of its steady-state voltage.

The capacitor and resistor voltage for any discharging time-constant interval can easily be determined using a calculator and the general equation (see Examples 24-3 and 24-4)

$$v_C = \frac{1}{e^x} \times V_A \qquad \text{volts} \tag{24-3}$$

The universal time-constant curves of Fig. 24-7 give the instantaneous value of v_C or i as a percentage of the initial or steady-state value with time given in time constants tau (τ).

With a fixed time constant and a square-wave input voltage, the voltage that the capacitor charges to will be determined by the pulse-width ON time.

With a fixed time constant and a square-wave input voltage, the voltage that the capacitor will discharge to will be determined by the pulse-width OFF time.

For a pulse width equal or greater than five time constants (5 τ), the capacitor will fully charge on the first pulse and fully discharge during the OFF-time portion of the pulse, with the resultant waveform as shown in Fig. 24-11(b) and (c).

For a pulse width less than five time constants, the capacitor will charge and discharge only partially during the input voltage pulse, with the resultant waveforms shown in Fig. 24-14.

With a long time constant and a narrow pulse width, the capacitor output voltage changes very little and the resistor's voltage waveform approaches that of the input voltage waveform shape.

When the output of a series *CR* circuit is taken across the capacitor, the circuit is called an integrator in terms of its time response.

When the output of a series *CR* circuit is taken across the resistor, the circuit is called a differentiator in terms of its time response.

The rate at which the current rises or decays is determined by the time constant of the series *LR* circuit and is given by

$$T = \frac{L}{R} \qquad \text{seconds} \tag{24-4}$$

One L/R time constant is the time interval that it takes the current to change 63% of its steady-state value in a series *LR* circuit.

One L/R decaying current time constant is the time interval that it takes the current to decay by 63% of its steady-state value. The current value at the end of the time-constant interval will then be 37% of its initial value at the beginning of that time-constant interval.

The rising current for any L/R time-constant interval can easily be determined using a calculator and the general equation

$$i_L = I_T\left(1 - \frac{1}{e^x}\right) \qquad \text{amperes} \tag{24-5}$$

Note I_T is the steady-state value for current and is calculated using Ohm's law.

The decaying current for any L/R time-constant interval can easily be determined using a calculator and the general equation

$$i_L = \frac{1}{e^x} \times I_T \qquad \text{amperes} \tag{24-6}$$

REVIEW QUESTIONS

24-1. Explain why there is a surge current when an uncharged capacitor is connected across a voltage source.

24-2. Why does the surge current diminish very quickly?

24-3. What component limits the surge current in a series *CR* circuit?

24-4. What is the voltage value of v_C and v_R at the instant the switch is closed in a series *CR* circuit of Fig. 24-3(a)?

24-5. What is the voltage value of v_C and v_R after the capacitor is fully charged in the series *CR* circuit of Fig. 24-3(a)?

24-6. Define a charging *CR* time constant.

24-7. A 5-μF capacitor and a 10-kΩ resistor are connected in series across a 20-V dc supply. Calculate:
(a) The time constant of the circuit.
(b) The capacitor voltage drop after one time-constant interval.
(c) The resistor voltage drop after two-constant intervals.
(d) The time required for the resistor voltage to reach approximately zero volts.

24-8. Use the universal time-constant graph to determine the voltage across the capacitor in the circuit of Question 24-7 after 75 ms.

24-9. Define a discharging time constant.

24-10. A 4-μF capacitor and 560-kΩ resistor are connected in series across a 110-V dc supply. Calculate:
(a) The time constant of the circuit.
Assume that the capacitor is fully charged. Calculate:
(b) The capacitor voltage after one discharge time-constant interval.
(c) The resistor voltage after two discharge time-constant intervals.
(d) The time required for the resistor voltage to reach approximately zero volts.

24-11. Use the universal time-constant graph to determine the voltage across the discharging capacitor after 3.36 s.

24-12. A series *CR* circuit has a square-wave input voltage with a pulse width that is equal to or greater than five time constants. Sketch the resultant voltage waveforms across the capacitor and resistor.

24-13. A series *CR* circuit has a square-wave input voltage with a narrow pulse width that is shorter than one time constant. Sketch the resultant voltage waveforms across the capacitor and resistor.

24-14. Draw a schematic diagram and the output voltage waveform of a *CR* integrator with a square-wave input voltage having a PW = 5τ.

24-15. Draw a schematic diagram and the output voltage waveform of a *CR* differentiator with a square-wave input voltage having a PW = 5τ.

24-16. A 4-H inductor and a 10-kΩ resistor are connected in series across a 55-V dc supply. Calculate:
(a) The time constant of the circuit.
(b) The inductor voltage drop after one time-constant interval.
(c) The resistor voltage drop after two time-constant intervals.
(d) The time required for the current to reach a steady-state value.

24-17. Draw a schematic diagram and the output voltage waveform of a *LR* integrator with a square-wave input voltage having a PW = 5τ.

24-18. Draw a schematic diagram and the output voltage waveform of a *LR* differentiator with a square-wave input voltage having a PW = 5τ.

SELF-EXAMINATION

Where applicable, choose the most suitable word, phrase, or value to complete each statement.

Reference *CR* Circuit:
Questions 24-1 to 24-20 refer to a *CR* circuit with a 25-μF capacitor and a 8-kΩ resistor that are series connected across a 20-V source.

24-1. Assume that the capacitor is not charged. What is the voltage drop across the capacitor at the instant the source is connected?

24-2. What is the voltage drop across the resistor at the instant the source is connected?

24-3. What is the value of the circuit current at the instant the source is connected?

24-4. What is the value of the circuit current after the capacitor is fully charged?

24-5. What is the voltage drop across the capacitor after it is fully charged?

24-6. What is the voltage drop across the resistor after it is fully charged?

24-7. The product of *CR* is termed the ________ ________ of the series *CR* circuit.

24-8. The symbol for time constant is ________, the Greek lowercase letter tau.

24-9. During a time interval equal to one time constant the charge on a capacitor will change by approximately ________ percent of its initial value.

24-10. What is the time constant of this circuit?

24-11. What is the capacitor voltage drop after one charge time-constant interval?

24-12. What is the resistor voltage drop after two charge time-constant intervals?

24-13. What is the time required for the capacitor voltage to reach approximately 20 V?

24-14. Use the universal time-constant graph to determine the voltage across the capacitor after 320 ms.

24-15. During a time interval equal to one time constant, the capacitor will discharge to approximately ________ percent of its steady-state value.

For the following questions, assume that the capacitor in the reference series *CR* circuit has been fully charged.

24-16. What is the capacitor voltage after two discharge time-constant intervals?

24-17. What is the resistor voltage after two discharge time-constant intervals?

24-18. Use the universal time-constant graph to determine the voltage across the discharging capacitor after 320 ms.

24-19. The input voltage to the reference *CR* circuit is now a square wave. What must the frequency of the pulse be in order to charge and discharge the capacitor completely?

24-20. What must the frequency of the pulse be when the average capacitor voltage is 10 V, while the resistor voltage has a similar waveform to the input voltage square wave?

Refer to Fig. 24-25 for Questions 24-21 to 24-24.

24-21. The *CR* circuit is called a *CR* ________.

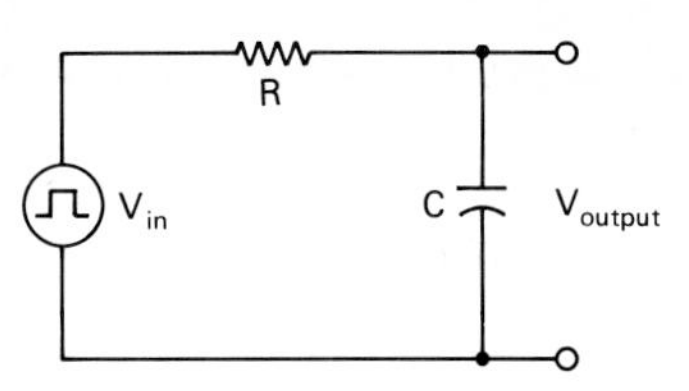

FIGURE 24-25 Schematic.

24-22. Sketch the output voltage waveform given that the input voltage pulse width is equal to 5τ and the *CR* circuit has a short time constant.

24-23. Where must the output voltage be taken if the circuit is to be considered as a differentiator?

24-24. Sketch the output voltage waveform of a *CR* differentiator with a short time constant given that the input voltage pulse width is equal to 5τ.

Questions 24-25 to 24-28 refer to a *LR* circuit with a 500-mH inductor and a 200-Ω resistor that are series connected across a 30-V source.

24-25. What is the time constant of the circuit?

24-26. What is the voltage drop across the inductor after one time-constant interval?

24-27. What is the voltage drop across the resistor after two time-constant intervals?

24-28. What is the time required for the current to reach a steady-state value?

24-29. Sketch the output voltage waveform of a *LR* integrator given that the input voltage pulse width is equal to 5τ.

24-30. Sketch the output voltage waveform of a *LR* differentiator given that the input voltage pulse width is equal to 5τ.

ANSWERS TO PRACTICE QUESTIONS

24-1

1. nonreactive
2. Ohm's law
3. zero
4. surge
5. resistor
6. source
7. source
8. zero
9. capacitor
10. heat
11. decay
12. time constant
13. product, microfarads, megohms
14. 0.05 s
15. 94.5 V
16. 150 V
17. 130 V
18. v_C = 20 V, v_R = 20 V
19. 35 ms
20. 80 ms

25

Capacitive Reactance

INTRODUCTION

With a pure resistive dc or ac circuit the opposition to current flow is resistance. We were able to calculate this resistance using Ohm's law. A capacitor in an ac circuit produces a counter emf due to the charging and discharging current. The effect of this counter emf is to oppose the current charge and discharge. Because the counter emf is frequency and capacitance dependent, its opposition to the current flow is termed reactance. Reactance can be calculated using the formula $1/2\pi fC$ (ohms). Reactance is also an Ohm's law value and therefore can be calculated using Ohm's law formula, $X_C = V/I$. The symbol for reactance is X, with the capital letter C as a subscript to indicate that it is capacitive reactance.

Any ac circuit that has capacitance will have reactance as a prime source of opposition to current flow. Some examples are tuned circuits, filter circuits, capacitor coupling, speaker crossover networks, AF and RF bypassing, and many others.

In this chapter we look at the reactance of a capacitor in a simple series and parallel circuit.

MAIN TOPICS IN CHAPTER 25

OBJECTIVES

After studying Chapter 25, the student should be able to

1. Master the fundamental concept of capacitive reactance.
2. Master the fundamental concept of why and how X_C changes with capacitance and frequency.
3. Solve X_C problems with a change in frequency and capacitance.

4. Solve series and parallel X_C circuit problems using Ohm's law.
5. Master the concept of a leading current in a pure capacitive circuit.

25-1 CAPACITANCE IN THE DC VERSUS AC CIRCUIT

Capacitance in the DC Circuit

In Chapter 24 we discovered the action of a capacitor when connected to a dc voltage. Initially, the current flows into the capacitor to charge it. After a time interval the capacitor becomes fully charged and no further current flows in the circuit. With the series CR circuit the time it took the capacitor to become fully charged was approximately five time constants.

Looking at the schematic diagram of Fig. 25-1(a) shows that the capacitor and neon lamp are initially (switch position 1) series connected to a 120-V dc source. Since the neon lamp requires very little current to make it glow, we will be able to see the effect of the charging current.

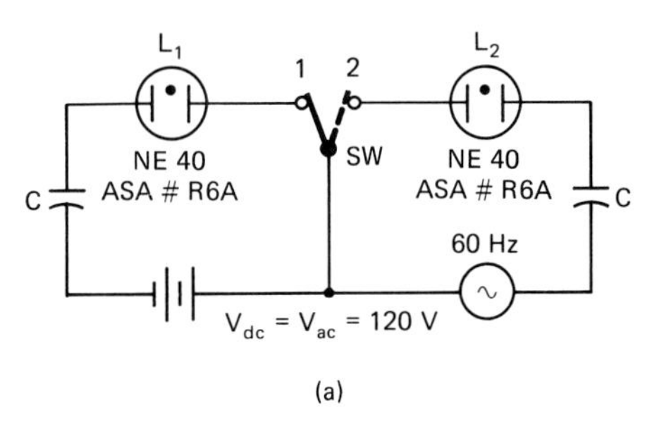

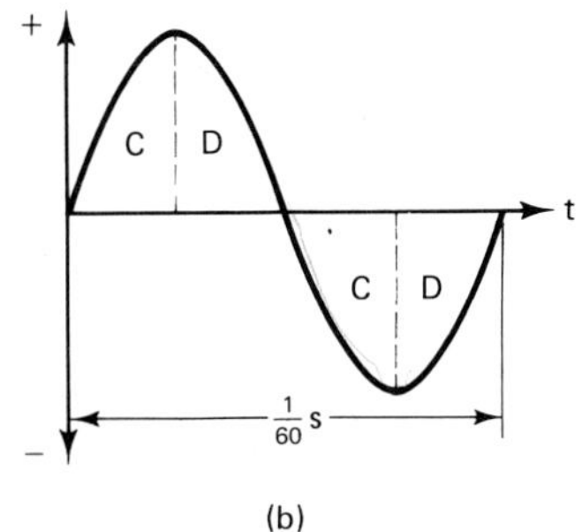

FIGURE 25-1 (a) Capacitor connected in series with a neon lamp to a 120-V ac/dc source. (b) Capacitor charges *C* twice and discharges *D* for every cycle of ac.

At the instant the switch is closed to position 1, the 120 V is across the neon lamp. This causes the lamp to ignite to full brilliance. Note that one electrode side of the neon lamp glows. If the polarity of the dc source is reversed, the glow switches to the other electrode side. The reason for this was explained in Chaper 19. As the capacitor starts to charge, it builds up a counter emf which reduces the voltage across the neon lamp. The neon lamp glow begins to fade and goes out as the capacitor approaches full charge.

The action of the neon lamp gives a visual indication of the capacitor's charging action. Note that once the capacitor is fully charged, no further current flows and the capacitor appears as an open circuit to the dc source. *The capacitor is said to block dc.*

Capacitance in the Ac Circuit

Toggling the switch to position 2 of the schematic in Fig. 25-1(a) places the series neon lamp and capacitor across the ac source. This time the ac charges the capacitor twice every cycle and discharges the capacitor twice every cycle [see Fig. 25-1(b)]. Both electrodes of the neon lamp glow to show that there is a continuous ac voltage and current in the circuit.

If the neon lamp is replaced with a 15-W incadescent lamp, it would glow dimly since the charge and discharge current is not great enough to light the lamp fully.

The brilliance of the lamp visually illustrates the effect of capacitance on the current in the ac circuit. Although the capacitor offers some opposition to current flow, *the capacitor is said to pass ac.*

25-2 OHM'S LAW AND CAPACITIVE REACTANCE, X_C

The 15-W lamp in the circuit schematic of Fig. 25-2 needs 125 mA for full brilliance. With a source of 120 V, 60 Hz, the lamp has approximately 64 mA flowing through it. This is enough current to light the lamp dimly.

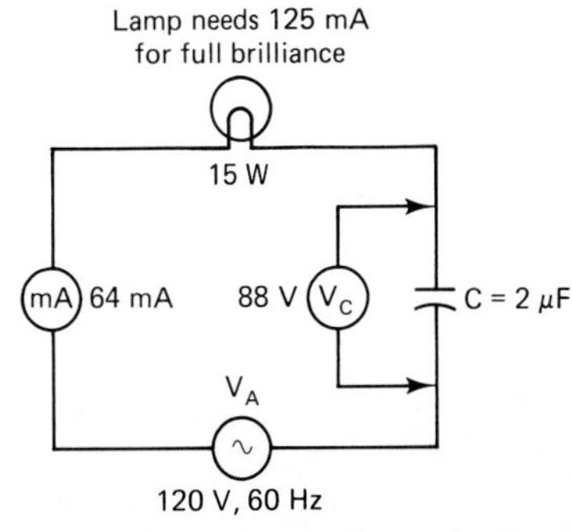

FIGURE 25-2 Circuit Schematic.

The lower value of current with the series capacitor indicates that an ac opposition is present in the circuit. Simplified, this opposition is due to the capacitor building up a counter emf that opposes a change in the charging current, therefore reduces the current. The counter emf is the reactive component that produces the opposition to the ac current flow. This opposition is termed capacitive reactance, X_C. Capacitive reactance is an Ohm's law value and can be calculated for any given voltage and current flowing in the capacitive circuit. In the schematic of Fig. 25-2 the reactance of the 2-μF capacitor is

$$X_C = \frac{V_C}{I_T} = \frac{88 \text{ V}}{64 \times 10^{-3} \text{ A}} = 1375 \ \Omega$$

EXAMPLE 25-1

Calculate the reactance of the 4-μF capacitor in the schematic of Fig. 25-3.

Solution: Using Ohm's law, we have

$$X_C = \frac{120 \text{ V}}{181 \times 10^{-3} \text{ A}} = 663 \ \Omega$$

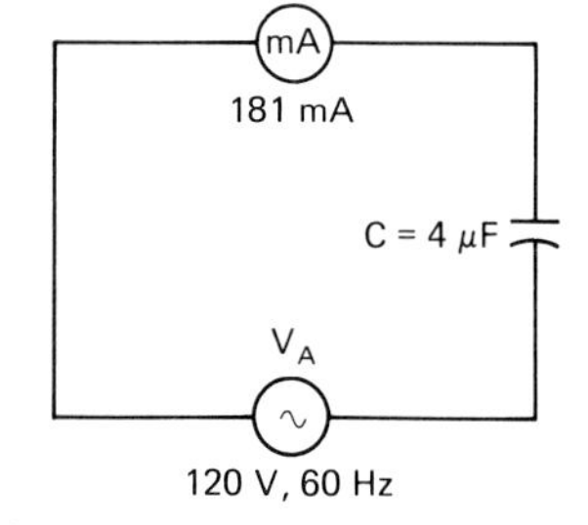

FIGURE 25-3 Circuit Schematic.

25-3 EFFECT OF CAPACITANCE ON X_C

Since the ac input voltage is constantly changing, the voltage across the capacitor must follow these sine-wave changes, as shown in Fig. 25-1(b). Regardless of the size of the capacitor, the capacitor voltage must then change at the same rate as the applied voltage. The amount of current required to charge and discharge the capacitor at a given rate will be directly proportional to capacitance. As capacitance increases the circuit, current increases, and vice versa.

Consider the capacitance in the schematic of Fig. 25-4(a) connected to a 120-V 60-Hz source. There are four current changes in $\frac{1}{60}$ of a second: that is, two current charges and two current discharges [see Fig. 25-1(b)] for one cycle.

For the 2-μF capacitor the peak charge during one cycle or $\frac{1}{60}$ of a second will be equal to

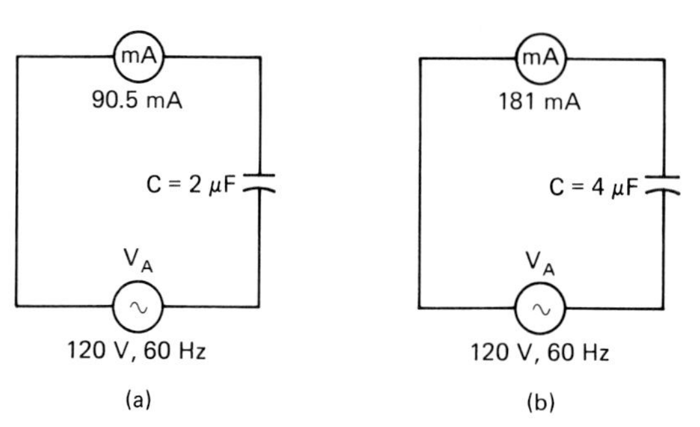

FIGURE 25-4 Current is directly proportional to capacitance.

$Q = CV = 2 \times 10^{-6} \text{ F} \times 120 \text{ V} \times \sqrt{2} = 339.4 \ \mu\text{C}$ or 0.339 mC.

The average current will be

$$\Delta I_{(\text{avg})} = \frac{\Delta Q}{\Delta t}(\text{avg}) = \frac{Q_P}{t/4} = \frac{4Q_P}{t}$$

and

$$\frac{\Delta Q}{\Delta t}(\text{avg}) = \frac{4 \times 0.339 \text{ mC}}{1/60 \text{ s}} = 81.4 \text{ mA}$$

Converting the average current value to rms gives

$$I_{(\text{rms})} = 1.11 \times I_{(\text{avg})} = 1.11 \times 81.4 \text{ mA} = 90.4 \text{ mA}$$

For the 4-μF capacitor shown in the schematic of Fig. 25-4(b), the circuit current will double since the capacitance is doubled with the same rate of charge and discharge.

The peak charge during one cycle or $\frac{1}{60}$ of a second will be equal to $Q = CV = 4 \times 10^{-6} \text{ F} \times 120 \text{ V} \times \sqrt{2} = 678.7 \ \mu\text{C}$ or 0.679 mC. The average current will be

$$I_{(\text{avg})} = \frac{\Delta Q}{\Delta t}(\text{avg}) = \frac{4Q_P}{t} = \frac{4 \times 0.679 \text{ mC}}{1/60 \text{ s}} = 163 \text{ mA}$$

Converting the average current value to rms gives

$$I_{(\text{rms})} = 1.11 \times I_{(\text{avg})} = 1.11 \times 163 \text{ mA} = 181 \text{ mA}$$

Ohm's law states that current is inversely proportional to resistance. Since the current doubled when capacitance was doubled, the reactance must have been halved, and vice versa. This means that reactance is inversely proportional to capacitance:

$$X_C \propto \frac{1}{C} \tag{25-1}$$

EXAMPLE 25-2

Calculate the current for the schematic of Fig. 25-3 with double the capacitance.

Solution: The reactance of the 4-μF capacitor at 60 Hz was found using Ohm's law and is 663 Ω. With double the capacitance, the reactance of the capacitor will be halved to 331.5 Ω. The current is then found using Ohm's law:

$$I = \frac{V_A}{X_C} = \frac{120 \text{ V}}{331.5 \ \Omega} = 362 \text{ mA}$$

Note that we could have deduced that the current would have doubled from the original value of 181 mA with the 4-μF capacitor since the reactance would be halved with double the capacitance (i.e., 8 μF).

25-4 EFFECT OF FREQUENCY ON X_C

The circuit current for the 2-μF capacitor connected across the 120-V 60-Hz source was found to be 90.4 mA. If the frequency of the source is doubled, the rate of current change will double for the comparable time interval of $\frac{1}{60}$ s, as shown in the waveform diagram of Fig. 25-5. Since the capacitor is being charged and discharged twice as fast with the 120-Hz source, the current will be twice as great, or 181 mA.

This can be shown by calculating the average current for eight changes in $\frac{1}{60}$ s, then finding the rms current as follows. The peak charge for the 2-μF capacitor was previously calculated to be 339.4 μC or 0.339 mA. Therefore, with double the frequency, the average current will be

$$I_{(\text{avg})} = \frac{\Delta Q}{\Delta t}(\text{avg}) = \frac{Q_P}{t/8} = \frac{8Q_P}{t}$$

and

$$I_{(\text{avg})} = \frac{\Delta Q}{\Delta t}(\text{avg}) = \frac{8 \times 0.339 \text{ mC}}{1/60 \text{ s}} = 163 \text{ mA}$$

Converting the average current value to rms gives

$$I_{(\text{rms})} = 1.11 \times I_{(\text{avg})} = 1.11 \times 163 \text{ mA} = 181 \text{ mA}$$

Ohm's law states that current is inversely proportional to resistance. Since the current doubled when frequency was doubled, the reactance must have been halved, and vice versa. This means that reactance is inversely proportional to frequency:

$$X_C \propto \frac{1}{f} \tag{25-2}$$

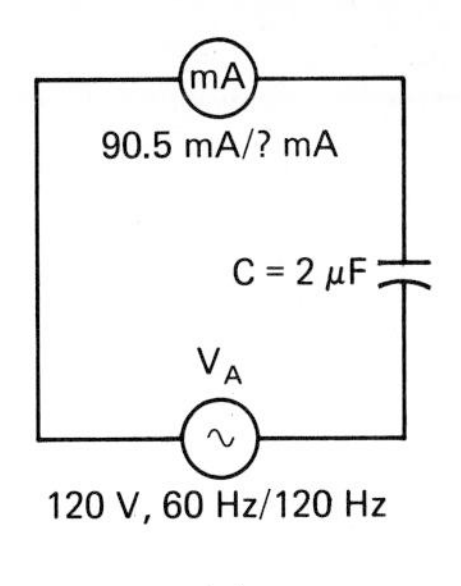

(a)

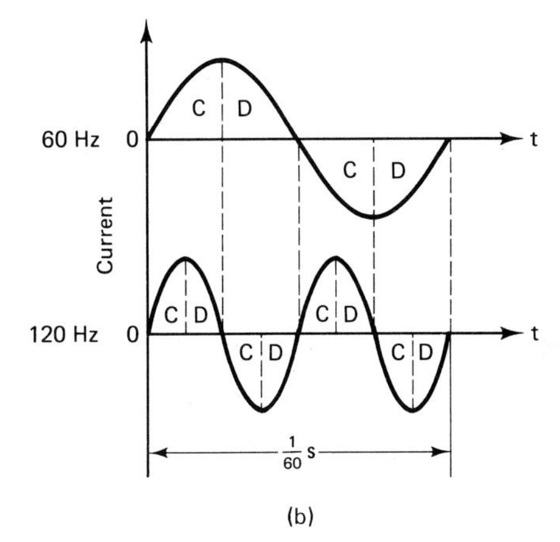

(b)

FIGURE 25-5 Current is directly proportional to frequency.

EXAMPLE 25-3

Calculate the current for the schematic of Fig. 25-5 if the frequency is doubled.

Solution: The reactance of the 2-μF capacitor with the 60-Hz source of Fig. 25-4(a) is $X_C = V/I = 1.33$ kΩ. With double the frequency, the reactance of the capacitor will be halved, to 665 Ω. The current is found using Ohm's law:

$$I = \frac{V_A}{X_C} = \frac{120 \text{ V}}{665 \ \Omega} = 180.5 \text{ mA} \quad \text{or} \quad \cong 181 \text{ mA}$$

Note: We could have deduced that the current would have doubled from the original value of 90.5 mA with the 60-Hz source since the reactance would be halved with the 120-Hz source.

25-5 EQUATION $X_C = 1/2\pi fC$ AND GRAPHICAL REPRESENTATION OF X_C VERSUS f

As long as the current of a capacitive circuit and the voltage across the capacitor are known, its reactance can be calculated using Ohm's law. An equation that is able to predict the reactance for any given frequency and capacitance is far more useful. This equation can be derived from the two previous equations, (25-1) and (25-2), which show that reactance is inversely proportional to frequency and capacitance:

$$X_C \propto \frac{1}{fC}$$

The frequency component in the equation is an ac sinusoidal waveform; therefore, the angular velocity $2\pi f$ for a sine wave must be considered. With the constant 2π the equation for capacitive reactance with frequency and capacitance becomes

$$X_C = \frac{1}{2\pi fC} \quad \text{ohms} \qquad (25\text{-}3)$$

where f is in hertz
C is in farads

Taking the inverse relation of 2π will give 0.159 rounded off to three significant figures; the zero is used to identify the decimal place. Substituting 0.159 for 2π in formula (25-3) gives

$$X_C = \frac{0.159}{fC} \quad \text{ohms} \qquad (25\text{-}4)$$

EXAMPLE 25-4

What is the reactance of the capacitor in the schematic of Fig. 25-4(a) if the frequency is increased to 240 Hz?

Solution: Using equation (25-4) gives

$$X_C = \frac{0.159}{fC} = \frac{0.159}{240\ \text{Hz} \times 2 \times 10^{-6}\ \text{F}} = 331\ \Omega$$

EXAMPLE 25-5

A certain capacitor has a reactance of 3180 Ω at 1 kHz. What is its capacitance?

Solution: Rearranging equation (25-4) gives

$$C = \frac{0.159}{fX_C} = \frac{0.159}{1000\ \text{Hz} \times 3180\ \Omega} = 0.05\ \mu\text{F}$$

Graphical Representation of X_C versus f

Although tuned circuits use a variable type of capacitor, most capacitors are fixed and the frequency of the circuit varies. Since frequency is the variable factor in an electronic circuit, it would be advantageous to have a visual graphical representation of capacitive reactance with a change in frequency. Equation (25-3) shows that reactance (X_C) is inversely proportional to frequency and capacitance. Keeping capacitance fixed, a plot of X_C versus f as graphed in Fig. 25-6 shows that X_C is an inverse function of frequency. (*Note:* By convention X_C is always considered to be a negative quantity, therefore is plotted in the negative quadrant.)

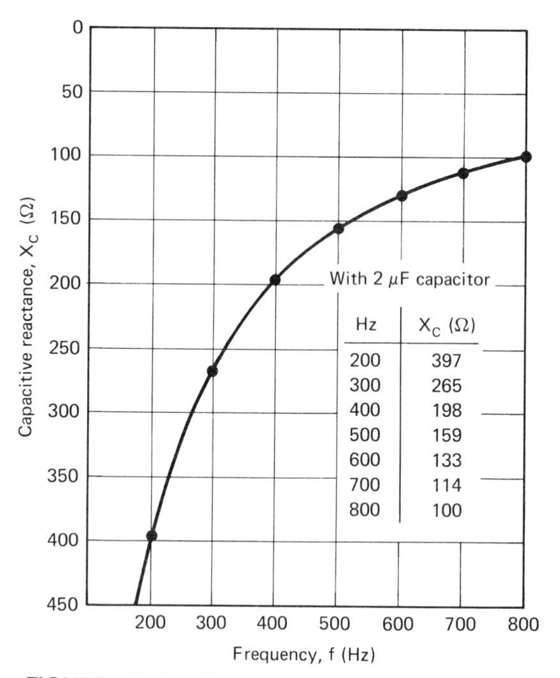

FIGURE 25-6 Graphical representation of capacitive reactance with frequency.

PRACTICE QUESTIONS 25-1

Choose the most suitable word or phrase to complete each statement.

1. A 1-μF capacitor is connected in series with a 20-kΩ resistor across a 10-V dc source. Calculate the circuit current after 1 s.
2. The same 1-μF capacitor has 45.3 mA flowing in the circuit when it is connected to 120 V, 60 Hz. Calculate its reactance.
3. Calculate the reactance of the capacitor in Question 2 if the frequency is halved.
4. Calculate the current flowing in the circuit of Question 2 if the capacitance is doubled.

5. Find the reactance of a 10-μF capacitor when it is connected across a 10-V 1-kHz source.
6. Capacitive reactance is ________ proportional to f and C.
7. The reactance of a certain capacitor is 159 Ω. What is its reactance if the source frequency is increased 10 times, to 10 kHz?
8. If the reactance of the capacitor in Question 7 increases to 1590 Ω, what must the new value of the source frequency be?
9. A 0.05-μF and a 0.1-μF capacitor are connected in series. The source frequency is 1 kHz. Calculate the net reactance of the circuit.
10. Calculate the net reactance for the two capacitors of Question 9 connected in parallel.

25-6 SERIES AND PARALLEL OHM'S LAW FOR X_C

Series Circuit

Since X_C is an Ohm's law opposition to ac current, the total reactance of series-connected capacitors can be found by the simple sum of the individual reactances:

$$X_{CT} = X_{C1} + X_{C2} + X_{C3} \quad \text{etc.} \qquad \text{ohms} \qquad (25\text{-}5)$$

EXAMPLE 25-6

Determine the total reactance of the capacitors in the circuit schematic of Fig. 25-7. What is the total capacitance of the circuit?

Solution: The total reactance can be found by summing the individual reactances. Using equation (25-5) gives

$$\begin{aligned} X_{CT} &= X_{C1} + X_{C2} + X_{C3} \\ &= 795\ \Omega + 318\ \Omega + 159\ \Omega \\ &= 1272\ \Omega \end{aligned}$$

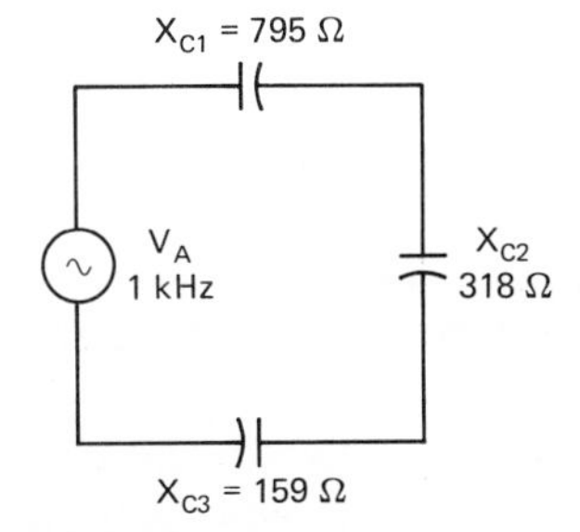

FIGURE 25-7 Circuit schematic.

The total capacitance of the circuit can be found by rearranging equation (25-4), which gives

$$C_T = \frac{0.159}{fX_{CT}} = \frac{0.159}{1000\ \text{Hz} \times 1272\ \Omega} = 0.1\ \mu\text{F}$$

EXAMPLE 25-7

From the circuit schematic of Fig. 25-8, determine:

(a) The total reactance of the circuit.
(b) The source voltage.

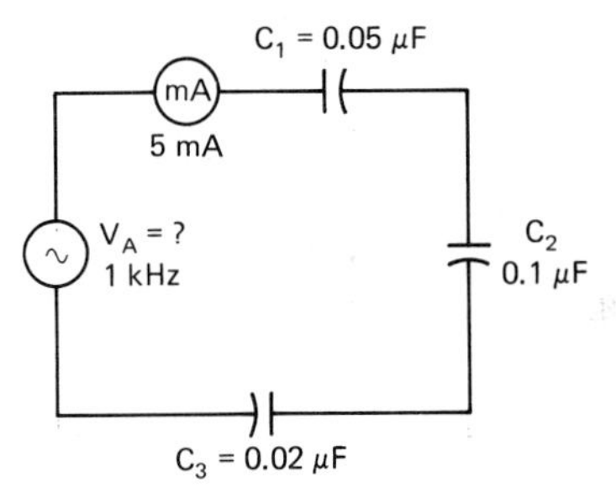

FIGURE 25-8 Circuit schematic.

Solution:

(a) Finding the total capacitance of the circuit first will enable us to calculate the total reactance using equation (25-3):

$$\begin{aligned} \frac{1}{C_T} &= \frac{1}{C_1} + \frac{1}{C_2} + \frac{1}{C_3} \\ &= \frac{1}{0.05\ \mu\text{F}} + \frac{1}{0.1\ \mu\text{F}} + \frac{1}{0.02\ \mu\text{F}} \\ &= 0.0125\ \mu\text{F} \cong 0.01\ \mu\text{F} \end{aligned}$$

$$\begin{aligned} X_{CT} &= \frac{0.159}{fC_T} = \frac{0.159}{1000\ \text{Hz} \times 0.01 \times 10^{-6}\ \text{F}} \\ &= 15.9\ \text{k}\Omega \end{aligned}$$

(b) Having found the total reactance to be 15.9 kΩ, and given 5 mA for the total current from the schematic diagram, the applied voltage can be found using Ohm's law:

$$\begin{aligned} V_A &= I_T X_{CT} = 5\ \text{mA} \times 15.9\ \text{k}\Omega = 79.5\ \text{V} \\ &\cong 80\ \text{V} \end{aligned}$$

Parallel Circuit

Similar to parallel resistance, the total reactance of a parallel circuit can be found using the reciprocal sum equation:

$$\frac{1}{X_{CT}} = \frac{1}{X_{C1}} + \frac{1}{X_{C2}} + \frac{1}{X_{C3}} \quad \text{etc.} \qquad \text{ohms} \quad (25\text{-}6)$$

If two parallel capacitors are taken at a time, their total reactance can be found using the product-over-sum equation:

$$X_{CT} = \frac{X_{C1} \times X_{C2}}{X_{C1} + X_{C2}} \qquad \text{ohms} \qquad (25\text{-}7)$$

EXAMPLE 25-8

Determine the total reactance of the capacitors in the circuit schematic of Fig. 25-9.

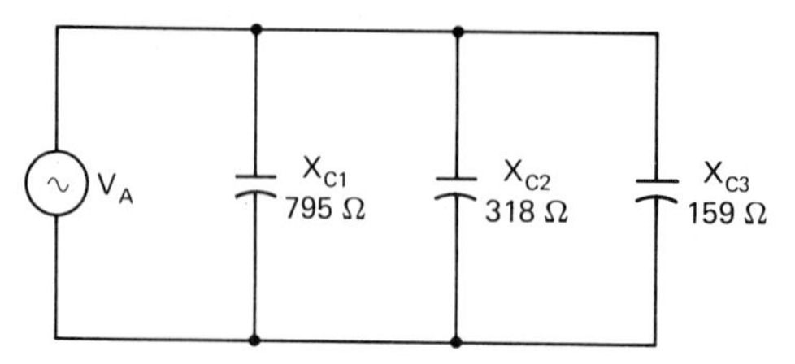

FIGURE 25-9 Circuit schematic.

Solution: The total reactance can be found by using the reciprocal sum equation (25-6) or product-over-sum equation (25-7) with two reactances at a time. Using equation (25-6) gives

$$\frac{1}{X_{CT}} = \frac{1}{X_{C1}} + \frac{1}{X_{C2}} + \frac{1}{X_{C3}}$$

$$= \frac{1}{795\ \Omega} + \frac{1}{318\ \Omega} + \frac{1}{159\ \Omega}$$

$$= 93.5\ \Omega$$

EXAMPLE 25-9

From the circuit schematic of Fig. 25-10, determine:

(a) The total reactance of the circuit.
(b) The source voltage.
(c) The total circuit current.

Solution:

(a) If we find the individual reactance of the capacitors, we can find the total reactance. Using equation (25-4) gives

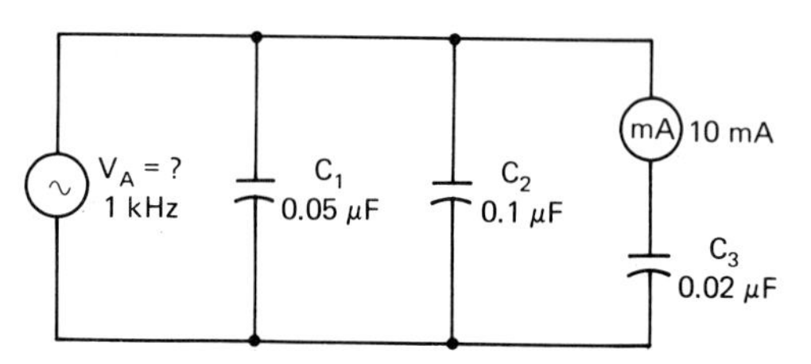

FIGURE 25-10 Circuit schematic.

$$X_{C1} = \frac{0.159}{fC} = \frac{0.159}{1000\text{ Hz} \times 0.05 \times 10^{-6}\text{ F}}$$

$$= 3180\ \Omega$$

$$X_{C2} = \frac{0.159}{fC} = \frac{0.159}{1000\text{ Hz} \times 0.1 \times 10^{-6}\text{ F}}$$

$$= 1590\ \Omega$$

$$X_{C3} = \frac{0.159}{fC} = \frac{0.159}{1000\text{ Hz} \times 0.02 \times 10^{-6}\text{ F}}$$

$$= 7950\ \Omega$$

The total reactance can be calculated by using the reciprocal sum equation (25-6):

$$\frac{1}{X_{CT}} = \frac{1}{X_{C1}} + \frac{1}{X_{C2}} + \frac{1}{X_{C3}}$$

$$= \frac{1}{3180} + \frac{1}{1590} + \frac{1}{7950}$$

$$= 935\ \Omega$$

(b) Since C_1, C_2, and C_3 are in parallel with the source, finding the voltage across X_{C3} will give this value.

$$V_A = V_{C3} = I_3 \times X_{C3} = 10\text{ mA} \times 7.95\text{ k}\Omega$$

$$\cong 80\text{ V}$$

(c) The total source current can be found using Ohm's law since we know the source voltage and the total reactance.

$$I_T = \frac{V_A}{X_{CT}} = \frac{80\text{ V}}{935\ \Omega} = 85.6\text{ mA}$$

25-7 PHASE RELATIONSHIP OF CAPACITIVE CURRENT AND V_C

The sinusoidal coulomb charge representative waveform as shown in Fig. 25-11(a) charges a certain capacitor. As a result of the coulomb charge, a voltage develops across the capacitor which follows the coulomb charge waveform shown in the representative waveform diagram of Fig. 25-11 (b).

An instantaneous maximum circuit current will flow when the coulomb charge and the capacitor voltage go through the zero value at 0°, 180°, and 360° of their cycle. This is because the capacitor is in an uncharged state at these values of v_C, and appears as a short circuit to the applied voltage.

When the coulomb charge and v_C reach their maximum value, there is no further change in the charge, so the capacitor instantaneously appears as an open circuit and the circuit current is zero.

In plotting the circuit current at the 0° of v_C, the rate of change $\Delta v_C/\Delta t$ is positive going, therefore will give a positive maximum value of circuit current. At the 180° v_C point, the current change is negative going, therefore will give a negative maximum value of circuit current. At the 360° v_C point, the voltage change is once more positive going, therefore will give a positive maximum value of circuit current, as shown in the waveform diagram of Fig. 25-11(b).

The capacitive voltage v_c can be superimposed over the capacitive current i_c as shown in Fig. 25-11(c). Then, the capacitive current can be shown as leading v_c by 90°.

Phasor Diagram

The waveforms in Fig. 25-11 is the method by which we can graphically show that the current leads the voltage by 90°. Once this concept is represented graphically, a phasor diagram is a much simpler method to show this relationship and is shown in Fig. 25-12.

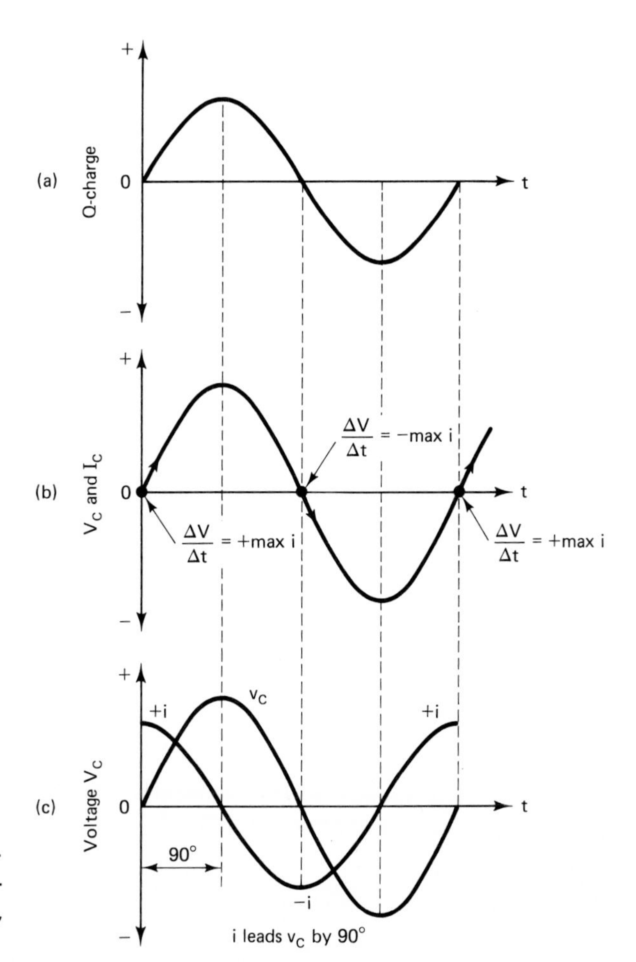

FIGURE 25-11 Waveform diagrams to represent the phase relationship of coulomb charge Q, voltage V_C and current I_C.

The voltage, V_C, waveform of Fig. 25-12 is used as the reference waveform so that the voltage phasor is drawn along the x axis and is the reference phasor. The current waveform is shown leading the voltage by 90° since the voltage is zero while the current is maximum, as shown in Fig. 25-12. As the phasor is considered to be rotating counterclockwise, the current is shown leading the voltage by 90°.

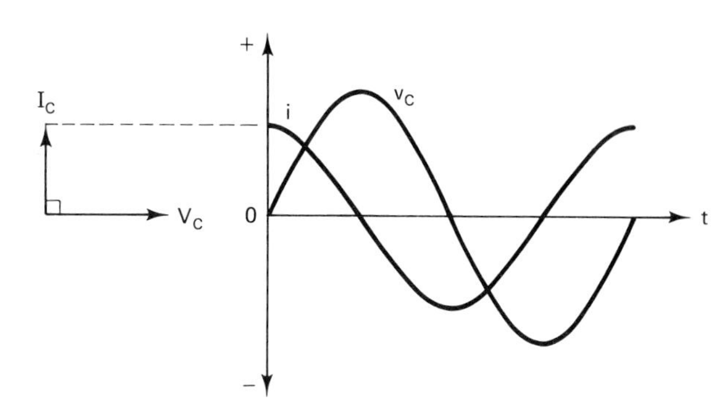

FIGURE 25-12 Waveform and phasor diagram to represent capacitive current I_C leading V_C by 90°.

SUMMARY

To dc a capacitor appears as an open circuit.

To ac a capacitor appears as a closed circuit.

Capacitive reactance (X_C) is an ac opposition to current flow and is measured in ohms.

Capacitive reactance can be calculated for any given voltage and current flowing in a capacitive circuit by applying Ohm's law, $X_C = V/I$.

X_C is inversely proportional to capacitance and frequency.

The equation to find X_C given the capacitance and frequency is

$$X_C = \frac{1}{2\pi f C} \quad \text{ohms} \tag{25-3}$$

A simplified equation to find X_C is

$$X_C = \frac{0.159}{fC} \quad \text{ohms} \tag{25-4}$$

The total reactance of series-connected capacitance can be found by summing the individual reactances:

$$X_{CT} = X_{C1} + X_{C2} + X_{C3} \quad \text{etc.} \quad \text{ohms} \tag{25-5}$$

The total reactance of a parallel capacitive circuit can be found by using the reciprocal sum equation:

$$\frac{1}{X_{CT}} = \frac{1}{X_{C1}} + \frac{1}{X_{C2}} + \frac{1}{X_{C3}} \quad \text{etc.} \quad \text{ohms} \tag{25-6}$$

If two parallel capacitors are taken at a time, their total reactance can be found using the product-over-sum equation:

$$X_{CT} = \frac{X_{C1} \times X_{C2}}{X_{C1} + X_{C2}} \quad \text{ohms} \tag{25-7}$$

In a pure capacitive ac circuit the current leads the applied voltage by 90°.

REVIEW QUESTIONS

25-1. Explain briefly why a capacitor appears as an open circuit when connected across a dc voltage.

25-2. Explain briefly why a capacitor appears as a closed circuit when connected across an ac voltage.

25-3. What reactive opposition establishes the capacitive reactance of a circuit?

25-4. A current of 20 mA flows in a capacitive circuit connected across 60 V ac. What is the reactance of the capacitor?

25-5. If the frequency of the source is halved, what will be the reactance of the capacitor in Question 25-4?

25-6. If the frequency of the source is doubled, what will be the reactance of the capacitor in Question 25-4?

25-7. What will be the reactance of the inductor in Question 25-4 if the capacitance is doubled?

25-8. A certain capacitor has a reactance of 361 Ω at 2 kHz. Calculate its capacitance.

25-9. Three capacitors, 0.02 μF, 120 μF, and 1200 pF, are connected in series. Find their total reactance if the source frequency is 5 kHz.

25-10. The three capacitors of Question 25-9 are connected in parallel. Find their total reactance.

25-11. From the circuit schematic of Fig. 25-13, calculate the voltage drop across each of the capacitors and the source voltage.

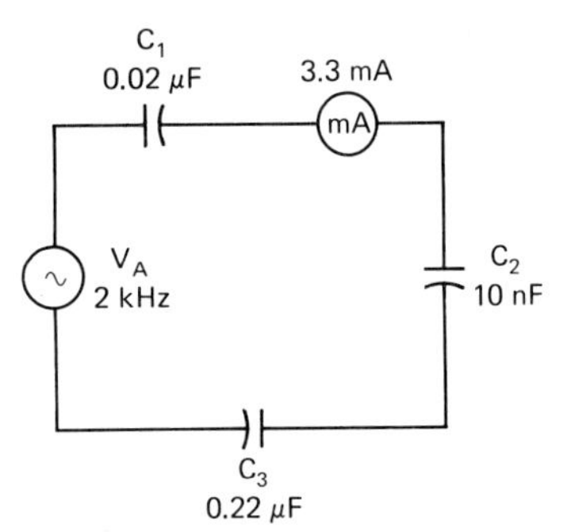

FIGURE 25-13 Circuit schematic.

25-12. From the circuit schematic of Fig. 25-14, calculate the source voltage and the total current in the circuit.

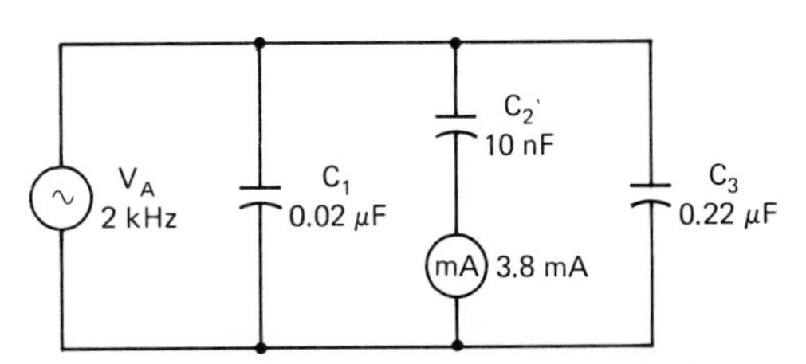

FIGURE 25-14 Circuit schematic.

25-13. What is the phase relationship of the applied voltage to the capacitive current? Which one leads? Draw a representative phasor diagram to show this.

SELF-EXAMINATION

Choose the most suitable word, phrase, or value to complete each statement.

25-1. What is the reactance of the capacitor in the circuit schematic of Fig. 25-15 at the instant the switch is closed?

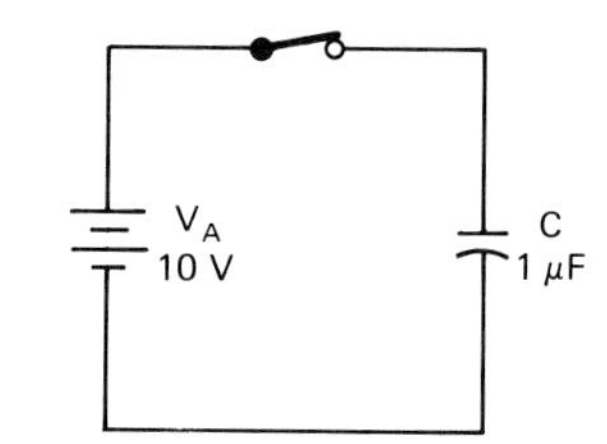

FIGURE 25-15 Circuit schematic.

25-2. Find the reactance of the capacitor in the circuit schematic of Fig. 25-15 if the source voltage is now 40 V ac with a frequency of 1 kHz.

25-3. What is the total reactance of two series-connected capacitors having X_{C1} of 600 Ω and X_{C2} of 900 Ω?

25-4. Find the total circuit current in the circuit schematic of Fig. 25-16.

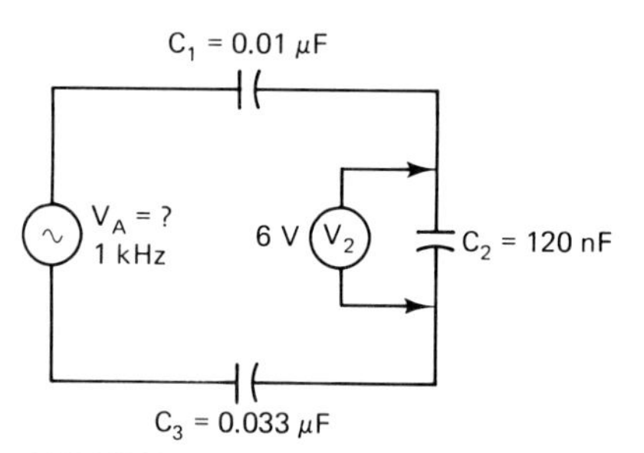

FIGURE 25-16 Circuit schematic.

25-5. What is the total capacitance of the circuit in Fig. 25-16?

25-6. What is the total reactance of the circuit in Fig. 25-16?

25-7. Find the applied voltage in the circuit schematic of Fig. 25-16.

25-8. Find the capacitance of C_3 in the circuit schematic of Fig. 25-17.

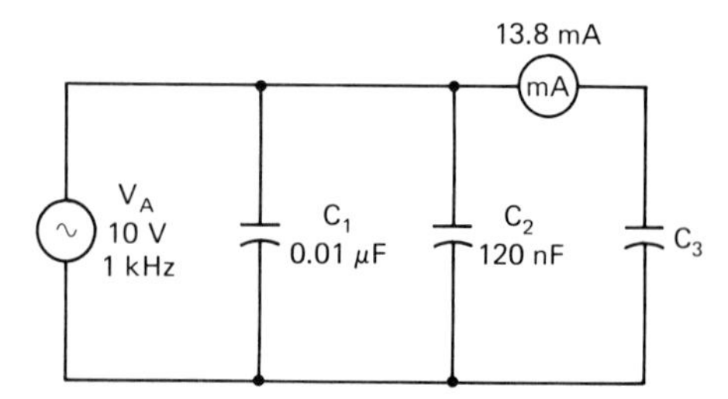

FIGURE 25-17 Circuit schematic.

25-9. Find the total reactance of the circuit in Fig. 25-17.

25-10. Find the total current in the circuit schematic of Fig. 25-17.

25-11. Find the total capacitance of the circuit in Fig. 25-17.

25-12. The voltage across a pure capacitive circuit (*lags, leads*) the current in the circuit by ________ degrees.

ANSWERS TO PRACTICE QUESTIONS

25-1

1. zero
2. $X_C = 2649\ \Omega$
3. $X_C = 5298\ \Omega$
4. $I = 90.6$ mA
5. $X_C = 15.9\ \Omega$
6. inversely
7. X_C decreases 10 times, or 15.9 Ω
8. f must decrease 10 times, or 1 kHz
9. $X_C = 4770\ \Omega$
10. $X_C = 1060\ \Omega$

26

Impedance in the Series and Parallel *RC* Circuit

INTRODUCTION

Some applications of *RC* circuits include filter circuits and *RC* coupling between stages of a transistor amplifier. Bypassing audio or RF voltages across a resistance is an application of a parallel *RC* circuit.

These *RC* circuit applications involve either a series or a parallel capacitance and resistance. In both cases the phase relationship (as in the case of *RL* circuits) of the ac voltages and currents is altered. The opposition to the current flow offered by the resistor remains the same even though the frequency of the current changes. Capacitance, like inductance, is reactive. Its opposition X_C changes with a change in frequency.

Combining the two types of oppositions, X_C and R, must be done by considering the phase relationship of the voltages and current in both the series and parallel *RC* circuits. This combined opposition to ac current flow is called impedance, Z.

In this chapter we examine impedance in a series and parallel *RC* circuit and the effect on impedance with a change in frequency and capacitance.

MAIN TOPICS IN CHAPTER 26

OBJECTIVES

After studying Chapter 26, the student should be able to

1. Master the current and voltage phase relationship concept for a series *RC* circuit.
2. Master the current and voltage phase relationship concept for a parallel *RC* circuit.
3. Apply phasor diagrams to series *RC* circuits and calculate voltage, current, and impedance.
4. Apply phasor diagrams to parallel *RC* circuits and calculate voltage, current, and impedance.

26-1 VOLTAGE AND CURRENT PHASE RELATIONSHIP FOR A PURE RESISTANCE VERSUS PURE CAPACITANCE

Ac current flowing through a resistor or resistance in an ac circuit does not produce a counter emf to react against the changing current. As in the case of dc, its opposition to ac current flow is due to the acceleration of conduction electrons that continually collide with the resistive material's atoms. The voltage required to produce a certain level of current flow through a resistor is its voltage drop or loss, or $I = V/R$. Since we are dealing with a sinusoidal ac voltage and current, the phase difference between the current flowing through the resistance and the voltage drop across the resistance will be zero degrees. The voltage and current are said to be in phase, as shown by the waveform diagram of Fig. 26-1(a).

The phasor diagram for the current and voltage across the resistance is drawn with the voltage phasor superimposed or "in line" with the current phasor, indicating a zero-degree phase angle [see Fig. 26-1(a)].

The ac charging and discharging current flowing into and out of a capacitor causes an opposing voltage (counter emf) to be set up across the capacitor. Since the counter emf is due to the amount of charge on the capacitor ($V = Q/C$), the charging current leads the voltage by 90°. Conversely, the voltage lags the current by 90°.

The phasor diagram for the 90° leading current or lagging voltage is shown in Fig. 26-1(b). Note that the voltage and current phasors are drawn to an arbitrary length. The lengths are not important except where a graphical solution is required. Each phasor is labeled with a capital letter, as shown.

26-2 VOLTAGE AND CURRENT PHASE RELATIONSHIP IN THE SERIES *RC* CIRCUIT

In analyzing the series *RC* circuit, we start with the basic series circuit characteristic—that current is common in all parts of the circuit. With this in mind, the current waveform shown at the top of the waveform diagram of Fig. 26-2(a) is used as our reference for all the other waveforms. The voltage waveform across the series resistor is in phase with the current through the resistor, as

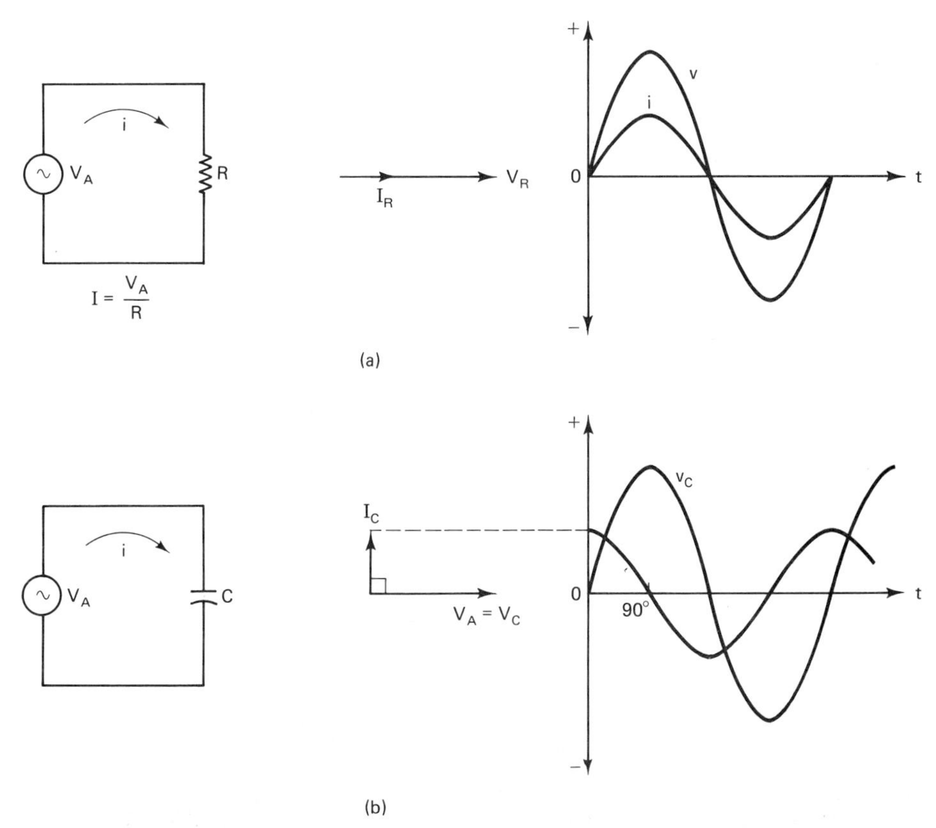

FIGURE 26-1 (a) Current is in phase with the applied voltage in the pure resistive circuit. (b) Current leads the applied voltage by 90° in the pure capacitive circuit.

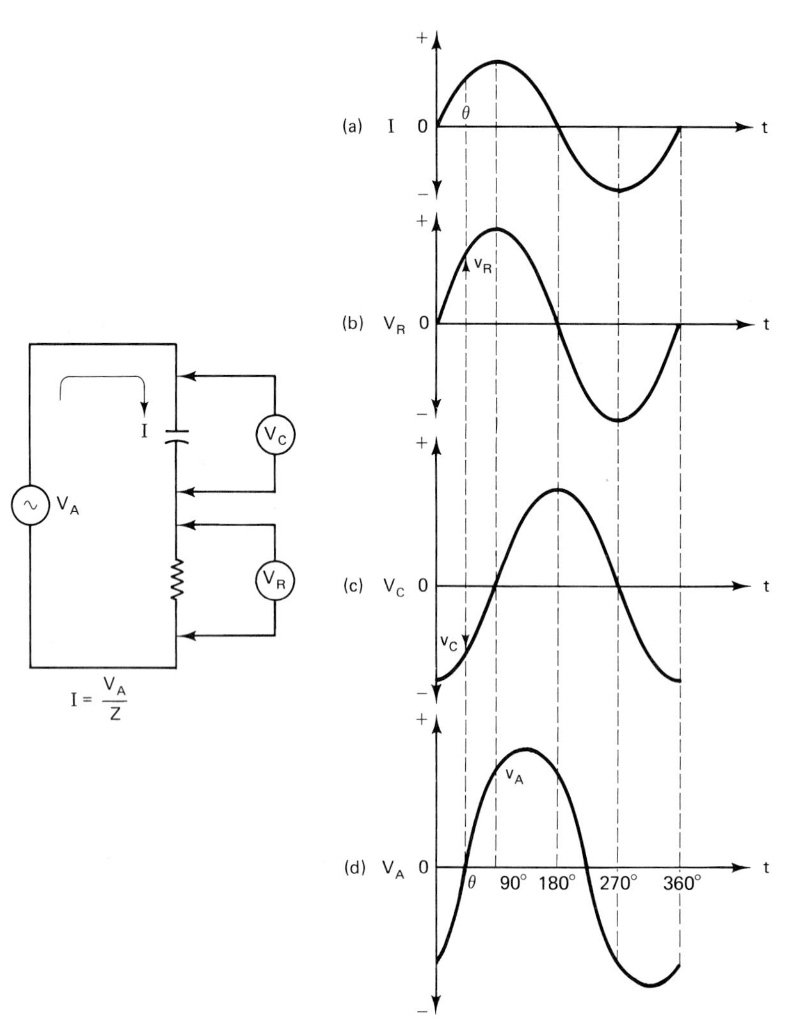

FIGURE 26-2 Series *RC* circuit and waveforms showing V_C is 90° out of phase with V_R. The applied voltage, V_A, is the resultant of summing the instantaneous values of v_R and v_C algebraically at each instant of time, or is the phasor sum of V_R and V_C.

shown by the waveform diagram of Fig. 26-2(b). Since the current through the capacitor will lead the voltage across the capacitor by 90°, V_C will be 90° out of phase with V_R, as shown by the waveform diagram of Fig. 26-2(c).

The applied voltage, V_A, is the *resultant of summing the instantaneous values* of v_R and v_C algebraically at each instant of time. The voltage drop V_R and V_C can be added as phasor quantities, shown in the phasor diagrams of Fig. 26-3. V_C and V_R form the two sides of a right-angle triangle and V_A the hypotenuse. Note V_C is shown in the negative quadrant (i.e., lagging the current).

The rms, peak, or peak-to-peak values (usually rms) for V_C and V_R are represented by the lengths of the two sides of the right-angle triangle. The length of the hypotenuse represents the rms, peak, or peak-to-peak value of the applied voltage. Note that in all cases the values must be consistent.

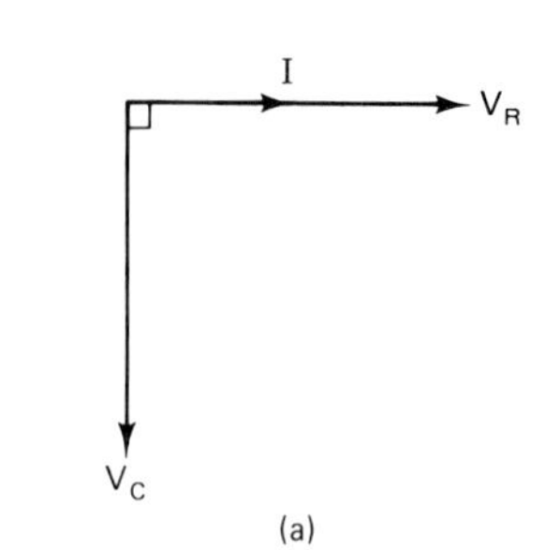

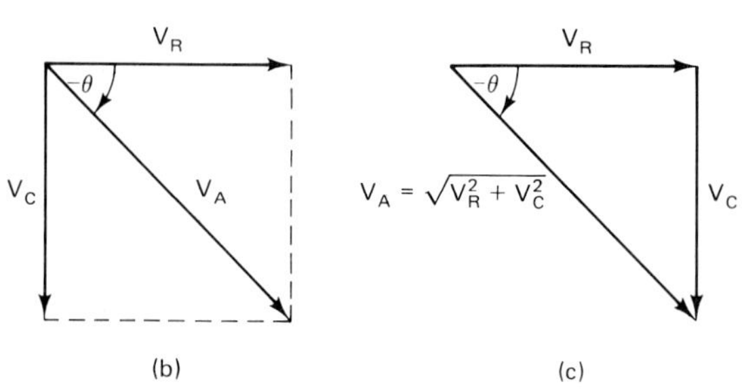

FIGURE 26-3 Voltage phasor diagram for the series *RC* circuit.

Voltages in a Series *RC* Circuit

V_R and V_C can be found by using Ohm's law, $V_R = IR$ and $V_C = IX_C$. The applied voltage, V_A, however, can be found by using the phasor sum—that is, by applying the Pythagorean theorem for a right-angle triangle shown in Fig. 26-3(c), to give

$$V_A = \sqrt{V_R^2 + V_C^2} \qquad \text{volts} \qquad (26\text{-}1)$$

V_A can also be found using Ohm's law provided that we know the current and the impedance, Z, which is the combined opposition of X_C and R with due regard to the phase relationship of the voltage and current. Therefore,

$$V_A = I_T Z \qquad \text{volts} \qquad (26\text{-}2)$$

Phase Angle between V_A and I

Since the current is in phase with V_R, we can find the phase angle by which the applied voltage lags the current using a trigonometric function. In this case the tangent (tan) lends itself as the most convenient, since V_R and V_C are adjacent and opposite sides to the angle θ [see Fig. 26-3(c)].

$$\tan\theta = \left(-\frac{V_C}{V_R}\right) \qquad (26\text{-}3)$$

or

$$\theta = \tan^{-1}\left(-\frac{V_C}{V_R}\right) \quad \text{or} \quad \arctan\left(-\frac{V_C}{V_R}\right) \qquad (26\text{-}3a)$$

Note: Since the current and V_R are referenced at 0° and the phasor rotates counterclockwise by convention, the angle θ will be negative. The negative sign then represents a lagging applied voltage with respect to the current and V_R.

EXAMPLE 26-1

Find (a) V_R, (b) V_C, (c) V_A, and (d) θ_I for the series *RC* circuit diagram of Fig. 26-4.

Solution: Using Ohm's law for both V_R and V_C since current is common gives

(a) $V_R = IR = 50\text{ mA} \times 1\text{ k}\Omega = 50\text{ V}$

(b) $V_C = IX_C = 50\text{ mA} \times 1\text{ k}\Omega = 50\text{ V}$

(c) Applying equation (26-1) to find V_A gives

$$V_A = \sqrt{V_R^2 + V_C^2} = \sqrt{(50\text{ V})^2 + (50\text{ V})^2} = 70.7\text{ V}$$

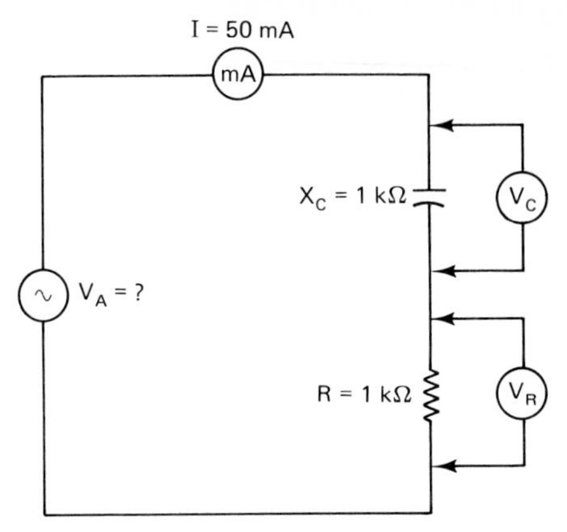

FIGURE 26-4 Series *RC* circuit.

(d) Applying equation (26-3a) gives

$$\theta = \tan^{-1} = \left(-\frac{V_C}{V_R}\right) = \tan^{-1}\left(-\frac{50\text{ V}}{50\text{ V}}\right) = \tan^{-1}(-1)$$

Therefore,

$$\theta = -45°$$

This means that the angle θ represents the phase angle by which V_A *lags* V_R, or −45°. Since V_R is in phase with the circuit current, the phase angle θ also represents the phase angle by which the current leads the applied voltage.

EXAMPLE 26-2

From the series *RC* circuit schematic of Fig. 26-5, find:

(a) The circuit current.
(b) The voltage drop V_C across the capacitor.
(c) The phase angle between the applied voltage and the current.
(d) The capacitance of the capacitor.

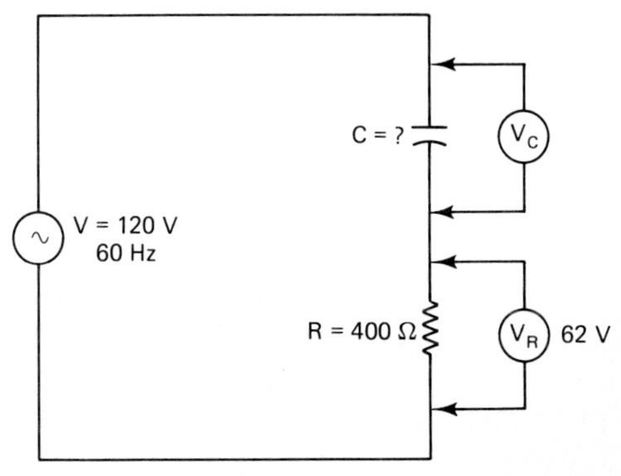

FIGURE 26-5 Series *RC* circuit schematic.

Solution:

(a) The current is common in a series circuit. Finding the current through the resistor will give the circuit current.

$$I_R = I_T = \frac{V_R}{R} = \frac{62\text{ V}}{400\ \Omega} = 155\text{ mA}$$

(b) Rearranging equation (26-1) will give V_C.

$$V_A = \sqrt{V_R^2 + V_C^2}$$

Therefore,

$$V_C = \sqrt{V_A^2 - V_R^2}$$
$$= \sqrt{(120\text{ V})^2 - (62\text{ V})^2} \cong 103\text{ V}$$

(c) The phase angle between V_A and the current can be found using equation (26-3) or (26-3a).

$$\theta = \tan^{-1}\left(-\frac{V_C}{V_R}\right) = \tan^{-1}\left(-\frac{103\text{ V}}{62\text{ V}}\right)$$
$$= \tan^{-1}(-1.66) = -59°$$

(d) The capacitance of the capacitor can be found by rearranging equation (25-3), $X_C = 1/2\pi fC$. But first we must find X_C using Ohm's law.

$$X_C = \frac{V_C}{I} = \frac{103\text{ V}}{155 \times 10^{-3}\text{ A}} = 665\ \Omega$$

and

$$C = \frac{1}{2\pi f X_C} = \frac{0.159}{60\text{ Hz} \times 665\ \Omega} = 4\ \mu\text{F}$$

26-3 IMPEDANCE IN THE SERIES *RC* CIRCUIT

The total opposition to ac current flow in a series *RC* circuit is called *impedance, Z*. Impedance can be found by using Ohm's law, provided that we know the total circuit current; therefore,

$$Z = \frac{V_A}{I_T} \quad \text{ohms} \qquad (26\text{-}4)$$

Capacitive reactance is a reactive component while resistance is a passive component. Their combined ac opposition in a series *RC* circuit is determined with due regard to the phase relationship of the current and voltages in the circuit.

The impedance phasor is developed from the voltage phasor of Fig. 26-6(a). With V_C transferred to form a closed triangle as shown in Fig. 26-6(b), IR, IX_C, and IZ are substituted for each of the phasor voltages. Since current is the same or common in all parts of a series circuit, the current of each voltage phasor can be dropped, leaving the impedance triangle as shown in Fig. 26-6(c). This can be shown by developing equation (26-1) as follows;

$$V_A = \sqrt{V_R^2 + V_C^2} \quad \text{volts} \qquad (26\text{-}1)$$

and

$$V_R = IR \qquad V_C = IX_C \qquad V_A = IZ$$

Substituting for each voltage gives

$$IZ = \sqrt{(IR)^2 + (IX_C)^2}$$
$$= \sqrt{I^2(R^2 + X_C^2)}$$
$$= I\sqrt{R^2 + X_C^2}$$

Therefore,

$$Z = \sqrt{R^2 + X_C^2} \quad \text{ohms} \qquad (26\text{-}5)$$

where Z is the impedance, in ohms
R is the resistance, in ohms
X_C is the capacitive reactance, in ohms

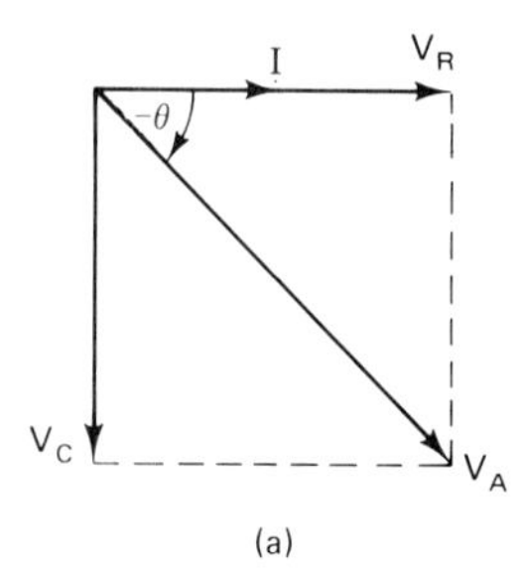

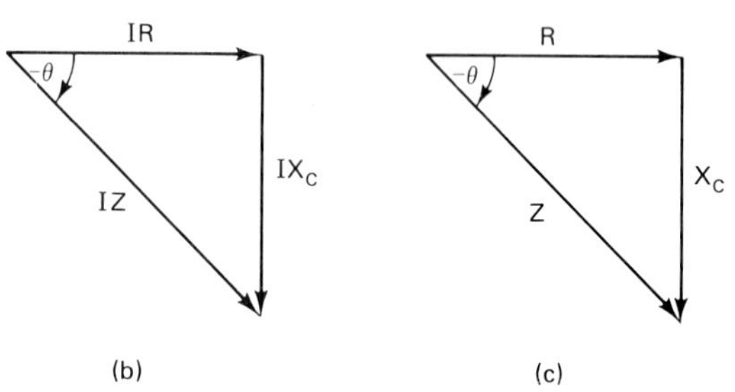

FIGURE 26-6 Voltage and impedance triangle for a series *RC* circuit. (a) Voltage phasor. (b) Equivalent triangle. (c) Impedance triangle.

Phase Angle between V_A and I

Since the impedance triangle corresponds to the voltage triangle, the phasor angle between V_A and I will correspond to the phase angle between Z and R. Using the tangent (tan) trigonometric function for the impedance triangle of Fig. 26-6(c) gives

$$\tan\theta = \theta_Z = \left(-\frac{X_C}{R}\right) \qquad (26\text{-}6)$$

or

$$\theta_Z = \tan^{-1}\left(-\frac{X_C}{R}\right) = \arctan\left(-\frac{X_C}{R}\right) \qquad (26\text{-}6a)$$

EXAMPLE 26-3

Use Fig. 26-4, the series RC circuit schematic, to find:

(a) The impedance of the circuit.
(b) The phase angle between V_A and the current.

Solution:

(a) The impedance of the circuit can be found using equation (26-5).

$$Z = \sqrt{R^2 + X_C^2}$$
$$= \sqrt{(1\text{ k}\Omega)^2 + (1\text{ k}\Omega)^2} = 1.41\text{ k}\Omega$$

(b) the phase between V_A and I can be found using equation (26-6a).

$$\theta_Z = \tan^{-1}\left(-\frac{X_C}{R}\right) = \tan^{-1}\left(-\frac{1\text{ k}\Omega}{1\text{ k}\Omega}\right)$$

Therefore,

$$\theta_I = -45°$$

Note that whenever X_C equals R, the phase angle between V_A and I will be $-45°$ (i.e., V_A lags I by 45°.

EXAMPLE 26-4

Use the example 26-2 solutions to find the following in the series RC, circuit schematic of Fig. 26-5.

(a) The impedance of the circuit.
(b) The phase angle between V_A and I.
(c) The total current in the circuit using Ohm's law equation.

Solution:

(a) The impedance of the circuit can be found using equation (26-4) or (26-5).

$$Z = \frac{V_A}{I_T} = \frac{120\text{ V}}{155\text{ mA}} \cong 776\ \Omega$$

or

$$Z = \sqrt{R^2 + X_C^2}$$
$$= \sqrt{(400\ \Omega)^2 + (665\ \Omega)^2} \cong 776\ \Omega$$

(b) The phase angle between V_A and I can be found using equation (26-6a).

$$\theta_Z = \tan^{-1}\left(-\frac{X_C}{R}\right) = \tan^{-1}\left(-\frac{665}{400}\right)$$

Therefore,

$$\theta = -59°$$

This shows that V_A lags I by 59°.

(c) The total current in the circuit can be found using Ohm's law.

$$I = \frac{V_A}{Z} = \frac{120\text{ V}}{776\ \Omega} = 155\text{ mA}$$

26-4 EFFECT OF FREQUENCY, CAPACITANCE, AND RESISTANCE ON IMPEDANCE IN THE SERIES *RC* CIRCUIT

Equation (25-3), $X_C = 1/2\pi fC$, shows us that X_C is inversely proportional to frequency and capacitance. This means that X_C is halved when frequency is doubled or when capacitance is doubled. Reactance X_C is doubled when frequency is halved, when capacitance is halved, and so on.

The impedance triangles of Fig. 26-7 show that as f or C increases, X_C and Z will decrease. If X_{C3} is much smaller than R, the phase θ_3, between the resistive voltage and lagging applied voltage approaches 0°. Most of the voltage drop will be across the resistance; therefore, Z_3 will be mostly resistive with a higher f or C.

When X_{C2} is equal to R, the phase angle θ_2 between the resistance voltage and lagging applied voltage will be 45°. The voltage drops across X_C and R will be equal to each other; therefore, Z_2 will be a combination of X_C and R.

When X_{C1} is much greater than R, the phase angle θ_1 between the resistive voltage and the lagging applied

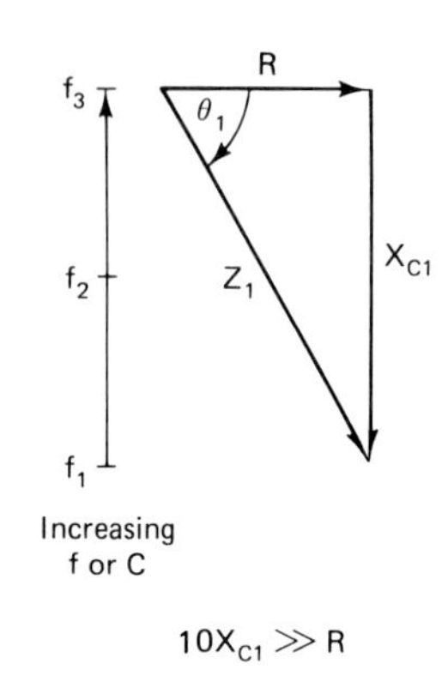

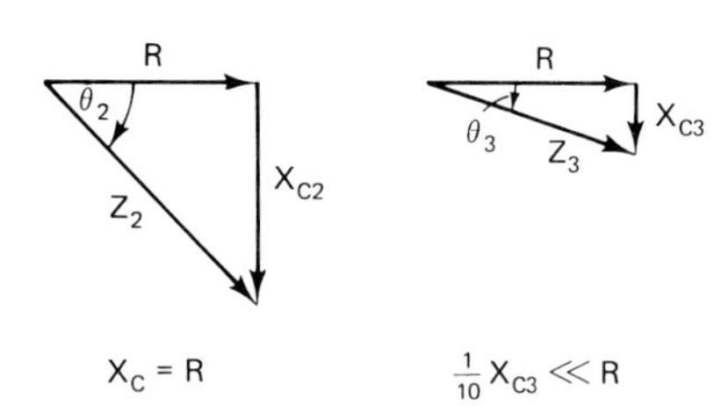

FIGURE 26-7 Effect of frequency and capacitance on impedance in the series *RC* circuit.

voltage approaches 90°. Most of the voltage drop will be across the capacitance; therefore, Z_1 will be mostly capacitive with a lower f or C.

The impedance triangles of Fig. 26-8 exemplify the effect of resistance on impedance. If R_1 is at least 10 times greater than X_C, the phase angle θ_1 approaches 0° and the impedance Z will be mostly resistive.

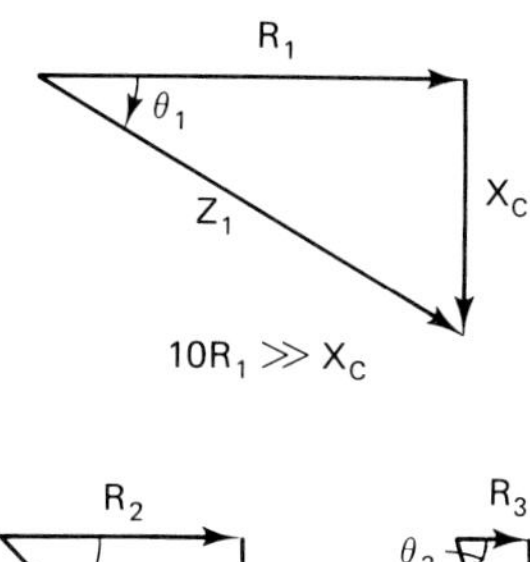

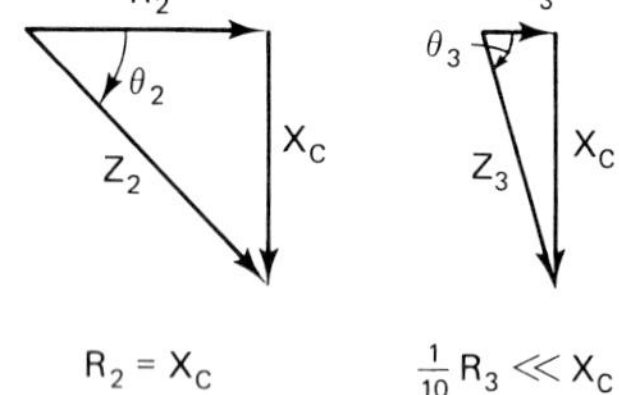

FIGURE 26-8 Effect of resistance on impedance in the series *RC* circuit.

When R_2 is equal to X_C, the phase angle θ_2 is −45°. The impedance Z_2 is a combination of R and X_C.

When R_3 is at least one-tenth smaller than X_C, the phase angle θ_3 approaches 90°, and the impedance Z_1 will be mostly capacitive.

PRACTICE QUESTIONS 26-1

Choose the most suitable word, phrase, or value to complete each statement.

1. An ac current flowing through a pure resistance is (*in phase, out of phase*) with its voltage drop by (*0°, 90°*).
2. An ac current flowing in a pure capacitive circuit is (*in phase, out of phase*) with its voltage drop by (*0°, 90°, 45°*).
3. The voltage across the capacitor will (*lag, lead*) the current in the capacitive circuit by ________ degrees.

Refer to Fig. 26-9, the series *RC* circuit schematic, for Questions 4 to 9.

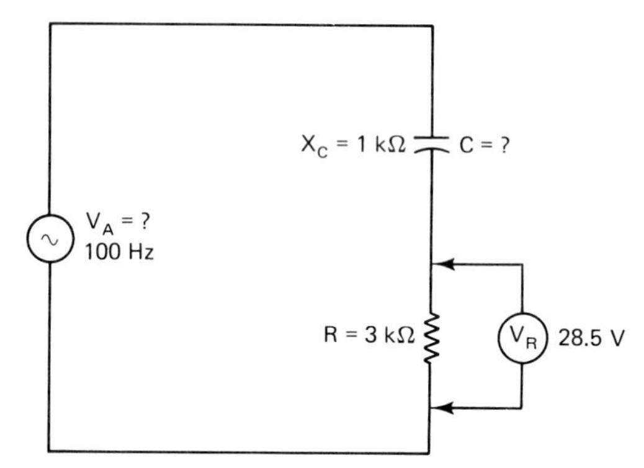

FIGURE 26-9 Series *RC* circuit schematic.

4. Find the total circuit current.
5. Find the voltage drop across the capacitor.
6. Find the applied voltage V_A.
7. Find the capacitance of the capacitor.
8. Find the impedance of the circuit.
9. Find the phase angle between the current and the applied voltage.

Refer to Fig. 26-10, the series *RC* circuit schematic, for Questions 10 to 17.

10. Find the voltage drop across the capacitor.
11. Find the voltage drop across the resistance.
12. Find the phase angle between the current and the applied voltage.
13. Find the capacitance of the capacitor.
14. What is the reactance of the capacitor if the frequency of the source is increased to 300 Hz?
15. What is the impedance of the circuit if the frequency of the source is increased to 300 Hz?

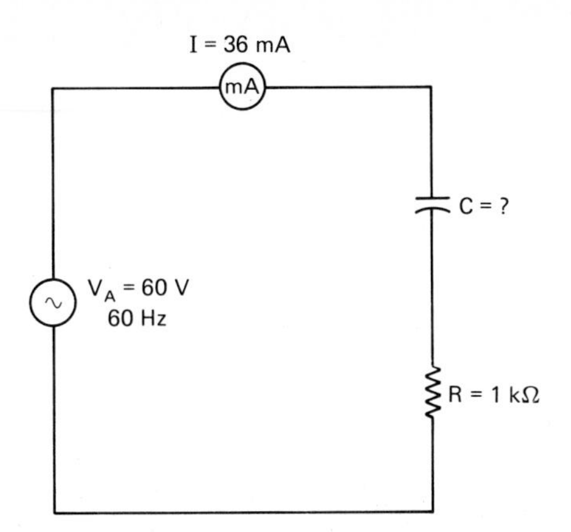

FIGURE 26-10 Series *RC* circuit schematic.

16. What is the phase angle between the current and the applied voltage of Question 15?

17. What type of impedance is mostly present in the circuit of Question 15?

26-5 CURRENT AND VOLTAGE PHASE RELATIONSHIP IN THE PARALLEL *RC* CIRCUIT

In analyzing the parallel *RC* circuit we start with the basic parallel circuit characteristics, that voltage is common to all parts of the circuit. With this in mind, the applied voltage waveform shown at the top of the waveform diagram of Fig. 26-11(a), is used as our reference for all other waveforms.

The current through the parallel resistor is in phase with the applied voltage, as shown by the waveform diagram of Fig. 26-11(b). Since the current through the capacitor will lead the applied voltage by 90°, I_C will be 90° out of phase with I_R as shown by the waveform diagram of Fig. 26-11(c).

The total circuit current, I_T, is the resultant of summing the *instantaneous values* of i_R and i_C algebraically at each instant of time as shown in Fig. 26-11(d).

The currents I_R and I_C can be added as phasor quantities, shown in the phasor diagram of Fig. 26-12. I_R and I_C form the two sides of a right-angle triangle and I_T the hypotenuse as shown in the current triangle of Fig. 26-12(c).

The rms, peak, or peak-to-peak value (usually rms) for I_R and I_C are represented by the length of the two sides of the right-angle triangle. The length of the hypotenuse represents the rms, peak, or peak-to-peak value of the total circuit current. Note that in all cases the values must be consistent.

Currents in a Parallel *RC* Circuit

I_R and I_C can be found by using Ohm's law, $I_R = V_A/R$ and $I_C = V_A/X_C$. The total circuit current, however, must be found by using the phasor sum or Pythagorean

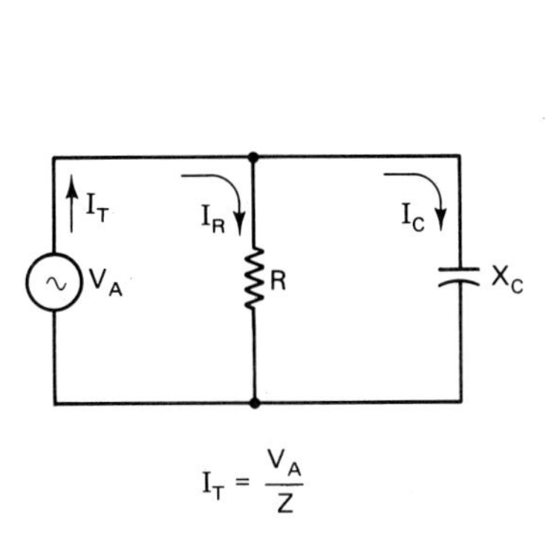

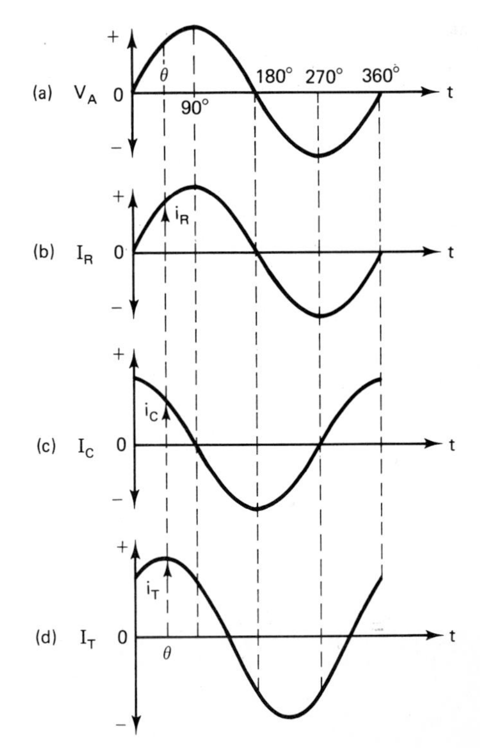

FIGURE 26-11 Parallel *RC* circuit and sine waves showing that I_C is 90° out of phase with I_R. The total circuit current I_T is the resultant of summing the instantaneous values of i_R and i_C algebraically at each instant of time, or is the phasor sum of I_R and I_C.

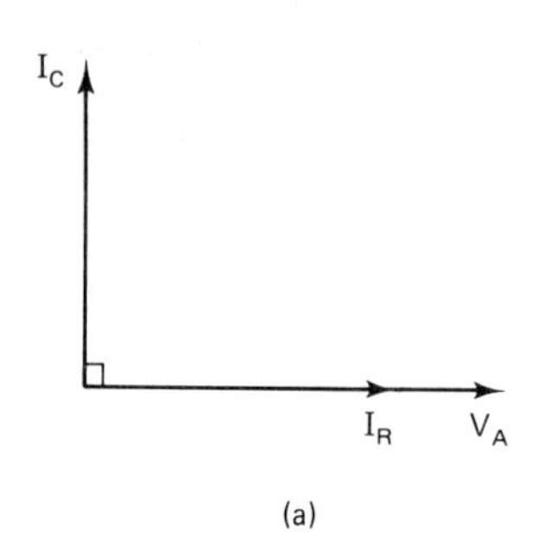

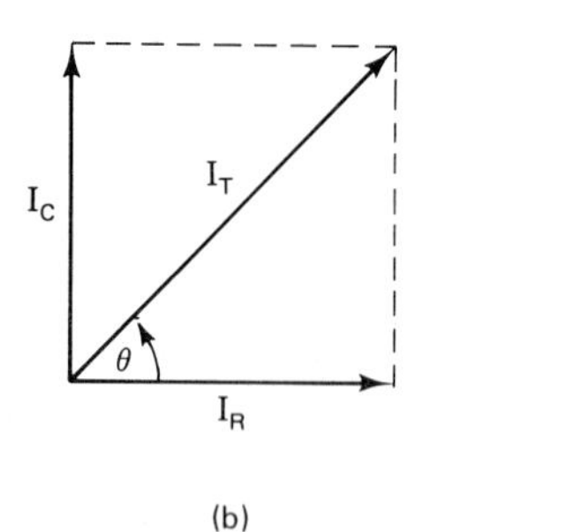

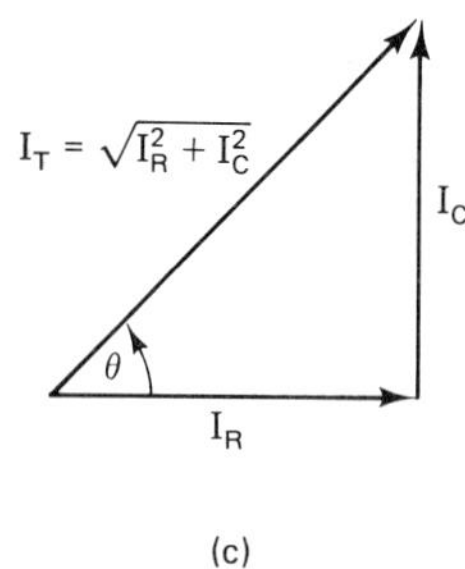

FIGURE 26-12 Current phasor diagram for the parallel *RC* circuit.

theorem for a right-angle triangle,

$$I_T = \sqrt{I_R^2 + I_C^2} \quad \text{amperes} \tag{26-7}$$

I_T can also be found by using Ohm's law provided that we know the current and the impedance, Z, for the circuit, or

$$I_T = \frac{V_A}{Z} \quad \text{amperes} \tag{26-8}$$

Phase Angle between I_T and V_A

Since the applied voltage is in phase with I_R, we can find the phase angle by which the total current leads the applied voltage using the tan function for the angle θ (see Fig. 26-12).

$$\tan \theta_I = \frac{I_C}{I_R} \tag{26-9}$$

or

$$\theta_I = \tan^{-1} \frac{I_C}{I_R} \tag{26-9a}$$

EXAMPLE 26-5

Find: (a) I_R, (b) V_A, and (c) C, in the parallel circuit of Fig. 26-13.

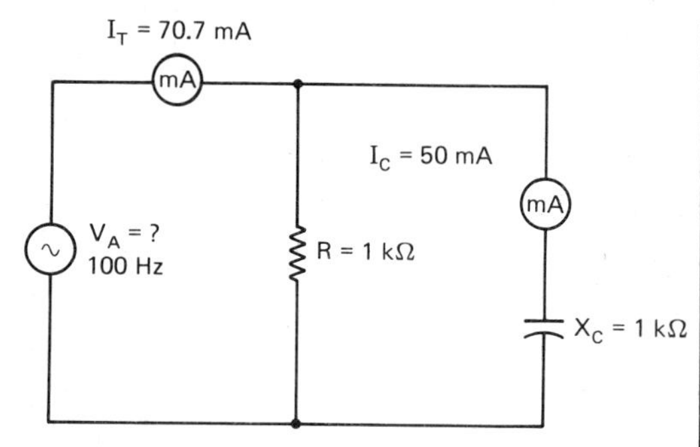

FIGURE 26-13 Parallel *RC* circuit schematic.

Solution:

(a) To find I_R, rearrange equation (26-7), or

$$I_R = \sqrt{I_T^2 - I_C^2} = \sqrt{(70\text{ mA})^2 - (50\text{ mA})^2}$$
$$= 50\text{ mA}$$

(b) Finding the voltage across the resistor or the capacitor will give V_A since they are connected in parallel.

$$V_C = V_A = I_C \times X_C = 50\text{ mA} \times 1\text{ k}\Omega = 50\text{ V}$$

(c) Since we know X_C, rearranging equation (25-3) will give the capacitance C,

$$C = \frac{1}{2\pi f X_C} = \frac{0.159}{100\text{ Hz} \times 1000\ \Omega} = 1.6\ \mu\text{F}$$

EXAMPLE 26-6

Find (a) I_R, (b) I_T, (c) θ_I, and (d) the capacitance C in the parallel *RC* schematic diagram of Fig. 26-14.

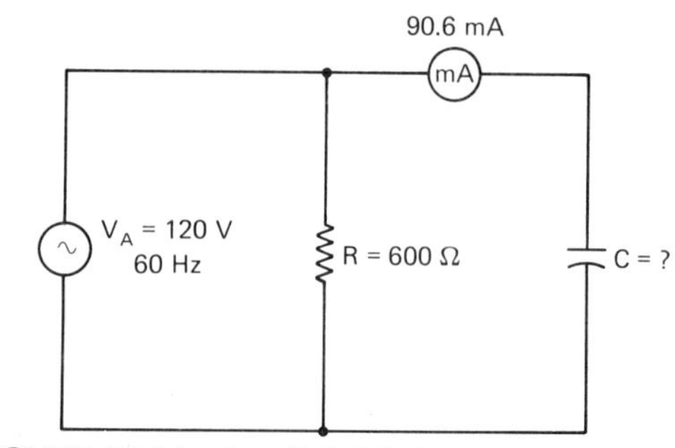

FIGURE 26-14 Parallel *RC* circuit schematic.

Solution:

(a) The current through the resistance can be found using Ohm's law.

$$I_R = \frac{V_A}{R} = \frac{120\text{ V}}{600\ \Omega} = 200\text{ mA}$$

(b) The total current can be found using equation (26-7).

$$I_T = \sqrt{I_R^2 + I_C^2} = \sqrt{(200 \text{ mA})^2 + (90.6 \text{ mA})^2}$$

$$= 220 \text{ mA}$$

(c) The phase angle between the leading current and the applied voltage can be found using equation (26-9) or (26-9a).

$$\theta_I = \tan^{-1}\frac{I_C}{I_R} = \arctan\frac{90.6 \text{ mA}}{200 \text{ mA}}$$

Therefore,

$$\theta_I \cong 24°$$

The small phase angle of 24° between the circuit current and V_A indicates that the current is mostly resistive. This is due to the much lower resistance value of R shunting X_C.

(d) First we must find the reactance of the capacitor by applying Ohm's law:

$$X_C = \frac{V_A}{I_C} = \frac{120 \text{ V}}{90.6 \text{ mA}} = 1325\ \Omega$$

Rearranging equation (25-3) will give the capacitance:

$$C = \frac{1}{2\pi f X_C} = \frac{1}{6.28 \times 60 \text{ Hz} \times 1325\ \Omega}$$

$$= 2\mu\text{F}$$

26-6 IMPEDANCE IN THE PARALLEL *RC* CIRCUIT

The total opposition to ac current flow in a parallel *RC* circuit is called impedance, Z. Impedance can be found by using Ohm's law, $Z = V_A/I_T$, provided that we know the total circuit current.

Since the capacitive current is out of phase with the resistive current, the total current in the circuit can be found using equation (26-7):

$$I_T = \sqrt{I_R^2 + I_C^2}$$

Removing the square root sign and substituting Ohm's law for each current gives

$$\left(\frac{V_A}{Z}\right)^2 = \left(\frac{V_A}{R}\right)^2 + \left(\frac{V_A}{X_C}\right)^2$$

Voltage is common in a parallel circuit, therefore we can drop V_A to give

$$\frac{1}{Z^2} = \frac{1}{R^2} + \frac{1}{X_C^2} \quad \text{ohms} \qquad (26\text{-}10a)$$

From equation (26-9a) the product-over-the-phasor-sum equation can be derived:

$$Z = \frac{\text{product}}{\text{sum}} = \frac{R \times X_C}{\sqrt{R^2 + X_C^2}} \quad \text{ohms} \qquad (26\text{-}10)$$

EXAMPLE 26-7

Find the impedance in the circuit schematic of Fig. 26-13.

Solution: Since the total circuit is given, the impedance of the circuit can be found by using Ohm's law.

$$Z = \frac{V_A}{I_T} = \frac{50 \text{ V}}{70.7 \text{ mA}} = 0.707 \text{ k}\Omega$$

Using the product-over-the-phasor-sum equation (26-10) gives

$$Z = \frac{R \times X_C}{\sqrt{R^2 + 1/X_C^2}} = \frac{1 \text{ k}\Omega \times 1 \text{ k}\Omega}{\sqrt{(1 \text{ k}\Omega)^2 + (1 \text{ k}\Omega)^2}}$$

$$= 0.707 \text{ k}\Omega$$

or using equation (26-10a) gives

$$\frac{1}{Z^2} = \frac{1}{R^2} + \frac{1}{X_C^2} = \frac{1}{(1 \text{ k}\Omega)^2} + \frac{1}{(1 \text{ k}\Omega)^2}$$

$$Z = \sqrt{1/2} \text{ k}\Omega = 0.707 \text{ k}\Omega$$

EXAMPLE 26-8

Find the impedance of the circuit in the schematic of Fig. 26-14.

Solution: Since the total current was found to be 220 mA from Example 26-6, the impedance can be found using Ohm's law.

$$Z = \frac{V_A}{I_T} = \frac{120 \text{ V}}{220 \times 10^{-3} \text{ A}} = 545\ \Omega$$

Using the product-over-the-phasor sum will also give Z. We must find X_C first since it is not given.

$$X_C = \frac{V_A}{I_C} = \frac{120 \text{ V}}{90.6 \times 10^{-3} \text{ A}} = 1325\ \Omega$$

Therefore,

$$Z = \frac{R \times X_C}{\sqrt{R^2 + X_C^2}} = \frac{600\ \Omega \times 1325\ \Omega}{\sqrt{(600\ \Omega)^2 + (1325\ \Omega)^2}}$$

$$= 547\ \Omega$$

Or using equation (26-10a) gives

$$\frac{1}{Z^2} = \frac{1}{R^2} + \frac{1}{X_C^2} = \frac{1}{(600\ \Omega)^2} + \frac{1}{(1325\ \Omega)^2}$$

$$Z = \sqrt{1/33 \times 10^{-7}} = 547\ \Omega$$

26-7 EFFECT OF FREQUENCY, CAPACITANCE, AND RESISTANCE ON IMPEDANCE IN THE PARALLEL *RC* CIRCUIT

The resistance connected in parallel remains fixed for any frequency changes, so that we need only consider X_C. Capacitive reactance is inversely proportional to frequency and will vary with frequency. If X_C increases, the charge and discharge current from the capacitor will decrease, and vice versa. The increase or decrease in I_C will affect I_T, which in turn will affect Z.

Consider the current triangles of Fig. 26-15. Given a parallel *RC* circuit where X_C is at least 10 times smaller than R, the charge and discharge current from the capacitor will be 10 times greater than I_R. As shown by the current triangle of Fig. 26-15(a), the total current or circuit current is mostly capacitive, I_C. The phase angle between the leading total current will approach 90°; therefore, Z will be approximately equal to X_C.

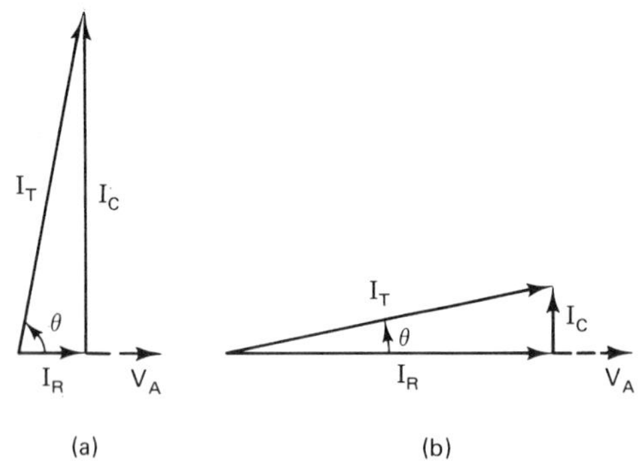

FIGURE 26-15 Current triangle for a parallel *RC* circuit where (a) X_C is 10 times smaller than R and (b) X_C is 10 times greater than R.

Conversely, if the frequency of the source is changed so that X_C is 10 times greater than R, the capacitor current will be 10 times less than I_R, as shown by the current triangle of Fig. 26-15(b). The phase angle between the leading total current and the applied voltage will approach 0°; therefore, Z will be approximately equal to R.

PRACTICE QUESTIONS 26-2

Choose the most suitable word, phrase, or value to complete each statement.

1. In a parallel *RC* circuit the current through the resistor is in phase with ________.
2. In a parallel *RC* circuit the charge and discharge current of the capacitor (*leads, lags*) the applied voltage by ________ degrees.

Refer to Fig. 26-16, the parallel *RC* circuit schematic, for Questions 3 to 6.

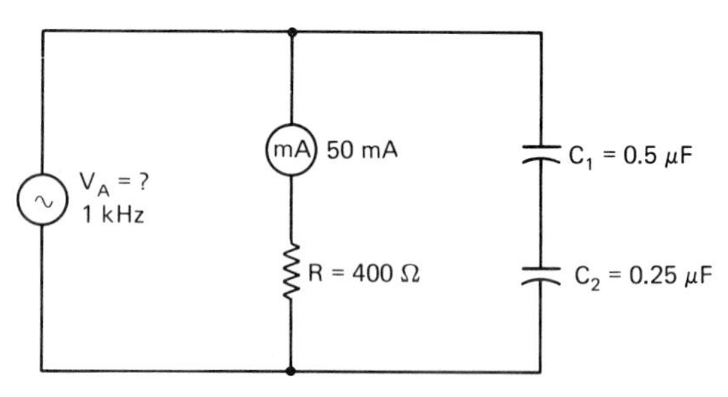

FIGURE 26-16 Parallel *RC* circuit schematic.

3. Find the total circuit current.
4. Find the applied voltage.
5. Find the phase angle between the total current and the applied voltage.
6. Find the impedance of the circuit.

Refer to Figure 26-17, the parallel *RC* circuit schematic, for Questions 7 to 11.

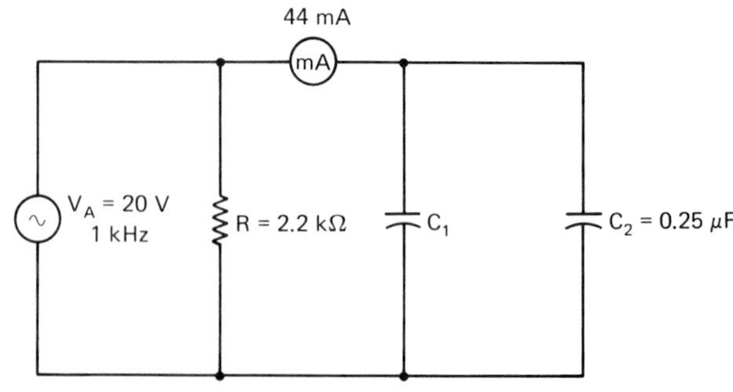

FIGURE 26-17 Parallel *RC* circuit schematic.

7. Find the current through the resistor.
8. Find the branch current for the capacitor C_1.
9. Find the branch current for the capacitor C_2.
10. Find the circuit current.

11. Find the phase angle between the current and the applied voltage.
12. Find the impedance of the circuit.
13. Find the capacitance of C_1.
14. What is the total capacitance in the circuit?

SUMMARY

An ac current flowing through a resistor will be in phase with the voltage drop it produces across the resistor.

An ac current flowing in a pure capacitive circuit will lead the applied voltage by 90°.

In a series RC circuit, V_R leads V_C by 90°.

In a series RC circuit, the phasor sum of the voltage drops is equal to the applied voltage:

$$V_A = \sqrt{V_R^2 + V_C^2} \quad \text{volts} \tag{26-1}$$

V_A can also be found by using Ohm's law provided that we know the impedance:

$$V_A = I_T Z \quad \text{volts} \tag{26-2}$$

The phase angle between the lagging applied voltage and the leading current in a series RC circuit can be found from the voltage triangle;

$$\tan \theta = \left(-\frac{V_C}{V_R}\right) \tag{26-3}$$

In a series RC circuit the total opposition to current flow is termed impedance, Z, and can be found using Ohm's law:

$$Z = \frac{V_A}{I_T} \quad \text{ohms} \tag{26-4}$$

The combined ac opposition of resistance and capacitive reactance in a series RC circuit is called impedance. The phasor sum of X_L and R gives,

$$Z = \sqrt{R^2 + X_C^2} \quad \text{ohms} \tag{26-5}$$

The phase angle between the applied voltage and the leading current can be found from the impedance triangle:

$$\tan \theta = \theta_Z = \left(-\frac{X_C}{R}\right) \tag{26-6}$$

Impedance is proportional to frequency since X_C is inversely proportional to frequency.

In a parallel RC circuit, I_C leads I_R by 90°.

In a parallel RC circuit, the phasor sum of the currents is equal to the total current:

$$I_T = \sqrt{I_R^2 + I_C^2} \quad \text{amperes} \tag{26-7}$$

I_T can also be found using Ohm's law provided that we know the impedance or,

$$I_T = \frac{V_A}{Z} \quad \text{amperes} \tag{26-8}$$

The phase angle between the leading current of the capacitor and the applied voltage can be found from the current triangle:

$$\tan \theta_I = \frac{I_C}{I_R} \tag{26-9}$$

In a parallel *RC* circuit the total opposition to current flow is termed impedance, Z, and can be found using Ohm's law:

$$Z = \frac{V_A}{I_T} \tag{26-4}$$

The product-over-the-phasor-sum equation for Z is

$$Z = \frac{R \times X_C}{\sqrt{R^2 + X_C^2}} \quad \text{ohms} \tag{26-10}$$

$$\text{or} \quad \frac{1}{Z^2} = \frac{1}{R^2} + \frac{1}{X_C^2} \quad \text{ohms} \tag{26-10a}$$

Since X_C is inversely proportional to frequency, impedance will also be proportional to frequency in a parallel *RC* circuit.

REVIEW QUESTIONS

26-1 Explain why there is a phase difference between the current and voltage of a capacitor, whereas there is no phase difference between voltage and current through a resistance.

26-2. Draw a phasor diagram to show the phase relationship of the applied voltage and current in a series *RC* circuit.

26-3. Draw a voltage triangle for a series *RC* circuit and state the equation used to solve for V_A.

26-4. From the voltage triangle, state the equation used to solve for the phase angle between the applied voltage and the current in a series *RC* circuit.

26-5. Draw an impedance triangle for a series *RC* circuit and state the equation used to solve for Z.

26-6. From the impedance triangle, give the equation used to solve for the phase angle between the applied voltage and the current in a series *RC* circuit.

26-7. Explain why impedance decreases in a series *RC* circuit with an increase in frequency.

26-8. At what approximate ratio of X_C to R is the phase angle between V_A and I nearly zero?

26-9. At what approximate ratio of X_C to R is the phase angle between V_A and I nearly 90°?

26-10. Draw a phasor diagram to show the phase relationship of the total circuit current and the branch currents of a parallel *RC* circuit.

26-11. Draw a current triangle for a parallel *RC* circuit and state the equation used to solve for I_T.

26-12. From the current triangle, state the equation used to solve for the phase angle between the total current and the applied voltage of a parallel *RC* circuit.

26-13. State two equations used to solve for Z in a parallel *RC* circuit.

26-14. What is the approximate phase relationship of V_A and I_T if the current through the capacitor is 10 times greater than through the resistor in a parallel *RC* circuit?

SELF-EXAMINATION

Choose the most suitable word, phrase, or value to complete each statement.

26-1. In a sinusoidal ac circuit having a pure resistance, the phase angle between the applied voltage and the current is ________ degrees.

26-2. In a sinusoidal ac circuit having a pure capacitance, the phase angle between the applied voltage and the current is ________ with the ________ leading.

Refer to Fig. 26-18, the series *RC* schematic, for Questions 26-3 to 26-8.

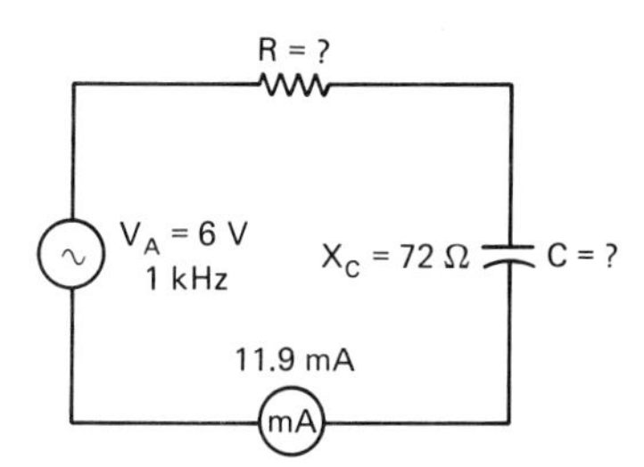

FIGURE 26-18 Series *RC* circuit schematic.

26-3. Find the impedance of the circuit.

26-4. Find the capacitance of the capacitor.

26-5. Find the voltage drop across X_C.

26-6. Find the voltage drop across R.

26-7. Find the resistance of the resistor.

26-8. Find the phase angle between the applied voltage and the current.

Refer to Fig. 26-19, the series *RC* schematic, for Questions 26-9 to 26-13.

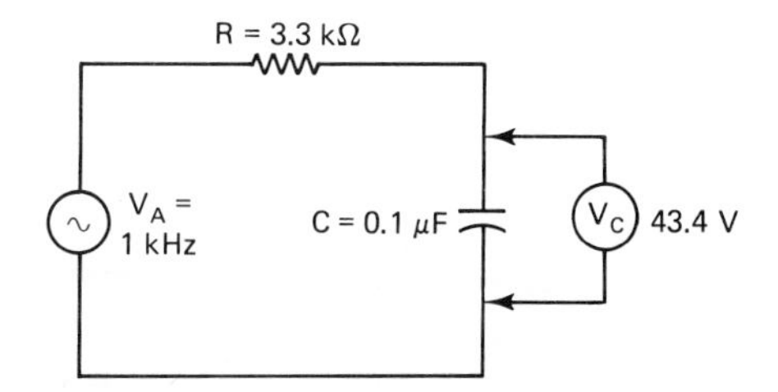

FIGURE 26-19 Series *RC* circuit schematic.

26-9. Find the reactance of the capacitor.
26-10. Find the impedance of the circuit.
26-11. Find the voltage drop across R.
26-12. Find the applied voltage.
26-13. Find the phase angle between the applied voltage and the current.

Refer to Fig. 26-20, the parallel RC schematic, for Questions 26-14 to 26-20.

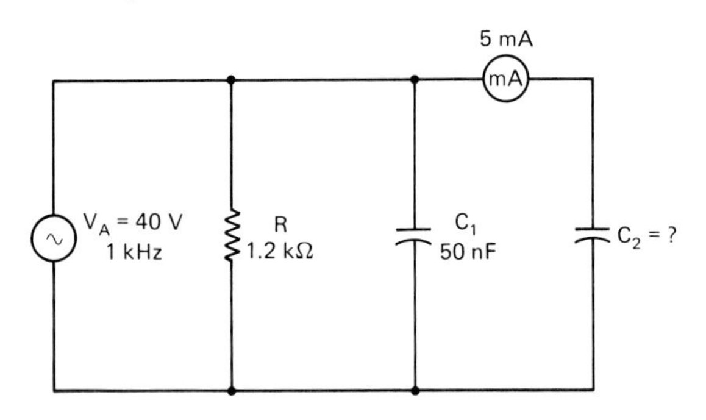

FIGURE 26-20 Parallel *RC* circuit schematic.

26-14. Find the current through R.
26-15. Find the branch current of C_1.
26-16. Find the total circuit current.
26-17. Find the impedance of the circuit.
26-18. Find the phase angle between the applied voltage and the current.
26-19. Find the capacitance of C_2.

ANSWERS TO PRACTICE QUESTIONS

26-1

1. in phase, 0°
2. out of phase, 90°
3. lag, 90°
4. 9.5 mA
5. 9.5 V
6. 30 V
7. 1.6 μF
8. 3162 Ω
9. 18.4°
10. 48 V
11. 36 V
12. 53°
13. 2 μF
14. 267 Ω
15. 1035 Ω
16. 15°
17. resistive

26-2

1. applied voltage
2. leads, 90
3. 54 mA
4. 20 V
5. 22.8°
6. 370 Ω
7. 9.09 mA
8. 12.6 mA
9. 31.4 mA
10. 45 mA
11. 78.6°
12. 444 Ω
13. 0.1 μF
14. 0.35 μF

27

Series Resonant Circuit

INTRODUCTION

In studying *LR* and *CR* circuits, we found that these circuits were applicable to timing and filtering applications. Another important application of *LR* and *CR* circuits is in band-pass and band-stop filters, discussed in Chapter 29.

Combining *LR* and *CR* circuits into either a series or a parallel configuration will give the most important application, a tuned *LCR* circuit. The series and parallel *LCR* tuned circuit is used in AM, FM, TV, and communication systems. It is here that the frequency must be "tuned in" in order to receive a specific channel. *LCR* tuned circuits serve to amplify and select the desired channel through an electrical process that involves resonance and band width.

In this chapter we study the series *LCR* circuit first, then study the series resonant circuit.

MAIN TOPICS IN CHAPTER 27

27-1 The Series *LCR* Circuit
27-2 Phase Relationship of I to V_L, V_C, V_R, and V_A
27-3 Impedance in the Series *LCR* Circuit
27-4 Series Resonance
27-5 Graphical Response Curves of X_L, X_C, and Z, I versus Frequency
27-6 Resonant Frequency Equation
27-7 Voltage Magnification
27-8 Tuned Circuit and Q Defined
27-9 Tuned Circuit Bandwidth
27-10 Plotting a General Response Curve
27-11 Reactance Chart

OBJECTIVES

After studying Chapter 27, the student should be able to

1. Master the fundamental concepts of current, voltage drops, and impedance of a series *LCR* circuit.
2. Solve series *LCR* current, voltage drop, and impedance problems.
3. Master the concept of current and voltage phase relationships about the circuit.
4. Master the concepts of series resonances, bandwidth, and the Q relationship.

5. Solve series resonant circuit current, voltage, and phase relationship; impedance, bandwidth; and Q problems.

27-1 THE SERIES *LCR* CIRCUIT

Having studied series *LR* and *CR* circuits, we are now ready to combine inductance, capacitance, and resistance into a series circuit as shown in the schematic of Fig. 27-1. Note that normally, the resistor in the series *LCR* circuit represents the value of the inductor's winding. In a case where a resistor is deliberately connected in series, its resistance plus the internal resistance of the coil is the net nonreactive resistance in the circuit.

As is the case for any series circuit, the current is the same in all parts of the circuit. This allows us to use the current as a reference point in developing the phase relationship of the voltage drops and the impedance in the series *LCR* circuit.

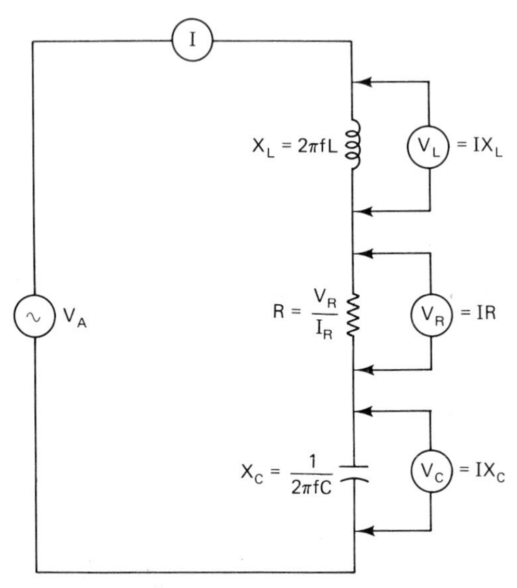

FIGURE 27-1 Series *LCR* circuit.

27-2 PHASE RELATIONSHIP OF *I* TO V_L, V_C, V_R, AND V_A

Phase Relationship of V_L and *I*

Recall that the induced counter emf produced by the inductor opposes the current change. This causes the current to lag the voltage across the inductor by 90°. Conversely, the *voltage* across the *inductor leads the current by 90°*, as shown by the representative phasor diagram of Fig. 27-2(a).

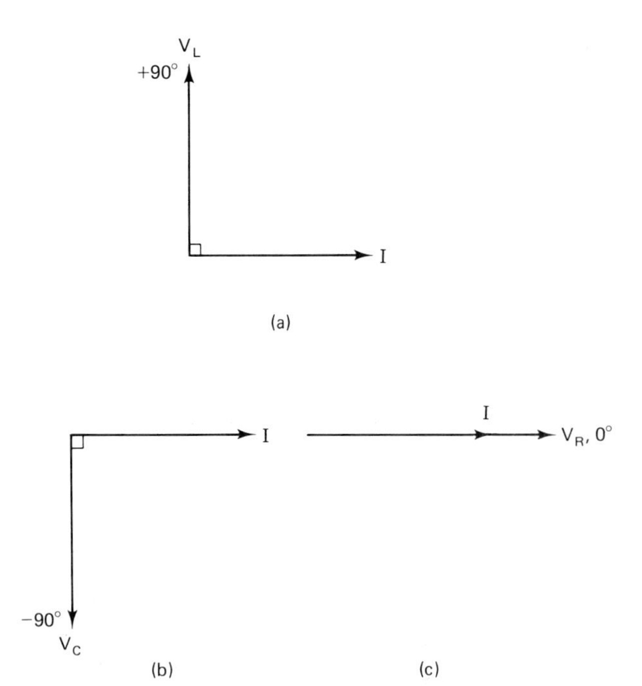

FIGURE 27-2 (a) V_L leads I by 90°. (b) V_C lags I by 90°. (c) V_R is in phase with I.

Phase Relationship of V_C and *I*

Again, recall that the current surges into a capacitor until it starts to build up a counter emf due to the coulomb charge on the capacitor. This current flow into the capacitor before the buildup of a voltage across the capacitor causes the current to lead the voltage by 90°. Conversely, the *voltage* across the capacitor *lags the current* by 90°, as shown by the representative phasor diagram of Fig. 27-2(b).

Phase Relationship of V_R and *I*

Since the resistance is a nonreactive component, the current will be in phase with the voltage across the resistor, as shown by the representative phasor diagram of Fig. 27-2(c).

Combining the phasor diagrams for the series *LCR* circuit will give the representative phasor diagram shown in Fig. 27-3(a). The phasor diagram shows that V_L leads I by 90° and V_C lags I by 90°. Most important, however, is the fact *that V_L is 180° out of phase with V_C and they cancel each other*.

The phasor diagram of Fig. 27-3(b) shows that a reactive voltage, V_X (either V_L or V_C) is left when V_C and V_L cancel each other. If both voltages were equal, V_X would be zero. In the phasor diagram of Fig. 27-3(b), V_X is inductive and must be summed with due consideration of the phase angle (θ) difference between V_X and V_R. The resultant voltage V_A is the phasor sum of V_X and V_R, as

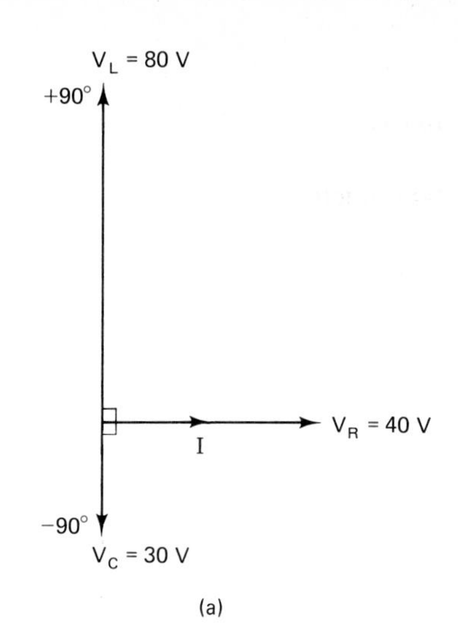

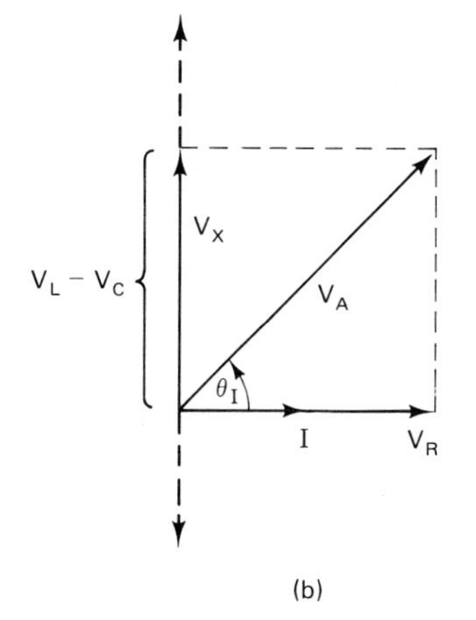

FIGURE 27-3 (a) Representative phasor diagram for current and voltages about a series *LCR* circuit. (b) V_L cancels V_C and leaves V_X. (c) V_A is the phasor sum of V_R and V_X in the voltage triangle.

shown in the voltage triangle of Fig. 27-3(c), and is

$$V_A = \sqrt{V_R^2 + (V_L - V_C)^2} \quad \text{volts} \qquad (27\text{-}1)$$

Phase Angle V_A to I

The phasor diagram of Fig. 27-3(b) shows that V_X will be inductive. The inductive voltage V_X is leading the current, which in turn causes V_A to lead the current. To find the phase angle between the leading applied voltage and the current, we can use the tan function equation

$$\tan\theta = \frac{V_L - V_C}{V_R} \qquad (27\text{-}2a)$$

Therefore,

$$\theta = \tan^{-1}\frac{V_X}{V_R} \qquad (27\text{-}2)$$

When V_C is greater than V_L, V_X will be capacitive and the applied voltage will lag the current; therefore, the angle θ is equal to

$$\tan\theta = -\left(\frac{V_L - V_C}{I_R}\right) \qquad (27\text{-}3a)$$

Therefore,

$$\theta = \tan^{-1}\left(-\frac{V_X}{V_R}\right) \qquad (27\text{-}3)$$

EXAMPLE 27-1

The current in a series *LCR* circuit was found to be 35 mA. If $X_L = 1\ \text{k}\Omega$, $X_C = 2\ \text{k}\Omega$, and $R = 100\ \Omega$, find:

(a) The voltages V_L, V_C, V_R, and the applied voltage.
(b) The phase angle between V_A and I.

Solution:

(a) The current is common in all parts of the series LCR circuit; therefore, the voltages V_L, V_C, and V_R can be found using Ohm's law.

$$V_L = IX_L = 35 \text{ mA} \times 1 \text{ k}\Omega = 35 \text{ V}$$

$$V_C = IX_C = 35 \text{ mA} \times 2 \text{ k}\Omega = 70 \text{ V}$$

$$V_R = IR = 35 \text{ mA} \times 100 \ \Omega = 3.5 \text{ V}$$

The applied voltage can be found using equation (27-1):

$$\begin{aligned} V_A &= \sqrt{V_R^2 + (V_L - V_C)^2} \\ &= \sqrt{(3.5 \text{ V})^2 + (35 \text{ V} - 70 \text{ V})^2} \\ &\cong 35 \text{ V} \end{aligned}$$

(b) Since $V_L - V_C = V_X$, and is capacitive, the current will be leading and V_X will be lagging; therefore, we use equation (27-3) to find the phase angle between V_A and I.

$$V_X = V_L - V_C = 35 \text{ V} - 70 \text{ V} = -35 \text{ V}$$

$$\theta = \tan^{-1}\left(-\frac{V_X}{V_R}\right) = \tan^{-1}\left(-\frac{35 \text{ V}}{3.5 \text{ V}}\right)$$

Therefore,

$$\theta = -84.3°$$

Note that the phase angle between V_A and I is negative. This indicates that the *applied voltage lags the current* in a capacitive circuit.

27-3 IMPEDANCE IN THE SERIES *LCR* CIRCUIT

The series LCR circuit presents three types of ac opposition to current flow, inductive, capacitative, and resistive. Combining the three types of oppositions requires that the phase angle between the current and voltages must be considered, as was the case in the RL and RC circuits.

Substituting the Ohm's law equation in each of the voltage phasors of Fig. 27-4(a) will show that the impedance phasor diagram of Fig. 27-4(b) is identical to the voltage phasor. This is because the current in the series LCR circuit is common to all the voltages and the ac opposition is directly proportional to the voltage.

Note that X_L cancels X_C, therefore, whichever is greater will be the predominate or "leftover" reactance (X), to be combined with the resistance as impedance in the circuit.

The impedance triangle of Fig. 27-4(c) shows that impedance, Z, is the phasor sum of X and R and is

$$Z = \sqrt{R^2 + (X_L - X_C)^2} \quad \text{ohms} \qquad (27\text{-}4)$$

When X_C is greater than X_L, the net reactance, $X = X_L - X_C$, will have a negative sign. The negative sign, however, only serves to identify that the net reactance is capacitive and is not to be considered in equation (27-4).

Impedance and Ohm's law

Ohm's law can be applied to find the current, the applied voltage, and the impedance of the series LCR circuit and is

$$Z = \frac{V_A}{I_T} \quad \text{ohms} \qquad (27\text{-}5)$$

Phase Angle (θ) between I and V_A

The phasor diagram of Fig. 27-4(b) shows that the phase angle between R and Z will be identical to the current and applied voltage phase angle. This is because R is representative of the current and Z is representative of the applied voltage. The phase angle between the applied voltage and the current can then be found by using the tan function in the impedance triangle of Fig. 27-4(c):

$$\tan \theta_Z = \theta_I = \frac{X_L - X_C}{R} \qquad (27\text{-}6a)$$

Therefore,

$$\theta_Z = \tan^{-1}\frac{X}{R} \qquad (27\text{-}6)$$

When X_C is greater than X_L, X will be capacitive and V_A will lag the current; therefore, the angle θ_Z is equal to

$$\tan \theta_Z = -\left(\frac{X_L - X_C}{R}\right) \qquad (27\text{-}7a)$$

Therefore,

$$\theta_Z = \tan^{-1}\left(-\frac{X}{R}\right) \qquad (27\text{-}7)$$

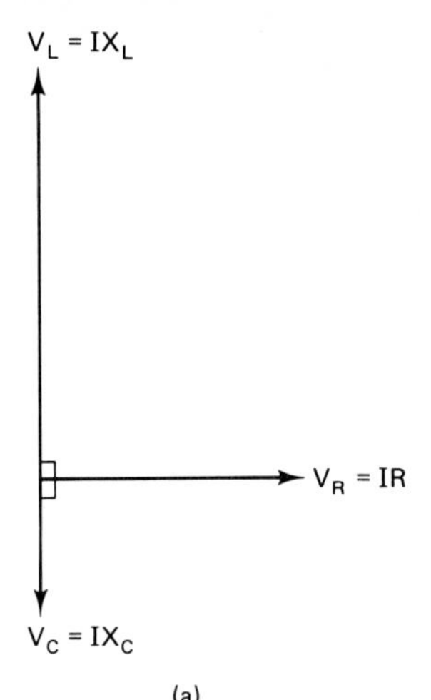

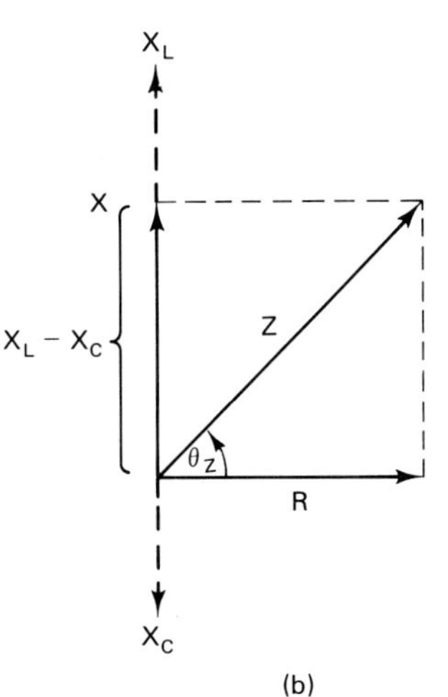

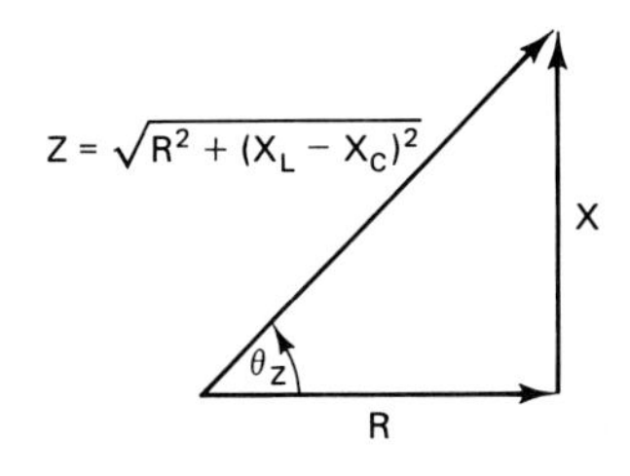

FIGURE 27-4 Representative phasor diagram for impedance.

EXAMPLE 27-2

A series-connected 0.01-μF capacitor and a 20-mH inductor having an internal resistance of 100 Ω are connected across a 10-V 5-kHz source.

(a) Find the impedance of the circuit.

(b) Find the phase angle between V_A and I.

Solution: The reactance of the capacitor and inductor must be found first. Using equation (21-3), the reactance of the inductor is

$$\begin{aligned} X_L &= 2\pi f L \\ &= 6.28 \times 5000 \text{ Hz} \times 0.02 \text{ H} \\ &= 628\ \Omega \end{aligned}$$

Using equation (25-3) gives the reactance of the capacitor.

$$\begin{aligned} X_C &= \frac{0.159}{fC} \\ &= \frac{0.159}{5 \times 10^3 \text{ Hz} \times 0.01 \times 10^{-6} \text{ F}} = 3180\ \Omega \end{aligned}$$

(a) The impedance of the circuit can now be found using equation (27-4).

$$\begin{aligned} Z &= \sqrt{R^2 + (X_L - X_C)^2} \\ &= \sqrt{(100\ \Omega)^2 + (628\ \Omega - 3180\ \Omega)^2} \\ &= 2554\ \Omega \end{aligned}$$

(b) In this case X_C cancels X_L and X_C is left.

$$\begin{aligned} X &= X_L - X_C \\ &= 628\ \Omega - 3180\ \Omega \\ &= -2552\ \Omega \end{aligned}$$

When X_C is greater than X_L, we use equation (27-7).

$$\theta_Z = \tan^{-1}\left(-\frac{X}{R}\right) = \tan^{-1}\left(-\frac{2552\ \Omega}{100\ \Omega}\right)$$

Therefore,

$$\theta_Z = -87.8°$$

The applied voltage lags the current by 87.8° in this predominately capacitive reactance circuit. Conversely, the current leads the applied voltage by 87.8°.

PRACTICE QUESTIONS 27-1

Choose the most suitable word, phrase, or value to complete each statement.

1. The voltage across a capacitor (*leads, lags*) the current by 90°.
2. The voltage across an inductor (*leads, lags*) the current by 90°.
3. The ________ is a nonreactive component in an ac circuit.
4. The voltage across the resistor will be (*90°, 180°, in phase*) with the current in a series *LCR* circuit.
5. In a series *LCR* circuit the voltage across the inductor is (*90°, 180°, 0°*) out of phase with the voltage across the capacitor.

Refer to the following circuit for Questions 6 to 12: A 10-nF capacitor and a 100-mH choke having an internal resistance of 278 Ω, and a 1 kΩ (external) resistor are connected in series across a 10-kHz source; the voltage drop across the 1-kΩ resistor is 3.9 V.

6. Find the reactance of the inductor.
7. Find the reactance of the capacitor.
8. Find the impedance of the circuit.
9. Find the voltage drop across the inductor.
10. Find the voltage drop across the capacitor.
11. Find the source voltage.
12. Find the phase angle between the current and the applied voltage.

27-4 SERIES RESONANCE

There are two types of resonance, mechanical and electrical. *Mechanical resonance* is identified with an object that vibrates at its natural rate or frequency. A prime example of this is a musical instrument. Reed instruments, such as the saxophone and clarinet, depend on the vibrations of a reed to produce their characteristic sound. Stringed instruments, such as the violin, guitar, and piano, depend on the vibrations of a number of stretched strings, such as piano wire, for their particular sound.

A startling example of mechanical resonance and its effect is that of some of the older bridges, which collapsed during heavy winds due mainly to vibrations started by the wind. Each vibration built upon the preceding vibration until the bridge broke up. Today, engineers have eliminated this problem by proper design utilizing wind tunnel testing.

Electrical resonance is identified with a circuit that is in tune with, or has the same natural frequency of oscillation as, the source voltage. Because of electrical resonance we are able to tune in radio, TV, and communication stations or channels. The circuit in the schematic of Fig. 27-5 is tuned to the resonant frequency of the source by adjusting either the inductance or the capacitance so that X_L will equal X_C.

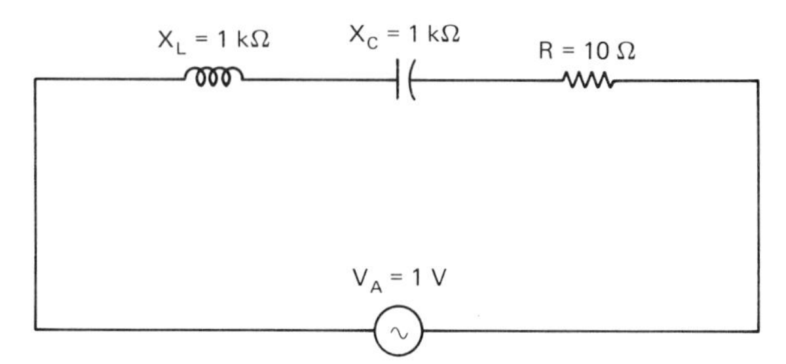

FIGURE 27-5 Series *LCR* circuit schematic.

Impedance at Resonance

Since X_L is 180° out of phase with X_C (see the representative phasor diagram, Fig. 27-4), they cancel and only R appears to be present to the source voltage. The impedance of the circuit in the schematic of Fig. 27-5 is equal to R or 10 Ω. This is shown by equation (27-4) as follows:

$$\begin{aligned} Z &= \sqrt{R^2 + (X_L - X_C)^2} \\ &= \sqrt{(10\ \Omega)^2 + (1000\ \Omega - 1000\ \Omega)^2} \\ &= 10\ \Omega \end{aligned}$$

At resonance

$$Z = R \qquad (27\text{-}8)$$

Circuit Current at Resonance

Since this is a series *LCR* circuit, the current will be the same in all parts of the circuit. The circuit current can be calculated by using Ohm's law and substituting R for Z. In the circuit of Fig. 27-5 the current is

$$I = \frac{V_A}{Z} = \frac{V_A}{R} = \frac{1\text{ V}}{10\ \Omega} = 100\text{ mA}$$

Note that at resonance the current in the circuit is due to the resistance and is large.

Voltage Drops at Resonance

Since V_L and V_C are 180° out of phase, they cancel each other and the source voltage appears to be across the resistor. This is shown by equation (27-1) and the circuit of Fig. 27-5 as follows:

$$V_L = IX_L = 100\text{ mA} \times 1\text{ k}\Omega = 100\text{ V}$$

$$V_C = IX_C = 100\text{ mA} \times 1\text{ k}\Omega = 100\text{ V}$$

$$V_R = IR = 100\text{ mA} \times 10\ \Omega = 1\text{ V}$$

Using equation (27-1) shows that the voltage across the resistor will be equal to the source voltage:

$$V_A = \sqrt{(V_R)^2 + (V_L - V_C)^2}$$

$$= \sqrt{(1\text{ V})^2 + (100\text{ V} - 100\text{ V})^2}$$

$$= 1\text{ V}$$

At resonance,

$$V_R = V_A \qquad (27\text{-}9)$$

27-5 GRAPHICAL RESPONSE CURVES OF X_L, X_C, AND Z, I VERSUS FREQUENCY

The phasor diagram of Fig. 27-4(b) shows that X_L is plotted as a positive quantity, while X_C is plotted as a negative quantity.

Figure 27-6 shows the variation of X_L and X_C with frequency. At the resonant frequency (f_r), X_L is equal to X_C and they cancel each other. The graph of Fig. 27-6 is an excellent visual representation to show that X_L is the predominate impedance above the resonant frequency, while X_C is predominate below the resonant frequency.

The response curve of Fig. 27-7(a) shows that the impedance of the series LCR circuit dips at the resonant frequency (f_r) and $Z = R$. Referring back to the graph of Fig. 27-6 will show the characteristic impedance above and below the resonant frequency.

The response curve of Fig. 27-7(b) shows that the current peaks or is maximum at the resonant frequency. This is because $Z = R$ and all the source voltage is across R with resultant increase or peak in current.

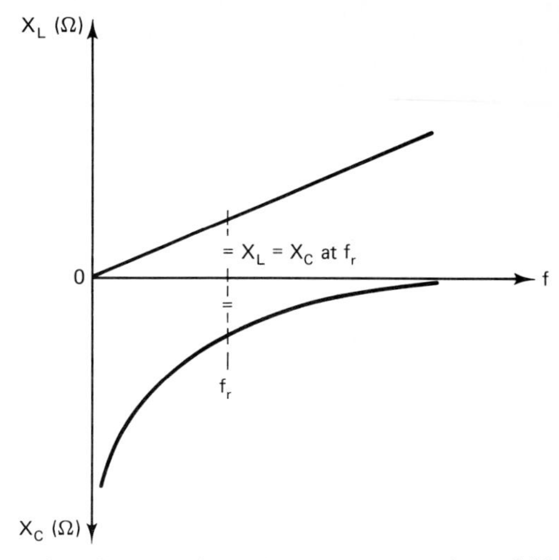

FIGURE 27-6 Graphical representation of X_L and X_C variation with frequency.

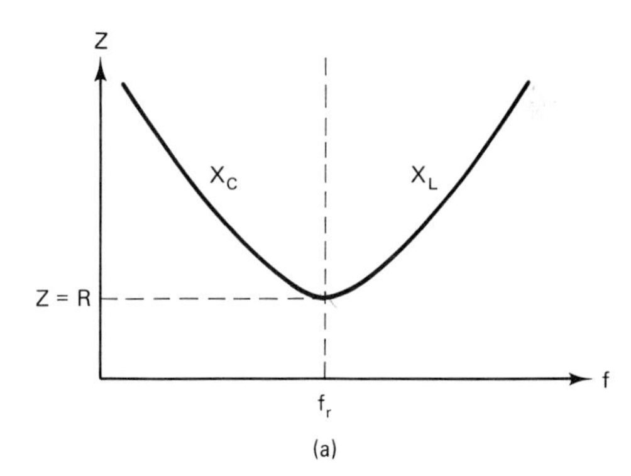

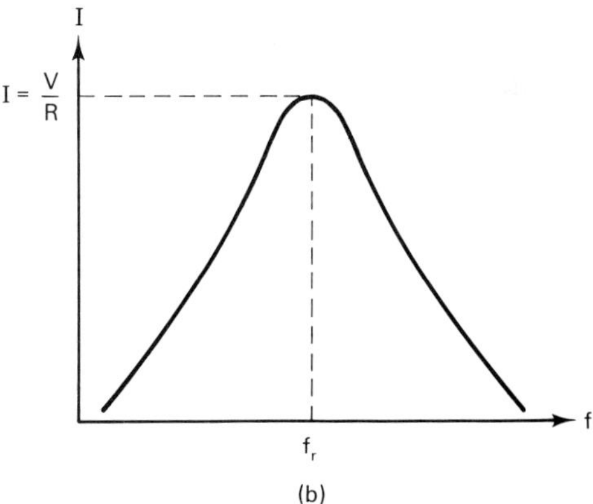

FIGURE 27-7 (a) Response curve of impedance variation with frequency. (b) Response curve of current variation with frequency.

27-6 RESONANT FREQUENCY EQUATION

Provided that we have the values of L and C, the resonant frequency can be calculated by using an equation. This resonant frequency equation stems from our previous knowledge that *at resonance*, $X_L = X_C$.

By substituting equation (25-3), $1/2\pi fC$, for X_C

and equation (21-3), $2\pi fL$, for X_L, we can derive the equation for the resonant frequency (f_r).

$$X_L = X_C$$

$$2\pi f_r L = \frac{1}{2\pi f_r C}$$

Rearranging the equation gives

$$f_r^2 = \frac{1}{4\pi^2 LC} \qquad \text{(27-10a)}$$

Taking the square root of each side of the equation gives

$$f_r = \frac{1}{2\pi\sqrt{LC}} \quad \text{hertz} \qquad \text{(27-10b)}$$

Substituting 0.159 for $1/2\pi$ gives

$$f_r = \frac{0.159}{\sqrt{LC}} \quad \text{hertz} \qquad \text{(27-10)}$$

where L is in henrys
C is in farads

EXAMPLE 27-3

An 18-mH choke and a resistance of 128 Ω is connected in series with a 22-nF capacitor across a 10-V ac source. Find:

(a) The resonant frequency of the circuit.
(b) The impedance at resonance.
(c) The circuit current at resonance.
(d) The voltage drop across R, X_L, and X_C.

Solution:

(a) The resonant frequency can be found by using equation (27-10).

$$f_r = \frac{0.159}{\sqrt{LC}}$$

$$= \frac{0.159}{\sqrt{18 \times 10^{-3}\ \text{H} \times 22 \times 10^{-9}\ \text{F}}}$$

$$= 7990\ \text{Hz} \quad \text{or} \quad \approx 8\ \text{kHz}$$

(b) At resonance, $Z = R = 128\ \Omega$.

(c) Circuit current can be found by using Ohm's law:

$$I = \frac{V_A}{Z} = \frac{V_A}{R} = \frac{10\ \text{V}}{128\ \Omega} = 78\ \text{mA}$$

(d) We must find X_L or X_C in order to calculate the voltage drops.

$$X_L = X_C = 2\pi fL$$

$$= 6.28 \times 8 \times 10^3\ \text{Hz} \times 18 \times 10^{-3}\ \text{H}$$

$$= 904\ \Omega$$

At resonance, $X_L = X_C$; therefore, $V_L = V_C$.

$$V_L = V_C = IX_L$$

$$= 78 \times 10^{-3}\ \text{A} \times 904\ \Omega = 70.5\ \text{V}$$

Finding C and L from f_r

In designing a tuned resonant circuit the inductance of the coil is normally fixed. Although it is desirable to obtain a high L/C ratio, the capacitance value that is required to resonate the circuit to a specific frequency can be calculated as follows.

Equation (27-10a) can be rearranged to find either C or L.

$$f_r^2 = \frac{1}{4\pi^2 LC} \quad \text{hertz} \qquad \text{(27-10a)}$$

Rearranging equation (27-10a) gives

$$C = \frac{1}{4\pi^2 f_r^2 L} \quad \text{farads} \qquad \text{(27-11)}$$

Rearranging equation (27-10a) to solve for L gives

$$L = \frac{1}{4\pi^2 f_r^2 C} \quad \text{henrys} \qquad \text{(27-12)}$$

EXAMPLE 27-4

What value of capacitance is required to resonate a 2.2-mH inductance to 250 kHz?

Solution: Use equation (27-11) to give

$$C = \frac{1}{4\pi^2 f_r^2 L}$$

$$= \frac{1}{4\pi^2 (250 \times 10^3\ \text{Hz})^2 \times 2.2 \times 10^{-3}\ \text{H}}$$

$$= 184\ \text{pF}$$

EXAMPLE 27-5

A tuned resonant circuit has a 500-pF capacitor in series with an unknown inductance. The circuit resonates at 450 kHz. Find the value of the inductance.

Solution: Use equation (27-12) to give

$$L = \frac{1}{4\pi^2 f_r^2 C}$$

$$= \frac{1}{4\pi^2(450 \times 10^3 \text{ Hz})^2 \times 500 \times 10^{-12} \text{ F}}$$

$$= 2.50 \times 10^{-4} \text{ H} = 250 \ \mu\text{H}$$

PRACTICE QUESTIONS 27-2

Choose the most suitable word, phrase, or value to complete each statement.

1. Electrical resonance is a condition where a circuit has the same natural frequency of oscillation as the ________ ________.
2. X_L cancels X_C because they are ________ degrees out of phase.
3. The impedance of a series *LCR* circuit at resonance is equal to ________.
4. At resonance the current in the series *LCR* circuit is due to the ________ and is very large.

Refer to the following circuit for Questions 5 to 11: A 22-nF capacitor, a 4.7-mH choke, and a 65-Ω resistor are connected in series across a voltage source.

5. Find the resonant frequency.
6. Calculate X_L at resonance.
7. Calculate X_C at resonance.
8. Find Z at resonance.
9. Given that the circuit current at resonance is 92 mA, find the value of the source voltage.
10. Find the voltage drop across X_L.
11. Find the voltage drop across the 65-Ω resistor.
12. What is the net voltage across L and C?

27-7 VOLTAGE MAGNIFICATION

The voltage drop across X_L, X_C, and R in the schematic shown in Table 27-1 was calculated for the resonant frequency and frequencies above and below resonance. Examining Table 27-1 shows that there is a current rise with resultant voltage rise across X_L and X_C at resonance. Below and above resonance the reactive voltage across the inductor or capacitor falls off quickly, in this case very rapidly.

Notice that the reactive voltage drop of 100 V is 100 times greater than the applied voltage of 1 V. This voltage magnification is due to the large current rise and large reactance at the resonant frequency. When the circuit is detuned off resonance, the circuit current falls off, with a resultant decrease in voltage drop across X_L and X_C.

The voltage magnification effect takes place at resonance and is dependent on the series resistance. At resonance X_L cancels X_C and only the resistance is left; therefore, the current rise will be limited by this series resistance and the applied voltage.

The voltage magnification property of a series *LCR* circuit at resonance is also called the circuit's *quality factor,* symbol Q. A circuit such as that shown in the schematic of Table 27-1 is termed a high-Q circuit. As

TABLE 27-1

Series *LCR* circuit calculations for the circuit shown

kHZ	X_L (Ω)	X_C (Ω)	Z (Ω)	I_T (mA)	V_R (mV)	V_L (V)	V_C (V)
28	879	1137	258	3.9	39	3.4	4.4
30	942	1062	120	8.3	83	7.8	8.8
f_r/32	1000	1000	10	100	1V	100	100
34	1068	937	131	7.6	76	8.1	7.1
36	1130	885	245	4.1	41	4.6	3.6

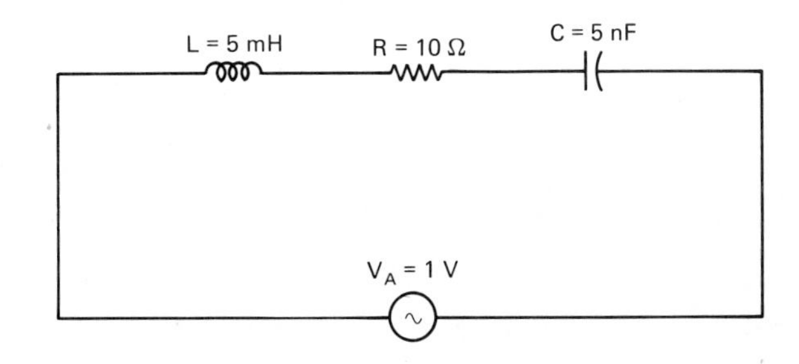

you probably are aware by now, the Q of a circuit is dependent on the series resistance. We define the Q of a circuit in Section 27-8.

27-8 TUNED CIRCUIT AND Q DEFINED

A principal application of both series and parallel resonant tuned circuits can be found in radio, TV, and communication receivers and transmitters. In this application a tuned radio-frequency (RF) transformer and a tuned intermediate-frequency (IF) transformer couple two or more stages of transistor amplification, as shown in the block diagram of Fig. 27-8. The primary and the secondary are normally tuned to the RF or IF frequency of the receiver. For AM receivers the IF is 455 kHz; for FM receivers, 10.7 MHz.

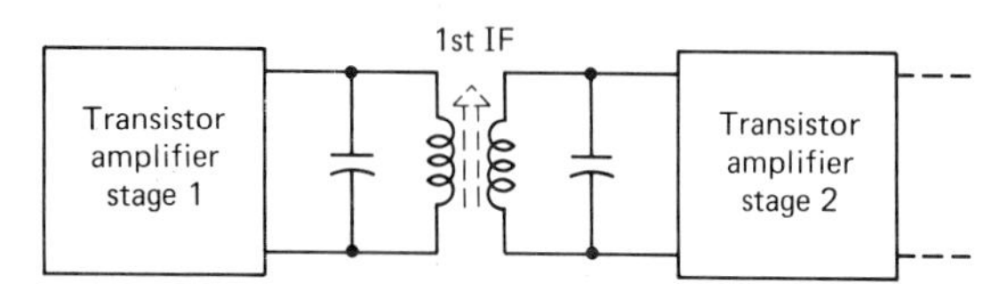

FIGURE 27-8 IF transformer couples transistor amplifier stage 1 to stage 2.

As shown in Fig. 27-9, the primary of these tuned IF transformers is connected across the output of a transistor amplifier. An ac voltage source symbol is substituted for the transistor voltage output to simplify the schematic. Since the ac output voltage of the transistor is in parallel with both L and C, *the primary characteristics is that of a parallel resonant circuit*. Tuning to resonance is accomplished by varying either L or C. Usually, C is fixed and a ferrite core is made variable, as shown by the dashed lines and the arrowhead in Figs. 27-8 and 27-9.

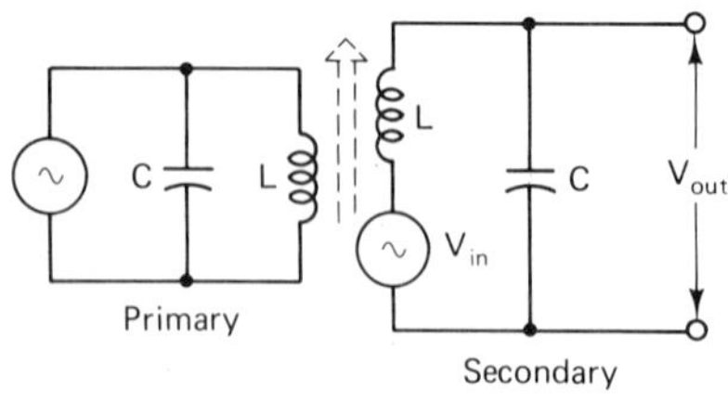

FIGURE 27-9 The primary of a tuned IF transformer is a parallel resonant circuit, while the secondary is a series resonant circuit.

The primary of the IF transformer produces the varying magnetic flux that links or couples the secondary. This induces a voltage (V_{in}) into the secondary which will be in series with L and C. *The characteristics of the secondary will be that of a series resonant circuit*. Note that the output is taken across the capacitor. This becomes the input voltage to the next transistor amplifier stage (2).

Q of a Series Resonant Circuit

The quality factor (Q) of a series resonant circuit is a measure of the voltage magnification or gain of the circuit and is a ratio of the output voltage to the input voltage. This ratio can be shown to be the ratio of X_L to r_s.

Consider the secondary circuit of the tuned transformer shown in Fig. 27-10. The secondary is a series resonant circuit with r_s representing the internal ac resistance of the coil winding. The induced input voltage (V_{in}) is in series with L, C, and r_s. The output is taken off the capacitor and is coupled to the input of the transistor amplifier.

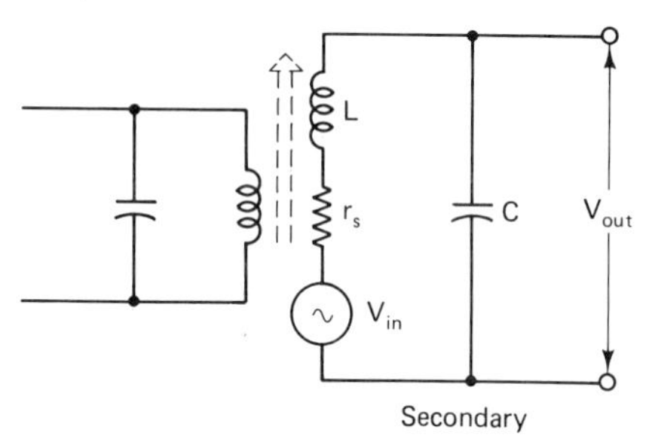

FIGURE 27-10 Secondary of a tuned IF transformer with the internal resistance of the coil r_S and the induced voltage v_{in} in series with L and C.

Now, voltage gain or magnification is a simple ratio of the output voltage to the input:

$$\text{voltage gain} = \frac{V_{out}}{V_{in}} \tag{27-13a}$$

If we substitute the applicable voltages in this simple ratio, we obtain

$$Q = \text{voltage gain} = \frac{V_C}{V_{in}} \tag{27-13b}$$

At resonance the induced input voltage will be across r_s; therefore, $V_{in} = I \times r_s$. Substituting $I \times X_C$ for V_C and $I \times r_s$ for V_{in}, then canceling I since the current is common in a series circuit will give

$$Q = \frac{I \times X_C}{I \times r_s} = \frac{X_C}{r_s} \tag{27-13c}$$

Although $X_C = X_L$ at resonance, the Q of a circuit is considered to be in terms of X_L since the series resistance is part of the coil; therefore,

$$Q = \frac{X_L}{r_s} \quad (27\text{-}13)$$

EXAMPLE 27-6

Consider the secondary circuit of the transformer in the schematic of Fig. 27-10. Inductance $L = 250\ \mu\text{H}$, $C = 200$ pF, and $r_s = 4\ \Omega$ with V_{in} of 100 mV.

(a) Find the voltage drop across X_L and X_C at resonance.
(b) Find the Q of the circuit at resonance.
(c) What is the Q of the circuit if the ac resistance of the coil is 12 Ω?

Solution:

(a) The resonant frequency of the circuit must be calculated before X_L and X_C can be found. Equation (27-10) will give the resonant frequency.

$$f_r = \frac{0.159}{\sqrt{LC}}$$

$$= \frac{0.159}{\sqrt{250 \times 10^{-6}\ \text{H} \times 200 \times 10^{-12}\ \text{F}}}$$

$$= 711\ \text{kHz}$$

At resonance $X_L = X_C$ therefore;

$$X_L = X_C = 2\pi fL$$

$$= 6.28 \times 711 \times 10^3\ \text{Hz} \times 250 \times 10^{-6}\ \text{H}$$

$$= 1116\ \Omega$$

The current at resonance is maximum when $Z = R$ and is found by using Ohm's law.

$$I = \frac{V_{in}}{Z} = \frac{V_{in}}{R} = \frac{100\ \text{mV}}{4\ \Omega} = 25\ \text{mA}$$

Current is common in a series circuit and $V_L \simeq V_C$ at resonance; therefore,

$$V_L = V_C = IX_L$$

$$= 25 \times 10^{-3}\ \text{A} \times 1116\ \Omega = 27.9\ \text{V}$$

(b) The Q of the circuit can be found by using either the ratio V_L/V_{rs} or X_L/r_s. Using $Q = V_L/V_{rs}$,

$$Q = \frac{27.9\ \text{V}}{100\ \text{mV}} = 279$$

Using $Q = X_L/r_s$,

$$Q = \frac{1116}{4\ \Omega} = 279$$

(c) If the ac resistance of the coil is increased to 12 Ω, the Q of the circuit will decrease:

$$Q = \frac{X_L}{r_s} = \frac{1116}{12\Omega} = 93$$

Q of a Series LCR Circuit versus Q of a Coil

Q of a Coil. The Q of the resonant circuit in Example 27-6 was shown to be 279 when the resistance of the coil was 4 Ω. If a coil was selected having an ac resistance of 12 Ω, the Q decreased to 93. This shows that to have a high-Q circuit, a coil having a low value of resistance must be used. A coil is considered to be a high-Q coil if it has a Q greater than 10 ($Q > 10$). *The Q of a series resonant circuit cannot be any greater than the Q of the inductor in the circuit.*

Q of a Circuit. If a series resistance is added into the circuit (other than changing the coil to increase the resistance), the Q of the circuit will be reduced. This is because the total current at resonance is reduced, with a resultant reduction in voltage magnification. The greater the resistance, the greater the reduction in current and Q of the circuit.

EXAMPLE 27-7

The Q of the coil and the circuit of Example 27-6 was 279. Find the new Q of the circuit if a resistance of 200 Ω is connected in series with L and C.

Solution: The Q of the circuit cannot be greater than 279. If a series resistance is added into the circuit, the effective resistance is increased and Q is decreased:

$$Q = \frac{X_L}{R + r_s} = \frac{1116\ \Omega}{200\ \Omega + 4\ \Omega} = 5.5$$

Using the ratio of V_L/V_{in} will also give the new Q. Since the current has been reduced, the voltage drop across L will be reduced, or

$$I = \frac{V}{Z} = \frac{V}{R} = \frac{100\ \text{mV}}{200\ \Omega + 4\ \Omega} \cong 0.5\ \text{mA}$$

and

$$V_L = IX_L = 0.5 \times 10^{-3}\ \text{A} \times 1116\ \Omega \cong 550\ \text{mV}$$

Therefore,

$$Q = \frac{V_L}{V_{in}} = \frac{550\ \text{mV}}{100\ \text{mV}} = 5.5$$

Effect of L/C Ratio on Q

With a higher inductance and smaller capacitance, X_L and X_C will be much higher at resonance. Since the Q of the circuit is dependent on the ratio of X_L to r_s, increasing the inductance of the coil will increase its reactance at resonance.

Practically, there are limitations to the amount of inductance and the minimum capacitance that can be used. Stray capacitance between the turns on the coil will limit the minimum capacitance. Besides stray capacitance, the turns on a coil are limited by the physical size of the coil as well as the resistance of the conductor due to its cross-sectional area. Typical Q values range up to 500 and higher.

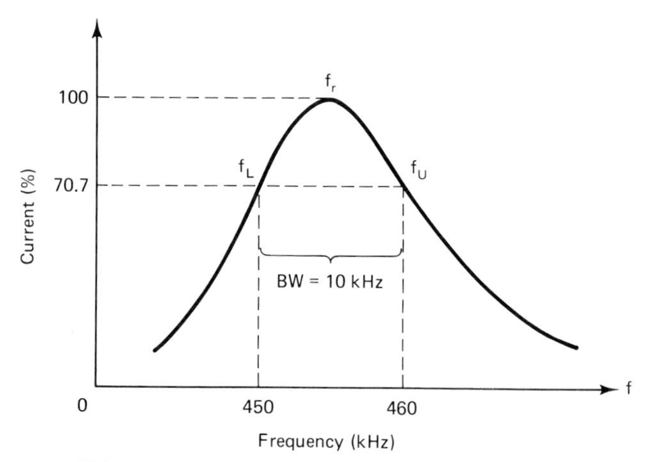

FIGURE 27-11 Bandwidth is measured between the 70.7% current or voltage response points, lower band frequency, f_L, and upper band frequency, f_U.

27-9 TUNED CIRCUIT BANDWIDTH

The bandwidth of a tuned circuit is of great importance in AM, FM, TV, and communication receivers and transmitters. For example, communication channels need a bandwidth of 2 kHz to ensure a proper quality of voice transmission and reception. For broadcasting an AM signal has a bandwidth of 10 kHz, while FM has a bandwidth of 200 kHz, to ensure the quality of audio transmission. The greatest bandwidth is required for a TV signal. Here a bandwidth of 6.0 MHz is required to ensure good picture and audio quality.

Bandwidth characteristic is most noticable in a TV set; the definition of the picture is greater in a TV set having a wide bandwidth and poorer in a TV set having a narrow bandwidth, since the eye is more critical than the ear.

Definition of Bandwidth

The frequency response curve of Fig. 27-11 shows that the current is maximum at resonance and falls off below and above the resonant frequency. The range of frequencies taken from the lower band frequency (f_L) to the upper band frequency (f_U) is called the bandwidth (BW) of the resonant circuit.

By definition the bandwidth of a tuned circuit is designated as that band of frequencies on either side of resonance (f_L and f_U) at which the current or voltage is 70.7% of or greater than its maximum resonant value. In equation format,

$$\text{BW} = f_U - f_L \quad \text{hertz} \tag{27-14}$$

For the response curve of Fig. 27-11 the bandwidth is

$$\text{BW} = f_U - f_L = 460 \text{ kHz} - 450 \text{ kHz} = 10 \text{ kHz}$$

The upper band frequency, f_U, and the lower band frequency, f_L, are also termed the *half-power* frequencies. This is because the power dissipation of the circuit resistance drops to one-half of its value at resonance.

This can be shown by using the resonant current of 25 mA in the circuit of Example 27-6. The power dissipated by the resistance at resonance will be

$$P = I^2R = (25 \times 10^{-3} \text{ A})^2 \times 4\ \Omega = 2.5 \text{ mW}$$

At the half-power frequencies the current falls to 70.7% of its resonant value; therefore, $I = 0.707 \times 25$ mA $= 17.7$ mA. The power dissipation at f_L and f_U will be

$$P = I^2R = (17.7 \times 10^{-3} \text{ A})^2 \times 4\ \Omega = 1.25 \text{ mW}$$

Effect of Q on Bandwidth and Selectivity

We have found that the Q of a resonant circuit cannot be greater than the Q of the coil. Conversely, connecting a series resistor into the circuit, and depending on its value, the Q of the resonant circuit can be decreased without changing the coil.

Bandwidth. Figure 27-12 shows the effect of Q on the bandwidth of a tuned circuit. For a low Q value in the range of 5, the bandwidth f_{L1} to f_{U1} at the 70.7% point will be very wide. This is the type of bandwidth required for TV video amplifiers. Note that video and FM-tuned IF amplifiers are stagger tuned to obtain the desired bandwidth. Stagger tuning is a method whereby a number of tuned IF amplifier stages have their resonant frequency stagger tuned (e.g., 41 MHz, 43 MHz, and 46 MHz).

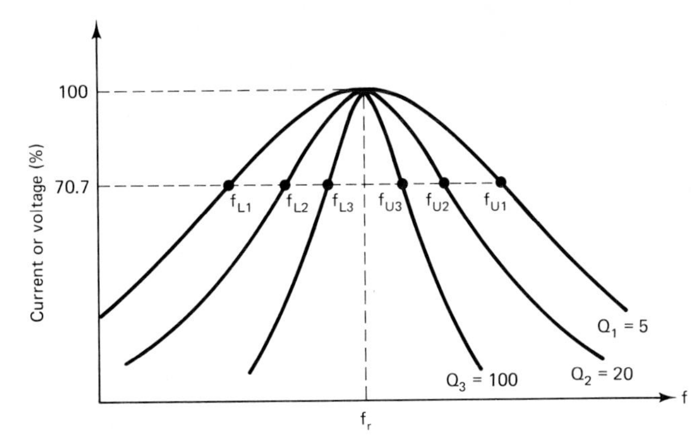

FIGURE 27-12 Bandwidth of a series *LC* resonant circuit for various *Q* values.

Increasing the Q to 20 narrows or decreases the BW. This type of bandwidth is suitable for FM and AM receiver-tuned IF amplifiers. A high-Q circuit will have a narrow bandwidth as shown by f_{L3} to f_{U3} in Fig. 27-12. Communication receivers and telephone channels require a narrow bandwidth in order to send and receive many voice channels simultaneously without interfering with one another.

A tuned or untuned circuit is considered to pass a range of frequencies at the 70.7% of maximum gain or Q point, termed bandwidth, and its gain is useful by the amplifier stage. Once the gain falls below this half-power or 70.7% point, its gain is too low for proper amplification by the following stage.

As shown previously, the bandwidth of a resonant circuit is dependent on the Q of the circuit. If the resonant frequency is increased, the bandwidth will increase, and vice versa. This is because the range in frequency below (f_L) and above (f_U) the resonant frequency has increased. Bandwidth is directly proportional to the resonant frequency f_r.

It was also shown that if the Q of the circuit is increased, the bandwidth was decreased, and vice versa. This means that the bandwidth is inversely proportional to Q. Putting f_r and Q in the relationship with bandwidth gives the equation

$$\text{BW} = \frac{f_r}{Q} \quad \text{hertz} \tag{27-15}$$

Since the resonant frequency is at the midpoint of the bandwidth frequency, the lower band frequency, f_L, can be found by subtracting one-half of the bandwidth frequency from the resonant frequency:

$$f_L = f_r - \frac{\text{BW}}{2} \quad \text{hertz} \tag{27-16}$$

Similarly, the upper band frequency, f_U, can be found by adding one-half of the bandwidth frequency to the resonant frequency:

$$f_U = f_r + \frac{\text{BW}}{2} \quad \text{hertz} \tag{27-17}$$

Selectivity. Since the tuned *LC* circuits and the amplifier stages reject frequencies below and above the half-power, or 70.7% of the gain point, the tuned circuit stage is said to have *selectivity*. A highly selective circuit requires that the Q of the circuit must be high, with resultant narrow bandwidth. Tuned stages of amplification having high selectivity must be used in voice communication receivers. These stages allow a specific communication channel to be accepted while rejecting accompanying channels for reception without interference. Conversely, TV video amplifier stages have low or poor selectivity since they must accept a wider range of frequencies at the high-power point.

Selectivity, then, is the ability of a tuned circuit or amplifier to reject frequencies above and below the half-power or 70.7%-of-gain point. Selectivity is determined by the Q and BW of the tuned circuit.

EXAMPLE 27-8

Find the bandwidth and the upper and lower band frequencies for the frequency response curves of Fig. 27-12 given that the resonant frequency is 455 kHz.

Solution: With a resonant frequency of 455 kHz, the bandwidth can be found using equation (27-15). For a Q of 5:

$$\text{BW} = \frac{f_r}{Q} = \frac{455\text{ kHz}}{5} = 91\text{ kHz}$$

$$f_L = f_r - \frac{\text{BW}}{2}$$

$$= 455 - \frac{91\text{ kHz}}{2} = 409.5\text{ kHz}$$

$$f_U = f_r + \frac{BW}{2}$$

$$= 455 + \frac{91\text{ kHz}}{2} = 500.5\text{ kHz}$$

For a Q of 20:

$$BW = \frac{f_r}{Q} = \frac{455\text{ kHz}}{20} = 22.75\text{ kHz}$$

$$f_L = f_r - \frac{BW}{2}$$

$$= 455 - \frac{22.75\text{ kHz}}{2} = 443.625\text{ kHz}$$

$$f_U = f_r + \frac{BW}{2}$$

$$= 455 + \frac{22.75\text{ kHz}}{2} = 466.375\text{ kHz}$$

For a Q of 100:

$$BW = \frac{f_r}{Q} = \frac{455\text{ kHz}}{100} = 4.55\text{ kHz}$$

$$f_L = f_r - \frac{BW}{2}$$

$$= 455 - \frac{4.55\text{ kHz}}{2} = 452.725\text{ kHz}$$

$$f_U = f_r + \frac{BW}{2}$$

$$= 455 + \frac{4.55\text{ kHz}}{2} = 457.275\text{ kHz}$$

The bandwidth calculations show that the bandwidth is inversely proportional to the Q of the circuit. When Q was increased from 5 to 20, a factor of 4, the BW was reduced by $\frac{1}{4}$; similarly, an increase of Q from 5 to 100 will decrease the BW by a factor of 20 to 4.55 kHz. Note that the BW must be divided by 2 in order to determine the lower and upper band frequencies, which are centered above and below the resonant frequency.

EXAMPLE 27-9

For resonant frequencies of 1 MHz and 10 MHz, find the bandwidth with a Q of 10 to show that the bandwidth is directly proportional to frequency.

Solution: For a Q of 10 and a resonant frequency of 1 MHz, using equation (27-16) gives a bandwidth of

$$BW = \frac{f_r}{Q} = \frac{1\text{ MHz}}{10} = 100\text{ kHz}$$

A Q of 10 and a resonant frequency of 10 MHz gives

$$BW = \frac{f_r}{Q} = \frac{10\text{ MHz}}{10} = 1000\text{ kHz}$$

The bandwidth calculations show that when the resonant frequency increases from 1 MHz to 10 MHz, a factor of 10, the BW will increase by a factor of 10, to 1000 kHz; and so on. This shows that the bandwidth is directly proportional to the resonant frequency.

27-10 PLOTTING A GENERAL RESPONSE CURVE

A series *LCR* resonant frequency response curve such as the curve shown in Fig. 27-11 can be plotted on graph paper by calculating X_L, X_C, Z, I, and V_C on either side of the resonant frequency (see Table 27-1). From these calculations, either a Z, I, or V_C response curve could be plotted and the bandwidth determined from the graph. This graphical method involves quite a few calculations and is lengthy. A technique can be used to simplify plotting a response curve. This technique uses resonant circuit theory based on one assumption: that for a small frequency change on either side of resonance, the Q of the circuit remains constant at the resonant value. With this assumption the relative output drops off (within 1% error) as follows:

Response gain assumptions

1. At f_r = 100% of gain
2. At $f_U - f_L$ = 1 BW = 70.7% of gain
3. At $f_U - f_L$ = 2 BW = 44.7% of gain
4. At $f_U - f_L$ = 3 BW = 32% of gain
5. At $f_U - f_L$ = 4 BW = 24% of gain

EXAMPLE 27-10

Plot a resonant frequency response curve for a series *LCR* circuit having an L of 235 μH with $r_s = 11.8\ \Omega$, and a C of 107 pF.

Solution: The resonant frequency, the Q, and the bandwidth must be found first.

$$f_r = \frac{0.159}{\sqrt{LC}}$$

$$= \frac{0.159}{\sqrt{235 \times 10^{-6}\ \text{H} \times 107 \times 10^{-12}\ \text{F}}}$$

$$\cong 1000\ \text{kHz or} \approx 1\ \text{MHz}$$

$$X_L = 2\pi f L$$

$$= 6.28 \times 1 \times 10^6\ \text{Hz} \times 235 \times 10^{-6}\ \text{H}$$

$$= 1476\ \Omega$$

$$Q = \frac{X_L}{r_s} = \frac{1476}{11.8} = 125$$

$$\text{BW} = \frac{f_r}{Q} = \frac{1000\ \text{kHz}}{125} = 8.0\ \text{kHz}$$

1. At the resonant frequency of 1000 kHz, the gain will be 100%.

2. *At one bandwidth* (1 BW), the gain at the lower and upper band frequencies is 70.7%, and

$$f_L = f_r - \frac{1\ \text{BW}}{2}$$

$$= \frac{8\ \text{kHz}}{2} - 1000\ \text{kHz} = 996\ \text{kHz}$$

$$f_U = f_r + \frac{1\ \text{BW}}{2}$$

$$= 1000 + \frac{8\ \text{kHz}}{2} = 1004\ \text{kHz}$$

3. *At two bandwidths* (2 BW), the gain at the lower and upper band frequencies is 44.7%, and

$$f_L = f_r - \frac{2\ \text{BW}}{2}$$

$$= 1000 + \frac{16\ \text{kHz}}{2} = 992\ \text{kHz}$$

$$f_U = f_r + \frac{2\ \text{BW}}{2}$$

$$= 1000 + \frac{16\ \text{kHz}}{2} = 1008\ \text{kHz}$$

4. *At three bandwidths* (3 BW), the gain at the lower and upper band frequencies is 32%, and

$$f_L = f_r - \frac{3\ \text{BW}}{2}$$

$$= 1000 - \frac{24\ \text{kHz}}{2} = 988\ \text{kHz}$$

$$f_U = f_r + \frac{3\ \text{BW}}{2}$$

$$= 1000 + \frac{24\ \text{kHz}}{2} = 1012\ \text{kHz}$$

5. *At four bandwidths* (4 BW), the gain at the lower and upper band frequencies is 24%, and

$$f_L = f_r - \frac{4\ \text{BW}}{2}$$

$$= 1000 - \frac{32\ \text{kHz}}{2} = 984\ \text{kHz}$$

$$f_U = f_r + \frac{4\ \text{BW}}{2}$$

$$= 1000 + \frac{32\ \text{kHz}}{2} = 1016\ \text{kHz}$$

The plotted graph in terms of percentage gain is shown in Fig. 27-13.

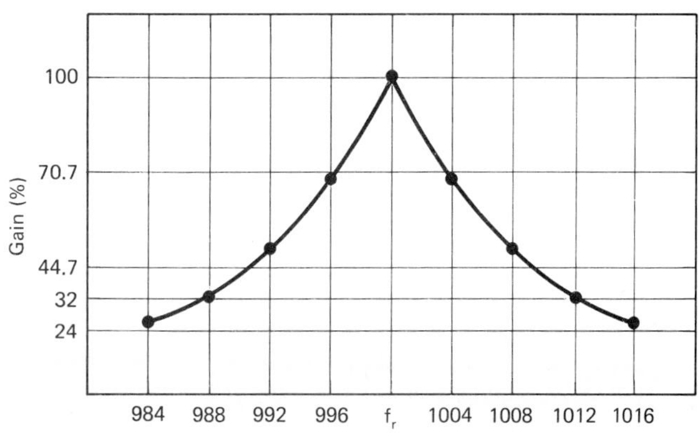

FIGURE 27-13 Frequency response curve for Example 27-10.

27-11 REACTANCE CHART

The reactance chart in Fig. 27-14 is useful to approximate values of X_L and X_C given the inductance or capacitance at a specified frequency. It is also useful to find combinations of L and C that would give a specific resonant frequency. Conversely, given a specific L and C, we can find the resonant frequency as well as X_L and X_C at resonance.

EXAMPLE 27-11

Use the reactance chart of Fig. 27-14 to find the resonant frequency, X_L and X_C, for the circuit in Example 27-3.

Solution: The circuit of Example 27-3 has $L = 18$ mH $= 0.02$ H, $C = 22$ nF $\cong 0.02$ μF. You will need to pick the closest diagonal line to 0.018 H or 0.02 H. Since the main division line is 0.01 H, the next diagonal line is 0.02 H, 0.03 H, and so on, up to 0.1 H. Similarly, you will need to pick out the 22 nF $\cong$ 0.02 μF diagonal line. Again the main division line is 0.01 μF, so that the next diagonal line is 0.02 μF, and so on down to 0.1 μF.

Now pick the intersection point of the 0.02-H and the 0.02-μF lines. Drop a vertical line from this intersection point to find the frequency of 8 kHz on the bottom scale. The horizontal line intersecting this point will give the reactance in ohms of X_C and X_L, in this case approximately 1 kΩ.

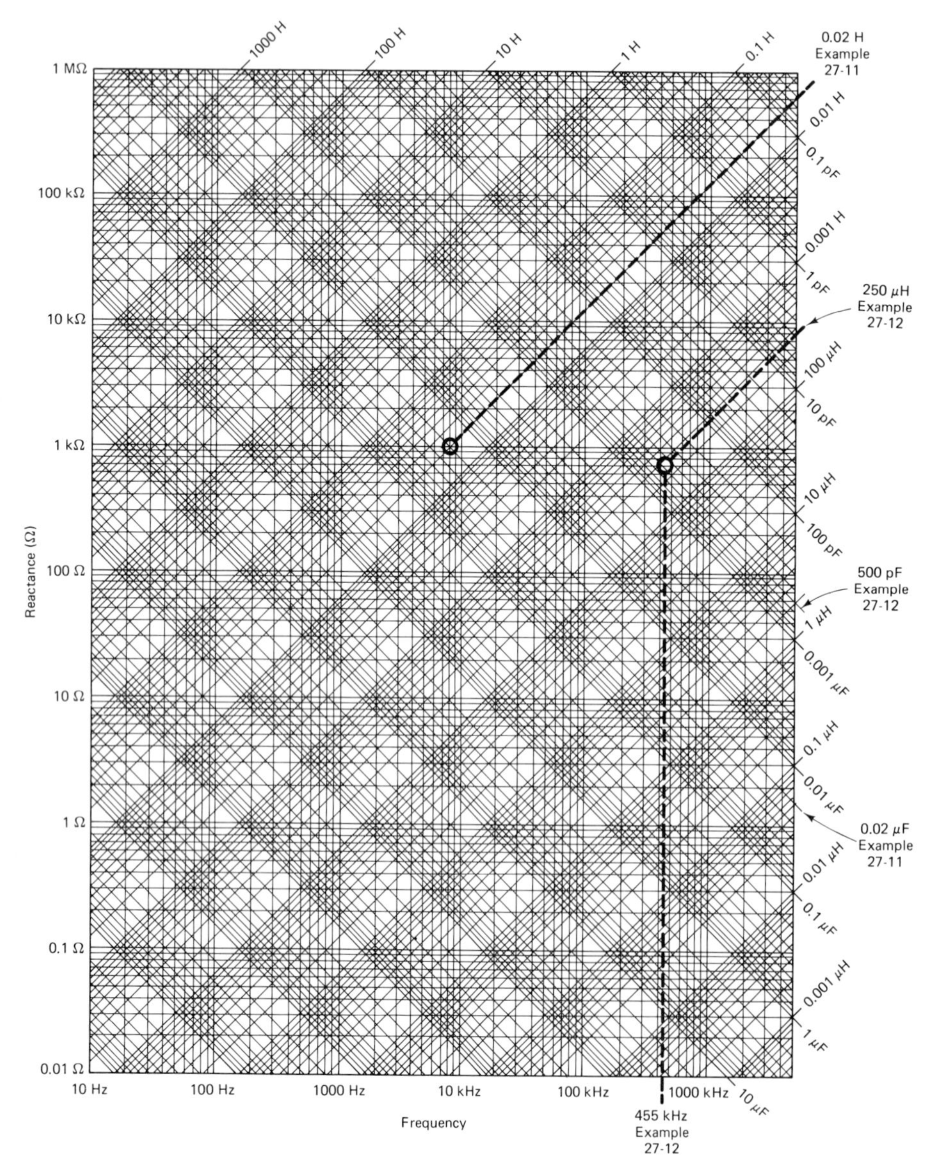

FIGURE 27-14 Reactance chart. Inductive and capacitive reactance over a wide range of operating frequencies.

EXAMPLE 27-12

Use the reactance chart to find the inductance required to resonate an LC circuit to 450 kHz with a 500-pF capacitor.

Solution: The inductance of the coil can be located at the intersection of the 500-pF diagonal line and the vertical 450-kHz line. The intersecting lines gives a value of 250 μH.

SUMMARY

The ac voltage across a capacitor lags the current by 90°.

The ac voltage across an inductor leads the current by 90°.

A resistor is nonreactive, so that its voltage drop is in phase with the current.

In a series LCR circuit V_L will be 180° out of phase with V_C and they cancel.

The phasor sum of a series V_L, V_C, and V_R is equal to

$$V_A = \sqrt{V_R^2 + (V_L - V_C)^2} \quad \text{volts} \tag{27-1}$$

The phase angle of V_A to I can be found using the equation

$$\theta = \tan^{-1} \frac{V_L - V_C}{V_R} = \frac{V_X}{V_R} \tag{27-2}$$

When V_C is greater than V_L, the phase angle V_A to I is negative to show that the V_A is lagging I:

$$\theta = \tan^{-1} \left(-\frac{V_L - V_C}{V_R} \right) = \left(-\frac{V_X}{V_R} \right) \tag{27-3}$$

In a series LCR circuit X_L cancels X_C and the impedance of the circuit is

$$Z = \sqrt{R^2 + (X_L - X_C)^2} \quad \text{ohms} \tag{27-4}$$

Impedance is an Ohm's law ac opposition and can be found using the equation

$$Z = \frac{V_A}{I_T} \quad \text{ohms} \tag{27-5}$$

The phase angle between Z and R is equal to the phase angle between V_A and I and is found using

$$\theta_Z = \theta_I = \tan^{-1} \frac{X}{R} \tag{27-6}$$

When X_C is greater than X_L, X will be capacitive and V_A will lag the current, therefore

$$\theta_z = \tan^{-1} \left(-\frac{X_L - X_C}{R} \right) = \left(-\frac{X}{R} \right) \tag{27-7}$$

Electrical resonance can be defined as a circuit that is in tune with or having the same natural frequency of oscillations as the source voltage. Impedance at resonance is

$$Z = R \qquad (27\text{-}8)$$

The voltage drops about an *LCR* circuit at resonance are

$$V_R = V_A \qquad (27\text{-}9)$$

and

$$V_C = V_L = IX_L$$

and are 180° out of phase.

To find the resonant frequency, we use the equation

$$f_r = \frac{0.159}{\sqrt{LC}} \quad \text{hertz} \qquad (27\text{-}10)$$

To find C we rearrange equation (27-11):

$$C = \frac{1}{4\pi^2 f_r^2 L} \quad \text{farads} \qquad (27\text{-}11)$$

To find L, we rearrange equation (27-11),

$$L = \frac{1}{4\pi^2 f_r^2 C} \quad \text{henrys} \qquad (27\text{-}12)$$

Voltage magnification takes place at resonance and is a large increase in voltage drop across X_L and X_C due to the increase in current and reactance at resonance.

The primary of a tuned IF transformer has parallel resonance characteristics.

The secondary of a tuned IF transformer has series resonance characteristics.

The quality factor (Q) of a series resonant circuit is a measure of the voltage magnification or gain of the circuit and is equal to

$$Q = \frac{X_L}{r_s} \qquad (27\text{-}13)$$

The Q of the series *LCR* circuit cannot be any greater than the Q of the inductor.

Q is dependent on the L/C ratio of the circuit.

By definition, the bandwidth of a tuned circuit is designated as that band of frequencies on either side of resonance (f_L and f_U) at which the current or voltage is 70.7% of its maximum resonant value.

$$\text{BW} = f_U - f_L \quad \text{hertz} \qquad (27\text{-}14)$$

The BW of a resonant circuit is dependent on the Q of the circuit.

$$\text{BW} = \frac{f_r}{Q} \quad \text{hertz} \qquad (27\text{-}15)$$

Selectivity is the ability of a tuned circuit or amplifier to reject frequencies above and below the half-power or 70.7% of gain point.

REVIEW QUESTIONS

27-1. Explain briefly why the voltage across a capacitor will lag the current by 90°.

27-2. Explain briefly why the voltage across an inductor leads the current by 90°.

27-3. An ac voltage drop across a resistor is in phase with the current. Explain.

27-4. What is the phase difference between V_L and V_C in a series *LCR* circuit?

27-5. The voltage drop across an X_L of 1 kΩ in a series *LCR* current was found to be 10 V. If X_C is 3 kΩ and *R* is 100 Ω, find:

(a) The voltage drop across X_C and *R*.
(b) The applied voltage V_A.
(c) The phase angle between V_A and *I*.
(d) The impedance of the circuit.
(e) If the frequency of the source is 1 kHz, find the inductance of the coil.
(f) If the frequency of the source is 1 kHz, find the capacitance of the capacitor.

27-6. The impedance of a series *LCR* circuit at a resonant frequency is equal to its pure resistance, that is, $Z = R$, equation (27-8). Explain.

27-7. Explain how the circuit of Question 27-5 can be brought into resonance without changing the frequency of the source.

27-8. A 250-μH coil having a resistance of 10 Ω is connected in series with a 500-pF capacitor across 100 mV. Find:

(a) The resonant frequency of the circuit.
(b) The impedance at resonance.
(c) The circuit current at resonance.
(d) The voltage across *R*, X_L, and X_C.

27-9. A tuned resonant circuit has a 10-mH coil connected in series with an unknown capacitor. The circuit resonates at 5 kHz. Find the capacitance of the capacitor.

27-10. Voltage magnification takes place at resonance. Explain.

27-11. Explain why the secondary characteristics of a tuned RF or IF transformer are that of a series resonant circuit.

27-12. The *Q* of a series resonant circuit cannot be any greater than the *Q* of the inductor in the circuit. Explain.

27-13. The *Q* of a series resonant circuit can be altered by changing the L/C ratio. Explain.

27-14. The *Q* of a series resonant circuit can be altered by increasing the series resistance. Explain.

27-15. What relationship is the *Q* of the circuit to the voltage magnification of V_C and V_L at resonance?

27-16. What is the bandwidth of a tuned circuit?

27-17. A greater or smaller *Q* will affect bandwidth. Explain.

27-18. An increase or decrease in frequency will affect bandwidth. Explain.

27-19. Find the bandwidth for a series *LCR* resonant circuit having an *L* of 120 μH, with $r_s = 24$ Ω and a *C* of 200 pF.

27-20. What are the upper and lower frequencies at the half-power point in the resonant circuit of Question 27-19?

27-21. How would you decrease the selectivity of the series *LCR* circuit in Question 27-19?

27-22. Plot the resonant frequency response curve for the series *LCR* circuit of Question 27-19 by using the response gain assumption.

27-23. Use the reactance chart of Fig. 27-14 to find the value of capacitance required to resonant a 1-mH coil at 50 kHz.

SELF-EXAMINATION

Choose the most suitable word or phrase to complete each statement. Refer to the schematic diagram of Fig. 27-15 for Questions 27-1 to 27-6.

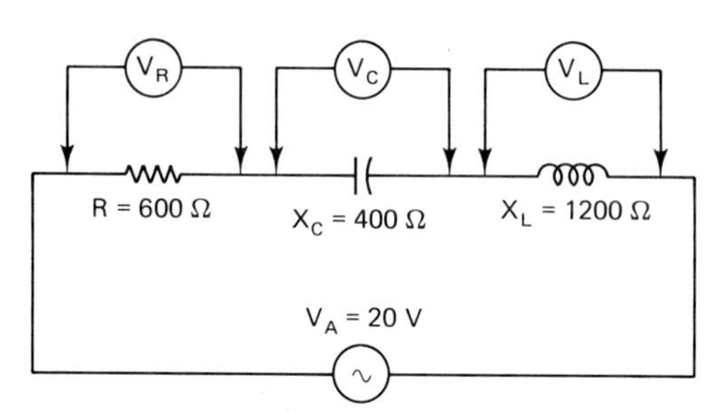

FIGURE 27-15 Schematic.

27-1. What is the impedance of the circuit?

27-2. The circuit or line current is ________ milliamperes.

27-3. What is the voltage drop V_R?

27-4. What is the voltage drop V_L?

27-5. What is the voltage drop V_C?

27-6. Determine the phase angle V_A to I.

27-7. A turned resonant circuit has a 200-μH inductance in series with an unknown capacitance. The circuit resonates at 200 kHz. Find the value of the capacitor.

Refer to the following circuit for Questions 27-8 to 27-14: A 200-pF capacitor and a 700-μH coil having an ac resistance of 40 Ω are connected in series across a voltage source.

27-8. Find the resonant frequency.

27-9. Calculate X_L at resonance.

27-10. Calculate X_C at resonance.

27-11. Find Z at resonance.

27-12. Given that the circuit current at resonance is 75 mA, find the value of the source voltage.

27-13. Find the voltage drop across X_L.

27-14. What is the net voltage across L and C?

27-15. The Q of a series resonant circuit is termed the ________ ________ of the circuit.

27-16. The primary characteristics of a tuned transformer are those of a ________ resonant circuit.

27-17. The secondary characteristics of a tuned transformer are those of a ________ resonant circuit.

27-18. The L/C ratio will determine the ________ of a series resonant circuit.

27-19. The Q of a series resonant circuit cannot be any greater than the Q of the ________.

27-20. The Q of a series resonant circuit can be reduced by adding a ________ in series with L and C.

27-21. The range of frequency at which the voltage or current is 70.7% of its maximum value or greater is termed the ________ ________ of a tuned circuit.

27-22. The upper band frequency, f_U, and the lower band frequency, f_L, are also termed the ________ ________ frequencies.

27-23. The selectivity of a tuned circuit must be high in order to obtain a (*narrow, wide, normal*) bandwidth.

27-24. ________ is the ability of a tuned circuit to reject frequencies above and below the half-power or 70.7%-of-gain point.

27-25. Selectivity is determined by the ________ and ________ of the tuned circuit.

Refer to the following circuit for Questions 27-27 to 27-31: A 25-nF capacitor and a 0.01-H inductance having an ac resistance of 100 Ω are found to be resonant across a 10-V 10-kHz source.

27-26. Find the Q of the circuit.

27-27. Find the bandwidth of the tuned circuit.

27-28. What is the lower band frequency at the half-power point?

27-29. What is the upper band frequency at the 70.7% point of maximum current or voltage?

27-30. What value of Q would increase the bandwidth to 3184 Hz?

27-31. What ac resistance value for the coil would be needed to decrease the bandwidth to 795 Hz?

ANSWERS TO PRACTICE QUESTIONS

27-1

1. lags
2. leads
3. resistor
4. in phase
5. 180°
6. $X_L = 2\pi fL = 6280\ \Omega$
7. $X_C = \dfrac{0.159}{fc} = 1590\ \Omega$
8. $Z = \sqrt{R^2 + (X_L - X_C)^2} = 4861\ \Omega$
9. $V = IX_L = 24.5$ V
10. $V = IX_C = 6.2$ V
11. $V_s = IZ \cong 19$ V
12. $\theta = \tan^{-1}\dfrac{X_L - X_C}{R}$; therefore, $\theta \cong 75°$

27-2

1. source voltage
2. 180°
3. resistance, R
4. resistance
5. $f_r = \dfrac{0.159}{\sqrt{LC}} = 15.66$ kHz
6. $X_L = 2\pi fL = 462\ \Omega$
7. $X_C = X_L = 462\ \Omega$
8. $Z = R = 65\ \Omega$
9. $V_s = IZ \cong 6$ V
10. $V_L = IX_L = 42.5$ V
11. $V_R = V_S = 6$ V
12. $V_X = V_L - V_C = 0$ V

28

Parallel Resonant Circuit

INTRODUCTION

In Chapter 27 we discovered a significant application of the series resonant circuit to be the secondary of a tuned IF or RF transformer. The primary of the same transformer becomes a parallel resonant tuned circuit since both L and C are connected in parallel with the source voltage, such as a transistor amplifier. The frequency is thus "tuned in" or resonated in both the primary and secondary of the IF or RF transformer and forms a high-Q, selective bandpass circuit. The parallel resonant circuit has an important application in a band-stop or rejector-type filter circuit.

In this chapter we study the parallel LCR and LC circuit first, then study the parallel resonant circuit.

MAIN TOPICS IN CHAPTER 28

OBJECTIVES

After studying Chapter 28, the student should be able to

1. Master the fundamental concepts of current, voltage, and impedance in a parallel LC circuit.
2. Solve parallel LC current and impedance problems.
3. Master the concept of current and voltage phase relationship in the parallel L and C circuit.
4. Master the concepts of parallel resonances, bandwidth, and Q relationships as well as the effect on Q and bandwidth with a damping resistor.
5. Solve parallel resonant circuit current and phase relationship, impedance, bandwidth, and Q problems.

28-1 PARALLEL *LCR* CIRCUIT

In Chapter 27 we discovered that the secondary of an intermediate-frequency (IF) or radio-frequency (RF) transformer was a tuned series LC resonant circuit, while the primary was tuned parallel LC resonant circuit. The parallel LCR circuit of Fig. 28-1 shows that L and C are connected in parallel; therefore, the source voltage is common to L, C, and R.

To analyze the parallel LCR circuit of Fig. 28-1, we must first recall and restate the basic phase relationship of current and voltage for an inductor, capacitor, and resistor connected across an ac voltage source.

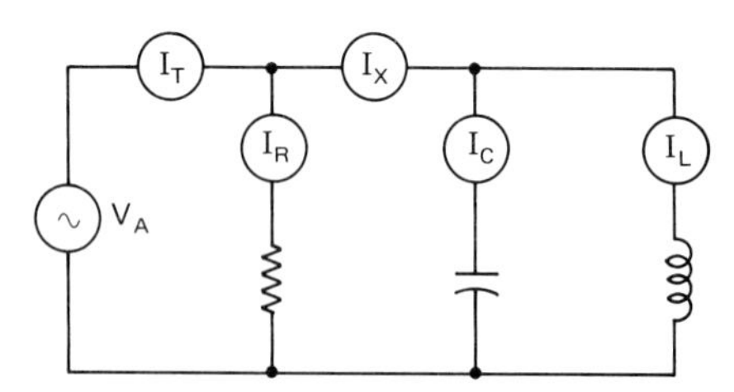

FIGURE 28-1 Parallel *LCR* circuit.

28-2 PHASE RELATIONSHIP OF I_C, I_L, AND I_R TO V_A

Phase Relationship of I_L to V_A

Although you have by now grasped the phase relationship of current to voltage in an inductor and capacitor, it is so important that it bears repeating. The induced emf produced by an inductor opposes the current change so that the current will lag the voltage across the inductor by 90°. In this case the voltage across the inductor is the applied voltage, so that the current lags the applied voltage. Conversely, the applied voltage *leads the inductive branch current by 90°*, as shown by the representative phasor diagram of Fig. 28-2(a).

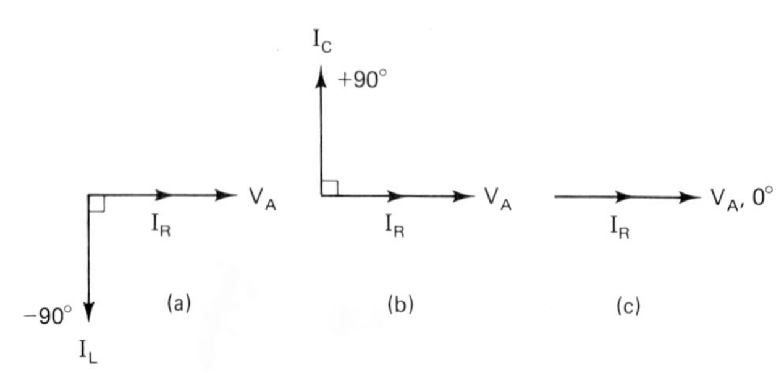

FIGURE 28-2 (a) I_L lags V_A by 90°. (b) I_C leads V_A by 90°. (c) I_R is in phase with V_A.

Phase Relationship of I_C to V_A

A voltage connected across an uncharged capacitor will cause a charging current surge which then builds up a coulomb charge. The coulomb charge, in turn, builds up a counter emf that opposes the charging current into the capacitor. This current flow into the capacitor before the buildup of a voltage across the capacitor causes the current to lead the voltage by 90°. Conversely, the applied voltage of a parallel capacitive branch circuit will lag the current in that branch circuit by 90°, as shown by the representative phasor diagram of Fig. 28-2(b).

Phase Relationship of I_R to V_A

Since the resistance is a nonreactive component, the current will be in phase with the voltage across the resistor, as shown by the representative phasor diagram of Fig. 28-2(c).

Combining the phasor diagram of the parallel LCR circuit will give the representative phasor diagram shown in Fig. 28-3(a). The phasor diagram shows that I_C leads the applied voltage V_A by 90°, while I_L lags the applied voltage by 90°. Most important, however, is the fact that I_C is 180° out of phase with I_L and they cancel each other. When I_C is equal to I_L, the total current flow from the source is due to the resistance.

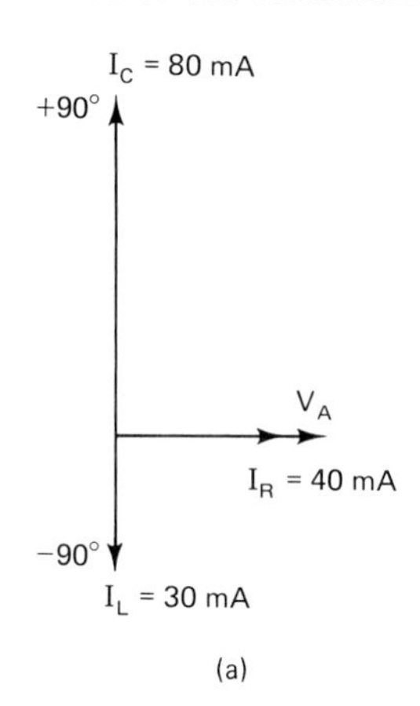

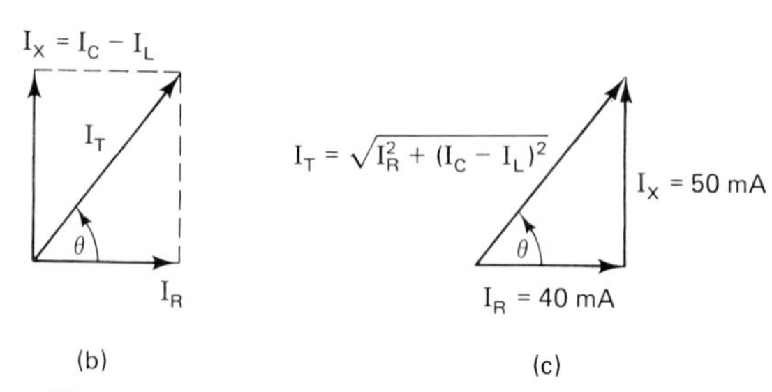

FIGURE 28-3 (a) Representative phasor diagram for current and applied voltage of a parallel *LCR* circuit. (b) I_L cancels I_C and leaves I_X. (c) I_T is the phasor sum of I_R and I_X in the current triangle.

The phasor diagram of Fig. 28-3(b) shows that a reactive current I_X (either I_C or I_L) is left when I_C and I_L are not equal to each other. If both I_C and I_L are equal, I_X

would be zero. In the phasor diagram of Fig. 28-3(b), I_X is capacitive and must be summed with due consideration of the phase angle θ difference between I_X and I_R. The resultant current I_T is the phasor sum of I_X and I_R, as shown in the current triangle of Fig. 28-3(c) and is

$$I_T = \sqrt{I_R^2 + (I_C - I_L)^2} \quad \text{amperes} \qquad (28\text{-}1)$$

Phase Angle I_T to V_A

The phasor diagram of Fig. 28-3(b) shows net current (I_T) to be leading V_A since I_X is capacitive. To find this phase angle, we can use the tan function equation,

$$\tan \theta = \frac{I_C - I_L}{I_R} \qquad (28\text{-}2a)$$

Therefore,

$$\theta_I = \tan^{-1} \frac{I_X}{I_R} \qquad (28\text{-}2)$$

When I_L is greater than I_C, I_X will be inductive and lagging V_A; therefore, the angle θ is equal to

$$\tan \theta_I = -\left(\frac{I_C - I_L}{I_R}\right) \qquad (28\text{-}3a)$$

Therefore,

$$\theta_I = \tan^{-1}\left(-\frac{I_X}{I_R}\right) \qquad (28\text{-}3)$$

EXAMPLE 28-1

The voltage across a parallel *LCR* circuit was found to be 500 mV. If $R = 100\ \Omega$, $X_L = 1\ \text{k}\Omega$, and $X_C = 2\ \text{k}\Omega$ find:

(a) The current through each branch and the total circuit current.

(b) The phase angle between V_A and I_T.

Solution:

(a) The voltage is common across all parts of the parallel *LCR* circuit; therefore, the current I_R, I_L, and I_C can be found using Ohm's law.

$$I_R = \frac{V_A}{R} = \frac{500\text{ mV}}{100} = 5\text{ mA}$$

$$I_L = \frac{V_A}{X_L} = \frac{500\text{ mV}}{1000} = 0.5\text{ mA}$$

$$I_C = \frac{V_A}{X_C} = \frac{500\text{ mV}}{2000} = 0.25\text{ mA}$$

The total current can be found using equation (28-1).

$$I_T = \sqrt{I_R^2 + (I_C - I_L)^2}$$

$$= \sqrt{(5\text{ mA})^2 + (0.25\text{ mA} - 0.5\text{ mA})^2}$$

$$= 5\text{ mA}$$

(b) Since, $I_C - I_L = I_X = -0.25$ mA and is very small compared to the resistive current of 5 mA, the total current in the circuit is mostly resistive and will be almost in phase with the applied voltage. We can use equation (28-3) to find the phase angle between I_T and V_A.

$$\theta_I = \tan^{-1}\left(-\frac{I_X}{I_R}\right) = -\frac{0.25\text{ mA}}{5\text{ mA}}$$

$$= -2.8^\circ \quad \text{(lagging)}$$

28-3 IMPEDANCE IN THE PARALLEL *LCR* CIRCUIT

The parallel *LCR* circuit presents three types of ac opposition to current flow: inductive, capacitive, and resistive. The net or total impedance can be found by using Ohm's law, with due consideration to the total current found using equation (28-1), so that

$$Z = \frac{V_A}{I_T} \quad \text{ohms} \qquad (28\text{-}4)$$

EXAMPLE 28-2

What is the impedance of the parallel *LCR* circuit given in Example 28-1?

Solution: Since the total current was calculated with due consideration of the phase differences, the Ohm's law equation (28-4) can be used to find the circuit impedance:

$$Z = \frac{V_A}{I_T} = \frac{500\text{ mV}}{5\text{ mA}} = 100\ \Omega$$

Note that when the reactance is at least 10 times greater than the parallel resistance, the circuit current will be mostly through the resistor. The impedance that the source "sees" will be mostly resistive and the circuit current will be nearly in phase with the applied voltage.

PRACTICE QUESTIONS 28-1

Choose the most suitable word or phrase to complete each statement.

1. In a parallel LCR circuit the ________ current will be in phase with the applied voltage.
2. In a parallel LCR circuit the ________ current will be lagging the applied voltage by 90°.
3. In a parallel LCR circuit the ________ current will be leading the applied voltage by 90°.

Given the following circuit, find each value for Questions 4 to 10: A 1-k ohm resistor, a 100-mH inductor, and a 0.01-μF capacitor are connected in parallel; the source voltage is 10 V at 10 kHz.

4. Find the resistor branch current.
5. Find the inductor branch current.
6. Find the capacitor branch current.
7. Find the total circuit current.
8. Find the phase angle between the total current and the applied voltage.
9. Is the total current in the circuit leading or lagging the applied voltage?
10. Find the impedance of the circuit using the Ohm's law equation.

28-4 PARALLEL *LC* NONRESONANT CIRCUIT

In the parallel LC circuit of Fig. 28-4, the inductor is considered to be ideal; that is, it has no internal resistance and the parallel resistance has been removed. This somewhat simplifies and approaches the actual parallel tuned resonant circuit. As in the case of any parallel circuit, the applied voltage at the source is common to both the inductance and the capacitance.

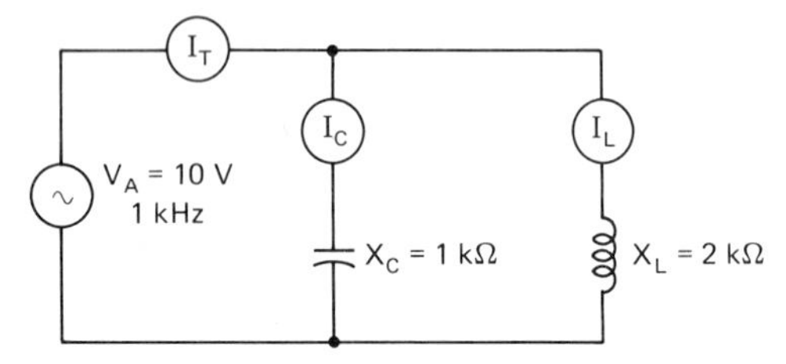

FIGURE 28-4 Parallel *LC* circuit.

The inductive branch current is 180° out of phase with the capacitive branch current [see the phasor diagram, Fig. 28-3(a)], so that I_L cancels I_C. If the inductive current is not equal to the capacitive current, the remaining current, I_X, will be the total current, I_T, flowing into and out of the source. In equation form, the net current of a parallel LC circuit can be found by the simple difference of I_C and I_L:

$$I_X = I_T = I_C - I_L \qquad \text{amperes} \qquad (28\text{-}5)$$

EXAMPLE 28-3

Find the total circuit current and impedance for the parallel LC circuit of Fig. 28-4.

Solution: The branch currents I_C and I_L can be found using Ohm's law, since V_A is common to both X_L and X_C.

$$I_L = \frac{V_A}{X_L} = \frac{10\text{ V}}{2\text{ k}\Omega} = 5\text{ mA}$$

$$I_C = \frac{V_A}{X_C} = \frac{10\text{ V}}{1\text{ k}\Omega} = 10\text{ mA}$$

The total circuit current can be found using equation (28-5):

$$I_T = I_C - I_L = 5\text{ mA} - 10\text{ mA} = -5\text{ mA}$$

The negative sign indicates that the circuit current is inductive and therefore will lag the applied voltage.

Once the circuit current is found, the impedance can be calculated by using the Ohm's law equation (28-4).

$$Z = \frac{V_A}{I_T} = \frac{10\text{ V}}{5\text{ mA}} = 2\text{ k}\Omega$$

For a parallel LC circuit in the example of a nonresonant circuit above, the impedance can also be found by the product-over-sum method, where the sum is actually the simple difference between X_L and X_C. Since I_C and I_L are 180° out of phase, they cancel, so X_L and X_C also cancel.

Product-Over-Difference Method for Impedance of a Nonresonant *LC* Circuit

$$Z = \frac{\text{product}}{\text{difference}} = \frac{X_L \times X_C}{X_L - X_C} \qquad \text{ohms} \qquad (28\text{-}6)$$

EXAMPLE 28-4

Find the impedance of the circuit in the schematic of Fig. 28-4 by using the product-over-difference method.

> ***Solution:*** Using the product-over-difference equation (28-6) for impedance gives
>
> $$Z = \frac{X_L X_C}{X_L - X_C} = \frac{2\text{ k}\Omega \times 1\text{ k}\Omega}{2\text{ k}\Omega - 1\text{ k}\Omega} = 2\text{ k}\Omega$$

28-5 PARALLEL *LC* TANK CIRCUIT AT RESONANCE

When X_L is equal to X_C in the parallel *LC* circuit of Fig. 28-5, the branch currents I_L and I_C will be equal. Since they are 180° out of phase with each other, they cancel. In effect, then, *X_L cancels X_C and the circuit is said to be resonant,* as was the case in the series resonant circuit. The frequency at which X_L equals X_C is called the resonant frequency, f_r. As in the case of the series resonant circuit, the resonant frequency can be determined using equation (27-11b):

$$f_r = \frac{1}{2\pi\sqrt{LC}} \quad \text{hertz} \qquad (27\text{-}11b)$$

FIGURE 28-5 Parallel resonant circuit.

LC Tank Circuit

Examine the parallel *LC* circuit of Fig. 28-6(a). At the instant the switch is closed, the capacitor appears as a short circuit to the source voltage since it is in an uncharged state. The capacitor charges up to its peak charge during the first quarter-cycle of the ac input voltage. At the same instant, the counter emf produced by the inductor because of the changing current makes it appear as an open circuit to the source as well as to the capacitor.

Once the charging current reaches its peak value, there is no further change in current, and the inductor's counter emf falls to zero. The inductor now appears as a short circuit [see Fig. 28-6(b)] to the capacitor, which proceeds to discharge its stored charge into the inductor. The stored electrostatic charge of the capacitor flows through the inductor and is transformed into a stored electromagnetic charge or energy in the inductor. The discharged capacitor now appears as a short circuit across the inductor. The inductor's magnetic field collapses and transforms its energy back to recharge the capacitor.

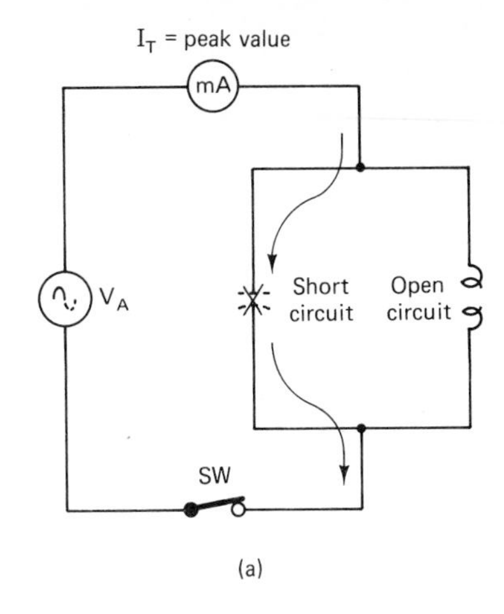

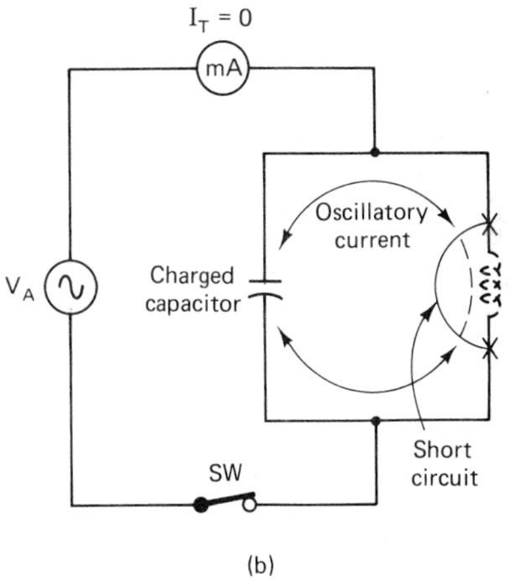

FIGURE 28-6 (a) Capacitor appears as a short circuit during first-quarter cycle of ac input voltage, while inductor appears as an open circuit. (b) The fully charged capacitor now discharges into the inductor, which appears as a short circuit across the capacitor.

This exchange of energy from the capacitor to the inductor and then back to the capacitor is termed the *oscillatory current.* Since the oscillatory current is contained in the parallel *LC* circuit, the circuit is termed a "tank" circuit.

The rate or frequency at which the tank circuit current oscillates is determined by the source frequency. Ideally, at resonance when X_L equals X_C, the oscillatory current given up by the capacitor will be equal to the current taken up by the inductor, and vice versa.

Tank Circuit Oscillatory Voltage

Recall that electrical resonance is defined as an *LC* circuit that is in tune with or has the same natural frequency of oscillation as the source voltage. In the tank circuit of Fig. 28-6, the source or applied voltage drives the charging current for the first quarter-cycle only. Once the capacitor is fully charged to peak value, the current oscillations persist only inside the LC tank circuit.

The voltage across the LC tank circuit due to the current oscillations will be 180° out of phase with V_A and cancels V_A as depicted in Fig. 28-7. Because the oscillatory tank circuit voltage opposes the source voltage, the source current is zero, and therefore by Ohm's law the impedance of the tank circuit is infinite.

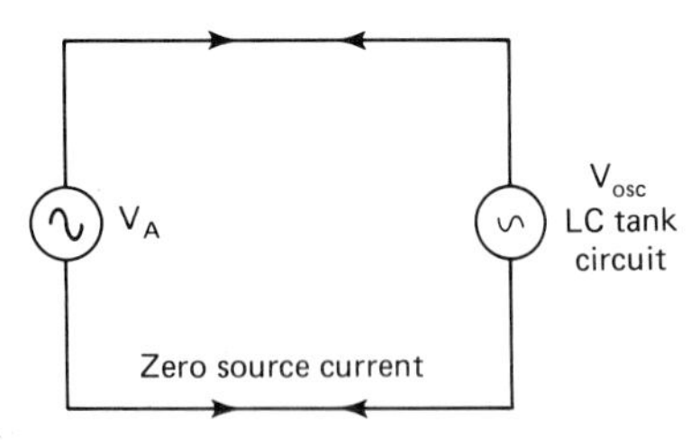

FIGURE 28-7 Oscillatory voltage of the tank circuit is 180° out of phase with V_A and cancels V_A.

28-6 PRACTICAL IMPEDANCE OF AN *LC* TANK CIRCUIT AT RESONANCE

In the practical tank circuit of Fig. 28-8, the series resistance r_s represents the internal resistance of the coil due to the wire size and length. The oscillatory current flowing through this resistance causes heat to be dissipated, producing a very small power loss. This power loss must be made up by the source current, which no longer will be zero. The capacitor losses are very low and can be ignored.

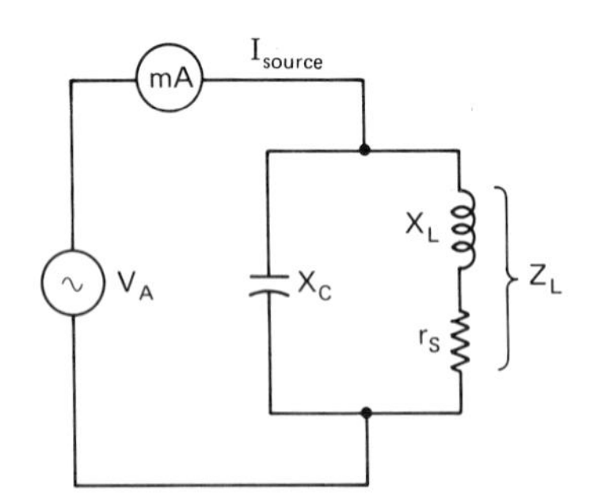

FIGURE 28-8 Practical tank circuit inductor has internal series resistance r_s due to its coil wire size and length.

At radio frequencies the RF current travels along the surface of a conductor. Only a small fraction of a conductor's cross-sectional area is effective, so the RF resistance is much higher than the ohmic (dc) resistance of the coil.

Practical Impedance of an *LC* Resonant Tank Circuit

Because of the power loss produced by the internal resistance of the coil, current flows from the source and the impedance is no longer infinite. The practical impedance of an LC tank circuit can be derived from the product-over-difference equation. Since r_s is considered to be in series with X_L,

$$Z_L = \sqrt{r_s^2 + X_L^2}$$

Using the product-over-difference equation for the parallel resonant tank circuit of Fig. 28-8, we obtain

$$Z = \frac{X_C \times \sqrt{r^2 + X_L^2}}{\sqrt{r_s^2 + (X_L - X_C)^2}} \qquad (28\text{-}7a)$$

The internal resistance, r_s, of a coil is very small compared to X_L at resonance; therefore, it can be ignored in the numerator of equation (28-7a), which leaves

$$Z = \frac{X_C X_L}{\sqrt{r_s^2 + (X_L - X_C)^2}} \qquad (28\text{-}7b)$$

Since this impedance applies only to a resonant circuit where X_L cancels X_C, $X_L - X_C$ = zero and drops out in the denominator of equation (28-7b) leaving

$$Z = \frac{X_C X_L}{r_s} \qquad (28\text{-}7c)$$

If we substitute the equation $1/2\pi fc$ for X_C and $2\pi fL$ for X_L, we obtain

$$Z = \frac{1/2\pi fC \times 2\pi fL}{r_s} = \frac{1/C \times L}{r_s}$$

Therefore,

$$Z = \frac{L}{Cr_s} \quad \text{ohms} \qquad (28\text{-}7)$$

where C is in farads
L is in henrys
r_s is in ohms and is the series resistance of the coil

Impedance Magnification

The circuit impedance at resonance will be Q times greater than X_L and the source current will be Q times smaller than the tank circuit current. This can be shown from equation (28-7c) by substituting X_L for X_C since they are equal at resonance, so that

$$Z = \frac{X_L X_L}{r_s} \qquad (28\text{-}8a)$$

Since X_L/r_s is equal to Q,

$$Z = QX_L \quad \text{ohms} \qquad (28\text{-}8)$$

EXAMPLE 28-5

A parallel resonant tank circuit consists of a coil having an inductance of 235 μH with $r_s = 8\ \Omega$, and a capacitor of 100 pF connected across a 3-V source. Find (a) f_r, (b) Q, (c) practical Z, (d) Z magnification, (e) I_T, and (f) I_L and I_C.

Solution:

(a) Using equation (27-11b) gives

$$f_r = \frac{1}{2\pi\sqrt{LC}}$$

$$= \frac{1}{6.28\sqrt{235 \times 10^{-6}\ \text{H} \times 100 \times 10^{-12}\ \text{F}}}$$

$$\cong 1038\ \text{kHz}$$

(b) To find Q we must first find X_L.

$$X_L = 2\pi fL$$

$$= 6.28 \times 1038 \times 10^3\ \text{Hz} \times 235 \times 10^{-6}\ \text{H}$$

$$= 1532\ \Omega$$

$$Q = \frac{X_L}{r_s} = \frac{1532\ \Omega}{8\ \Omega} = 192$$

(c) $$Z = \frac{L}{Cr_s} = \frac{235 \times 10^{-6}\ \text{H}}{100 \times 10^{-12}\ \text{F} \times 8\ \Omega}$$

$$= 294\ \text{k}\Omega$$

(d) $Z = Q \times X_L = 192 \times 1532\ \Omega = 294\ \text{k}\Omega$

(e) $$I_T = \frac{V_A}{Z} = \frac{3\ \text{V}}{294 \times 10^3\ \Omega} = 10.2\ \mu\text{A}$$

(f) $$I_L = I_C \text{ at resonance} = \frac{V_A}{X_L} = \frac{3\ \text{V}}{1532\ \Omega}$$

$$= 1.96\ \text{mA}$$

28-7 PRACTICAL RESPONSE CURVES FOR *Z* AND *I* VERSUS FREQUENCY

The graphical representation of Z and I_T of Fig. 28-9 shows that although very large, the impedance will have some finite value compared to the infinite impedance of the ideal resonant tank circuit.

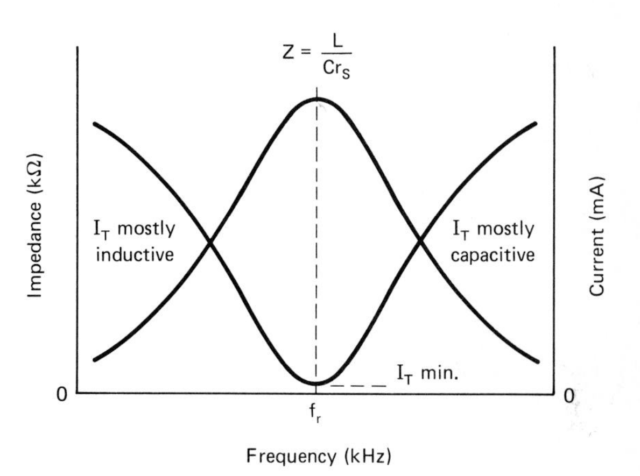

FIGURE 28-9 Practical response curves for Z and I_T versus frequency.

Since the impedance is very large at resonance, the source current will be very small. Below resonance the impedance will be mostly inductive and will be small; therefore, the current will be mostly inductive. Above resonance the impedance will be mostly capacitive and will be small; therefore, the current will be mostly capacitive.

28-8 CURRENT GAIN OR *Q* AND BANDWIDTH

For the series *LCR* resonant circuit, the current was common, so we had voltage magnification or a voltage gain given by the ratio of X_L to r_s.

In the parallel *LC* resonant tank circuit the voltage is common; therefore, we will have a current gain in the tank circuit. This can be shown by using the general output-over-input ratio to show the gain or Q as follows:

$$Q = \frac{\text{output}}{\text{input}} = \frac{I_L}{I_Z} \qquad (28\text{-}9)$$

The tank circuit Q cannot be any greater than the Q of the coil; therefore, the Q of a parallel *LC* resonant tank circuit can also be determined by using equation (27-14), $Q = X_L/r_s$.

EXAMPLE 28-6

Find the current gain of the parallel resonant circuit as given in Example 28-5. Compare the Q of the circuit to the calculated current gain of the tank circuit.

Solution: In Example 28-5 the tank circuit current was found to be 1.96 mA and the source total current was found to be 10.2 μA. Using equation (28-9) gives the current gain of the tank circuit.

$$Q = \frac{I_L}{I_Z} = \frac{1.96 \times 10^{-3}\ \text{A}}{10.2 \times 10^{-6}\ \text{A}} = 192$$

The Q of the circuit using X_L/r_s was found to be 192, which compares with the current gain method.

Bandwidth of a Parallel *LC* Resonant Circuit

The frequency response curve of Fig. 28-10 shows that the impedance is maximum at resonance and falls off below and above the resonant frequency. The range of frequencies taken from the lower band frequency (f_L) to the upper band frequency (f_U) is called the bandwidth (BW) of the resonant circuit.

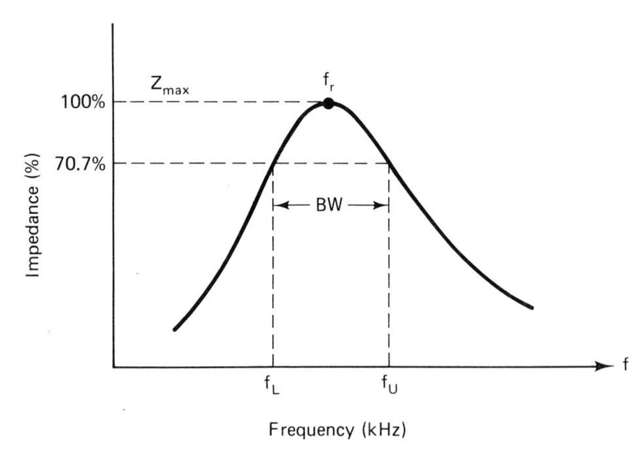

FIGURE 28-10 Bandwidth is measured between the 70.7% impedance response points, lower band frequency, f_L, and upper band frequency, f_U.

By definition the bandwidth of a tuned circuit is designated as that band of frequencies on either side of resonance (f_L and f_U) at which the Q of the circuit, or impedance in the case of the parallel resonant circuit, is 70.7% or more of its maximum value. Equation (27-15) is applicable, and is

$$\text{BW} = f_U - f_L \qquad \text{hertz} \qquad (27\text{-}15)$$

Since the resonant frequency is at the midpoint of the bandwidth frequency, the lower band frequency, f_L, can be found by subtracting one-half of the bandwidth frequency from the resonant frequency:

$$f_L = f_r - \frac{\text{BW}}{2} \qquad \text{hertz} \qquad (27\text{-}17)$$

Similarly, the upper band frequency, f_U, can be found by adding one-half of the bandwidth frequency to the resonant frequency:

$$f_U = f_r + \frac{\text{BW}}{2} \qquad \text{hertz} \qquad (27\text{-}18)$$

Measuring Bandwidth. Note that the impedance decreases to 70.7% of its maximum value at the upper and lower band frequencies. This means that the source current will increase to 1/0.707, which is equal to 1.414 times its minimum resonant value.

If the source current is monitored by a sensitive FET ac milliammeter, the upper and lower frequencies can be determined from the milliammeter readings, which will increase by 1.414 times on either side of the resonant frequency. Equation (27-15) can then be applied to determine the bandwidth.

Theoretical Bandwidth. As shown in the graphical response curve of Fig. 28-11, the Q of the parallel resonant circuit will determine the bandwidth. If the Q of the circuit is increased, the bandwidth is decreased, and vice versa. Since Q is equal to X_L/r_s, equation (27-16) can be applied once the resonant frequency is calculated to give the bandwidth:

$$\text{BW} = \frac{f_r}{Q} \qquad \text{hertz} \qquad (27\text{-}16)$$

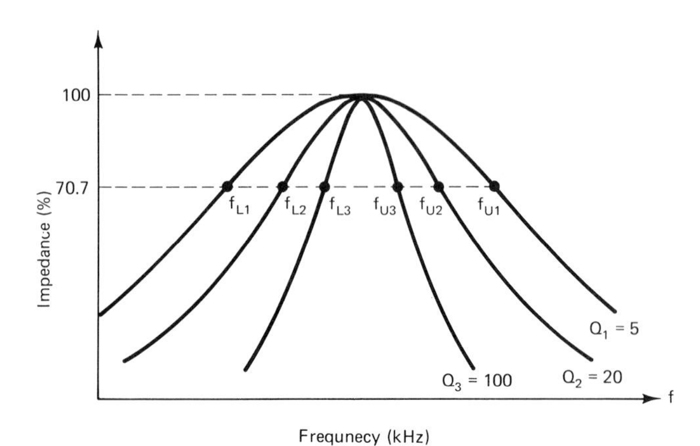

FIGURE 28-11 Bandwidth of a parallel *LC* resonant circuit for various Q values.

EXAMPLE 28-7

Determine the bandwidth of the tuned circuit given in Example 28-5 for the Q values shown in Fig. 28-11.

Solution: The resonant frequency of the parallel tuned circuit in Example 28-5 was found to be 1038 kHz. The bandwidth for a Q of 5 is

$$\text{BW} = \frac{f_r}{Q} = \frac{1038 \text{ kHz}}{5} = 207.6 \text{ kHz}$$

For a Q of 20,

$$\text{BW} = \frac{f_r}{Q} = \frac{1038 \text{ kHz}}{20} = 51.9 \text{ kHz}$$

For a Q of 100,

$$\text{BW} = \frac{f_r}{Q} = \frac{1038 \text{ kHz}}{100} = 10.38 \text{ kHz}$$

EXAMPLE 28-8

What is the upper band frequency and the lower band frequency for a Q of 20 in the circuit of Example 28-5?

Solution: The upper and lower band frequencies are found using equations (27-18) and (27-17).

$$f_U = f_r + \frac{\text{BW}}{2}$$

$$= 1038 \text{ kHz} + \frac{51.9 \text{ kHz}}{2} = 1064 \text{ kHz}$$

$$f_L = f_r - \frac{\text{BW}}{2}$$

$$= 1038 \text{ kHz} - \frac{51.9 \text{ kHz}}{2} = 1012 \text{ kHz}$$

PRACTICE QUESTIONS 28-2

Choose the most suitable word, phrase, or value to complete each statement.

1. In a parallel LC circuit the inductive current is ________ degrees out of phase with the capacitive branch current.
2. The net source current of a parallel LC circuit can be found by the simple ________ of I_L and I_C.

Refer to the parallel LC schematic diagram of Fig. 28-12 for Questions 3 to 17.

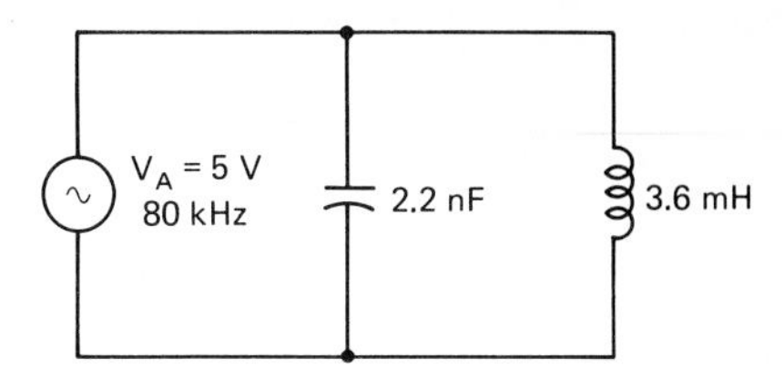

FIGURE 28-12 Parallel LC schematic.

3. Find the capacitive branch current.
4. Find the inductive branch current.
5. Find the total source current.
6. Determine the impedance of the circuit.
7. What is the impedance of the circuit using the product-over-difference equation?
8. What value of capacitive is required to bring the circuit to resonance?
9. What is the practical impedance of the now resonant circuit given that r_s is equal to 100 Ω?
10. Determine the Q of the resonant circuit.
11. Determine the circuit impedance using the Q or impedance magnification of the tank circuit.
12. Find the total source current of the resonant circuit.
13. What is the tank circuit current at resonance?
14. What is the current gain of the tank circuit?
15. Determine the bandwidth of the circuit.
16. Determine the lower band frequency for the bandwidth of the resonant circuit.
17. Determine the upper band frequency for the bandwidth of the resonant circuit.

28-9 DAMPING A RESONANT TANK CIRCUIT

The Q of a practical parallel resonant circuit, as we have seen, can be very high, giving a narrow bandwidth.

In tuned circuit applications such as those used in TV and FM, it is required that the tuned circuit bandwidth be wide in order to accept a greater range of frequencies. Increasing the bandwidth of the tuned circuits ensures that the upper and lower and frequencies of the received signal frequency are amplified and faithfully reproduced by either audio amplifiers or video amplifiers. Thus the quality of the reception is determined to a great extent from the bandwidth of the tuned circuits in the RF and IF sections of an FM or TV receiver.

For this purpose a resistor, R_p, is shunted across a certain stage or stages of a resonant circuit as shown in the schematic of Fig. 28-13. The effect of the shunt re-

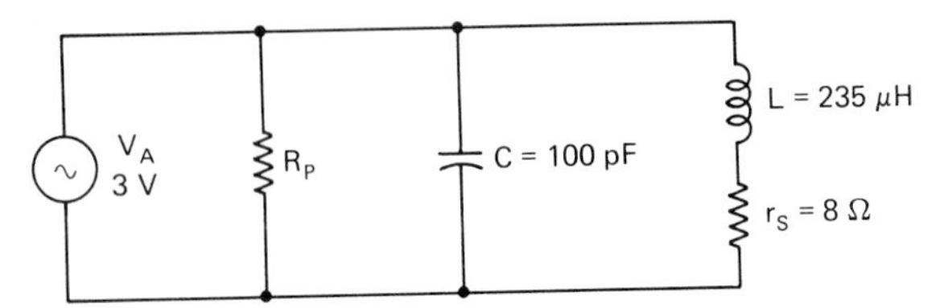

FIGURE 28-13 A parallel or shunt resistor lowers or dampens the Q of an LC tank circuit.

sistor, R_p, is to reduce the Q of the tuned circuit. This effect is termed *damping* the resonant tuned circuit, which widens or increases the bandwidth as shown in Fig. 28-11.

Damping can be deliberate or it can be due to the type of load resistance that the tank circuit is driving. An actual resistor placed across the tuned circuit in combination with stagger tuning increases the bandwidth to a desired value.

Bandwidth with a Parallel Resistance, R_p

The parallel resistor, R_p, reduces the Q of the resonant circuit. Once the Q is reduced, the impedance, Z_T, is also reduced increasing the source current, I_T, at resonance.

Rearranging equation (28-8) will show that the Q of the circuit is dependent on the impedance:

$$Z = QX_L \quad \text{ohms} \qquad (28\text{-}8)$$

Therefore,

$$Q = \frac{Z}{X_L} \quad \text{ohms} \qquad (28\text{-}10)$$

Shunting a resistor, R_p, across the tuned circuit means that R_p is in parallel with Z_T. Applying R_p in equation (28-10) gives

$$Q = \frac{Z \parallel R_p}{X_L} \qquad (28\text{-}11)$$

EXAMPLE 28-9

The resonant frequency for the tuned circuit of Fig. 28-13 was found to be 1038 kHz, $Z_T = 294\ \text{k}\Omega$, and X_L was 1532 Ω. Find the bandwidth with the following conditions:

(a) BW without R_p.
(b) BW with R_p of 220 Ω.
(c) BW with R_p of 2200 Ω.
(d) BW with R_p of 22 kΩ.

Solution: Using equation (28-11) and the product-over-sum equation to approximate the parallel impedance, the Q of the circuit is first determined; then the bandwidth can be found from equation (27-16):

$$\text{BW} = \frac{f_r}{Q}$$

(a) $$Q = \frac{X_L}{r_s} = \frac{1532\ \Omega}{8\ \Omega} = 192$$

Therefore,

$$\text{BW} = \frac{f_r}{Q} = \frac{1038\ \text{kHz}}{192} = 5.4\ \text{kHz}$$

The BW without R_p is 5.4 kHz.

(b) With an R_p of 220 Ω,

$$Q = \frac{Z \parallel R_p}{X_L} = \frac{294\ \text{k}\Omega \parallel 220\ \Omega}{1532\ \Omega}$$

$$\cong \frac{220\ \Omega}{1532\ \Omega} = 0.14$$

$$\text{BW} = \frac{f_r}{Q} = \frac{1038\ \text{kHz}}{0.14} = 7414\ \text{kHz}$$

This shows that the Q has been drastically reduced, thereby flattening the bandwidth over the complete frequency range from the resonant frequency down to dc, the lower band frequency, and 4745 kHz, the upper band frequency.

(c) With an R_p of 2200 Ω,

$$Q = \frac{Z \parallel R_p}{X_L} = \frac{294\ \text{k}\Omega \parallel 2200\ \Omega}{1532\ \Omega}$$

$$\cong \frac{2200\ \Omega}{1532\ \Omega} = 1.4$$

$$\text{BW} = \frac{f_r}{Q} = \frac{1038\ \text{kHz}}{1.4} = 741.4\ \text{kHz}$$

The lower band frequency is now 667.3 kHz, while the upper band frequency is 1408.7 kHz. We now have a little bit of gain with a Q of 1.4, and the frequency response curve is not as flat across the frequency spectrum.

(d) With an R_p of 22 kΩ,

$$Q = \frac{Z \parallel R_p}{X_L} = \frac{294\ \text{k}\Omega \parallel 22\ \text{k}\Omega}{1532\ \Omega} \cong \frac{20\ \text{k}\Omega}{1532\ \Omega} = 13$$

$$BW = \frac{f_r}{Q} = \frac{1038 \text{ kHz}}{13} = 79.8 \text{ kHz}$$

The frequency response curve is now much sharper (i.e., has greater selectivity). The lower band frequency is now 998.1 kHz and the upper band frequency is 1077.9 kHz.

The three examples of parallel resistance damping show the effect of R_p on Q and bandwidth of a resonant tank circuit. This is the preferred method of damping since the parallel resistance maintains a symmetrical resonant frequency response curve.

28-10 APPLICATIONS OF BANDWIDTH TO TUNED CIRCUITS

As mentioned previously, some of the most important applications to bandwidth are in AM, FM, and TV reception. To appreciate the significance of bandwidth, we must understand the need or requirement for such a bandwidth. The audio-frequency range of hearing for human beings is approximately from 30 Hz to 20 kHz. Dogs, bats, and other species are able to discern sound at frequencies beyond this range.

AM, FM, and TV stations broadcast what is called a *carrier frequency* that is modulated by the sound and/or video information. As an example, 1400 on the dial is the carrier frequency of a particular radio station. The sound modulations produce upper and lower band frequencies that depend on the frequency range of the audio being broadcast, which could be up to 20 kHz.

In practice, the AM broadcast frequency band range is 550 to 1600 kHz. To accommodate, without interference, the many broadcasting stations within this frequency spectrum, each broadcast carrier is limited to a bandwidth of 10 kHz. Note that the total spectrum space required by each AM station will be determined by guard bands on either side of the carrier and the power output alloted to a particular station by government regulations.

The FM broadcast frequency spectrum is 88 to 108 MHz. This means that the bandwidth of the broadcast carrier can be increased. In fact, the bandwidth of an FM radio station carrier is 150 kHz and each station is separated by guard bands of 25 kHz (see Fig. 28-14). As you might have deduced by now, the fidelity or quality of the audio being broadcast by the FM station is limited only by the limitations of transducers, amplifiers, and tuned circuit bandwidth in the transmitter and receiver system.

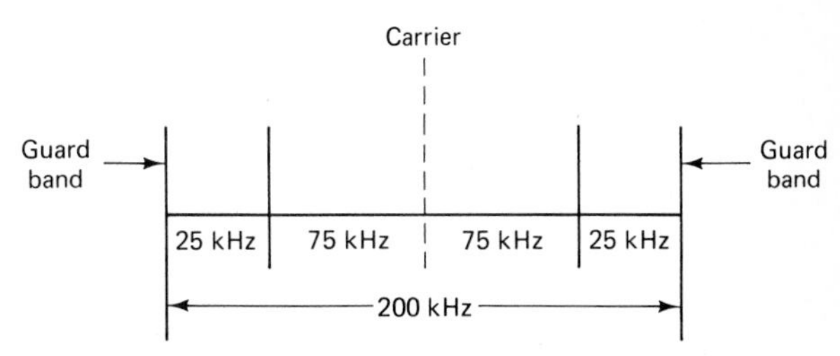

FIGURE 28-14 FM radio station frequency spectrum requirement.

The TV broadcast frequency spectrum ranges from 54 to 890 MHz. The bandwidth requirement for the video and FM sound carrier is 6 MHz. This bandwidth ensures that a good-quality resolution of the picture is faithfully reproduced on the TV screen.

The graphical frequency response curve for the TV carrier, video, and audio is shown in Fig. 28-15. Note that the video bandwidth is approximately 4 MHz. The lower side band is suppressed and limited to 1.25 MHz. The audio carrier is 4.5 MHz from the video carrier with a bandwidth of 50 kHz. The remaining 25 kHz is a guard band to separate the video from the audio lower band frequency.

These examples of bandwidth to AM, FM, and TV only serve to give a broad overview. A more detailed explanation of stagger tuning methods to obtain a broad bandpass can be obtained in telecommunication or TV textbooks.

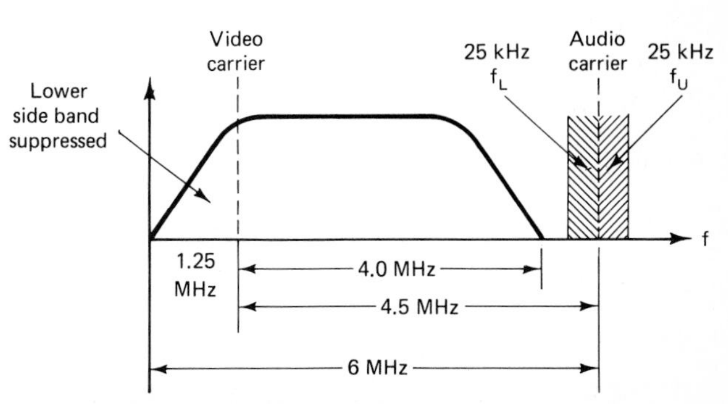

FIGURE 28-15 TV station frequency spectrum requirement.

SUMMARY

Current lags the voltage across an inductor by 90°.

Current leads the voltage across a capacitor by 90°.

I_C will be 180° out of phase with I_L in a parallel L_C circuit and they cancel.

The total current in a parallel *LCR* circuit is the phasor sum of I_X and I_R and is

$$I_T = \sqrt{I_R^2 + (I_C - I_L)^2} \quad \text{amperes} \tag{28-1}$$

The phase angle between the total current and the reactive current is found using the equation

$$\theta_I = \tan^{-1}\frac{I_X}{I_R} \tag{28-2}$$

where $I_X = I_C - I_L$.

When I_L is greater than I_C, I_X will be inductive and lagging V_A; therefore, the angle θ is equal to

$$\tan \theta_I = -\left(\frac{I_C - I_L}{I_e}\right) \tag{28-3a}$$

The net or total impedance of a parallel *LCR* circuit can be found using Ohm's law:

$$Z = \frac{V_A}{I_T} \quad \text{ohms} \tag{28-4}$$

In an ideal parallel *LC* circuit the coil has no internal resistance. When $X_L = X_C$, I_C cancels I_L, so that no source current flows.

The impedance of an ideal parallel nonresonant circuit can be found by using the product-over-difference equation:

$$Z = \frac{\text{product}}{\text{difference}} = \frac{X_L \times X_C}{X_L - X_C} \quad \text{ohms} \tag{28-6}$$

When $X_L = X_C$, the parallel *LC* circuit is said to be resonant, and the resonant frequency can be found using equation (27-11b):

$$f_r = \frac{1}{2\pi\sqrt{LC}} \quad \text{hertz} \tag{27-11b}$$

The exchange of energy from the capacitor to the inductor in a parallel *LC* circuit is termed the oscillatory current. This *LC* circuit is called a "tank" circuit.

The practical impedance of an *LC* tank circuit can be found using the equation

$$Z = \frac{L}{Cr_s} \quad \text{ohms} \tag{28-7}$$

At resonance the impedance will be Q times greater than X_L:

$$Z = QX_L \quad \text{ohms} \tag{28-8}$$

The Q of a parallel LC resonant circuit can be found using the equation

$$Q = \frac{I_L}{I_Z} \qquad (28\text{-}9)$$

$$Q = \frac{X_L}{r_s} \qquad (27\text{-}14)$$

The bandwidth of a tuned circuit is designated as that band of frequencies on either side of resonance (f_L and f_U) at which the Q of the circuit or impedance is 70.7% or more of its maximum value.

A theoretical bandwidth be calculated using the equation

$$\text{BW} = \frac{f_r}{Q} \quad \text{hertz} \qquad (27\text{-}16)$$

The lower band frequency, f_L, can be found using the equation

$$f_L = f_r - \frac{\text{BW}}{2} \quad \text{hertz} \qquad (27\text{-}17)$$

The upper band frequency, f_U, can be found using the equation

$$f_U = f_r + \frac{\text{BW}}{2} \quad \text{hertz} \qquad (27\text{-}18)$$

A parallel resistor across a resonant LC tank circuit reduces the Q of the circuit. This is shown by the equation

$$Q = \frac{Z \parallel R_p}{X_L} \qquad (28\text{-}11)$$

REVIEW QUESTIONS

28-1. Explain briefly why the capacitive current is 180° out of phase with the inductive current in a parallel LC circuit.

28-2. Explain why Ohm's law can be used to find the impedance in a parallel LCR circuit.

28-3. What type of impedance does the source "see" when the reactance is greater than 10 times the resistance in a parallel LCR circuit?

28-4. What is the impedance of a parallel LCR circuit when X_L is equal to X_C? Why?

28-5. What would be the phase angle between the current and the applied voltage in a parallel LCR circuit when $X_L = X_C$?

28-6. What is the term used to describe the frequency at which $X_L = X_C$?

28-7. Give the equation to find the frequency at which $X_L = X_C$.

28-8. Explain briefly why the current oscillates in the LC resonant tank circuit.

28-9. What is the phase difference between the applied voltage and the tank circuit at resonance? Explain.

28-10. In an ideal tank circuit the circuit impedance is infinite. Explain why this is not the case in a practical circuit.

28-11. Give the equation to find the impedance in a practical resonant tank circuit.

28-12. Explain why impedance is magnified Q times X_L in a resonant tank circuit.

28-13. What is the Q of a parallel resonant LC circuit in terms of the tank circuit current and the source current?

28-14. Define the bandwidth of a tuned resonant circuit.

28-15. Explain why the Q of a resonant tuned circuit will determine its bandwidth.

28-16. What are upper and lower band frequencies?

28-17. A 250-μH coil having a resistance of 35 Ω is connected in parallel with a 500-pF capacitor and a 500-mV source. Find:
- **(a)** The resonant frequency of the tank circuit.
- **(b)** The bandwidth of the tank circuit.
- **(c)** The upper and lower band frequencies.
- **(d)** The practical impedance of the circuit.
- **(e)** The tank circuit current.
- **(f)** The source supply current.

28-18. Find the bandwidth for the circuit of Question 28-17 given that a damping resistor of 3 kΩ is connected across the tank circuit.

28-19. Explain why the bandwidth of a tuned circuit amplifier is important in transmission and reception of (a) FM and (b) TV.

SELF-EXAMINATION

Choose the most suitable word, phrase, or value to complete each statement.

28-1. The induced emf produced by an inductor opposes the current change, so the voltage across the inductor will (*lead, lag*) the current by ________ degrees.

28-2. The voltage across a capacitor is due to its coulomb charge, so the voltage across the capacitor will (*lead, lag*) the current by degrees.

28-3. In a parallel LC circuit the capacitive current will be ________ degrees out of phase with the inductive current.

28-4. When I_C is equal to I_L, the total current flow from the source is due to the ________ in a parallel LCR circuit.

Refer to the following circuit for Questions 28-5 to 28-11: A parallel LCR circuit has a 100-mH, a 1-nF, and a 10-kΩ resistor connected across a 10-V, 20-kHz source.

28-5. Find the total source current.

28-6. Find the phase angle between the source current and the applied voltage.

28-7. Is the source current leading or lagging the applied voltage?

28-8. What is the impedance of this circuit?

28-9. What is the impedance of the circuit with the resistance removed? What type of impedance is left?

28-10. Calculate the value of capacitance required to resonate this circuit.

28-11. What is the impedance of the LCR circuit at resonance?

28-12. The term ________ circuit is applied to a parallel resonant LC circuit because of the oscillatory current contained in that circuit.

28-13. The circuit impedance at resonance will be Q times greater than __________.

28-14. The source current will be Q times smaller than the __________ circuit current.

Refer to the following circuit for Questions 28-15 to 28-22: A parallel LC circuit has a 127-μH coil, with an internal resistance of 36 Ω and a 200-pF capacitor connected across a 3-V, 1-MHz source.

28-15. Find the Q of the circuit.

28-16. Calculate the practical impedance of the circuit.

28-17. Calculate the bandwidth of the circuit.

28-18. Find the lower band frequency.

28-19. Find the upper band frequency.

28-20. Calculate the tank circuit current.

28-21. Calculate the source current.

28-22. What value of parallel resistor would be required to "damp" the Q of the circuit to a value of 10?

28-23. The bandwidth of an AM broadcasting signal frequency is __________ kHz.

28-24. The bandwidth of an FM broadcasting signal frequency is __________ kHz.

28-25. The bandwidth requirement for the video and FM sound carrier of a TV signal frequency is __________ MHz.

ANSWERS TO PRACTICE QUESTIONS

28-1

1. resistive
2. inductive
3. capacitive
4. 10 mA
5. 1.6 mA
6. 6.3 mA
7. 11 mA
8. 25°
9. leading
10. 909 Ω

28-2

1. 180
2. difference
3. 5.5 mA
4. 2.8 mA
5. 2.7 mA
6. 1.85 kΩ
7. 1.8 kΩ
8. 1.1 nF
9. 32.7 kΩ
10. 18
11. 32.5 kΩ
12. 153 μA
13. 2.8 mA
14. 18
15. 4444 Hz
16. 77,777 Hz
17. 82,222 Hz

29

Filters

INTRODUCTION

An analogy to electronic filters can be made for mechanical filters. The mechanical filter separates desired particles from undesired particles. In an electronic filter the desired frequency of input voltage is accepted and separated from the undesired frequency of input voltage or voltages.

Most electronic devices or systems contain voltages that have a dc level, as well as an audio frequency (AF) or radio frequency (RF) or a combination of all three voltage levels.

A typical example of this is in the AF detector or demodulator circuit of an AM, FM, or TV receiver. Here the AF voltage must be separated from the RF voltage while maintaining a dc level.

Another typical example is the filter circuit of a full-wave or half-wave power rectifier ciruit. Here the ripple voltage must be filtered from the dc voltage to obtain a pure dc voltage. Since capacitance and inductance are frequency dependent, filter circuits contain capacitors and inductors.

In this chapter we study basic types of filters in order to develop a thorough understanding of filter principles.

MAIN TOPICS IN CHAPTER 29

OBJECTIVES

After studying Chapter 29, the student should be able to

1. Master the basic concepts of low-pass, high-pass, bandpass, resonant bandpass, and reject filters.
2. Solve for cutoff frequency, attenuation, and bandwidth with upper and lower band frequencies.
3. Apply filter circuit principles to bypass capacitor, coupling capacitor, and power supply *RC* filters.

29-1 TYPES OF FILTERS

Filter circuits are used in many electronic systems and devices. These circuits utilize various configurations of resistors, inductors, and capacitors to obtain a frequency bandpass and/or band rejection characteristic between the input voltage and the filter circuit output voltage.

There are four types or categories of filter circuits that are functionally descriptive: (1) low pass, (2) high pass, (3) bandpass, and (4) band reject. A functional description of each is presented before we examine the actual circuit and its characteristics.

Low-Pass Filter

The configuration of capacitance and/or inductance and resistance in the low-pass filter block diagram of Fig. 29-1(a) is such that it passes only the low-frequency portion of the input voltage, V_{in}.

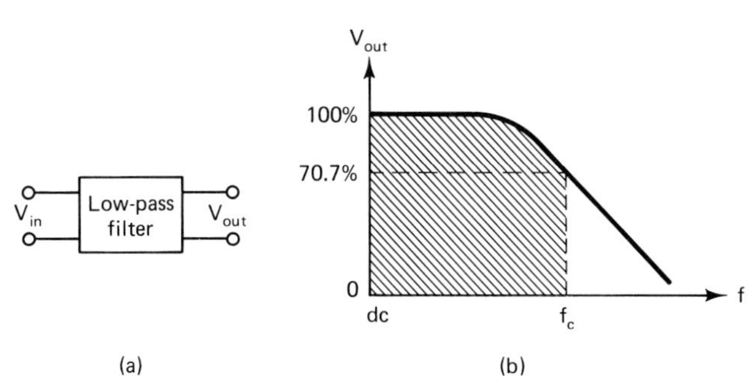

FIGURE 29-1 (a) Block diagram of low-pass filter. (b) Graphical frequency response curve for a low-pass filter.

The output voltage V_{out} will drop to 70.7% of its maximum or 100% value at the cutoff frequency, f_c, of the circuit as shown in the frequency response curve of Fig. 29-1(b).

The term *low-pass filter* applies to a configuration of L or C and R that passes a voltage having low frequencies while *attenuating* (or reducing) the higher-frequency voltage to 70.7% or less of the maximum input value.

High-Pass Filter

The configuration of capacitance and/or inductance and resistance in the high-pass filter block diagram of Fig. 29-2(a) is such that it passes only the high-frequency portion of the input voltage, V_{in}.

The output voltage, V_{out}, will drop to 70.7% of its maximum or 100% value at the cutoff frequency, f_c, of the circuit, as shown in the frequency response curve of Fig. 29-2(b).

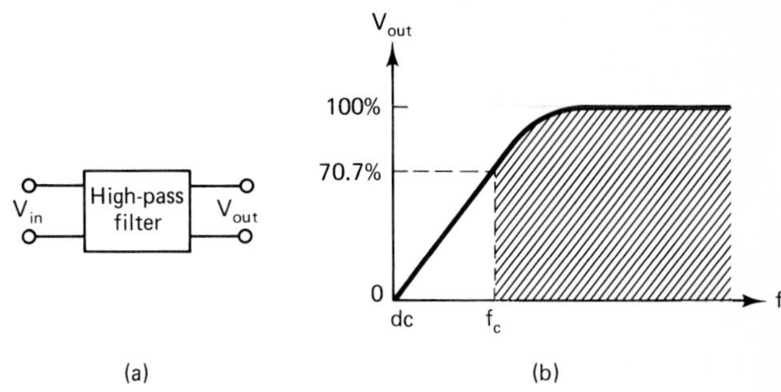

FIGURE 29-2 (a) Block diagram of high-pass filter. (b) Graphical frequency response curve for a high-pass filter.

The term *high-pass filter* applies to a configuration of L or C and R that passes a voltage having high frequencies while *attenuating* (or reducing) the lower-frequency voltage to 70.7% or less of the maximum value.

Bandpass Filter

The configuration of capacitance, inductance, or both and/or resistance in the bandpass filter block diagram of Fig. 29-3(a) is such that it passes a range of frequencies termed bandwidth and rejects the lower and upper frequency portions of the input voltage, V_{in}.

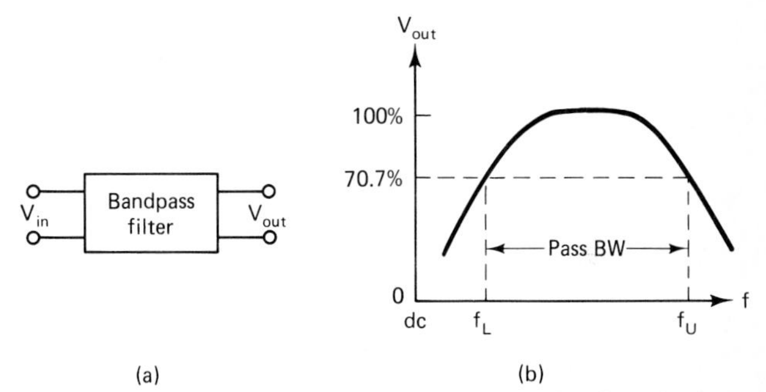

FIGURE 29-3 (a) Block diagram of a bandpass filter. (b) Graphical frequency curve for a bandpass filter.

The output voltage, V_{out}, will drop to 70.7% of its maximum or 100% value at the lower cutoff frequency, f_L, and at the upper cutoff frequency, f_U, as shown in the frequency response curve of Fig. 29-3(b).

The term *bandpass filter* applies to a configuration of L or C and R that passes a bandwidth range of frequencies while attenuating (or reducing) frequencies below 70.7% of the maximum output voltage points, f_L and f_U.

Band-Reject Filter

The configuration of capacitance inductance, or both, and/or resistance in the band-reject filter block diagram of Fig. 29-4(a) is such that it rejects a range of frequencies termed bandwidth and accepts the lower and upper frequency portions of the input voltage, V_{in}.

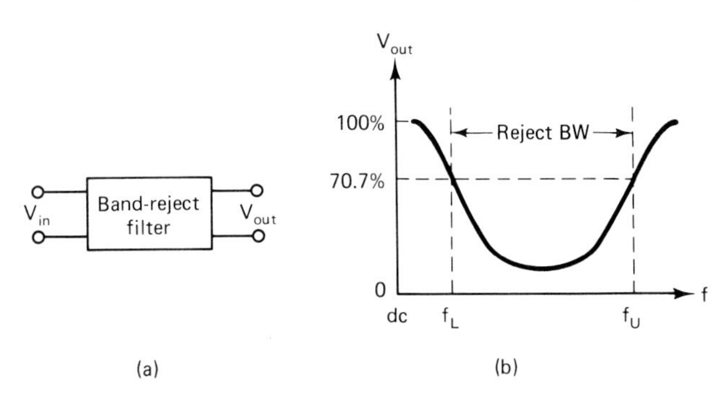

FIGURE 29-4 (a) Block diagram of band-reject filter. (b) Graphical frequency response curve for a band-reject filter.

The output voltage, V_{out}, will drop to 70.7% of its maximum or 100% value at the cutoff frequency, f_L, and at the upper cutoff frequency, f_U, as shown in the frequency response curve of Fig. 29-4(b).

The term *band-reject filter* applies to a configuration of L or C and R that rejects or attenuates a range of frequencies below 70.7% of the maximum output voltage points, f_L and f_U.

Having defined the four catagories of filters, we can now focus our attention on the actual LCR circuits that make up these filters.

29-2 LOW-PASS FILTERS

Capacitor to Ground, *RC* Low-Pass Filter

A very common circuit configuration of a low-pass filter is the capacitor-to-ground, shown in Fig. 29-5(a). At the low frequency end of V_{in}, the capacitor appears as a very high reactance and the output voltage will be approximately equal to the input voltage, V_{in}. As the frequency of the input voltage increases, the capacitor reactance decreases and it appears as a short circuit to the source voltage, as shown in Fig. 29-5(b). The output voltage is shorted out by the capacitor and is dependent on the value of X_C. If X_C approaches zero ohms, the output voltage will be approximately zero volts.

V_{out} and Attenuation Ratio. Since the capacitor and resistor form a basic series voltage divider across the input voltage, the output voltage can be shown as

$$V_{out} = IX_C$$

Since the current is common, we substitute V_{in}/Z for I,

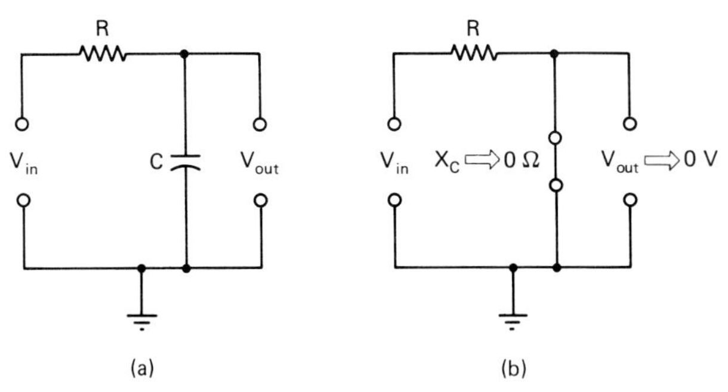

FIGURE 29-5 (a) *RC* low-pass filter. (b) Capacitor appears as a short circuit beyond the cutoff frequency f_C.

$$\frac{V_{in}}{Z} = \frac{V_{in}}{\sqrt{R^2 + X_C^2}}$$

so that

$$V_{out} = \frac{V_{in} \times X_C}{\sqrt{R^2 + X_C^2}} \quad \text{volts} \tag{29-1a}$$

Rearranging equation (29-1a) gives

$$V_{out} = \frac{X_C}{\sqrt{R^2 + X_C^2}} \times V_{in} \quad \text{volts} \tag{29-1}$$

The filter attenuation ratio can be determined from equation (29-1):

$$\frac{V_{out}}{V_{in}} = \frac{X_C}{\sqrt{R^2 + X_C^2}} \tag{29-2}$$

Cutoff Frequency. The cutoff frequency (f_c) will be reached when the output voltage falls to 70.7% of the input voltage. This can be shown using equation (29-1). If $X_C = R$, we can substitute R for X_C to give

$$V_{out} = \frac{R}{\sqrt{R^2 + X_C^2}} \times V_{in}$$
$$= \frac{R}{\sqrt{2R^2}} \times V_{in}$$
$$= \frac{R}{R\sqrt{2}} \times V_{in} = \frac{V_{in}}{\sqrt{2}}$$

Therefore,

$$V_{out} = V_{in} \times 0.707$$

The cutoff frequency, f_c, can be calculated by substituting the value of R for X_C in the equation $X_C = 1/2\pi f_c C$, since $X_C = R$, and rearranging the equation to give

$$f_c = \frac{1}{2\pi RC} \quad \text{hertz} \tag{29-3}$$

EXAMPLE 29-1

The low-pass filter circuit of Fig. 29-5(a) has a resistance of 5 kΩ and a 0.05-μF capacitor to ground. If the input voltage is 700 mV, calculate:

(a) The cutoff frequency.
(b) The output voltage when $X_C = R$.
(c) The attenuation ratio.

Solution:

(a) The cutoff frequency can be calculated using equation (29-3):

$$f_c = \frac{1}{2\pi RC}$$
$$= \frac{1}{6.28 \times 5 \times 10^3\ \Omega \times 0.05 \times 10^{-6}\ \text{F}}$$
$$= 637\ \text{Hz}$$

(b) When $X_C = R$, the output voltage falls to the 70.7% level of the input voltage:

$$V_{out} = V_{in} \times 0.707 = 700\ \text{mV} \times 0.707$$
$$= 495\ \text{mV}$$

(c) The attenuation ratio is given by equation (29-2):

$$\frac{V_{out}}{V_{in}} = \frac{495\ \text{mV}}{700\ \text{mV}} = 0.707$$

Resistor to Ground, *LR* Low-Pass Filter

The *LR* configuration of Fig. 29-6(a) is another type of low-pass filter. At the low-frequency end of V_{in}, the inductor appears as a low reactance and the output voltage will be approximately equal to the input voltage, V_{in}. As the frequency of the input voltage increases, the reactance of the inductor increases and it appears as an open circuit to the source voltage, as shown in Fig. 29-6(b). The output voltage will actually be dependent on the value of X_L. If X_L is very high, the output voltage will approach zero volts.

V_{out} and Attenuation Ratio. Since the inductor and resistor form a basic series voltage divider across the input voltage, the output voltage can be shown by the voltage-divider equation similar to equation (29-1a):

$$V_{out} = \frac{R}{\sqrt{R^2 + X_L^2}} \times V_{in} \quad \text{volts} \qquad (29\text{-}4)$$

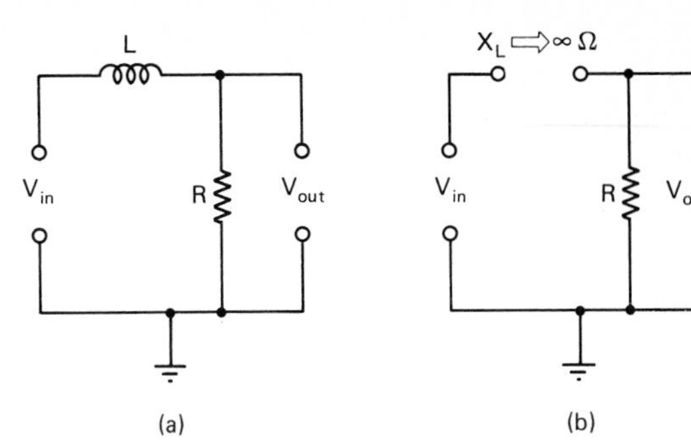

FIGURE 29-6 (a) *LR* low-pass filter. (b) Inductor appears as an open circuit beyond the cutoff frequency f_C.

The filter attenuation ratio can be determined from equation (29-4):

$$\frac{V_{out}}{V_{in}} = \frac{R}{\sqrt{R^2 + X_L^2}} \qquad (29\text{-}5)$$

Cutoff Frequency. The cutoff frequency will be reached when the output voltage falls to 70.7% of the input voltage. This can be shown using equation (29-4) and is similar to the *RC* derivation. If $X_L = R$, we can substitute R for X_L to give

$$V_{out} = \frac{R}{\sqrt{R^2 + X_L^2}} \times V_{in} = \frac{R}{\sqrt{2R^2}}$$
$$= \frac{R}{R\sqrt{2}} \times V_{in} = \frac{V_{in}}{\sqrt{2}}$$

Therefore,

$$V_{out} = V_{in} \times 0.707$$

The cutoff frequency can be calculated by substituting the value of R for X_L in the equation $X_L = 2\pi fL$, since $X_L = R$, and rearranging the equation to give

$$f_c = \frac{R}{2\pi L} \quad \text{hertz} \qquad (29\text{-}6)$$

EXAMPLE 29-2

The low-pass filter circuit of Fig. 29-6(a) has an inductance of 2 H and a 1.5-kΩ resistance to ground. If the input voltage is 3 V, calculate:

(a) The cutoff frequency.
(b) The output voltage when $X_L = R$.
(c) The attenuation ratio.

Solution:

(a) The cutoff frequency can be calculated by using equation (29-6):

$$f_c = \frac{R}{2\pi L} = \frac{1500\ \Omega}{6.28 \times 2\text{H}} \cong 119\ \text{Hz}$$

(b) When $X_L = R$, the output voltage falls to the 70.7% level of the input voltage, or

$$V_{out} = V_{in} \times 0.707 = 3\ \text{V} \times 0.707 = 2.1\ \text{V}$$

(c) The attenuation ratio is given by equation (29-5):

$$\frac{V_{out}}{V_{in}} = \frac{2.1\ \text{V}}{3\ \text{V}} = 0.707$$

Capacitor to Ground, *LC* Low-Pass Filter

The *L*-type (inverted) configuration of low-pass filter, shown in Fig. 29-7, is used in dc power supply filtering and filtering spurious or noise signal voltage on dc or ac voltages.

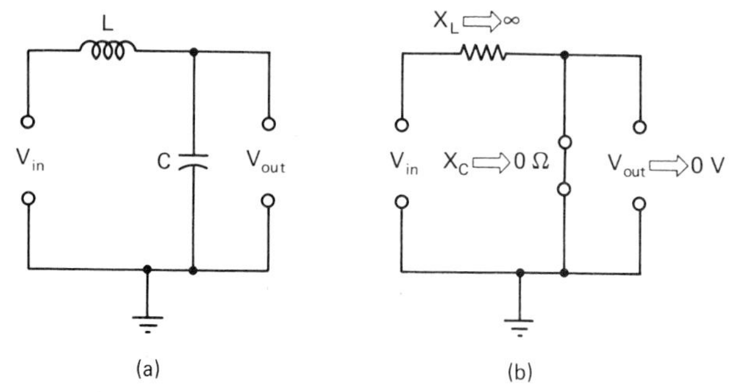

FIGURE 29-7 (a) L-type configuration low-pass filter. (b) With a higher frequency X_L approaches infinite reactance while X_C approaches zero reactance and the output voltage approaches zero volts.

At low frequencies and dc, the inductor's reactance approaches zero while the capacitive reactance is very high. Since there will be very little voltage drop across the inductor, most or nearly all of the input voltage will appear across the capacitive reactance as an output voltage.

With a higher frequency the inductive reactance increases while the capacitive reactance decreases. Now the capacitor appears as a short circuit and the output voltage taken across the capacitor approaches zero volts, as shown in Fig. 29-7(b).

This configuration of low-pass filter will readily pass dc from input to output since X_L is almost zero ohms to dc. Thus any ac spurious or noise voltage will be "choked off" by the inductor as well as shorted to ground by the capacitor.

V_{out} and Attenuation Ratio. Normally, the resistive component of the inductor is less than one-tenth of the inductive reactance, so that we can use the voltage-divider method to find the output voltage and the attenuation ratio:

$$V_{out} = \frac{X_C}{X_L - X_C} \times V_{in} \quad \text{volts} \qquad (29\text{-}7)$$

The filter attenuation ratio can be determined from equation (29-7).

$$\frac{V_{out}}{V_{in}} = \frac{X_C}{X_L - X_C} \qquad (29\text{-}8)$$

EXAMPLE 29-3

The low-pass circuit of Fig. 29-7(a) is used to filter the ripple voltage of a dc rectifier circuit. The dc input voltage has an ac ripple voltage content of 5.19 V at 120 Hz. If the choke is 7 H and the capacitor is 10 μF, calculate the ac ripple content on the dc output voltage.

Solution: First we must calculate the reactance for the inductor and the capacitor.

$$X_L = 2\pi fL = 5275\ \Omega$$

$$X_C = \frac{1}{2\pi fC} = 133\ \Omega$$

Using equation (29-8) will give the ripple voltage content at the output of the filter:

$$\frac{V_{out}}{V_{in}} = \frac{X_C}{X_L - X_C}$$

Therefore,

$$\frac{V_{out}}{5.19\ \text{V}} = \frac{133\ \Omega}{5275\ \Omega - 133\ \Omega}$$

$$V_{out} = \frac{133\ \Omega}{5142\ \Omega} \times 5.19\ \text{V} = 134\ \text{mV}$$

29-3 HIGH-PASS FILTERS

Inductor to Ground, *RL* High-Pass Filter

We have seen that a high-pass filter allows only frequencies beyond the cutoff frequency, f_c, to pass while attenuating the lower frequencies. A basic *RL* high-pass filter circuit is shown in Fig. 29-8(a).

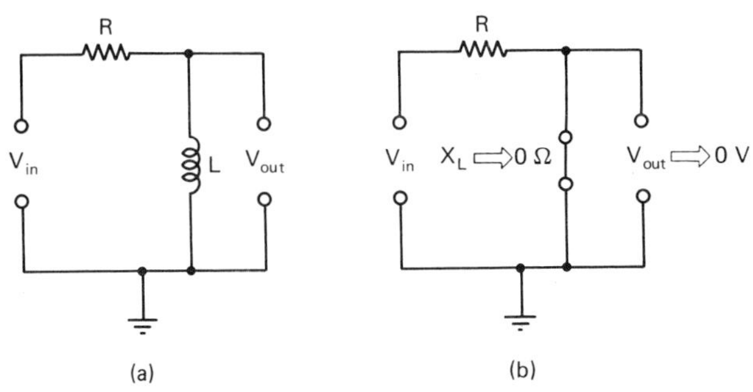

FIGURE 29-8 (a) L-type configuration RL high-pass filter. (b) With lower frequency X_L approaches zero reactance and output voltage approaches zero volts.

At low frequencies the reactance of the inductor approaches zero ohms; therefore, the output voltage approaches zero volts, as illustrated in Fig. 29-8(b). At higher frequencies the reactance of the inductor increases so that the output voltage across X_L will be determined by the ratio of X_L to R.

V_{out} and Attenuation Ratio. The resistive component of the inductor is normally less than one-tenth of X_L, so that we can use the voltage-divider method to find the output voltage and the attenuation ratio:

$$V_{out} = \frac{X_L}{\sqrt{R^2 + X_L^2}} \times V_{in} \quad \text{volts} \qquad (29\text{-}9)$$

The filter attenuation ratio can be determined from equation (29-9).

$$\frac{V_{out}}{V_{in}} = \frac{X_L}{\sqrt{R^2 + X_L^2}} \qquad (29\text{-}10)$$

The cutoff frequency can be calculated by substituting the value of R for X_L in the equation $X_L = 2\pi fL$, since $X_L = R$ to give equation (29-6), or

$$f_c = \frac{R}{2\pi L} \quad \text{hertz}$$

Resistor to Ground, *CR* High-Pass Filter

The high-pass *CR* filter of Fig. 29-9(a) is commonly used as a coupling capacitor, C_c, between audio amplifier stages. It serves to block dc bias voltages on the transistor while passing the ac voltage to the following transistor or amplifier stage.

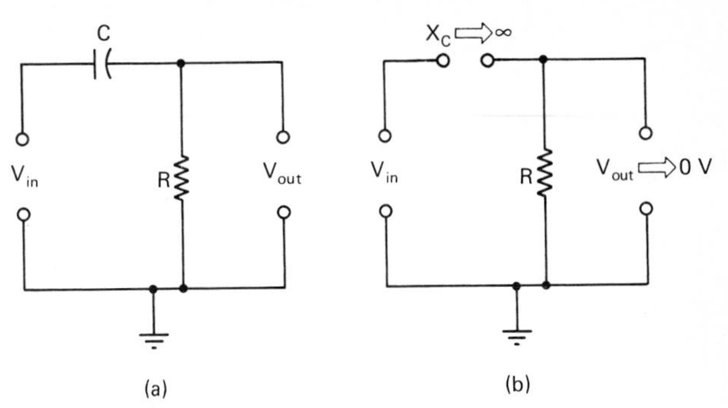

FIGURE 29-9 (a) L-type configuration CR high-pass filter. (b) With lower frequency X_C approaches infinite ohms and output voltage approaches zero volts.

At lower frequencies the capacitive reactance approaches infinite ohms and appears as an open circuit, as illustrated in Fig. 29-9(b). With the capacitor effectively open, the output voltage approaches zero volts.

V_{out} and Attenuation Ratio. As in the previous filter circuits, the voltage-divider method for the series *CR* circuit can be used if the phase difference of R and X_C are considered.

$$V_{out} = \frac{R}{\sqrt{R^2 + X_C^2}} \times V_{in} \quad \text{volts} \qquad (29\text{-}11)$$

The filter attenuation ratio can be determined from equation (29-11);

$$\frac{V_{out}}{V_{in}} = \frac{R}{\sqrt{R^2 + X_C^2}} \qquad (29\text{-}12)$$

The cutoff frequency can be calculated using equation (29-3).

EXAMPLE 29-4

The high-pass circuit of Fig. 29-9(a) couples a transistor amplifier stage. Calculate the cutoff frequency if $C = 0.1\ \mu$F and $R = 47$ kΩ.

Solution: Using equation (29-3) gives

$$f_c = \frac{1}{2\pi RC}$$
$$= \frac{1}{6.28 \times 47 \times 10^3\ \Omega \times 0.1 \times 10^{-6}\ \text{F}}$$
$$= 34\ \text{Hz}$$

This *CR* filter will pass a voltage having a frequency of 34 Hz or greater.

PRACTICE QUESTIONS 29-1

Choose the most suitable word or phrase to complete each statement.

1. A low-pass filter passes a voltage with only the _________ frequencies while attenuating the _________ frequencies.
2. A high-pass filter passes a voltage with only the _________ frequencies while attenuating the _________ frequencies.
3. A _________ filter passes a range of frequencies termed bandwidth and rejects the lower and upper frequency portion of the input voltage.
4. A _________-_________ filter rejects a range of frequencies termed bandwidth and accepts the lower and upper frequency portions of the input voltage.

Refer to the following circuit description for Questions 5 to 8: The low-pass filter circuit of Fig. 29-5(a) has a resistance of 4 kΩ and a 0.1-μF capacitor to ground; the input voltage is 1.3 V at 800 Hz.

5. What is the output voltage for the given circuit?
6. Find the attenuation ratio.
7. Calculate the cutoff frequency of the circuit.
8. What would V_{out} be at f_c?

Refer to the following circuit description for Questions 9 and 10: The high-pass circuit of Fig. 29-9(a) couples a transistor amplifier stage that has an input resistance of 10 kΩ.

9. Find the value of C if the lowest frequency to be passed by the filter is 30 Hz.
10. What would be the input voltage to the next stage if the input to the coupling circuit were 3 V at 30 Hz?

29-4 RESONANT BANDPASS FILTERS

There are two types of resonant bandpass filters, the series resonant and the parallel resonant bandpass circuits. Let's examine the series resonant bandpass circuit first.

Series Bandpass Filter

Figure 29-10(a) shows the schematic for a series resonant bandpass circuit. At the resonant frequency the series LC circuit will have minimum impedance and maximum current. The output voltage across R will then be maximum at the resonant frequency.

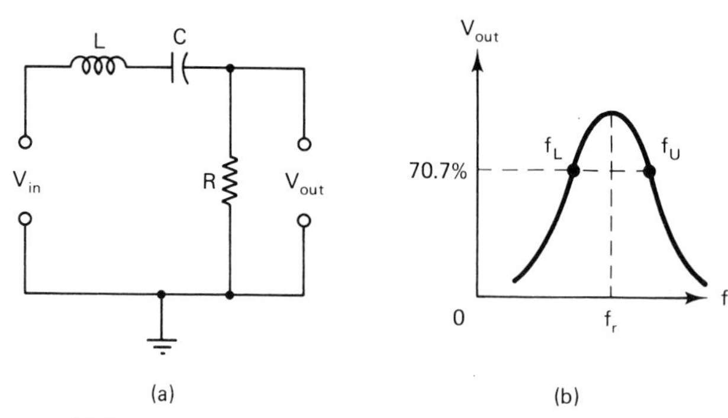

FIGURE 29-10 (a) Series resonant bandpass filter. (b) Output voltage across R has bandpass characteristics dependent upon circuit Q at resonance.

The bandpass characteristics f_L and f_U, shown in the response curve of Fig. 29-10(b), depend on the Q of the circuit.

EXAMPLE 29-5

The series LC resonant bandpass circuit of Fig. 29-10(a) has a 10-mH coil connected in series with a 2.2-nF capacitor. If the coil resistance is 100 Ω, determine (a) the resonant frequency, (b) the bandwidth, and (c) the output voltage across R of 1 kΩ if the input voltage is 5 V.

Solution:

(a) The resonant frequency can be found by using equation (29-11):

$$f_r = \frac{0.159}{\sqrt{LC}} = \frac{0.159}{\sqrt{10 \text{ mH} \times 2.2 \text{ nF}}} = 33.9 \text{ kHz}$$

(b) We need to know the Q of the circuit in order to find the bandwidth; therefore,

$$Q = \frac{X_L}{r_s} = \frac{2129\ \Omega}{100\ \Omega} = 21.3$$

Therefore,

$$\text{BW} = \frac{f_r}{Q} = \frac{33.9 \text{ kHz}}{21.3} = 1.59 \text{ kHz}$$

(c) The output voltage is across R. Since the impedance of the series resonant circuit is equal to the coil's internal resistance of 100 Ω, the voltage-divider method can be used to find the output voltage:

$$V_{out} = \frac{R}{R + r_s} \times V_{in}$$
$$= \frac{1000\ \Omega}{100\ \Omega + 1000} \times 5\text{ V} = 4.5\text{ V}$$

Parallel Band-Pass Filter

The parallel resonant circuit of Fig. 29-11(a) is another type of bandpass filter circuit. The LC tank circuit is part of the voltage divider across the source, V_{in}. At resonance the tank circuit impedance is maximum and is much greater than R, so most of the input voltage will be across the tank circuit. Since the output is taken off the tank circuit, the output voltage, V_{out}, will be maximum at resonance.

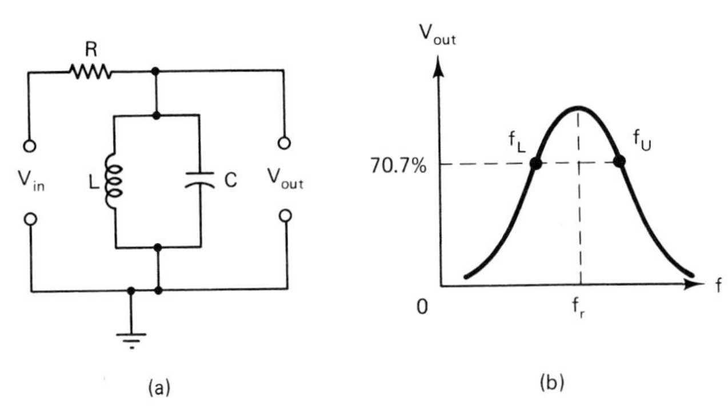

FIGURE 29-11 (a) Parallel resonant bandpass filter. (b) Output voltage across LC has bandpass characteristics dependent upon circuit Q at resonance

The bandpass characteristics, f_L and f_U, shown in the response curve of Fig. 29-11(b) depend on the Q of the tank circuit.

29-5 RESONANT BAND-REJECT FILTERS

Series Band-Reject Filter

The series LC circuit of Fig. 29-12(a) is part of the voltage divider across the source, V_{in}. At resonance the LC circuit impedance will be minimum and equal to the internal resistance of the coil. If R is at least 10 times greater than the resistance of the coil, most of the input voltage will be across R. The output voltage taken across the series LC circuit will be approximately zero volts.

As in previous circuits, the band-reject characteristics, f_L and f_U, shown in the response curve of Fig. 29-12(b), depend on the Q of the series LC circuit.

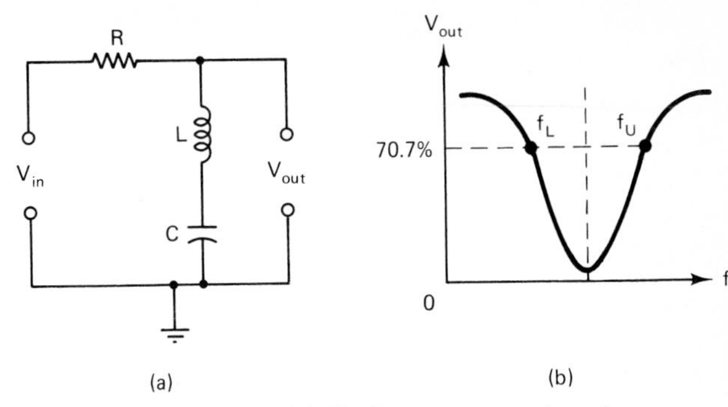

FIGURE 29-12 (a) Series resonant band-reject filter. (b) Output voltage across LC has band-reject characteristics dependent upon circuit Q at resonance.

EXAMPLE 29-6

The series resonant band-reject circuit of Fig. 29-12(a) has a 3-mH coil connected in series with a 10-nF capacitor. If the coil resistance is 10 Ω, determine:

(a) The resonant frequency.
(b) f_L and f_U.
(c) The output voltage at f_L and f_U if R is equal to 100 Ω and the input voltage is 5 V.

Solution:

(a) The resonant frequency can be found using equation (27-11):

$$f_r = \frac{0.159}{\sqrt{LC}} = \frac{0.159}{\sqrt{3\text{ mH} \times 10\text{ nF}}} = 29\text{ kHz}$$

(b) The Q and BW must be determined in order to find f_L and f_U.

$$X_L = 2\pi fL$$
$$= 6.28 \times 29 \times 10^3\text{ Hz} \times 3 \times 10^{-3}\text{ H}$$
$$= 546\ \Omega$$

$$Q = \frac{X_L}{r_s} = \frac{546}{10} = 54.6$$

$$\text{BW} = \frac{f_r}{Q} = \frac{29\text{ kHz}}{54.6} = 531\text{ Hz}$$

$$f_L = f_r - \frac{\text{BW}}{2} = 29\text{ kHz} - \frac{0.531\text{ kHz}}{2}$$
$$= 28.735\text{ kHz}$$

$$f_U = f_r + \frac{\text{BW}}{2} = 29\text{ kHz} + \frac{0.531\text{ kHz}}{2}$$
$$= 29.266\text{ kHz}$$

(c) The output voltage at f_L and f_U can be found using the voltage-divider equation, but the impedance of the series LC circuit must be recalculated.

$$Z = \sqrt{(r_s)^2 + (X_L - X_C)^2}$$

and

$$X_L = 2\pi f_L L$$
$$= 6.28 \times 28.735 \text{ kHz} \times 3 \text{ mH} = 541\ \Omega$$

$$X_C = \frac{1}{2\pi f_L C} = \frac{0.159}{28.735 \text{ kHz} \times 10 \text{ nF}} = 553\ \Omega$$

Therefore,

$$Z = \sqrt{(10\ \Omega)^2 + (541 - 553\ \Omega)^2} = 15.6\ \Omega$$

V_{out} at f_L will be the same at f_U, or

$$V_{out} = \frac{15.6\ \Omega}{15.6\ \Omega + 100\ \Omega} \times 5 \text{ V} = 0.67 \text{ V}$$

Parallel Band-Reject Filter

The parallel LC tank circuit of Fig. 29-13(a) is part of the voltage divider across the source, V_{in}. At resonance the tank circuit impedance will be maximum and most of the input voltage will be dropped across the LC circuit. The output voltage is therefore very nearly zero at resonance. Below and above the resonant frequency, the impedance of the tank circuit decreases and the output voltage increases.

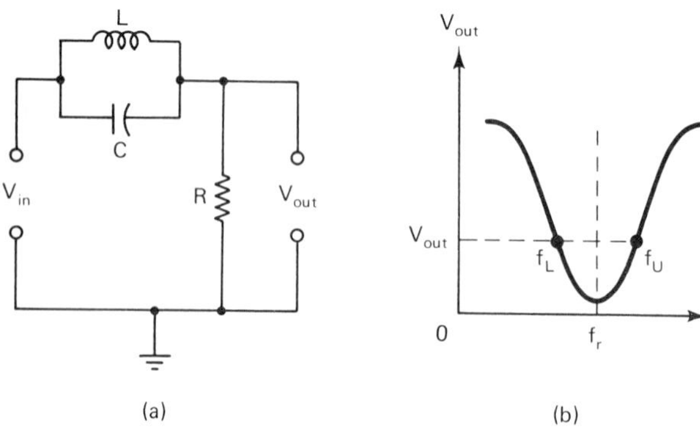

FIGURE 29-13 (a) Parallel resonant band-reject filter. (b) Output voltage across R has band-reject characteristics dependent upon circuit Q at resonance.

At low frequencies X_C approaches infinite resistance while X_L approaches zero resistance; therefore, the output voltage will be very nearly equal to the input voltage. Similarly, at higher frequencies above resonance, X_C approaches zero ohms while X_L approaches infinite resistance, and the output voltage will very nearly equal the input voltage.

EXAMPLE 29-7

The parallel resonant band-reject circuit of Fig. 29-13(a) has a 3-mH coil with an internal resistance of 10 Ω. If the parallel capacitor is 10 nF, determine the output voltage at resonance given that the input voltage is 5 V and the load resistance is 2 kΩ.

Solution: The impedance of the LC tank circuit must be found first.

$$Z = \frac{L}{Cr_s} = \frac{3 \times 10^{-3} \text{ H}}{10 \times 10^{-9} \text{ F} \times 10\ \Omega} = 30 \text{ k}\Omega$$

The voltage-divider equation can be used to approximate the output voltage at resonance:

$$V_{out} = \frac{R}{R + Z_{tank}} \times V_{in}$$
$$= \frac{2000\ \Omega}{2000\ \Omega + 30{,}000\ \Omega} \times 5 \text{ V} = 313 \text{ mV}$$

29-6 CASCADED FILTERS

The filter circuits of Fig. 29-14 are commonly used in dc power rectifier circuits to filter the ac ripple voltage. They provide much greater attenuation of a voltage with unwanted frequencies.

The T-type filter configuration shown in Fig. 29-14(a) is most suited to signal voltage sources that have a low source impedance. The series inductor increases the effective source impedance so that any high-frequency voltage input is attenuated by the combined effect of X_{L1} and X_C. Further attenuation is provided by the additional inductance X_{L2}. In this case the inductance of L_1 and L_2 can be halved and still provide superior attenuation characteristics to the single L-type filter of Fig. 29-7(a).

The π-type arrangement of L and C shown in Fig. 29-14(b) has similar characteristics to that of the T-type. This type of filter is most common in power rectifier filter circuits, where the 120-Hz ripple voltage must be filtered to obtain a pure dc.

The L-, π-, and T-type of filters are known as constant-K filters. It is desirable that the filter circuit input impedance be matched to the source impedance while the output impedance of the filter is matched to the load. This can be accomplished by choosing the value of C and L to make the product of X_L and X_C constant at all frequencies.

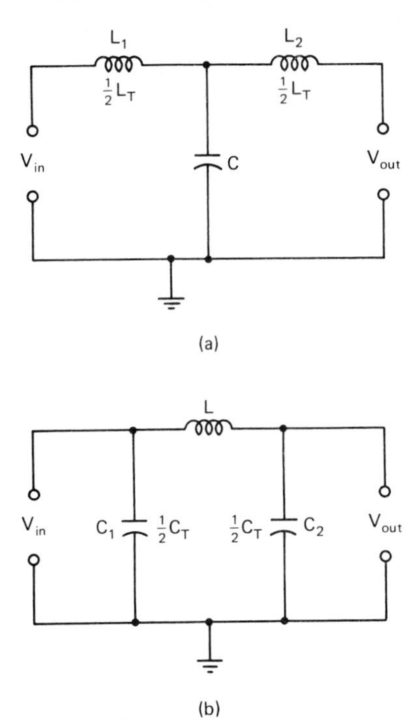

FIGURE 29-14 (a) T-type configuration low-pass filter. (b) Π-type low-pass filter.

As an example, if X_L is 800 Ω and X_C is chosen to be 400 Ω, the initial K value is 800 × 400, or 320,000. If the frequency is doubled, X_L is doubled to 1600 Ω while X_C is halved to 200 Ω, but their K value remains the same (i.e., 320,000).

The terminating resistance value of a constant-K filter can be found by the following equations, where R is the terminating resistance or the image impedance value.

$$R = \sqrt{X_L X_C} \quad \text{ohms} \qquad (29\text{-}13)$$

$$R = \sqrt{\frac{L}{C}} \quad \text{ohms} \qquad (29\text{-}14)$$

If the series components in the L-, T-, and π-type filters are capacitive and the shunt components are inductive, the filter circuits will be high-pass types. As in the case of the low-pass L, T, and π filters, all three basic types have equal series and shunt component total values.

29-7 BYPASS CAPACITORS

An application of the capacitor-to-ground low-pass filter is the so-called bypass capacitor. Electronic circuits such as audio frequency, AF amplifier stages or radio frequency, and RF amplifier stages have dc as well as AF and/or RF voltages present at certain points in the circuit. These time-varying voltages interfere with the dc biasing voltage on certain parts of the circuit and must be shunted while keeping the dc voltage intact.

Figure 29-15 shows a circuit that has a bypass capacitor across the emitter resistor, R_E, of a transistor amplifier. The bypass capacitor has a very low reactance to the *lowest frequency* being bypassed, so point A or the emitter is effectively grounded while the dc voltage across the emitter resistor is left undisturbed.

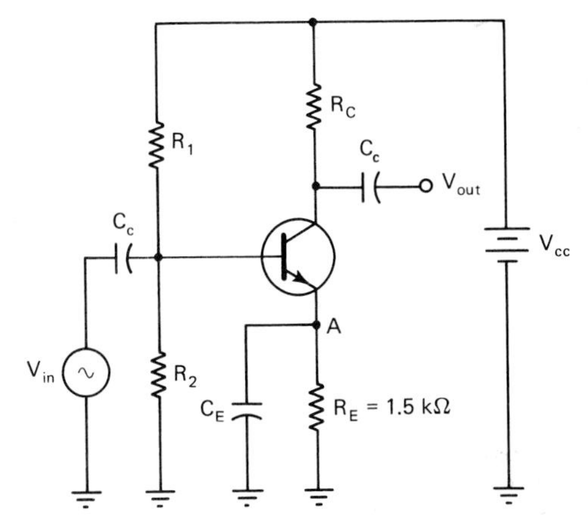

FIGURE 29-15 An audio-frequency transistor amplifier stage with emitter resistor bypassed to the AF voltage.

As a general rule, the AF or RF voltage will be effectively shorted to ground if the reactance of the capacitor is at least one-tenth of the shunt resistor value. By using this general rule, the value of the capacitance can then be calculated for the lowest frequency that is to be bypassed.

EXAMPLE 29-8

Find the value of C_E in Fig. 29-15 if the lowest frequency to be bypassed is 30 Hz.

Solution: Using the general rule that the reactance of the bypass capacitor must be one-tenth of the shunt resistance for effective bypassing gives

$$X_C = \frac{1}{10} \times 1.5\ \text{k}\Omega = 150\ \Omega$$

Therefore,

$$C = \frac{1}{2\pi f X_C} = \frac{1}{6.28 \times 30\ \text{Hz} \times 150\ \Omega} = 35\ \mu\text{F}$$

29-8 COUPLING CAPACITORS

The capacitors in the schematic of Fig. 29-15 labeled C_C are termed coupling capacitors. They couple the AF input or output voltage of the amplifier stage without disturbing the dc bias on that stage or the preceding stage.

Ideally, the capacitor appears as an open circuit to dc and a short circuit to the AF or RF voltage. Practically, for AF amplifiers, the capacitor behaves as a high-pass filter circuit and the cutoff frequency depends on the value of resistance of impedance that the capacitor "looks" into. With a higher-frequency voltage, the capacitor approaches the ideal condition of a short circuit to the AF, as shown in Fig. 29-16.

As in the case of the bypass capacitor, the value of the coupling capacitor is chosen so that its reactance, X_C, is at least one-tenth of R_L at the lowest frequency.

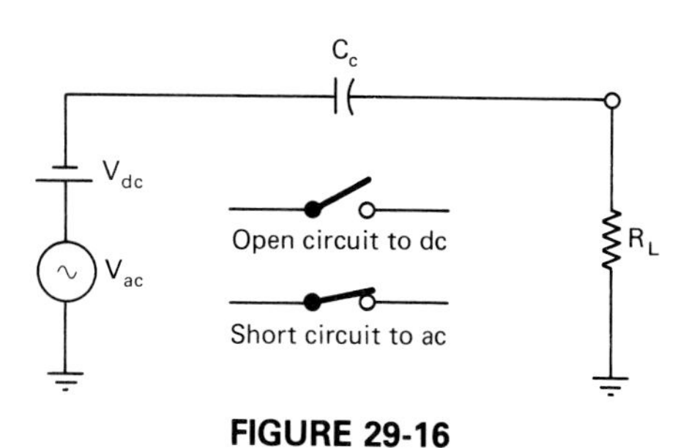

FIGURE 29-16

SUMMARY

A *low-pass filter* applies to a configuration of L or C and R that passes a voltage having low frequencies while attenuating higher frequencies to 70.7% or less of the maximum input value.

A *high-pass filter* applies to a configuration of L or C and R that passes a voltage having high frequencies while attenuating lower frequencies to 70.7% or less of the maximum input value.

A *bandpass filter* applies to a configuration of capacitance, inductance, or both and/or resistance that passes a range of frequencies termed bandwidth and rejects the lower and upper frequency portions of the input voltage, V_{in}.

A *band-reject filter* applies to a configuration of capacitance, inductance, or both, and/or resistance that rejects a range of frequencies termed bandwidth and accepts the lower and upper frequency portions of the input voltage, V_{in}.

The *RC, capacitor-to-ground* circuit of Fig. 29-5 is a low-pass filter because the shunt capacitive reactance approaches zero at higher frequencies of V_{in} and shorts out these frequencies.

The *output voltage* for an *RC*, capacitor-to-ground, low-pass filter circuit is

$$V_{out} = \frac{X_C}{\sqrt{R^2 + X_C^2}} \times V_{in} \qquad \text{volts} \qquad (29\text{-}1)$$

The *attenuation ratio* for an *RC*, capacitor-to-ground, low-pass filter circuit is

$$\frac{V_{out}}{V_{in}} = \frac{X_C}{\sqrt{R^2 + X_C^2}} \qquad (29\text{-}2)$$

The cutoff frequency, f_c, for an *RC* filter circuit can be determined by

$$f_c = \frac{1}{2\pi RC} \qquad \text{hertz} \qquad (29\text{-}3)$$

The *LR, resistor-to-ground* circuit of Fig. 26-6 is a low-pass filter because the series inductive reactance is very low at low frequencies and appears as an open circuit at the high frequencies.

The *output voltage* for an *LR*, resistor-to-ground, low-pass filter circuit is

$$V_{out} = \frac{R}{\sqrt{R^2 + X_L^2}} \times V_{in} \quad \text{volts} \qquad (29\text{-}4)$$

The attenuation ratio for an *LR*, resistor-to-ground, low-pass filter circuit is

$$\frac{V_{out}}{V_{in}} = \frac{R}{\sqrt{R^2 + X_L^2}} \qquad (29\text{-}5)$$

The *LC*, *capacitor-to-ground,* inverted *L*-type circuit of Fig. 29-7 is a low-pass filter because the series inductive reactance is very low at low frequencies while the capacitive reactance is very high, and nearly all of the input voltage will appear across the capacitive reactance as output voltage.

The output voltage for an *LC*, capacitor-to-ground, low-pass filter circuit is

$$V_{out} = \frac{X_C}{X_L - X_C} \times V_{in} \quad \text{volts} \qquad (29\text{-}7)$$

The *RL*, *inductor-to-ground* circuit of Fig. 29-8 is a high-pass filter because at low frequencies the reactance of the inductor approaches zero ohms; therefore, the output voltage is effectively shorted and approaches zero.

The *output voltage* for an *RL*, inductor-to-ground, high-pass filter circuit is

$$V_{out} = \frac{X_L}{\sqrt{R^2 + X_L^2}} \times V_{in} \quad \text{volts} \qquad (29\text{-}9)$$

The *attenuation ratio* for an *RL*, inductor-to-ground, high-pass filter circuit is

$$\frac{V_{out}}{V_{in}} = \frac{X_L}{\sqrt{R^2 + X_L^2}} \qquad (29\text{-}10)$$

The *CR*, *resistor-to-ground* circuit of Fig. 29-9 is a high-pass filter because at lower frequencies the capacitive reactance approaches infinite ohms; therefore, the output voltage is effectively open circuited and approaches zero.

The *output voltage* for a *CR*, resistor-to-ground, high-pass filter circuit is

$$V_{out} = \frac{R}{\sqrt{R^2 + X_C^2}} \times V_{in} \quad \text{volts} \qquad (29\text{-}11)$$

The *attenuation ratio* for a *CR*, resistor-to-ground, high-pass filter circuit is

$$\frac{V_{out}}{V_{in}} = \frac{R}{\sqrt{R^2 + X_C^2}} \qquad (29\text{-}12)$$

The *series LC resonant bandpass* filter circuit of Fig. 29-10 passes a range of frequencies that depend on the *Q* of the circuit.

The *parallel LC resonant bandpass* filter circuit of Fig. 29-11 passes a range of frequencies that depend on the *Q* of the circuit.

The *series LC resonant band-reject* filter circuit of Fig. 29-12 rejects a range of frequencies that depend on the *Q* of the circuit.

The *parallel LC resonant band-reject* circuit of Fig. 29-13 rejects a range of frequencies that depend on the *Q* of the circuit.

The *L-, π, and T-type* cascaded filters are known as constant-*K* filters.

In a *constant-K filter* the values of *L* and *C* are chosen so that the product of X_L and X_C is constant at all frequencies. As a result, the input and output impedances of the filter are matched to the source and load, respectively.

The *terminating resistance* of a constant-*K* filter can be found by

$$R = \sqrt{X_L X_C} \quad \text{ohms} \tag{29-13}$$

$$R = \sqrt{\frac{L}{C}} \quad \text{ohms} \tag{29-14}$$

The *reactance of a bypass capacitor* must be at least one-tenth of the shunt resistance at the lowest frequency for effective bypassing.

The reactance of a coupling capacitor must be at least one-tenth of the load resistance at the lowest frequency for effective coupling.

REVIEW QUESTIONS

29-1. **(a)** Draw a schematic diagram of an *RC*, capacitor-to-ground filter circuit.
(b) Describe briefly why it is termed a low-pass filter.
(c) Assume that $R = 10\ \text{k}\Omega$ and $C = 0.05\ \mu\text{F}$. Find the cutoff frequency for the circuit.
(d) What is the output voltage if V_{in} is 5 V at 636 Hz?
(e) What is the filter attenuation ratio for the conditions in the circuit part (d)?

29-2. **(a)** Draw a schematic diagram of an *LR*, resistor-to-ground filter circuit.
(b) Describe briefly why it is termed a low-pass filter.
(c) Assume that $R = 10\ \text{k}\Omega$ and $L = 3$ H. Find the cutoff frequency for the circuit.
(d) What is the output voltage if V_{in} is 5 V at 1061 Hz?
(e) What is the filter attenuation ratio for the conditions in the circuit of part (d)?

29-3 **(a)** Draw a schematic diagram of an *LC*, capacitor-to-ground filter circuit.
(b) Describe briefly why it is termed a low-pass filter.
(c) Assume that $L = 7$ H and $C = 20\ \mu\text{F}$. Find the output voltage of the filter if the ac ripple voltage has a frequency of 60 Hz and is 100 V peak-to-peak.

29-4. **(a)** Draw a schematic diagram of an *RL*, inductor-to-ground filter circuit.
(b) Describe briefly why it is termed a high-pass filter.

29-5. **(a)** Draw a schematic diagram of a *CR*, resistor-to-ground filter.
(b) Describe briefly why it is termed a high-pass filter.
(c) Assume that $C = 20\mu\text{F}$ and $R = 5\ \text{k}\Omega$. Find the cutoff frequency for the circuit.
(d) What will be the value of the output voltage at the cutoff frequency if the input is 5 V?

29-6. **(a)** Draw a schematic diagram of a series resonant bandpass filter.
(b) Explain briefly why the *Q* of the series *LC* resonant circuit will determine the bandpass capability of the filter.

29-7. (a) Draw a schematic diagram of a parallel resonant band-pass filter.
(b) Explain briefly why it is termed a bandpass filter.
(c) Given that the inductance of the parallel bandpass circuit is 4.7 mH with an internal resistance of 65 Ω connected across a capacitor of 800 pF, find the upper and lower bandpass frequencies.

29-8. (a) Draw a schematic diagram of a series resonant band-reject filter.
(b) Explain briefly why it is termed a series resonant band-reject filter.

29-9. (a) Draw a schematic diagram of a parallel resonant band-reject filter.
(b) Explain briefly why it is termed a parallel band-reject filter.

29-10. Explain briefly why the T- or π-type *LC* filter is most suited to signal voltage sources that have a low source impedance.

29-11. What is constant-*K* filter?

29-12. What is the function of a bypass capacitor?

29-13. Explain the function of a coupling capacitor.

29-14. The value of the coupling or bypass capacitor is chosen so that its reactance, X_C, is at least one-tenth of R_L at the lowest frequency. Explain briefly.

SELF-EXAMINATION

Choose the most suitable word or phrase to complete each statement.

29-1. The term "low-pass filter" applies to a configuration of *L* or *C* and *R* that passes a voltage having __________ frequencies while attenuating the __________ frequencies.

29-2. The term "high-pass filter" applies to a configuration of *L* or *C* and *R* that passes a voltage having __________ frequencies while attenuating the __________ frequencies.

29-3. A __________ filter passes a range of frequencies termed bandwidth and rejects the lower and upper frequency portions of the input voltage.

29-4. A band-reject filter rejects a range of frequencies termed bandwidth and __________ the lower and upper frequency portions of the input voltage.

Refer to Fig. 29-17 for Questions 29-5 to 29-8.

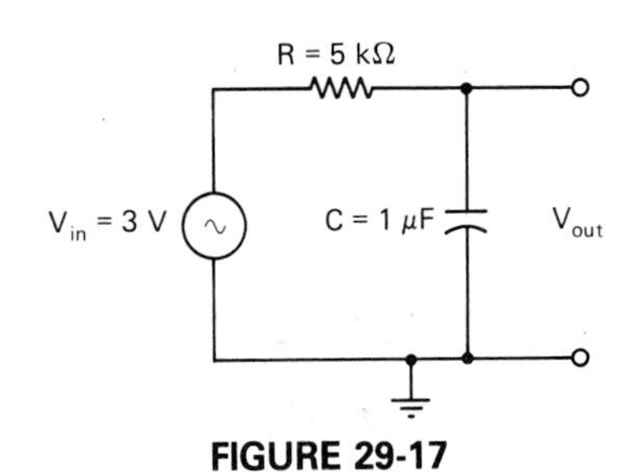

FIGURE 29-17

29-5. What type of filter circuit is represented in the schematic of Fig. 29-17?

29-6. Find the cutoff frequency for the filter.

29-7. What is the output voltage if the frequency of V_{in} is 128 Hz?

29-8. What is the output voltage at the cutoff frequency?

Refer to Fig. 29-18 for Questions 29-9 to 29-12.

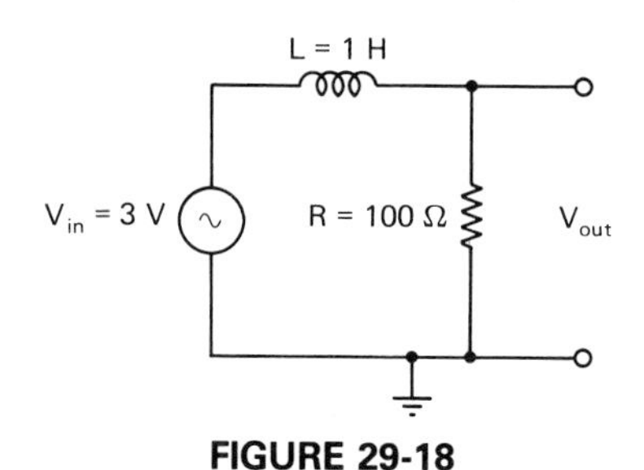

FIGURE 29-18

29-9. What type of filter circuit is represented in the schematic of Fig. 29-18?
29-10. Find the cutoff frequency for the circuit.
29-11. What is the output voltage if the frequency of V_{in} is 60 Hz?
29-12. What is the output voltage at the cutoff frequncy?

Refer to Fig. 29-19 for Questions 29-13 to 29-16.

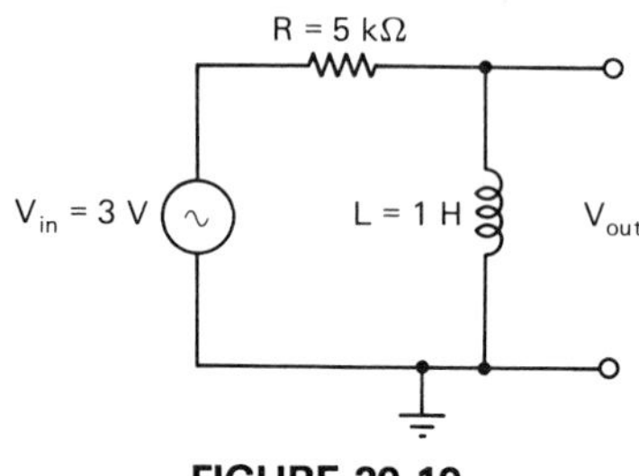

FIGURE 29-19

29-13. What type of filter circuit is represented in the schematic of Fig. 29-19?
29-14. Find the cutoff frequency for the filter.
29-15. What is the filter attenuation ratio at 398 Hz?
29-16. What is the output voltage if the frequency of V_{in} is 398 Hz?

Refer to Fig. 29-20 for Questions 29-17 to 29-20.

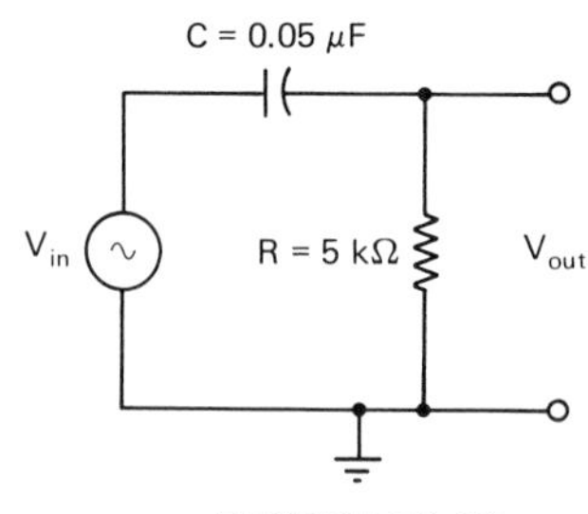

FIGURE 29-20

29-17. What type of filter circuit is represented in the schematic of Fig. 29-20?
29-18. Find the cutoff frequency for the filter.
29-19. What is the filter attenuation ratio at 212 Hz?
29-20. What is the output voltage if the frequency of V_{in} is 212 Hz?

Refer to Fig. 29-21 for Questions 29-21 to 29-23.

29-21. What type of filter circuit is represented in the schematic of Fig. 29-21?
29-22. Find the bandwidth of the filter circuit.

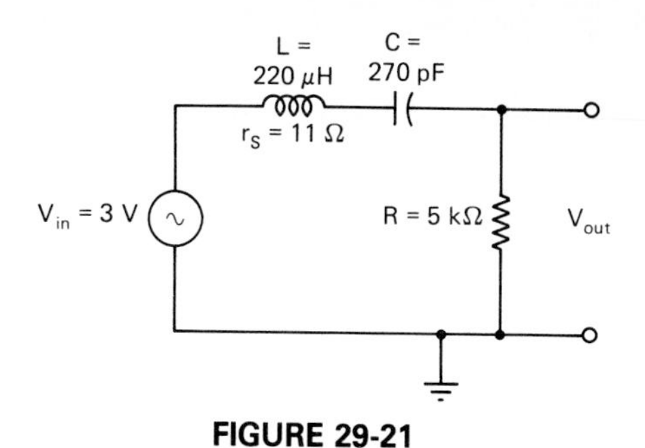

FIGURE 29-21

29-23. Find the output voltage at the resonant frequency.

Refer to Fig. 29-22 for Questions 29-24 to 29-27.

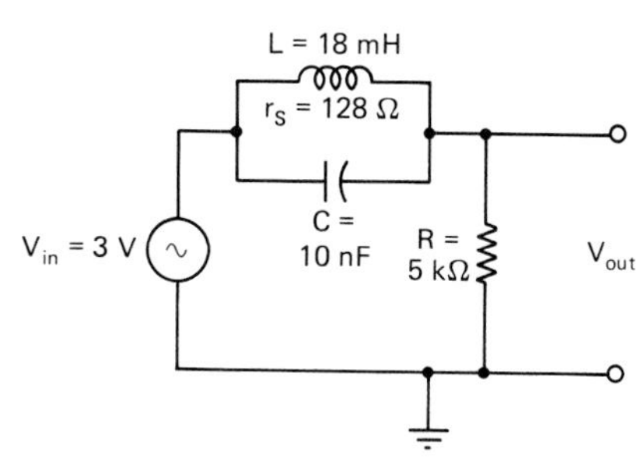

FIGURE 29-22

29-24. What type of filter circuit is represented in the schematic of Fig. 29-22?

29-25. Find the bandwidth of the filter circuit.

29-26. What is the lower band frequency that this circuit will reject? The upper band frequency?

29-27. What is the output voltage at the resonant frequency?

29-28. Three types of constant-*K* filters are ________, ________, and ________.

29-29. In a constant-*K* filter the input impedance is matched to the source impedance while the output impedance is matched to the load. This is done by choosing the value of *L* and *C* to make the ________ of X_L and X_C constant.

29-30. For effective bypassing of an ac signal voltage the reactance of the bypass capacitor must not be greater than $\frac{1}{10}$ of the resistance at the ________ frequency to be bypassed.

29-31. A coupling capacitor couples the ac signal voltage from one amplifier stage to another and appears as a ________-pass filter circuit.

29-32. A coupling capacitor appears as an open circuit to the ________ ________ voltage on the transistor amplifier.

ANSWERS TO PRACTICE QUESTIONS

29-1

1. low, high
2. high, low
3. bandpass
4. band-reject
5. 0.65 V
6. 0.5 to 1
7. 398 Hz
8. 0.92 V
9. 0.5 μF
10. 2.1 V

30

Ac Power

INTRODUCTION

In a dc circuit the heat released by a device or component is the real power, or product of *VI* dissipated. A capacitor in an ac circuit will store energy in the form of an electric field which it returns to the source. Conversely, an inductor will store energy in the form of an electromagnetic field which it returns to the source. The true power that is drawn from the source by an RC or *RL* circuit is no longer the product of *V* and *I* because of this return of energy. The capacitor and inductor are termed wattless components.

In this chapter we study the effects on the actual or real power dissipated by a pure resistance, capacitance, inductance, and combined *RC* and *RL* circuit. We learn how to correct for apparent power.

MAIN TOPICS IN CHAPTER 30

OBJECTIVES

After studying Chapter 30, the student should be able to

1. Master the concepts of power in ac resistive, capacitive, inductive, and *RL* circuits.
2. Solve and correct for power factor in an ac circuit.

30-1 AC POWER IN A RESISTIVE CIRCUIT

The voltage and current waveforms of Fig. 30-1(a) are in phase in a pure resistive circuit because their instantaneous values v and i reach their positive and negative peaks simultaneously. The power being supplied by the source and dissipated by the resistor will also vary since the instantaneous power is the product of the instantaneous voltage and current, $p = v \times i$.

Plotting the product of the instantaneous v_R and i values of Fig. 30-1(a) will give the power waveform of Fig. 31-1(b). The maximum or peak power takes place when both instantaneous values of voltage and current reach their peak values. Conversely, the power reaches a minimum or zero value when the instantaneous voltage and current are zero, as shown in Fig. 30-1(b).

During the negative ac alternation, the power pulse is repeated since the current and voltage direction

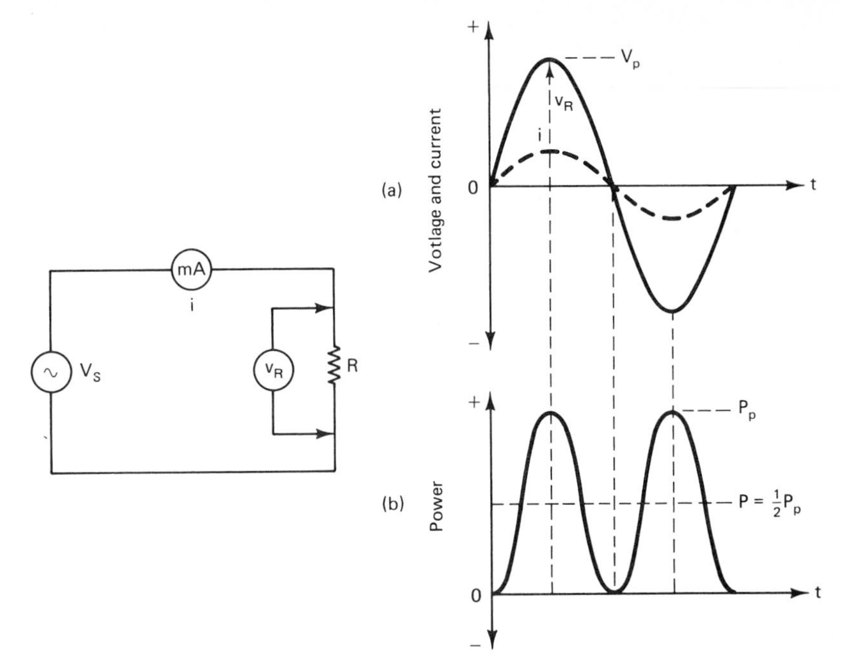

FIGURE 30-1 Power in a pure resistive circuit. The power waveform is not a sine wave since power varies as the square of the current ($P = I^2R$).

does not make any difference to the power that must be supplied by the source and resulting heat dissipated by the resistor.

Since the power is continually varying, the power that is actually dissipated by the resistor will be an *average power*. The average power, P_{avg}, can be obtained by finding the instantaneous power at various points along the ac cycle and finding the average of these values. Adding these values and dividing by the total number of values will give an approximate value for the average power value. The exact average value of ac power is actually one-half of the peak power, as shown in Fig. 30-1(b). In equation form,

$$P_{avg} = \frac{1}{2} V_p \times I_p \quad \text{watts} \qquad (30\text{-}1a)$$

Substituting V_p/R for I_p gives

$$P_{avg} = \frac{1}{2} \times \frac{V_p^2}{R} \quad \text{watts} \qquad (30\text{-}1)$$

EXAMPLE 30-1

What is the average power dissipated by a 100-Ω resistor with 100 V peak applied across it?

Solution: Using equation (30-1) gives

$$P_{avg} = \frac{1}{2} \times \frac{(100V_p)^2}{100\ \Omega} = 50\ \text{W}$$

Since dc power can be given by $(V_{dc})^2/R$, the average or ac power will provide the same heating effect as dc if the voltage and current are in rms values, where the rms value = 0.707 × V_p; therefore,

$$P_{avg} = \frac{V^2}{R} = \frac{(0.707 \times V_p) \times (0.707 \times V_p)}{R}$$

$$= \frac{1}{2} \times \frac{V_p^2}{R}$$

The power dissipated by the resistor in Example 30-1 can now be shown to be equivalent to dc by converting the voltage to rms:

$$V_{rms} = 0.707 \times V_p = 0.707 \times 100\ V_p = 70.7\ \text{V}$$

Therefore,

$$P = \frac{V^2}{R} = \frac{(70.7\ \text{V})^2}{100\ \Omega} = 50\ \text{W}$$

In a pure resistive ac circuit the actual or *true power* that is dissipated by a resistive component will be equivalent to dc and can be determined from either power equation, VI, I^2R, or V^2/R, where rms values for voltage and current are used.

The true power is the power dissipated by the resistor in the form of heat energy. Unlike the capacitor and inductor, this energy is not stored and is never returned to the source. Furthermore, the resistance of the resistor is not affected by frequency, so its power dissipation is unaffected by a change in the source frequency. The symbol for true power is P_T. The three dc power equations then apply to the true power dissipated by a resistor in an ac circuit:

$$P_T\,(\text{true power}) = V \times I \qquad \text{watts} \qquad (30\text{-}2)$$

$$P_T\,(\text{true power}) = I^2R \qquad \text{watts} \qquad (30\text{-}3)$$

$$P_T\,(\text{true power}) = \frac{V^2}{R} \qquad \text{watts} \qquad (30\text{-}4)$$

30-2 AC POWER IN A CAPACITIVE CIRCUIT

The voltage and current waveform of Fig. 30-2(a) shows the current leading the voltage, V_C, by 90° in a pure capacitive ac circuit. As in the case of the resistive circuit, the power being supplied by the source to charge the capacitor during one-quarter of a cycle will depend on the product of the instantaneous voltage and current, $P = v_C \times i$.

Plotting the product of the instantaneous v_C and i values of Fig. 30-2(a) will give the power waveform of Fig. 30-2(b). Energy from the source is supplied to the capacitor only during one-quarter of a cycle, when the capacitor is being charged. Once charged to the peak of the voltage source, the capacitor's stored energy is returned to the source during the next quarter-cycle, when the capacitor discharges. Thus there are two positive and two negative power peaks.

The negative power peaks indicate that energy is being returned to the source and that the resultant power supplied by the source is zero. Since there are two positive and two negative power peaks during one cycle of capacitor voltage, the *average power is zero.*

Although practically, the capacitor does have a very high internal resistance, the power dissipated by a capacitor due to its internal resistance can be considered to be negligible. Then the actual or *true power,* P_T, dissipated is zero, and the capacitor is termed a *wattless component.*

Reactive Power in a Capacitive Circuit

In an ac circuit at any time interval the capacitor is either being charged by the source current and voltage or is discharging back into the source. The product of this rms capacitor voltage and the circuit current is called the *reactive power.*

Reactive power, P_X, represents the amount of power being transferred back and forth between the source and the capacitor. The three dc power equations for the resistive circuit are used to find the reactive power. However, to distinguish the true power, the unit of reactive power is specified in *volt-amperes reactive* (var). Thus for a capacitive circuit,

$$P_X = V_C \times I_C \qquad \text{volt-amperes reactive} \qquad (30\text{-}5)$$

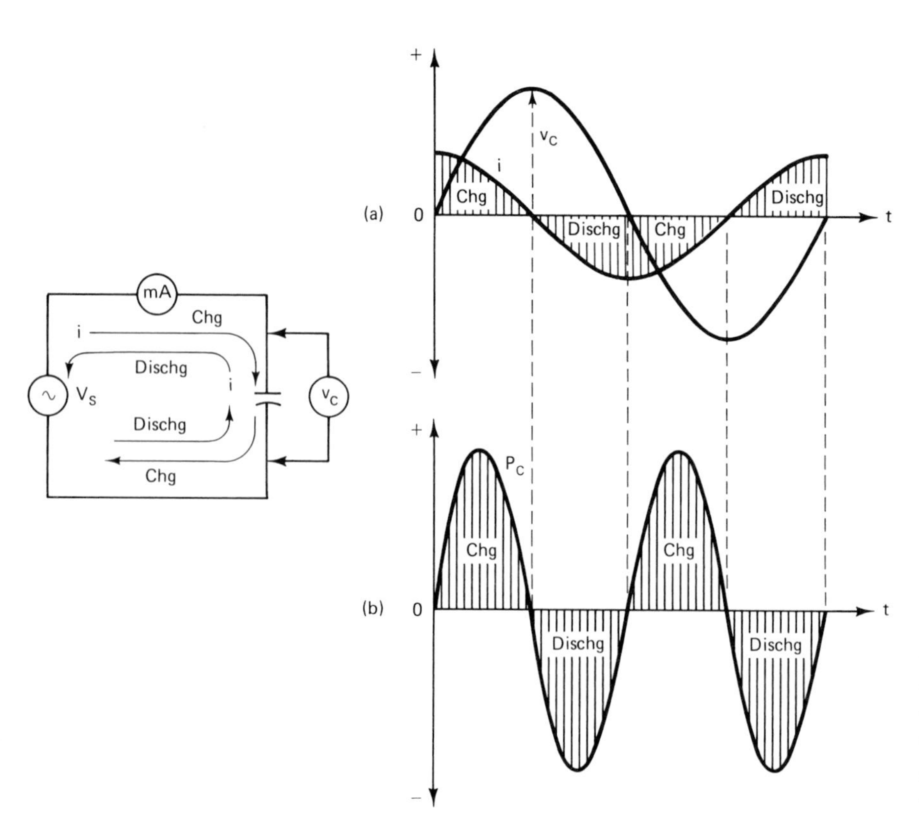

FIGURE 30-2 Power in a pure capacitive circuit. Positive values of power indicate that the capacitor is being charged. Negative values of power indicate that the capacitor is discharging back to the source.

$$P_X = I_C^2 \times X_C \quad \text{volt-amperes reactive} \quad (30\text{-}6)$$

$$P_X = \frac{V_C^2}{X_C} \quad \text{volt-amperes reactive} \quad (30\text{-}7)$$

EXAMPLE 30-2

A 10-μF capacitor is connected across a 120-V 60-Hz line.

(a) What is the true power dissipated by the capacitor?
(b) What is the reactive power of the circuit?

Solution:

(a) Practically, the leakage resistance of a capacitor can be ignored and its true power dissipation is zero.
(b) To find the reactive power, we must first find the reactance of the capacitor:

$$X_C = \frac{1}{2\pi fC} = 265\ \Omega$$

The reactive power can be found by using equation (30-7):

$$P_X = \frac{V^2}{X_C} = \frac{(120\ \text{V})^2}{265\ \Omega} = 54.3\ \text{var}$$

30-3 AC POWER IN AN *RC* CIRCUIT

Power in the Series *RC* Circuit

The *true power* in the series *RC* circuit of Fig. 30-3 will be the actual power dissipated by the resistor since the resistor has a nonreactive resistance to the frequency of the source current.

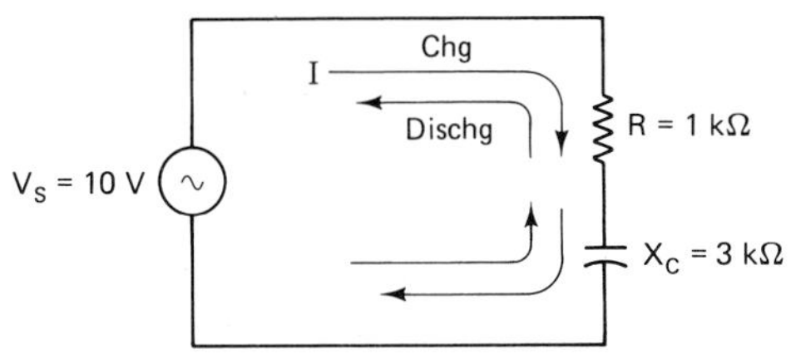

FIGURE 30-3 Series *RC* circuit.

Since current is common in a series circuit, the charging and discharging action of the capacitor or reactive power (var) will affect the total energy delivered to the *RC* circuit by the source.

Apparent Power

The product of $V_S I_T$, the power supplied by the source, is no longer the true power of the circuit. $V_S I_T$ represents the amount of power that is *apparently* being supplied to the circuit due to both the true power drawn by the resistor and the reactive power drawn by the capacitor.

The letter symbol for apparent power is P_A and the unit is volt-amperes, VA. The product *VI* equation for apparent power is then shown as

$$P_A = V_S I_T \quad \text{volt-amperes} \quad (30\text{-}8)$$

Since the total circuit current is dependent on the impedance of the circuit, or $I_T = V_S/Z$, the alternative equations for apparent power can be used provided that the values are in rms units:

$$P_A = I^2 Z \quad \text{volt-amperes} \quad (30\text{-}9)$$

$$P_A = \frac{V^2}{Z} \quad \text{volt-amperes} \quad (30\text{-}10)$$

EXAMPLE 30-3

Find (a) the true power, (b) the reactive power, and (c) the apparent power in the series *RC* circuit of Fig. 30-3.

Solution: The impedance and the circuit current must be found first.

$$Z = \sqrt{R^2 + X_C^2} = \sqrt{(1\ \text{k}\Omega)^2 + (3\ \text{k}\Omega)^2}$$
$$= 3.16\ \text{k}\Omega$$

and

$$I_T = \frac{V_S}{Z} = \frac{10\ \text{V}}{3.16\ \text{k}\Omega} = 3.2\ \text{mA}$$

(a) Equation (30-3) gives true power:

$$P_T = I^2 R = (3.2\ \text{mA})^2 \times 1\ \text{k}\Omega = 10.2\ \text{mW}$$

(b) Equation (30-6) gives reactive power:

$$P_X = I^2 X_C = (3.2\ \text{mA})^2 \times 3\ \text{k}\Omega = 30.7\ \text{mvar}$$

(c) Equation (30-9) gives apparent power:

$$P_A = I^2 Z = (3.2\ \text{mA})^2 \times 3.16\ \text{k}\Omega = 32.4\ \text{mVA}$$

The Power Triangle

The impedance phasor of Fig. 30-4(a) is representative of the resistance and capacitive reactance in the series *RC* circuit. The phase angle, θ, between *R* and *Z* is also the angle by which the current leads the source voltage.

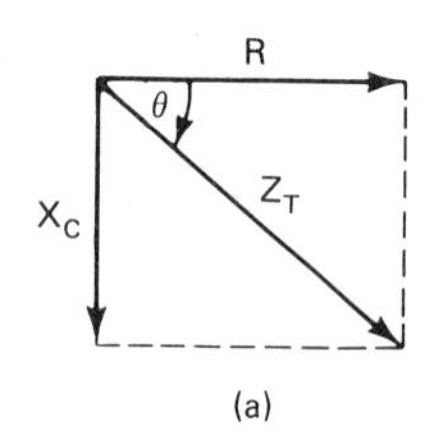

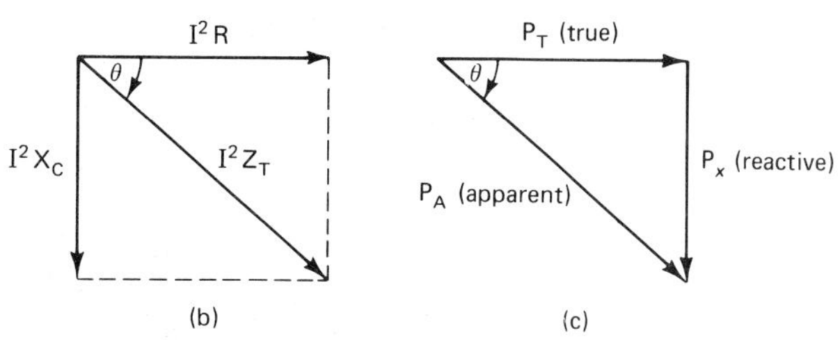

FIGURE 30-4 Power triangle for an *RC* impedance network, $P_A = \sqrt{P_T^2 + P_X^2}$.

Respective power, P_T, P_X, and P_A, can be represented by the same phasor provided that the phasor magnitude of each power is representative of the current squared, I^2, as shown in the phasor diagram of Fig. 30-4(b). From this phasor diagram we derive the power triangle shown in Fig. 30-4(c).

Using the rules for a right-angle triangle, the relationship for P_A and P_T can be stated as follows:

$$P_A = \sqrt{P_T^2 + P_X^2} \quad \text{volt-amperes} \quad (30\text{-}11)$$

$$P_T = P_A \cos\theta \quad \text{watts} \quad (30\text{-}12)$$

EXAMPLE 30-4

Find the true power in the series *RC* circuit of Fig. 30-3 given that the current leads the applied voltage by 45°.

Solution: When the phase angle between the applied voltage and the current is 45°, R will be equal to X_C. In this case $R = X_C = 1\ \text{k}\Omega$.

The apparent power can be found using equation (30-9), but we must find the total impedance and current first.

$$Z = \sqrt{R^2 + X_C^2} = \sqrt{(1\ \text{k}\Omega)^2 + (1\ \text{k}\Omega)^2}$$
$$= 1.414\ \text{k}\Omega$$

$$I = \frac{V}{Z} = \frac{10\ \text{V}}{1.414\ \text{k}\Omega} = 7.07\ \text{mA}$$

Therefore,

$$P_A = I^2Z = (7.07\ \text{mA})^2 \times 1.414\ \text{k}\Omega$$
$$= 70.7\ \text{mVA}$$

Using equation (30-12), we can find the true power,

$$P_T = P_A \cos\theta = 70.7\ \text{mVA} \times \cos 45° = 50\ \text{mW}$$

Power in the Parallel *RC* Circuit

As was the case in the series *RC* circuit, the *true power* dissipated by the resistor will be lost to the surrounding environment. The true power can be found by using the V^2/R equation provided that the values are rms.

$$P_T = \frac{V^2}{R} \quad \text{or} \quad I_R^2 R \quad \text{watts} \quad (30\text{-}13)$$

The capacitor stores its energy when it is being charged and returns the same amount of energy to the source on the discharge cycle. Thus the capacitor draws a reactive power from the source which can be found from

$$P_X = \frac{V^2}{X_C} \quad \text{or} \quad I_C^2 X_C \quad \text{volt-amperes reactive} \quad (30\text{-}14)$$

The total amount of power that is apparently being supplied by the source due to both the true power drawn by the resistor and the reactive power drawn by the capacitor or the apparent power can be found from

$$P_A = \frac{V^2}{Z} \quad \text{or} \quad I_T^2 Z \quad \text{volt-amperes} \quad (30\text{-}15)$$

EXAMPLE 30-5

Find (a) the true power, (b) the reactive power, and (c) the apparent power in the parallel *RC* circuit of Fig. 30-5.

Solution:

(a) Using equation (30-13) gives the true power dissipated by the resistor.

$$P_T = \frac{V^2}{R} = \frac{(10\ \text{V})^2}{1\ \text{k}\Omega} = 100\ \text{mW}$$

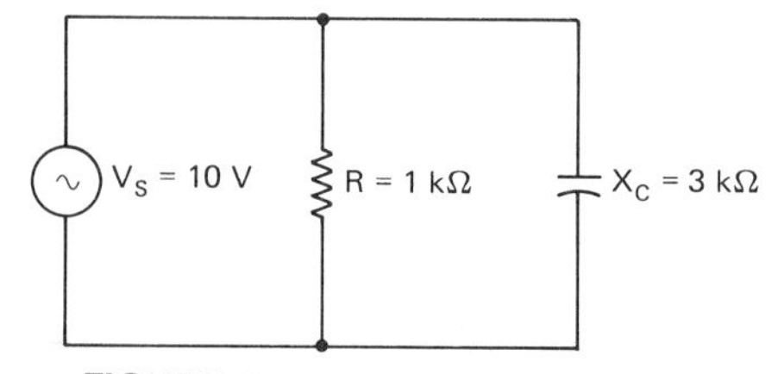

FIGURE 30-5 Parallel *RC* circuit.

(b) Using equation (30-14) gives the reactive power drawn by the capacitor.

$$P_X = \frac{V^2}{X_C} = \frac{(10\ \text{V})^2}{3\ \text{k}\Omega} = 33.3\ \text{mvar}$$

(c) Using equation (30-15) gives the apparent power supplied by the source. Calculating for impedance first gives

$$Z = \frac{R \times X_C}{\sqrt{R^2 + X_C^2}} = \frac{1\ \text{k}\Omega \times 3\ \text{k}\Omega}{\sqrt{(1\ \text{k}\Omega)^2 + (3\ \text{k}\Omega)^2}}$$

$$= 0.949\ \text{k}\Omega$$

$$P_A = \frac{V^2}{Z} = \frac{(10\ \text{V})^2}{0.949\ \text{k}\Omega} \cong 105\ \text{mVA}$$

30-4 AC POWER IN AN INDUCTIVE CIRCUIT

The voltage and current waveform of Fig. 30-6 shows the current lagging the voltage V_L by 90° in a pure inductive ac circuit. The power being supplied by the source to build up the magnetic flux is taken during one-quarter of a cycle and will depend on the product of the instantaneous voltage and current, $P = v_L \times i$.

Plotting the product of the instantaneous v_L and i values of Fig. 30-6(a) will give the power waveform of Fig. 30-6(b). Energy is returned to the source during the collapse or fall of the magnetic flux. During the next quarter of a cycle, current flows from the source to build up the magnetic flux and energy is stored in the inductor in the form of this magnetic flux. Thus there are two positive and two negative power peaks.

The negative power peaks indicate that energy is being returned to the source and the resultant power supplied by the source is zero. Since there are two positive and two negative power peaks during one cycle of inductor current, the *average power is zero*.

Practically, the inductor has some internal resistance in the winding so that some small amount of power will be dissipated continuously. For our discussion and analysis we approximate and consider the inductor to be ideal; therefore, its true power is zero, and the inductor is a wattless component.

Reactive Power in an Inductive Circuit

In an ac circuit at any instant the source is either supplying current to build up the magnetic flux [flux rising; see Fig. 30-6(a)], or the magnetic flux is falling and returning a current back to the source. The product of this rms

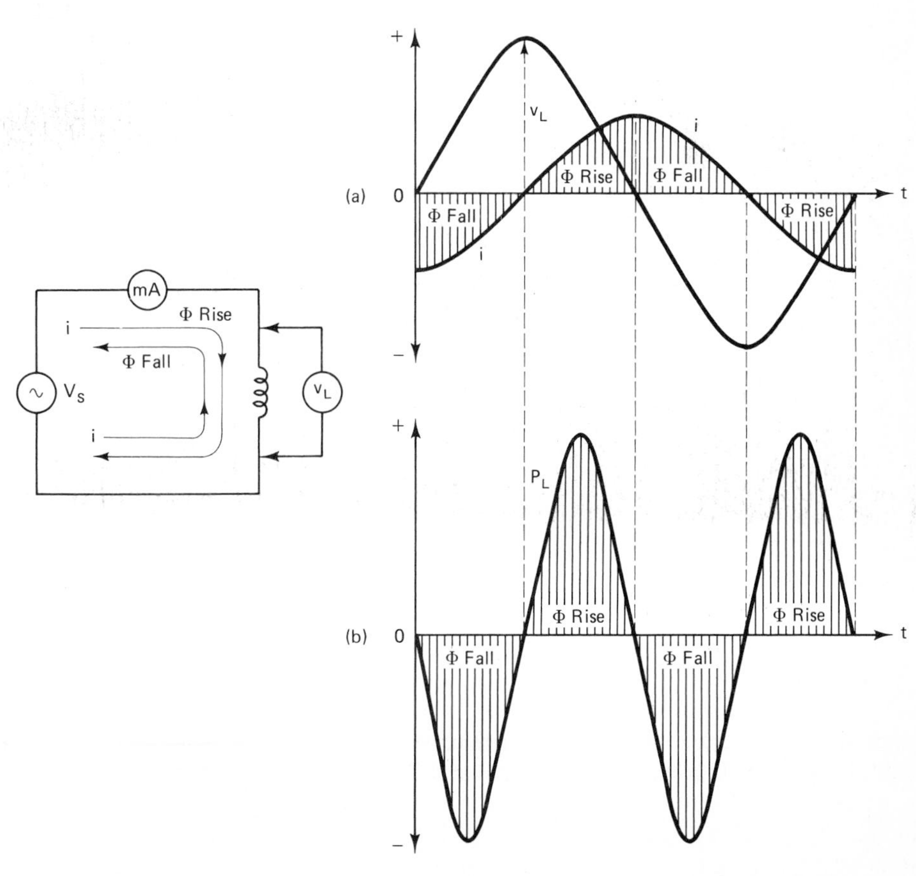

FIGURE 30-6 Power in a pure inductive circuit. Negative values of power indicate that the inductor's energy is being returned to the source. Positive power values indicate that energy is being stored by the inductor.

inductive voltage and the circuit current is the reactive power.

As was the case in the capacitive circuit, reactive power, P_X, represents the amount of power being transferred back and forth between the source and the inductor. The three dc power equations are used to find the reactive power. The unit of reactive power for the inductor is identical to the capacitive circuit and is the volt-ampere reactive (var). Thus for an inductive circuit,

$$P_X = V_L \times I_L \quad \text{volt-amperes reactive} \quad (30\text{-}16)$$

$$P_X = I_L^2 \times X_L \quad \text{volt-amperes reactive} \quad (30\text{-}17)$$

$$P_X = \frac{V_L^2}{X_L} \quad \text{volt-amperes reactive} \quad (30\text{-}18)$$

EXAMPLE 30-6

A 10-H choke is connected across a 24-V, 120-Hz line.

(a) What is the true power dissipated by the choke?
(b) What is the reactive power of the circuit?

Solution:

(a) Practically, the internal resistance of the choke winding will dissipate some true power; therefore, the true power dissipation will be slightly greater than zero.
(b) To find the reactive power, we must first find the reactance of the inductor:

$$X_L = 2\pi f L = 7536\ \Omega$$

The reactive power can be found by using equation (30-18):

$$P_X = \frac{V_L^2}{X_L} = \frac{(24\ \text{V})^2}{7536\ \Omega} = 76.4\ \text{mVA}$$

30-5 AC POWER IN A SERIES *RL* CIRCUIT

The *true power* in the series *RL* circuit of Fig. 30-7 will be the actual power dissipated by the resistor since the resistor has a nonreactive resistance to the frequency of the source current.

Since current is common is a series circuit, the energy being stored in the magnetic field of the inductor and then being returned back to the source by the collapsing magnetic field will be part of the total energy circulating in the circuit. This reactive power (var) will affect the total energy delivered to the *RL* circuit by the source.

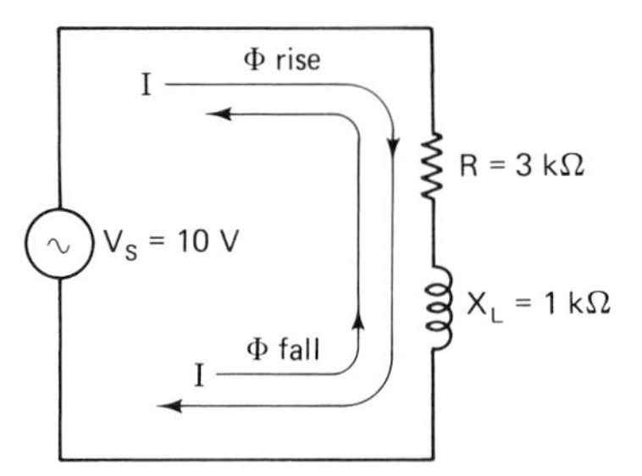

FIGURE 30-7 Series *RL* circuit.

Apparent Power

The product of $V_S I_T$, the power supplied by the source, is no longer the true power of the circuit. $V_S I_T$ represents the amount of power that is apparently being supplied to the circuit due to both the true power drawn by the resistor and the reactive power drawn by the capacitor.

As in the case of the *RC* circuit, the apparent power is found using the *VI* product and can be found using equation (30-8) as well as equations (30-9) and (30-10).

EXAMPLE 30-7

Find (a) the true power, (b) the reactive power, and (c) the apparent power in the series *RL* circuit of Fig. 30-7.

Solution: The impedance and the circuit current must be found first.

$$Z = \sqrt{R^2 + X_L^2}$$
$$= \sqrt{(3\ \text{k}\Omega)^2 + (1\ \ \text{k}\Omega)^2} = 3.16\ \text{k}\Omega$$

and

$$I_T = \frac{V_S}{Z} = \frac{10\ \text{V}}{3.16\ \text{k}\Omega} = 3.2\ \text{mA}$$

(a) Equation (30-3) gives true power:

$$P_T = I^2 R = (3.2\ \text{mA})^2 \times 3\ \text{k}\Omega = 30.7\ \text{mW}$$

(b) Equation (30-17) gives reactive power:

$$P_X = I^2 X_L = (3.2\ \text{mA})^2 \times 1\ \text{k}\Omega = 10.2\ \text{mVA}$$

(c) Equation (30-9) gives apparent power:

$$P_A = I^2 Z = (3.2\ \text{mA})^2 \times 3.16\ \text{k}\Omega = 32.4\ \text{mVA}$$

The Power Triangle

The impedance phasor of Fig. 30-8(a) is representative of the resistance and inductive reactance in the series *RL* circuit. The phase angle, θ, between *R* and *Z* is also the angle by which the current lags the source voltage.

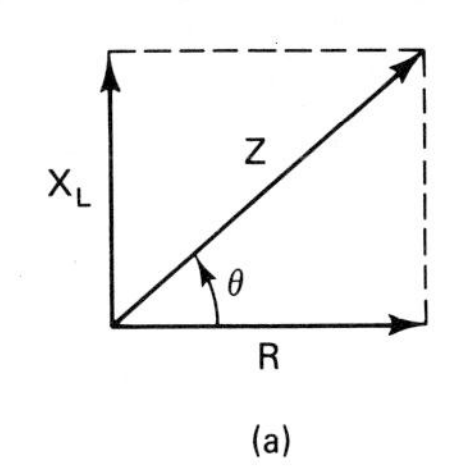

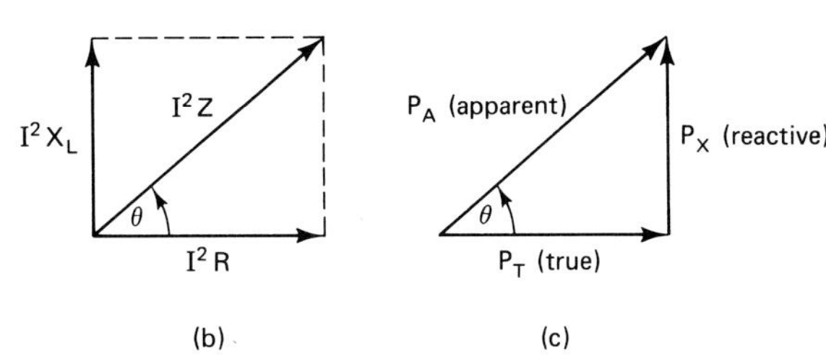

FIGURE 30-8 Power triangle for an *RL* impedance network.

Respective power, P_T, P_X, and P_A, can be represented by the same phasor provided that the phasor magnitude of each power is representative of the current squared, I^2, as shown in the phasor diagram of Fig. 30-8(b). From this phasor diagram we derive the power triangle shown in Fig. 30-8(c).

Using the rules for a right-angle triangle, the relationship for P_A and P_T can be stated as in the case of the *RC* circuit [see equations (30-11) and (30-12)].

EXAMPLE 30-8

Find the true power in the series *RL* circuit of Fig. 30-7 given that R is representative of the inductor's internal resistance and is equal to 100 Ω. The apparent power is 100 mVA.

Solution: Equation (30-12), $P_T = P_A \cos\theta$, can be used if the angle θ is found.

$$\theta = \arctan\frac{X_L}{R} = \arctan\frac{1000\ \Omega}{100\ \Omega}$$

Therefore,

$$\theta = 84.3°$$

and

$$P_T = P_A \cos\theta = 100\ \text{mVA} \times \cos 84.3°$$
$$= 9.9\ \text{mW} \quad \text{or} \quad \text{approx. } 10\ \text{mW}$$

Power in the Parallel *RL* Circuit

In the parallel *RL* circuit of Fig. 30-9, the true power dissipated by the resistor can be found using one of the three power terms $V_R I_R$, $I_R^2 R$, or V_R^2/R. The reactive power drawn by the inductor can be found using equation (30-18) or equation (30-17).

$$P_X = \frac{V^2}{X_L} \quad \text{or} \quad I_L^2 X_L$$

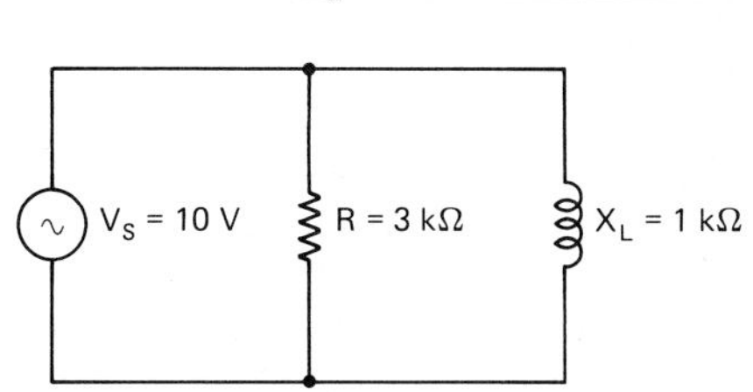

FIGURE 30-9 Parallel *RL* circuit.

EXAMPLE 30-9

Find (a) the true power, (b) the reactive power, and (c) the apparent power in the parallel *RL* circuit of Fig. 30-9.

Solution:

(a) Using equation (30-13) gives the true power dissipated by the resistor.

$$P_T = \frac{V^2}{R} = \frac{(10\ \text{V})^2}{3\ \text{k}\Omega} = 33.3\ \text{mW}$$

(b) Using equation (30-18) gives the reactive power drawn by the inductor.

$$P_X = \frac{V^2}{X_L} = \frac{(10\ \text{V})^2}{1\ \text{k}\Omega} = 100\ \text{mvar}$$

(c) Using equation (30-15) gives the apparent power supplied by the source. Calculating for impedance gives

$$Z = \frac{R \times X_L}{\sqrt{R^2 + X_L^2}} = \frac{3\ \text{k}\Omega \times 1\ \text{k}\Omega}{\sqrt{(3\ \text{k}\Omega)^2 + (1\ \text{k}\Omega)^2}}$$
$$= 0.949\ \text{k}\Omega$$

$$P_A = \frac{V^2}{Z} = \frac{(10\ \text{V})^2}{0.949\ \text{k}\Omega} \cong 105\ \text{mVA}$$

PRACTICE QUESTIONS 30-1

Choose the most suitable word, phrase, or value to complete each statement.

1. The power dissipated by a resistor is in the form of ________ energy and is lost to the surrounding ________.
2. Since the ac power is continually varying, the power that is dissipated by the resistor will be ________ power.

3. The exact average value of ac power is one-half of the ________ power.
4. The power dissipated by a resistor in an ac circuit is termed the ________ power.
5. The average power is ________ in a capacitive ac circuit.
6. The product of the rms capacitor voltage and the circuit current is called the ________ power.
7. The power delivered by the source to an RC circuit is called the ________ power.

Refer to the following circuit description for Questions 8 to 11: A 2-μF capacitor is connected in series with a 1-kΩ resistor to a 6-V 100-Hz source.

8. Find the true power of the circuit.
9. Find the reactive power of the circuit.
10. Find the apparent power of the circuit.
11. Find the phase angle between the leading current and source voltage.

Refer to the following circuit description for Questions 12 to 14: A 2-H choke is connected in parallel with a 1-kΩ resistor to a 6-V 100-Hz source.

12. Find the true power of the circuit.
13. Find the reactive power of the circuit.
14. Find the apparent power of the circuit.

30-6 POWER FACTOR

In a pure resistive circuit the true power dissipated by the resistor will be equal to the power supplied by the source. If the apparent power supplied by the source is the same as the true power dissipated by the load, the source current is in phase with the voltage. The phase angle in the power triangle of Fig. 30-8(c) would be zero degrees. This means that the true power dissipated in a combined reactive-resistive circuit cannot be greater than the apparent power supplied by the source to the circuit.

The ratio of true power to apparent power is called the *power factor* (PF) of the circuit or load and is

$$\text{PF} = \frac{P_T}{P_A} \quad \text{(no unit)} \qquad (30\text{-}19\text{a})$$

Since power factor is a ratio, it does not have a unit. More commonly, it is stated in terms of percentage; therefore,

$$\text{PF} = \frac{P_T}{P_A} \times 100 \quad \text{percentage} \qquad (30\text{-}19)$$

When PF is 100% the true power will be equal to the apparent power, so the circuit will appear to be totally resistive. If PF is less than 100%, the apparent power will be greater than the true power and the circuit will appear to have a combined reactive power (var) and true power (watts).

The power triangle of Fig. 30-8(c) shows that the cosine of the angle θ is P_T/P_A; therefore,

$$\text{PF} = \frac{P_T}{P_A} = \cos\theta \times 100 \quad \text{percentage} \qquad (30\text{-}20)$$

If the angle θ is 90°, it indicates that the circuit load is either totally inductive (+90°) [see Fig. 30-8(c)] or totally capacitive (−90°) [see Fig. 30-4(c)]; then the power factor is zero since cos 90° equals zero.

When the angle θ is 0°, the cosine of 0° is 1; therefore, the power factor is unity or 100%. An inductive load is said to have a *lagging power factor* since the load current will lag the source voltage. Conversely, a capacitive load is said to have a *leading power factor* since the current leads the applied voltage.

EXAMPLE 30-10

An ac motor or inductive load draws 5.6A from the 120-V source. If a wattmeter reads a true power of 420 W, find:

(a) The reactive power.
(b) The power factor of the circuit.
(c) The angle by which the current lags the source voltage.

Solution:

(a) Using the rules for a right-angle triangle, the reactive power in the power triangle of Fig. 30-8(c) can be found provided that we calculate the apparent power first. Equation (30-8) will give P_A:

$$P_A = I_S V_S$$
$$= 5.6\text{ A} \times 120\text{ V} = 672\text{ VA}$$
$$P_X = \sqrt{P_A^2 - P_T^2}$$
$$= \sqrt{(672\text{ VA})^2 - (420\text{ W})^2} = 525\text{ var}$$

(b) $$\text{PF} = \frac{P_T}{P_A} = \frac{420\text{ W}}{672\text{ VA}} \times 100 = 62.5\%$$

(c) $\cos\theta = \dfrac{P_T}{P_A}$ $\theta = 51.3°$

The current lags the applied voltage by 51.3° and the circuit has a lagging power factor.

30-7 CORRECTING FOR POWER FACTOR

In electronic circuits our main concern is to ensure that a maximum transfer of power takes place from the source to the load. This is accomplished by ensuring that the load impedance matches or is approximately equal to the source impedance (see Section 6-8). The power factor in the electronic circuit is normally of no importance.

Industrial plants have a high component of motors, which make up a highly inductive load across the hydroelectric power line source. This means that the reactive power drawn by the motors is transferred continually back and forth from the motors to the source. Because of this reactive power, the apparent power delivered by the source is going to be much greater than the actual or true power used by the motors.

The larger apparent power drawn from hydroelectric power lines means that the size of the equipment, such as transformers and line conductors, must be increased to accommodate the increase in current.

Ideally, then, a hydroelectric power utility would like to see a power factor of unity or 1. This avoids the excessive cost of increasing the transformer and line capacity, which reduces the cost of supplying power.

Power factor correction is accomplished by adding a capacitor or capacitance across the line, as shown in Fig. 30-10. If you compare Fig. 30-2(b), the capacitive power curve, to Fig. 30-6(b), the inductive power curve, you will notice that the peak power of both curves is 180° out of phase. This means that when the inductor flux is falling and returning the power to the source in the motor circuit of Fig. 30-10, the capacitor is charging and accepting power from the inductor. In effect, the inductor returns the power to the capacitor rather than to the source. Once the capacitor is charged to the peak voltage value, it returns its reactive power to the inductor. If the capacitive power (var) is equal to the inductive power (var), the reactive power cancels and only the true power is delivered by the source.

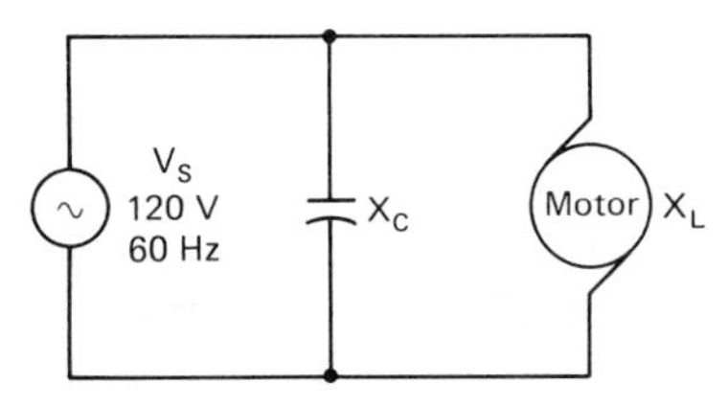

FIGURE 30-10 Power factor correction with a capacitor across the source.

EXAMPLE 30-11

Consider the motor load in the circuit schematic of Fig. 30-10. Without a capacitor the source current was 9.4 A. With the capacitor the current decreased to 5.8 A. Assume a unity power factor when the capacitor is connected into the circuit.

(a) Find the value of the capacitor.
(b) What was the power factor of the circuit before the capacitor was added?
(c) Find the lagging current phase angle before PF correction.

Solution:

(a) Considering that unity PF is assumed when the capacitor was connected into the circuit, the true power will be equal to the apparent power. Using the power triangle of Fig. 30-8(c) will give the reactive power. Without the capacitor:

$$P_A = V_S I_S = 120\text{ V} \times 9.4\text{ A} = 1128\text{ VA}$$

With the capacitor: Assuming a unity power factor,

$$P_A = P_T = V_S I_S = 120\text{ V} \times 5.8\text{ A} = 696\text{ W}$$

The power triangle of Fig. 30-8(c) will give the reactive power:

$$P_X = \sqrt{P_A^2 - P_T^2}$$
$$= \sqrt{(1128\text{ VA})^2 - (696\text{ W})^2} = 888\text{ var}$$

From $P_X = V^2/X_L$,

$$X_L = \frac{V^2}{P_X} = \frac{(120\text{ V})^2}{888\text{ var}} = 16.2\ \Omega$$

With a unity power factor correction X_C must equal X_L; therefore,

$$X_C = X_L = 16.2\ \Omega$$

and

$$C = \frac{1}{2\pi f X_C} = \frac{1}{6.28 \times 60\text{ Hz} \times 16.2\ \Omega}$$
$$= 164\ \mu\text{F}$$

(b) Equation (30-19) will give the power factor of the circuit before the capacitor was added: or,

$$\text{PF} = \frac{P_T}{P_A} \times 100 = \frac{696\text{ W}}{1128\text{ VA}} \times 100 = 62\%$$

(c) The lagging current phase angle before power factor correction can be found using equation (30-20):

$$\cos\theta = \frac{P_T}{P_A} = \frac{696 \text{ W}}{1128 \text{ VA}} = 0.62$$

Therefore, $\theta = 52°$.

PRACTICE QUESTIONS 30-2

Choose the most suitable word or phrase to complete each statement.

1. The ratio of true power to the apparent power is called the ________ ________ of the circuit or load.

2. When PF is 100%, the ________ power will be equal to the apparent power.

3. If PF is less than 100%, the ________ power will be less than the apparent power.

4. An inductive load is said to have a ________ power factor since the load current will (*lag, lead*) the source voltage.

5. A capacitive load is said to have a ________ power factor since the load current will (*lag, lead*) the source voltage.

Refer to the following circuit description for Questions 6 to 9: An ammeter connected in a 220-V 60-Hz motor circuit reads 9.6 A; a wattmeter connected to the circuit reads 1500 W (keep in mind that the wattmeter reads true power).

6. Find the apparent power in the circuit.

7. Find the reactive power in the circuit.

8. Find the percentage power factor of the circuit.

9. Find the angle by which the current lags the source voltage.

10. The size of the hydro equipment and line conductors must be (*decreased, increased*) to accommodate a load with a PF lower than 100%.

11. Power factor correction is accomplished by connecting a ________ across the source.

12. What must be the value of a capacitor connected across the line to increase the PF to 100% in the circuit of Questions 6 to 9?

SUMMARY

Ac power actually dissipated by a resistor will be an average power.

$$P_{\text{avg}} = \frac{1}{2} V_p \times I_p \quad \text{watts} \qquad (30\text{-}1a)$$

Also

$$P_{\text{avg}} = \frac{1}{2} \times \frac{V_P^2}{R} \text{ watts} \qquad (30\text{-}1)$$

The true power is the power dissipated by the resistor in the form of heat energy. The true power can be found by

$$P_T = V \times I \quad \text{watts} \qquad (30\text{-}2)$$

$$P_T = I^2 R \quad \text{watts} \qquad (30\text{-}3)$$

$$P_T = \frac{V^2}{R} \quad \text{watts} \qquad (30\text{-}4)$$

A capacitor is a wattless component because it does not dissipate power. It charges and then discharges, returning its energy to the source.

The unit of reactive power is the volt-ampere reactive (var).

The reactive power for a capacitive circuit is

$$P_X = V_C I_C \quad \text{volt-amperes reactive} \tag{30-5}$$

$$P_X = I_C^2 X_C \quad \text{volt-amperes reactive} \tag{30-6}$$

$$P_X = \frac{V_C^2}{X_C} \quad \text{volt-amperes reactive} \tag{30-7}$$

In a series or parallel *RC* circuit, the product of $V_S I_T$ represents the amount of power that is apparently being supplied by the source and is termed apparent power, P_A.

$$P_A = V_S I_T \quad \text{volt-amperes} \tag{30-8}$$

$$P_A = I^2 Z \quad \text{volt-amperes} \tag{30-9}$$

$$P_A = \frac{V^2}{Z} \quad \text{volt-amperes} \tag{30-10}$$

Apparent power can also be found by the phasor sum:

$$P_A = \sqrt{P_T^2 + P_X^2} \quad \text{volt-amperes} \tag{30-11}$$

From the power triangle shown in Fig. 30-4(c),

$$P_T = P_A \cos\theta \quad \text{watts} \tag{30-12}$$

In a parallel *RC* circuit the true power is dissipated by the resistor and is found

$$P_T = \frac{V^2}{R} \text{ or } I^2 R \quad \text{watts} \tag{30-13}$$

The reactive power in a parallel *RC* circuit is found by

$$P_X = \frac{V^2}{X_C} \text{ or } I^2 X_C \quad \text{volt-amperes reactive} \tag{30-14}$$

The apparent power in a parallel *RC* circuit is found by

$$P_A = \frac{V^2}{Z} \text{ or } I_T^2 Z \quad \text{volt-amperes} \tag{30-15}$$

An inductor is a wattless component because it does not dissipate power. Its stored energy, in the form of magnetic flux, is returned to the source.

The reactive power for an inductive circuit is

$$P_X = V_L \times I_L \quad \text{volt-amperes reactive} \tag{30-16}$$

$$P_X = I_L^2 \times X_L \quad \text{volt-amperes reactive} \tag{30-17}$$

$$P_X = \frac{V^2}{X_L} \quad \text{volt-amperes reactive} \tag{30-18}$$

The apparent power in a series *RL* circuit can be found using equations (30-8) to (30-12).

In a parallel *RL* circuit the true power is dissipated by the resistor and is found by

$$P_T = \frac{V^2}{R} \quad \text{or} \quad I^2R \qquad \text{watts} \tag{30-13}$$

The reactive power in a parallel *RL* circuit can be found using equation (30-15). The ratio of true power to the apparent power, called the power factor of the circuit, is

$$\text{PF} = \frac{P_T}{P_A} \qquad \text{(no unit)} \tag{30-19a}$$

Stated in terms of percentage,

$$\text{PF} = \frac{P_T}{P_A} \times 100 \qquad \text{percentage} \tag{30-19}$$

With unity power factor (100%) the true power will be equal to the apparent power. The power triangle of Fig. 30-8(c) shows that the cosine of the angle θ is P_T/P_A; therefore,

$$\text{PF} = \frac{P_T}{P_A} = \cos\theta \times 100 \qquad \text{percentage} \tag{30-20}$$

Adding a capacitance across the line will correct the power factor of a circuit. If the capacitive power is equal to the inductive power, the reacitve power cancels and only true power is delivered by the source.

REVIEW QUESTIONS

30-1. What is the exact average value of ac power in terms of peak power?

30-2. Find the average power dissipated by a 3.3-kΩ resistor with 100 V peak applied.

30-3. Explain why the ac power of a resistive circuit is termed the true power.

30-4. Why is a capacitor termed a wattless component in an ac circuit?

30-5. A 100-μF capacitor is connected across a 120-V 60-Hz line.
- **(a)** What is the true power dissipated by the capacitor?
- **(b)** What is the reactive power of the circuit?

30-6. Explain why the ac power of an *RC* or *RL* circuit is termed the apparent power.

30-7. A 1-kΩ resistor and a 0.05-μF capacitor are connected in series across a 10-V, 10-kHz source. Find (**a**) the true power, (**b**) the reactive power, and (**c**) the apparent power of the circuit.

30-8. The same 1-kΩ resistor and 0.05-μF capacitor are connected in parallel to a 10-V 10-kHz source. Find (**a**) the true power, (**b**) the reactive power, and (**c**) the apparent power of the circuit.

30-9. Why is the inductor termed a wattless component in an ac circuit?

30-10. A 1-kΩ resistor and a 30-mH choke are connected in series across a 10-V 10-kHz source. Find (**a**) the true power, (**b**) the reactive power, and (**c**) the apparent power of the circuit.

30-11. Draw a power triangle for Question 30-10 and prove that the true power, $P_T = P_A \cos\theta$.

30-12. Explain why the power factor of a circuit cannot be greater than 1.

30-13. What is a leading power factor? A lagging power factor?

30-14. What remedy is used to correct for power factor? Explain why the PF is corrected.

30-15. An industrial plant with a large component of motor-driven machinery is encouraged to correct for power factor. Explain.

30-16. A certain motor load connected across a 220-V 60-Hz source draws a current of 106 A. The power factor of the load is 60%:
(**a**) What is the true power in the circuit?
(**b**) What is the reactive power in the circuit?

30-17. What must the value of a capacitor or a capacitor bank be in order to bring the power factor of the motor load in Question 30-16 to 100% or unity?

SELF-EXAMINATION

Choose the most suitable word, phrase, or value to complete each statement.

30-1. The exact average value of ac power is one-half of the ________ power.

30-2. The actual power dissipated by a resistor will be the (*peak, peak-to-peak, average*) power.

30-3. In a pure resistive ac circuit the actual power dissipated by a resistor is termed the ________ power.

30-4. In a capacitive ac circuit the power being transfered back and forth between the source and the capacitor is termed the ________ power and is specified in ________ units.

Refer to the following circuit description for Questions 30-5 to 30-8: A 0.1-μF capacitor is connected in series with a 1-kΩ resistor to a 10 V 1-kHz source.

30-5. Find the true power of the circuit.

30-6. Find the reactive power of the circuit.

30-7. Find the apparent power in the circuit.

30-8. Find the phase angle between the leading current and source voltage.

Refer to the following circuit description for Questions 30-9 to 30-11: A 600-mH choke is connected in parallel with a 1-kΩ resistor to a 10-V 1 kHz source.

30-9. Find the true power of the circuit.

30-10. Find the reactive power of the circuit.

30-11. Find the apparent power of the circuit.

30-12. The power factor of an *RC* or *RL* circuit is the ratio of ________ power to ________ power.

30-13. When the true power is equal to the apparent power, the PF is ________ percent.

Refer to the following circuit description for Questions 30-14 to 30-18: A wattmeter connected into a 60-Hz, 220-V ac inductive circuit reads 3.6 kW of true power; the circuit current is 22 A.

30-14. Find the apparent power in the circuit.

30-15. Find the reactive power in the circuit.

30-16. Find the percentage power factor of the circuit.

30-17. Find the angle by which the current lags the source voltage.

30-18. What must be the value of a capacitor connected across the line to increase the PF to 100% in this circuit?

ANSWERS TO PRACTICE QUESTIONS

30-1

1. heat, environment
2. average
3. peak
4. true
5. zero
6. reactive
7. apparent
8. 22 mW
9. 17.6 mvar
10. 28.2 mVA
11. 38.5°
12. 36 mW
13. 28.7 mvar
14. 46 mVA

30-2

1. power factor
2. true
3. true
4. lagging, lag
5. leading, lead
6. 2112 VA
7. 1487 var
8. 71%
9. 44.7°
10. increased
11. capacitor
12. 80 μF

31

Ac Meters

INTRODUCTION

The permanent-magnet moving-coil (PMMC) meter is particularly suited to measure dc current and voltage. In an ac circuit the meter pointer is unable to respond to the constant polarity reversals, with the result that the pointer simply vibrates at low frequencies and reads zero at higher frequencies.

Ac meters such as the electrodynamometer and the moving-iron vane can be used on dc as well as ac. Because the meter pointer deflection responds to I^2 values of current, the scale can be calibrated in rms values. The electrodynamometer is particularly well suited as a wattmeter to measure the true power of an ac circuit.

In this chapter we discover how the permanent-magnet moving-coil meter movement can be converted to read ac. An overview of the electrodynamometer and the moving-iron vane meter is given as well as the electrodynamometer wattmeter.

MAIN TOPICS IN CHAPTER 31

31-1 Rectifier-Type Ac Meter
31-2 Dynamometer-Type Meter
31-3 Repulsion-Iron Vane Meter
31-4 Electrodynamometer Wattmeter

OBJECTIVE

After studying Chapter 31, the student should be able to

1. Obtain an overview of an ac analog volt-ammeter and wattmeter.

31-1 RECTIFIER-TYPE AC METER

The permanent-magnet moving-coil type of meter movement studied in Chapter 11 is not able to measure ac voltages or currents. Applying ac to the meter movement would drive the pointer upscale half of the time (on positive peaks), and with an ac voltage reversal the pointer would be driven down-scale the other half of the time. The net result is that the meter pointer vibrates noticeably at lower frequencies and is unresponsive at higher frequencies.

Since the meter movement operates strictly on dc, the ac must be converted or *rectified* to dc. This is accomplished by means of the full-wave rectifier circuit shown in Fig. 31-1(a).

The full-wave bridge rectifier circuit provides a pulsating dc voltage output as shown in Fig. 31-1(b). Diodes D_3 and D_4 conduct on the positive ac input

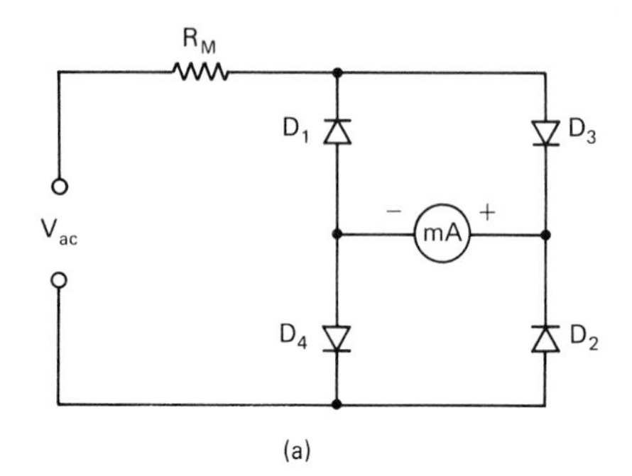

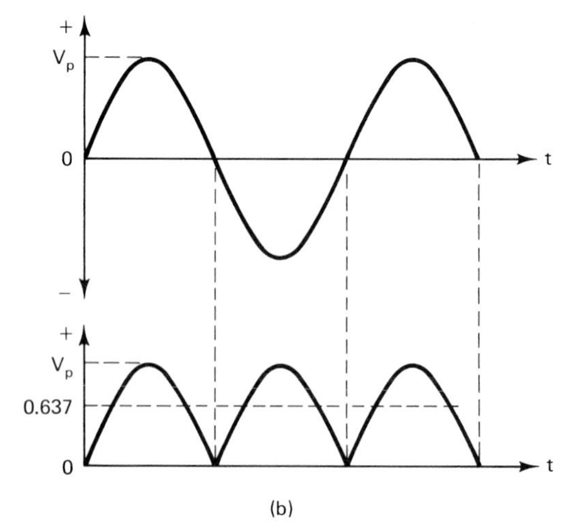

FIGURE 31-1 (a) Full-wave bridge rectifier circuit. (b) The ac voltage is rectified into pulsating dc and meter will measure the average of the peak value or 0.637 × peak value.

voltage to produce a positive dc pulse. Diodes D_1 and D_2 conduct on the negative ac input voltage to produce another positive dc pulse. Since there are two pulses for every cycle, the dc pulses are of very short duration. Thus the meter movement responds only to the average of these pulses, which is 0.637 of the peak value. For an ac *peak* value of 100 V, the pointer would be responding to 63.7 V on the scale. The scale of the meter must be calibrated in rms values since ac voltage and current are normally in rms units.

Calibrating the Ac Voltage Scale

To calibrate the dc meter scale-to-ac rms voltage with the full-wave rectifier output, it is necessary to convert from average value to rms value. Converting from 0.637 to 0.707 of the peak value gives the form factor $0.707/0.637 = 1.11$. Thus any average dc reading on the scale must be multiplied by 1.11 to convert it from average value to rms value, and the scale is calibrated accordingly.

As an example, the actual voltage driving the dc meter pointer to full scale on the 100-V ac range would be $0.637 \times V_{peak}$ and $V_{avg} = 0.637 \times 100\text{ V} \times 1.414 = 90\text{ V}$. To calibrate the 100 V range to rms, the multiplier resistance, R_M, must be designed for 90 V average dc value to obtain full-scale calibration. In equation form,

$$R_M = \frac{V_{FS(avg)}}{I_{FS}} \quad \text{ohms} \tag{31-1}$$

EXAMPLE 31-1

Assume that a meter movement of 0–1 mA dc with an internal resistance of 100 Ω is to be used to measure an ac voltage of 100 V. The multiplier resistor, R_M, can be calculated using equation (31-1) once the ac average voltage value is found:

$$V_{FS(avg)} = 100\text{ V} \times 1.414 \times 0.637 = 90\text{ V(avg)}$$

$$R_M = \frac{V_{FS(avg)}}{I_{FS}} = \frac{90\text{ V}}{1\text{ mA}} = 90\text{ k}\Omega$$

Therefore,

$$R_M = 90\text{ k}\Omega - 100\text{ }\Omega = 89.9\text{ k}\Omega$$

Ac Milliammeter

The bridge rectifier circuit can also serve to measure ac current. Placing the shunt resistor, R_S, across the bridge circuit as shown in Fig. 31-2 will bypass the circuit current except for the full-scale requirement of the meter movement. Since the meter pointer responds to the average of the peak current value, the scale of the meter must be calibrated in terms of rms values.

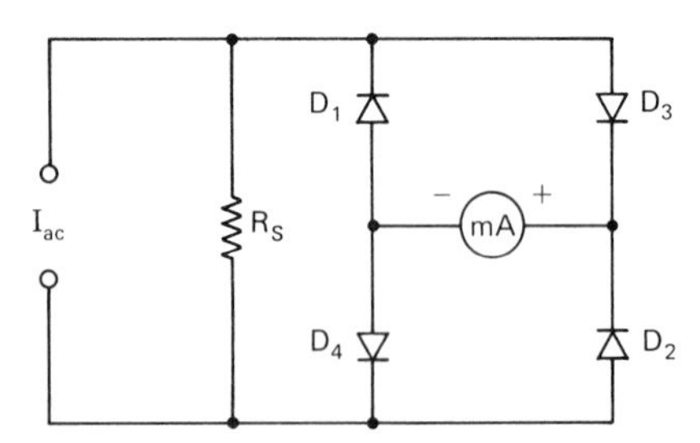

FIGURE 31-2 Full-wave bridge rectifier circuit for an ac milliammeter.

Calibrating the Ac Current Scale

As in the case of the voltmeter the rectifier output must be converted from peak value to average value and the full-scale current reading calibrated accordingly.

If we use a 0-1 mA movement with an internal resistance of 100 Ω, the actual current driving the dc meter pointer to full scale on, say, the 100-mA ac range, would only be $0.637 \times I_{peak}$ or $I_{avg} = 0.637 \times 1 \text{ mA} \times 1.414 = 0.9$ mA.

Since 1 mA is required to drive the meter pointer to FS, the shunt resistor, R_S, must be designed to give a full-scale reading with the average current of 0.9 mA. The shunt resistor, R_S, value can be calculated from $R_S = V_M/I_S$.

Since the scale is to be calibrated in rms values, I_S must be converted to average value, and

$$R_S = \frac{V_{M(FS)}}{I_{S(avg)}} \quad \text{ohms} \qquad (31\text{-}2)$$

EXAMPLE 31-2

Assume that the 0-1 mA meter movement in Example 31-1, with an internal resistance of 100 Ω, is to measure an ac current of 100 mA. The shunt resistance value, R_S, can be calculated using equation (31-2) once the ac average current value is found.

$$I_S = I_T - I_M = 100 \text{ mA} - 1 \text{ mA}$$
$$= 99 \text{ mA (rms)}$$
$$I_{S(avg)} = 99 \text{ mA} \times 1.414 \times 0.637$$
$$= 89.17 \text{ mA (avg)}$$
$$R_S = \frac{V_{M(FS)}}{I_{S(avg)}} = \frac{1 \text{ mA} \times 100\ \Omega}{89.17 \text{ mA}} = 1.121\ \Omega$$

31-2 DYNAMOMETER-TYPE METER

Although the dynamometer is used extensively as an ac voltmeter, ammeter, and wattmeter, it *can* also be used for dc measurements.

The dynamometer operates on the principle of a dynamo (i.e., a motor). It has two fixed coils and a moving coil that converts electrical energy into mechanical motion, as illustrated in Fig. 31-3(a).

Current flowing through the field coils produces a magnetic flux that is series aiding. The moving coil, however, is wound in the opposite direction so that the series current produces opposing poles and the pointer is deflected upscale or positively, as shown in Fig. 31-3(a).

With an ac reversal of current the magnetic polarity of the field coils reverses. However, the moving-coil flux polarity also reverses to cause a positive or upscale deflection of the pointer, as shown in Fig. 31-3(b).

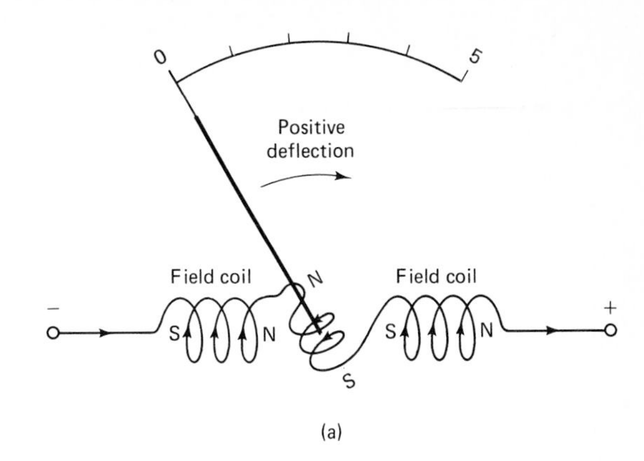

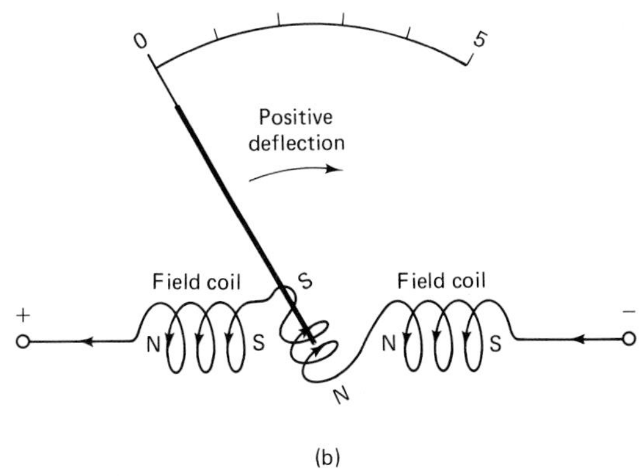

FIGURE 31-3 (a) Dynamometer voltmeter/ammeter. Current flowing through coils produces repulsion of moving coil and positive deflection of pointer. (b) Reversal of current produces repulsion of moving coil and positive deflection of pointer.

Because the field coil winding must produce a strong magnetic flux, the current through the winding is fairly large. Current requirements for full-scale deflection ranges up to hundreds of milliamperes compared to, say, 50 μA for a dc d'Arsonval meter movement. For this reason the dynamometer sensitivity is poor.

Dynamometer-type meters are calibrated to operate within a narrow frequency range, such as 50 to 60 Hz. This is because the inductive reactance of the coils is directly proportional to frequency. Current through the coils decreases with an increase in frequency.

There are many turns of fine wire on the field windings when the dynamometer is used as a voltmeter. As an ammeter the field windings have fewer turns of heavier wire.

The strength of the magnetic flux of the field coil and moving coil will be directly proportional to the current. As the current through the coils doubles, the magnetic field strength doubles. This causes the meter pointer to move four times as far for each time the current doubles, resulting in a current-squared (I^2) meter scale.

Since rms or I^2 values are equivalent to dc, the dynamometer scale calibration is identical for both ac and dc ranges. Step-down transformers are used where high voltages or higher ac current values are to be measured.

31-3 REPULSION IRON-VANE METER

The principle of the repulsion iron-vane meter movement is based on magnetic repulsion between two soft iron vanes that are mounted within a current-carrying coil. One vane is free to move while the other vane is fixed, as depicted in Fig. 31-4(a).

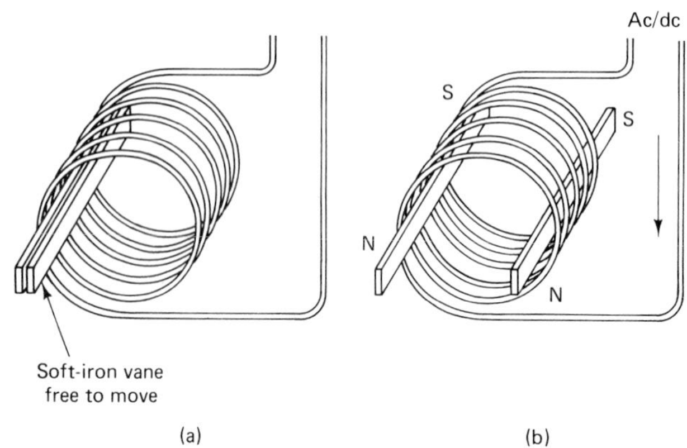

FIGURE 31-4 (a) Two soft-iron vanes mounted within a coil. One vane is free to move. (b) With an ac or dc current flow through the coil the vanes are magnetized having like poles adjacent. The repulsion displaces the movable iron vane away from the fixed vane.

When either ac or dc current flows through the coil, the iron vanes will be similarly magnetized, having like poles adjacent. Since like poles repel, the free vane will be repelled from the fixed vane as depicted in Fig. 31-4(b). This allows the meter movement to be used for measuring ac or dc current and voltage.

Figure 31-5(a) shows the typical construction of a concentric vane type of meter movement. Shaping the vanes allows more linear calibration along the major portion of the current-squared (rms) scale. The shaped vanes also help to reduce errors due to dc polarity reversal and residual magnetism.

A more sensitive repulsion iron-vane type of meter movement is shown in Fig. 31-5(b). The radial vane movement has a hinged vane that is attached to the pointer shaft as shown in Fig. 31-5(b). Maximum repulsion occurs when the current value is low, which increases the sensitivity of the meter movement as well as improving the linearity of the scale.

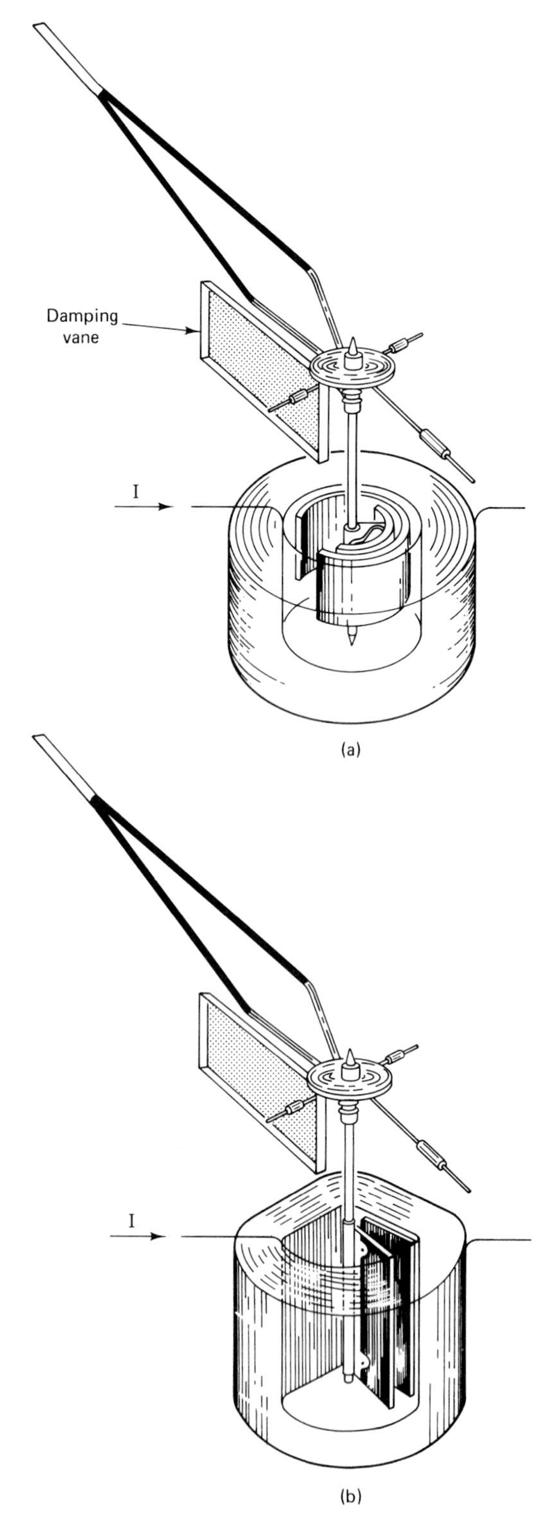

FIGURE 31-5 (a) Concentric iron-vane meter movement. (b) Radial iron-vane meter movement. (Weston Instruments, Newark, N.J.)

The aluminum damping vane attached to the meter shaft below the pointer [see Fig. 31-5(b)] rotates in a close-fitting chamber. This serves to bring the pointer to rest quickly.

31-4 ELECTRODYNAMOMETER WATTMETER

The electrodynamometer, moving-coil type of meter movement responds equally well to ac or dc. Its major application is as an ac/dc wattmeter.

Other coil arrangements expand its use in a variety of functions, such as polyphase wattmeter, measurement of power factor, measurement of reactive power, and other functions.

Since the electrodynamometer meter movement responds only to the current squared (I^2) values, it will *measure only the true power* in the circuit when used as an ac wattmeter. Figure 31-6 illustrates the coil arrangement of an ac single-phase wattmeter. The ac wattmeter is specifically calibrated to measure power from the 60-Hz power line, since the reactance of the fixed and movable coils increases with frequency.

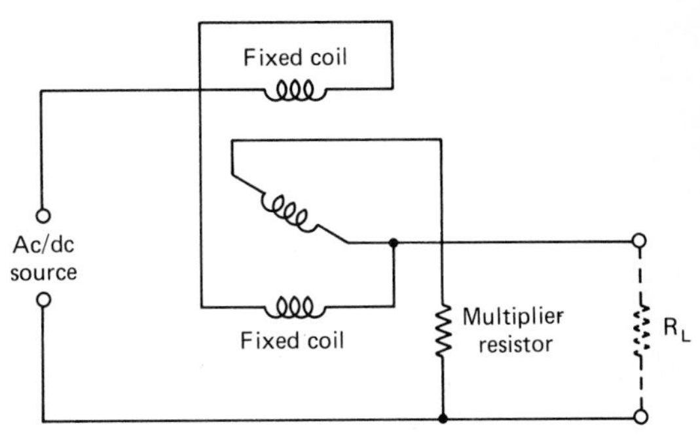

FIGURE 31-6 The electrodynamometer meter movement used as a wattmeter.

SUMMARY

The permanent-magnet moving-coil type of meter operates only on dc.

To measure ac current or voltage, the permanent-magnet moving-coil meter movement must be connected to a full-wave rectifier circuit.

The full-wave rectifier circuit rectifies ac to pulsating dc.

The meter movement responds to the average of the rectified dc pulses, which is 0.637 of the peak ac value.

The average value being read by the PMMC meter must be multiplied by 1.11 to convert it from average value to rms value and the scale calibrated accordingly.

To calibrate the ac voltmeter range the multiplier resistor is calculated to give full-scale deflection with an average value of dc volts:

$$V_{avg} = 0.637 \times V_{peak}$$

$$R_M = \frac{V_{FS(avg)}}{I_{FS}} \quad \text{ohms} \tag{31-1}$$

To calibrate the ac milliammeter range, the shunt resistance is calculated to give full-scale deflection with an average value of dc current:

$$I_{avg} = 0.637 \times I_{peak}$$

$$R_S = \frac{V_{M(FS)}}{I_{S(avg)}} \quad \text{ohms} \tag{31-2}$$

The dynamometer type of meter movement operates on ac and dc.

A dynamometer meter movement requires several hundreds of milliamperes for full-scale deflection; therefore, its sensitivity is poor.

The repulsion iron-vane meter movement operates on ac and dc and is more sensitive than the electrodynamometer meter.

The principle of the repulsion iron-vane meter movement is based on magnetic repulsion between two soft-iron vanes that are mounted within a current-carrying coil.

The electrodynamometer wattmeter measures only the true power in an ac circuit.

REVIEW QUESTIONS

31-1. To measure ac current or voltage with a permanent-magnet moving-coil meter movement, the ac must be rectified to dc. Explain.

31-2. To what value of pulsating dc does the permanent-magnet moving-coil meter movement respond when connected across the output of a full-wave rectifier?

31-3. Calculate the value of a series voltmeter multiplier resistor given a 50-μA PMMC meter movement which is to be used to measure 150 V ac.

31-4. For the same meter movement as in Question 31-3, calculate the value of a shunt resistance given that the meter movement internal resistance is 2200 Ω and it is to be used to measure 15 mA ac.

31-5. Explain briefly why the pointer of a dynamometer type of meter movement is always deflected positively or upscale regardless of the polarity of dc or ac.

31-6. Explain why the sensitivity of the dynamometer movement is poor.

31-7. What causes the movable vane in a repulsion iron-vane ac meter movement to be repelled from the fixed vane?

31-8. Explain why the repulsion iron-vane meter movement can be used on either ac or dc.

31-9. Why is the dynamometer type of ac meter movement used extensively as a wattmeter?

SELF-EXAMINATION

Choose the most suitable word, phrase, or value to complete each statement.

31-1. A permanent-magnet moving-coil meter movement pointer will not respond when operated on __________.

31-2. To operate a PMMC movement on ac, the ac must be rectified to dc by means of a __________ __________ rectifier.

Refer to the following meter movement for Questions 31-3 and 31-4: A 50 μA PMMC meter movement has an internal resistance of 2000 Ω and is connected across a full-wave bridge circuit rectifier.

31-3. Calculate the multiplier resistor value required to measure 250 V ac.

31-4. Calculate the shunt resistance value required to measure a current of 200 mA ac.

31-5. The electrodynamometer type of meter movement has two fixed coils which produce a series (*opposing, canceling, aiding*) magnetic field.

31-6. The moving coil is wound so that the series current produces a magnetic field that (*opposes, cancels, aids*) the magnetic polarity of the field coils.

31-7. The meter sensitivity of a dynamometer type of meter movement is much (*better, poorer*) than the PMMC meter movement.

31-8. Dynamometer-type meters are calibrated to operate within the narrow frequency range ________ to ________ Hz.

31-9. The meter scale of a dynamometer-type meter movement is calibrated in ________ values; therefore, it can be used on dc as well.

31-10. In the repulsion type of meter movement, a fixed soft-iron vane and a movable soft-iron vane are mounted within a ________.

31-11. The repulsion iron-vane meter movement can be used for measuring ________ or ________ current and voltage.

31-12. The calibrated scale of a repulsion iron-vane meter movement allows measurements to be made on dc or ________ values of ac.

31-13. The sensitivity of a repulsion iron-vane meter movement is (*less, greater, the same*) compared to the electrodynamometer type of meter movement.

31-14. The major application of an electrodynamometer type of meter movement is as a ________.

31-15. The electrodynamometer type of wattmeter measures only (*apparent power, true power, reactive power*).

32

Complex Numbers for Ac Circuits

INTRODUCTION

Series–parallel *RC*, *RL*, and *RLC* circuits up to this point were solved by applying the right-angle triangle theory to phasor quantities. For complex ac circuits involving resistance capacitive reactance and inductive reactance, the magnitude and phase of circuit parameters can be more systematically solved using the complex number system. The complex number system allows us to add, subtract, multiply, and divide phasor quantities that are easier to handle using either the polar or rectangular format of phasor analysis.

In this chapter we examine both the polar and rectangular forms of complex ac circuit analysis as well as applying complex numbers to series, parallel, and series–parallel *RLC* circuits.

MAIN TOPICS IN CHAPTER 32

OBJECTIVES

After studying Chapter 32, the student should be able to

1. Master the fundamental concepts of complex numbers.
2. Apply complex numbers to phasor quantities in rectangular and polar forms.
3. Convert from polar to rectangular and vice versa.
4. Solve series, parallel, and series–parallel reactive-resistant circuits by applying complex number theory.

32-1 COMPLEX NUMBERS

The Cartesian Plane

As shown in Fig. 32-1(a), the Cartesian mathematical plane has two coordinates, the horizontal x axis and the vertical y axis. Numbers or graduated units to the right of the zero along the x axis are positive, while the numbers to the left of the zero are negative.

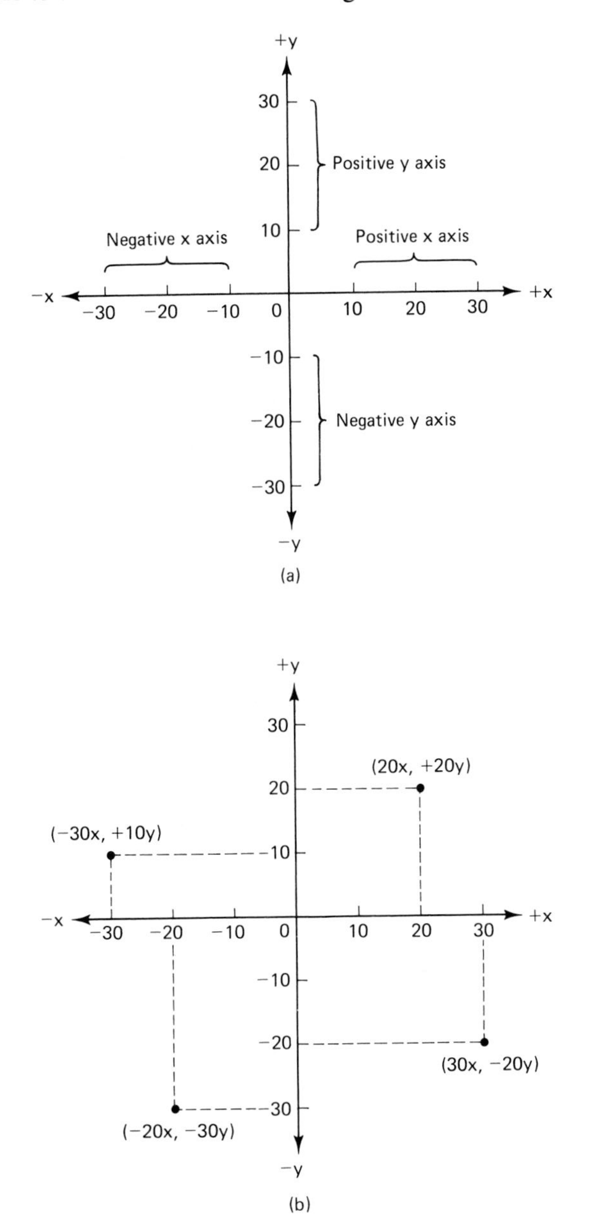

FIGURE 32-1 (a) The Cartesian x-y plane. (b) Coordinate points on the Cartesian plane.

To locate or plot a point in any one of the four areas (quadrants) within the plane, we need another set of coordinates along the y axis, as shown in Fig. 32-1(b). Numbers or graduations above the x axis are positive, while the numbers or graduations below the x axis are negative. To locate or plot a point within one of the four planes, it is only necessary to give its coordinates as shown in Fig. 32-1(b).

The Complex Plane

The term "complex" stems from the fact that the numbers along the coordinates are two-dimensional. The numbers along the horizontal (x) axis are termed the *real numbers*, while the numbers along the y axis are termed *imaginary numbers*. Real numbers to the right of the zero or origin are positive, while the real numbers to the left of the zero are negative, as shown in Fig. 32-2(a).

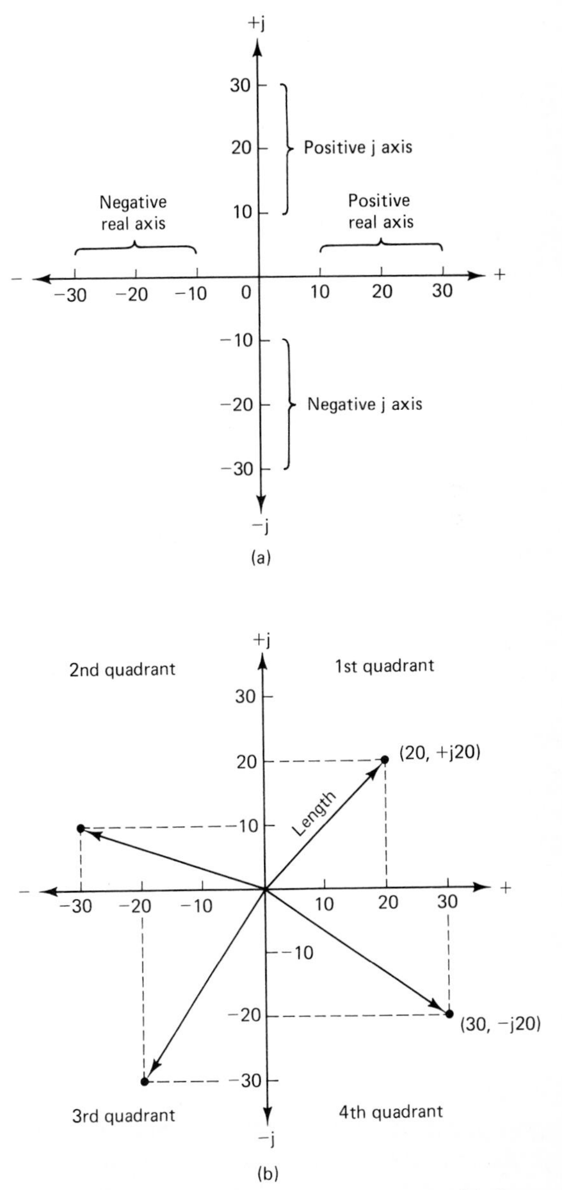

FIGURE 32-2 (a) The complex plane. (b) Coordinate points on the complex plane.

Since dc circuits have a constant or steady current and voltage with a fixed polarity, dc circuit values are stated in real numbers only, along the positive real axis. If the polarity is reversed, the circuit values can be stated in terms of negative real numbers along the negative real axis.

In ac circuits the alternating current and voltage produces the same heating effect or power loss in a resistor as dc. Furthermore, frequency has no effect on this power loss; therefore, pure ac resistance values are stated in real numbers as in the case of dc along the positive real axis shown in Fig. 32-2(a).

If X_L or X_C is introduced into the ac circuit, the current lags or leads the voltage, and the circuit values can no longer be represented by real numbers. Inductive or capacitive reactance is effectively in a 90° direction from the resistive apposition. Therefore, the positive or negative imaginary numbers along the y axis shown as $\pm j$ in Fig. 32-2(a) are representative of the reactance *with respect to the resistive quantity*.

The Y Axis, j Operator. The y axis in the complex plane of Fig. 32-2(a) is labeled j. In electrical math the $\pm j$ is attached to a number to show that it is 90° out of phase with a reference value. *This prefix is known as the j operator*.

Because an inductive voltage leads the current, inductive reactance is considered to be a positive quantity; therefore, the prefix $+j$ is used before an X_L value. Similarly, capacitive reactance is considered as a negative quantity so that the prefix $-j$ is used before an X_C value.

Adding resistance and reactance in series will give the coordinate point or impedance on the complex plane of Fig. 32-2(b). The coordinate point in the first quadrant shows that the series impedance is made up of 20 units of resistance (or 20 Ω) and 20 units of inductive reactance ($+j20$), which is 90° out of phase and leading the resistive component. The coordinate point in the fourth quadrant shows that the series impedance is made up of 30 units of resistance (or 30 Ω) and 20 units of capacitive reactance ($-j20$), which is 90° out of phase and lagging the resistive component.

Any 90° out-of-phase quantity in an electrical circuit will have the j prefix. The j operator applies only to a quantity that is leading or lagging by 90°, or it must be a multiple of 90° through one cycle of 360°. There are four conditions to consider when using the j operator shown in Fig. 32-3.

The representative arcs of a counterclockwise rotating vector of Fig. 32-3 show the three resultant j values with respect to the reference direction of 0°. Rotating a vector by 90° CCW is the same as multiplying it by j.

Figure 32-3(a) represents a quantity that is in phase and does not have a j prefix.

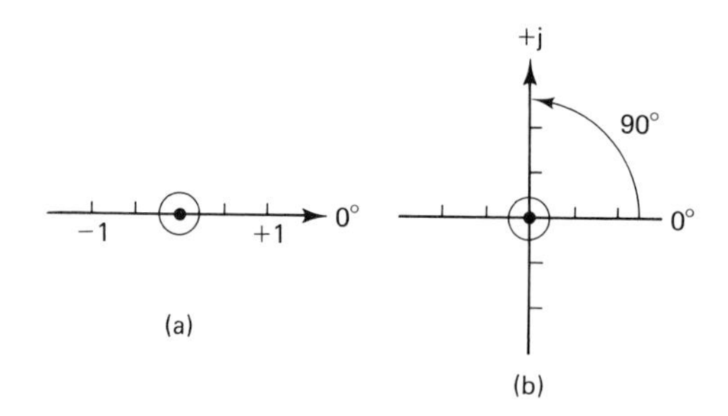

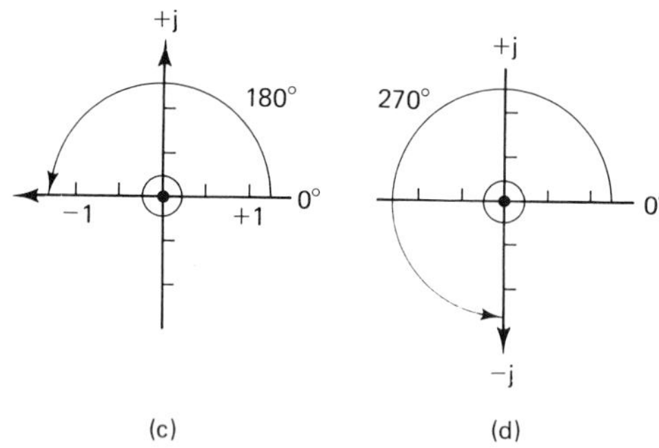

FIGURE 32-3 Arcs used to represent the three positions of a vector rotating counterclockwise and resultant j prefix. (a) 0° = in phase. (b) 90° = $+j$. (c) 180° = $j^2 = -1$. (d) 270° = $j^3 = -j$.

Figure 32-3(b) shows a vector quantity that is rotated by 90°. This quantity leads the reference quantity by 90° and is assigned a $+j$ prefix. The j is referred to as an operator since it operates or modifies a number by rotating it through a positive angle of 90°.

Figure 32-3(c) shows a vector quantity that is leading (or lagging) the reference quantity by 180° and is assigned a prefix of j^2. Note that a real number given a coordinate value of 1 that is rotated 180° from the reference axis becomes a j value of -1. In effect, the vector went through two 90° rotations; therefore, $j \times j = j^2 = -1$, then $j = \sqrt{-1}$. $j = \sqrt{-1}$ is called "imaginary." This is because although -1 can exist, its square root must be imagined. No real number multiplied by itself can have a minus sign; therefore, $j = \sqrt{-1}$ is termed *imaginary*.

Figure 32-3(*d*) Shows a vector quantity that is 90° lagging from the reference quantity. This is the same as the CCW position of 270°. This vector value is equivalent to j^3 or $j2 + j$, which is equivalent to $-1 \times j$ or $-j$. The lagging position is thus assigned the prefix $-j$ to a voltage, current, or impedance in a reactive circuit.

Phasor Quantities in Rectangular Form. Earlier we noted that the x axis of the complex plane was referenced to the pure ac resistive component and stated in terms of a real number. Reactive values are effectively 90° out of phase with the resistive values; therefore, they are represented along the $\pm j$ axis of the complex plane as shown in Fig. 32-4.

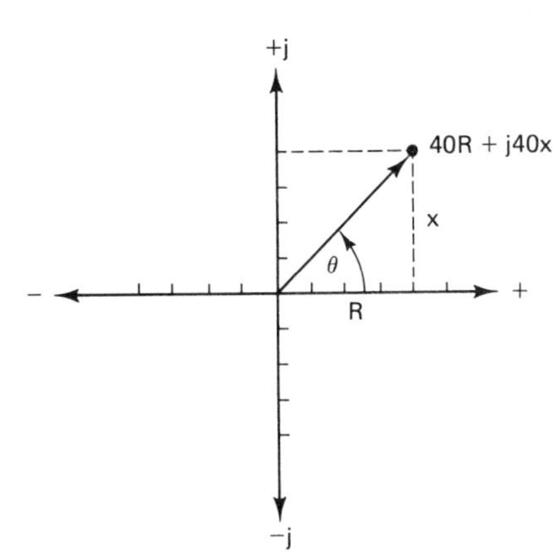

FIGURE 32-4 Rectangular coordinates for two out-of-phase quantities.

A phasor quantity in rectangular form is the sum of the real coordinates and the j coordinates. The coordinate point in phasor format is identified by drawing an arrow from the origin to the coordinate point as shown in Fig. 32-4.

Phasor quantities expressed in rectangular form provide a means of easily adding and substracting other rectangular quantities. The rectangular quantity can then be converted to polar form.

Phasor Quantities in Polar Form. The term *polar* is derived from the concept that the phasor or vector is considered to start from a central point on the reference line and extends outward from that point at a specified angle as shown by the examples of Fig. 32-5.

Since the phasor magnitude is represented by the hypoteneuse of a right-angle triangle, its magnitude can easily be found by applying the Pythagorean theorem as shown in Fig. 32-6. *The polar form of phasor quantity is in terms of the magnitude and phase angle referenced to the positive real axis.*

Quantities expressed in polar form provide a means of easily multiplying and dividing electrical quantities. The resulting answers give the magnitude and phase angle, which you are able to read from a meter and an oscilloscope or other instrument.

32-2 SERIES CIRCUIT IMPEDANCE IN COMPLEX FORM

The 25 Ω represents a pure resistance in the ac circuit schematic of Fig. 32-7(a). Since its resistance does not change with frequency, it is said to be nonreactive;

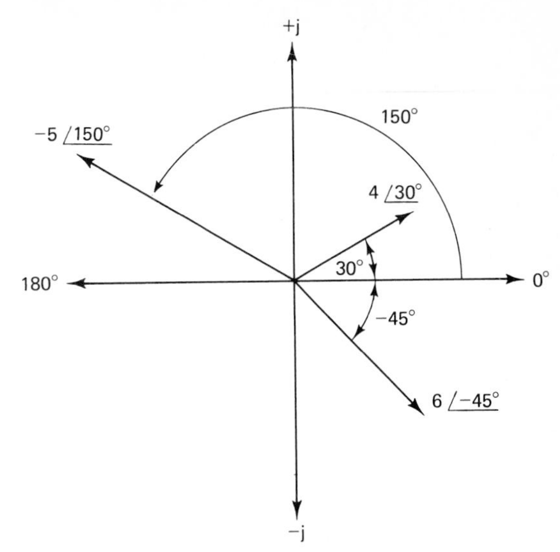

FIGURE 32-5 Phasor magnitude and phase angle specified in polar form.

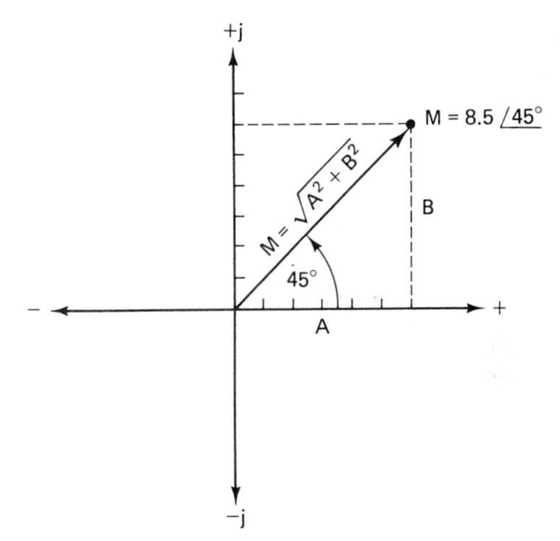

FIGURE 32-6 Polar coordinates for two out-of-phase quantities.

therefore, its quantity R is along the real axis corresponding to 0°.

In rectangular form R is written simply as R or 25 Ω. In polar form it is shown as $R\angle 0$ Ω, or $25\angle 0°$ Ω.

For the capacitive reactance of the circuit in Fig. 32-7(b), we know that the voltage lags the current by 90°; therefore, X_C lags R by 90°. The capacitive reactance quantity is shown along the imaginary or $-j$ axis.

In rectangular form X_C is shown as $-jX_C$, or $-j500$ Ω. In polar form it is shown as $X_C\angle -90°$ Ω or $500\angle -90°$ Ω.

In the inductive reactance circuit of Fig. 32-7(c) we know that the voltage leads the current by 90°; therefore, X_L leads R by 90°. The inductive reactance quantity is shown along the imaginary or $+j$ axis.

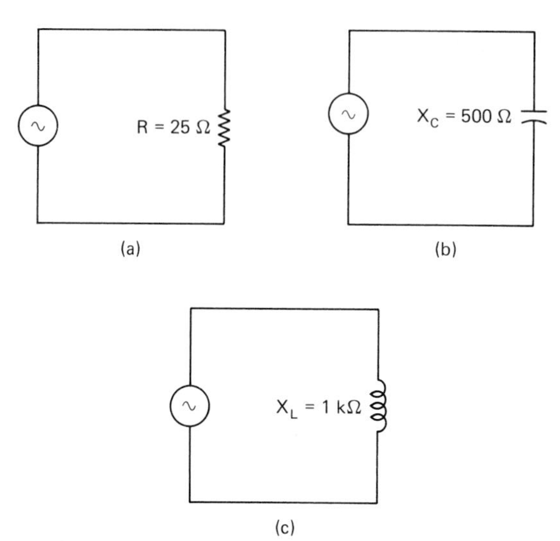

FIGURE 32-7 Resistive and reactive circuits.

In rectangular form X_L is shown as $+jX_L$, or $+j1$ kΩ. In polar form it is shown as $X_L\underline{/90^\circ}$ Ω, or 1 kΩ$\underline{/90^\circ}$ Ω.

Series *RL* Circuit

Adding the resistance along the real axis to the inductive reactance along the $+j$ axis will give the coordinate point of the impedance in rectangular form as shown in Figure 32-8 and is given as

$$Z = R + jX_L \qquad \text{ohms} \qquad (32\text{-}1)$$

Note, that the real component always represents resistance, while the j component always represents the reactance.

The polar form states the magnitude or impedance quantity as well as the leading phase angle. Applying the Pythagorean theorem to the impedance triangle of Fig. 32-8 will give the magnitude of Z:

$$Z = \sqrt{R^2 + X_L^2} \qquad \text{ohms} \qquad (32\text{-}2a)$$

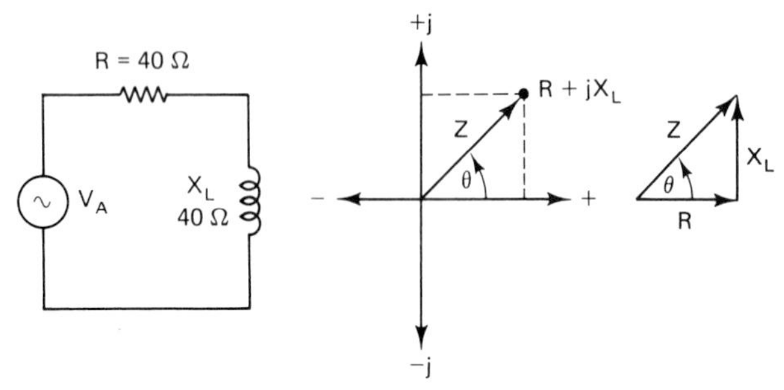

FIGURE 32-8 Series *RL* circuit impedance in rectangular and polar form.

Applying the tan trig function to the impedance triangle of Fig. 32-8 will give the phase angle:

$$\theta = \arctan \frac{X_L}{R} \qquad (32\text{-}2b)$$

Combining equations (32-2a) and (32-2b) will give the impedance in polar form:

$$Z = \underline{/\pm\theta^\circ} = \sqrt{R^2 + X_L^2}\,\underline{/\arctan \frac{X_L}{R}} \qquad \text{ohms} \qquad (32\text{-}2)$$

EXAMPLE 32-1

Express the magnitude of the impedance in rectangular and polar forms for the *RL* circuit of Fig. 32-8.

Solution: Adding the resistance of 40 Ω along the real axis to the inductive reactance of 40 Ω along the $+j$ axis gives the impedance in rectangular form as given by equation (32-1):

$$Z = R + jX_L = 40\ \Omega + j40\ \Omega$$

To obtain the impedance in polar form, we solve for impedance by applying equation (32-2a) and then calculate the phase angle by applying equation (32-2b).

$$Z = \sqrt{R^2 + X_L^2} = \sqrt{40^2 + 40^2} = 56.6\ \Omega$$

$$\theta = \arctan \frac{X_L}{R} = \frac{40\ \Omega}{40\ \Omega}$$

Therefore,

$$\theta = 45^\circ$$

The impedance in polar form then is

$$Z = 56.6\underline{/45^\circ}\ \Omega$$

Series *RC* Circuit

Adding the resistance along the real axis to the capacitive reactance along the $-j$ axis will give the coordinate point of the impedance in rectangular form as shown in Fig. 32-9 and is given as

$$Z = R - jX_C \qquad \text{ohms} \qquad (32\text{-}3)$$

Note that the real component always represents resistance, while the j component always represents the reactance.

The polar form states the magnitude or impedance quantity as well as the lagging phase angle. Applying

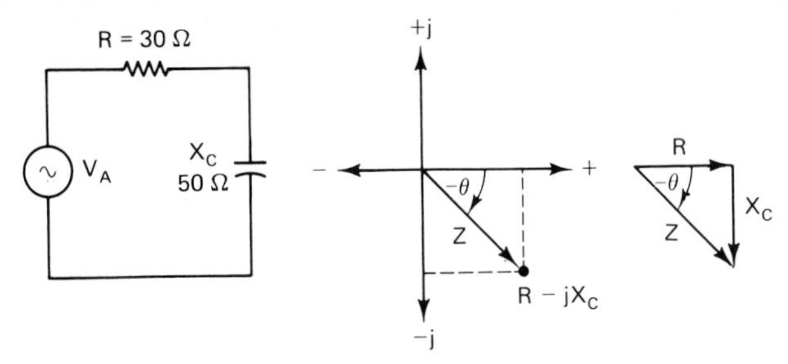

FIGURE 32-9 Series *RC* circuit impedance in rectangular and polar form.

the Pythagorean theorem to the impedance triangle of Fig. 32-8 will give the magnitude of Z:

$$Z = \sqrt{R^2 + X_C^2} \quad \text{ohms} \qquad (32\text{-}4a)$$

Applying the tan trigonometric function to the impedance triangle of Fig. 32-9 will give the phase angle:

$$\theta = -\arctan \frac{X_C}{R} \qquad (32\text{-}4b)$$

Combining equations (32-4a) and 32-4b) will give the impedance in polar form:

$$Z\underline{/\pm\theta°} = \sqrt{R^2 + X_C^2}\underline{\Big/ -\arctan \frac{X_C}{R}} \quad \text{ohms} \qquad (32\text{-}4)$$

EXAMPLE 32-2

Express the magnitude of the impedance in rectangular and polar form for the *RC* circuit of Fig. 32-9.

Solution: Adding the resistance of 30 Ω along the real axis to the inductive reactance of 50 Ω along the $-j$ axis gives the impedance in rectangular form as given by equation (32-3):

$$Z = R - jX_C = 30\ \Omega - j50\ \Omega$$

To obtain the impedance in polar form, we solve for impedance by applying equation (32-4a), then calculating the phase angle by applying equation (32-4b).

$$Z = \sqrt{R^2 + X_C^2} = \sqrt{30^2 + 50^2} = 58.3\ \Omega$$

$$\theta = -\arctan \frac{X_C}{R} = \frac{50\ \Omega}{30\ \Omega}$$

Therefore,

$$\theta = -59°$$

The impedance in polar form then is

$$Z = 58.3\underline{/-59°}\ \Omega$$

32-3 RECTANGULAR-TO-POLAR CONVERSION

So far we have dealt with simple series circuits in both polar and rectangular forms. As we progress into more complex circuits, multiplication and division calculations are easier to perform when the phasor quantities are in polar form. Addition and subtraction of phasor quantities, on the other hand, are easier to perform when in rectangular form; therefore, we must be able to convert from one form to another.

Rectangular coordinates only give the point on the complex plane to which the magnitude reaches, as shown in Fig. 32-10(a). To convert these coordinates into polar form requires only that we apply the Pythagorean theorem as shown in Fig. 32-10(b). This gives the relationship of the magnitude (M) as

$$M = \sqrt{A^2 + B^2} \qquad (32\text{-}5a)$$

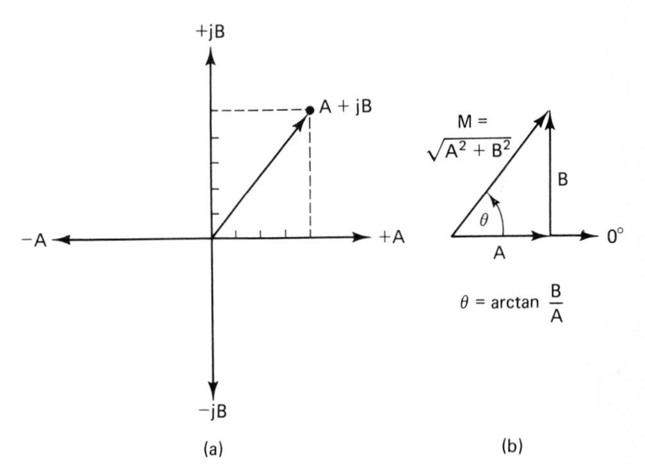

FIGURE 32-10 (a) Rectangular form. (b) Polar form.

The angle θ is the trigonometric function of the right-angle triangle in Fig. 32-10(b) and is

$$\theta = \arctan \frac{B}{A} \qquad (32\text{-}5b)$$

Combining equations (32-5a) and (32-5b) gives the general equation for converting rectangular to polar form and is

$$M = \sqrt{A^2 + B^2} \angle \pm\arctan\frac{B}{A} = M\angle\pm\theta \qquad (32\text{-}5)$$

where M is the general term for the phasor magnitude.

EXAMPLE 32-3

Convert the following complex numbers from rectangular to polar form.

(a) $40 + j40$
(b) $30 - j50$
(c) $-30 + j40$

Solution:

(a) The magnitude of the phasor with rectangular coordinates $40 + j40$ is given by equation (32-5a):

$$M = \sqrt{A^2 + B^2} = \sqrt{40^2 + 40^2} = 56.6$$

The phasor angle θ is given by equation (32-5b):

$$\theta = \arctan\frac{B}{A} = \frac{40}{40}$$

Therefore,

$$\theta = 45°$$

The polar expression for this phasor is

$$M = 56.6\angle 45°$$

(b) The magnitude of the phasor with reactangular coordinates $30 - j50$ is

$$M = \sqrt{30^2 + (-50)^2} = 58.3$$

The phase angle θ is

$$\theta = -\arctan\frac{50}{30}$$

Therefore,

$$\theta = -59°$$

The polar expression for this phasor is

$$M = 58.3\angle -59°$$

(c) The magnitude of the phasor with rectangular coordinates $-30 + j40$ is

$$M = \sqrt{(-30)^2 + 40^2} = 50$$

The phasor angle θ is

$$\theta = 180° - \left(\arctan\frac{40}{30}\right)$$

Therefore,

$$\theta = 126.8°$$

Note: The polar magnitude is always positive, while the angle can be given as a positive or a negative value.

The polar expression for this equation is

$$M = 50\angle 126.80°$$

32-4 POLAR-TO-RECTANGULAR CONVERSION

The polar form gives the magnitude and phase angle so we need only to apply right-angle triangle trigonometry to convert from polar form to rectangular. This is shown in the triangle of Fig. 32-11. Side A or the real number, can be found from

$$A = M\cos\theta \qquad (32\text{-}6)$$

Side B, the j operator number, can be found from

$$B = M\sin\theta \qquad (32\text{-}7)$$

Putting equations (32-6) and (32-7) into a general conversion formula gives

$$A + jB = M\cos\theta + jM\sin\theta \qquad (32\text{-}8)$$

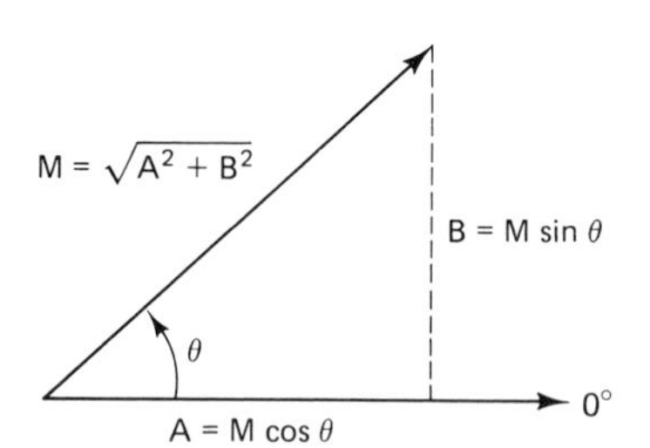

FIGURE 32-11 Complex numbers triangle.

EXAMPLE 32-4

Convert the following polar quantities to rectangular form.

(a) $20\angle 60°$
(b) $30\angle -53°$
(c) $50\angle 125°$

Solution:

(a) Equation (32-6) will give the real number of the rectagular coordinates.

$$A = M \cos \theta = 20 \cos 60° = 20 \times 0.5 = 10$$

The j operator number can be found using equation (32-7).

$$B = M \sin \theta = 20 \sin 60° = 20 \times 0.866 = 17.3$$

The rectangular expression is

$$10 + j17.3$$

(b) Using equation (32-6) gives the real number of the rectangular coordinates.

$$A = M \cos \theta = 30 \cos (-53°)$$
$$= 30 \times 0.60 = 18$$

The j operator number can be found using equation (32-7).

$$B = M \sin \theta = 30 \sin (-53°)$$
$$= 30 \times 0.799 = -24$$

The rectangular expression is

$$18 - j24$$

(c) Equation (32-6) gives the real number of the rectangular coordinates.

$$A = M \cos \theta = 50 \cos 125° = 50 \times (-0.574)$$
$$= -28.68$$

The j operator number can be found using equation (32-7).

$$B = M \sin \theta = 50 \sin 125° = 50 \times 0.189 = 41$$

The rectangular expression is

$$28.68 + j41$$

Note: The real number is always referenced to 0°, therefore cannot be a leading or lagging quantity.

32-5 COMPLEX NUMBERS ARITHMETIC

In solving more complex circuits, we must be able to express the various equations in either polar or rectangular form or both. Polar expressions can be divided and multiplied easily, while rectangular expression can be subtracted and added simply. As an example, using both polar and rectangular forms is the most simple method of solving series–parallel impedance by applying the product-over-sum equation.

An understanding of the basic arithmetic involved with polar and rectangular expressions must be grasped before we start solving more complex circuits.

Rectangular Addition

Rule: Add the real parts separately, then add the j parts.

EXAMPLE 32-5

Add 3 kΩ + j3 kΩ and 1 kΩ + j1 kΩ

Solution:

$$(3\text{ k}\Omega + j3\text{ k}\Omega) + (1\text{ k}\Omega + j1\text{ k}\Omega)$$
$$= (3\text{ k}\Omega + 1\text{ k}\Omega) + j(3\text{ k}\Omega + 1\text{ k}\Omega)$$
$$= 4\text{ k}\Omega + j4\text{ k}\Omega$$

or

$$\begin{array}{r} 3\text{ k}\Omega + j3\text{ k}\Omega \\ + \underline{1\text{ k}\Omega + j1\text{ k}\Omega} \\ 4\text{ k}\Omega + j4\text{ k}\Omega \end{array}$$

EXAMPLE 32-6

Add 2 kΩ + j6 kΩ and 3 kΩ − j2 kΩ.

Solution

$$(2\text{ k}\Omega + j6\text{ k}\Omega) + (3\text{ k}\Omega - j2\text{ k}\Omega)$$
$$= (2\text{ k}\Omega + 3\text{ k}\Omega) + j(6\text{ k}\Omega - 2\text{ k}\Omega)$$
$$= 5\text{ k}\Omega + j4\text{ k}\Omega$$

or

$$\begin{array}{r} 2\text{ k}\Omega + j6\text{ k}\Omega \\ + \underline{3\text{ k}\Omega - j2\text{ k}\Omega} \\ 5\text{ k}\Omega + j4\text{ k}\Omega \end{array}$$

Rectangular Subtraction

Rule: Subtract the real parts separately, then subtract the j parts.

EXAMPLE 32-7

Subtract 3 kΩ + j4 kΩ from 6 kΩ + j6 kΩ.

Solution

$$(3\text{ k}\Omega + j4\text{ k}\Omega) - (6\text{ k}\Omega + j6\text{ k}\Omega)$$
$$= (6\text{ k}\Omega - 3\text{ k}\Omega) + j(6\text{ k}\Omega - 4\text{ k}\Omega)$$
$$= 3\text{ k}\Omega + j2\text{ k}\Omega$$

or

$$\begin{array}{r} 6\text{ k}\Omega + j6\text{ k}\Omega \\ -\ 3\text{ k}\Omega - j4\text{ k}\Omega \\ \hline 3\text{ k}\Omega + j2\text{ k}\Omega \end{array} \quad \text{(change sign and add)}$$

EXAMPLE 32-8

Subtract 2 kΩ − j4 kΩ from 4 kΩ + j6 kΩ.

Solution

$$(2\text{ k}\Omega - j4\text{ k}\Omega) - (4\text{ k}\Omega + j6\text{ k}\Omega)$$
$$= (4\text{ k}\Omega - 2\text{ k}\Omega) + j[6\text{ k}\Omega - (-4\text{ k}\Omega)]$$
$$= 2\text{ k}\Omega + j10\text{ k}\Omega$$

or

$$\begin{array}{r} 4\text{ k}\Omega + j6\text{ k}\Omega \\ -\ 2\text{ k}\Omega + j4\text{ k}\Omega \\ \hline 2\text{ k}\Omega + j10\text{ k}\Omega \end{array} \quad \text{(change sign and add)}$$

Polar Multiplication

Rule: Multiply the magnitudes, and add the phase angles algebraically.

EXAMPLE 32-9

Multiply $30\angle 60^\circ$ by $10\angle 20^\circ$.

Solution

$$(30\angle 60^\circ)(10\angle 20^\circ) = (30 \times 10)\angle 60^\circ + 20^\circ$$
$$= 300\angle 80^\circ$$

EXAMPLE 32-10

Multiply $50\angle 30^\circ$ by $15\angle -55^\circ$.

Solution

$$(50\angle 30^\circ)(15\angle 20^\circ) = (50 \times 15)\angle 30^\circ + (-55^\circ)$$
$$= 750\angle -25^\circ$$

Polar Division

Rule: Divide the magnitude, change the sign of the denomintor phase angle, and add it algebraically to the numerator phase angle.

EXAMPLE 32-11

Divide $400\angle 60^\circ$ by $10\angle 30^\circ$.

Solution

$$\frac{400\angle 60^\circ}{10\angle 30^\circ} = \frac{400}{10}\angle 60^\circ - 30^\circ = 40\angle 30^\circ$$

EXAMPLE 32-12

Divide $75\angle 45^\circ$ by $25\angle -30^\circ$.

Solution

$$\frac{75\angle 45^\circ}{25\angle -30^\circ} = \frac{75}{25}\angle 45 + 30^\circ = 3\angle 75^\circ$$

PRACTICE QUESTIONS 32-1

Refer to the series *RL* circuit of Fig. 32-12 for Questions 1 and 2.

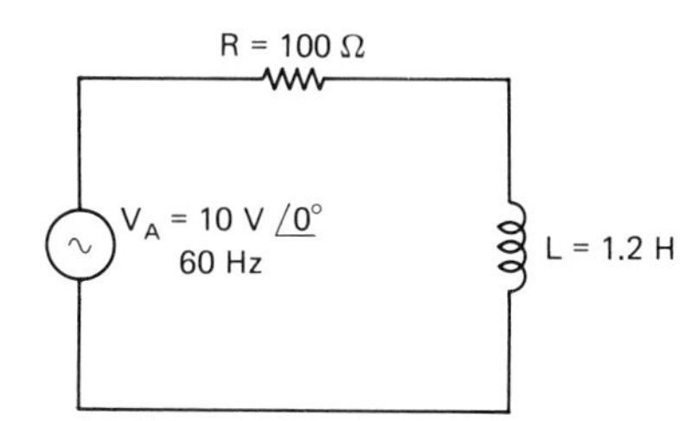

FIGURE 32-12 Series *RL* circuit.

1. Find and express the impedance of the circuit in polar form.
2. Express the impedance of the circuit in rectangular form.

Refer to the series *RC* circuit of Fig. 32-13 for Questions 3 and 4.

3. Find and express the impedance of the circuit in polar form.
4. Express the impedance of the circuit in rectangular form.
5. Convert 40 + j20 into polar form.
6. Convert 30 − j40 into polar form.

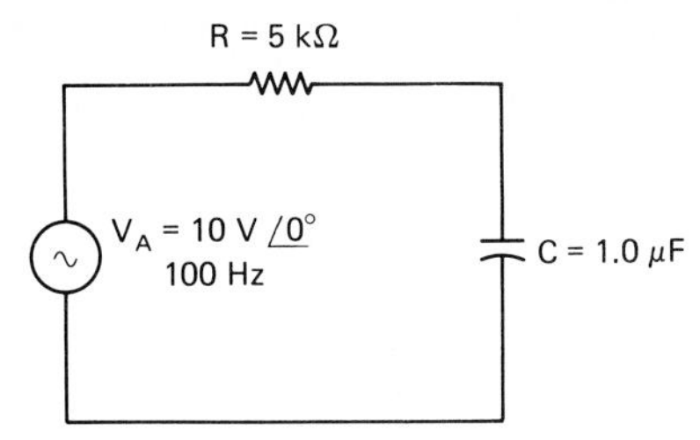

FIGURE 32-13 Series *RC* circuit.

7. Convert $70\angle -60^\circ$ into rectangular form.
8. Add $15 + j30$ to $6 - j10$.
9. Subtract $10 - j20$ from $100 + j80$.
10. Multiply $40\angle -53^\circ$ by $30\angle 60^\circ$.
11. Divide $85\angle 75^\circ$ by $25\angle -30^\circ$.

32-6 PARALLEL CIRCUIT IMPEDANCE IN COMPLEX FORM

We can now apply both polar and rectangular expressions to show that the impedance is the product-over-the-phasor sum for the parallel *RL* or *RC* circuits of Figs. 32-14 and 32-15 as follows:

$$Z = \frac{(R\angle 0^\circ)\,(X_L\angle 90^\circ)}{R + jX_L} \qquad \text{(32-9a)}$$

Converting the rectangular expression to polar form in the denominator [equation (32-2)] and multiplying the polar units in the numerator gives

$$Z = \frac{R \times X_L\angle 0^\circ + 90^\circ}{\sqrt{R^2 + X_L^2}\angle \arctan (X_L/R)} \qquad \text{(32-9b)}$$

Changing the sign of the angle $\arctan (X_L/R)$ in the denominator and adding it to the phase angle of the numerator gives

$$Z\angle \pm\theta^\circ = \frac{R \times X_L}{\sqrt{R^2 + X_L^2}}\angle 90^\circ - \arctan \frac{X_L}{R} \quad \text{ohms} \qquad \text{(32-9)}$$

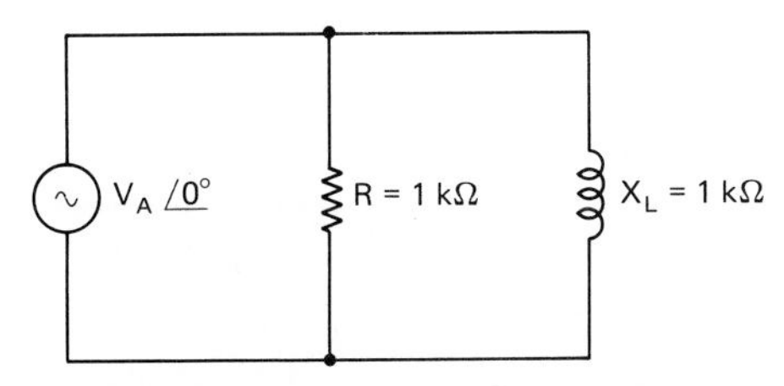

FIGURE 32-14 Parallel *RL* circuit.

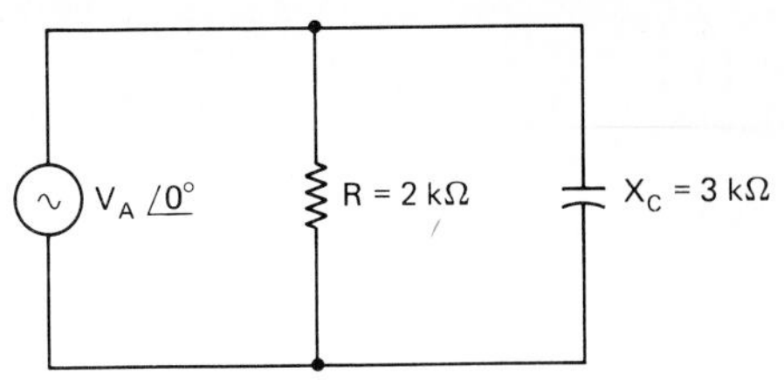

FIGURE 32-15 Parallel *RC* circuit.

EXAMPLE 32-13

Express the impedance of the parallel *RL* circuit in Fig. 32-14 in both polar and rectangular forms.

Solution: Using equation (32-9) will give the impedance in polar form.

$$Z\angle \pm\theta^\circ = \frac{R \times X_L}{\sqrt{R^2 + X_L^2}}\angle 90^\circ - \arctan \frac{X_L}{R}$$

$$= \frac{1\text{ k}\Omega \times 1\text{ k}\Omega}{\sqrt{(1\text{ k}\Omega)^2 + (1\text{ k}\Omega)^2}}\angle 90^\circ - \arctan \frac{1\text{ k}\Omega}{1\text{ k}\Omega}$$

$$= 707\ \Omega\angle 90^\circ - 45^\circ = 707\angle 45^\circ\ \Omega$$

Therefore, the impedance of the circuit is 707 Ω and V_A leads the current by 45°. To obtain the impedance in rectangular form, we must convert from polar to rectangular form by using the general conversion equation (32-8).

$$A \pm jB = M \cos \theta \pm jM \sin \theta$$

The resistive component is found:

$$R = 707\ \Omega \cos 45^\circ = 500\ \Omega$$

The inductive reactance is the j component and is found:

$$X_L = 707 \sin 45^\circ = 500\ \Omega$$

The impedance of the *RL* circuit in Fig. 32-14 in rectangular form is

$$Z = 500 + j500\ \Omega$$

For the parallel *RC* circuit of Fig. 32-15 the product-over-phasor-sum equation is similarly derived.

$$Z = \frac{(R\angle 0^\circ)\,(X_C\angle 90^\circ)}{R - jX_C} \qquad \text{(32-10a)}$$

$$= \frac{R \times X_C(0^\circ - 90^\circ)}{\sqrt{R^2 + X_C^2}\angle -\arctan (X_C/R)} \qquad \text{(32-10b)}$$

Changing the sign of the angle $-$ arctan (X_C/R) in the denominator and adding it to the phase angle of the numerator gives

$$Z\underline{/\pm\theta^\circ} = \frac{R \times X_C}{\sqrt{R^2 + X_C^2}}\underline{/-90^\circ + \arctan\frac{X_C}{R}} \quad \text{ohms} \tag{32-10}$$

EXAMPLE 32-14

Express the impedance of the parallel RC circuit in Fig. 32-15 in both polar and rectangular forms.

Solution: Using equation (32-10) will give the impedance in polar form.

$$Z = \underline{/\pm\theta^\circ}$$

$$= \frac{R \times X_C}{\sqrt{R^2 + X_C^2}}\underline{/-90^\circ + \arctan\frac{X_C}{R}} \quad \text{ohms}$$

$$= \frac{2\text{ k}\Omega \times 3\text{ k}\Omega}{\sqrt{(2\text{ k}\Omega)^2 + (3\text{ k}\Omega)^2}}\underline{/-90^\circ + \arctan\frac{3\text{ k}\Omega}{2\text{ k}\Omega}}$$

$$Z = 1664\underline{/-90^\circ + 56.3^\circ} = 1664\underline{/-33.7^\circ}\ \Omega$$

Therefore, the impedance of the circuit is 1664 Ω and V_A lags the current by 33.7°. To obtain the impedance in rectangular form, we must convert from polar to rectangular form by using the general conversion equation (32-8).

$$A + jB = M\cos\theta \pm jM\sin\theta$$

The resistive component is found:

$$R = 1664\cos(-33.7^\circ) = 1384\ \Omega$$

The capacitive reactance is the j component and is found:

$$X_C = 1664\sin(-33.7^\circ) = -923\ \Omega$$

The impedance of the RC circuit in Fig. 32-15 in rectangular form is

$$Z = 1384 - j923\ \Omega$$

32-7 COMPLEX NUMBER APPLICATION TO THE SERIES *LCR* CIRCUIT IMPEDANCE

In any series LCR circuit the inductive voltage is 180° out of phase with the capacitive voltage, so they cancel each other. Because of the inductive leading and capacitive lagging voltage, X_L and X_C are 180° out of phase and also cancel each other. The impedance of the LCR circuit is then dependent on the resistance and the net reactance $X_L - X_C$.

The total impedance for the series LCR circuit of Fig. 32-16 in polar form is given by equation (32-11), while equation (32-12) gives the total impedance in rectangular form.

$$Z\underline{/\pm 0^\circ} = \sqrt{R^2 + (X_L - X_C)^2}\underline{/\arctan\frac{X_L - X_C}{R}} \quad \text{ohms} \tag{32-11}$$

$$Z = R + jX_L - jX_C \quad \text{ohms} \tag{32-12}$$

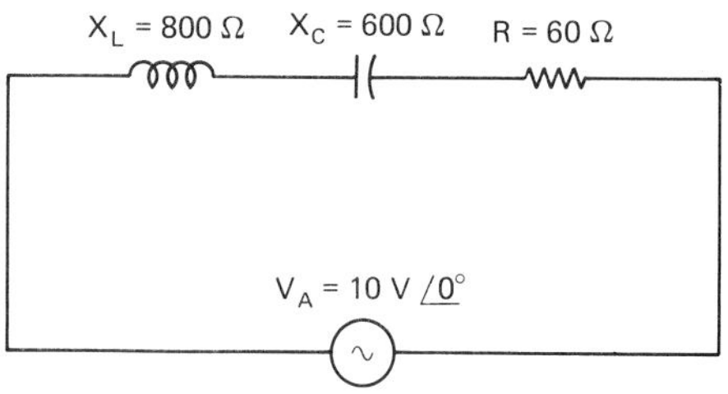

X_L = 800 Ω X_C = 600 Ω   R = 60 Ω

V_A = 10 V /0°

FIGURE 32-16 Series *LCR* circuit.

Voltages V_A, V_R, V_L, and V_C Relationship

In previous series LCR phasor diagram analysis the current and resistance was referenced along the 0° axis of the phasor. Polar expressions are easily adaptable to multiplication and division for Ohm's law application, so we will now *reference the applied voltage along the 0° axis. The phase angles for V_L, V_C, and V_R will be with respect to V_A.*

EXAMPLE 32-15

(a) Find and express in polar form the circuit current and voltage drops across X_L, X_C, and R in the series LCR circuit of Fig. 32-16.

(b) Draw a phasor diagram of the voltage with V_A as the reference voltage.

Solution:

(a) The circuit current can be found by applying Ohm's law once the impedance is known. Equation (32-11) will give the impedance in polar form.

$$Z\underline{/\pm\theta^\circ} = \sqrt{R^2(X_L - X_C)^2}\underline{/\arctan\frac{X_L - X_C}{R}}$$

(where, $X_L - X_C = 800\ \Omega - 600\ \Omega = 200\ \Omega$)

$$= \sqrt{(60\ \Omega)^2 + (200\ \Omega)^2}\underline{/\arctan\frac{200\ \Omega}{60\ \Omega}}$$

$$Z = 209\underline{/73^\circ}\ \Omega$$

Applying Ohm's law will give the circuit current.

$$I = \frac{V_A}{Z} = \frac{10 \text{ V}\angle 0^\circ \text{ V}}{209\angle 73^\circ \ \Omega} = 47.8\angle -73^\circ \text{ mA}$$

Applying Ohm's law will give the voltage drops V_L, V_C, and V_R.

$$V_L = IX_L$$
$$= (47.8 \times 10^{-3}\angle -73^\circ \text{ A})(800\angle 90^\circ \ \Omega)$$
$$= 38.2\angle 17^\circ \text{ V}$$
$$V_C = IX_C = (47.8 \times 10^{-3}\angle -73^\circ \text{ A})(600\angle -90^\circ \ \Omega)$$
$$= 28.7\angle -163^\circ \text{ V}$$
$$V_R = IR = (47.8 \times 10^{-3} \angle -73^\circ \text{ A})(60\angle 0^\circ \ \Omega$$
$$= 2.87\angle -73 \text{ V}$$

(b) In the phasor diagram of Fig. 32-17 the applied voltage is referenced along the X or 0° axis. The phasor diagram shows that V_L is leading V_R by 90°, while V_C is lagging V_R by 90°. Furthermore, it shows that V_L is 180° out of phase with V_C. The circuit current is in phase with V_R and is lagging V_A, indicating a predominately inductive circuit.

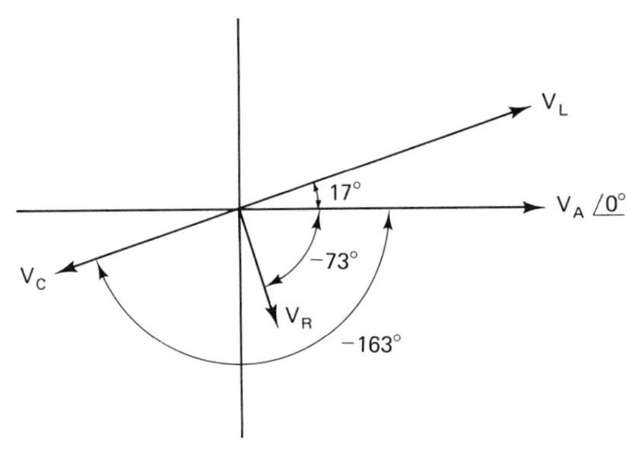

FIGURE 32-17 Phasor diagram of voltage for the circuit of Fig. 32-16.

32-8 COMPLEX NUMBER APPLICATION TO THE PARALLEL *LCR* CIRCUIT

In a parallel *LCR* circuit the inductive current is 180° out of phase with the capacitive current, and to the source they appear to cancel each other. Because of the 180° phase difference between I_L and I_C, the circuit current is dependent on the resistive current and the net reactive current $I_C - I_L$. *Note that the capacitive current leads V_A and the resistive current by 90°; therefore the capacitive current is shown as $+I_C$ while the 90° lagging inductive current is shown as $-I_L$.*

The total current for the parallel *LCR* circuit of Fig. 32-18 in rectangular form can be shown as

$$I_T = I_R + j(I_C - I_L) \quad \text{amperes} \quad (32\text{-}13)$$

Converting the rectangular form into polar form will give the total current expressed in polar form:

$$I_T = \sqrt{I_R^2 + (I_C - I_L)^2}\Big/ \arctan \frac{I_C - I_L}{I_R} \quad \text{amperes} \quad (32\text{-}14)$$

Once the total current is determined, the circuit impedance can be found by using Ohm's law. The impedance of the parallel LCR circuit expressed in polar form is

$$Z = \frac{V_A\angle \theta^\circ}{I_T \Big/ \arctan \dfrac{I_C - I_L}{I_R}} \quad \text{ohms} \quad (32\text{-}15)$$

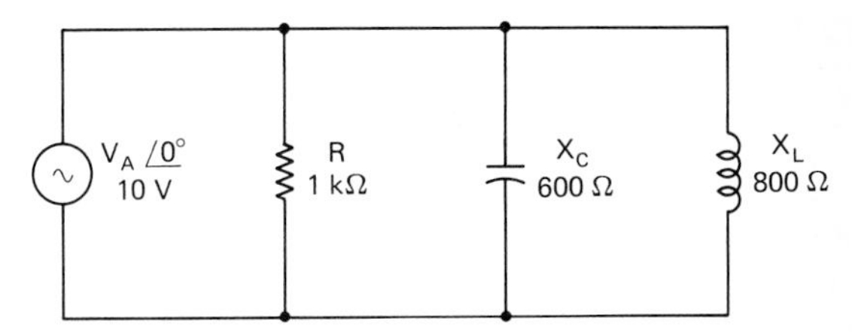

FIGURE 32-18 Parallel *LCR* circuit.

EXAMPLE 32-16

(a) Find and express in polar form each branch current in the parallel *LCR* circuit of Fig. 32-18.

(b) Express the total current in rectangular and polar forms.

(c) Find and express the impedance in polar form.

(d) Draw a phasor diagram of the branch currents with V_A as reference.

Solution:

(a) Each branch current can be found by using Ohm's law.

$$I_R = \frac{V_A}{R} = \frac{10\angle 0^\circ \text{ V}}{1 \text{ k}\angle 0^\circ \ \Omega} = 10\angle 0^\circ \text{ mA}$$

$$I_C = \frac{V_A}{X_C} = \frac{10\angle 0^\circ \text{ V}}{600\angle -90^\circ \ \Omega} = 16.7\angle 90^\circ \text{ mA}$$

$$I_L = \frac{V_A}{X_L} = \frac{10\angle 0^\circ \text{ V}}{800\angle 90^\circ \ \Omega} = 12.5\angle -90^\circ \text{ mA}$$

(b) The total current expressed in rectangular form then is

$$I_T = 10 \text{ mA} + j16.7 \text{ mA} - j12.5 \text{ mA}$$
$$= 10 \text{ mA} + j4.2 \text{ mA}$$

In polar form the total current can be found by using equation (32-14):

$$I_T = \sqrt{I_R^2 + (I_C - I_L)^2} \Big/ \arctan \frac{I_C - I_L}{I_R}$$

$$I_T = \sqrt{(10 \text{ mA})^2 + (4.2 \text{ mA})^2} \Big/ \arctan \frac{4.2 \text{ mA}}{10 \text{ mA}}$$
$$= 10.8\angle 23^\circ \text{ mA}$$

(c) The impedance can be found by using Ohm's law.

$$Z = \frac{V_A}{I_T} = \frac{10\angle 0^\circ \text{ V}}{10.8\angle 23^\circ \text{ mA}} = 926\angle -23^\circ \ \Omega$$

(d) The phasor diagram of the branch currents is shown in Fig. 32-19. Since I_R is in phase with the source voltage, its phasor is drawn along the 0° axis. I_C leads I_R by 90° and I_L lags I_R by 90°. The total current leads the resistive current and source voltage by approximately 23°.

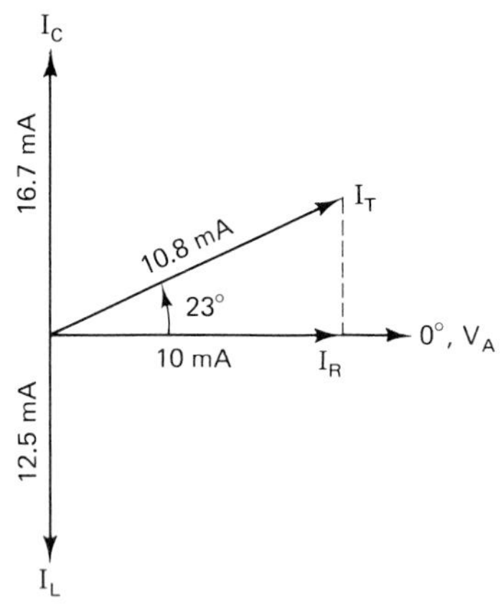

FIGURE 32-19 Phasor diagram for the parallel *LCR* circuit of Fig. 32-18.

PRACTICE QUESTIONS 32-2

Refer to the parallel *RL* circuit of Fig. 32-20 for Questions 1 and 2.

1. Find and express the impedance of the circuit in polar form.
2. Convert the polar form impedance into rectangular form.

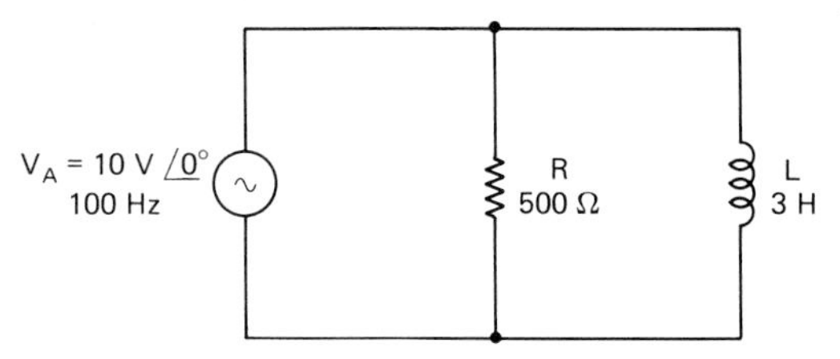

FIGURE 32-20 Parallel *RL* circuit.

Refer to the series *LCR* circuit of Fig. 32-21 for Questions 3 to 9.

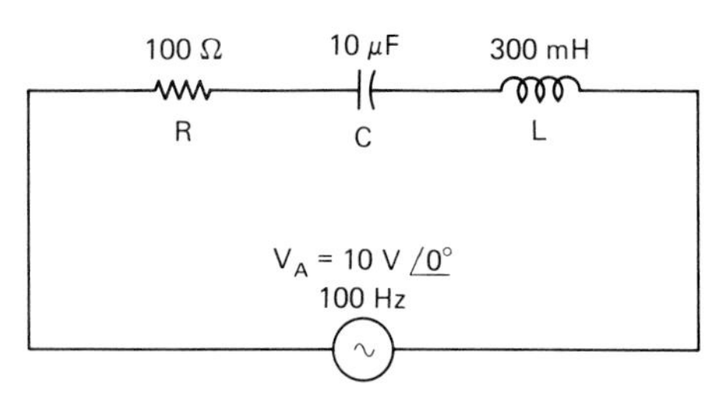

FIGURE 32-21 Series *LCR* circuit.

3. Find and express the impedance of the circuit in polar form.
4. Express the impedance of the circuit in rectangular form.
5. Find and express the circuit current in polar form.
6. Find and express V_R in polar form.
7. Find and express V_L in polar form.
8. Find and express V_C in polar form.
9. Draw a voltage phasor diagram for the circuit of Fig. 32-21.

Refer to the parallel *LCR* circuit of Fig. 32-22 for Questions 10 to 16.

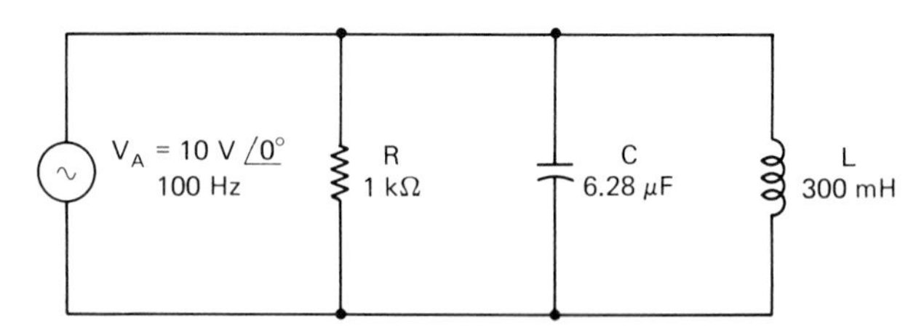

FIGURE 32-22 Parallel *LCR* circuit.

10. Find and express the resistive current of the circuit in polar form.
11. Find and express the capacitive current of the circuit in polar form.
12. Find and express the inductive current of the circuit in polar form.
13. Find and express the total circuit current in polar form.
14. Express the total current in rectangular form.

15. Find and express the impedance of the circuit in polar form.

16. Draw a current phasor diagram for the circuit of Fig. 32-22.

32-9 CONDUCTANCE, SUSCEPTANCE, AND ADMITTANCE IN A PARALLEL *LCR* CIRCUIT

The "sum of the reciprocal" of each branch resistance is another method of finding the impedance of a parallel *LCR* circuit. Recall that in a parallel circuit the total resistance was found by summing the conductance (reciprocal of resistance) of each resistive branch.

In a similar manner the reciprocal of each R, X_L, and X_C can be summed to give the net conductance; then taking the reciprocal of the net conductance will give the net impedance of the circuit.

The total impedance in polar form of a parallel *LCR* circuit can be found by applying the sum of the reciprocal equation.

$$\frac{1}{Z} = \frac{1}{R\angle 0^\circ} + \frac{1}{X_L\angle 90^\circ} + \frac{1}{X_C\angle -90^\circ} \quad \text{ohms} \qquad (32\text{-}16)$$

EXAMPLE 32-17

Find the impedance in polar form for the circuit of Fig. 32-22.

Solution: Calculating for X_C and X_L, then applying equation (32-16), gives

$$X_C = \frac{1}{2\pi fC} = \frac{0.159}{100 \text{ Hz} \times 6.28 \times 10^{-6} \text{ F}}$$

$$= 253 \angle -90^\circ \ \Omega$$

$$X_L = 2\pi fL = 6.28 \times 100 \text{ Hz} \times 300 \times 10^{-3} \text{ H}$$

$$= 188 \angle +90^\circ \ \Omega$$

$$\frac{1}{Z} = \frac{1}{1000 \angle 0^\circ \ \Omega} + \frac{1}{188 \angle +90^\circ \ \Omega} + \frac{1}{253 \angle -90^\circ \ \Omega}$$

Taking the inverse of each value and applying the rule for division of polar number phase angles, we obtain

$$\frac{1}{Z} = 0.001 \angle 0^\circ \text{ S} + 0.0053 \angle -90^\circ \text{ S} + 0.00395 \angle +90 \text{ S}$$

Converting each item to rectangular form gives

$$\frac{1}{Z} = 0.001 \text{ S} - j0.0053 \text{ S} + j0.00395 \text{ S}$$

$$= 0.001 \text{ S} - j0.00135 \text{ S}$$

Converting Z to polar form

where, $\sqrt{(0.001\ S)^2 + (0.001325 \text{ S})^2}$

$$= 0.00168 \text{ S}$$

$$Z = \frac{1}{(0.00168 \text{ S} \angle \arctan - \dfrac{0.00135 \text{ S}}{0.001 \text{ S}}}$$

$$= 595 \angle 53.5^\circ \ \Omega$$

For the parallel complex circuit the reciprocal reactance is termed *susceptance, B*. A subscript C indicates capacitive susceptance, while L indicates inductive susceptance. As in the case of conductance, the unit of susceptance is the siemens. In equation form,

$$G = \frac{1}{R\angle 0^\circ} \quad \text{siemens} \qquad (32\text{-}17a)$$

$$B_L = \frac{1}{X_L \angle +90^\circ} = B_L \angle -90^\circ = -jB_L \quad \text{siemens} \qquad (32\text{-}17b)$$

$$B_C = \frac{1}{X_C \angle -90^\circ} = B_C \angle +90^\circ = jB_C \quad \text{siemens} \qquad (32\text{-}17c)$$

Note that the inductive susceptance is a negative j quantity, while capacitive susceptance becomes a positive j quantity. This is because of the mathematical division or inverse process, as shown in equations (32-17b) and (32-17c).

For a parallel *LCR* circuit the phasor sum of conductance, G, and susceptance, B, will give the *admittance*, Y, of the circuit. The reciprocal of the net admittance will then give the total impedance of the circuit.

$$Y = \frac{1}{Z \angle \pm\theta^\circ} \quad \text{siemens} \qquad (32\text{-}17d)$$

$$Z = \frac{1}{Y \angle \theta^\circ} \quad \text{ohms} \qquad (32\text{-}17e)$$

The complex sum of the conductance is the total admittance of the circuit and is expressed in rectangular form as

$$Y \angle \pm\theta^\circ = G + jB_C - jB_L \quad \text{siemens} \qquad (32\text{-}18)$$

The admittance can be found simply by applying the phasor sum of the conductance and susceptance:

$$Y \underline{/\pm\theta^\circ} = \sqrt{G^2 + (B_C - B_L)^2} \Big/ \arctan \frac{B}{G} \quad \text{siemens} \tag{32-19}$$

EXAMPLE 32-18

Find the admittance and impedance of the parallel *LCR* circuit of Fig. 32-23.

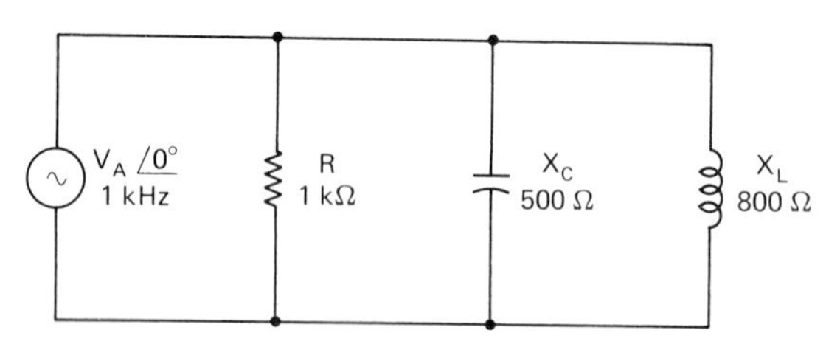

FIGURE 32-23 Parallel *LCR* circuit.

Solution: Equation (32-17a) gives the conductance, G.

$$G = \frac{1}{R\underline{/0^\circ}} = \frac{1}{1000\underline{/0^\circ}} = 0.001\underline{/0^\circ}\text{ S}$$

Equation (32-17b) gives the susceptance of the inductive branch.

$$B_L = \frac{1}{X_L\underline{/+90^\circ}} = \frac{1}{800\underline{/90^\circ}} = 0.00125\underline{/-90^\circ}\text{ S}$$

Equation (32-17c) gives the susceptance of the capacitive branch.

$$B_C = \frac{1}{X_C\underline{/-90^\circ}} = \frac{1}{500\underline{/-90^\circ}} = 0.002\underline{/+90^\circ}\text{ S}$$

The admittance of the circuit can now be expressed in rectangular form by applying equation (32-18):

$$\begin{aligned} Y\underline{/\pm\theta^\circ} &= G + jB_C - jB_L \\ &= 0.001\text{ S} + j0.002\text{ S} - j0.00125\text{ S} \\ &= 0.001\text{ S} + j0.00075\text{ S} \end{aligned}$$

Applying equation (32-19) will give the admittance of the circuit.

$$Y = \sqrt{G^2 + (B_C - B_L)^2} \Big/ \arctan \frac{B}{G} \quad \text{siemens}$$

where, $\sqrt{(0.001\text{ S})^2 + (0.00075\text{ S})^2} = 0.00125\text{ S}$

$$= 0.00125\text{ S} \Big/ \arctan \frac{0.00075\text{ S}}{0.001\text{ S}}$$

$$= 0.00125\underline{/37^\circ}\text{ S}$$

The impedance of the circuit is then found by applying equation (32-17e)

$$Z = \frac{1}{Y\underline{/\pm\theta^\circ}} \quad \text{ohms}$$

$$= \frac{1}{0.00125\text{ S}\underline{/37^\circ}} = 800\underline{/-37^\circ}\ \Omega$$

Note that the negative phase angle shows that the circuit Z is more capacitive. This is because the reactance of 500 Ω is much lower than the resistance or inductive reactance. In a parallel circuit the lowest R or X will have the greatest current and therefore the greatest effect on the impedance.

32-10 CONVERTING A COMPLEX CIRCUIT INTO AN EQUIVALENT CIRCUIT

Solving series–parallel complex circuits involves reducing a group of parallel components into a series equivalent Z and θ, then adding this impedance and phase angle to other series components in order to compute a final equivalent circuit Z and θ. Conversely, a series impedance and phase angle can be converted to an equivalent parallel Z and θ. Note that in both cases the Z and θ is a theoretical equivalency as seen by the source and are not the same circuit components.

EXAMPLE 32-19

Convert the parallel *LCR* circuit of Fig. 32-24 into its equivalent series circuit.

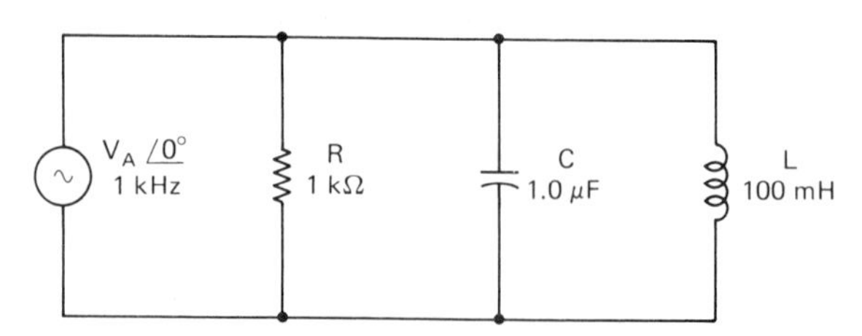

FIGURE 32-24 Parallel *LCR* circuit.

Solution: The capacitive reactance and inductive reactance must be found first. Next the impedance is calculated and converted into polar form. The polar equation is then converted into rectangular form to give the two equivalent components of the circuit.

$$X_C = \frac{1}{2\pi f C} = \frac{0.159}{1000\text{ Hz} \times 1.0 \times 10^{-6}\text{ F}}$$

$$= 159\ \Omega$$

$$X_L = 2\pi fL$$

$$= 6.28 \times 1 \times 10^3\ \text{Hz} \times 100 \times 10^{-3}\ \text{H}$$

$$= 628\ \Omega$$

Using equation (32-16) will give the impedance of the circuit.

$$\frac{1}{Z} = \frac{1}{R\angle 0^\circ} + \frac{1}{X_L\angle 90^\circ} + \frac{1}{X_C\angle -90^\circ}$$

$$= \frac{1}{500\angle 0^\circ} + \frac{1}{628\angle 90^\circ} + \frac{1}{159\angle -90^\circ}$$

$$= 0.002\angle 0^\circ\ \text{S} + 0.00159\angle -90^\circ\ \text{S} + 0.00629\angle 90^\circ\ \text{S}$$

$$= 0.002\ \text{S} - j0.00159\ \text{S} + j0.00629\ \text{S}$$

$$= 0.002\ \text{S} + j0.0047\ \text{S}$$

Converting Z to polar form

where, $\sqrt{(0.0025\ \text{S})^2 + (0.0047\ \text{S})^2} = 0.0051\ \text{S}$

$$Z = \frac{1}{0.0051\ \text{S}\angle \arctan \dfrac{0.0047\ \text{S}}{0.002\ \text{S}}}$$

$$= 196\angle -67^\circ\ \Omega$$

We can use the general equation (32-8) to convert from polar to rectangular form:

$$A \pm jB = M \cos\theta \pm jM \sin\theta$$

$$= 196 \cos(-67^\circ) \pm j196 \sin(-67^\circ)$$

$$= 76.6 - j180\ \Omega$$

The series equivalent circuit has a resistance of 76.6 Ω and a capacitive reactance of 180 Ω; solving for capacitance gives the equivalent circuit components as shown in Fig. 32-25.

Solving for capacitance gives

$$C = \frac{1}{2\pi f X_C} = \frac{0.159}{1 \times 10^3\ \text{Hz} \times 180\ \Omega} = 0.9\ \mu\text{F}$$

The series circuit of Fig. 32-25 will appear to the source as having the same impedance and leading current as the parallel *LCR* circuit.

Having converted a parallel *LCR* circuit into a series equivalent circuit, we will now convert the series *LCR* circuit into an equivalent parallel circuit.

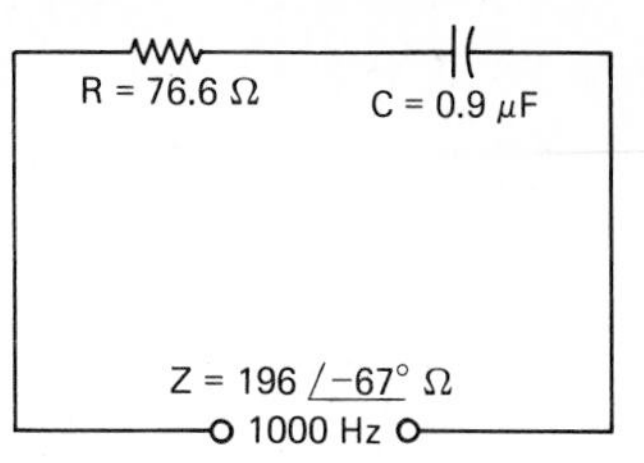

FIGURE 32-25 Series equivalent circuit of Fig. 32-24, parallel *LCR* circuit.

EXAMPLE 32-20

Convert the series *LCR* circuit of Fig. 32-26(a) into a parallel equivalent circuit.

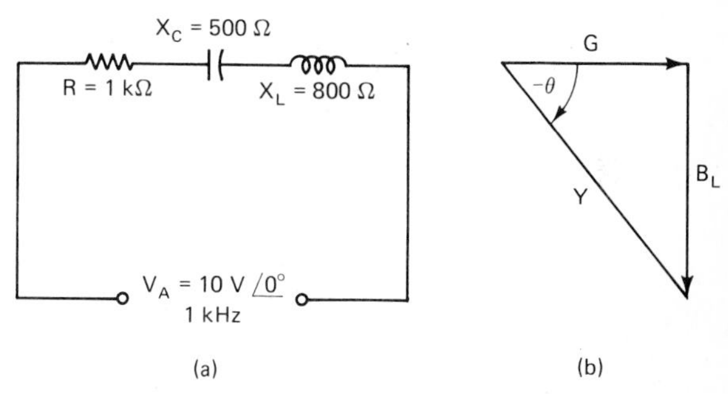

FIGURE 32-26 (a) Series *LCR* circuit. (b) Admittance triangle.

Solution: The impedance and phase angle of the series *LCR* circuit must be found first. The total impedance and phase angle can be found by applying equation (32-11).

$$Z = \sqrt{R^2 + (X_L - X_C)^2}\angle \arctan \frac{X_L - X_C}{R} \quad \text{ohms}$$

where, $\sqrt{(1000\ \Omega)^2 + (800\ \Omega - 500\ \Omega)^2}$

$$= 1044\ \Omega$$

$$= 1044\angle \arctan \frac{800 - 500\ \Omega}{1000\ \Omega}$$

$$= 1044\angle 16.7^\circ\ \Omega$$

This means that the circuit has an impedance of 1044 Ω and the applied voltage leads the current by 16.7°.

Next we can apply the general conversion formula (32-8) to find the conductance, G, and B_L.

$$A \pm jB = M \cos\theta \pm jM \sin\theta$$

Therefore,

$$Y\angle\theta^\circ = Y \cos\theta \pm jY \sin\theta$$

From the admittance triangle [see Fig. 32-26(b)] and given the phase angle and impedance, we can easily find Y, G, and B_L as follows:

$$Y\angle\theta° = \frac{1}{Z\angle\theta°} = \frac{1}{1044\angle 16.7°}$$

$$= 9.579 \times 10^{-4}\angle -16.7° \text{ S}$$

$$\cos\theta = \frac{G}{Y} \qquad \text{therefore, } G = Y\cos\theta$$

$$G = 9.579 \times 10^{-4}\text{ S} \times \cos(-16.7°)$$

$$= 9.175 \times 10^{-4}\text{ S}$$

$$\sin\theta = \frac{B_L}{Y} \qquad \text{therefore, } B_L = Y\sin\theta$$

$$B_L = 9.579 \times 10^{-4}\text{ S} \times \sin(-16.7°)$$

$$= 2.752 \times 10^{-4}\text{ S}$$

The equivalent component values in parallel are

$$R = \frac{1}{G} = \frac{1}{9.175 \times 10^{-4}\text{ S}} = 1090\ \Omega$$

$$X_L = \frac{1}{B_L} = \frac{1}{2.752 \times 10^{-4}\text{ S}} = 3634\ \Omega$$

The equivalent parallel inductance is

$$L = \frac{X_L}{2\pi f} = \frac{3634\ \Omega}{6.28 \times 1000\text{ Hz}} = 579\text{ mH}$$

A 1-kHz source would see the same impedance in the parallel equivalent circuit (Fig. 32-27) as the series LCR circuit with the circuit current lagging the applied voltage by 16.7° or approximately 17°. Comparable phasor diagrams are shown in Fig. 32-28(a) for the series circuit and Fig. 32-28(b) for the equivalent parallel circuit.

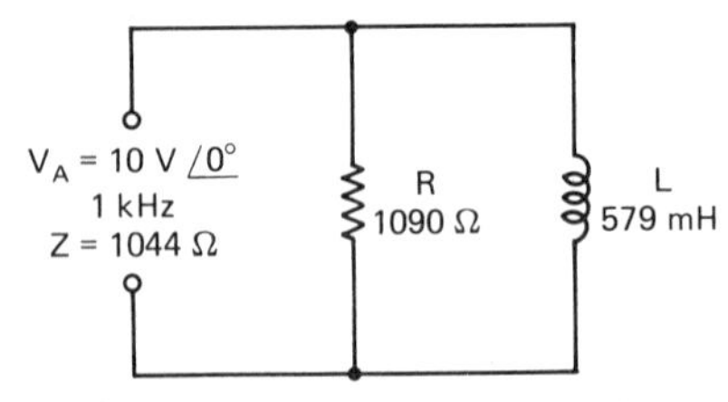

FIGURE 32-27 Equivalent parallel circuit of Fig. 32-26, Series *LCR* circuit.

32-11 COMPLEX NUMBER APPLICATION TO SERIES–PARALLEL *LCR* CIRCUITS

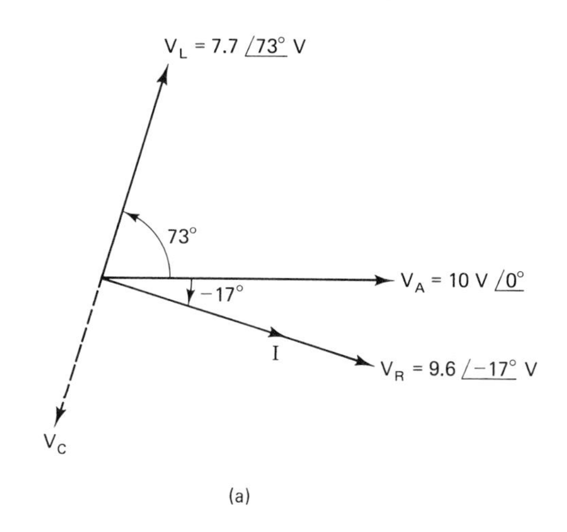

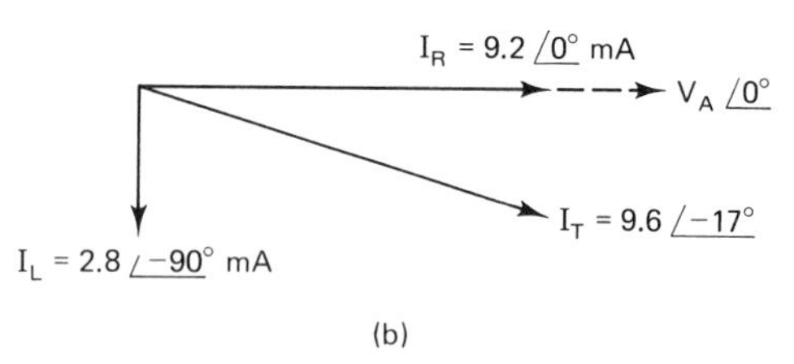

FIGURE 32-28 (a) Phasor diagram of series *LCR* circuit of Fig. 32-26. (b) Phasor diagram of parallel equivalent circuit of Fig. 32-26.

The circuit parameters of series–parallel complex circuits can be solved by finding the total impedance of the circuit. We can then find the total current by using a convenient value of applied voltage.

The total impedance of a series–parallel circuit can be reduced into either two or more series impedances or two or more parallel branch impedances, depending on the circuit configuration.

Combining Series Impedance

If the circuit is reduced to a number of series impedances the total impedance can be found by converting each impedance into its rectangular expression, then adding the real and j numbers to give the total impedance expressed in rectangular form, and by conversion into polar form.

EXAMPLE 32-21

Find the total impedance and the voltage drops in polar form, for the circuit of Fig. 32-29(a).

Solution: First reduce the parallel LR component to an equivalent impedance of Fig. 32-29(b) by applying equation (32-9).

$$Z_1 = \frac{R \times X_L}{\sqrt{R^2 + X_L^2}}\Big/ 90° - \arctan\frac{X_L}{R} \quad \text{ohms}$$

where,

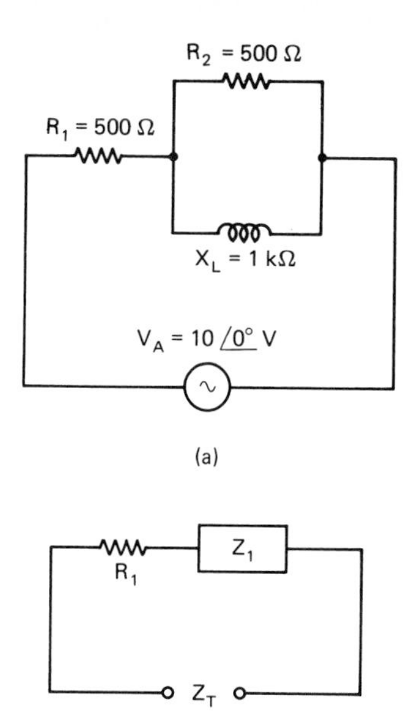

FIGURE 32-29 (a) Series–parallel complex circuit. (b) Equivalent series impedance.

$$\frac{R \times X_L}{\sqrt{R^2 + X_L^2}} = \frac{500\ \Omega \times 1000\ \Omega}{\sqrt{(500\ \Omega)^2 + (1000\ \Omega)^2}} = 447\ \Omega$$

$$= 447 \Big/ 90^\circ - \arctan \frac{1000\ \Omega}{500\ \Omega}$$

$$= 447\underline{/26.6^\circ}\ \Omega$$

The impedance of the parallel components, Fig. 32-29(b), can also be found by the reciprocal sum or conductance method.

$$\frac{1}{Z_1} = \frac{1}{500\underline{/0^\circ}} + \frac{1}{1000\underline{/90^\circ}}$$

$$= 0.002\underline{/0^\circ} + 0.001\underline{/-90^\circ}$$

$$= 0.002\ \text{S} - j0.001\ \text{S}$$

Converting Z to polar form where,

$$\frac{1}{Z_1} = \sqrt{(0.002\ \text{S})^2 + (0.001\ \text{S})^2} = 0.0022\ \text{S}$$

$$Z_1 = \frac{1}{0.0022\ \text{S}} \Big/ \arctan\left(-\frac{0.001\ \text{S}}{0.002\ \text{S}}\right)$$

$$= 447\underline{/26.6^\circ}\ \Omega$$

The parallel impedance can now be converted to equivalent series impedance by using the general equation (32-8).

$$A \pm jB = M \cos\theta \pm jM \sin\theta$$

$$Z_1 = 447 \cos 26.6^\circ \pm j447 \sin 26.6^\circ$$

$$= 400 + j200\ \Omega$$

The total impedance of the circuit is equal to the sum of R_1 and Z_2:

$$Z_T = R_1\underline{/\theta^\circ} + Z_1\underline{/\theta^\circ}$$

The total impedance in rectangular form of the circuit now becomes

$$Z_T = 500 + 400 + j200\ \Omega$$

$$= 900 + j200\ \Omega$$

The total impedance in polar form is then

$$Z_T = \sqrt{(900\ \Omega)^2 + (200\ \Omega)^2} \Big/ \arctan \frac{200\ \Omega}{900\ \Omega}$$

$$= 922\underline{/12.5^\circ}\Omega$$

Having found the total impedance, the circuit current can be found by applying Ohm's law.

$$I_T = \frac{V_A}{Z_T} = \frac{10\underline{/0^\circ}\ \text{V}}{922\underline{/12.5^\circ}} = 10.8\underline{/-12.5^\circ}\ \text{mA}$$

$$V_{R1} = I_T \times R_1 = 10.8\underline{/-12.5^\circ}\ \text{mA} \times 0.5\ \text{k}\Omega$$

$$= 5.4\underline{/-12.5^\circ}\ \text{V}$$

$$V_{R2} = I_T \times Z_1$$

$$= 10.8\underline{/-12.5^\circ}\ \text{mA} \times 0.447\ \text{k}\Omega\underline{/26.6^\circ}$$

and

$$V_{R2} = V_{XL} = 4.8\underline{/14.1^\circ}\ \text{V}$$

EXAMPLE 32-22

Find the voltage drop across the series RC impedance and the parallel RL impedance in the circuit of Fig. 32-30(a), in polar form.

Solution: Reducing the series RC component into Z_1 and the parallel RL components into Z_2 as shown in Fig. 32-30(b) will allow us to find the total impedance. The total circuit current and the voltage across each impedance can then be found by applying Ohm's law. An alternative and much quicker solution would be to apply the voltage-divider method to solve for the voltage drops across each Z. We shall look at both methods of solution.

Finding the Total Impedance. The series impedance, Z_1, in rectangular form is

$$Z_1 = 200 - j1000\ \Omega$$

Converting the rectangular expression for Z_1 into polar form gives

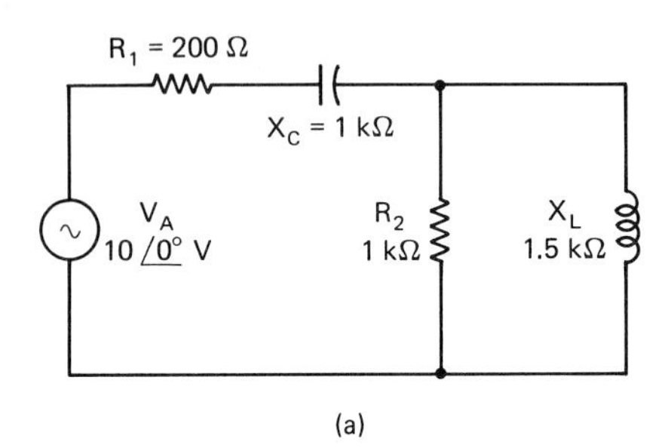

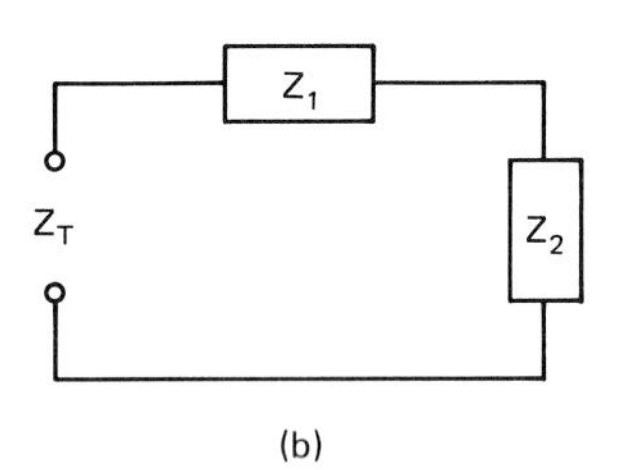

FIGURE 32-30 (a) Series–parallel complex circuit. (b) Equivalent series impedance.

Z_1

$$= \sqrt{(200\ \Omega)^2 + (1000\ \Omega)^2} \angle \arctan\left(-\frac{1000\ \Omega}{200\ \Omega}\right)$$

$$= 1020\angle -78.7°\ \Omega$$

The parallel impedance Z_2 can be found by applying equation (32-9).

$$Z_2 = \frac{R \times X_L}{\sqrt{R^2 + X_L^2}} \angle 90° - \arctan\frac{X_L}{R} \quad \text{ohms}$$

where,

$$\frac{R \times X_L}{\sqrt{R^2 + X_L^2}} = \frac{1\ \text{k}\Omega \times 1.5\ \text{k}\Omega}{\sqrt{(1\ \text{k}\Omega)^2 + (1.5\ \text{k}\Omega)^2}} = 832\ \Omega$$

$$= 832 \angle 90° - \arctan\frac{1.5\ \text{k}\Omega}{1.0\ \text{k}\Omega}$$

$$= 832\angle 33.7°\ \Omega$$

Converting Z_2 into its rectangular expression will allow us to find Z_T by the simple rectangular sum of Z_1 and Z_2. Applying the general equation (32-8) will give the rectangular expression for Z_2.

$$Z_2 = 832\cos(33.7°) + j832\sin(33.7°)$$

$$= 692 + j462\ \Omega$$

The total impedance of the circuit is the sum of Z_1 and Z_2.

$$Z_T = Z_1 + Z_2 = 200 - j1000 + 692 + j462\ \Omega$$

$$= 892 - j538\ \Omega$$

Converting Z_T to polar form gives

$Z_T\angle\theta°$

$$= \sqrt{(892\ \Omega)^2 + (538\ \Omega)^2} \angle -90° + \arctan\frac{538\ \Omega}{892\ \Omega}$$

$$= 1042\angle -59°\ \Omega$$

Ohm's law method to find voltage drops:

$$I_T = \frac{V_A\angle 0°}{Z_T\angle\theta} = \frac{10\angle 0°\ \text{V}}{1042\angle -59°\ \Omega} = 9.6\angle 59°\ \text{mA}$$

$$V_{Z1} = I_T \times Z_1 = 9.6\angle 59°\ \text{mA} \times 1.02\angle -78.7°\ \text{k}\Omega$$
$$= 9.8\angle -19.7°$$

$$V_{Z2} = I_T \times Z_2 = 9.6\angle 59°\ \text{mA} \times 0.832\angle 33.7°\ \text{k}\Omega$$
$$= 8.0\angle 92.7°\ \text{V}$$

Voltage-divider method: Applying the voltage-divider method to find the voltage drop across Z_1 and Z_2 gives

$$V_{Z1} = \frac{Z_1}{Z_T} \times V_A = \frac{1020\angle -78.7°\ \Omega}{1042\angle -59°\ \Omega} \times 10\angle 0°\ \text{V}$$
$$= 9.8\angle -19.7°\ \text{V}$$

$$V_{Z2} = \frac{Z_2}{Z_T} \times V_A = \frac{832\angle 33.7°\ \Omega}{1042\angle -59°\ \Omega} \times 10\angle 0°\ \text{V}$$
$$= 8.0\angle 92.7°\ \text{V}$$

The 9.8 V across the first series impedance lags the applied voltage by approximately 20°, while the parallel voltage across Z_2 leads the applied voltage by approximately 93°.

EXAMPLE 32-23

Find the total impedance of the complex circuit in Fig. 32-31(a).

Solution: First find the branch impedances Z_1 and Z_2 as represented in Fig. 32-31(b); then use product-over-sum method to find the total impedance.

RL Branch Impedance (Z_1)

$$Z_1 = 100 + j500\ \Omega$$

Converting to polar form gives

$Z_1\angle\theta°$

$$= \sqrt{(100\ \Omega)^2 + (500\ \Omega)^2} \angle \arctan\frac{X_L}{R} \quad \text{ohms}$$

$$= 510\angle 78.7°\ \Omega$$

LCR Branch Impedance (Z_2)

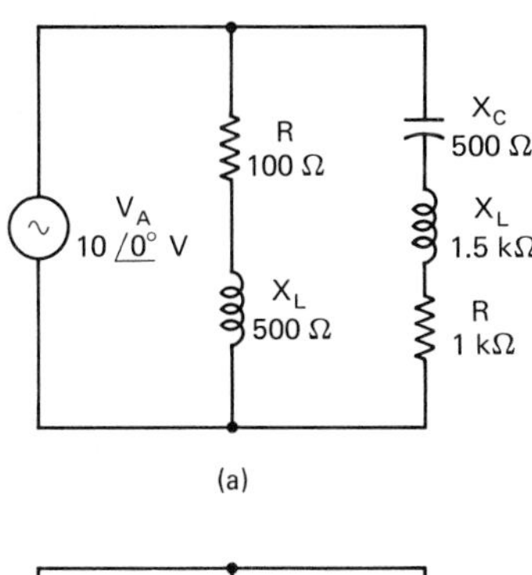

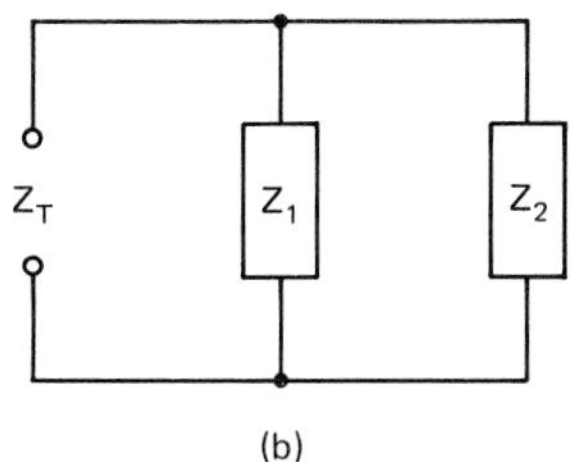

FIGURE 32-31 (a) Series–parallel complex circuit. (b) Equivalent parallel impedance.

$$Z_2 = 1\text{ k}\Omega + j1.5\text{ k}\Omega - j0.5\text{ k}\Omega$$
$$= 1\text{ k}\Omega + j1\text{ k}\Omega$$

Converting to polar form gives

$$Z_2\underline{/\theta^\circ} = \sqrt{(1\text{ k}\Omega)^2 + (1\text{ k}\Omega)^2}\,\Big/\arctan\frac{1\text{ k}\Omega}{1\text{ k}\Omega}$$
$$= 1414\underline{/45^\circ}\ \Omega$$

Once the branch impedances are found, each branch current can be calculated by applying Ohm's law.

Total Impedance

The product-over-sum method can be used to find Z_T.

$$Z_T = \frac{Z_1 \times Z_2}{Z_1 + Z_2} = \frac{510\underline{/78.7^\circ}\ \Omega \times 1414\underline{/45^\circ}\ \Omega}{100 + j500 + 1000 + j1000\ \Omega}$$
$$= \frac{721{,}140\underline{/123.7^\circ}\ \Omega}{1100 + j1500\ \Omega}$$

Converting the denominator to polar form for easier division gives

$$Z_T\underline{/\theta^\circ} = \frac{721{,}140\underline{/123.7^\circ}\ \Omega}{1860\underline{/53.7^\circ}\ \Omega} = 388\underline{/70^\circ}\ \Omega$$

EXAMPLE 32-24

(a) Find the total impedance of the circuit in Fig. 32-32.

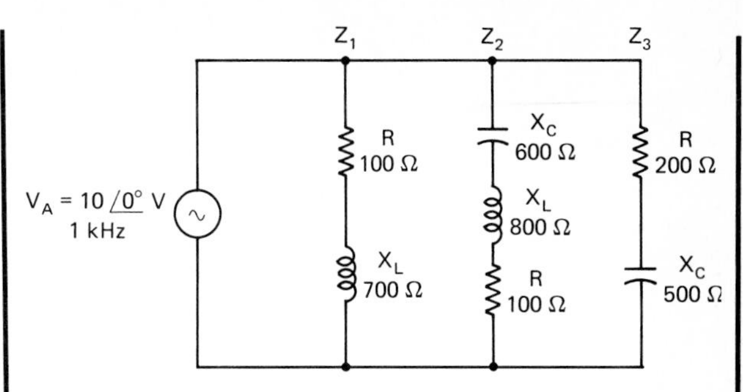

FIGURE 32-32 Complex *LCR* circuit.

(b) What is its equivalent series components?

Solution: Two methods can be applied to solve for the total impedance of the circuit in Fig. 32-32.

1. Product-over-sum method of branch impedance taking two branches at a time.
2. Ohm's law by first finding the total circuit current.

Product-Over-Sum Method.

(a) The impedance of each branch circuit is converted from rectangular form into polar form.

$$Z_1 = 100 + j700\ \Omega = 707\underline{/81.9^\circ}\ \Omega$$
$$Z_2 = 100 + j800 - j600 = 224\underline{/63.4^\circ}\ \Omega$$
$$Z_3 = 200 - j500\ \Omega = 539\underline{/-68.2^\circ}\ \Omega$$

Taking the product-over-sum of two branches at one time gives

$$Z_T' = \frac{Z_1 \times Z_2}{Z_1 + Z_2} = \frac{707\underline{/81.9^\circ}\ \Omega \times 224\underline{/63.4^\circ}\ \Omega}{100 + j700 + 100 + j200\ \Omega}$$
$$= \frac{158{,}368\underline{/145.3^\circ}\ \Omega}{200 + j900\ \Omega} = \frac{158{,}368\underline{/145.3^\circ}\ \Omega}{922\underline{/77.5^\circ}\ \Omega}$$
$$= 172\underline{/67.8^\circ}\ \Omega$$

$$Z_T = \frac{Z_3 \times Z_T'}{Z_3 + Z_T'} = \frac{539\underline{/-68.2^\circ} \times 172\underline{/67.8^\circ}\ \Omega}{200 - j500 + 65 + j159\ \Omega}$$
$$= \frac{92{,}708\underline{/-0.4^\circ}\ \Omega}{265 - j341\ \Omega} = \frac{92{,}708\underline{/-0.4^\circ}\ \Omega}{432\underline{/-52.1^\circ}\ \Omega}$$
$$Z_T\underline{/\theta^\circ} = 215\underline{/51.7^\circ}\ \Omega$$

(b) The equivalent series circuit components can be found by converting the total impedance into rectangular form. Applying the general equation (32-8) gives

$$Z_T = 215 \cos(51.7°) + j215 \sin(51.7°)$$
$$= 133 + j169\ \Omega$$
$$L = \frac{X_L}{2\pi f} = \frac{169\ \Omega}{6.28 \times 10^3\ \text{Hz}} \cong 27\ \text{mH}$$

The series equivalent circuit consists of a 133-Ω resistance in series with a 27-mH choke.

Total Circuit Current Method. Find each branch current in polar form.

$$I_1 = \frac{V_A}{Z_1} = \frac{10\angle 0°\ \text{V}}{707\angle 81.9°\ \Omega} = 14.1\angle -81.9°\ \text{mA}$$
$$I_2 = \frac{V_A}{Z_2} = \frac{10\angle 0°\ \text{V}}{224\angle 63.4°\ \Omega} = 44.6\angle -63.4°\ \text{mA}$$
$$I_3 = \frac{V_A}{Z_3} = \frac{10\angle 0°\ \text{V}}{539\angle -68.2\ \Omega} = 18.6\angle 68.2°\ \text{mA}$$

To find the total current, we can sum the branch currents by converting each branch current into its rectangular form for ease of addition.

$$I_1 = 14.1 \cos(-81.9°) \pm 14.1 \sin(-81.9°)\ \text{mA}$$
$$= 1.99 - j14\ \text{mA}$$
$$I_2 = 44.6 \cos(-63.4°) \pm 44.6 \sin(-63.4°)$$
$$= 20 - j39.9\ \text{mA}$$
$$I_3 = 18.6 \cos(68.2°) \pm 18.6 \sin(68.2°)$$
$$= 6.9 + j17.3\ \text{mA}$$
$$I_T = 1.99 - j14 + 20 - j39.9 + 6.9 + j17.3$$
$$= 28.9 - j36.6\ \text{mA}$$

In polar form the total current is

$$I_T = \sqrt{(28.9\ \text{mA})^2 + (36.6\ \text{mA})^2} \angle \arctan\left(-\frac{36.6}{28.9}\right)$$
$$= 46.6\angle -51.7°\ \text{mA}$$
$$Z_T = \frac{V_A\angle 0°}{I_T\angle\theta} = \frac{10\angle 0°\ \text{V}}{46.6\angle -51.7°\ \text{mA}} = 215\angle 51.7°\ \Omega$$

PRACTICE QUESTIONS 32-3

Refer to the series–parallel *LCR* circuit of Fig. 32-33 for Questions 1 to 6.

1. Find the impedance of the series components R_1 and X_L in polar form.
2. Find the impedance of the parallel components R_2 and X_C in polar form.

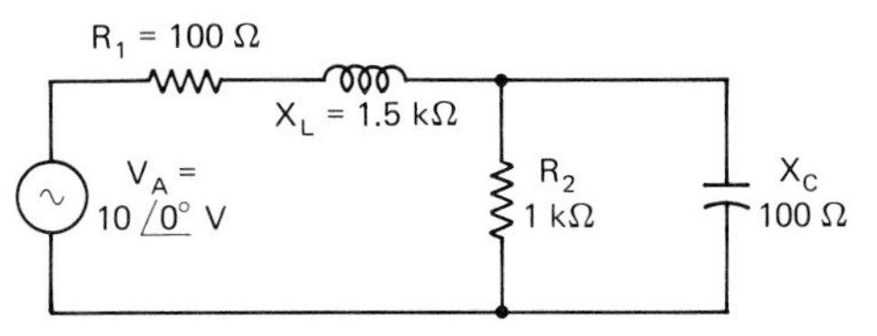

FIGURE 32-33 Series–parallel *LCR* circuit.

3. Express the total impedance in rectangular form.
4. Express the total impedance in polar form.
5. Find the voltage drop across the capacitance.
6. Find the voltage drop across the series impedance.

Refer to the complex *LCR* circuit of Fig. 32-34 for Questions 7 to 19.

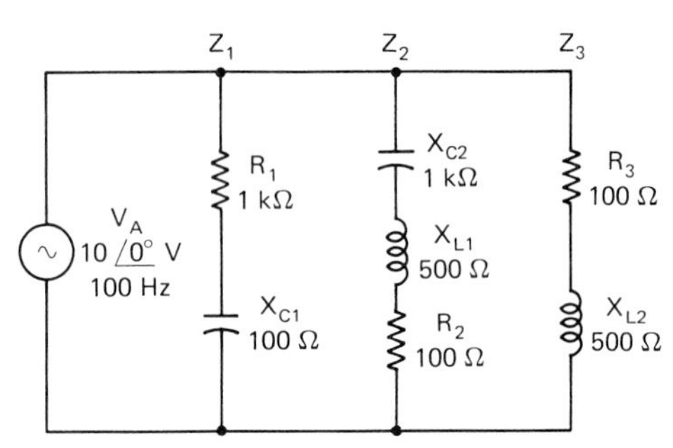

FIGURE 32-34 Complex *LCR* circuit.

7. Find the impedance of branch Z_1 in polar form.
8. Find the impedance of branch Z_2 in polar form.
9. Find and express the current in branch Z_1 in polar form.
10. Express the current in branch Z_1 in rectangular form.
11. Find and express the current in branch Z_2 in polar form.
12. Express the current in branch Z_2 in rectangular form.
13. Find and express the current in branch Z_3 in polar form.
14. Express the current in branch Z_3 in rectangular form.
15. Express the total circuit current in rectangular form.
16. Express the total current in polar form.
17. Find and express the total impedance in polar form.
18. What is the equivalent series resistance for this circuit?
19. What is the equivalent reactive component for this circuit?

SUMMARY

Numbers along the horizontal (x) axis are termed the *real numbers*.

Numbers along the vertical (y) axis are termed imaginary numbers shown as $\pm j$ and represent the reactance with respect to the resistance.

A vector quantity that is rotated CCW by 90° from the reference direction of 0° is given a $+j$ prefix and is a leading position.

A vector quantity that is rotated CCW by 270° from the reference direction of 0° is given a $-j$ prefix and is a lagging position.

A phasor quantity in rectangular form is the sum of the real coordinates and the j coordinates.

Phasor quantities in rectangular form provide a means of easily adding and subtracting other rectangular quantities.

The polar form of a phasor quantity is in terms of the magnitude and phase angle referenced to the positive or real axis.

Quantities expressed in polar form provide a means of easily multiplying and dividing electrical quantities.

Resistance in rectangular form is written as R or its value. In polar form resistance is shown as $R\angle 0^\circ\ \Omega$.

Inductive reactance is considered to be a leading quantity, and therefore is written $+j$ in rectangular form. In polar form it is shown as $X_L = \angle 90^\circ\ \Omega$.

Capacitive reactance is considered to be a lagging quantity, and therefore is written as $-j$ in rectangular form. In polar form it is shown as $X_C = \angle -90^\circ\ \Omega$.

Impedance of a series RL circuit is given in rectangular form by equation,

$$Z = R + jX_L \qquad \text{ohms} \tag{32-1}$$

Impedance of a series RL circuit is given in polar form by equation,

$$Z = \sqrt{R^2 + X_L^2}\angle \arctan\frac{X_L}{R} \qquad \text{ohms} \tag{32-2}$$

Impedance of a series RC circuit is given in rectangular form by equation,

$$Z = R - jX_C \qquad \text{ohms} \tag{32-3}$$

Impedance of a series RC circuit is given in polar form by equation,

$$Z = \sqrt{R^2 + X_C^2}\angle -\arctan\frac{X_C}{R} \qquad \text{ohms} \tag{32-4}$$

The general equation for converting rectangular to polar form is given by equation,

$$A \pm jB = \sqrt{A^2 + B^2}\angle \pm\arctan\frac{B}{A} = M\angle \pm\theta^\circ \tag{32-5}$$

where M is the general term for the phasor magnitude.

The general equation for converting polar to rectangular form is given by equation,

$$A \pm jB = M\cos\theta \pm jM\sin\theta \tag{32-8}$$

Rectangular addition rule: Add the real parts separately, then add the j parts.

Rectangular subtraction rule: Subtract the real parts separately, then subtract the j parts.

Polar multiplication rule: Multiply the magnitudes, and add the phase angles algebraically.

Polar division rule: Divide the magnitude. Change the sign of the denominator phase angle and add it algebraically to the numerator phase angle.

The impedance of a parallel RL circuit is given by equation,

$$Z = \frac{R \times X_L}{\sqrt{R^2 + X_L^2}} \angle 90 - \arctan \frac{X_L}{R} \quad \text{ohms} \tag{32-9}$$

The impedance of a parallel RC circuit is given by equation,

$$Z = \frac{R \times X_C}{\sqrt{R^2 + X_C^2}} \angle -90 + \arctan \frac{X_C}{R} \quad \text{ohms} \tag{32-10}$$

The total impedance for a series LCR circuit in rectangular form is given by equation,

$$Z = R + jX_L - jX_C \quad \text{ohms} \tag{32-12}$$

In polar form,

$$Z = \sqrt{R^2 + (X_L - X_C)^2} \angle \arctan \frac{X_L - X_C}{R} \quad \text{ohms} \tag{32-11}$$

The total current in a parallel LCR circuit in rectangular form is

$$I_T = I_R + j(I_C - I_L) \quad \text{amperes} \tag{32-13}$$

The total current in a parallel LCR circuit in polar form is

$$I_T = \sqrt{I_R^2 + (I_C - I_L)^2} \angle \arctan \frac{I_C - I_L}{R} \quad \text{amperes} \tag{32-14}$$

$$Z = \frac{V_A \angle \theta^\circ}{I_T \angle \arctan \frac{I_C - I_L}{R}} \quad \text{ohms} \tag{32-15}$$

The total impedance in polar form of a parallel LCR circuit can be found by applying the sum of the reciprocal equation (32-16).

$$\frac{1}{Z} = \frac{1}{R\angle 0^\circ} + \frac{1}{X_L\angle 90^\circ} + \frac{1}{X_C\angle -90^\circ} \quad \text{ohms} \tag{32-16}$$

The complex sum of the conductance is the total admittance of the parallel LCR circuit and is expressed in rectangular form as

$$Y\angle \pm\theta = G + jB_C - jB_L \quad \text{siemens} \tag{32-18}$$

The admittance of a parallel LCR circuit in polar form is

$$Y\angle \pm\theta = \sqrt{G^2 + (B_C - B_L)^2} \angle \arctan \frac{B}{G} \quad \text{siemens} \tag{32-19}$$

REVIEW QUESTIONS

32-1. What term is applied to the numbers along the (**a**) horizontal and (**b**) vertical axis of a complex plane?

32-2. Explain why the reactive component is always represented by the prefix $\pm j$ operator.

32-3. What is the advantage of expressing phasor quantities in rectangular form?

32-4. In what two quantities is the polar form of a phasor expressed?

32-5. What are two advantages of expressing phasor quantities in polar form?

32-6. Express the impedance of a series *RL* circuit in rectangular form.

32-7. Show the equation to give the impedance of a series *RL* circuit in polar form.

32-8. Express the impedance of a series *RC* circuit in rectangular form.

32-9. Show the equation to give the impedance of a series *RC* circuit in polar form.

32-10. Convert the following complex numbers from rectangular to polar form.
(**a**) $100 + j300$ (**b**) $60 - j150$

32-11. Convert the following polar quantities to rectangular form.
(**a**) $400\underline{/35^\circ}$ (**b**) $300\underline{/-45^\circ}$

Refer to Fig. 32-35 for Questions 32-12 to 32-19.

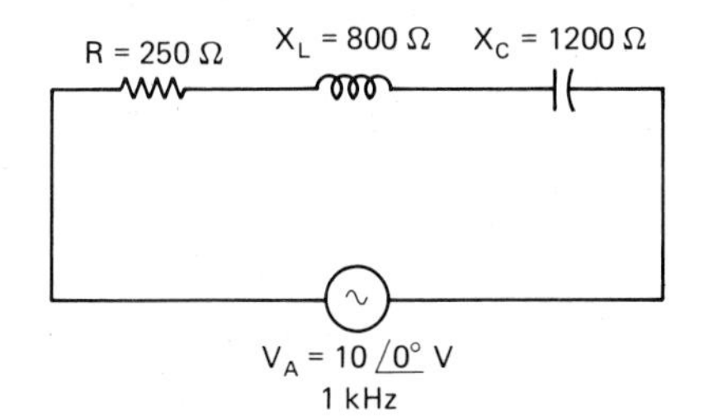

FIGURE 32-35 Series *LCR* circuit.

32-12. Express the impedance of the series *LCR* circuit in rectangular form.

32-13. Find and express the impedance of the series *LCR* circuit in polar form.

32-14. Find the total circuit current.

32-15. Find the voltage drop across the resistance.

32-16. Find the voltage drop across the inductive reactance.

32-17. Find the voltage drop across the capacitive reactance.

32-18. Draw a phasor diagram of the applied voltage and the voltage drops. Include the circuit current.

32-19. Convert the series *LCR* circuit of Fig. 32-35 into a parallel equivalent circuit.

Refer to Fig. 32-36 for Questions 32-20 to 32-32.

32-20. Find and express the impedance of branch 1 in polar form.

32-21. Find and express the impedance of branch 2 in polar form.

32-22. Find and express the impedance of branch 3 in polar form.

32-23. Find the current in branch 1.

32-24. Find the current in branch 2.

32-25. Find the current in branch 3.

32-26. Find and express the current of branch 1 in rectangular form.

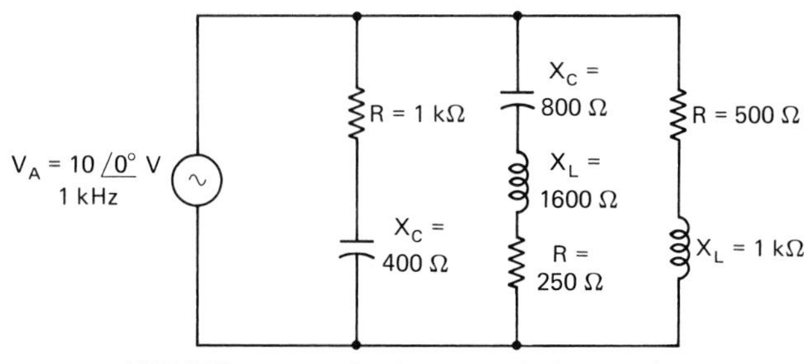

FIGURE 32-36 Series–parallel complex circuit.

32-27. Find and express the current of branch 2 in rectangular form.
32-28. Find and express the current of branch 3 in rectangular form.
32-29. Find and express the total circuit current in rectangular form.
32-30. Find and express the total circuit current in polar form.
32-31. Find the total impedance of the circuit.
32-32. Convert the parallel circuit of Fig. 32-36 into a series equivalent circuit.

SELF-EXAMINATION

Choose the most suitable word, phrase, or value to complete each statement.

32-1. Numbers along the horizontal (x) axis of a complex plane are termed __________ numbers.
32-2. Numbers along the vertical (y) axis of a complex plane are termed __________ numbers.
32-3. A phasor quantity in rectangular form is the sum of the __________ part and the __________ part.
32-4. The polar form of a phasor quantity is in terms of the __________ and __________ referenced to the positive real axis.
32-5. Express the impedance of the *RL* circuit in Fig. 32-37 in rectangular form.

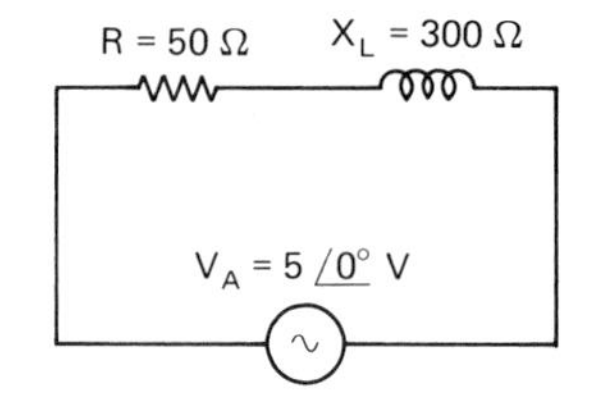

FIGURE 32-37 Series *RL* circuit.

32-6. Find and express the impedance of the *RL* circuit in Fig. 32-37 in polar form.
32-7. Express the impedance of the *RC* circuit in Fig. 32-38 in rectangular form.
32-8. Find and express the impedance of the *RC* circuit in Fig. 32-38 in polar form.
32-9. Convert the magnitude of the impedance phasor $150 + j450$ into polar form.
32-10. Convert the magnitude of the impedance phasor $456\angle -65°$ into rectangular form.

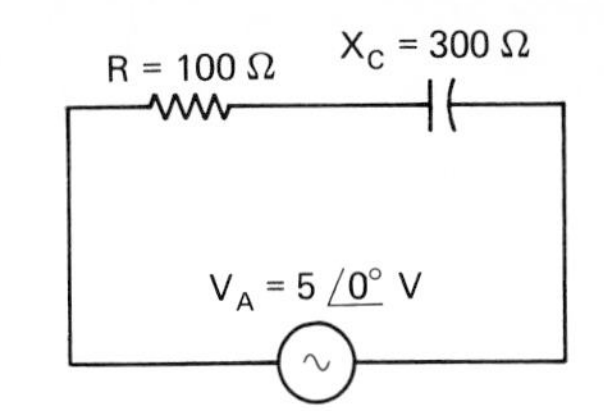

FIGURE 32-38 Series *RC* circuit.

Refer to Fig. 32-39 for Questions 32-11 and 32-12.

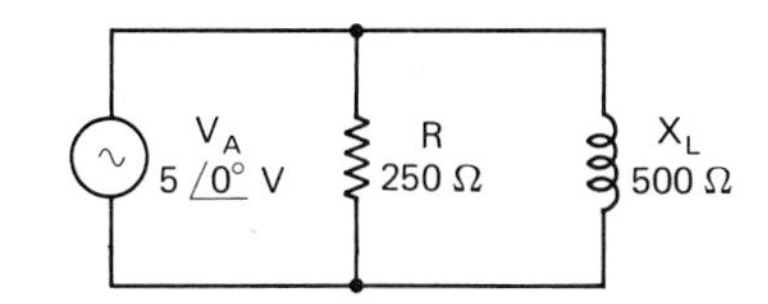

FIGURE 32-39 Parallel *RL* circuit.

32-11. Find and express the impedance of the *RL* circuit in polar form.

32-12. Convert the impedance from polar to rectangular form.

Refer to Fig. 32-40 for Questions 32-13 and 32-14.

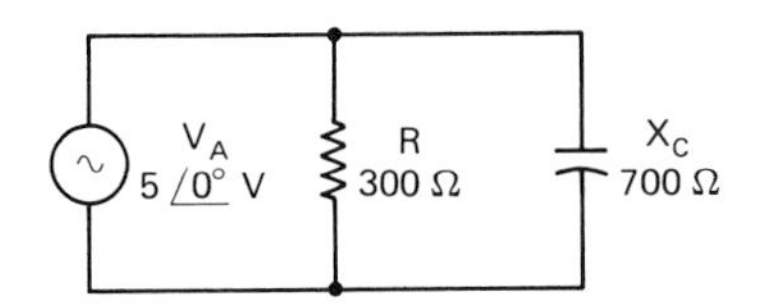

FIGURE 32-40 Parallel *RC* circuit.

32-13. Find and express the impedance of the *RC* circuit in polar form.

32-14. Convert the impedance from polar to rectangular form.

Refer to Fig. 32-41 for Questions 32-15 to 32-19.

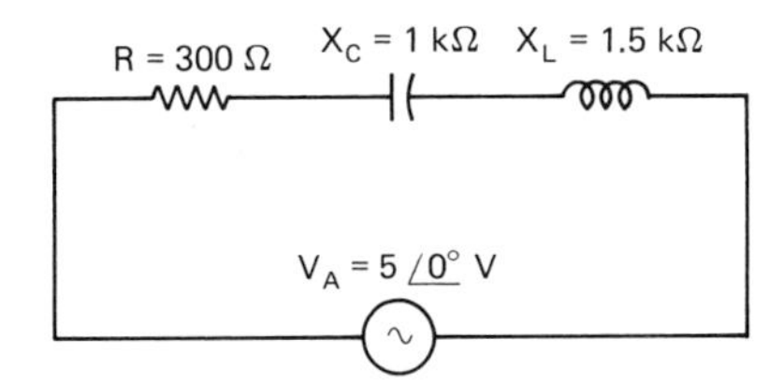

FIGURE 32-41 Series *LCR* circuit.

32-15. Find and express the impedance of the *LCR* circuit in polar form.

32-16. The source current (*lags, leads*) the applied voltage by ________ degrees.

32-17. Find and express the resistor voltage drop in polar form.

32-18. Find and express the inductor voltage drop in polar form.

32-19. Find and express the capacitor voltage drop in polar form.

Refer to Fig. 32-42 for Questions 32-20 to 32-28.

32-20. Find and express the resistor current in polar form.

32-21. Find and express the inductor current in polar form.

32-22. Find and express the capacitor current in polar form.

32-23. Express the total circuit current in rectangular form.

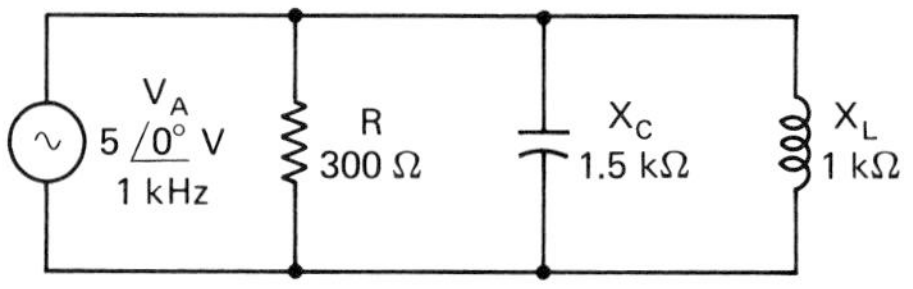

FIGURE 32-42 Parallel *LCR* circuit.

32-24. Find and express the total circuit current in polar form.

32-25. Find and express the circuit impedance in polar form.

32-26. Find the admittance of the parallel *LCR* circuit.

32-27. Use the circuit admittance to find the impedance of the parallel *LCR* circuit.

32-28. Convert the parallel *LCR* circuit into its equivalent series components.

Refer to Fig. 32-43 for Questions 32-29 to 32.33.

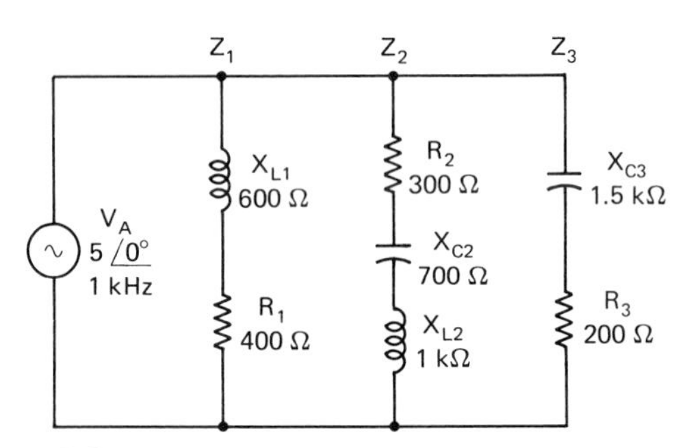

FIGURE 32-43 Complex *LCR* circuit.

32-29. Find and express the impedance of branch one in polar form.

32-30. Find and express the impedance of branch two in polar form.

32-31. Find and express the impedance of branch three in polar form.

32-32. Find the impedance of the circuit.

32-33. Find the equivalent series circuit components for the parallel *LCR* complex circuit of Fig. 32-43.

ANSWERS TO PRACTICE QUESTIONS

32-1

1. $Z = 463\angle 77.5^\circ\ \Omega$
2. $Z = 100 + j452\ \Omega$
3. $Z = 5247\angle -17.6^\circ\ \Omega$
4. $Z = 5000 - j1592\ \Omega$
5. $44.7\angle 26.6^\circ$
6. $50\angle -53^\circ$
7. $35 - j61$
8. $21 + j20$
9. $90 + j100$
10. $1200\angle 7^\circ$
11. $3.4\angle +105^\circ$

32-2

1. $Z = 483\underline{/15°}\ \Omega$
2. $466 + j125\ \Omega$
3. $Z = 104\underline{/16°}\ \Omega$
4. $Z = 100 + j188 - j159\ \Omega$
5. $I_T = 96.2\underline{/-16°}$ mA
6. $V_R = 9.6\underline{/-16°}$ V
7. $V_L = 18\underline{/74°}$ V
8. $V_C = 15.3\underline{/-106}$ V
9. 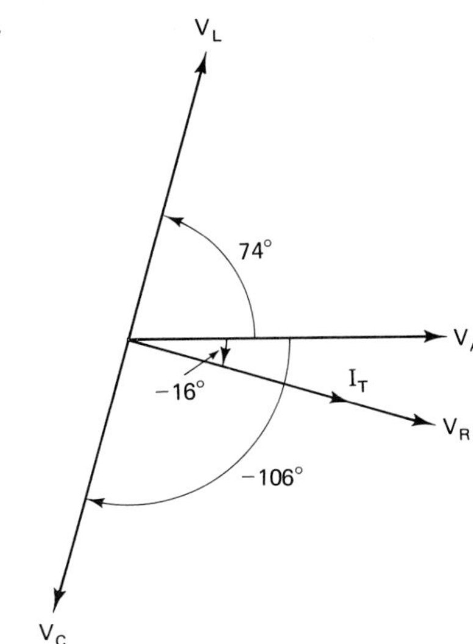

10. $10\underline{/0°}$ mA
11. $39.4\underline{/90°}$ mA
12. $53.2\underline{/-90°}$ mA
13. $I_T\underline{/\theta°} = 17\underline{/-54°}$ mA
14. $I_T = 10 + j39.4 - j53.2$ mA; therefore $I_T = 10 - j13.8$ mA
15. $Z = 588\underline{/54°}\ \Omega$
16. 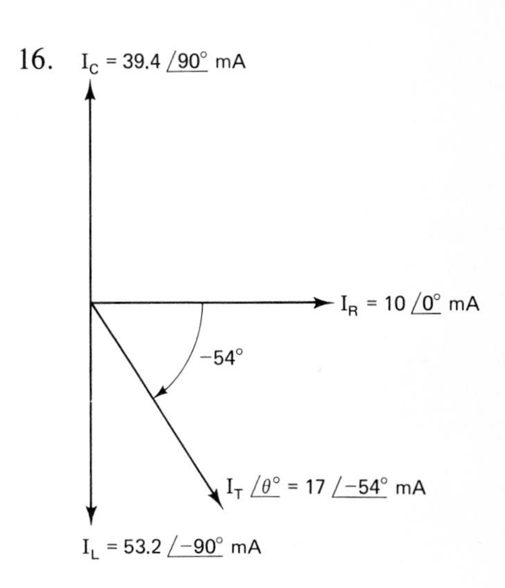

32-3

1. $Z_1 = 1503\underline{/86.2°}\ \Omega$
2. $Z_2 = 99.5\underline{/-84.3°}\ \Omega$
3. $Z_T = 110 + j1400\ \Omega$
4. $Z_T = 1404\underline{/85.5°}\ \Omega$
5. $V_C \cong 706\underline{/-170°}$ mV
6. $V_{Z1} = 10.7\underline{/0.7°}$ V
7. $Z_1 = 1005\underline{/-5.7°}\ \Omega$
8. $Z_2 = 510\underline{/-78.7°}\ \Omega$
9. $I_{Z1} = 9.95\underline{/5.7°}$ mA
10. $I_{Z1} = 9.9 + j0.99$ mA
11. $I_{Z2} = 19.6\underline{/78.7°}$ mA
12. $I_{Z2} = 3.8 + j19.2$ mA
13. $I_{Z3} = 19.6\underline{/-78.7°}$ mA
14. $I_{Z3} = 3.8 - j19.2$ mA
15. $I_T = 17.5 + j0.99$ mA
16. $I_T = 17.5\underline{/3.2°}$ mA
17. $Z = 571\underline{/-3.2°}\ \Omega$
18. $R = 570\ \Omega$
19. $X = -j32\ \Omega$; therefore, $C \cong 50\ \mu$F

Glossary

Absolute permeability (μ): the specific magnetic conductivity per unit length and cross-sectional area of a material.

Ac generator: a rotating electrical machine that converts mechanical energy into electrical energy in the form of an ac current and voltage brought out through slip rings.

Admittance (y): the reciprocal of impedance.

Air core: a coil or transformer winding wound on an insulating former having air as its core.

Alkaline cell: a cell that uses an alkaline electrolyte.

Alternating current or voltage (ac): normally applied to a sinusoidal waveform of ac current or voltage.

Alternation: one-half (180°) of an ac cycle.

Alternator: *see* Ac generator.

Ammeter: an analog or digital meter used to measure current.

Ampere-hour: the current rating of a cell for the specified time of 1 hour.

Ampere-turns: the magnetizing force of a current-carrying coil.

Ampere-turns per metre: Magnetic field intensity (H) per unit length of 1 metre.

Analog: refers to a device that operates by angular rotation.

Anode: the electrode that collects electrons.

Apparent power: the product of voltage and current in a reactive circuit.

Armature: the rotating part of a generator or motor.

Atomic number: indicates the number of shell electrons about an atom.

Attenuation factor: a factor by which a filter circuit reduces the input voltage.

Autotransformer: a transformer with one continuous winding and tapped connections or slider to give step-up or step-down voltages.

Average ac value: the average of the instantaneous values for 180° of a sinusoidal waveform.

Average power: the average of the instantaneous power values for 360° of a sinusoidal waveform.

AWG American wire gauge, used to determine the gauge number of a wire.

Ayrton shunt: A series–parallel combination of shunt resistors that provide different current ranges with a meter movement.

Balanced bridge: the null point on the galvanometer

of a bridge circuit.

Bandpass filter: a filter circuit that accepts a range of frequencies and rejects all other frequencies.

Band-reject filter: a filter circuit that rejects a range of frequencies and accepts all other frequencies.

Bandwidth (BW): a range of frequencies between the lower and upper half-power frequencies.

Battery: a number of series–parallel interconnected cells.

***B–H* magnetization curve:** a plot of flux density *B* versus field intensity *H* for a magnetic core.

Bleeder current: an independent current flow through a voltage-divider network that stablilizes the output voltage with load current.

Bohr model: a planetary model of an atom having a nucleus composed of neutrons and protons which are surrounded by orbiting electrons.

Breakdown: normally, the voltage point at which a large amount of current flows through the component.

Bridge circuit: a voltage-divider circuit where the voltage drops of two branches can be balanced to provide zero voltage between the two branches. Also designated a Wheatstone bridge.

Brushes: as applied to motors and generators, graphite connectors that are spring-mounted and brush against slip rings or a commutator to provide an electrical connection.

Bus: a printed-circuit strip or wire to which all the common connections of an electrical circuit or system are connected.

Bypass capacitor: a capacitor that provides a parallel or shunt path to ac.

Capacitance (*C*): the ability to store an electric charge.

Capacitor: a discrete component that is designed to have capacitance.

Carbon-zinc cell: a cell that uses zinc as the negative electrode and carbon rod as the positive electrode.

Cascaded filter: two or more series–parallel interconnected filter circuits.

Cathode: the electrode that emits electrons.

Cell: a discrete component that converts one form of energy such as chemical or solar into electrical energy.

Ceramic capacitor: a capacitor having ceramic as its dielectric.

Chassis ground: metal frame of electronic equipment that is connected to the grounded side of the power line.

Choke: a coil having inductance.

Circuit breaker: an automatic current overload protection switch that can also be operated manually.

Circular mil: the cross-sectional area of a conductor having a diameter of 10^{-3} inch equal to 1 mil.

Closed circuit: a complete circuit.

Coil: number of turns of insulated wire.

Color code: standardized color band method to identify resistor, capacitor, inductor values, and ratings.

Complex number: a number statement containing a real number and an imaginary number $(3 + j6)$.

Complex plane: a coordinate x- and y-axis system with the real numbers along the x axis and the imaginary numbers along the y axis.

Commutator: an arrangement of copper segments on a rotating armature used to connect the armature coils to the brushes.

Condenser: another term for capacitor.

Conductance (*G*): measure of the ease of current flow in a circuit; the reciprocal of resistance.

Continuity: normally, a wire conductor that is continuous (i.e., without a break or open).

Conventional current flow: by convention a positive ion, or charge in motion, whose direction is from the positive electrode of the voltage source through the external circuit, then continuing to the negative electrode.

Core loss: a power loss due to hystersis and eddy currents in a ferromagnetic core.

Coulomb (C): an electric charge of 6.25×10^{18} electrons.

Counter emf (counter electromotive force): an opposing out-of-phase voltage.

Coupling capacitor: a capacitor that provides an ac path from one device to another while blocking dc.

***CR* differentiator:** *see* Differentiator circuit.

***CR* integrator:** *see* Integrator.

CR time constant: the time in seconds for a series *RC* capacitor to reach a charge of 63% (symbol τ).

CRT: cathode-ray tube; used in an oscilloscope or TV picture tube.

Current divider: two or more resistive parallel branches.

Cycle: a complete voltage or current variation between two successive points having the same polarity and value.

D'Arsonval movement: a basic permanent-magnet moving-coil dc meter.

Damping: reducing the Q of a tuned circuit, which increases bandwidth.

Delta (Δ) circuit: three-phase transformer windings or components connected in series that form a closed loop. Also equivalent to pi (π) circuit.

Diamagnetic: a material that weakly opposes being magnetized.

Dielectric: an insulating material (eg., the insulator

between the plates of a capacitor).

Dielectric constant (K): a comparative ratio or percentage between air and an insulator to show a measure of its ability to concentrate an electric field.

Dielectric strength: a measure per unit thickness of an insulator to withstand a maximum voltage without arc over through the insulator.

Differentiator circuit: an *RC* circuit where the output is taken across the series resistor. An *RL* circuit where the output is taken across the inductor.

Diode: a semiconductor component or a vacuum tube that behaves electrically like a switch.

Discrete component: a capacitor, inductor, and resistor or similarly packaged component.

DMM: digital multimeter.

Domain: small internal region(s) of molecular magnetic fields of a material; usually refers to a ferromagnetic material.

Domain wall: boundary between one magnetic domain and another magnetic domain; the area where the direction of magnetization reverses.

DPDT: double-pole, double-throw switch.

DPST: double-pole, single-throw switch.

Dynamometer: a type of ac ammeter or voltmeter specifically used to measure voltage or current on a 60-Hz line.

Eddy currents: a self-induced circulating current in a conducting core due to the changing magnetic flux.

Effective value: an ac value having the same heating effect as an equivalent dc value. Also called rms value.

Efficiency: ratio of output over input times 100 to give percentage efficiency.

Electric field: invisible lines of force about a charged body that "flow" into an unlike charge but are repelled by a similar or like charge.

Electricity: static or dynamic energy inherent in nature which when harnessed can produce heat, light, mechanical work, sound, and computation.

Electrode: a metal conductor through which an electrical current enters or leaves a device.

Electrolyte: a paste or liquid solution that causes the dissociation of positive and negative ions, thus conducting an electric current.

Electrolytic capacitor: a capacitor having an electrolyte as its dielectric. Usually polarized with the electrolyte as the true cathode, which is negative.

Electromagnet: a coil of insulated copper wire wound on an air-core or soft-iron core that produces a magnetic field because of the current flowing through its turns.

Electron: a negatively charged particle orbiting in shells about the nucleus of an atom.

Electron flow: a negative charge, or current in motion, whose direction is from the negative electrode of the voltage source through the external circuit, then continuing to the positive electrode.

Emf: electromotive force; an electrical term meaning an electrical pressure or voltage.

Farad (F): the unit of capacitance.

Faraday's law: applied to electromagnetic induction. The induced voltage in a conductor is directly proportional to the rate at which the magnetic flux is cut by the conductor.

Ferrite core: powdered iron combined with a binder and molded into a core.

Ferromagnetic: a material that exhibits magnetic properties.

Field: applied to lines of force surrounding a magnet or electromagnet and a charge or a charged electrostatic body.

Field coil: usually, the stationary coil windings that set up the magnetic field of a generator or motor.

Field-effect transistor (FET): a transistor designed to control its drain current by means of the electric field set up between the gate and the channel semiconductor.

Field intensity: *see* magnetic field intensity.

Filter: a circuit that accepts and/or rejects an input voltage within its designed cutoff frequency range.

Flux (Φ): as applied to magnetic or electric field lines of force.

Flux density: the amount of flux per unit cross-sectional area.

Free electron: a valence electron that requires little energy to break away from its orbital path and moves at random through the atoms of the material.

Frequency (f): as applied to a periodic waveform, it is the number of cycles the wave goes through every second. Unit is the hertz (Hz).

Full-wave rectifier: a circuit configuration utilizing diodes that converts each alternation of an ac voltage cycle into pulsating dc.

Fuse: a device with a lead alloy link that melts due to the heat generated by an excessive current and opens the circuit.

Galvanometer: an extremely sensitive ammeter capable of measuring current in the range of picoamperes.

Generator: ususally refers to a rotary machine that converts mechanical energy into either ac or dc voltage and current. Also refers to an electronic device that produces a source of periodic voltage and current waveform(s).

Giga (G): standard international prefix for 10^9 units.

Ground: a common reference point in an electrical circuit used as a return path for current and a reference point for voltage. It can be either the metal chassis or a common bus. Also refers to the common earth connection of ac electric utility power lines.

Ground wire: a bare or green insulated wire that is connected to the hydroelectric utility's earth ground.

Half-power frequency: the frequency at which the output voltage or current falls to 70.7% of its maximum value.

Harmonic: an even or odd multiple of a fundamental frequency.

henry (H): a unit of inductance.

Hertz (Hz): a unit of frequency equal to cycles per second.

Horse power (hp): a unit of mechanical work used in the English system where 1 hp = 550 ft-1b/s and is equal to 746 watts.

Hot wire: the voltage or potential conductor of a source. Especially referred to in the hydroelectric power lines. Also called the live wire.

Hysteresis: a characteristic of a ferromagnetic material; the time lag between the application of a magnetizing force and the time it takes the material to become magnetized in a specific direction.

Imaginary number: a reactive quantity referenced at 90° to the resistive quantity given a $-j$ prefix for X_C and $+j$ for X_L.

Impedance (Z): the total ac opposition in a circuit containing any combination of *RLC*.

Induced: *as applied to a magnetic field or an electric field,* it is the magnetization or charge produced at a distance or without physical contact by a magnetic and an electric flux, respectively.

As applied to an electromagnetic circuit, it is the voltage and current within a coil or within a coupled coil produced without physical contact and as a result of the changing flux or motion of the coil through the flux.

Inductance (L): the ability of a coil or inductor to induce a voltage within itself or into another coil as a result of a changing magnetic flux.

Induction: *see* Induced.

Induction motor: a motor in which the stator coils produce a rotating magnetic field which induces current in the shorted turns of the rotor, with resultant repelling and rotating action.

Inductor: any length of wire coiled or straight with a changing current flowing through it.

Instantaneous value: applied to a periodic wave, it is the value at any given instant of time.

Insulator: a nonconductor of electricity.

Integrated circuit (IC): a microminiaturized packaged circuit or system composed of transistors, diodes, resistors, and capacitors.

Integrator: the output voltage of a series *LR* circuit is taken across the resistor and is proportional to the mathematical integral of the input. In an *RC* circuit the output voltage is taken across the capacitor.

Internal resistance: any electrical device, such as a battery, power source, amplifier, or transformer, exhibits some resistance within itself.

Inversely proportional: a reciprocal function; an increase in one variable will decrease the other variable, and vice versa.

Ion: an atom that has a deficiency or excess of electrons.

Iron-vane meter: an ac meter that has a deflecting iron vane and a stationary coil to measure ac voltage and current on a 60-Hz line.

***j*-operator:** a prefix to a number, $+j$ or $-j$, to show that the quantity is 90° out of phase with a reference value.

Joule (J): unit of work or energy.

Kilo (k): standard international prefix for 10^3 units.

Kirchhoff's laws: a set of algebraic laws to show the distribution of voltage and current in any ac or dc circuit.

Lagging: refers to the delay in time or phase of an instantaneous waveform value as compared to another instantaneous waveform value at the same time.

Laminations: thin sheets of silicon iron that are stacked together to form a core for transformers, inductors, relays, and motors.

Leading: refers to the advance in time or phase of one instantaneous waveform value as compared to another instantaneous waveform value at the same time.

LeClanche cell: a carbon-zinc primary cell.

Lenz's law: a statement that the induced voltage and current will always be in such a direction as to oppose the action creating it.

Linear relation: a graphical straight-line relation where one variable quantity is directly proportional to another variable quantity.

Line cord: two or three insulated conductors that are enclosed in an insulating sheath or cable used to connect portable equipment to the 120/240-V ac line.

Lines of force: imaginary lines that identify the direction of magnetic or electric flux to support theoretical and measurable concepts.

Line voltage: usually refers to the hydroelectric 120 V.

Line wire: a wire connecting the source to the device.

Load: any device that converts the effect of ac or dc current flow into another form of energy, such as heat, light, or motor action.

Loading effect: the increased source current and decrease in source voltage due to the parallel resistance of a load.

Loop: a closed current path in a circuit.

Magnet: a bar or U-shaped magnetized ferromagnetic material that has a north and a south magnetic pole.

Magnetic: a material that exhibits the properties of magnetism.

Magnetic circuit: a complete ferromagnetic enclosed path for magnetic flux.

Magnetic field: the total lines of force or flux set up about the region of a magnetized material or magnet.

Magnetic field intensity (H): a measure of the magnetic field strength set up by the number of ampere-turns per metre unit length of a magnetic circuit; its unit is amperes per metre (A/m).

Magnetic flux: magnetic lines of force.

Magnetic induction: see Induced.

Magneto: an ac generator that uses permanent magnets for its magnetic field.

Magnetomotive force (mmf): the magnetizing force produced by a current flowing through the turns of a coil. Its SI unit is the ampere (A); its metric unit is ampere-turns (At).

Magnitude: value of any electrical quantity without regard for the phase angle.

Maximum power transfer: the optimum circuit condition where the internal resistance of a dc source, or the impedance of an ac source, is equal to the load resistance or impedance and maximum power from input to output takes place.

Maxwell: one magnetic line of force (cgs system).

Mega (M): standard international prefix for 10^6 units.

Mesh current: assumed direction of current in a closed loop applied to Kirchhoff's current law.

Micro (μ): standard international prefix for 10^{-6} units.

Milli (m): standard international prefix for 10^{-3} units.

Monostable: always reverts to one stable state, as in a multivibrator.

Multimeter: a portable multifunction (volts, ohms, amperes) measuring meter.

Multiplier resistance: a series resistor used to extend the voltage range of a multimeter.

Multivibrator: an oscillator that generates timed square-wave pulses for logic circuits.

Mutual inductance (L_m): ability of a primary coil to induce a voltage into a secondary coil because the changing magnetic flux is common to both.

Nano (n): standard international prefix for 10^{-9} units.

Neutral: the grounded conductor of the electric utility system identified by its white insulation.

Node: a common junction where two or more currents enter or leave.

Nonlinear relation: a graphical curve relation where one variable quantity exhibits a nonuniform proportionality to another variable quantity.

Nucleus: the innermost part of an atom, which consists of protons and neutrons.

Null: a balanced bridge condition where the galvanometer reads zero or approximately zero.

Ohm (Ω): unit of resistance.

Ohmmeter: a meter that measures resistance.

Ohms per volt: sensitivity rating for a meter movement used as a voltmeter.

Open circuit: an incomplete path for current flow.

Parallel circuit: two or more loads connected across one source.

Paramagnetic material: a material that will establish a weak magnetic field in the opposite direction of the magnetizing force.

Peak-to-peak value (p-p): the value between two opposite polarity peak values.

Peak value: the maximum amplitude of a periodic wave.

Period: the length of time it takes a periodic wave to complete one cycle.

Periodic waveform: an amplitude-varying waveform that repeats itself every cycle.

Permeability (μ): relative measure of a material's ability to concentrate magnetic lines of force within itself.

Phase angle: a measure in degrees of the displacement between two periodic waveforms with one waveform used as a reference.

Phasor: a vector diagram rotating CCW by convention.

Pico (p): standard international prefix for 10^{-12} units.

Polar form: format of complex numbers that specifies magnitude and its phase angle.

Polarity: designates two opposite conditions of electrical charges (i.e., positive and negative) or between two magnetic poles, a north and a south pole.

Polarized: a preset condition of connections, such as an electrolytic capacitor or a U-ground outlet.

Potential difference: voltage difference between two points in a circuit.

Potentiometer: a variable three-terminal resistor normally used as a voltage divider.

Power: rate of doing work. Electrical unit is the watt.

Power factor: ratio of true power in watts to reactive power in volt-amperes.

Primary cell: a cell that is not rechargeable.

Primary winding: a transformer winding that is connected to the source voltage.

Proton: a positive particle of electricity in the nucleus of an atom.

Pulsating dc: amplitude pulses of voltage or current that vary in one direction only, either positively or negatively.

Quality factor (Q): figure of quality or merit of a coil or resonant circuit, or a ratio of voltage or current magnification in resonant circuits. It is the ratio of reactive power to true power in a coil.

Radian: angular measure of 57.3°.

Reactance (X): Ohm's law ac opposition to current flow through either a capacitor or an inductor.

Reactive power (var): the power that is alternately drawn from the source by a reactive component, then returned to the source.

Real number: in the complex number system it represents the magnitude of a pure resistance.

Real power: power dissipated by a pure resistance in an ac circuit.

Reciprocal: an inverse relation as one magnitude increases the other decreases.

Rectangular form: format of complex numbers that specifies the phasor sum of a real number and an imaginary j number ($A + jB$).

Rectifier: an electronic circuit that converts alternating current to direct current.

Relative permeability (μ_r): permeability of a material as compared to free space (air).

Relay: remote electromagnetic-operated switch.

Reluctance ($\mathcal{R}$): opposition offered by a magnetic circuit to the magnetic flux.

Resistance (R): the opposition offered to current flow in a circuit.

Resistivity: the resistance of a material per unit volume and temperature.

Resonance: a circuit that is in tune with or having the same natural frequency of oscillations as the source voltage.

Retentivity: ability of a material to retain some of its magnetic properties after the magnetizing force is removed.

Rheostat: a two-terminal variable resistance.

Rms value: an ac value that is equivalent to a dc value.

Rotor: the rotating part of a motor or generator.

Saturation: as applied to electromagnetism, a further increase in the magnetizing force will not appreciably increase the strength of the flux.

Secondary cell: a cell that can be recharged.

Secondary winding: the transformer winding that is connected to a load.

Selectivity: ability of a tuned circuit to accept a band of frequencies and reject all other frequencies, which is determined by its bandwidth (BW).

Self-inductance (L): ability of a coil to produce a counter emf within itself due to the changing current through the coil.

Series circuit: two or more loads connected to form one current path.

Short circuit: a zero- or low-resistance path for current across two points of a circuit.

Shunt: connected in parallel, such as an ammeter shunt for a meter movement.

Siemens (S): unit of conductance, susceptance, and admittance. It is equal to the reciprocal of resistance.

Sine function: trigonometric function of an angle; the ratio of the opposite side to the hypotenuse.

Sine wave: a periodic wave in which the instantaneous values vary in proportion to the sine function of their phase angle.

Sinusoidal: having a waveform shape of a sine wave.

Slip rings: a continuous conducting ring(s) in contact with a carbon brush to provide a current path into or from the coils of a rotating rotor to the external circuit.

Solenoid: an iron-core electromagnet used for relays and other electromagnetic device.

Source of power: refers to dc or ac voltage and current sources.

SPDT: single-pole, double-throw switch.

Specific gravity: the ratio of weight of a substance to that of an equal volume of water.

Specific resistance (ρ): the resistance of a cubic centimetre of a material.

SPST: single-pole, single-throw switch.

Square wave: a periodic wave whose amplitude resembles a square top and its ON time is equal to its OFF time.

Static: stationary or not in motion and nonvarying.

Static electricity: electrical charge that is not in motion and is nonvarying.

Stator: the fixed or stationary coil(s) of a motor or generator.

Superposition theorem: a method of solving a network with multiple sources by systematically finding the current for each component with one source connected at a time, then superimposing the currents and adding them algebraically.

Susceptance (*B*): a measure of the ease of current flow through a reactive component; the reciprocal of reactance.

Switch: a device that opens or closes the path for current in a circuit.

Tangent: trigonometric function of an angle; the ratio of the opposite side to the adjacent side.

Tank circuit: the parallel *LC* resonant circuit.

Tantalum: a base-oxidizable material used in electrolytic capacitors.

Taper: the rate at which the resistance of a potentiometer varies with percentage rotation of its shaft.

Temperature coefficient: a number used to specify the change in resistance per degree rise in temperature.

Tesla (T): unit of flux density, equal to 10^8 lines of force per square metre.

Thermistor: a resistor having a negative temperature coefficient of resistance.

Thévenin's theorem: a rule by which a complex circuit is reduced to one voltage source with an equivalent series resistance.

Three-phase voltage: three sinusoidal voltages, each having a phase difference of 120 electrical degrees.

Time constant (τ): time required for the voltage to rise by 63% from its initial value across a capacitor, or the current to rise by 63% from its initial value through an inductor. It is the product of *R* and *C* and ratio of *L*/*R* in seconds.

Tolerance: specifies the maximum and the minimum deviation from a component's nominal value.

Transformer: a device having a primary and secondary coil(s) which steps up or steps down an ac voltage.

Transient: a short-duration pulse of current or voltage.

Transistor: a three-terminal semiconductor device, either bipolor or field effect; amplifies current and/or voltage.

Tuned circuit: a series or parallel *LC* circuit having its capacitance or inductance adjustable in order to bring it into a resonant condition.

Turns ratio: the ratio of the number of turns on a transformer primary compared to the secondary.

U-ground receptacle: an ac 120-V outlet having an extra U-shaped terminal which is grounded to the hydroelectric system.

Unity power factor: a circuit condition where $X_L = X_C$ so that the current and applied voltage are in phase and the dissipated power is 100% resistive or unity (1).

Valence electron: an electron that has its orbit in the outermost shell of an atom.

Variac: a single-winding transformer constructed so that its output voltage can be varied in small increments.

Vector: a line drawn to scale representing magnitude and direction.

Volt (V): unity of electrical potential difference, emf, pressure, or tension.

Voltage divider: resistors in series or a potentiometer connected across a voltage source to provide a lower output voltage.

Volt-ampere (VA): unit of apparent power in an ac circuit.

Volt-ampere reactive (var): unit of reactive power in an ac circuit.

Voltmeter loading: the shunting affect of a voltmeter multiplier resistance causing the circuit current and voltage to be inaccurate.

Voltmeter sensitivity (*S*): the ohms per volt rating of a meter movement, equal to 1 V divided by the full-scale current.

VOM: Volt-ohm-milliammeter.

Watt (W): unit of electrical power.

Wattmeter: measures dc power or true ac power.

Waveform: the shape of the combined voltage or current instantaneous values plotted against time.

Wave trap: a parallel resonant *LC* circuit that rejects a band of frequencies.

Weber (Wb): unit of magnetic flux; equal to 10^8 lines of force.

Wheatstone bridge: *see* Bridge circuit.

Wire gauge: a gauge having a number of slotted holes which are used to measure the size of a conductor, as in AWG.

Work: a process of converting one form of energy into another form; the standard unit of work is the joule.

Wye (Y) circuit: three-phase transformer windings or components connected with one end to a common point and the other ends connected to the line or load; equivalent to a (T) circuit.

Y: symbol for admittance.

Z: symbol for impedance.

Zero-ohms adjust: a variable resistance to allow the ohmmeter pointer to be set to zero ohms.

Index

A

R

S